STUDENT'S SOLUTIONS MANUAL

PART ONE

ARDIS • BORZELLINO • BUCHANAN • KOUBA
MOGILL • NELSON

UNIVERSITY CALCULUS

Joel Hass

University of California, Davis

Maurice D. Weir

Naval Postgraduate School

George B. Thomas, Jr.

Massachusetts Institute of Technology

PEARSON
Addison
Wesley

Boston San Francisco New York
London Toronto Sydney Tokyo Singapore Madrid
Mexico City Munich Paris Cape Town Hong Kong Montreal

ISBN   0-321-38849-6

8 9 BR 09 08

PREFACE TO THE STUDENT

The Student's Solutions Manual contains the solutions to all of the odd-numbered exercise in the 11th Edition of THOMAS' UNIVERSITY CALCULUS by Maurice Weir, Joel Hass and Frank Giordano, excluding the Computer Algebra System (CAS) exercises. We have worked each solution to ensure that it

- conforms exactly to the methods, procedures and steps presented in the text

- is mathematically correct

- includes all of the steps necessary so you can follow the logical argument and algebra

- includes a graph or figure whenever called for by the exercise, or if needed to help with the explanation

- is formatted in an appropriate style to aid in its understanding

How to use a solution's manual

- solve the assigned problem yourself

- if you get stuck along the way, refer to the solution in the manual as an aid but continue to solve the problem on your own

- if you cannot continue, reread the textbook section, or work through that section in the Student Study Guide, or consult your instructor

- if your answer is correct by your solution procedure seems to differ from the one in the manual, and you are unsure your method is correct, consult your instructor

- if your answer is incorrect and you cannot find your error, consult your instructor

Acknowledgments

Solutions Writers
    William Ardis, Collin County Community College-Preston Ridge Campus
    Joseph Borzellino, California Polytechnic State University
    Linda Buchanan, Howard College
    Duane Kouba, University of California-Davis
    Tim Mogill
    Patricia Nelson, University of Wisconsin-La Crosse

Accuracy Checkers
    Karl Kattchee, University of Wisconsin-La Crosse
    Marie Vanisko, California State University, Stanislaus
    Tom Weigleitner, VISTA Information Technologies

Thanks to Rachel Reeve, Christine O'Brien, Sheila Spinney, Elka Block, and Joe Vetere for all their guidance and help at every step.

TABLE OF CONTENTS

CHAPTER 1 FUNCTIONS

1.1 FUNCTIONS AND THEIR GRAPHS

1. domain $= (-\infty, \infty)$; range $= [1, \infty)$

3. domain $= (0, \infty)$; y in range $\Rightarrow y = \frac{1}{\sqrt{t}}$, $t > 0 \Rightarrow y^2 = \frac{1}{t}$ and $y > 0 \Rightarrow$ y can be any positive real number
   $\Rightarrow$ range $= (0, \infty)$.

5. (a) Not the graph of a function of x since it fails the vertical line test.
   (b) Is the graph of a function of x since any vertical line intersects the graph at most once.

7. base $= x$; $(\text{height})^2 + \left(\frac{x}{2}\right)^2 = x^2 \Rightarrow$ height $= \frac{\sqrt{3}}{2} x$; area is $a(x) = \frac{1}{2} (\text{base})(\text{height}) = \frac{1}{2} (x) \left(\frac{\sqrt{3}}{2}x\right) = \frac{\sqrt{3}}{4} x^2$;
   perimeter is $p(x) = x + x + x = 3x$.

9. Let $D = $ diagonal of a face of the cube and $\ell = $ the length of an edge. Then $\ell^2 + D^2 = d^2$ and $D^2 = 2\ell^2 \Rightarrow 3\ell^2 = d^2$
   $\Rightarrow \ell = \frac{d}{\sqrt{3}}$. The surface area is $6\ell^2 = \frac{6d^2}{3} = 2d^2$ and the volume is $\ell^3 = \left(\frac{d^2}{3}\right)^{3/2} = \frac{d^3}{3\sqrt{3}}$.

11. The domain is $(-\infty, \infty)$.

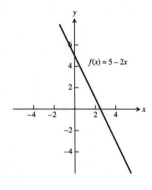

13. The domain is $(-\infty, \infty)$.

15. The domain is $(-\infty, 0) \cup (0, \infty)$.

17. Neither graph passes the vertical line test
    (a)

    (b)

19.

x	0	1	2
y	0	1	0

21. $G(x) = \begin{cases} 3 - x, & x \le 1 \\ 2x, & x > 1 \end{cases}$

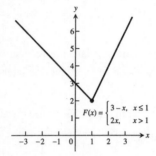

23. (a) Line through $(0, 0)$ and $(1, 1)$: $y = x$

Line through $(1, 1)$ and $(2, 0)$: $y = -x + 2$

$f(x) = \begin{cases} x, & 0 \le x \le 1 \\ -x + 2, & 1 < x \le 2 \end{cases}$

(b)　$f(x) = \begin{cases} 2, & 0 \le x < 1 \\ 0, & 1 \le x < 2 \\ 2, & 2 \le x < 3 \\ 0, & 3 \le x \le 4 \end{cases}$

25. (a) Line through $(-1, 1)$ and $(0, 0)$: $y = -x$

Line through $(0, 1)$ and $(1, 1)$: $y = 1$

Line through $(1, 1)$ and $(3, 0)$: $m = \frac{0-1}{3-1} = \frac{-1}{2} = -\frac{1}{2}$, so $y = -\frac{1}{2}(x - 1) + 1 = -\frac{1}{2}x + \frac{3}{2}$

$f(x) = \begin{cases} -x & -1 \le x < 0 \\ 1 & 0 < x \le 1 \\ -\frac{1}{2}x + \frac{3}{2} & 1 < x < 3 \end{cases}$

(b) Line through $(-2, -1)$ and $(0, 0)$: $y = \frac{1}{2}x$

Line through $(0, 2)$ and $(1, 0)$: $y = -2x + 2$

Line through $(1, -1)$ and $(3, -1)$: $y = -1$

$f(x) = \begin{cases} \frac{1}{2}x & -2 \le x \le 0 \\ -2x + 2 & 0 < x \le 1 \\ -1 & 1 < x \le 3 \end{cases}$

27. (a) $\lfloor x \rfloor = 0$ for $x \in [0, 1)$　　　　　　　　(b) $\lceil x \rceil = 0$ for $x \in (-1, 0]$

29. For any real number x, $n \le x \le n + 1$, where n is an integer. Now: $n \le x \le n + 1 \Rightarrow -(n + 1) \le -x \le -n$. By definition: $\lceil -x \rceil = -n$ and $\lfloor x \rfloor = n \Rightarrow -\lfloor x \rfloor = -n$. So $\lceil -x \rceil = -\lfloor x \rfloor$ for all $x \in \Re$.

31. Symmetric about the origin

    Dec: $-\infty < x < \infty$

    Inc: nowhere

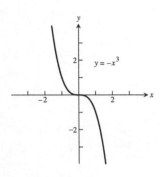

33. Symmetric about the origin

    Dec: nowhere

    Inc: $-\infty < x < 0$

    $0 < x < \infty$

35. Symmetric about the y-axis

    Dec: $-\infty < x \le 0$

    Inc: $0 < x < \infty$

37. Symmetric about the origin

    Dec: nowhere

    Inc: $-\infty < x < \infty$

39. No symmetry

    Dec: $0 \le x < \infty$

    Inc: nowhere

41. Symmetric about the y-axis

    Dec: $-\infty < x \le 0$

    Inc: $0 < x < \infty$

43. Since a horizontal line not through the origin is symmetric with respect to the y-axis, but not with respect to the origin, the function is even.

45. Since $f(x) = x^2 + 1 = (-x)^2 + 1 = -f(x)$. The function is even.

47. Since $g(x) = x^3 + x$, $g(-x) = -x^3 - x = -(x^3 + x) = -g(x)$. So the function is odd.

49. $g(x) = \frac{1}{x^2 - 1} = \frac{1}{(-x)^2 - 1} = g(-x)$. Thus the function is even.

51. $h(t) = \frac{1}{t - 1}$; $h(-t) = \frac{1}{-t - 1}$; $-h(t) = \frac{1}{1 - t}$. Since $h(t) \ne -h(t)$ and $h(t) \ne h(-t)$, the function is neither even nor odd.

53. $h(t) = 2t + 1$, $h(-t) = -2t + 1$. So $h(t) \ne h(-t)$. $-h(t) = -2t - 1$, so $h(t) \ne -h(t)$. The function is neither even nor odd.

55. $v = f(x) = x(14 - 2x)(22 - 2x) = 4x^3 - 72x^2 + 308x; \; 0 < x < 7.$

57. (a) Graph h because it is an even function and rises less rapidly than does Graph g.

    (b) Graph f because it is an odd function.

    (c) Graph g because it is an even function and rises more rapidly than does Graph h.

59. (a) From the graph, $\frac{x}{2} > 1 + \frac{4}{x} \; \Rightarrow \; x \in (-2, 0) \cup (4, \infty)$

    (b) $\frac{x}{2} > 1 + \frac{4}{x} \; \Rightarrow \; \frac{x}{2} - 1 - \frac{4}{x} > 0$

    $x > 0: \; \frac{x}{2} - 1 - \frac{4}{x} > 0 \; \Rightarrow \; \frac{x^2 - 2x - 8}{2x} > 0 \; \Rightarrow \; \frac{(x-4)(x+2)}{2x} > 0$
    $\Rightarrow \; x > 4$ since x is positive;

    $x < 0: \; \frac{x}{2} - 1 - \frac{4}{x} > 0 \; \Rightarrow \; \frac{x^2 - 2x - 8}{2x} < 0 \; \Rightarrow \; \frac{(x-4)(x+2)}{2x} < 0$
    $\Rightarrow \; x < -2$ since x is negative;

    sign of $(x - 4)(x + 2)$

    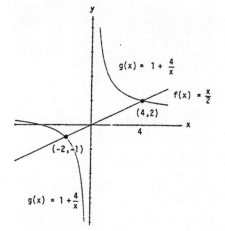

    Solution interval: $(-2, 0) \cup (4, \infty)$

61. A curve symmetric about the x-axis will not pass the vertical line test because the points (x, y) and $(x, -y)$ lie on the same vertical line. The graph of the function $y = f(x) = 0$ is the x-axis, a horizontal line for which there is a single y-value, 0, for any x.

1.2 COMBINING FUNCTIONS; SHIFTING AND SCALING GRAPHS

1. $D_f: \; -\infty < x < \infty, D_g: \; x \geq 1 \; \Rightarrow \; D_{f+g} = D_{fg}: \; x \geq 1. \; R_f: \; -\infty < y < \infty, R_g: \; y \geq 0, R_{f+g}: \; y \geq 1, R_{fg}: \; y \geq 0$

3. $D_f: \; -\infty < x < \infty, D_g: \; -\infty < x < \infty, \; D_{f/g}: \; -\infty < x < \infty, D_{g/f}: \; -\infty < x < \infty, \; R_f: \; y = 2, R_g: \; y \geq 1,$
   $R_{f/g}: \; 0 < y \leq 2, R_{g/f}: \; \frac{1}{2} \leq y < \infty$

5. (a) 2

   (b) 22

   (c) $x^2 + 2$

   (d) $(x + 5)^2 - 3 = x^2 + 10x + 22$

   (e) 5

   (f) -2

   (g) $x + 10$

   (h) $(x^2 - 3)^2 - 3 = x^4 - 6x^2 + 6$

7. (a) $\frac{4}{x^2} - 5$

   (b) $\frac{4}{x^2} - 5$

   (c) $\left(\frac{4}{x} - 5\right)^2$

   (d) $\frac{1}{(4x - 5)^2}$

   (e) $\frac{1}{4x^2 - 5}$

   (f) $\frac{1}{(4x - 5)^2}$

9. (a) $(f \circ g)(x)$

   (b) $(j \circ g)(x)$

   (c) $(g \circ g)(x)$

   (d) $(j \circ j)(x)$

   (e) $(g \circ h \circ f)(x)$

   (f) $(h \circ j \circ f)(x)$

11.

	$g(x)$	$f(x)$	$(f \circ g)(x)$
(a)	$x - 7$	$\sqrt{x}$	$\sqrt{x - 7}$
(b)	$x + 2$	$3x$	$3(x + 2) = 3x + 6$
(c)	x^2	$\sqrt{x - 5}$	$\sqrt{x^2 - 5}$
(d)	$\frac{x}{x - 1}$	$\frac{x}{x - 1}$	$\frac{\frac{x}{x-1}}{\frac{x}{x-1} - 1} = \frac{x}{x - (x-1)} = x$
(e)	$\frac{1}{x - 1}$	$1 + \frac{1}{x}$	x
(f)	$\frac{1}{x}$	$\frac{1}{x}$	x

13. (a) $f(g(x)) = \sqrt{\frac{1}{x} + 1} = \sqrt{\frac{1+x}{x}}$

$g(f(x)) = \frac{1}{\sqrt{x+1}}$

(b) Domain (f∘g): $(0, \infty)$, domain (g∘f): $(-1, \infty)$

(c) Range (f∘g): $(1, \infty)$, range (g∘f): $(0, \infty)$

15. (a) $y = -(x + 7)^2$ (b) $y = -(x - 4)^2$

17. (a) Position 4   (b) Position 1   (c) Position 2   (d) Position 3

19.

21.

23.

25.

27.

29.

31.

33.

35.

37.

39.

41.

43.

45.

47.

49. (a) domain: $[0, 2]$; range: $[2, 3]$

(b) domain: $[0, 2]$; range: $[-1, 0]$

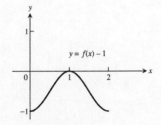

(c) domain: $[0, 2]$; range: $[0, 2]$

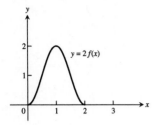

(d) domain: $[0, 2]$; range: $[-1, 0]$

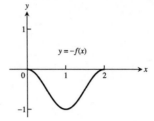

(e) domain: $[-2, 0]$; range: $[0, 1]$

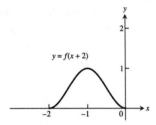

(f) domain: $[1, 3]$; range: $[0, 1]$

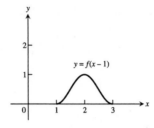

(g) domain: $[-2, 0]$; range: $[0, 1]$

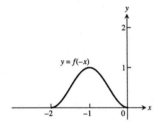

(h) domain: $[-1, 1]$; range: $[0, 1]$

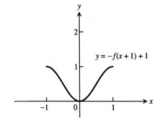

51. $y = 3x^2 - 3$

53. $y = \frac{1}{2}\left(1 + \frac{1}{x^2}\right) = \frac{1}{2} + \frac{1}{2x^2}$

55. $y = \sqrt{4x + 1}$

57. $y = \sqrt{4 - \left(\frac{x}{2}\right)^2} = \frac{1}{2}\sqrt{16 - x^2}$

59. $y = 1 - (3x)^3 = 1 - 27x^3$

61. Let $y = -\sqrt{2x + 1} = f(x)$ and let $g(x) = x^{1/2}$, $h(x) = \left(x + \frac{1}{2}\right)^{1/2}$, $i(x) = \sqrt{2}\left(x + \frac{1}{2}\right)^{1/2}$, and

$j(x) = -\left[\sqrt{2}\left(x + \frac{1}{2}\right)^{1/2}\right] = f(x)$. The graph of $h(x)$ is the graph of $g(x)$ shifted left $\frac{1}{2}$ unit; the graph of $i(x)$ is the graph

of $h(x)$ stretched vertically by a factor of $\sqrt{2}$; and the graph of $j(x) = f(x)$ is the graph of $i(x)$ reflected across the x-axis.

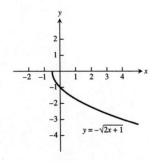

63. $y = f(x) = x^3$. Shift $f(x)$ one unit right followed by a shift two units up to get $g(x) = (x - 1)^3 + 2$.

65. Compress the graph of $f(x) = \frac{1}{x}$ horizontally by a factor of 2 to get $g(x) = \frac{1}{2x}$. Then shift $g(x)$ vertically down 1 unit to get $h(x) = \frac{1}{2x} - 1$.

67. Reflect the graph of $y = f(x) = \sqrt[3]{x}$ across the x-axis to get $g(x) = -\sqrt[3]{x}$.

69.

71. $9x^2 + 25y^2 = 225 \Rightarrow \frac{x^2}{5^2} + \frac{y^2}{3^2} = 1$

73. $3x^2 + (y-2)^2 = 3 \Rightarrow \frac{x^2}{1^2} + \frac{(y-2)^2}{(\sqrt{3})^2} = 1$

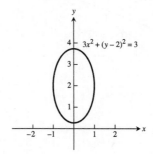

75. $3(x-1)^2 + 2(y+2)^2 = 6$

$\Rightarrow \frac{(x-1)^2}{(\sqrt{2})^2} + \frac{[y-(-2)]^2}{(\sqrt{3})^2} = 1$

77. $\frac{x^2}{16} + \frac{y^2}{9} = 1$ has its center at $(0, 0)$. Shiftinig 4 units left and 3 units up gives the center at $(h, k) = (-4, 3)$. So the

equation is $\frac{[x-(-4)]^2}{4^2} + \frac{(y-3)^2}{3^2} = 1 \Rightarrow \frac{(x+4)^2}{4^2} + \frac{(y-3)^2}{3^2} = 1$. Center, C, is $(-4, 3)$, and major axis, $\overline{AB}$, is the segment from $(-8, 3)$ to $(0, 3)$.

79. (a) $(fg)(-x) = f(-x)g(-x) = f(x)(-g(x)) = -(fg)(x)$, odd

(b) $\left(\frac{f}{g}\right)(-x) = \frac{f(-x)}{g(-x)} = \frac{f(x)}{-g(x)} = -\left(\frac{f}{g}\right)(x)$, odd

(c) $\left(\frac{g}{f}\right)(-x) = \frac{g(-x)}{f(-x)} = \frac{-g(x)}{f(x)} = -\left(\frac{g}{f}\right)(x)$, odd

(d) $f^2(-x) = f(-x)f(-x) = f(x)f(x) = f^2(x)$, even

(e) $g^2(-x) = (g(-x))^2 = (-g(x))^2 = g^2(x)$, even

(f) $(f \circ g)(-x) = f(g(-x)) = f(-g(x)) = f(g(x)) = (f \circ g)(x)$, even

(g) $(g \circ f)(-x) = g(f(-x)) = g(f(x)) = (g \circ f)(x)$, even

(h) $(f \circ f)(-x) = f(f(-x)) = f(f(x)) = (f \circ f)(x)$, even

(i) $(g \circ g)(-x) = g(g(-x)) = g(-g(x)) = -g(g(x)) = -(g \circ g)(x)$, odd

81. (a)

(b)

(c)

(d)

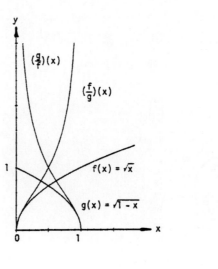

1.3 TRIGONOMETRIC FUNCTIONS

1. (a) $s = r\theta = (10)\left(\frac{4\pi}{5}\right) = 8\pi$ m

(b) $s = r\theta = (10)(110°)\left(\frac{\pi}{180°}\right) = \frac{110\pi}{18} = \frac{55\pi}{9}$ m

3. $\theta = 80° \Rightarrow \theta = 80°\left(\frac{\pi}{180°}\right) = \frac{4\pi}{9} \Rightarrow s = (6)\left(\frac{4\pi}{9}\right) = 8.4$ in. (since the diameter $= 12$ in. $\Rightarrow$ radius $= 6$ in.)

5.

θ	$-\pi$	$-\frac{2\pi}{3}$	0	$\frac{\pi}{2}$	$\frac{3\pi}{4}$
$\sin\theta$	0	$-\frac{\sqrt{3}}{2}$	0	1	$\frac{1}{\sqrt{2}}$
$\cos\theta$	-1	$-\frac{1}{2}$	1	0	$-\frac{1}{\sqrt{2}}$
$\tan\theta$	0	$\sqrt{3}$	0	und.	-1
$\cot\theta$	und.	$\frac{1}{\sqrt{3}}$	und.	0	-1
$\sec\theta$	-1	-2	1	und.	$-\sqrt{2}$
$\csc\theta$	und.	$-\frac{2}{\sqrt{3}}$	und.	1	$\sqrt{2}$

7. $\cos x = -\frac{4}{5}, \tan x = -\frac{3}{4}$

9. $\sin x = -\frac{\sqrt{8}}{3}, \tan x = -\sqrt{8}$

11. $\sin x = -\frac{1}{\sqrt{5}}, \cos x = -\frac{2}{\sqrt{5}}$

13.

period $= \pi$

15.

period $= 2$

17.

period $= 6$

19.

period $= 2\pi$

21.

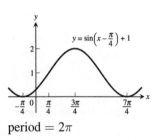

period $= 2\pi$

23. period $= \frac{\pi}{2}$, symmetric about the origin

25. period $= 4$, symmetric about the y-axis

27. (a) Cos x and sec x are positive in QI and QIV and
negative in QII and QIII. Sec x is undefined when
cos x is 0. The range of sec x is $(-\infty, -1] \cup [1, \infty)$;
the range of cos x is $[-1, 1]$.

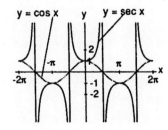

(b) Sin x and csc x are positive in QI and QII and negative in QIII and QIV. Csc x is undefined when sin x is 0. The range of csc x is $(-\infty, -1] \cup [1, \infty)$; the range of sin x is $[-1, 1]$.

29. D: $-\infty < x < \infty$; R: $y = -1, 0, 1$

31. $\cos\left(x - \frac{\pi}{2}\right) = \cos x \cos\left(-\frac{\pi}{2}\right) - \sin x \sin\left(-\frac{\pi}{2}\right) = (\cos x)(0) - (\sin x)(-1) = \sin x$

33. $\sin\left(x + \frac{\pi}{2}\right) = \sin x \cos\left(\frac{\pi}{2}\right) + \cos x \sin\left(\frac{\pi}{2}\right) = (\sin x)(0) + (\cos x)(1) = \cos x$

35. $\cos(A - B) = \cos(A + (-B)) = \cos A \cos(-B) - \sin A \sin(-B) = \cos A \cos B - \sin A(-\sin B)$
    $= \cos A \cos B + \sin A \sin B$

37. If $B = A$, $A - B = 0 \Rightarrow \cos(A - B) = \cos 0 = 1$. Also $\cos(A - B) = \cos(A - A) = \cos A \cos A + \sin A \sin A$
    $= \cos^2 A + \sin^2 A$. Therefore, $\cos^2 A + \sin^2 A = 1$.

39. $\cos(\pi + x) = \cos \pi \cos x - \sin \pi \sin x = (-1)(\cos x) - (0)(\sin x) = -\cos x$

41. $\sin\left(\frac{3\pi}{2} - x\right) = \sin\left(\frac{3\pi}{2}\right)\cos(-x) + \cos\left(\frac{3\pi}{2}\right)\sin(-x) = (-1)(\cos x) + (0)(\sin(-x)) = -\cos x$

43. $\sin \frac{7\pi}{12} = \sin\left(\frac{\pi}{4} + \frac{\pi}{3}\right) = \sin\frac{\pi}{4}\cos\frac{\pi}{3} + \cos\frac{\pi}{4}\sin\frac{\pi}{3} = \left(\frac{\sqrt{2}}{2}\right)\left(\frac{1}{2}\right) + \left(\frac{\sqrt{2}}{2}\right)\left(\frac{\sqrt{3}}{2}\right) = \frac{\sqrt{6}+\sqrt{2}}{4}$

45. $\cos \frac{\pi}{12} = \cos\left(\frac{\pi}{3} - \frac{\pi}{4}\right) = \cos\frac{\pi}{3}\cos\left(-\frac{\pi}{4}\right) - \sin\frac{\pi}{3}\sin\left(-\frac{\pi}{4}\right) = \left(\frac{1}{2}\right)\left(\frac{\sqrt{2}}{2}\right) - \left(\frac{\sqrt{3}}{2}\right)\left(-\frac{\sqrt{2}}{2}\right) = \frac{1+\sqrt{3}}{2\sqrt{2}}$

47. $\cos^2 \frac{\pi}{8} = \frac{1+\cos\left(\frac{2\pi}{8}\right)}{2} = \frac{1+\frac{\sqrt{2}}{2}}{2} = \frac{2+\sqrt{2}}{4}$

49. $\sin^2 \frac{\pi}{12} = \frac{1-\cos\left(\frac{2\pi}{12}\right)}{2} = \frac{1-\frac{\sqrt{3}}{2}}{2} = \frac{2-\sqrt{3}}{4}$

51. $\tan(A + B) = \frac{\sin(A+B)}{\cos(A+B)} = \frac{\sin A \cos B + \cos A \cos B}{\cos A \cos B - \sin A \sin B} = \frac{\frac{\sin A \cos B}{\cos A \cos B} + \frac{\cos A \sin B}{\cos A \cos B}}{\frac{\cos A \cos B}{\cos A \cos B} - \frac{\sin A \sin B}{\cos A \cos B}} = \frac{\tan A + \tan B}{1 - \tan A \tan B}$

53. According to the figure in the text, we have the following: By the law of cosines, $c^2 = a^2 + b^2 - 2ab \cos \theta$
    $= 1^2 + 1^2 - 2\cos(A - B) = 2 - 2\cos(A - B)$. By distance formula, $c^2 = (\cos A - \cos B)^2 + (\sin A - \sin B)^2$
    $= \cos^2 A - 2\cos A \cos B + \cos^2 B + \sin^2 A - 2\sin A \sin B + \sin^2 B = 2 - 2(\cos A \cos B + \sin A \sin B)$. Thus
    $c^2 = 2 - 2\cos(A - B) = 2 - 2(\cos A \cos B + \sin A \sin B) \Rightarrow \cos(A - B) = \cos A \cos B + \sin A \sin B$.

55. $c^2 = a^2 + b^2 - 2ab \cos C = 2^2 + 3^2 - 2(2)(3)\cos(60°) = 4 + 9 - 12\cos(60°) = 13 - 12\left(\frac{1}{2}\right) = 7$.
    Thus, $c = \sqrt{7} \approx 2.65$.

57. From the figures in the text, we see that $\sin B = \frac{h}{c}$. If C is an acute angle, then $\sin C = \frac{h}{b}$. On the other hand, if C is obtuse (as in the figure on the right), then $\sin C = \sin(\pi - C) = \frac{h}{b}$. Thus, in either case,
$$h = b\sin C = c\sin B \;\Rightarrow\; ah = ab\sin C = ac\sin B.$$
By the law of cosines, $\cos C = \frac{a^2+b^2-c^2}{2ab}$ and $\cos B = \frac{a^2+c^2-b^2}{2ac}$. Moreover, since the sum of the interior angles of a triangle is π, we have $\sin A = \sin(\pi - (B+C)) = \sin(B+C) = \sin B \cos C + \cos B \sin C$
$$= \left(\frac{h}{c}\right)\left[\frac{a^2+b^2-c^2}{2ab}\right] + \left[\frac{a^2+c^2-b^2}{2ac}\right]\left(\frac{h}{b}\right) = \left(\frac{h}{2abc}\right)(2a^2 + b^2 - c^2 + c^2 - b^2) = \frac{ah}{bc} \;\Rightarrow\; ah = bc\sin A.$$
Combining our results we have $ah = ab\sin C$, $ah = ac\sin B$, and $ah = bc\sin A$. Dividing by abc gives
$$\underbrace{\frac{h}{bc} = \frac{\sin A}{a} = \frac{\sin C}{c} = \frac{\sin B}{b}}_{\text{law of sines}}.$$

59. From the figure at the right and the law of cosines,

$$b^2 = a^2 + 2^2 - 2(2a)\cos B$$
$$= a^2 + 4 - 4a\left(\tfrac{1}{2}\right) = a^2 - 2a + 4.$$
Applying the law of sines to the figure, $\frac{\sin A}{a} = \frac{\sin B}{b}$
$$\Rightarrow \frac{\sqrt{2}/2}{a} = \frac{\sqrt{3}/2}{b} \;\Rightarrow\; b = \sqrt{\tfrac{3}{2}}\,a.$$ Thus, combining results,
$$a^2 - 2a + 4 = b^2 = \tfrac{3}{2}a^2 \;\Rightarrow\; 0 = \tfrac{1}{2}a^2 + 2a - 4$$
$$\Rightarrow 0 = a^2 + 4a - 8.$$ From the quadratic formula and the fact that $a > 0$, we have
$$a = \frac{-4 + \sqrt{4^2 - 4(1)(-8)}}{2} = \frac{4\sqrt{3}-4}{2} \simeq 1.464.$$

61. $A = 2, B = 2\pi, C = -\pi, D = -1$

63. $A = -\frac{2}{\pi}, B = 4, C = 0, D = \frac{1}{\pi}$

1.4 EXPONENTIAL FUNCTIONS

1.

3.

5.

7.

9.

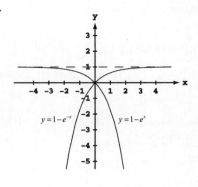

11. $16^2 \cdot 16^{-1.75} = 16^{2+(-1.75)} = 16^{0.25} = 16^{1/4} = 2$

13. $\frac{4^{4.2}}{4^{3.7}} = 4^{4.2-3.7} = 4^{0.5} = 4^{1/2} = 2$

15. $\left(25^{1/8}\right)^4 = 25^{4/8} = 25^{1/2} = 5$

17. $2^{\sqrt{3}} \cdot 7^{\sqrt{3}} = (2 \cdot 7)^{\sqrt{3}} = 14^{\sqrt{3}}$

19. $\left(\frac{2}{\sqrt{2}}\right)^4 = \frac{2^4}{(2^{1/2})^4} = \frac{16}{2^2} = 4$

21. Domain: $(-\infty, \infty)$; y in range $\Rightarrow y = \frac{1}{2+e^x}$. As x increases, e^x becomes infinitely large and y becomes a smaller and smaller positive real number. As x decreases, e^x becomes a smaller and smaller positive real number, $y < \frac{1}{2}$, and y gets arbitrarily close to $\frac{1}{2} \Rightarrow$ Range: $\left(0, \frac{1}{2}\right)$.

23. Domain: $(-\infty, \infty)$; y in range $\Rightarrow y = \sqrt{1 + 3^{-t}}$. Since the values of 3^{-t} are $(0, \infty) \Rightarrow$ Range: $(1, \infty)$.

25.

[-6, 6] by [-2, 6]

$x \approx 2.3219$

27.

[-6, 6] by [-3, 5]

$x \approx -0.6309$

29. Let t be the number of years. Solving $500{,}000(1.0375)^t = 1{,}000{,}000$ graphically, we find that $t \approx 18.828$. The population will reach 1 million in about 19 years.

31. (a) $A(t) = 6.6\left(\frac{1}{2}\right)^{t/14}$

(b) Solving $A(t) = 1$ graphically, we find that $t \approx 38$. There will be 1 gram remaining after about 38.1145 days.

33. Let A be the amount of the initial investment, and let t be the number of years. We wish to solve $A(1.0625)^t = 2A$, which is equivalent to $1.0625^t = 2$. Solving graphically, we find that $t \approx 11.433$. It will take about 11.433 years. (If the interest is credited at the end of each year, it will take 12 years.)

35. After t hours, the population is $P(t) = 2^{t/0.5}$, or equivalently, $P(t) = 2^{2t}$. After 24 hours, the population is $P(24) = 2^{48} \approx 2.815 \times 10^{14}$ bacteria.

1.5 INVERSE FUNCTIONS AND LOGARITHMS

1. Yes one-to-one, the graph passes the horizontal test.

3. Not one-to-one since (for example) the horizontal line $y = 2$ intersects the graph twice.

5. Yes one-to-one, the graph passes the horizontal test

7. Domain: $0 < x \le 1$, Range: $0 \le y$

9. Domain: $-1 \le x \le 1$, Range: $-\frac{\pi}{2} \le y \le \frac{\pi}{2}$

11. (a) The graph is symmetric about $y = x$.

(b) $y = \sqrt{1 - x^2} \Rightarrow y^2 = 1 - x^2 \Rightarrow x^2 = 1 - y^2 \Rightarrow x = \sqrt{1 - y^2} \Rightarrow y = \sqrt{1 - x^2} = f^{-1}(x)$

13. Step 1: $y = x^2 + 1 \Rightarrow x^2 = y - 1 \Rightarrow x = \sqrt{y - 1}$
    Step 2: $y = \sqrt{x - 1} = f^{-1}(x)$

15. Step 1: $y = x^3 - 1 \Rightarrow x^3 = y + 1 \Rightarrow x = (y + 1)^{1/3}$
    Step 2: $y = \sqrt[3]{x + 1} = f^{-1}(x)$

17. Step 1: $y = (x + 1)^2 \Rightarrow \sqrt{y} = x + 1,$ since $x \geq -1 \Rightarrow x = \sqrt{y} - 1$
    Step 2: $y = \sqrt{x} - 1 = f^{-1}(x)$

19. Step 1: $y = x^5 \Rightarrow x = y^{1/5}$
    Step 2: $y = \sqrt[5]{x} = f^{-1}(x);$
    Domain and Range of f^{-1}: all reals;
    $f(f^{-1}(x)) = \left(x^{1/5}\right)^5 = x$ and $f^{-1}(f(x)) = \left(x^5\right)^{1/5} = x$

21. Step 1: $y = x^3 + 1 \Rightarrow x^3 = y - 1 \Rightarrow x = (y - 1)^{1/3}$
    Step 2: $y = \sqrt[3]{x - 1} = f^{-1}(x);$
    Domain and Range of f^{-1}: all reals;
    $f(f^{-1}(x)) = \left((x - 1)^{1/3}\right)^3 + 1 = (x - 1) + 1 = x$ and $f^{-1}(f(x)) = \left((x^3 + 1) - 1\right)^{1/3} = \left(x^3\right)^{1/3} = x$

23. Step 1: $y = \frac{1}{x^2} \Rightarrow x^2 = \frac{1}{y} \Rightarrow x = \frac{1}{\sqrt{y}}$
    Step 2: $y = \frac{1}{\sqrt{x}} = f^{-1}(x)$
    Domain of f^{-1}: $x > 0$, Range of f^{-1}: $y > 0$;
    $f(f^{-1}(x)) = \frac{1}{\left(\frac{1}{\sqrt{x}}\right)^2} = \frac{1}{\left(\frac{1}{x}\right)} = x$ and $f^{-1}(f(x)) = \frac{1}{\sqrt{\frac{1}{x^2}}} = \frac{1}{\left(\frac{1}{x}\right)} = x$ since $x > 0$

25. (a) $\ln 0.75 = \ln \frac{3}{4} = \ln 3 - \ln 4 = \ln 3 - \ln 2^2 = \ln 3 - 2 \ln 2$
    (b) $\ln \frac{4}{9} = \ln 4 - \ln 9 = \ln 2^2 - \ln 3^2 = 2 \ln 2 - 2 \ln 3$
    (c) $\ln \frac{1}{2} = \ln 1 - \ln 2 = -\ln 2$ \qquad (d) $\ln \sqrt[3]{9} = \frac{1}{3} \ln 9 = \frac{1}{3} \ln 3^2 = \frac{2}{3} \ln 3$
    (e) $\ln 3\sqrt{2} = \ln 3 + \ln 2^{1/2} = \ln 3 + \frac{1}{2} \ln 2$
    (f) $\ln \sqrt{13.5} = \frac{1}{2} \ln 13.5 = \frac{1}{2} \ln \frac{27}{2} = \frac{1}{2} (\ln 3^3 - \ln 2) = \frac{1}{2} (3 \ln 3 - \ln 2)$

27. (a) $\ln \sin \theta - \ln \left(\frac{\sin \theta}{5}\right) = \ln \left(\frac{\sin \theta}{\left(\frac{\sin \theta}{5}\right)}\right) = \ln 5$ \qquad (b) $\ln (3x^2 - 9x) + \ln \left(\frac{1}{3x}\right) = \ln \left(\frac{3x^2 - 9x}{3x}\right) = \ln (x - 3)$

    (c) $\frac{1}{2} \ln (4t^4) - \ln 2 = \ln \sqrt{4t^4} - \ln 2 = \ln 2t^2 - \ln 2 = \ln \left(\frac{2t^2}{2}\right) = \ln (t^2)$

29. (a) $e^{\ln 7.2} = 7.2$ \qquad (b) $e^{-\ln x^2} = \frac{1}{e^{\ln x^2}} = \frac{1}{x^2}$ \qquad (c) $e^{\ln x - \ln y} = e^{\ln(x/y)} = \frac{x}{y}$

31. (a) $2 \ln \sqrt{e} = 2 \ln e^{1/2} = (2) \left(\frac{1}{2}\right) \ln e = 1$ \qquad (b) $\ln (\ln e^e) = \ln (e \ln e) = \ln e = 1$
    (c) $\ln e^{(-x^2 - y^2)} = (-x^2 - y^2) \ln e = -x^2 - y^2$

33. $\ln y = 2t + 4 \Rightarrow e^{\ln y} = e^{2t + 4} \Rightarrow y = e^{2t + 4}$

35. $\ln (y - 40) = 5t \Rightarrow e^{\ln (y - 40)} = e^{5t} \Rightarrow y - 40 = e^{5t} \Rightarrow y = e^{5t} + 40$

37. $\ln(y-1) - \ln 2 = x + \ln x \Rightarrow \ln(y-1) - \ln 2 - \ln x = x \Rightarrow \ln\left(\frac{y-1}{2x}\right) = x \Rightarrow e^{\ln\left(\frac{y-1}{2x}\right)} = e^x \Rightarrow \frac{y-1}{2x} = e^x$
$\Rightarrow y - 1 = 2xe^x \Rightarrow y = 2xe^x + 1$

39. (a) $e^{2k} = 4 \Rightarrow \ln e^{2k} = \ln 4 \Rightarrow 2k \ln e = \ln 2^2 \Rightarrow 2k = 2 \ln 2 \Rightarrow k = \ln 2$
   (b) $100e^{10k} = 200 \Rightarrow e^{10k} = 2 \Rightarrow \ln e^{10k} = \ln 2 \Rightarrow 10k \ln e = \ln 2 \Rightarrow 10k = \ln 2 \Rightarrow k = \frac{\ln 2}{10}$
   (c) $e^{k/1000} = a \Rightarrow \ln e^{k/1000} = \ln a \Rightarrow \frac{k}{1000} \ln e = \ln a \Rightarrow \frac{k}{1000} = \ln a \Rightarrow k = 1000 \ln a$

41. (a) $e^{-0.3t} = 27 \Rightarrow \ln e^{-0.3t} = \ln 3^3 \Rightarrow (-0.3t) \ln e = 3 \ln 3 \Rightarrow -0.3t = 3 \ln 3 \Rightarrow t = -10 \ln 3$
   (b) $e^{kt} = \frac{1}{2} \Rightarrow \ln e^{kt} = \ln 2^{-1} = kt \ln e = -\ln 2 \Rightarrow t = -\frac{\ln 2}{k}$
   (c) $e^{(\ln 0.2)t} = 0.4 \Rightarrow \left(e^{\ln 0.2}\right)^t = 0.4 \Rightarrow 0.2^t = 0.4 \Rightarrow \ln 0.2^t = \ln 0.4 \Rightarrow t \ln 0.2 = \ln 0.4 \Rightarrow t = \frac{\ln 0.4}{\ln 0.2}$

43. $e^{\sqrt{t}} = x^2 \Rightarrow \ln e^{\sqrt{t}} = \ln x^2 \Rightarrow \sqrt{t} = 2 \ln x \Rightarrow t = 4(\ln x)^2$

45. (a) $5^{\log_5 7} = 7$ (b) $8^{\log_8 \sqrt{2}} = \sqrt{2}$ (c) $1.3^{\log_{1.3} 75} = 75$
   (d) $\log_4 16 = \log_4 4^2 = 2 \log_4 4 = 2 \cdot 1 = 2$ (e) $\log_3 \sqrt{3} = \log_3 3^{1/2} = \frac{1}{2} \log_3 3 = \frac{1}{2} \cdot 1 = \frac{1}{2} = 0.5$
   (f) $\log_4 \left(\frac{1}{4}\right) = \log_4 4^{-1} = -1 \log_4 4 = -1 \cdot 1 = -1$

47. (a) Let $z = \log_4 x \Rightarrow 4^z = x \Rightarrow 2^{2z} = x \Rightarrow (2^z)^2 = x \Rightarrow 2^z = \sqrt{x}$
   (b) Let $z = \log_3 x \Rightarrow 3^z = x \Rightarrow (3^z)^2 = x^2 \Rightarrow 3^{2z} = x^2 \Rightarrow 9^z = x^2$
   (c) $\log_2 \left(e^{(\ln 2) \sin x}\right) = \log_2 2^{\sin x} = \sin x$

49. (a) $\frac{\log_2 x}{\log_3 x} = \frac{\ln x}{\ln 2} \div \frac{\ln x}{\ln 3} = \frac{\ln x}{\ln 2} \cdot \frac{\ln 3}{\ln x} = \frac{\ln 3}{\ln 2}$ (b) $\frac{\log_2 x}{\log_8 x} = \frac{\ln x}{\ln 2} \div \frac{\ln x}{\ln 8} = \frac{\ln x}{\ln 2} \cdot \frac{\ln 8}{\ln x} = \frac{3 \ln 2}{\ln 2} = 3$
   (c) $\frac{\log_x a}{\log_{x^2} a} = \frac{\ln a}{\ln x} \div \frac{\ln a}{\ln x^2} = \frac{\ln a}{\ln x} \cdot \frac{\ln x^2}{\ln a} = \frac{2 \ln x}{\ln x} = 2$

51. (a) $-\frac{\pi}{6}$ (b) $\frac{\pi}{4}$ (c) $-\frac{\pi}{3}$

53. (a) $\arccos(-1) = \pi$ since $\cos(\pi) = -1$ and $0 \le \pi \le \pi$.
   (b) $\arccos(0) = \frac{\pi}{2}$ since $\cos\left(\frac{\pi}{2}\right) = 0$ and $0 \le \frac{\pi}{2} \le \pi$.

55. The function g(x) is also one-to-one. The reasoning: f(x) is one-to-one means that if $x_1 \ne x_2$ then $f(x_1) \ne f(x_2)$, so $-f(x_1) \ne -f(x_2)$ and therefore $g(x_1) \ne g(x_2)$. Therefore g(x) is one-to-one as well.

57. The composite is one-to-one also. The reasoning: If $x_1 \ne x_2$ then $g(x_1) \ne g(x_2)$ because g is one-to-one. Since $g(x_1) \ne g(x_2)$, we also have $f(g(x_1)) \ne f(g(x_2))$ because f is one-to-one. Thus, $f \circ g$ is one-to-one because $x_1 \ne x_2 \Rightarrow f(g(x_1)) \ne f(g(x_2))$.

59. (a) $y = \frac{100}{1+2^{-x}} \rightarrow 1 + 2^{-x} = \frac{100}{y} \rightarrow 2^{-x} = \frac{100}{y} - 1 \rightarrow \log_2(2^{-x}) = \log_2\left(\frac{100}{y} - 1\right) \rightarrow -x = \log_2\left(\frac{100}{y} - 1\right)$
   $x = -\log_2\left(\frac{100}{y} - 1\right) = -\log_2\left(\frac{100-y}{y}\right) = \log_2\left(\frac{y}{100-y}\right)$.
   Interchange x and y: $y = \log_2\left(\frac{x}{100-x}\right) \rightarrow f^{-1}(x) = \log_2\left(\frac{x}{100-x}\right)$
   Verify.
   $(f \circ f^{-1})(x) = f\left(\log_2\left(\frac{x}{100-x}\right)\right) = \frac{100}{1+2^{-\log_2\left(\frac{x}{100-x}\right)}} = \frac{100}{1+2^{\log_2\left(\frac{100-x}{x}\right)}} = \frac{100}{1+\frac{100-x}{x}} = \frac{100x}{x+100-x} = \frac{100x}{100} = x$
   $(f^{-1} \circ f)(x) = f^{-1}\left(\frac{100}{1+2^{-x}}\right) = \log_2\left(\frac{\frac{100}{1+2^{-x}}}{100-\frac{100}{1+2^{-x}}}\right) = \log_2\left(\frac{100}{100(1+2^{-x})-100}\right) = \log_2\left(\frac{1}{2^{-x}}\right) = \log_2(2^x) = x$

(b) $y = \frac{50}{1+1.1^{-x}} \rightarrow 1 + 1.1^{-x} = \frac{50}{y} \rightarrow 1.1^{-x} = \frac{50}{y} - 1 \rightarrow \log_{1.1}\left(1.1^{-x}\right) = \log_{1.1}\left(\frac{50}{y} - 1\right) \rightarrow -x = \log_{1.1}\left(\frac{50}{y} - 1\right)$

$x = -\log_{1.1}\left(\frac{50}{y} - 1\right) = -\log_{1.1}\left(\frac{50-y}{y}\right) = \log_{1.1}\left(\frac{y}{50-y}\right).$

Interchange x and y: $y = \log_{1.1}\left(\frac{x}{50-x}\right) \rightarrow f^{-1}(x) = \log_{1.1}\left(\frac{x}{50-x}\right)$

Verify.

$(f \circ f^{-1})(x) = f\left(\log_{1.1}\left(\frac{x}{50-x}\right)\right) = \frac{50}{1+1.1^{-\log_{1.1}\left(\frac{x}{50-x}\right)}} = \frac{50}{1+1.1^{\log_{1.1}\left(\frac{50-x}{x}\right)}} = \frac{50}{1+\frac{50-x}{x}} = \frac{50x}{x+50-x} = \frac{50x}{50} = x$

$(f^{-1} \circ f)(x) = f^{-1}\left(\frac{50}{1+1.1^{-x}}\right) = \log_{1.1}\left(\frac{\frac{50}{1+1.1^{-x}}}{50-\frac{50}{1+1.1^{-x}}}\right) = \log_{1.1}\left(\frac{50}{50(1+1.1^{-x})-50}\right) = \log_{1.1}\left(\frac{1}{1.1^{-x}}\right) = \log_{1.1}(1.1^x) = x$

61. (a) Amount $= 8\left(\frac{1}{2}\right)^{t/12}$

(b) $8\left(\frac{1}{2}\right)^{t/12} = 1 \rightarrow \left(\frac{1}{2}\right)^{t/12} = \frac{1}{8} \rightarrow \left(\frac{1}{2}\right)^{t/12} = \left(\frac{1}{2}\right)^3 \rightarrow \frac{t}{12} = 3 \rightarrow t = 36$

There will be 1 gram remaining after 36 hours.

63. $375,000(1.0225)^t = 1,000,000 \rightarrow 1.0225^t = \frac{8}{3} \rightarrow \ln(1.0225^t) = \ln\left(\frac{8}{3}\right) \rightarrow t\ln(1.0225) = \ln\left(\frac{8}{3}\right) \rightarrow t = \frac{\ln\left(\frac{8}{3}\right)}{\ln(1.0225)} \approx 44.081$

It will take about 44.081 years.

65. From zooming in on the graph at the right, we estimate
the third root to be x ≈ −0.76666

67. (a) Begin with $y = \ln x$ and reduce the y-value by $3 \Rightarrow y = \ln x - 3$.

(b) Begin with $y = \ln x$ and replace x with $x - 1 \Rightarrow y = \ln(x - 1)$.

(c) Begin with $y = \ln x$, replace x with $x + 1$, and increase the y-value by $3 \Rightarrow y = \ln(x + 1) + 3$.

(d) Begin with $y = \ln x$, reduce the y-value by 4, and replace x with $x - 2 \Rightarrow y = \ln(x - 2) - 4$.

(e) Begin with $y = \ln x$ and replace x with $-x \Rightarrow y = \ln(-x)$.

(f) Begin with $y = \ln x$ and switch x and $y \Rightarrow x = \ln y$ or $y = e^x$.

69. There are a few cases to consider. First, if $x = 0$ then $\cos(\sin^{-1}(x)) = \cos(\sin^{-1}(0)) = \cos(0) = 1 = \sqrt{1-0}$

$= \sqrt{1-x^2}$. Note that if $x \neq 0$ then either $x > 0$ or $x < 0$. However, since $\sin(x)$ is an odd function, so is $\sin^{-1}(x)$ odd. Now $\cos(x)$ is even so that $\cos(\sin^{-1}(x)) = \cos(-\sin^{-1}(-x)) = \cos(\sin^{-1}(-x))$. Thus, it suffices to assume only that $x > 0$; the case $x < 0$ follows. So consider the given right triangle with $\theta = \sin^{-1}(x)$. We have $\sin\theta = x = \frac{x}{1}$

$= \frac{\text{opposite}}{\text{hypotenuse}}$. Thus $\cos(\sin^{-1}(x)) = \cos\theta = \frac{\text{adjacent}}{\text{hypotenuse}} = \sqrt{1-x^2}$.

1.6 GRAPHING WITH CALCULATORS AND COMPUTERS

1-4.   The most appropriate viewing window displays the maxima, minima, intercepts, and end behavior of the graphs and has little unused space.

1. d.

3. d.

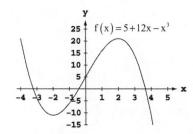

5-30.   For any display there are many appropriate display widows. The graphs given as answers in Exercises 5−30 are not unique in appearance.

5. $[-2, 5]$ by $[-15, 40]$

7. $[-2, 6]$ by $[-250, 50]$

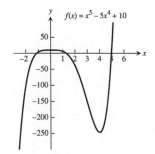

9. $[-4, 4]$ by $[-5, 5]$

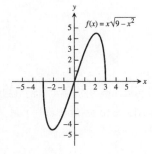

11. $[-2, 6]$ by $[-5, 4]$

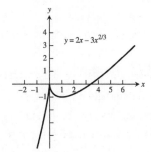

13. $[-1, 6]$ by $[-1, 4]$

15. $[-3, 3]$ by $[0, 10]$

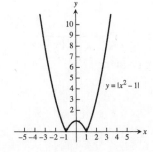

17. $[-5, 1]$ by $[-5, 5]$

19. $[-4, 4]$ by $[0, 3]$

21. $[-10, 10]$ by $[-6, 6]$

23. $[-6, 10]$ by $[-6, 6]$

25. $[-0.03, 0.03]$ by $[-1.25, 1.25]$

27. $[-300, 300]$ by $[-1.25, 1.25]$

29. $[-0.25, 0.25]$ by $[-0.3, 0.3]$

31. $x^2 + 2x = 4 + 4y - y^2 \Rightarrow y = 2 \pm \sqrt{-x^2 - 2x + 8}$.
    The lower half is produced by graphing
    $y = 2 - \sqrt{-x^2 - 2x + 8}$.

33.

35.

37.

39.

41.

43.

45.

47.

49.

51.

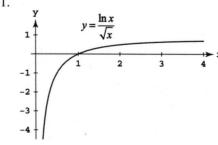

NOTES:

CHAPTER 2 LIMITS AND CONTINUITY

2.1 RATES OF CHANGE AND TANGENTS TO CURVES

1. (a) $\frac{\Delta f}{\Delta x} = \frac{f(3) - f(2)}{3 - 2} = \frac{28 - 9}{1} = 19$ (b) $\frac{\Delta f}{\Delta x} = \frac{f(1) - f(-1)}{1 - (-1)} = \frac{2 - 0}{2} = 1$

3. (a) $\frac{\Delta h}{\Delta t} = \frac{h\left(\frac{3\pi}{4}\right) - h\left(\frac{\pi}{4}\right)}{\frac{3\pi}{4} - \frac{\pi}{4}} = \frac{-1 - 1}{\frac{\pi}{2}} = -\frac{4}{\pi}$ (b) $\frac{\Delta h}{\Delta t} = \frac{h\left(\frac{\pi}{2}\right) - h\left(\frac{\pi}{6}\right)}{\frac{\pi}{2} - \frac{\pi}{6}} = \frac{0 - \sqrt{3}}{\frac{\pi}{3}} = \frac{-3\sqrt{3}}{\pi}$

5. $\frac{\Delta R}{\Delta \theta} = \frac{R(2) - R(0)}{2 - 0} = \frac{\sqrt{8+1} - \sqrt{1}}{2} = \frac{3 - 1}{2} = 1$

7. (a)

Q	Slope of PQ $= \frac{\Delta p}{\Delta t}$
$Q_1(10, 225)$	$\frac{650 - 225}{20 - 10} = 42.5$ m/sec
$Q_2(14, 375)$	$\frac{650 - 375}{20 - 14} = 45.83$ m/sec
$Q_3(16.5, 475)$	$\frac{650 - 475}{20 - 16.5} = 50.00$ m/sec
$Q_4(18, 550)$	$\frac{650 - 550}{20 - 18} = 50.00$ m/sec

   (b) At $t = 20$, the sportscar was traveling approximately 50 m/sec or 180 km/h.

9. (a) $\frac{\Delta y}{\Delta x} = \frac{((2+h)^2 - 3) - (2^2 - 3)}{h} = \frac{4 + 4h + h^2 - 3 - 1}{h} = \frac{4h + h^2}{h} = 4 + 4h$. As $h \to 0, 4 + 4h \to 4 \Rightarrow$ at $P(2, 1)$ the slope is 4.
   (b) $y - 1 = 4(x - 2) \Rightarrow y - 1 = 4x - 8 \Rightarrow y = 4x - 7$

11. (a) $\frac{\Delta y}{\Delta x} = \frac{((2+h)^2 - 2(2+h) - 3) - (2^2 - 2(2) - 3)}{h} = \frac{4 + 4h + h^2 - 4 - 2h - 3 - (-3)}{h} = \frac{2h + h^2}{h} = 2 + h$. As $h \to 0, 2 + h \to 2 \Rightarrow$ at $P(2, -3)$ the slope is 2.
   (b) $y - (-3) = 2(x - 2) \Rightarrow y + 3 = 2x - 4 \Rightarrow y = 2x - 7$.

13. (a) $\frac{\Delta y}{\Delta x} = \frac{(2+h)^3 - 2^3}{h} = \frac{8 + 12h + 4h^2 + h^3 - 8}{h} = \frac{12h + 4h^2 + h^3}{h} = 12 + 4h + h^2$. As $h \to 0, 12 + 4h + h^2 \to 12, \Rightarrow$ at $P(2, 8)$ the slope is 12.
   (b) $y - 8 = 12(x - 2) \Rightarrow y - 8 = 12x - 24 \Rightarrow y = 12x - 16$.

15. (a) $\frac{\Delta y}{\Delta x} = \frac{(1+h)^3 - 12(1+h) - (1^3 - 12(1))}{h} = \frac{1 + 3h + 3h^2 + h^3 - 12 - 12h - (-11)}{h} = \frac{-9h + 3h^2 + h^3}{h} = -9 + 3h + h^2$. As $h \to 0,$
   $-9 + 3h + h^2 \to -9 \Rightarrow$ at $P(1, -11)$ the slope is -9.
   (b) $y - (-11) = (-9)(x - 1) \Rightarrow y + 11 = -9x + 9 \Rightarrow y = -9x - 2$.

17. (a)

   (b) $\frac{\Delta p}{\Delta t} = \frac{174 - 62}{2004 - 2002} = \frac{112}{2} = 56$ thousand dollars per year
   (c) The average rate of change from 2001 to 2002 is $\frac{\Delta p}{\Delta t} = \frac{62 - 27}{20022 - 2001} = 35$ thousand dollars per year.

   The average rate of change from 2002 to 2003 is $\frac{\Delta p}{\Delta t} = \frac{111 - 62}{2003 - 2002} = 49$ thousand dollars per year.

   So, the rate at which profits were changing in 2002 is approximatley $\frac{1}{2}(35 + 49) = 42$ thousand dollars per year.

19. (a) $\frac{\Delta g}{\Delta x} = \frac{g(2) - g(1)}{2 - 1} = \frac{\sqrt{2} - 1}{2 - 1} \approx 0.414213$ $\frac{\Delta g}{\Delta x} = \frac{g(1.5) - g(1)}{1.5 - 1} = \frac{\sqrt{1.5} - 1}{0.5} \approx 0.449489$

   $\frac{\Delta g}{\Delta x} = \frac{g(1 + h) - g(1)}{(1 + h) - 1} = \frac{\sqrt{1 + h} - 1}{h}$

   (b) $g(x) = \sqrt{x}$

$1 + h$	1.1	1.01	1.001	1.0001	1.00001	1.000001
$\sqrt{1 + h}$	1.04880	1.004987	1.0004998	1.0000499	1.000005	1.0000005
$\left(\sqrt{1 + h} - 1\right) / h$	0.4880	0.4987	0.4998	0.499	0.5	0.5

   (c) The rate of change of g(x) at x = 1 is 0.5.

   (d) The calculator gives $\lim\limits_{h \to 0} \frac{\sqrt{1 + h} - 1}{h} = \frac{1}{2}$.

2.2 LIMIT OF A FUNCTION AND LIMIT LAWS

1. (a) Does not exist. As x approaches 1 from the right, g(x) approaches 0. As x approaches 1 from the left, g(x) approaches 1. There is no single number L that all the values g(x) get arbitrarily close to as x → 1.

   (b) 1

   (c) 0

3. (a) True   (b) True   (c) False

   (d) False   (e) False   (f) True

5. $\lim\limits_{x \to 0} \frac{x}{|x|}$ does not exist because $\frac{x}{|x|} = \frac{x}{x} = 1$ if $x > 0$ and $\frac{x}{|x|} = \frac{x}{-x} = -1$ if $x < 0$. As x approaches 0 from the left, $\frac{x}{|x|}$ approaches −1. As x approaches 0 from the right, $\frac{x}{|x|}$ approaches 1. There is no single number L that all the function values get arbitrarily close to as x → 0.

7. Nothing can be said about f(x) because the existence of a limit as x → x_0 does not depend on how the function is defined at x_0. In order for a limit to exist, f(x) must be arbitrarily close to a single real number L when x is close enough to x_0. That is, the existence of a limit depends on the values of f(x) for x <u>near</u> x_0, not on the definition of f(x) at x_0 itself.

9. No, the definition does not require that f be defined at x = 1 in order for a limiting value to exist there. If f(1) is defined, it can be any real number, so we can conclude nothing about f(1) from $\lim\limits_{x \to 1} f(x) = 5$.

11. $\lim\limits_{x \to -7} (2x + 5) = 2(-7) + 5 = -14 + 5 = -9$

13. $\lim\limits_{x \to 2} (-x^2 + 5x - 2) = -(2)^2 + 5(2) - 2 = -4 + 10 - 2 = 4$

15. $\lim\limits_{t \to 6} 8(t - 5)(t - 7) = 8(6 - 5)(6 - 7) = -8$ 17. $\lim\limits_{x \to 2} \frac{x + 3}{x + 6} = \frac{2 + 3}{2 + 6} = \frac{5}{8}$

19. $\lim\limits_{y \to -5} \frac{y^2}{5 - y} = \frac{(-5)^2}{5 - (-5)} = \frac{25}{10} = \frac{5}{2}$

21. $\lim\limits_{x \to -1} 3(2x - 1)^2 = 3(2(-1) - 1)^2 = 3(-3)^2 = 27$

23. $\lim\limits_{y \to -3} (5 - y)^{4/3} = [5 - (-3)]^{4/3} = (8)^{4/3} = \left((8)^{1/3}\right)^4 = 2^4 = 16$

25. $\displaystyle\lim_{h \to 0} \frac{3}{\sqrt{3h+1}+1} = \frac{3}{\sqrt{3(0)+1}+1} = \frac{3}{\sqrt{1}+1} = \frac{3}{2}$

27. $\displaystyle\lim_{h \to 0} \frac{\sqrt{3h+1}-1}{h} = \lim_{h \to 0} \frac{\sqrt{3h+1}-1}{h} \cdot \frac{\sqrt{3h+1}+1}{\sqrt{3h+1}+1} = \lim_{h \to 0} \frac{(3h+1)-1}{h\left(\sqrt{3h+1}+1\right)} = \lim_{h \to 0} \frac{3h}{h\left(\sqrt{3h+1}+1\right)} = \lim_{h \to 0} \frac{3}{\sqrt{3h+1}+1}$

   $= \frac{3}{\sqrt{1}+1} = \frac{3}{2}$

29. $\displaystyle\lim_{x \to 5} \frac{x-5}{x^2-25} = \lim_{x \to 5} \frac{x-5}{(x+5)(x-5)} = \lim_{x \to 5} \frac{1}{x+5} = \frac{1}{5+5} = \frac{1}{10}$

31. $\displaystyle\lim_{x \to -5} \frac{x^2+3x-10}{x+5} = \lim_{x \to -5} \frac{(x+5)(x-2)}{x+5} = \lim_{x \to -5} (x-2) = -5 - 2 = -7$

33. $\displaystyle\lim_{t \to 1} \frac{t^2+t-2}{t^2-1} = \lim_{t \to 1} \frac{(t+2)(t-1)}{(t-1)(t+1)} = \lim_{t \to 1} \frac{t+2}{t+1} = \frac{1+2}{1+1} = \frac{3}{2}$

35. $\displaystyle\lim_{x \to -2} \frac{-2x-4}{x^3+2x^2} = \lim_{x \to -2} \frac{-2(x+2)}{x^2(x+2)} = \lim_{x \to -2} \frac{-2}{x^2} = \frac{-2}{4} = -\frac{1}{2}$

37. $\displaystyle\lim_{u \to 1} \frac{u^4-1}{u^3-1} = \lim_{u \to 1} \frac{(u^2+1)(u+1)(u-1)}{(u^2+u+1)(u-1)} = \lim_{u \to 1} \frac{(u^2+1)(u+1)}{u^2+u+1} = \frac{(1+1)(1+1)}{1+1+1} = \frac{4}{3}$

39. $\displaystyle\lim_{x \to 9} \frac{\sqrt{x}-3}{x-9} = \lim_{x \to 9} \frac{\sqrt{x}-3}{(\sqrt{x}-3)(\sqrt{x}+3)} = \lim_{x \to 9} \frac{1}{\sqrt{x}+3} = \frac{1}{\sqrt{9}+3} = \frac{1}{6}$

41. $\displaystyle\lim_{x \to 1} \frac{x-1}{\sqrt{x+3}-2} = \lim_{x \to 1} \frac{(x-1)\left(\sqrt{x+3}+2\right)}{(\sqrt{x+3}-2)(\sqrt{x+3}+2)} = \lim_{x \to 1} \frac{(x-1)\left(\sqrt{x+3}+2\right)}{(x+3)-4} = \lim_{x \to 1} \left(\sqrt{x+3}+2\right)$

   $= \sqrt{4}+2 = 4$

43. $\displaystyle\lim_{x \to 2} \frac{\sqrt{x^2+12}-4}{x-2} = \lim_{x \to 2} \frac{\left(\sqrt{x^2+12}-4\right)\left(\sqrt{x^2+12}+4\right)}{(x-2)\left(\sqrt{x^2+12}+4\right)} = \lim_{x \to 2} \frac{(x^2+12)-16}{(x-2)\left(\sqrt{x^2+12}+4\right)}$

   $= \lim_{x \to 2} \frac{(x-2)(x+2)}{(x-2)\left(\sqrt{x^2+12}+4\right)} = \lim_{x \to 2} \frac{x+2}{\sqrt{x^2+12}+4} = \frac{4}{\sqrt{16}+4} = \frac{1}{2}$

45. $\displaystyle\lim_{x \to -3} \frac{2-\sqrt{x^2-5}}{x+3} = \lim_{x \to -3} \frac{\left(2-\sqrt{x^2-5}\right)\left(2+\sqrt{x^2-5}\right)}{(x+3)\left(2+\sqrt{x^2-5}\right)} = \lim_{x \to -3} \frac{4-(x^2-5)}{(x+3)\left(2+\sqrt{x^2-5}\right)}$

   $= \lim_{x \to -3} \frac{9-x^2}{(x+3)\left(2+\sqrt{x^2-5}\right)} = \lim_{x \to -3} \frac{(3-x)(3+x)}{(x+3)\left(2+\sqrt{x^2-5}\right)} = \lim_{x \to -3} \frac{3-x}{2+\sqrt{x^2-5}} = \frac{6}{2+\sqrt{4}} = \frac{3}{2}$

47. $\displaystyle\lim_{x \to 0} (2\sin x - 1) = 2\sin 0 - 1 = 0 - 1 = -1$

49. $\displaystyle\lim_{x \to 0} \sec x = \lim_{x \to 0} \frac{1}{\cos x} = \frac{1}{\cos 0} = \frac{1}{1} = 1$

51. $\displaystyle\lim_{x \to 0} \frac{1+x+\sin x}{3\cos x} = \frac{1+0+\sin 0}{3\cos 0} = \frac{1+0+0}{3} = \frac{1}{3}$

53. $\displaystyle\lim_{x \to 0} \sqrt{x+1}\,\cos^{1/3} x = \lim_{x \to 0} \sqrt{x+1} \cdot \sqrt[3]{\cos x} = \sqrt{0+1} \cdot \sqrt[3]{\cos 0} = \sqrt{1} \cdot \sqrt[3]{1} = 1 \cdot 1 = 1$

55. (a) quotient rule

   (b) difference and power rules

   (c) sum and constant multiple rules

57. (a) $\lim\limits_{x \to c} f(x)\,g(x) = \left[\lim\limits_{x \to c} f(x)\right]\left[\lim\limits_{x \to c} g(x)\right] = (5)(-2) = -10$

(b) $\lim\limits_{x \to c} 2f(x)\,g(x) = 2\left[\lim\limits_{x \to c} f(x)\right]\left[\lim\limits_{x \to c} g(x)\right] = 2(5)(-2) = -20$

(c) $\lim\limits_{x \to c} [f(x) + 3g(x)] = \lim\limits_{x \to c} f(x) + 3\lim\limits_{x \to c} g(x) = 5 + 3(-2) = -1$

(d) $\lim\limits_{x \to c} \frac{f(x)}{f(x) - g(x)} = \frac{\lim\limits_{x \to c} f(x)}{\lim\limits_{x \to c} f(x) - \lim\limits_{x \to c} g(x)} = \frac{5}{5 - (-2)} = \frac{5}{7}$

59. (a) $\lim\limits_{x \to b} [f(x) + g(x)] = \lim\limits_{x \to b} f(x) + \lim\limits_{x \to b} g(x) = 7 + (-3) = 4$

(b) $\lim\limits_{x \to b} f(x) \cdot g(x) = \left[\lim\limits_{x \to b} f(x)\right]\left[\lim\limits_{x \to b} g(x)\right] = (7)(-3) = -21$

(c) $\lim\limits_{x \to b} 4g(x) = \left[\lim\limits_{x \to b} 4\right]\left[\lim\limits_{x \to b} g(x)\right] = (4)(-3) = -12$

(d) $\lim\limits_{x \to b} f(x)/g(x) = \lim\limits_{x \to b} f(x) / \lim\limits_{x \to b} g(x) = \frac{7}{-3} = -\frac{7}{3}$

61. $\lim\limits_{h \to 0} \frac{(1+h)^2 - 1^2}{h} = \lim\limits_{h \to 0} \frac{1 + 2h + h^2 - 1}{h} = \lim\limits_{h \to 0} \frac{h(2+h)}{h} = \lim\limits_{h \to 0} (2 + h) = 2$

63. $\lim\limits_{h \to 0} \frac{[3(2+h) - 4] - [3(2) - 4]}{h} = \lim\limits_{h \to 0} \frac{3h}{h} = 3$

65. $\lim\limits_{h \to 0} \frac{\sqrt{7+h} - \sqrt{7}}{h} = \lim\limits_{h \to 0} \frac{\left(\sqrt{7+h} - \sqrt{7}\right)\left(\sqrt{7+h} + \sqrt{7}\right)}{h\left(\sqrt{7+h} + \sqrt{7}\right)} = \lim\limits_{h \to 0} \frac{(7+h) - 7}{h\left(\sqrt{7+h} + \sqrt{7}\right)}$

$= \lim\limits_{h \to 0} \frac{h}{h\left(\sqrt{7+h} + \sqrt{7}\right)} = \lim\limits_{h \to 0} \frac{1}{\sqrt{7+h} + \sqrt{7}} = \frac{1}{2\sqrt{7}}$

67. $\lim\limits_{x \to 0} \sqrt{5 - 2x^2} = \sqrt{5 - 2(0)^2} = \sqrt{5}$ and $\lim\limits_{x \to 0} \sqrt{5 - x^2} = \sqrt{5 - (0)^2} = \sqrt{5}$; by the sandwich theorem,

$\lim\limits_{x \to 0} f(x) = \sqrt{5}$

69. (a) $\lim\limits_{x \to 0} \left(1 - \frac{x^2}{6}\right) = 1 - \frac{0}{6} = 1$ and $\lim\limits_{x \to 0} 1 = 1$; by the sandwich theorem, $\lim\limits_{x \to 0} \frac{x \sin x}{2 - 2\cos x} = 1$

(b) For $x \neq 0$, $y = (x \sin x)/(2 - 2\cos x)$

lies between the other two graphs in the

figure, and the graphs converge as $x \to 0$.

71. (a) $f(x) = (x^2 - 9)/(x + 3)$

x	−3.1	−3.01	−3.001	−3.0001	−3.00001	−3.000001
f(x)	−6.1	−6.01	−6.001	−6.0001	−6.00001	−6.000001

x	−2.9	−2.99	−2.999	−2.9999	−2.99999	−2.999999
f(x)	−5.9	−5.99	−5.999	−5.9999	−5.99999	−5.999999

The estimate is $\lim\limits_{x \to -3} f(x) = -6$.

(b)

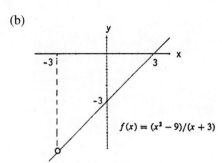

$f(x) = (x^2 - 9)/(x + 3)$

(c) $f(x) = \frac{x^2 - 9}{x + 3} = \frac{(x + 3)(x - 3)}{x + 3} = x - 3$ if $x \neq -3$, and $\lim\limits_{x \to -3} (x - 3) = -3 - 3 = -6$.

73. (a) $G(x) = (x + 6) / (x^2 + 4x - 12)$

x	−5.9	−5.99	−5.999	−5.9999	−5.99999	−5.999999
G(x)	−.126582	−.1251564	−.1250156	−.1250015	−.1250001	−.1250000

x	−6.1	−6.01	−6.001	−6.0001	−6.00001	−6.000001
G(x)	−.123456	−.124843	−.124984	−.124998	−.124999	−.124999

(b)

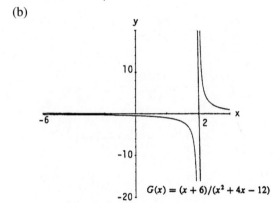

$G(x) = (x + 6)/(x^2 + 4x - 12)$

(c) $G(x) = \frac{x + 6}{(x^2 + 4x - 12)} = \frac{x + 6}{(x + 6)(x - 2)} = \frac{1}{x - 2}$ if $x \neq -6$, and $\lim\limits_{x \to -6} \frac{1}{x - 2} = \frac{1}{-6 - 2} = -\frac{1}{8} = -0.125$.

75. (a) $f(x) = (x^2 - 1) / (|x| - 1)$

x	−1.1	−1.01	−1.001	−1.0001	−1.00001	−1.000001
f(x)	2.1	2.01	2.001	2.0001	2.00001	2.000001

x	−.9	−.99	−.999	−.9999	−.99999	−.999999
f(x)	1.9	1.99	1.999	1.9999	1.99999	1.999999

(b)

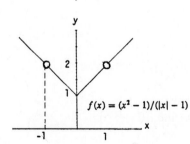

$f(x) = (x^2 - 1)/(|x| - 1)$

(c) $f(x) = \frac{x^2 - 1}{|x| - 1} = \begin{cases} \frac{(x + 1)(x - 1)}{x - 1} = x + 1, & x \geq 0 \text{ and } x \neq 1 \\ \frac{(x + 1)(x - 1)}{-(x + 1)} = 1 - x, & x < 0 \text{ and } x \neq -1 \end{cases}$, and $\lim\limits_{x \to -1} (1 - x) = 1 - (-1) = 2$.

77. (a) $g(\theta) = (\sin \theta)/\theta$

θ	.1	.01	.001	.0001	.00001	.000001
$g(\theta)$	.998334	.999983	.999999	.999999	.999999	.999999

θ	$-.1$	$-.01$	$-.001$	$-.0001$	$-.00001$	$-.000001$
$g(\theta)$	.998334	.999983	.999999	.999999	.999999	.999999

$\lim_{\theta \to 0} g(\theta) = 1$

(b)

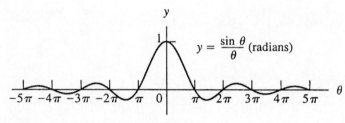

NOT TO SCALE

79. (a) $f(x) = x^{1/(1-x)}$

x	.9	.99	.999	.9999	.99999	.999999
f(x)	.348678	.366032	.367695	.367861	.367877	.367879

x	1.1	1.01	1.001	1.0001	1.00001	1.000001
f(x)	.385543	.369711	.368063	.367897	.367881	.367878

$\lim_{x \to 1} f(x) \approx 0.36788$

(b)

Graph is NOT TO SCALE. Also the intersection of the axes is not the origin: the axes intersect at the point $(1, 2.71820)$.

81. $\lim_{x \to c} f(x)$ exists at those points c where $\lim_{x \to c} x^4 = \lim_{x \to c} x^2$. Thus, $c^4 = c^2 \Rightarrow c^2 (1 - c^2) = 0$

$\Rightarrow c = 0, 1,$ or -1. Moreover, $\lim_{x \to 0} f(x) = \lim_{x \to 0} x^2 = 0$ and $\lim_{x \to -1} f(x) = \lim_{x \to 1} f(x) = 1$.

83. $1 = \lim_{x \to 4} \frac{f(x)-5}{x-2} = \frac{\lim_{x \to 4} f(x) - \lim_{x \to 4} 5}{\lim_{x \to 4} x - \lim_{x \to 4} 2} = \frac{\lim_{x \to 4} f(x) - 5}{4 - 2} \Rightarrow \lim_{x \to 4} f(x) - 5 = 2(1) \Rightarrow \lim_{x \to 4} f(x) = 2 + 5 = 7.$

85. (a) $0 = 3 \cdot 0 = \left[\lim_{x \to 2} \frac{f(x)-5}{x-2} \right] \left[\lim_{x \to 2} (x-2) \right] = \lim_{x \to 2} \left[\left(\frac{f(x)-5}{x-2} \right) (x-2) \right] = \lim_{x \to 2} [f(x) - 5] = \lim_{x \to 2} f(x) - 5$

$\Rightarrow \lim_{x \to 2} f(x) = 5.$

(b) $0 = 4 \cdot 0 = \left[\lim_{x \to 2} \frac{f(x)-5}{x-2} \right] \left[\lim_{x \to 2} (x-2) \right] \Rightarrow \lim_{x \to 2} f(x) = 5$ as in part (a).

87. (a) $\lim\limits_{x \to 0} x \sin \frac{1}{x} = 0$

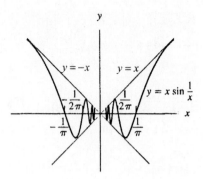

(b) $-1 \le \sin \frac{1}{x} \le 1$ for $x \ne 0$:

$x > 0 \Rightarrow -x \le x \sin \frac{1}{x} \le x \Rightarrow \lim\limits_{x \to 0} x \sin \frac{1}{x} = 0$ by the sandwich theorem;

$x < 0 \Rightarrow -x \ge x \sin \frac{1}{x} \ge x \Rightarrow \lim\limits_{x \to 0} x \sin \frac{1}{x} = 0$ by the sandwich theorem.

2.3 THE PRECISE DEFINITION OF A LIMIT

1.

Step 1: $|x - 5| < \delta \Rightarrow -\delta < x - 5 < \delta \Rightarrow -\delta + 5 < x < \delta + 5$

Step 2: $\delta + 5 = 7 \Rightarrow \delta = 2$, or $-\delta + 5 = 1 \Rightarrow \delta = 4$.

The value of δ which assures $|x - 5| < \delta \Rightarrow 1 < x < 7$ is the smaller value, $\delta = 2$.

3.

Step 1: $|x - (-3)| < \delta \Rightarrow -\delta < x + 3 < \delta \Rightarrow -\delta - 3 < x < \delta - 3$

Step 2: $-\delta - 3 = -\frac{7}{2} \Rightarrow \delta = \frac{1}{2}$, or $\delta - 3 = -\frac{1}{2} \Rightarrow \delta = \frac{5}{2}$.

The value of δ which assures $|x - (-3)| < \delta \Rightarrow -\frac{7}{2} < x < -\frac{1}{2}$ is the smaller value, $\delta = \frac{1}{2}$.

5.

Step 1: $\left|x - \frac{1}{2}\right| < \delta \Rightarrow -\delta < x - \frac{1}{2} < \delta \Rightarrow -\delta + \frac{1}{2} < x < \delta + \frac{1}{2}$

Step 2: $-\delta + \frac{1}{2} = \frac{4}{9} \Rightarrow \delta = \frac{1}{18}$, or $\delta + \frac{1}{2} = \frac{4}{7} \Rightarrow \delta = \frac{1}{14}$.

The value of δ which assures $\left|x - \frac{1}{2}\right| < \delta \Rightarrow \frac{4}{9} < x < \frac{4}{7}$ is the smaller value, $\delta = \frac{1}{18}$.

7. Step 1: $|x - 5| < \delta \Rightarrow -\delta < x - 5 < \delta \Rightarrow -\delta + 5 < x < \delta + 5$

Step 2: From the graph, $-\delta + 5 = 4.9 \Rightarrow \delta = 0.1$, or $\delta + 5 = 5.1 \Rightarrow \delta = 0.1$; thus $\delta = 0.1$ in either case.

9. Step 1: $|x - 1| < \delta \Rightarrow -\delta < x - 1 < \delta \Rightarrow -\delta + 1 < x < \delta + 1$

Step 2: From the graph, $-\delta + 1 = \frac{9}{16} \Rightarrow \delta = \frac{7}{16}$, or $\delta + 1 = \frac{25}{16} \Rightarrow \delta = \frac{9}{16}$; thus $\delta = \frac{7}{16}$.

11. Step 1: $|x - 2| < \delta \Rightarrow -\delta < x - 2 < \delta \Rightarrow -\delta + 2 < x < \delta + 2$

Step 2: From the graph, $-\delta + 2 = \sqrt{3} \Rightarrow \delta = 2 - \sqrt{3} \approx 0.2679$, or $\delta + 2 = \sqrt{5} \Rightarrow \delta = \sqrt{5} - 2 \approx 0.2361$; thus $\delta = \sqrt{5} - 2$.

13. Step 1: $|x - (-1)| < \delta \Rightarrow -\delta < x + 1 < \delta \Rightarrow -\delta - 1 < x < \delta - 1$

Step 2: From the graph, $-\delta - 1 = -\frac{16}{9} \Rightarrow \delta = \frac{7}{9} \approx 0.77$, or $\delta - 1 = -\frac{16}{25} \Rightarrow \frac{9}{25} = 0.36$; thus $\delta = \frac{9}{25} = 0.36$.

15. Step 1: $|(x + 1) - 5| < 0.01 \Rightarrow |x - 4| < 0.01 \Rightarrow -0.01 < x - 4 < 0.01 \Rightarrow 3.99 < x < 4.01$

Step 2: $|x - 4| < \delta \Rightarrow -\delta < x - 4 < \delta \Rightarrow -\delta + 4 < x < \delta + 4 \Rightarrow \delta = 0.01$.

17. Step 1: $\left|\sqrt{x+1}-1\right| < 0.1 \Rightarrow -0.1 < \sqrt{x+1}-1 < 0.1 \Rightarrow 0.9 < \sqrt{x+1} < 1.1 \Rightarrow 0.81 < x+1 < 1.21$
$\Rightarrow -0.19 < x < 0.21$

    Step 2: $|x - 0| < \delta \Rightarrow -\delta < x < \delta.$ Then, $-\delta = -0.19 \Rightarrow \delta = 0.19$ or $\delta = 0.21$; thus, $\delta = 0.19.$

19. Step 1: $\left|\sqrt{19-x}-3\right| < 1 \Rightarrow -1 < \sqrt{19-x}-3 < 1 \Rightarrow 2 < \sqrt{19-x} < 4 \Rightarrow 4 < 19-x < 16$
$\Rightarrow -4 > x - 19 > -16 \Rightarrow 15 > x > 3$ or $3 < x < 15$

    Step 2: $|x - 10| < \delta \Rightarrow -\delta < x - 10 < \delta \Rightarrow -\delta + 10 < x < \delta + 10.$
Then $-\delta + 10 = 3 \Rightarrow \delta = 7,$ or $\delta + 10 = 15 \Rightarrow \delta = 5$; thus $\delta = 5.$

21. Step 1: $\left|\frac{1}{x} - \frac{1}{4}\right| < 0.05 \Rightarrow -0.05 < \frac{1}{x} - \frac{1}{4} < 0.05 \Rightarrow 0.2 < \frac{1}{x} < 0.3 \Rightarrow \frac{10}{2} > x > \frac{10}{3}$ or $\frac{10}{3} < x < 5.$

    Step 2: $|x - 4| < \delta \Rightarrow -\delta < x - 4 < \delta \Rightarrow -\delta + 4 < x < \delta + 4.$
Then $-\delta + 4 = \frac{10}{3}$ or $\delta = \frac{2}{3},$ or $\delta + 4 = 5$ or $\delta = 1$; thus $\delta = \frac{2}{3}.$

23. Step 1: $|x^2 - 4| < 0.5 \Rightarrow -0.5 < x^2 - 4 < 0.5 \Rightarrow 3.5 < x^2 < 4.5 \Rightarrow \sqrt{3.5} < |x| < \sqrt{4.5} \Rightarrow -\sqrt{4.5} < x < -\sqrt{3.5},$
for x near $-2.$

    Step 2: $|x - (-2)| < \delta \Rightarrow -\delta < x + 2 < \delta \Rightarrow -\delta - 2 < x < \delta - 2.$
Then $-\delta - 2 = -\sqrt{4.5} \Rightarrow \delta = \sqrt{4.5} - 2 \approx 0.1213,$ or $\delta - 2 = -\sqrt{3.5} \Rightarrow \delta = 2 - \sqrt{3.5} \approx 0.1292$;
thus $\delta = \sqrt{4.5} - 2 \approx 0.12.$

25. Step 1: $|(x^2 - 5) - 11| < 1 \Rightarrow |x^2 - 16| < 1 \Rightarrow -1 < x^2 - 16 < 1 \Rightarrow 15 < x^2 < 17 \Rightarrow \sqrt{15} < x < \sqrt{17}.$

    Step 2: $|x - 4| < \delta \Rightarrow -\delta < x - 4 < \delta \Rightarrow -\delta + 4 < x < \delta + 4.$
Then $-\delta + 4 = \sqrt{15} \Rightarrow \delta = 4 - \sqrt{15} \approx 0.1270,$ or $\delta + 4 = \sqrt{17} \Rightarrow \delta = \sqrt{17} - 4 \approx 0.1231$;
thus $\delta = \sqrt{17} - 4 \approx 0.12.$

27. Step 1: $|mx - 2m| < 0.03 \Rightarrow -0.03 < mx - 2m < 0.03 \Rightarrow -0.03 + 2m < mx < 0.03 + 2m \Rightarrow$
$2 - \frac{0.03}{m} < x < 2 + \frac{0.03}{m}.$

    Step 2: $|x - 2| < \delta \Rightarrow -\delta < x - 2 < \delta \Rightarrow -\delta + 2 < x < \delta + 2.$
Then $-\delta + 2 = 2 - \frac{0.03}{m} \Rightarrow \delta = \frac{0.03}{m},$ or $\delta + 2 = 2 + \frac{0.03}{m} \Rightarrow \delta = \frac{0.03}{m}.$ In either case, $\delta = \frac{0.03}{m}.$

29. Step 1: $\left|(mx + b) - \left(\frac{m}{2} + b\right)\right| < c \Rightarrow -c < mx - \frac{m}{2} < c \Rightarrow -c + \frac{m}{2} < mx < c + \frac{m}{2} \Rightarrow \frac{1}{2} - \frac{c}{m} < x < \frac{1}{2} + \frac{c}{m}.$

    Step 2: $\left|x - \frac{1}{2}\right| < \delta \Rightarrow -\delta < x - \frac{1}{2} < \delta \Rightarrow -\delta + \frac{1}{2} < x < \delta + \frac{1}{2}.$
Then $-\delta + \frac{1}{2} = \frac{1}{2} - \frac{c}{m} \Rightarrow \delta = \frac{c}{m},$ or $\delta + \frac{1}{2} = \frac{1}{2} + \frac{c}{m} \Rightarrow \delta = \frac{c}{m}.$ In either case, $\delta = \frac{c}{m}.$

31. $\lim\limits_{x \to 3} (3 - 2x) = 3 - 2(3) = -3$

    Step 1: $|(3 - 2x) - (-3)| < 0.02 \Rightarrow -0.02 < 6 - 2x < 0.02 \Rightarrow -6.02 < -2x < -5.98 \Rightarrow 3.01 > x > 2.99$ or
$2.99 < x < 3.01.$

    Step 2: $0 < |x - 3| < \delta \Rightarrow -\delta < x - 3 < \delta \Rightarrow -\delta + 3 < x < \delta + 3.$
Then $-\delta + 3 = 2.99 \Rightarrow \delta = 0.01,$ or $\delta + 3 = 3.01 \Rightarrow \delta = 0.01$; thus $\delta = 0.01.$

33. $\lim\limits_{x \to 2} \frac{x^2 - 4}{x - 2} = \lim\limits_{x \to 2} \frac{(x+2)(x-2)}{(x-2)} = \lim\limits_{x \to 2} (x + 2) = 2 + 2 = 4, x \neq 2$

    Step 1: $\left|\left(\frac{x^2 - 4}{x - 2}\right) - 4\right| < 0.05 \Rightarrow -0.05 < \frac{(x+2)(x-2)}{(x-2)} - 4 < 0.05 \Rightarrow 3.95 < x + 2 < 4.05, x \neq 2$
$\Rightarrow 1.95 < x < 2.05, x \neq 2.$

    Step 2: $|x - 2| < \delta \Rightarrow -\delta < x - 2 < \delta \Rightarrow -\delta + 2 < x < \delta + 2.$
Then $-\delta + 2 = 1.95 \Rightarrow \delta = 0.05,$ or $\delta + 2 = 2.05 \Rightarrow \delta = 0.05$; thus $\delta = 0.05.$

35. $\lim_{x \to -3} \sqrt{1-5x} = \sqrt{1-5(-3)} = \sqrt{16} = 4$

Step 1: $\left|\sqrt{1-5x} - 4\right| < 0.5 \Rightarrow -0.5 < \sqrt{1-5x} - 4 < 0.5 \Rightarrow 3.5 < \sqrt{1-5x} < 4.5 \Rightarrow 12.25 < 1 - 5x < 20.25$

$\Rightarrow 11.25 < -5x < 19.25 \Rightarrow -3.85 < x < -2.25.$

Step 2: $|x - (-3)| < \delta \Rightarrow -\delta < x + 3 < \delta \Rightarrow -\delta - 3 < x < \delta - 3.$

Then $-\delta - 3 = -3.85 \Rightarrow \delta = 0.85$, or $\delta - 3 = -2.25 \Rightarrow 0.75$; thus $\delta = 0.75.$

37. Step 1: $|(9 - x) - 5| < \epsilon \Rightarrow -\epsilon < 4 - x < \epsilon \Rightarrow -\epsilon - 4 < -x < \epsilon - 4 \Rightarrow \epsilon + 4 > x > 4 - \epsilon \Rightarrow 4 - \epsilon < x < 4 + \epsilon.$

Step 2: $|x - 4| < \delta \Rightarrow -\delta < x - 4 < \delta \Rightarrow -\delta + 4 < x < \delta + 4.$

Then $-\delta + 4 = -\epsilon + 4 \Rightarrow \delta = \epsilon$, or $\delta + 4 = \epsilon + 4 \Rightarrow \delta = \epsilon.$ Thus choose $\delta = \epsilon.$

39. Step 1: $\left|\sqrt{x-5} - 2\right| < \epsilon \Rightarrow -\epsilon < \sqrt{x-5} - 2 < \epsilon \Rightarrow 2 - \epsilon < \sqrt{x-5} < 2 + \epsilon \Rightarrow (2 - \epsilon)^2 < x - 5 < (2 + \epsilon)^2$

$\Rightarrow (2 - \epsilon)^2 + 5 < x < (2 + \epsilon)^2 + 5.$

Step 2: $|x - 9| < \delta \Rightarrow -\delta < x - 9 < \delta \Rightarrow -\delta + 9 < x < \delta + 9.$

Then $-\delta + 9 = \epsilon^2 - 4\epsilon + 9 \Rightarrow \delta = 4\epsilon - \epsilon^2$, or $\delta + 9 = \epsilon^2 + 4\epsilon + 9 \Rightarrow \delta = 4\epsilon + \epsilon^2.$ Thus choose the smaller distance, $\delta = 4\epsilon - \epsilon^2.$

41. Step 1: For $x \neq 1$, $|x^2 - 1| < \epsilon \Rightarrow -\epsilon < x^2 - 1 < \epsilon \Rightarrow 1 - \epsilon < x^2 < 1 + \epsilon \Rightarrow \sqrt{1 - \epsilon} < |x| < \sqrt{1 + \epsilon}$

$\Rightarrow \sqrt{1 - \epsilon} < x < \sqrt{1 + \epsilon}$ near $x = 1.$

Step 2: $|x - 1| < \delta \Rightarrow -\delta < x - 1 < \delta \Rightarrow -\delta + 1 < x < \delta + 1.$

Then $-\delta + 1 = \sqrt{1 - \epsilon} \Rightarrow \delta = 1 - \sqrt{1 - \epsilon}$, or $\delta + 1 = \sqrt{1 + \epsilon} \Rightarrow \delta = \sqrt{1 + \epsilon} - 1.$ Choose

$\delta = \min\left\{1 - \sqrt{1 - \epsilon}, \sqrt{1 + \epsilon} - 1\right\}$, that is, the smaller of the two distances.

43. Step 1: $\left|\frac{1}{x} - 1\right| < \epsilon \Rightarrow -\epsilon < \frac{1}{x} - 1 < \epsilon \Rightarrow 1 - \epsilon < \frac{1}{x} < 1 + \epsilon \Rightarrow \frac{1}{1+\epsilon} < x < \frac{1}{1-\epsilon}.$

Step 2: $|x - 1| < \delta \Rightarrow -\delta < x - 1 < \delta \Rightarrow 1 - \delta < x < 1 + \delta.$

Then $1 - \delta = \frac{1}{1+\epsilon} \Rightarrow \delta = 1 - \frac{1}{1+\epsilon} = \frac{\epsilon}{1+\epsilon}$, or $1 + \delta = \frac{1}{1-\epsilon} \Rightarrow \delta = \frac{1}{1-\epsilon} - 1 = \frac{\epsilon}{1-\epsilon}.$

Choose $\delta = \frac{\epsilon}{1+\epsilon}$, the smaller of the two distances.

45. Step 1: $\left|\left(\frac{x^2-9}{x+3}\right) - (-6)\right| < \epsilon \Rightarrow -\epsilon < (x - 3) + 6 < \epsilon, x \neq -3 \Rightarrow -\epsilon < x + 3 < \epsilon \Rightarrow -\epsilon - 3 < x < \epsilon - 3.$

Step 2: $|x - (-3)| < \delta \Rightarrow -\delta < x + 3 < \delta \Rightarrow -\delta - 3 < x < \delta - 3.$

Then $-\delta - 3 = -\epsilon - 3 \Rightarrow \delta = \epsilon$, or $\delta - 3 = \epsilon - 3 \Rightarrow \delta = \epsilon.$ Choose $\delta = \epsilon.$

47. Step 1: $x < 1$: $|(4 - 2x) - 2| < \epsilon \Rightarrow 0 < 2 - 2x < \epsilon$ since $x < 1.$ Thus, $1 - \frac{\epsilon}{2} < x < 0$;

$x \geq 1$: $|(6x - 4) - 2| < \epsilon \Rightarrow 0 \leq 6x - 6 < \epsilon$ since $x \geq 1.$ Thus, $1 \leq x < 1 + \frac{\epsilon}{6}.$

Step 2: $|x - 1| < \delta \Rightarrow -\delta < x - 1 < \delta \Rightarrow 1 - \delta < x < 1 + \delta.$

Then $1 - \delta = 1 - \frac{\epsilon}{2} \Rightarrow \delta = \frac{\epsilon}{2}$, or $1 + \delta = 1 + \frac{\epsilon}{6} \Rightarrow \delta = \frac{\epsilon}{6}.$ Choose $\delta = \frac{\epsilon}{6}.$

49. By the figure, $-x \leq x \sin \frac{1}{x} \leq x$ for all $x > 0$ and $-x \geq x \sin \frac{1}{x} \geq x$ for $x < 0.$ Since $\lim_{x \to 0} (-x) = \lim_{x \to 0} x = 0$,

then by the sandwich theorem, in either case, $\lim_{x \to 0} x \sin \frac{1}{x} = 0.$

51. As x approaches the value 0, the values of $g(x)$ approach k. Thus for every number $\epsilon > 0$, there exists a $\delta > 0$ such that $0 < |x - 0| < \delta \Rightarrow |g(x) - k| < \epsilon.$

53. Let $f(x) = x^2.$ The function values do get closer to -1 as x approaches 0, but $\lim_{x \to 0} f(x) = 0$, not $-1.$ The function $f(x) = x^2$ never gets <u>arbitrarily</u> <u>close</u> to -1 for x near 0.

55. $|A - 9| \leq 0.01 \Rightarrow -0.01 \leq \pi \left(\frac{x}{2}\right)^2 - 9 \leq 0.01 \Rightarrow 8.99 \leq \frac{\pi x^2}{4} \leq 9.01 \Rightarrow \frac{4}{\pi}(8.99) \leq x^2 \leq \frac{4}{\pi}(9.01)$

$\Rightarrow 2\sqrt{\frac{8.99}{\pi}} \leq x \leq 2\sqrt{\frac{9.01}{\pi}}$ or $3.384 \leq x \leq 3.387$. To be safe, the left endpoint was rounded up and the right endpoint was rounded down.

57. (a) $-\delta < x - 1 < 0 \Rightarrow 1 - \delta < x < 1 \Rightarrow f(x) = x$. Then $|f(x) - 2| = |x - 2| = 2 - x > 2 - 1 = 1$. That is, $|f(x) - 2| \geq 1 \geq \frac{1}{2}$ no matter how small δ is taken when $1 - \delta < x < 1 \Rightarrow \lim_{x \to 1} f(x) \neq 2$.

(b) $0 < x - 1 < \delta \Rightarrow 1 < x < 1 + \delta \Rightarrow f(x) = x + 1$. Then $|f(x) - 1| = |(x + 1) - 1| = |x| = x > 1$. That is, $|f(x) - 1| \geq 1$ no matter how small δ is taken when $1 < x < 1 + \delta \Rightarrow \lim_{x \to 1} f(x) \neq 1$.

(c) $-\delta < x - 1 < 0 \Rightarrow 1 - \delta < x < 1 \Rightarrow f(x) = x$. Then $|f(x) - 1.5| = |x - 1.5| = 1.5 - x > 1.5 - 1 = 0.5$. Also, $0 < x - 1 < \delta \Rightarrow 1 < x < 1 + \delta \Rightarrow f(x) = x + 1$. Then $|f(x) - 1.5| = |(x + 1) - 1.5| = |x - 0.5|$ $= x - 0.5 > 1 - 0.5 = 0.5$. Thus, no matter how small δ is taken, there exists a value of x such that $-\delta < x - 1 < \delta$ but $|f(x) - 1.5| \geq \frac{1}{2} \Rightarrow \lim_{x \to 1} f(x) \neq 1.5$.

59. (a) For $3 - \delta < x < 3 \Rightarrow f(x) > 4.8 \Rightarrow |f(x) - 4| \geq 0.8$. Thus for $\epsilon < 0.8$, $|f(x) - 4| \geq \epsilon$ whenever $3 - \delta < x < 3$ no matter how small we choose $\delta > 0 \Rightarrow \lim_{x \to 3} f(x) \neq 4$.

(b) For $3 < x < 3 + \delta \Rightarrow f(x) < 3 \Rightarrow |f(x) - 4.8| \geq 1.8$. Thus for $\epsilon < 1.8$, $|f(x) - 4.8| \geq \epsilon$ whenever $3 < x < 3 + \delta$ no matter how small we choose $\delta > 0 \Rightarrow \lim_{x \to 3} f(x) \neq 4.8$.

(c) For $3 - \delta < x < 3 \Rightarrow f(x) > 4.8 \Rightarrow |f(x) - 3| \geq 1.8$. Again, for $\epsilon < 1.8$, $|f(x) - 3| \geq \epsilon$ whenever $3 - \delta < x < 3$ no matter how small we choose $\delta > 0 \Rightarrow \lim_{x \to 3} f(x) \neq 3$.

2.4 ONE-SIDED LIMITS AND LIMITS AT INFINITY

1. (a) True   (b) True   (c) False   (d) True
   (e) True   (f) True   (g) False   (h) False
   (i) False   (j) False   (k) True   (l) False

3. (a) $\lim_{x \to 2^+} f(x) = \frac{2}{2} + 1 = 2$, $\lim_{x \to 2^-} f(x) = 3 - 2 = 1$

(b) No, $\lim_{x \to 2} f(x)$ does not exist because $\lim_{x \to 2^+} f(x) \neq \lim_{x \to 2^-} f(x)$

(c) $\lim_{x \to 4^-} f(x) = \frac{4}{2} + 1 = 3$, $\lim_{x \to 4^+} f(x) = \frac{4}{2} + 1 = 3$

(d) Yes, $\lim_{x \to 4} f(x) = 3$ because $3 = \lim_{x \to 4^-} f(x) = \lim_{x \to 4^+} f(x)$

5. (a) No, $\lim_{x \to 0^+} f(x)$ does not exist since $\sin\left(\frac{1}{x}\right)$ does not approach any single value as x approaches 0

(b) $\lim_{x \to 0^-} f(x) = \lim_{x \to 0^-} 0 = 0$

(c) $\lim_{x \to 0} f(x)$ does not exist because $\lim_{x \to 0^+} f(x)$ does not exist

7. (a)

(b) $\lim_{x \to 1^-} f(x) = 1 = \lim_{x \to 1^+} f(x)$

(c) Yes, $\lim_{x \to 1} f(x) = 1$ since the right-hand and left-hand limits exist and equal 1

9. (a) domain: $0 \le x \le 2$

range: $0 < y \le 1$ and $y = 2$

(b) $\lim\limits_{x \to c} f(x)$ exists for c belonging to

$(0, 1) \cup (1, 2)$

(c) $x = 2$

(d) $x = 0$

$y = \begin{cases} \sqrt{1-x^2}, & 0 \le x < 1 \\ 1, & 1 \le x < 2 \\ 2, & x = 2 \end{cases}$

11. $\lim\limits_{x \to -0.5^-} \sqrt{\frac{x+2}{x-1}} = \sqrt{\frac{-0.5+2}{-0.5+1}} = \sqrt{\frac{3/2}{1/2}} = \sqrt{3}$

13. $\lim\limits_{x \to -2^+} \left(\frac{x}{x+1}\right)\left(\frac{2x+5}{x^2+x}\right) = \left(\frac{-2}{-2+1}\right)\left(\frac{2(-2)+5}{(-2)^2+(-2)}\right) = (2)\left(\frac{1}{2}\right) = 1$

15. $\lim\limits_{h \to 0^+} \frac{\sqrt{h^2+4h+5}-\sqrt{5}}{h} = \lim\limits_{h \to 0^+} \left(\frac{\sqrt{h^2+4h+5}-\sqrt{5}}{h}\right)\left(\frac{\sqrt{h^2+4h+5}+\sqrt{5}}{\sqrt{h^2+4h+5}+\sqrt{5}}\right)$

$= \lim\limits_{h \to 0^+} \frac{(h^2+4h+5)-5}{h\left(\sqrt{h^2+4h+5}+\sqrt{5}\right)} = \lim\limits_{h \to 0^+} \frac{h(h+4)}{h\left(\sqrt{h^2+4h+5}+\sqrt{5}\right)} = \frac{0+4}{\sqrt{5}+\sqrt{5}} = \frac{2}{\sqrt{5}}$

17. (a) $\lim\limits_{x \to -2^+} (x+3)\frac{|x+2|}{x+2} = \lim\limits_{x \to -2^+} (x+3)\frac{(x+2)}{(x+2)}$ ($|x+2| = x+2$ for $x > -2$)

$= \lim\limits_{x \to -2^+} (x+3) = (-2)+3 = 1$

(b) $\lim\limits_{x \to -2^-} (x+3)\frac{|x+2|}{x+2} = \lim\limits_{x \to -2^-} (x+3)\left[\frac{-(x+2)}{(x+2)}\right]$ ($|x+2| = -(x+2)$ for $x < -2$)

$= \lim\limits_{x \to -2^-} (x+3)(-1) = -(-2+3) = -1$

19. (a) $\lim\limits_{\theta \to 3^+} \frac{|\theta|}{\theta} = \frac{3}{3} = 1$ (b) $\lim\limits_{\theta \to 3^-} \frac{|\theta|}{\theta} = \frac{2}{3}$

21. $\lim\limits_{\theta \to 0} \frac{\sin\sqrt{2\theta}}{\sqrt{2\theta}} = \lim\limits_{x \to 0} \frac{\sin x}{x} = 1$ (where $x = \sqrt{2\theta}$)

23. $\lim\limits_{y \to 0} \frac{\sin 3y}{4y} = \frac{1}{4}\lim\limits_{y \to 0} \frac{3\sin 3y}{3y} = \frac{3}{4}\lim\limits_{y \to 0} \frac{\sin 3y}{3y} = \frac{3}{4}\lim\limits_{\theta \to 0} \frac{\sin \theta}{\theta} = \frac{3}{4}$ (where $\theta = 3y$)

25. $\lim\limits_{x \to 0} \frac{\tan 2x}{x} = \lim\limits_{x \to 0} \frac{\left(\frac{\sin 2x}{\cos 2x}\right)}{x} = \lim\limits_{x \to 0} \frac{\sin 2x}{x\cos 2x} = \left(\lim\limits_{x \to 0} \frac{1}{\cos 2x}\right)\left(\lim\limits_{x \to 0} \frac{2\sin 2x}{2x}\right) = 1 \cdot 2 = 2$

27. $\lim\limits_{x \to 0} \frac{x \csc 2x}{\cos 5x} = \lim\limits_{x \to 0} \left(\frac{x}{\sin 2x} \cdot \frac{1}{\cos 5x}\right) = \left(\frac{1}{2}\lim\limits_{x \to 0} \frac{2x}{\sin 2x}\right)\left(\lim\limits_{x \to 0} \frac{1}{\cos 5x}\right) = \left(\frac{1}{2} \cdot 1\right)(1) = \frac{1}{2}$

29. $\lim\limits_{x \to 0} \frac{x+x\cos x}{\sin x \cos x} = \lim\limits_{x \to 0} \left(\frac{x}{\sin x \cos x} + \frac{x\cos x}{\sin x \cos x}\right) = \lim\limits_{x \to 0} \left(\frac{x}{\sin x} \cdot \frac{1}{\cos x}\right) + \lim\limits_{x \to 0} \frac{x}{\sin x}$

$= \lim\limits_{x \to 0} \left(\frac{1}{\frac{\sin x}{x}}\right) \cdot \lim\limits_{x \to 0} \left(\frac{1}{\cos x}\right) + \lim\limits_{x \to 0} \left(\frac{1}{\frac{\sin x}{x}}\right) = (1)(1) + 1 = 2$

31. $\lim\limits_{t \to 0} \frac{\sin(1-\cos t)}{1-\cos t} = \lim\limits_{\theta \to 0} \frac{\sin \theta}{\theta} = 1$ since $\theta = 1-\cos t \to 0$ as $t \to 0$

33. $\lim\limits_{\theta \to 0} \frac{\sin \theta}{\sin 2\theta} = \lim\limits_{\theta \to 0} \left(\frac{\sin \theta}{\sin 2\theta} \cdot \frac{2\theta}{2\theta}\right) = \frac{1}{2}\lim\limits_{\theta \to 0} \left(\frac{\sin \theta}{\theta} \cdot \frac{2\theta}{\sin 2\theta}\right) = \frac{1}{2} \cdot 1 \cdot 1 = \frac{1}{2}$

35. $\lim\limits_{x \to 0} \frac{\tan 3x}{\sin 8x} = \lim\limits_{x \to 0} \left(\frac{\sin 3x}{\cos 3x} \cdot \frac{1}{\sin 8x}\right) = \lim\limits_{x \to 0} \left(\frac{\sin 3x}{\cos 3x} \cdot \frac{1}{\sin 8x} \cdot \frac{8x}{3x} \cdot \frac{3}{8}\right)$

$= \frac{3}{8}\lim\limits_{x \to 0} \left(\frac{1}{\cos 3x}\right)\left(\frac{\sin 3x}{3x}\right)\left(\frac{8x}{\sin 8x}\right) = \frac{3}{8} \cdot 1 \cdot 1 \cdot 1 = \frac{3}{8}$

Note: In these exercises we use the result $\lim\limits_{x \to \pm\infty} \frac{1}{x^{m/n}} = 0$ whenever $\frac{m}{n} > 0$. This result follows immediately from Example 6 and the power rule in Theorem 8: $\lim\limits_{x \to \pm\infty}\left(\frac{1}{x^{m/n}}\right) = \lim\limits_{x \to \pm\infty}\left(\frac{1}{x}\right)^{m/n} = \left(\lim\limits_{x \to \pm\infty}\frac{1}{x}\right)^{m/n} = 0^{m/n} = 0.$

37. (a) -3　　　　　　　　　　　　　　　　　　(b) -3

39. (a) $\frac{1}{2}$　　　　　　　　　　　　　　　　　(b) $\frac{1}{2}$

41. (a) $-\frac{5}{3}$　　　　　　　　　　　　　　　　(b) $-\frac{5}{3}$

43. $-\frac{1}{x} \le \frac{\sin 2x}{x} \le \frac{1}{x} \Rightarrow \lim\limits_{x \to \infty} \frac{\sin 2x}{x} = 0$ by the Sandwich Theorem

45. $\lim\limits_{t \to \infty} \frac{2 - t + \sin t}{t + \cos t} = \lim\limits_{t \to \infty} \frac{\frac{2}{t} - 1 + \left(\frac{\sin t}{t}\right)}{1 + \left(\frac{\cos t}{t}\right)} = \frac{0 - 1 + 0}{1 + 0} = -1$

47. $-\frac{1}{e^x} \le \frac{\sin x}{e^x} \le \frac{1}{e^x} \Rightarrow \lim\limits_{x \to \infty} e^{-x}\sin x = 0$ by the Sandwich Theorem.

49. $\lim\limits_{x \to -\infty} \frac{e^x - e^{-x}}{e^x + e^{-x}} = \lim\limits_{x \to -\infty} \frac{e^x - \frac{1}{e^x}}{e^x + \frac{1}{e^x}} = \lim\limits_{x \to -\infty} \frac{\frac{e^{2x} - 1}{e^x}}{\frac{e^{2x} + 1}{e^x}} = \lim\limits_{x \to -\infty} \frac{e^{2x} - 1}{e^{2x} + 1} = \frac{0 - 1}{0 + 1} = -1$

51. (a) $\lim\limits_{x \to \infty} \frac{2x + 3}{5x + 7} = \lim\limits_{x \to \infty} \frac{2 + \frac{3}{x}}{5 + \frac{7}{x}} = \frac{2}{5}$　　　　　(b) $\frac{2}{5}$ (same process as part (a))

53. (a) $\lim\limits_{x \to \infty} \frac{x + 1}{x^2 + 3} = \lim\limits_{x \to \infty} \frac{\frac{1}{x} + \frac{1}{x^2}}{1 + \frac{3}{x^2}} = 0$　　　　　(b) 0 (same process as part (a))

55. (a) $\lim\limits_{x \to \infty} \frac{7x^3}{x^3 - 3x^2 + 6x} = \lim\limits_{x \to \infty} \frac{7}{1 - \frac{3}{x} + \frac{6}{x^2}} = 7$　　　　　(b) 7 (same process as part (a))

57. (a) $\lim\limits_{x \to \infty} \frac{10x^5 + x^4 + 31}{x^6} = \lim\limits_{x \to \infty} \frac{\frac{10}{x} + \frac{1}{x^2} + \frac{31}{x^6}}{1} = 0$
　　(b) 0 (same process as part (a))

59. (a) $\lim\limits_{x \to \infty} \frac{-2x^3 - 2x + 3}{3x^3 + 3x^2 - 5x} = \lim\limits_{x \to \infty} \frac{-2 - \frac{2}{x^2} + \frac{3}{x^3}}{3 + \frac{3}{x} - \frac{5}{x^2}} = -\frac{2}{3}$
　　(b) $-\frac{2}{3}$ (same process as part (a))

61. $\lim\limits_{x \to \infty} \frac{2\sqrt{x} + x^{-1}}{3x - 7} = \lim\limits_{x \to \infty} \frac{\left(\frac{2}{x^{1/2}}\right) + \left(\frac{1}{x^2}\right)}{3 - \frac{7}{x}} = 0$

63. $\lim\limits_{x \to -\infty} \frac{\sqrt[3]{x} - \sqrt[5]{x}}{\sqrt[3]{x} + \sqrt[5]{x}} = \lim\limits_{x \to -\infty} \frac{1 - x^{(1/5) - (1/3)}}{1 + x^{(1/5) - (1/3)}} = \lim\limits_{x \to -\infty} \frac{1 - \left(\frac{1}{x^{2/15}}\right)}{1 + \left(\frac{1}{x^{2/15}}\right)} = 1$

65. $\lim\limits_{x \to \infty} \frac{2x^{5/3} - x^{1/3} + 7}{x^{8/5} + 3x + \sqrt{x}} = \lim\limits_{x \to \infty} \frac{2x^{1/15} - \frac{1}{x^{19/15}} + \frac{7}{x^{8/5}}}{1 + \frac{3}{x^{3/5}} + \frac{1}{x^{11/10}}} = \infty$

67. Yes. If $\lim\limits_{x \to a^+} f(x) = L = \lim\limits_{x \to a^-} f(x)$, then $\lim\limits_{x \to a} f(x) = L$. If $\lim\limits_{x \to a^+} f(x) \ne \lim\limits_{x \to a^-} f(x)$, then $\lim\limits_{x \to a} f(x)$ does not exist.

69. If f is an odd function of x, then $f(-x) = -f(x)$. Given $\lim\limits_{x \to 0^+} f(x) = 3$, then $\lim\limits_{x \to 0^-} f(x) = -3$.

71. Yes. If $\lim\limits_{x \to \infty} \frac{f(x)}{g(x)} = 2$ then the ratio of the polynomials' leading coefficients is 2, so $\lim\limits_{x \to -\infty} \frac{f(x)}{g(x)} = 2$ as well.

73. At most 1 horizontal asymptote: If $\lim\limits_{x \to \infty} \frac{f(x)}{g(x)} = L$, then the ratio of the polynomials' leading coefficients is L, so $\lim\limits_{x \to -\infty} \frac{f(x)}{g(x)} = L$ as well.

75. For any $\epsilon > 0$, take N = 1. Then for all $x > N$ we have that $|f(x) - k| = |k - k| = 0 < \epsilon$.

77. $I = (5, 5 + \delta) \Rightarrow 5 < x < 5 + \delta$. Also, $\sqrt{x - 5} < \epsilon \Rightarrow x - 5 < \epsilon^2 \Rightarrow x < 5 + \epsilon^2$. Choose $\delta = \epsilon^2$
    $\Rightarrow \lim\limits_{x \to 5^+} \sqrt{x - 5} = 0$.

79. As $x \to 0^-$ the number x is always negative. Thus, $\left|\frac{x}{|x|} - (-1)\right| < \epsilon \Rightarrow \left|\frac{x}{-x} + 1\right| < \epsilon \Rightarrow 0 < \epsilon$ which is always
    true independent of the value of x. Hence we can choose any $\delta > 0$ with $-\delta < x < 0 \Rightarrow \lim\limits_{x \to 0^-} \frac{x}{|x|} = -1$.

81. (a) $\lim\limits_{x \to 400^+} \lfloor x \rfloor = 400$. Just observe that if $400 < x < 401$, then $\lfloor x \rfloor = 400$. Thus if we choose $\delta = 1$, we have for any
        number $\epsilon > 0$ that $400 < x < 400 + \delta \Rightarrow |\lfloor x \rfloor - 400| = |400 - 400| = 0 < \epsilon$.
    (b) $\lim\limits_{x \to 400^-} \lfloor x \rfloor = 399$. Just observe that if $399 < x < 400$ then $\lfloor x \rfloor = 399$. Thus if we choose $\delta = 1$, we have for any
        number $\epsilon > 0$ that $400 - \delta < x < 400 \Rightarrow |\lfloor x \rfloor - 399| = |399 - 399| = 0 < \epsilon$.
    (c) Since $\lim\limits_{x \to 400^+} \lfloor x \rfloor \neq \lim\limits_{x \to 400^-} \lfloor x \rfloor$ we conclude that $\lim\limits_{x \to 400} \lfloor x \rfloor$ does not exist.

83. $\lim\limits_{x \to \pm\infty} x \sin \frac{1}{x} = \lim\limits_{\theta \to 0} \frac{1}{\theta} \sin \theta = 1, \ \left(\theta = \frac{1}{x}\right)$

85. $\lim\limits_{x \to \pm\infty} \frac{3x + 4}{2x - 5} = \lim\limits_{x \to \pm\infty} \frac{3 + \frac{4}{x}}{2 - \frac{5}{x}} = \lim\limits_{t \to 0} \frac{3 + 4t}{2 - 5t} = \frac{3}{2}, \ \left(t = \frac{1}{x}\right)$

87. $\lim\limits_{x \to \pm\infty} \left(3 + \frac{2}{x}\right)\left(\cos \frac{1}{x}\right) = \lim\limits_{\theta \to 0}(3 + 2\theta)(\cos \theta) = (3)(1) = 3, \ \left(\theta = \frac{1}{x}\right)$

2.5 INFINITE LIMITS AND VERTICAL ASYMPTOTES

1. $\lim\limits_{x \to 0^+} \frac{1}{3x} = \infty$ $\left(\frac{\text{positive}}{\text{positive}}\right)$

3. $\lim\limits_{x \to 2^-} \frac{3}{x - 2} = -\infty$ $\left(\frac{\text{positive}}{\text{negative}}\right)$

5. $\lim\limits_{x \to -8^+} \frac{2x}{x + 8} = -\infty$ $\left(\frac{\text{negative}}{\text{positive}}\right)$

7. $\lim\limits_{x \to 7} \frac{4}{(x - 7)^2} = \infty$ $\left(\frac{\text{positive}}{\text{positive}}\right)$

9. (a) $\lim\limits_{x \to 0^+} \frac{2}{3x^{1/3}} = \infty$

    (b) $\lim\limits_{x \to 0^-} \frac{2}{3x^{1/3}} = -\infty$

11. $\lim\limits_{x \to 0} \frac{4}{x^{2/5}} = \lim\limits_{x \to 0} \frac{4}{(x^{1/5})^2} = \infty$

13. $\lim\limits_{x \to \left(\frac{\pi}{2}\right)^-} \tan x = \infty$

15. $\lim\limits_{\theta \to 0^-}(1 + \csc \theta) = -\infty$

17. (a) $\lim\limits_{x \to 2^+} \frac{1}{x^2 - 4} = \lim\limits_{x \to 2^+} \frac{1}{(x + 2)(x - 2)} = \infty$ $\left(\frac{1}{\text{positive·positive}}\right)$

    (b) $\lim\limits_{x \to 2^-} \frac{1}{x^2 - 4} = \lim\limits_{x \to 2^-} \frac{1}{(x + 2)(x - 2)} = -\infty$ $\left(\frac{1}{\text{positive·negative}}\right)$

    (c) $\lim\limits_{x \to -2^+} \frac{1}{x^2 - 4} = \lim\limits_{x \to -2^+} \frac{1}{(x + 2)(x - 2)} = -\infty$ $\left(\frac{1}{\text{positive·negative}}\right)$

    (d) $\lim\limits_{x \to -2^-} \frac{1}{x^2 - 4} = \lim\limits_{x \to -2^-} \frac{1}{(x + 2)(x - 2)} = \infty$ $\left(\frac{1}{\text{negative·negative}}\right)$

19. (a) $\displaystyle\lim_{x \to 0^+} \frac{x^2}{2} - \frac{1}{x} = 0 + \lim_{x \to 0^+} \frac{1}{-x} = -\infty$ $\left(\frac{1}{\text{negative}}\right)$

(b) $\displaystyle\lim_{x \to 0^-} \frac{x^2}{2} - \frac{1}{x} = 0 + \lim_{x \to 0^-} \frac{1}{-x} = \infty$ $\left(\frac{1}{\text{positive}}\right)$

(c) $\displaystyle\lim_{x \to \sqrt[3]{2}} \frac{x^2}{2} - \frac{1}{x} = \frac{2^{2/3}}{2} - \frac{1}{2^{1/3}} = 2^{-1/3} - 2^{-1/3} = 0$

(d) $\displaystyle\lim_{x \to -1} \frac{x^2}{2} - \frac{1}{x} = \frac{1}{2} - \left(\frac{1}{-1}\right) = \frac{3}{2}$

21. (a) $\displaystyle\lim_{x \to 0^+} \frac{x^2 - 3x + 2}{x^3 - 2x^2} = \lim_{x \to 0^+} \frac{(x-2)(x-1)}{x^2(x-2)} = -\infty$ $\left(\frac{\text{negative·negative}}{\text{positive·negative}}\right)$

(b) $\displaystyle\lim_{x \to 2^+} \frac{x^2 - 3x + 2}{x^3 - 2x^2} = \lim_{x \to 2^+} \frac{(x-2)(x-1)}{x^2(x-2)} = \lim_{x \to 2^+} \frac{x-1}{x^2} = \frac{1}{4}, x \neq 2$

(c) $\displaystyle\lim_{x \to 2^-} \frac{x^2 - 3x + 2}{x^3 - 2x^2} = \lim_{x \to 2^-} \frac{(x-2)(x-1)}{x^2(x-2)} = \lim_{x \to 2^-} \frac{x-1}{x^2} = \frac{1}{4}, x \neq 2$

(d) $\displaystyle\lim_{x \to 2} \frac{x^2 - 3x + 2}{x^3 - 2x^2} = \lim_{x \to 2} \frac{(x-2)(x-1)}{x^2(x-2)} = \lim_{x \to 2} \frac{x-1}{x^2} = \frac{1}{4}, x \neq 2$

(e) $\displaystyle\lim_{x \to 0} \frac{x^2 - 3x + 2}{x^3 - 2x^2} = \lim_{x \to 0} \frac{(x-2)(x-1)}{x^2(x-2)} = -\infty$ $\left(\frac{\text{negative·negative}}{\text{positive·negative}}\right)$

23. (a) $\displaystyle\lim_{t \to 0^+} \left[2 - \frac{3}{t^{1/3}}\right] = -\infty$ (b) $\displaystyle\lim_{t \to 0^-} \left[2 - \frac{3}{t^{1/3}}\right] = \infty$

25. (a) $\displaystyle\lim_{x \to 0^+} \left[\frac{1}{x^{2/3}} + \frac{2}{(x-1)^{2/3}}\right] = \infty$ (b) $\displaystyle\lim_{x \to 0^-} \left[\frac{1}{x^{2/3}} + \frac{2}{(x-1)^{2/3}}\right] = \infty$

(c) $\displaystyle\lim_{x \to 1^+} \left[\frac{1}{x^{2/3}} + \frac{2}{(x-1)^{2/3}}\right] = \infty$ (d) $\displaystyle\lim_{x \to 1^-} \left[\frac{1}{x^{2/3}} + \frac{2}{(x-1)^{2/3}}\right] = \infty$

27. $y = \frac{1}{x-1}$ 29. $y = \frac{1}{2x+4}$

31. $y = \frac{x+3}{x+2} = 1 + \frac{1}{x+2}$ 33. $y = \frac{x^2}{x-1} = x + 1 + \frac{1}{x-1}$

35. $y = \frac{x^2 - 4}{x - 1} = x + 1 - \frac{3}{x - 1}$

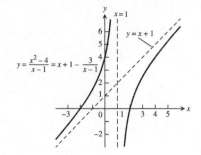

37. $y = \frac{x^2 - 1}{x} = x - \frac{1}{x}$

39. Here is one possibility.

41. Here is one possibility.

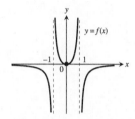

43. Here is one possibility.

45. Here is one possibility.

47. For every real number $-B < 0$, we must find a $\delta > 0$ such that for all x, $0 < |x - 0| < \delta \Rightarrow \frac{-1}{x^2} < -B$. Now,

   $-\frac{1}{x^2} < -B < 0 \Leftrightarrow \frac{1}{x^2} > B > 0 \Leftrightarrow x^2 < \frac{1}{B} \Leftrightarrow |x| < \frac{1}{\sqrt{B}}$. Choose $\delta = \frac{1}{\sqrt{B}}$, then $0 < |x| < \delta \Rightarrow |x| < \frac{1}{\sqrt{B}}$

   $\Rightarrow \frac{-1}{x^2} < -B$ so that $\lim\limits_{x \to 0} -\frac{1}{x^2} = -\infty$.

49. For every real number $-B < 0$, we must find a $\delta > 0$ such that for all x, $0 < |x - 3| < \delta \Rightarrow \frac{-2}{(x - 3)^2} < -B$.

   Now, $\frac{-2}{(x-3)^2} < -B < 0 \Leftrightarrow \frac{2}{(x-3)^2} > B > 0 \Leftrightarrow \frac{(x-3)^2}{2} < \frac{1}{B} \Leftrightarrow (x-3)^2 < \frac{2}{B} \Leftrightarrow 0 < |x - 3| < \sqrt{\frac{2}{B}}$. Choose

   $\delta = \sqrt{\frac{2}{B}}$, then $0 < |x - 3| < \delta \Rightarrow \frac{-2}{(x-3)^2} < -B < 0$ so that $\lim\limits_{x \to 3} \frac{-2}{(x-3)^2} = -\infty$.

51. (a) We say that f(x) approaches infinity as x approaches x_0 from the left, and write $\lim\limits_{x \to x_0^-} f(x) = \infty$, if

   for every positive number B, there exists a corresponding number $\delta > 0$ such that for all x,
   $x_0 - \delta < x < x_0 \Rightarrow f(x) > B$.

   (b) We say that f(x) approaches minus infinity as x approaches x_0 from the right, and write $\lim\limits_{x \to x_0^+} f(x) = -\infty$,

   if for every positive number B (or negative number $-B$) there exists a corresponding number $\delta > 0$ such
   that for all x, $x_0 < x < x_0 + \delta \Rightarrow f(x) < -B$.

(c) We say that f(x) approaches minus infinity as x approaches x_0 from the left, and write $\lim\limits_{x \to x_0^-} f(x) = -\infty$,

if for every positive number B (or negative number $-B$) there exists a corresponding number $\delta > 0$ such that for all x, $x_0 - \delta < x < x_0 \Rightarrow f(x) < -B$.

53. For $B > 0$, $\frac{1}{x} < -B < 0 \Leftrightarrow -\frac{1}{x} > B > 0 \Leftrightarrow -x < \frac{1}{B} \Leftrightarrow -\frac{1}{B} < x$. Choose $\delta = \frac{1}{B}$. Then $-\delta < x < 0$
$\Rightarrow -\frac{1}{B} < x \Rightarrow \frac{1}{x} < -B$ so that $\lim\limits_{x \to 0^-} \frac{1}{x} = -\infty$.

55. For $B > 0$, $\frac{1}{x-2} > B \Leftrightarrow 0 < x - 2 < \frac{1}{B}$. Choose $\delta = \frac{1}{B}$. Then $2 < x < 2 + \delta \Rightarrow 0 < x - 2 < \delta \Rightarrow 0 < x - 2 < \frac{1}{B}$
$\Rightarrow \frac{1}{x-2} > B > 0$ so that $\lim\limits_{x \to 2^+} \frac{1}{x-2} = \infty$.

57. $y = \sec x + \frac{1}{x}$

59. $y = \tan x + \frac{1}{x^2}$

61. $y = \dfrac{x}{\sqrt{4 - x^2}}$

63. $y = x^{2/3} + \frac{1}{x^{1/3}}$

65. The graph of $y = f\left(\frac{1}{x}\right) = \frac{1}{x}e^{1/x}$ is shown.
$\lim\limits_{x \to \infty} f(x) = \lim\limits_{x \to 0^+} f\left(\frac{1}{x}\right) = \infty$
$\lim\limits_{x \to -\infty} f(x) = \lim\limits_{x \to 0^-} f\left(\frac{1}{x}\right) = 0$

[-4, 4] by [-1, 3]

67. The graph of $y = f\left(\frac{1}{x}\right) = x \ln\left|\frac{1}{x}\right|$ is shown.
$\lim\limits_{x \to \infty} f(x) = \lim\limits_{x \to 0^+} f\left(\frac{1}{x}\right) = 0$
$\lim\limits_{x \to -\infty} f(x) = \lim\limits_{x \to 0^-} f\left(\frac{1}{x}\right) = 0$

[-3, 3] by [-2, 2]

69. (a) $y \to \infty$ (see accompanying graph)

(b) $y \to \infty$ (see accompanying graph)

(c) cusps at $x = \pm 1$ (see accompanying graph)

$y = \frac{3}{2}\left(x - \frac{1}{x}\right)^{2/3}$

2.6 CONTINUITY

1. No, discontinuous at $x = 2$, not defined at $x = 2$

3. Continuous on $[-1, 3]$

5. (a) Yes

(b) Yes, $\lim\limits_{x \to -1^+} f(x) = 0$

(c) Yes

(d) Yes

7. (a) No

(b) No

9. $f(2) = 0$, since $\lim\limits_{x \to 2^-} f(x) = -2(2) + 4 = 0 = \lim\limits_{x \to 2^+} f(x)$

11. Nonremovable discontinuity at $x = 1$ because $\lim\limits_{x \to 1} f(x)$ fails to exist ($\lim\limits_{x \to 1^-} f(x) = 1$ and $\lim\limits_{x \to 1^+} f(x) = 0$).
    Removable discontinuity at $x = 0$ by assigning the number $\lim\limits_{x \to 0} f(x) = 0$ to be the value of $f(0)$ rather than $f(0) = 1$.

13. Discontinuous only when $x - 2 = 0 \Rightarrow x = 2$

15. Discontinuous only when $x^2 - 4x + 3 = 0 \Rightarrow (x - 3)(x - 1) = 0 \Rightarrow x = 3$ or $x = 1$

17. Continuous everywhere. ($|x - 1| + \sin x$ defined for all x; limits exist and are equal to function values.)

19. Discontinuous only at $x = 0$

21. Discontinuous when $2x$ is an integer multiple of π, i.e., $2x = n\pi$, n an integer $\Rightarrow x = \frac{n\pi}{2}$, n an integer, but continuous at all other x.

23. Discontinuous at odd integer multiples of $\frac{\pi}{2}$, i.e., $x = (2n - 1)\frac{\pi}{2}$, n an integer, but continuous at all other x.

25. Discontinuous when $2x + 3 < 0$ or $x < -\frac{3}{2} \Rightarrow$ continuous on the interval $\left[-\frac{3}{2}, \infty\right)$.

27. Continuous everywhere: $(2x - 1)^{1/3}$ is defined for all x; limits exist and are equal to function values.

29. $\lim\limits_{x \to \pi} \sin(x - \sin x) = \sin(\pi - \sin \pi) = \sin(\pi - 0) = \sin \pi = 0$, and function continuous at $x = \pi$.

31. $\lim\limits_{y \to 1} \sec(y \sec^2 y - \tan^2 y - 1) = \lim\limits_{y \to 1} \sec(y \sec^2 y - \sec^2 y) = \lim\limits_{y \to 1} \sec((y - 1)\sec^2 y) = \sec((1 - 1)\sec^2 1)$
    $= \sec 0 = 1$, and function continuous at $y = 1$.

33. $\lim\limits_{x \to 0^+} \sin\left(\frac{\pi}{2}e^{\sqrt{x}}\right) = \sin\left(\frac{\pi}{2}e^0\right) = \sin\left(\frac{\pi}{2}\right) = 1$, and function is continuous at x = 0.

35. $g(x) = \frac{x^2 - 9}{x - 3} = \frac{(x+3)(x-3)}{(x-3)} = x + 3, x \neq 3 \Rightarrow g(3) = \lim\limits_{x \to 3}(x + 3) = 6$

37. $f(s) = \frac{s^3 - 1}{s^2 - 1} = \frac{(s^2 + s + 1)(s - 1)}{(s+1)(s-1)} = \frac{s^2 + s + 1}{s + 1}, s \neq 1 \Rightarrow f(1) = \lim\limits_{s \to 1}\left(\frac{s^2 + s + 1}{s + 1}\right) = \frac{3}{2}$

39. As defined, $\lim\limits_{x \to 3^-} f(x) = (3)^2 - 1 = 8$ and $\lim\limits_{x \to 3^+} (2a)(3) = 6a$. For f(x) to be continuous we must have
    $6a = 8 \Rightarrow a = \frac{4}{3}$.

41. The function can be extended: $f(0) \approx 2.3$.

43. The function cannot be extended to be continuous at x = 0. If f(0) = 1, it will be continuous from the right. Or if f(0) = −1, it will be continuous from the left.

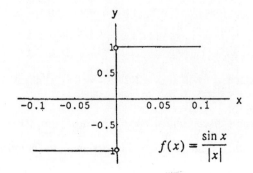

45. f(x) is continuous on [0, 1] and f(0) < 0, f(1) > 0 ⇒ by the Intermediate Value Theorem f(x) takes on every value between f(0) and f(1) ⇒ the equation f(x) = 0 has at least one solution between x = 0 and x = 1.

47. Let $f(x) = x^3 - 15x + 1$, which is continuous on $[-4, 4]$. Then $f(-4) = -3$, $f(-1) = 15$, $f(1) = -13$, and $f(4) = 5$. By the Intermediate Value Theorem, f(x) = 0 for some x in each of the intervals $-4 < x < -1$, $-1 < x < 1$, and $1 < x < 4$. That is, $x^3 - 15x + 1 = 0$ has three solutions in $[-4, 4]$. Since a polynomial of degree 3 can have at most 3 solutions, these are the only solutions.

49. Answers may vary. Note that f is continuous for every value of x.
    (a) $f(0) = 10$, $f(1) = 1^3 - 8(1) + 10 = 3$. Since $3 < \pi < 10$, by the Intermediate Value Theorem, there exists a c so that $0 < c < 1$ and $f(c) = \pi$.
    (b) $f(0) = 10$, $f(-4) = (-4)^3 - 8(-4) + 10 = -22$. Since $-22 < -\sqrt{3} < 10$, by the Intermediate Value Theorem, there exists a c so that $-4 < c < 0$ and $f(c) = -\sqrt{3}$.
    (c) $f(0) = 10$, $f(1000) = (1000)^3 - 8(1000) + 10 = 999,992,010$. Since $10 < 5,000,000 < 999,992,010$, by the Intermediate Value Theorem, there exists a c so that $0 < c < 1000$ and $f(c) = 5,000,000$.

51. Answers may vary. For example, $f(x) = \frac{\sin(x-2)}{x-2}$ is discontinuous at x = 2 because it is not defined there. However, the discontinuity can be removed because f has a limit (namely 1) as x → 2.

53. (a) Suppose x_0 is rational $\Rightarrow f(x_0) = 1$. Choose $\epsilon = \frac{1}{2}$. For any $\delta > 0$ there is an irrational number x (actually infinitely many) in the interval $(x_0 - \delta, x_0 + \delta) \Rightarrow f(x) = 0$. Then $0 < |x - x_0| < \delta$ but $|f(x) - f(x_0)|$ $= 1 > \frac{1}{2} = \epsilon$, so $\lim\limits_{x \to x_0} f(x)$ fails to exist $\Rightarrow$ f is discontinuous at x_0 rational.

   On the other hand, x_0 irrational $\Rightarrow f(x_0) = 0$ and there is a rational number x in $(x_0 - \delta, x_0 + \delta) \Rightarrow f(x)$ $= 1$. Again $\lim\limits_{x \to x_0} f(x)$ fails to exist $\Rightarrow$ f is discontinuous at x_0 irrational. That is, f is discontinuous at every point.

   (b) f is neither right-continuous nor left-continuous at any point x_0 because in every interval $(x_0 - \delta, x_0)$ or $(x_0, x_0 + \delta)$ there exist both rational and irrational real numbers. Thus neither limits $\lim\limits_{x \to x_0^-} f(x)$ and $\lim\limits_{x \to x_0^+} f(x)$ exist by the same arguments used in part (a).

55. No. For instance, if $f(x) = 0$, $g(x) = \lceil x \rceil$, then $h(x) = 0 (\lceil x \rceil) = 0$ is continuous at $x = 0$ and $g(x)$ is not.

57. Yes, because of the Intermediate Value Theorem. If f(a) and f(b) did have different signs then f would have to equal zero at some point between a and b since f is continuous on [a, b].

59. If $f(0) = 0$ or $f(1) = 1$, we are done (i.e., $c = 0$ or $c = 1$ in those cases). Then let $f(0) = a > 0$ and $f(1) = b < 1$ because $0 \le f(x) \le 1$. Define $g(x) = f(x) - x \Rightarrow$ g is continuous on [0, 1]. Moreover, $g(0) = f(0) - 0 = a > 0$ and $g(1) = f(1) - 1 = b - 1 < 0 \Rightarrow$ by the Intermediate Value Theorem there is a number c in $(0, 1)$ such that $g(c) = 0 \Rightarrow f(c) - c = 0$ or $f(c) = c$.

61. By Exercises 52 in Section 2.3, we have $\lim\limits_{x \to c} f(x) = L \Leftrightarrow \lim\limits_{h \to 0} f(c + h) = L$.

   Thus, $f(x)$ is continuous at $x = c \Leftrightarrow \lim\limits_{x \to c} f(x) = f(c) \Leftrightarrow \lim\limits_{h \to 0} f(c + h) = f(c)$.

63. $x \approx 1.8794, -1.5321, -0.3473$ 65. $x \approx 1.7549$

67. $x \approx 3.5156$ 69. $x \approx 0.7391$

2.7 TANGENTS AND DERIVATIVES AT A POINT

1. P_1: $m_1 = 1$, P_2: $m_2 = 5$ 3. P_1: $m_1 = \frac{5}{2}$, P_2: $m_2 = -\frac{1}{2}$

5. $m = \lim\limits_{h \to 0} \frac{[3 - (-1 + h)^2] - (3 - (-1)^2)}{h}$

   $= \lim\limits_{h \to 0} \frac{3 - (1 - 2h + h^2) - 2}{h} = \lim\limits_{h \to 0} \frac{h(2 - h)}{h} = 2$;
   at $(-1, 2)$: $y = 2 + 2(x - (-1)) \Rightarrow y = 2x + 4$, tangent line

7. $m = \lim\limits_{h \to 0} \dfrac{2\sqrt{1+h} - 2\sqrt{1}}{h} = \lim\limits_{h \to 0} \dfrac{2\sqrt{1+h} - 2}{h} \cdot \dfrac{2\sqrt{1+h} + 2}{2\sqrt{1+h} + 2}$

$= \lim\limits_{h \to 0} \dfrac{4(1+h) - 4}{2h\left(\sqrt{1+h} + 1\right)} = \lim\limits_{h \to 0} \dfrac{2}{\sqrt{1+h} + 1} = 1;$

at $(1, 2)$: $y = 2 + 1(x - 1) \Rightarrow y = x + 1$, tangent line

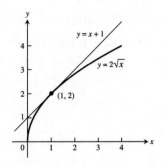

9. $m = \lim\limits_{h \to 0} \dfrac{(-2+h)^3 + 1 - ((-2)^3 + 1)}{h} = \lim\limits_{h \to 0} \dfrac{-8 + 12h - 6h^2 + h^3 + 8}{h}$

$= \lim\limits_{h \to 0} \left(12 - 6h + h^2\right) = 12;$

at $(-2, -7)$: $y = -7 + 12(x - (-2)) \Rightarrow y = 12x + 17$,
tangent line

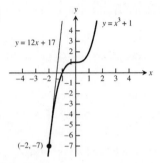

11. $m = \lim\limits_{h \to 0} \dfrac{[(2+h)^2 + 1] - 5}{h} = \lim\limits_{h \to 0} \dfrac{(5 + 4h + h^2) - 5}{h} = \lim\limits_{h \to 0} \dfrac{h(4+h)}{h} = 4;$

at $(2, 5)$: $y - 5 = 4(x - 2)$, tangent line

13. $m = \lim\limits_{h \to 0} \dfrac{\frac{3+h}{(3+h)-2} - 3}{h} = \lim\limits_{h \to 0} \dfrac{(3+h) - 3(h+1)}{h(h+1)} = \lim\limits_{h \to 0} \dfrac{-2h}{h(h+1)} = -2;$

at $(3, 3)$: $y - 3 = -2(x - 3)$, tangent line

15. $m = \lim\limits_{h \to 0} \dfrac{((2+h)^3 - (2+h)) - 6}{h} = \lim\limits_{h \to 0} \dfrac{(6 + 11h + 6h^2 + h^3) - 6}{h} = \lim\limits_{h \to 0} \dfrac{h(11 + 6h + h^2)}{h} = 11;$

at $(2, 6)$: $y - 6 = 11(t - 2)$, tangent line

17. $m = \lim\limits_{h \to 0} \dfrac{\sqrt{4+h} - 2}{h} = \lim\limits_{h \to 0} \dfrac{\sqrt{4+h} - 2}{h} \cdot \dfrac{\sqrt{4+h} + 2}{\sqrt{4+h} + 2} = \lim\limits_{h \to 0} \dfrac{(4+h) - 4}{h\left(\sqrt{4+h} + 2\right)} = \lim\limits_{h \to 0} \dfrac{h}{h\left(\sqrt{4+h} + 2\right)} = \dfrac{1}{\sqrt{4} + 2}$

$= \frac{1}{4}$; at $(4, 2)$: $y - 2 = \frac{1}{4}(x - 4)$, tangent line

19. At $x = -1$, $y = 5 \Rightarrow m = \lim\limits_{h \to 0} \dfrac{5(-1+h)^2 - 5}{h} = \lim\limits_{h \to 0} \dfrac{5(1 - 2h + h^2) - 5}{h} = \lim\limits_{h \to 0} \dfrac{5h(-2+h)}{h} = -10$, slope

21. At $x = 3$, $y = \frac{1}{2} \Rightarrow m = \lim\limits_{h \to 0} \dfrac{\frac{1}{(3+h)-1} - \frac{1}{2}}{h} = \lim\limits_{h \to 0} \dfrac{2 - (2+h)}{2h(2+h)} = \lim\limits_{h \to 0} \dfrac{-h}{2h(2+h)} = -\frac{1}{4}$, slope

23. At a horizontal tangent the slope $m = 0 \Rightarrow 0 = m = \lim\limits_{h \to 0} \dfrac{[(x+h)^2 + 4(x+h) - 1] - (x^2 + 4x - 1)}{h}$

$= \lim\limits_{h \to 0} \dfrac{(x^2 + 2xh + h^2 + 4x + 4h - 1) - (x^2 + 4x - 1)}{h} = \lim\limits_{h \to 0} \dfrac{(2xh + h^2 + 4h)}{h} = \lim\limits_{h \to 0} (2x + h + 4) = 2x + 4;$

$2x + 4 = 0 \Rightarrow x = -2$. Then $f(-2) = 4 - 8 - 1 = -5 \Rightarrow (-2, -5)$ is the point on the graph where there is a horizontal tangent.

25. $-1 = m = \lim\limits_{h \to 0} \dfrac{\frac{1}{(x+h)-1} - \frac{1}{x-1}}{h} = \lim\limits_{h \to 0} \dfrac{(x-1) - (x+h-1)}{h(x-1)(x+h-1)} = \lim\limits_{h \to 0} \dfrac{-h}{h(x-1)(x+h-1)} = -\dfrac{1}{(x-1)^2}$

$\Rightarrow (x-1)^2 = 1 \Rightarrow x^2 - 2x = 0 \Rightarrow x(x-2) = 0 \Rightarrow x = 0$ or $x = 2$. If $x = 0$, then $y = -1$ and $m = -1$

$\Rightarrow y = -1 - (x - 0) = -(x + 1)$. If $x = 2$, then $y = 1$ and $m = -1 \Rightarrow y = 1 - (x - 2) = -(x - 3)$.

27. $\lim\limits_{h \to 0} \frac{f(2+h)-f(2)}{h} = \lim\limits_{h \to 0} \frac{(100-4.9(2+h)^2)-(100-4.9(2)^2)}{h} = \lim\limits_{h \to 0} \frac{-4.9(4+4h+h^2)+4.9(4)}{h}$

$= \lim\limits_{h \to 0} (-19.6 - 4.9h) = -19.6.$ The minus sign indicates the object is falling <u>downward</u> at a speed of

19.6 m/sec.

29. $\lim\limits_{h \to 0} \frac{f(3+h)-f(3)}{h} = \lim\limits_{h \to 0} \frac{\pi(3+h)^2 - \pi(3)^2}{h} = \lim\limits_{h \to 0} \frac{\pi[9+6h+h^2-9]}{h} = \lim\limits_{h \to 0} \pi(6+h) = 6\pi$

31. $\lim\limits_{h \to 0} \frac{(m(x_0+h)+b)-(mx_0+b)}{h} = \lim\limits_{h \to 0} \frac{mx_0+mh+b-mx_0-b}{h} = \lim\limits_{h \to 0} \frac{mh}{h} = \lim\limits_{h \to 0} m = m$

$y - (mx_0 + b) = m(x - x_0) \to y - mx_0 - b = mx - mx_0 \to y = mx + b$

33. Slope at origin $= \lim\limits_{h \to 0} \frac{f(0+h)-f(0)}{h} = \lim\limits_{h \to 0} \frac{h^2 \sin\left(\frac{1}{h}\right)}{h} = \lim\limits_{h \to 0} h \sin\left(\frac{1}{h}\right) = 0 \Rightarrow$ yes, f(x) does have a tangent at

the origin with slope 0.

35. $\lim\limits_{h \to 0^-} \frac{f(0+h)-f(0)}{h} = \lim\limits_{h \to 0^-} \frac{-1-0}{h} = \infty$, and $\lim\limits_{h \to 0^+} \frac{f(0+h)-f(0)}{h} = \lim\limits_{h \to 0^+} \frac{1-0}{h} = \infty$. Therefore,

$\lim\limits_{h \to 0} \frac{f(0+h)-f(0)}{h} = \infty \Rightarrow$ yes, the graph of f has a vertical tangent at the origin.

37. (a) The graph appears to have a cusp at $x = 0$.

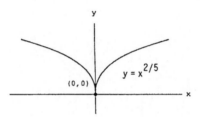

(b) $\lim\limits_{h \to 0^-} \frac{f(0+h)-f(0)}{h} = \lim\limits_{h \to 0^-} \frac{h^{2/5}-0}{h} = \lim\limits_{h \to 0^-} \frac{1}{h^{3/5}} = -\infty$ and $\lim\limits_{h \to 0^+} \frac{1}{h^{3/5}} = \infty \Rightarrow$ limit does not exist

$\Rightarrow$ the graph of $y = x^{2/5}$ does not have a vertical tangent at $x = 0$.

39. (a) The graph appears to have a vertical tangent at $x = 0$.

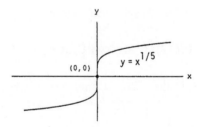

(b) $\lim\limits_{h \to 0} \frac{f(0+h)-f(0)}{h} = \lim\limits_{h \to 0} \frac{h^{1/5}-0}{h} = \lim\limits_{h \to 0} \frac{1}{h^{4/5}} = \infty \Rightarrow y = x^{1/5}$ has a vertical tangent at $x = 0$.

41. (a) The graph appears to have a cusp at $x = 0$.

(b) $\lim\limits_{h \to 0^-} \frac{f(0+h)-f(0)}{h} = \lim\limits_{h \to 0^-} \frac{4h^{2/5}-2h}{h} = \lim\limits_{h \to 0^-} \frac{4}{h^{3/5}} - 2 = -\infty$ and $\lim\limits_{h \to 0^+} \frac{4}{h^{3/5}} - 2 = \infty$

$\Rightarrow$ limit does not exist $\Rightarrow$ the graph of $y = 4x^{2/5} - 2x$ does not have a vertical tangent at $x = 0$.

43. (a) The graph appears to have a vertical tangent at $x = 1$ and a cusp at $x = 0$.

$y = x^{2/3} - (x-1)^{1/3}$

(b) $x = 1$: $\lim\limits_{h \to 0} \frac{(1+h)^{2/3} - (1+h-1)^{1/3} - 1}{h} = \lim\limits_{h \to 0} \frac{(1+h)^{2/3} - h^{1/3} - 1}{h} = -\infty$

$\Rightarrow y = x^{2/3} - (x-1)^{1/3}$ has a vertical tangent at $x = 1$;

$x = 0$: $\lim\limits_{h \to 0} \frac{f(0+h) - f(0)}{h} = \lim\limits_{h \to 0} \frac{h^{2/3} - (h-1)^{1/3} - (-1)^{1/3}}{h} = \lim\limits_{h \to 0} \left[\frac{1}{h^{1/3}} - \frac{(h-1)^{1/3}}{h} + \frac{1}{h} \right]$

does not exist $\Rightarrow y = x^{2/3} - (x-1)^{1/3}$ does not have a vertical tangent at $x = 0$.

45. (a) The graph appears to have a vertical tangent at $x = 0$.

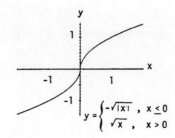

$y = \begin{cases} -\sqrt{|x|}, & x \le 0 \\ \sqrt{x}, & x > 0 \end{cases}$

(b) $\lim\limits_{h \to 0^+} \frac{f(0+h) - f(0)}{h} = \lim\limits_{x \to 0^+} \frac{\sqrt{h} - 0}{h} = \lim\limits_{h \to 0} \frac{1}{\sqrt{h}} = \infty$;

$\lim\limits_{h \to 0^-} \frac{f(0+h) - f(0)}{h} = \lim\limits_{h \to 0^-} \frac{-\sqrt{|h|} - 0}{h} = \lim\limits_{h \to 0^-} \frac{-\sqrt{|h|}}{-|h|} = \lim\limits_{h \to 0^-} \frac{1}{\sqrt{|h|}} = \infty$

$\Rightarrow y$ has a vertical tangent at $x = 0$.

CHAPTER 2 PRACTICE EXERCISES

1. At $x = -1$: $\lim\limits_{x \to -1^-} f(x) = \lim\limits_{x \to -1^+} f(x) = 1$

$\Rightarrow \lim\limits_{x \to -1} f(x) = 1 = f(-1)$

$\Rightarrow f$ is continuous at $x = -1$.

At $x = 0$: $\lim\limits_{x \to 0^-} f(x) = \lim\limits_{x \to 0^+} f(x) = 0 \Rightarrow \lim\limits_{x \to 0} f(x) = 0$.

But $f(0) = 1 \ne \lim\limits_{x \to 0} f(x)$

$\Rightarrow f$ is discontinuous at $x = 0$.

If we define $f(0) = 0$, then the discontinuity at $x = 0$ is removable.

At $x = 1$: $\lim\limits_{x \to 1^-} f(x) = -1$ and $\lim\limits_{x \to 1^+} f(x) = 1$

$\Rightarrow \lim\limits_{x \to 1} f(x)$ does not exist

$\Rightarrow f$ is discontinuous at $x = 1$.

$y = f(x)$

3. (a) $\lim\limits_{t \to t_0} (3f(t)) = 3 \lim\limits_{t \to t_0} f(t) = 3(-7) = -21$

(b) $\lim\limits_{t \to t_0} (f(t))^2 = \left(\lim\limits_{t \to t_0} f(t) \right)^2 = (-7)^2 = 49$

(c) $\lim\limits_{t \to t_0} (f(t) \cdot g(t)) = \lim\limits_{t \to t_0} f(t) \cdot \lim\limits_{t \to t_0} g(t) = (-7)(0) = 0$

(d) $\lim\limits_{t \to t_0} \frac{f(t)}{g(t)-7} = \frac{\lim\limits_{t \to t_0} f(t)}{\lim\limits_{t \to t_0}(g(t)-7)} = \frac{\lim\limits_{t \to t_0} f(t)}{\lim\limits_{t \to t_0} g(t) - \lim\limits_{t \to t_0} 7} = \frac{-7}{0-7} = 1$

(e) $\lim\limits_{t \to t_0} \cos(g(t)) = \cos\left(\lim\limits_{t \to t_0} g(t)\right) = \cos 0 = 1$

(f) $\lim\limits_{t \to t_0} |f(t)| = \left|\lim\limits_{t \to t_0} f(t)\right| = |-7| = 7$

(g) $\lim\limits_{t \to t_0}(f(t)+g(t)) = \lim\limits_{t \to t_0} f(t) + \lim\limits_{t \to t_0} g(t) = -7 + 0 = -7$

(h) $\lim\limits_{t \to t_0}\left(\frac{1}{f(t)}\right) = \frac{1}{\lim\limits_{t \to t_0} f(t)} = \frac{1}{-7} = -\frac{1}{7}$

5. Since $\lim\limits_{x \to 0} x = 0$ we must have that $\lim\limits_{x \to 0}(4 - g(x)) = 0$. Otherwise, if $\lim\limits_{x \to 0}(4 - g(x))$ is a finite positive number, we would have $\lim\limits_{x \to 0^-}\left[\frac{4-g(x)}{x}\right] = -\infty$ and $\lim\limits_{x \to 0^+}\left[\frac{4-g(x)}{x}\right] = \infty$ so the limit could not equal 1 as $x \to 0$. Similar reasoning holds if $\lim\limits_{x \to 0}(4 - g(x))$ is a finite negative number. We conclude that $\lim\limits_{x \to 0} g(x) = 4$.

7. (a) $\lim\limits_{x \to c} f(x) = \lim\limits_{x \to c} x^{1/3} = c^{1/3} = f(c)$ for every real number $c \Rightarrow f$ is continuous on $(-\infty, \infty)$.

(b) $\lim\limits_{x \to c} g(x) = \lim\limits_{x \to c} x^{3/4} = c^{3/4} = g(c)$ for every nonnegative real number $c \Rightarrow g$ is continuous on $[0, \infty)$.

(c) $\lim\limits_{x \to c} h(x) = \lim\limits_{x \to c} x^{-2/3} = \frac{1}{c^{2/3}} = h(c)$ for every nonzero real number $c \Rightarrow h$ is continuous on $(-\infty, 0)$ and $(-\infty, \infty)$.

(d) $\lim\limits_{x \to c} k(x) = \lim\limits_{x \to c} x^{-1/6} = \frac{1}{c^{1/6}} = k(c)$ for every positive real number $c \Rightarrow k$ is continuous on $(0, \infty)$

9. (a) $\lim\limits_{x \to 0} \frac{x^2 - 4x + 4}{x^3 + 5x^2 - 14x} = \lim\limits_{x \to 0} \frac{(x-2)(x-2)}{x(x+7)(x-2)} = \lim\limits_{x \to 0} \frac{x-2}{x(x+7)}$, $x \neq 2$; the limit does not exist because $\lim\limits_{x \to 0^-} \frac{x-2}{x(x+7)} = \infty$ and $\lim\limits_{x \to 0^+} \frac{x-2}{x(x+7)} = -\infty$

(b) $\lim\limits_{x \to 2} \frac{x^2 - 4x + 4}{x^3 + 5x^2 - 14x} = \lim\limits_{x \to 2} \frac{(x-2)(x-2)}{x(x+7)(x-2)} = \lim\limits_{x \to 2} \frac{x-2}{x(x+7)}$, $x \neq 2$, and $\lim\limits_{x \to 2} \frac{x-2}{x(x+7)} = \frac{0}{2(9)} = 0$

11. $\lim\limits_{x \to 1} \frac{1 - \sqrt{x}}{1 - x} = \lim\limits_{x \to 1} \frac{1 - \sqrt{x}}{(1 - \sqrt{x})(1 + \sqrt{x})} = \lim\limits_{x \to 1} \frac{1}{1 + \sqrt{x}} = \frac{1}{2}$

13. $\lim\limits_{h \to 0} \frac{(x+h)^2 - x^2}{h} = \lim\limits_{h \to 0} \frac{(x^2 + 2hx + h^2) - x^2}{h} = \lim\limits_{h \to 0}(2x + h) = 2x$

15. $\lim\limits_{x \to 0} \frac{\frac{1}{2+x} - \frac{1}{2}}{x} = \lim\limits_{x \to 0} \frac{2 - (2+x)}{2x(2+x)} = \lim\limits_{x \to 0} \frac{-1}{4 + 2x} = -\frac{1}{4}$

17. $\lim\limits_{x \to 0} \frac{\tan 2x}{\tan \pi x} = \lim\limits_{x \to 0} \frac{\sin 2x}{\cos 2x} \cdot \frac{\cos \pi x}{\sin \pi x} = \lim\limits_{x \to 0} \left(\frac{\sin 2x}{2x}\right)\left(\frac{\cos \pi x}{\cos 2x}\right)\left(\frac{\pi x}{\sin \pi x}\right)\left(\frac{2x}{\pi x}\right) = 1 \cdot 1 \cdot 1 \cdot \frac{2}{\pi} = \frac{2}{\pi}$

19. $\lim\limits_{x \to \pi} \sin\left(\frac{x}{2} + \sin x\right) = \sin\left(\frac{\pi}{2} + \sin \pi\right) = \sin\left(\frac{\pi}{2}\right) = 1$

21. Let $x = t - 3 \Rightarrow \lim\limits_{t \to 3^+} \ln(t - 3) = \lim\limits_{x \to 0^+} \ln x = -\infty$

23. $-1 \leq \cos\left(\frac{\pi}{\theta}\right) \leq 1 \Rightarrow e^{-1} \leq e^{\cos(\pi/\theta)} \leq e^1 \Rightarrow \sqrt{\theta} e^{-1} \leq \sqrt{\theta} e^{\cos(\pi/\theta)} \leq \sqrt{\theta} e \Rightarrow \lim\limits_{\theta \to 0^+} \sqrt{\theta} e^{\cos(\pi/\theta)} = 0$ by the Sandwich Theorem

25. $\lim\limits_{x \to 0^+} [4 g(x)]^{1/3} = 2 \Rightarrow \left[\lim\limits_{x \to 0^+} 4 g(x)\right]^{1/3} = 2 \Rightarrow \lim\limits_{x \to 0^+} 4 g(x) = 8$, since $2^3 = 8$. Then $\lim\limits_{x \to 0^+} g(x) = 2$.

27. $\lim\limits_{x \to 1} \frac{3x^2 + 1}{g(x)} = \infty \Rightarrow \lim\limits_{x \to 1} g(x) = 0$ since $\lim\limits_{x \to 1}(3x^2 + 1) = 4$

29. $\lim\limits_{x \to \infty} \frac{2x + 3}{5x + 7} = \lim\limits_{x \to \infty} \frac{2 + \frac{3}{x}}{5 + \frac{7}{x}} = \frac{2 + 0}{5 + 0} = \frac{2}{5}$

31. $\lim\limits_{x \to -\infty} \frac{x^2 - 4x + 8}{3x^3} = \lim\limits_{x \to -\infty} \left(\frac{1}{3x} - \frac{4}{3x^2} + \frac{8}{3x^3} \right) = 0 - 0 + 0 = 0$

33. $\lim\limits_{x \to -\infty} \frac{x^2 - 7x}{x + 1} = \lim\limits_{x \to -\infty} \frac{x - 7}{1 + \frac{1}{x}} = -\infty$

35. $\lim\limits_{x \to \infty} \frac{\sin x}{\lfloor x \rfloor} \leq \lim\limits_{x \to \infty} \frac{1}{\lfloor x \rfloor} = 0$ since int $x \to \infty$ as $x \to \infty \Rightarrow \lim\limits_{x \to \infty} \frac{\sin x}{\lfloor x \rfloor} = 0$.

37. $\lim\limits_{x \to \infty} \frac{x + \sin x + 2\sqrt{x}}{x + \sin x} = \lim\limits_{x \to \infty} \frac{1 + \frac{\sin x}{x} + \frac{2}{\sqrt{x}}}{1 + \frac{\sin x}{x}} = \frac{1 + 0 + 0}{1 + 0} = 1$

39. $\lim\limits_{x \to \infty} e^{1/x} \cos\left(\frac{1}{x}\right) = e^0 \cdot \cos(0) = 1 \cdot 1 = 1$

41. $\lim\limits_{x \to -\infty} \tan^{-1} x = -\frac{\pi}{2}$

43. At $x = -1$: $\lim\limits_{x \to -1^-} f(x) = \lim\limits_{x \to -1^-} \frac{x(x^2 - 1)}{|x^2 - 1|}$

    $= \lim\limits_{x \to -1^-} \frac{x(x^2 - 1)}{x^2 - 1} = \lim\limits_{x \to -1^-} x = -1$, and

    $\lim\limits_{x \to -1^+} f(x) = \lim\limits_{x \to -1^+} \frac{x(x^2 - 1)}{|x^2 - 1|} = \lim\limits_{x \to -1^+} \frac{x(x^2 - 1)}{-(x^2 - 1)}$

    $= \lim\limits_{x \to -1} (-x) = -(-1) = 1$. Since

    $\lim\limits_{x \to -1^-} f(x) \neq \lim\limits_{x \to -1^+} f(x)$

    $\Rightarrow \lim\limits_{x \to -1} f(x)$ does not exist, the function f <u>cannot</u> be

    extended to a continuous function at $x = -1$.

    At $x = 1$: $\lim\limits_{x \to 1^-} f(x) = \lim\limits_{x \to 1^-} \frac{x(x^2 - 1)}{|x^2 - 1|} = \lim\limits_{x \to 1^-} \frac{x(x^2 - 1)}{-(x^2 - 1)} = \lim\limits_{x \to 1^-} (-x) = -1$, and

    $\lim\limits_{x \to 1^+} f(x) = \lim\limits_{x \to 1^+} \frac{x(x^2 - 1)}{|x^2 - 1|} = \lim\limits_{x \to 1^+} \frac{x(x^2 - 1)}{x^2 - 1} = \lim\limits_{x \to 1^+} x = 1$. Again $\lim\limits_{x \to 1} f(x)$ does not exist so f

    <u>cannot</u> be extended to a continuous function at $x = 1$ either.

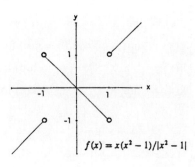

$f(x) = x(x^2 - 1)/|x^2 - 1|$

45. Yes, f does have a continuous extension to $a = 1$:

    define $f(1) = \lim\limits_{x \to 1} \frac{x - 1}{x - \sqrt[4]{x}} = \frac{4}{3}$.

$f(x) = \dfrac{x - 1}{x - \sqrt[4]{x}}$, $a = 1$

47. From the graph we see that $\lim\limits_{t \to 0^-} h(t) \neq \lim\limits_{t \to 0^+} h(t)$

    so h <u>cannot</u> be extended to a continuous function

    at $a = 0$.

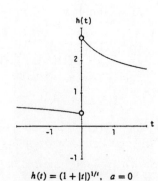

$h(t) = (1 + |t|)^{1/t}$, $a = 0$

49. (a) $f(-1) = -1$ and $f(2) = 5 \Rightarrow$ f has a root between -1 and 2 by the Intermediate Value Theorem.

(b), (c) root is 1.32471795724

CHAPTER 2 ADDITIONAL AND ADVANCED EXERCISES

1. (a)

x	0.1	0.01	0.001	0.0001	0.00001
x^x	0.7943	0.9550	0.9931	0.9991	0.9999

Apparently, $\lim\limits_{x \to 0^+} x^x = 1$

(b)

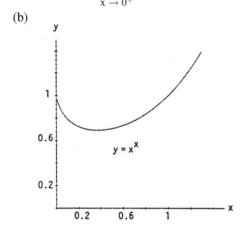

3. $\lim\limits_{v \to c^-} L = \lim\limits_{v \to c^-} L_0 \sqrt{1 - \frac{v^2}{c^2}} = L_0 \sqrt{1 - \frac{\lim\limits_{v \to c^-} v^2}{c^2}} = L_0 \sqrt{1 - \frac{c^2}{c^2}} = 0$

The left-hand limit was needed because the function L is undefined if $v > c$ (the rocket cannot move faster than the speed of light).

5. $|10 + (t - 70) \times 10^{-4} - 10| < 0.0005 \Rightarrow |(t - 70) \times 10^{-4}| < 0.0005 \Rightarrow -0.0005 < (t - 70) \times 10^{-4} < 0.0005$
$\Rightarrow -5 < t - 70 < 5 \Rightarrow 65° < t < 75° \Rightarrow$ Within 5° F.

7. Show $\lim\limits_{x \to 1} f(x) = \lim\limits_{x \to 1} (x^2 - 7) = -6 = f(1)$.

Step 1: $|(x^2 - 7) + 6| < \epsilon \Rightarrow -\epsilon < x^2 - 1 < \epsilon \Rightarrow 1 - \epsilon < x^2 < 1 + \epsilon \Rightarrow \sqrt{1 - \epsilon} < x < \sqrt{1 + \epsilon}$.

Step 2: $|x - 1| < \delta \Rightarrow -\delta < x - 1 < \delta \Rightarrow -\delta + 1 < x < \delta + 1$.

Then $-\delta + 1 = \sqrt{1 - \epsilon}$ or $\delta + 1 = \sqrt{1 + \epsilon}$. Choose $\delta = \min\left\{1 - \sqrt{1 - \epsilon}, \sqrt{1 + \epsilon} - 1\right\}$, then

$0 < |x - 1| < \delta \Rightarrow |(x^2 - 7) - 6| < \epsilon$ and $\lim\limits_{x \to 1} f(x) = -6$. By the continuity test, f(x) is continuous at $x = 1$.

9. Show $\lim\limits_{x \to 2} h(x) = \lim\limits_{x \to 2} \sqrt{2x - 3} = 1 = h(2)$.

Step 1: $\left|\sqrt{2x - 3} - 1\right| < \epsilon \Rightarrow -\epsilon < \sqrt{2x - 3} - 1 < \epsilon \Rightarrow 1 - \epsilon < \sqrt{2x - 3} < 1 + \epsilon \Rightarrow \frac{(1 - \epsilon)^2 + 3}{2} < x < \frac{(1 + \epsilon)^2 + 3}{2}$.

Step 2: $|x - 2| < \delta \Rightarrow -\delta < x - 2 < \delta$ or $-\delta + 2 < x < \delta + 2$.

Then $-\delta + 2 = \frac{(1 - \epsilon)^2 + 3}{2} \Rightarrow \delta = 2 - \frac{(1 - \epsilon)^2 + 3}{2} = \frac{1 - (1 - \epsilon)^2}{2} = \epsilon - \frac{\epsilon^2}{2}$, or $\delta + 2 = \frac{(1 + \epsilon)^2 + 3}{2}$

$\Rightarrow \delta = \frac{(1 + \epsilon)^2 + 3}{2} - 2 = \frac{(1 + \epsilon)^2 - 1}{2} = \epsilon + \frac{\epsilon^2}{2}$. Choose $\delta = \epsilon - \frac{\epsilon^2}{2}$, the smaller of the two values. Then,

$0 < |x - 2| < \delta \Rightarrow \left|\sqrt{2x - 3} - 1\right| < \epsilon$, so $\lim\limits_{x \to 2} \sqrt{2x - 3} = 1$. By the continuity test, h(x) is continuous at $x = 2$.

11. Suppose L_1 and L_2 are two different limits. Without loss of generality assume $L_2 > L_1$. Let $\epsilon = \frac{1}{3}(L_2 - L_1)$.

Since $\lim\limits_{x \to x_0} f(x) = L_1$ there is a $\delta_1 > 0$ such that $0 < |x - x_0| < \delta_1 \Rightarrow |f(x) - L_1| < \epsilon \Rightarrow -\epsilon < f(x) - L_1 < \epsilon$

$\Rightarrow -\frac{1}{3}(L_2 - L_1) + L_1 < f(x) < \frac{1}{3}(L_2 - L_1) + L_1 \Rightarrow 4L_1 - L_2 < 3f(x) < 2L_1 + L_2$. Likewise, $\lim\limits_{x \to x_0} f(x) = L_2$

so there is a δ_2 such that $0 < |x - x_0| < \delta_2 \Rightarrow |f(x) - L_2| < \epsilon \Rightarrow -\epsilon < f(x) - L_2 < \epsilon$

$\Rightarrow -\frac{1}{3}(L_2 - L_1) + L_2 < f(x) < \frac{1}{3}(L_2 - L_1) + L_2 \Rightarrow 2L_2 + L_1 < 3f(x) < 4L_2 - L_1$

$\Rightarrow L_1 - 4L_2 < -3f(x) < -2L_2 - L_1$. If $\delta = \min\{\delta_1, \delta_2\}$ both inequalities must hold for $0 < |x - x_0| < \delta$:

$\left.\begin{array}{c} 4L_1 - L_2 < \ \ 3f(x) < \ \ 2L_1 + L_2 \\ L_1 - 4L_2 < -3f(x) < -2L_2 - L_1 \end{array}\right\} \Rightarrow 5(L_1 - L_2) < 0 < L_1 - L_2$. That is, $L_1 - L_2 < 0 \ \underline{\text{and}} \ L_1 - L_2 > 0$,

a contradiction.

13. (a) Since $x \to 0^+, 0 < x^3 < x < 1 \Rightarrow (x^3 - x) \to 0^- \Rightarrow \lim_{x \to 0^+} f(x^3 - x) = \lim_{y \to 0^-} f(y) = B$ where $y = x^3 - x$.

   (b) Since $x \to 0^-, -1 < x < x^3 < 0 \Rightarrow (x^3 - x) \to 0^+ \Rightarrow \lim_{x \to 0^-} f(x^3 - x) = \lim_{y \to 0^+} f(y) = A$ where $y = x^3 - x$.

   (c) Since $x \to 0^+, 0 < x^4 < x^2 < 1 \Rightarrow (x^2 - x^4) \to 0^+ \Rightarrow \lim_{x \to 0^+} f(x^2 - x^4) = \lim_{y \to 0^+} f(y) = A$ where $y = x^2 - x^4$.

   (d) Since $x \to 0^-, -1 < x < 0 \Rightarrow 0 < x^4 < x^2 < 1 \Rightarrow (x^2 - x^4) \to 0^+ \Rightarrow \lim_{x \to 0^+} f(x^2 - x^4) = A$ as in part (c).

15. Show $\lim_{x \to -1} f(x) = \lim_{x \to -1} \frac{x^2 - 1}{x + 1} = \lim_{x \to -1} \frac{(x+1)(x-1)}{(x+1)} = -2, x \neq -1$.

   Define the continuous extension of $f(x)$ as $F(x) = \begin{cases} \frac{x^2-1}{x+1}, & x \neq -1 \\ -2, & x = -1 \end{cases}$. We now prove the limit of $f(x)$ as $x \to -1$

   exists and has the correct value.

   Step 1: $\left|\frac{x^2-1}{x+1} - (-2)\right| < \epsilon \Rightarrow -\epsilon < \frac{(x+1)(x-1)}{(x+1)} + 2 < \epsilon \Rightarrow -\epsilon < (x-1) + 2 < \epsilon, x \neq -1 \Rightarrow -\epsilon - 1 < x < \epsilon - 1$.

   Step 2: $|x - (-1)| < \delta \Rightarrow -\delta < x + 1 < \delta \Rightarrow -\delta - 1 < x < \delta - 1$.

   Then $-\delta - 1 = -\epsilon - 1 \Rightarrow \delta = \epsilon$, or $\delta - 1 = \epsilon - 1 \Rightarrow \delta = \epsilon$. Choose $\delta = \epsilon$. Then $0 < |x - (-1)| < \delta$

   $\Rightarrow \left|\frac{x^2-1}{x+1} - (-2)\right| < \epsilon \Rightarrow \lim_{x \to -1} F(x) = -2$. Since the conditions of the continuity test are met by $F(x)$, then $f(x)$ has a

   continuous extension to $F(x)$ at $x = -1$.

17. (a) Let $\epsilon > 0$ be given. If x is rational, then $f(x) = x \Rightarrow |f(x) - 0| = |x - 0| < \epsilon \Leftrightarrow |x - 0| < \epsilon$; i.e., choose

   $\delta = \epsilon$. Then $|x - 0| < \delta \Rightarrow |f(x) - 0| < \epsilon$ for x rational. If x is irrational, then $f(x) = 0 \Rightarrow |f(x) - 0| < \epsilon$

   $\Leftrightarrow 0 < \epsilon$ which is true no matter how close irrational x is to 0, so again we can choose $\delta = \epsilon$. In either case,

   given $\epsilon > 0$ there is a $\delta = \epsilon > 0$ such that $0 < |x - 0| < \delta \Rightarrow |f(x) - 0| < \epsilon$. Therefore, f is continuous at $x = 0$.

   (b) Choose $x = c > 0$. Then within any interval $(c - \delta, c + \delta)$ there are both rational and irrational numbers.

   If c is rational, pick $\epsilon = \frac{c}{2}$. No matter how small we choose $\delta > 0$ there is an irrational number x in

   $(c - \delta, c + \delta) \Rightarrow |f(x) - f(c)| = |0 - c| = c > \frac{c}{2} = \epsilon$. That is, f is not continuous at any rational $c > 0$. On

   the other hand, suppose c is irrational $\Rightarrow f(c) = 0$. Again pick $\epsilon = \frac{c}{2}$. No matter how small we choose $\delta > 0$

   there is a rational number x in $(c - \delta, c + \delta)$ with $|x - c| < \frac{c}{2} = \epsilon \Leftrightarrow \frac{c}{2} < x < \frac{3c}{2}$. Then $|f(x) - f(c)| = |x - 0|$

   $= |x| > \frac{c}{2} = \epsilon \Rightarrow$ f is not continuous at any irrational $c > 0$.

   If $x = c < 0$, repeat the argument picking $\epsilon = \frac{|c|}{2} = \frac{-c}{2}$. Therefore f fails to be continuous at any

   nonzero value $x = c$.

19. Yes. Let R be the radius of the equator (earth) and suppose at a fixed instant of time we label noon as the zero

   point, 0, on the equator $\Rightarrow 0 + \pi R$ represents the midnight point (at the same exact time). Suppose x_1 is a point

   on the equator "just after" noon $\Rightarrow x_1 + \pi R$ is simultaneously "just after" midnight. It seems reasonable that

   the temperature T at a point just after noon is hotter than it would be at the diametrically opposite point just

   after midnight: That is, $T(x_1) - T(x_1 + \pi R) > 0$. At exactly the same moment in time pick x_2 to be a point

   just before midnight $\Rightarrow x_2 + \pi R$ is just before noon. Then $T(x_2) - T(x_2 + \pi R) < 0$. Assuming the temperature

   function T is continuous along the equator (which is reasonable), the Intermediate Value Theorem says there

   is a point c between 0 (noon) and πR (simultaneously midnight) such that $T(c) - T(c + \pi R) = 0$; i.e.,

   there is always a pair of antipodal points on the earth's equator where the temperatures are the same.

21. (a) $\lim\limits_{a \to 0} r_+(a) = \lim\limits_{a \to 0} \dfrac{-1+\sqrt{1+a}}{a} = \lim\limits_{a \to 0} \left(\dfrac{-1+\sqrt{1+a}}{a}\right)\left(\dfrac{-1-\sqrt{1+a}}{-1-\sqrt{1+a}}\right)$

$= \lim\limits_{a \to 0} \dfrac{1-(1+a)}{a\left(-1-\sqrt{1+a}\right)} = \dfrac{-1}{-1-\sqrt{1+0}} = \dfrac{1}{2}$

$\lim\limits_{a \to -1^+} r_+(a) = \lim\limits_{a \to -1^+} \dfrac{1-(1+a)}{a\left(-1-\sqrt{1+a}\right)} = \lim\limits_{a \to -1} \dfrac{-a}{a\left(-1-\sqrt{1+a}\right)} = \dfrac{-1}{-1-\sqrt{0}} = 1$

(b) $\lim\limits_{a \to 0^-} r_-(a) = \lim\limits_{a \to 0^-} \dfrac{-1-\sqrt{1+a}}{a} = \lim\limits_{a \to 0^-} \left(\dfrac{-1-\sqrt{1+a}}{a}\right)\left(\dfrac{-1+\sqrt{1+a}}{-1+\sqrt{1+a}}\right)$

$= \lim\limits_{a \to 0^-} \dfrac{1-(1+a)}{a\left(-1+\sqrt{1+a}\right)} = \lim\limits_{a \to 0^-} \dfrac{-a}{a\left(-1+\sqrt{1+a}\right)} = \lim\limits_{a \to 0^-} \dfrac{-1}{-1+\sqrt{1+a}} = \infty$ (because the

denominator is always negative); $\lim\limits_{a \to 0^+} r_-(a) = \lim\limits_{a \to 0^+} \dfrac{-1}{-1+\sqrt{1+a}} = -\infty$ (because the denominator

is always positive). Therefore, $\lim\limits_{a \to 0} r_-(a)$ does not exist.

$\lim\limits_{a \to -1^+} r_-(a) = \lim\limits_{a \to -1^+} \dfrac{-1-\sqrt{1+a}}{a} = \lim\limits_{a \to -1^+} \dfrac{-1}{-1+\sqrt{1+a}} = 1$

(c)

Graph not to scale

(d)

$f(x) = ax^2 + 2x - 1$

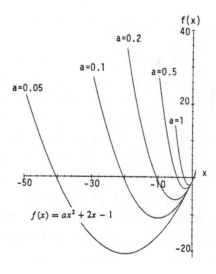

$f(x) = ax^2 + 2x - 1$

23. (a) The function f is bounded on D if $f(x) \geq M$ and $f(x) \leq N$ for all x in D. This means $M \leq f(x) \leq N$ for all x in D. Choose B to be max $\{|M|, |N|\}$. Then $|f(x)| \leq B$. On the other hand, if $|f(x)| \leq B$, then $-B \leq f(x) \leq B \Rightarrow f(x) \geq -B$ and $f(x) \leq B \Rightarrow f(x)$ is bounded on D with $N = B$ an upper bound and $M = -B$ a lower bound.

(b) Assume $f(x) \leq N$ for all x and that $L > N$. Let $\epsilon = \dfrac{L-N}{2}$. Since $\lim\limits_{x \to x_0} f(x) = L$ there is a $\delta > 0$ such that

$0 < |x - x_0| < \delta \Rightarrow |f(x) - L| < \epsilon \Leftrightarrow L - \epsilon < f(x) < L + \epsilon \Leftrightarrow L - \dfrac{L-N}{2} < f(x) < L + \dfrac{L-N}{2}$

$\Leftrightarrow \dfrac{L+N}{2} < f(x) < \dfrac{3L-N}{2}$. But $L > N \Rightarrow \dfrac{L+N}{2} > N \Rightarrow N < f(x)$ contrary to the boundedness assumption

$f(x) \leq N$. This contradiction proves $L \leq N$.

(c) Assume $M \leq f(x)$ for all x and that $L < M$. Let $\epsilon = \dfrac{M-L}{2}$. As in part (b), $0 < |x - x_0| < \delta$

$\Rightarrow L - \dfrac{M-L}{2} < f(x) < L + \dfrac{M-L}{2} \Leftrightarrow \dfrac{3L-M}{2} < f(x) < \dfrac{M+L}{2} < M$, a contradiction.

25. $\lim\limits_{x \to 0} = \frac{\sin(1-\cos x)}{x} = \lim\limits_{x \to 0} \frac{\sin(1-\cos x)}{1-\cos x} \cdot \frac{1-\cos x}{x} \cdot \frac{1+\cos x}{1+\cos x} = \lim\limits_{x \to 0} \frac{\sin(1-\cos x)}{1-\cos x} \cdot \lim\limits_{x \to 0} \frac{1-\cos^2 x}{x(1+\cos x)} = 1 \cdot \lim\limits_{x \to 0} \frac{\sin^2 x}{x(1+\cos x)}$

       $= \lim\limits_{x \to 0} \frac{\sin x}{x} \cdot \frac{\sin x}{1+\cos x} = 1 \cdot \left(\frac{0}{2}\right) = 0.$

27. $\lim\limits_{x \to 0} \frac{\sin(\sin x)}{x} = \lim\limits_{x \to 0} \frac{\sin(\sin x)}{\sin x} \cdot \frac{\sin x}{x} = \lim\limits_{x \to 0} \frac{\sin(\sin x)}{\sin x} \cdot \lim\limits_{x \to 0} \frac{\sin x}{x} = 1 \cdot 1 = 1.$

29. $\lim\limits_{x \to 2} \frac{\sin(x^2-4)}{x-2} = \lim\limits_{x \to 2} \frac{\sin(x^2-4)}{x^2-4} \cdot (x+2) = \lim\limits_{x \to 2} \frac{\sin(x^2-4)}{x^2-4} \cdot \lim\limits_{x \to 2} (x+2) = 1 \cdot 4 = 4$

CHAPTER 3 DIFFERENTIATION

3.1 THE DERIVATIVE AS A FUNCTION

1. Step 1: $f(x) = 4 - x^2$ and $f(x + h) = 4 - (x + h)^2$

   Step 2: $\frac{f(x+h)-f(x)}{h} = \frac{[4-(x+h)^2]-(4-x^2)}{h} = \frac{(4-x^2-2xh-h^2)-4+x^2}{h} = \frac{-2xh-h^2}{h} = \frac{h(-2x-h)}{h} = -2x - h$

   Step 3: $f'(x) = \lim\limits_{h \to 0} (-2x - h) = -2x; f'(-3) = 6, f'(0) = 0, f'(1) = -2$

3. Step 1: $g(t) = \frac{1}{t^2}$ and $g(t + h) = \frac{1}{(t+h)^2}$

   Step 2: $\frac{g(t+h)-g(t)}{h} = \frac{\frac{1}{(t+h)^2} - \frac{1}{t^2}}{h} = \frac{\left(\frac{t^2-(t+h)^2}{(t+h)^2 \cdot t^2}\right)}{h} = \frac{t^2-(t^2+2th+h^2)}{(t+h)^2 \cdot t^2 \cdot h} = \frac{-2th-h^2}{(t+h)^2 t^2 h} = \frac{h(-2t-h)}{(t+h)^2 t^2 h} = \frac{-2t-h}{(t+h)^2 t^2}$

   Step 3: $g'(t) = \lim\limits_{h \to 0} \frac{-2t-h}{(t+h)^2 t^2} = \frac{-2t}{t^2 \cdot t^2} = \frac{-2}{t^3}$; $g'(-1) = 2, g'(2) = -\frac{1}{4}, g'\left(\sqrt{3}\right) = -\frac{2}{3\sqrt{3}}$

5. Step 1: $p(\theta) = \sqrt{3\theta}$ and $p(\theta + h) = \sqrt{3(\theta + h)}$

   Step 2: $\frac{p(\theta+h)-p(\theta)}{h} = \frac{\sqrt{3(\theta+h)}-\sqrt{3\theta}}{h} = \frac{\left(\sqrt{3\theta+3h}-\sqrt{3\theta}\right)}{h} \cdot \frac{\left(\sqrt{3\theta+3h}+\sqrt{3\theta}\right)}{\left(\sqrt{3\theta+3h}+\sqrt{3\theta}\right)} = \frac{(3\theta+3h)-3\theta}{h\left(\sqrt{3\theta+3h}+\sqrt{3\theta}\right)}$

   $= \frac{3h}{h\left(\sqrt{3\theta+3h}+\sqrt{3\theta}\right)} = \frac{3}{\sqrt{3\theta+3h}+\sqrt{3\theta}}$

   Step 3: $p'(\theta) = \lim\limits_{h \to 0} \frac{3}{\sqrt{3\theta+3h}+\sqrt{3\theta}} = \frac{3}{\sqrt{3\theta}+\sqrt{3\theta}} = \frac{3}{2\sqrt{3\theta}}$; $p'(1) = \frac{3}{2\sqrt{3}}, p'(3) = \frac{1}{2}, p'\left(\frac{2}{3}\right) = \frac{3}{2\sqrt{2}}$

7. $y = f(x) = 2x^3$ and $f(x + h) = 2(x + h)^3 \Rightarrow \frac{dy}{dx} = \lim\limits_{h \to 0} \frac{2(x+h)^3 - 2x^3}{h} = \lim\limits_{h \to 0} \frac{2(x^3+3x^2h+3xh^2+h^3)-2x^3}{h}$

   $= \lim\limits_{h \to 0} \frac{6x^2h+6xh^2+2h^3}{h} = \lim\limits_{h \to 0} \frac{h(6x^2+6xh+2h^2)}{h} = \lim\limits_{h \to 0} (6x^2 + 6xh + 2h^2) = 6x^2$

9. $s = r(t) = \frac{t}{2t+1}$ and $r(t + h) = \frac{t+h}{2(t+h)+1} \Rightarrow \frac{ds}{dt} = \lim\limits_{h \to 0} \frac{\left(\frac{t+h}{2(t+h)+1}\right) - \left(\frac{t}{2t+1}\right)}{h}$

   $= \lim\limits_{h \to 0} \frac{\left(\frac{(t+h)(2t+1)-t(2t+2h+1)}{(2t+2h+1)(2t+1)}\right)}{h} = \lim\limits_{h \to 0} \frac{(t+h)(2t+1)-t(2t+2h+1)}{(2t+2h+1)(2t+1)h} = \lim\limits_{h \to 0} \frac{2t^2+t+2ht+h-2t^2-2ht-t}{(2t+2h+1)(2t+1)h}$

   $= \lim\limits_{h \to 0} \frac{h}{(2t+2h+1)(2t+1)h} = \lim\limits_{h \to 0} \frac{1}{(2t+2h+1)(2t+1)} = \frac{1}{(2t+1)(2t+1)} = \frac{1}{(2t+1)^2}$

11. $p = f(q) = \frac{1}{\sqrt{q+1}}$ and $f(q + h) = \frac{1}{\sqrt{(q+h)+1}} \Rightarrow \frac{dp}{dq} = \lim\limits_{h \to 0} \frac{\left(\frac{1}{\sqrt{(q+h)+1}}\right) - \left(\frac{1}{\sqrt{q+1}}\right)}{h}$

   $= \lim\limits_{h \to 0} \frac{\left(\frac{\sqrt{q+1}-\sqrt{q+h+1}}{\sqrt{q+h+1}\sqrt{q+1}}\right)}{h} = \lim\limits_{h \to 0} \frac{\sqrt{q+1}-\sqrt{q+h+1}}{h\sqrt{q+h+1}\sqrt{q+1}} = \lim\limits_{h \to 0} \frac{(\sqrt{q+1}-\sqrt{q+h+1})}{h\sqrt{q+h+1}\sqrt{q+1}} \cdot \frac{(\sqrt{q+1}+\sqrt{q+h+1})}{(\sqrt{q+1}+\sqrt{q+h+1})}$

   $= \lim\limits_{h \to 0} \frac{(q+1)-(q+h+1)}{h\sqrt{q+h+1}\sqrt{q+1}\left(\sqrt{q+1}+\sqrt{q+h+1}\right)} = \lim\limits_{h \to 0} \frac{-h}{h\sqrt{q+h+1}\sqrt{q+1}\left(\sqrt{q+1}+\sqrt{q+h+1}\right)}$

   $= \lim\limits_{h \to 0} \frac{-1}{\sqrt{q+h+1}\sqrt{q+1}\left(\sqrt{q+1}+\sqrt{q+h+1}\right)} = \frac{-1}{\sqrt{q+1}\sqrt{q+1}\left(\sqrt{q+1}+\sqrt{q+1}\right)} = \frac{-1}{2(q+1)\sqrt{q+1}}$

13. $f(x) = x + \frac{9}{x}$ and $f(x + h) = (x + h) + \frac{9}{(x+h)} \Rightarrow \frac{f(x+h)-f(x)}{h} = \frac{\left[(x+h)+\frac{9}{(x+h)}\right]-\left[x+\frac{9}{x}\right]}{h}$

   $= \frac{x(x+h)^2+9x-x^2(x+h)-9(x+h)}{x(x+h)h} = \frac{x^3+2x^2h+xh^2+9x-x^3-x^2h-9x-9h}{x(x+h)h} = \frac{x^2h+xh^2-9h}{x(x+h)h}$

   $= \frac{h(x^2+xh-9)}{x(x+h)h} = \frac{x^2+xh-9}{x(x+h)}$; $f'(x) = \lim\limits_{h \to 0} \frac{x^2+xh-9}{x(x+h)} = \frac{x^2-9}{x^2} = 1 - \frac{9}{x^2}$; $m = f'(-3) = 0$

15. $\dfrac{ds}{dt} = \lim\limits_{h \to 0} \dfrac{[(t+h)^3 - (t+h)^2] - (t^3 - t^2)}{h} = \lim\limits_{h \to 0} \dfrac{(t^3 + 3t^2h + 3th^2 + h^3) - (t^2 + 2th + h^2) - t^3 + t^2}{h}$

$= \lim\limits_{h \to 0} \dfrac{3t^2h + 3th^2 + h^3 - 2th - h^2}{h} = \lim\limits_{h \to 0} \dfrac{h(3t^2 + 3th + h^2 - 2t - h)}{h} = \lim\limits_{h \to 0} (3t^2 + 3th + h^2 - 2t - h)$

$= 3t^2 - 2t; \ m = \dfrac{ds}{dt}\Big|_{t=-1} = 5$

17. $f(x) = \dfrac{8}{\sqrt{x-2}}$ and $f(x+h) = \dfrac{8}{\sqrt{(x+h)-2}} \ \Rightarrow \ \dfrac{f(x+h)-f(x)}{h} = \dfrac{\frac{8}{\sqrt{(x+h)-2}} - \frac{8}{\sqrt{x-2}}}{h}$

$= \dfrac{8\left(\sqrt{x-2} - \sqrt{x+h-2}\right)}{h\sqrt{x+h-2}\sqrt{x-2}} \cdot \dfrac{\left(\sqrt{x-2} + \sqrt{x+h-2}\right)}{\left(\sqrt{x-2} + \sqrt{x+h-2}\right)} = \dfrac{8[(x-2)-(x+h-2)]}{h\sqrt{x+h-2}\sqrt{x-2}\left(\sqrt{x-2}+\sqrt{x+h-2}\right)}$

$= \dfrac{-8h}{h\sqrt{x+h-2}\sqrt{x-2}\left(\sqrt{x-2}+\sqrt{x+h-2}\right)} \ \Rightarrow \ f'(x) = \lim\limits_{h \to 0} \dfrac{-8}{\sqrt{x+h-2}\sqrt{x-2}\left(\sqrt{x-2}+\sqrt{x+h-2}\right)}$

$= \dfrac{-8}{\sqrt{x-2}\sqrt{x-2}\left(\sqrt{x-2}+\sqrt{x-2}\right)} = \dfrac{-4}{(x-2)\sqrt{x-2}} \ ; \ m = f'(6) = \dfrac{-4}{4\sqrt{4}} = -\dfrac{1}{2} \ \Rightarrow \ $ the equation of the tangent

line at $(6, 4)$ is $y - 4 = -\dfrac{1}{2}(x - 6) \Rightarrow y = -\dfrac{1}{2}x + 3 + 4 \Rightarrow y = -\dfrac{1}{2}x + 7.$

19. $s = f(t) = 1 - 3t^2$ and $f(t+h) = 1 - 3(t+h)^2 = 1 - 3t^2 - 6th - 3h^2 \ \Rightarrow \ \dfrac{ds}{dt} = \lim\limits_{h \to 0} \dfrac{f(t+h)-f(t)}{h}$

$= \lim\limits_{h \to 0} \dfrac{(1 - 3t^2 - 6th - 3h^2) - (1 - 3t^2)}{h} = \lim\limits_{h \to 0} (-6t - 3h) = -6t \ \Rightarrow \ \dfrac{ds}{dt}\Big|_{t=-1} = 6$

21. $r = f(\theta) = \dfrac{2}{\sqrt{4-\theta}}$ and $f(\theta + h) = \dfrac{2}{\sqrt{4-(\theta + h)}} \ \Rightarrow \ \dfrac{dr}{d\theta} = \lim\limits_{h \to 0} \dfrac{f(\theta+h)-f(\theta)}{h} = \lim\limits_{h \to 0} \dfrac{\frac{2}{\sqrt{4-\theta-h}} - \frac{2}{\sqrt{4-\theta}}}{h}$

$= \lim\limits_{h \to 0} \dfrac{2\sqrt{4-\theta} - 2\sqrt{4-\theta-h}}{h\sqrt{4-\theta}\sqrt{4-\theta-h}} = \lim\limits_{h \to 0} \dfrac{2\sqrt{4-\theta} - 2\sqrt{4-\theta-h}}{h\sqrt{4-\theta}\sqrt{4-\theta-h}} \cdot \dfrac{\left(2\sqrt{4-\theta}+2\sqrt{4-\theta-h}\right)}{\left(2\sqrt{4-\theta}+2\sqrt{4-\theta-h}\right)}$

$= \lim\limits_{h \to 0} \dfrac{4(4-\theta) - 4(4-\theta-h)}{2h\sqrt{4-\theta}\sqrt{4-\theta-h}\left(\sqrt{4-\theta}+\sqrt{4-\theta-h}\right)} = \lim\limits_{h \to 0} \dfrac{2}{\sqrt{4-\theta}\sqrt{4-\theta-h}\left(\sqrt{4-\theta}+\sqrt{4-\theta-h}\right)}$

$= \dfrac{2}{(4-\theta)\left(2\sqrt{4-\theta}\right)} = \dfrac{1}{(4-\theta)\sqrt{4-\theta}} \ \Rightarrow \ \dfrac{dr}{d\theta}\Big|_{\theta=0} = \dfrac{1}{8}$

23. $f'(x) = \lim\limits_{z \to x} \dfrac{f(z) - f(x)}{z - x} = \lim\limits_{z \to x} \dfrac{\frac{1}{z+2} - \frac{1}{x+2}}{z - x} = \lim\limits_{z \to x} \dfrac{(x+2)-(z+2)}{(z-x)(z+2)(x+2)} = \lim\limits_{z \to x} \dfrac{x - z}{(z-x)(z+2)(x+2)} = \lim\limits_{z \to x} \dfrac{-1}{(z+2)(x+2)}$

$= \dfrac{-1}{(x+2)^2}$

25. $g'(x) = \lim\limits_{z \to x} \dfrac{g(z) - g(x)}{z - x} = \lim\limits_{z \to x} \dfrac{\frac{z}{z-1} - \frac{x}{x-1}}{z - x} = \lim\limits_{z \to x} \dfrac{z(x-1) - x(z-1)}{(z-x)(z-1)(x-1)} = \lim\limits_{z \to x} \dfrac{-z + x}{(z-x)(z-1)(x-1)} = \lim\limits_{z \to x} \dfrac{-1}{(z-1)(x-1)}$

$= \dfrac{-1}{(x-1)^2}$

27. Note that as x increases, the slope of the tangent line to the curve is first negative, then zero (when $x = 0$), then positive $\Rightarrow$ the slope is always increasing which matches (b).

29. $f_3(x)$ is an oscillating function like the cosine. Everywhere that the graph of f_3 has a horizontal tangent we expect f_3' to be zero, and (d) matches this condition.

31. (a) f' is not defined at $x = 0, 1, 4$. At these points, the left-hand and right-hand derivatives do not agree.

For example, $\lim\limits_{x \to 0^-} \dfrac{f(x) - f(0)}{x - 0} = $ slope of line joining $(-4, 0)$ and $(0, 2) = \dfrac{1}{2}$ but $\lim\limits_{x \to 0^+} \dfrac{f(x) - f(0)}{x - 0} = $ slope of

line joining $(0, 2)$ and $(1, -2) = -4$. Since these values are not equal, $f'(0) = \lim\limits_{x \to 0} \dfrac{f(x) - f(0)}{x - 0}$ does not exist.

(b)

33. Left-hand derivative: For $h < 0$, $f(0 + h) = f(h) = h^2$ (using $y = x^2$ curve) $\Rightarrow \lim\limits_{h \to 0^-} \frac{f(0+h) - f(0)}{h}$

$= \lim\limits_{h \to 0^-} \frac{h^2 - 0}{h} = \lim\limits_{h \to 0^-} h = 0$;

Right-hand derivative: For $h > 0$, $f(0 + h) = f(h) = h$ (using $y = x$ curve) $\Rightarrow \lim\limits_{h \to 0^+} \frac{f(0+h) - f(0)}{h}$

$= \lim\limits_{h \to 0^+} \frac{h - 0}{h} = \lim\limits_{h \to 0^+} 1 = 1$;

Then $\lim\limits_{h \to 0^-} \frac{f(0+h) - f(0)}{h} \neq \lim\limits_{h \to 0^+} \frac{f(0+h) - f(0)}{h} \Rightarrow$ the derivative $f'(0)$ does not exist.

35. Left-hand derivative: When $h < 0$, $1 + h < 1 \Rightarrow f(1 + h) = \sqrt{1 + h} \Rightarrow \lim\limits_{h \to 0^-} \frac{f(1+h) - f(1)}{h}$

$= \lim\limits_{h \to 0^-} \frac{\sqrt{1+h} - 1}{h} = \lim\limits_{h \to 0^-} \frac{\left(\sqrt{1+h} - 1\right)}{h} \cdot \frac{\left(\sqrt{1+h} + 1\right)}{\left(\sqrt{1+h} + 1\right)} = \lim\limits_{h \to 0^-} \frac{(1+h) - 1}{h\left(\sqrt{1+h} + 1\right)} = \lim\limits_{h \to 0^-} \frac{1}{\sqrt{1+h} + 1} = \frac{1}{2}$;

Right-hand derivative: When $h > 0$, $1 + h > 1 \Rightarrow f(1 + h) = 2(1 + h) - 1 = 2h + 1 \Rightarrow \lim\limits_{h \to 0^+} \frac{f(1+h) - f(1)}{h}$

$= \lim\limits_{h \to 0^+} \frac{(2h+1) - 1}{h} = \lim\limits_{h \to 0^+} 2 = 2$;

Then $\lim\limits_{h \to 0^-} \frac{f(1+h) - f(1)}{h} \neq \lim\limits_{h \to 0^+} \frac{f(1+h) - f(1)}{h} \Rightarrow$ the derivative $f'(1)$ does not exist.

37. (a) The function is differentiable on its domain $-3 \leq x \leq 2$ (it is smooth)

    (b) none

    (c) none

39. (a) The function is differentiable on $-3 \leq x < 0$ and $0 < x \leq 3$

    (b) none

    (c) The function is neither continuous nor differentiable at $x = 0$ since $\lim\limits_{x \to 0^-} f(x) \neq \lim\limits_{x \to 0^+} f(x)$

41. (a) f is differentiable on $-1 \leq x < 0$ and $0 < x \leq 2$

    (b) f is continuous but not differentiable at $x = 0$: $\lim\limits_{x \to 0} f(x) = 0$ exists but there is a cusp at $x = 0$, so

    $f'(0) = \lim\limits_{h \to 0} \frac{f(0+h) - f(0)}{h}$ does not exist

    (c) none

43. $y' = \lim\limits_{h \to 0} \frac{(2(x+h)^2 - 13(x+h) + 5) - (2x^2 - 13x + 5)}{h} = \lim\limits_{h \to 0} \frac{2x^2 + 4xh + 2h^2 - 13x - 13h + 5 - 2x^2 + 13x - 5}{h}$

$= \lim\limits_{h \to 0} \frac{4xh + 2h^2 - 13h}{h} = \lim\limits_{h \to 0} (4x + 2h - 13) = 4x - 13$, slope at x. The slope is -1 when $4x - 13 = -1$

$\Rightarrow 4x = 12 \Rightarrow x = 3 \Rightarrow y = 2 \cdot 3^2 - 13 \cdot 3 + 5 = -16$. Thus the tangent line is $y + 16 = (-1)(x - 3)$

$\Rightarrow y = -x - 13$ and the point of tangency is $(3, -16)$.

45. (a) Suppose $|f(x)| \leq x^2$ for $-1 \leq x \leq 1$. Then $|f(0)| \leq 0^2 \Rightarrow f(0) = 0$. Then $f'(0) = \lim\limits_{h \to 0} \frac{f(0+h) - f(0)}{h}$

    $= \lim\limits_{h \to 0} \frac{f(h) - 0}{h} = \lim\limits_{h \to 0} \frac{f(h)}{h}$. For $|h| \leq 1$, $-h^2 \leq f(h) \leq h^2 \Rightarrow -h \leq \frac{f(h)}{h} \leq h \Rightarrow f'(0) = \lim\limits_{h \to 0} \frac{f(h)}{h} = 0$

    by the Sandwich Theorem for limits.

(b) Note that for $x \neq 0$, $|f(x)| = \left|x^2 \sin \frac{1}{x}\right| = |x^2| \, |\sin x| \leq |x^2| \cdot 1 = x^2$ (since $-1 \leq \sin x \leq 1$). By part (a),
f is differentiable at $x = 0$ and $f'(0) = 0$.

47. Weierstrass's nowhere differentiable continuous function.

$$g(x) = \cos(\pi x) + \left(\frac{2}{3}\right)^1 \cos(9\pi x) + \left(\frac{2}{3}\right)^2 \cos(9^2\pi x) + \left(\frac{2}{3}\right)^3 \cos(9^3\pi x)$$

$$+ \cdots + \left(\frac{2}{3}\right)^7 \cos(9^7\pi x)$$

3.2 DIFFERENTIATION RULES FOR POLYNOMIALS, EXPONENTIALS, PRODUCTS, AND QUOTIENTS

1. $y = -x^2 + 3 \Rightarrow \frac{dy}{dx} = \frac{d}{dx}\left(-x^2\right) + \frac{d}{dx}\left(3\right) = -2x + 0 = -2x \Rightarrow \frac{d^2y}{dx^2} = -2$

3. $s = 5t^3 - 3t^5 \Rightarrow \frac{ds}{dt} = \frac{d}{dt}\left(5t^3\right) - \frac{d}{dt}\left(3t^5\right) = 15t^2 - 15t^4 \Rightarrow \frac{d^2s}{dt^2} = \frac{d}{dt}\left(15t^2\right) - \frac{d}{dt}\left(15t^4\right) = 30t - 60t^3$

5. $y = \frac{4x^3}{3} - x + 2e^x \Rightarrow \frac{dy}{dx} = 4x^2 - 1 + 2e^x \Rightarrow \frac{d^2y}{dx^2} = 8x + 2e^x$

7. $w = 3z^{-2} - z^{-1} \Rightarrow \frac{dw}{dz} = -6z^{-3} + z^{-2} = \frac{-6}{z^3} + \frac{1}{z^2} \Rightarrow \frac{d^2w}{dz^2} = 18z^{-4} - 2z^{-3} = \frac{18}{z^4} - \frac{2}{z^3}$

9. $y = 6x^2 - 10x - 5x^{-2} \Rightarrow \frac{dy}{dx} = 12x - 10 + 10x^{-3} = 12x - 10 + \frac{10}{x^3} \Rightarrow \frac{d^2y}{dx^2} = 12 - 0 - 30x^{-4} = 12 - \frac{30}{x^4}$

11. $r = \frac{1}{3}s^{-2} - \frac{5}{2}s^{-1} \Rightarrow \frac{dr}{ds} = -\frac{2}{3}s^{-3} + \frac{5}{2}s^{-2} = \frac{-2}{3s^3} + \frac{5}{2s^2} \Rightarrow \frac{d^2r}{ds^2} = 2s^{-4} - 5s^{-3} = \frac{2}{s^4} - \frac{5}{s^3}$

13. (a) $y = \left(3 - x^2\right)\left(x^3 - x + 1\right) \Rightarrow y' = \left(3 - x^2\right) \cdot \frac{d}{dx}\left(x^3 - x + 1\right) + \left(x^3 - x + 1\right) \cdot \frac{d}{dx}\left(3 - x^2\right)$
$= \left(3 - x^2\right)\left(3x^2 - 1\right) + \left(x^3 - x + 1\right)\left(-2x\right) = -5x^4 + 12x^2 - 2x - 3$
  (b) $y = -x^5 + 4x^3 - x^2 - 3x + 3 \Rightarrow y' = -5x^4 + 12x^2 - 2x - 3$

15. (a) $y = \left(x^2 + 1\right)\left(x + 5 + \frac{1}{x}\right) \Rightarrow y' = \left(x^2 + 1\right) \cdot \frac{d}{dx}\left(x + 5 + \frac{1}{x}\right) + \left(x + 5 + \frac{1}{x}\right) \cdot \frac{d}{dx}\left(x^2 + 1\right)$
$= \left(x^2 + 1\right)\left(1 - x^{-2}\right) + \left(x + 5 + x^{-1}\right)\left(2x\right) = \left(x^2 - 1 + 1 - x^{-2}\right) + \left(2x^2 + 10x + 2\right) = 3x^2 + 10x + 2 - \frac{1}{x^2}$
  (b) $y = x^3 + 5x^2 + 2x + 5 + \frac{1}{x} \Rightarrow y' = 3x^2 + 10x + 2 - \frac{1}{x^2}$

17. $y = \frac{2x+5}{3x-2}$; use the quotient rule: $u = 2x + 5$ and $v = 3x - 2 \Rightarrow u' = 2$ and $v' = 3 \Rightarrow y' = \frac{vu' - uv'}{v^2}$
$= \frac{(3x-2)(2) - (2x+5)(3)}{(3x-2)^2} = \frac{6x - 4 - 6x - 15}{(3x-2)^2} = \frac{-19}{(3x-2)^2}$

19. $g(x) = \frac{x^2-4}{x+0.5}$; use the quotient rule: $u = x^2 - 4$ and $v = x + 0.5$ $\Rightarrow$ $u' = 2x$ and $v' = 1$ $\Rightarrow$ $g'(x) = \frac{vu' - uv'}{v^2}$

$= \frac{(x+0.5)(2x) - (x^2-4)(1)}{(x+0.5)^2} = \frac{2x^2 + x - x^2 + 4}{(x+0.5)^2} = \frac{x^2 + x + 4}{(x+0.5)^2}$

21. $v = (1-t)(1+t^2)^{-1} = \frac{1-t}{1+t^2}$ $\Rightarrow$ $\frac{dv}{dt} = \frac{(1+t^2)(-1) - (1-t)(2t)}{(1+t^2)^2} = \frac{-1-t^2-2t+2t^2}{(1+t^2)^2} = \frac{t^2-2t-1}{(1+t^2)^2}$

23. $y = 2e^{-x} \Rightarrow y' = 2e^{-x} \cdot (-1) = -2e^{-x}$

25. $v = \frac{1+x-4\sqrt{x}}{x}$ $\Rightarrow$ $v' = \frac{x\left(1 - \frac{2}{\sqrt{x}}\right) - (1+x-4\sqrt{x})}{x^2} = \frac{2\sqrt{x}-1}{x^2}$

27. $y = x^3 e^x \Rightarrow y' = x^3 \cdot e^x + 3x^2 \cdot e^x = (x^3 + 3x^2)e^x$

29. $y = x^{9/4} \Rightarrow y' = \frac{9}{4}x^{5/4}$

31. $s = 2t^{3/2} + 3e^2 \Rightarrow s' = 3t^{1/2} + 0 \Rightarrow s' = 3t^{1/2}$

33. $y = \sqrt[7]{x^2} - x^e = x^{2/7} - x^e \Rightarrow y' = \frac{2}{7}x^{-5/7} - e\,x^{e-1} \Rightarrow y' = \frac{2}{7x^{5/7}} - e\,x^{e-1}$

35. $r = \frac{e^s}{s} \Rightarrow r' = \frac{s \cdot e^s - e^s(1)}{s^2} \Rightarrow r' = \frac{e^s(s-1)}{s^2}$

37. $y = \frac{1}{2}x^4 - \frac{3}{2}x^2 - x \Rightarrow y' = 2x^3 - 3x - 1 \Rightarrow y'' = 6x^2 - 3 \Rightarrow y''' = 12x \Rightarrow y^{(4)} = 12 \Rightarrow y^{(n)} = 0$ for all $n \geq 5$

39. $y = \frac{x^3+7}{x} = x^2 + 7x^{-1} \Rightarrow \frac{dy}{dx} = 2x - 7x^{-2} = 2x - \frac{7}{x^2} \Rightarrow \frac{d^2y}{dx^2} = 2 + 14x^{-3} = 2 + \frac{14}{x^3}$

41. $r = \frac{(\theta-1)(\theta^2+\theta+1)}{\theta^3} = \frac{\theta^3-1}{\theta^3} = 1 - \frac{1}{\theta^3} = 1 - \theta^{-3} \Rightarrow \frac{dr}{d\theta} = 0 + 3\theta^{-4} = 3\theta^{-4} = \frac{3}{\theta^4} \Rightarrow \frac{d^2r}{d\theta^2} = -12\theta^{-5} = \frac{-12}{\theta^5}$

43. $w = 3z^2 e^z \Rightarrow w' = 3z^2 \cdot e^z + 6z \cdot e^z = (3z^2 + 6z)e^z$ and $w'' = (3z^2 + 6z)e^z + (6z+6)e^z = (3z^2 + 12z + 6)e^z$

45. $p = \left(\frac{q^2+3}{12q}\right)\left(\frac{q^4-1}{q^3}\right) = \frac{q^6-q^2+3q^4-3}{12q^4} = \frac{1}{12}q^2 - \frac{1}{12}q^{-2} + \frac{1}{4} - \frac{1}{4}q^{-4} \Rightarrow \frac{dp}{dq} = \frac{1}{6}q + \frac{1}{6}q^{-3} + q^{-5} = \frac{1}{6}q + \frac{1}{6q^3} + \frac{1}{q^5}$

$\Rightarrow \frac{d^2p}{dq^2} = \frac{1}{6} - \frac{1}{2}q^{-4} - 5q^{-6} = \frac{1}{6} - \frac{1}{2q^4} - \frac{5}{q^6}$

47. $u(0) = 5$, $u'(0) = -3$, $v(0) = -1$, $v'(0) = 2$

(a) $\frac{d}{dx}(uv) = uv' + vu' \Rightarrow \frac{d}{dx}(uv)\big|_{x=0} = u(0)v'(0) + v(0)u'(0) = 5 \cdot 2 + (-1)(-3) = 13$

(b) $\frac{d}{dx}\left(\frac{u}{v}\right) = \frac{vu' - uv'}{v^2} \Rightarrow \frac{d}{dx}\left(\frac{u}{v}\right)\big|_{x=0} = \frac{v(0)u'(0) - u(0)v'(0)}{(v(0))^2} = \frac{(-1)(-3) - (5)(2)}{(-1)^2} = -7$

(c) $\frac{d}{dx}\left(\frac{v}{u}\right) = \frac{uv' - vu'}{u^2} \Rightarrow \frac{d}{dx}\left(\frac{v}{u}\right)\big|_{x=0} = \frac{u(0)v'(0) - v(0)u'(0)}{(u(0))^2} = \frac{(5)(2) - (-1)(-3)}{(5)^2} = \frac{7}{25}$

(d) $\frac{d}{dx}(7v - 2u) = 7v' - 2u' \Rightarrow \frac{d}{dx}(7v-2u)\big|_{x=0} = 7v'(0) - 2u'(0) = 7 \cdot 2 - 2(-3) = 20$

49. $y = x^3 - 4x + 1$. Note that $(2, 1)$ is on the curve: $1 = 2^3 - 4(2) + 1$

(a) Slope of the tangent at (x, y) is $y' = 3x^2 - 4$ $\Rightarrow$ slope of the tangent at $(2, 1)$ is $y'(2) = 3(2)^2 - 4 = 8$. Thus the slope of the line perpendicular to the tangent at $(2, 1)$ is $-\frac{1}{8}$ $\Rightarrow$ the equation of the line perpendicular to the tangent line at $(2, 1)$ is $y - 1 = -\frac{1}{8}(x - 2)$ or $y = -\frac{x}{8} + \frac{5}{4}$.

(b) The slope of the curve at x is $m = 3x^2 - 4$ and the smallest value for m is -4 when $x = 0$ and $y = 1$.

(c) We want the slope of the curve to be 8 $\Rightarrow$ y$'$ = 8 $\Rightarrow$ 3x^2 − 4 = 8 $\Rightarrow$ 3x^2 = 12 $\Rightarrow$ x^2 = 4 $\Rightarrow$ x = ± 2. When
x = 2, y = 1 and the tangent line has equation y − 1 = 8(x − 2) or y = 8x − 15; when x = −2,
y = (−2)^3 − 4(−2) + 1 = 1, and the tangent line has equation y − 1 = 8(x + 2) or y = 8x + 17.

51. y = $\frac{4x}{x^2+1}$ $\Rightarrow$ $\frac{dy}{dx}$ = $\frac{(x^2+1)(4)-(4x)(2x)}{(x^2+1)^2}$ = $\frac{4x^2+4-8x^2}{(x^2+1)^2}$ = $\frac{4(-x^2+1)}{(x^2+1)^2}$. When x = 0, y = 0 and y$'$ = $\frac{4(0+1)}{1}$
= 4, so the tangent to the curve at (0, 0) is the line y = 4x. When x = 1, y = 2 $\Rightarrow$ y$'$ = 0, so the tangent to the
curve at (1, 2) is the line y = 2.

53. y = ax^2 + bx + c passes through (0, 0) $\Rightarrow$ 0 = a(0) + b(0) + c $\Rightarrow$ c = 0; y = ax^2 + bx passes through (1, 2)
$\Rightarrow$ 2 = a + b; y$'$ = 2ax + b and since the curve is tangent to y = x at the origin, its slope is 1 at x = 0
$\Rightarrow$ y$'$ = 1 when x = 0 $\Rightarrow$ 1 = 2a(0) + b $\Rightarrow$ b = 1. Then a + b = 2 $\Rightarrow$ a = 1. In summary a = b = 1 and c = 0 so
the curve is y = x^2 + x.

55. (a) y = x^3 − x $\Rightarrow$ y$'$ = 3x^2 − 1. When x = −1, y = 0 and y$'$ = 2 $\Rightarrow$ the tangent line to the curve at (−1, 0) is
y = 2(x + 1) or y = 2x + 2.
(b)

(c) $\left.\begin{matrix} y = x^3 - x \\ y = 2x + 2 \end{matrix}\right\}$ $\Rightarrow$ x^3 − x = 2x + 2 $\Rightarrow$ x^3 − 3x − 2 = (x − 2)(x + 1)^2 = 0 $\Rightarrow$ x = 2 or x = −1. Since
y = 2(2) + 2 = 6; the other intersection point is (2, 6)

57. P(x) = a$_n$x^n + a$_{n-1}$x^{n-1} + ⋯ + a$_2$x^2 + a$_1$x + a$_0$ $\Rightarrow$ P$'$(x) = na$_n$x^{n-1} + (n − 1)a$_{n-1}$x^{n-2} + ⋯ + 2a$_2$x + a$_1$

59. Let c be a constant $\Rightarrow$ $\frac{dc}{dx}$ = 0 $\Rightarrow$ $\frac{d}{dx}$(u · c) = u · $\frac{dc}{dx}$ + c · $\frac{du}{dx}$ = u · 0 + c $\frac{du}{dx}$ = c $\frac{du}{dx}$. Thus when one of the
functions is a constant, the Product Rule is just the Constant Multiple Rule $\Rightarrow$ the Constant Multiple Rule is
a special case of the Product Rule.

61. (a) $\frac{d}{dx}$ (uvw) = $\frac{d}{dx}$ ((uv) · w) = (uv) $\frac{dw}{dx}$ + w · $\frac{d}{dx}$ (uv) = uv $\frac{dw}{dx}$ + w $\left(u \frac{dv}{dx} + v \frac{du}{dx}\right)$ = uv $\frac{dw}{dx}$ + wu $\frac{dv}{dx}$ + wv $\frac{du}{dx}$
= uvw$'$ + uv$'$w + u$'$vw
(b) $\frac{d}{dx}$ (u$_1$u$_2$u$_3$u$_4$) = $\frac{d}{dx}$ ((u$_1$u$_2$u$_3$) u$_4$) = (u$_1$u$_2$u$_3$) $\frac{du_4}{dx}$ + u$_4$ $\frac{d}{dx}$ (u$_1$u$_2$u$_3$) $\Rightarrow$ $\frac{d}{dx}$ (u$_1$u$_2$u$_3$u$_4$)
= u$_1$u$_2$u$_3$ $\frac{du_4}{dx}$ + u$_4$ $\left(u_1u_2 \frac{du_3}{dx} + u_3u_1 \frac{du_2}{dx} + u_3u_2 \frac{du_1}{dx}\right)$ (using (a) above)
$\Rightarrow$ $\frac{d}{dx}$ (u$_1$u$_2$u$_3$u$_4$) = u$_1$u$_2$u$_3$ $\frac{du_4}{dx}$ + u$_1$u$_2$u$_4$ $\frac{du_3}{dx}$ + u$_1$u$_3$u$_4$ $\frac{du_2}{dx}$ + u$_2$u$_3$u$_4$ $\frac{du_1}{dx}$
= u$_1$u$_2$u$_3$u$'_4$ + u$_1$u$_2$u$'_3$u$_4$ + u$_1$u$'_2$u$_3$u$_4$ + u$'_1$u$_2$u$_3$u$_4$
(c) Generalizing (a) and (b) above, $\frac{d}{dx}$ (u$_1$ ⋯ u$_n$) = u$_1$u$_2$ ⋯ u$_{n-1}$u$'_n$ + u$_1$u$_2$ ⋯ u$_{n-2}$u$'_{n-1}$u$_n$ + … + u$'_1$u$_2$ ⋯ u$_n$

63. P = $\frac{nRT}{V-nb}$ − $\frac{an^2}{V^2}$. We are holding T constant, and a, b, n, R are also constant so their derivatives are zero
$\Rightarrow$ $\frac{dP}{dV}$ = $\frac{(V-nb)\cdot 0 - (nRT)(1)}{(V-nb)^2}$ − $\frac{V^2(0) - (an^2)(2V)}{(V^2)^2}$ = $\frac{-nRT}{(V-nb)^2}$ + $\frac{2an^2}{V^3}$

3.3 THE DERIVATIVE AS A RATE OF CHANGE

1. $s = t^2 - 3t + 2, 0 \leq t \leq 2$
   (a) displacement $= \Delta s = s(2) - s(0) = 0\text{m} - 2\text{m} = -2$ m, $v_{av} = \frac{\Delta s}{\Delta t} = \frac{-2}{2} = -1$ m/sec
   (b) $v = \frac{ds}{dt} = 2t - 3 \Rightarrow |v(0)| = |-3| = 3$ m/sec and $|v(2)| = 1$ m/sec;
       $a = \frac{d^2s}{dt^2} = 2 \Rightarrow a(0) = 2$ m/sec^2 and $a(2) = 2$ m/sec^2
   (c) $v = 0 \Rightarrow 2t - 3 = 0 \Rightarrow t = \frac{3}{2}$. v is negative in the interval $0 < t < \frac{3}{2}$ and v is positive when $\frac{3}{2} < t < 2 \Rightarrow$ the body
       changes direction at $t = \frac{3}{2}$.

3. $s = -t^3 + 3t^2 - 3t, 0 \leq t \leq 3$
   (a) displacement $= \Delta s = s(3) - s(0) = -9$ m, $v_{av} = \frac{\Delta s}{\Delta t} = \frac{-9}{3} = -3$ m/sec
   (b) $v = \frac{ds}{dt} = -3t^2 + 6t - 3 \Rightarrow |v(0)| = |-3| = 3$ m/sec and $|v(3)| = |-12| = 12$ m/sec; $a = \frac{d^2s}{dt^2} = -6t + 6$
       $\Rightarrow a(0) = 6$ m/sec^2 and $a(3) = -12$ m/sec^2
   (c) $v = 0 \Rightarrow -3t^2 + 6t - 3 = 0 \Rightarrow t^2 - 2t + 1 = 0 \Rightarrow (t - 1)^2 = 0 \Rightarrow t = 1$. For all other values of t in the
       interval the velocity v is negative (the graph of $v = -3t^2 + 6t - 3$ is a parabola with vertex at $t = 1$ which
       opens downward $\Rightarrow$ the body never changes direction).

5. $s = \frac{25}{t^2} - \frac{5}{t}, 1 \leq t \leq 5$
   (a) $\Delta s = s(5) - s(1) = -20$ m, $v_{av} = \frac{-20}{4} = -5$ m/sec
   (b) $v = \frac{-50}{t^3} + \frac{5}{t^2} \Rightarrow |v(1)| = 45$ m/sec and $|v(5)| = \frac{1}{5}$ m/sec; $a = \frac{150}{t^4} - \frac{10}{t^3} \Rightarrow a(1) = 140$ m/sec^2 and $a(5) = \frac{4}{25}$ m/sec^2
   (c) $v = 0 \Rightarrow \frac{-50 + 5t}{t^3} = 0 \Rightarrow -50 + 5t = 0 \Rightarrow t = 10 \Rightarrow$ the body does not change direction in the interval

7. $s = t^3 - 6t^2 + 9t$ and let the positive direction be to the right on the s-axis.
   (a) $v = 3t^2 - 12t + 9$ so that $v = 0 \Rightarrow t^2 - 4t + 3 = (t - 3)(t - 1) = 0 \Rightarrow t = 1$ or 3; $a = 6t - 12 \Rightarrow a(1) = -6$ m/sec^2
       and $a(3) = 6$ m/sec^2. Thus the body is motionless but being accelerated left when $t = 1$, and motionless but being
       accelerated right when $t = 3$.
   (b) $a = 0 \Rightarrow 6t - 12 = 0 \Rightarrow t = 2$ with speed $|v(2)| = |12 - 24 + 9| = 3$ m/sec
   (c) The body moves to the right or forward on $0 \leq t < 1$, and to the left or backward on $1 < t < 2$. The positions are
       $s(0) = 0, s(1) = 4$ and $s(2) = 2 \Rightarrow$ total distance $= |s(1) - s(0)| + |s(2) - s(1)| = |4| + |-2| = 6$ m.

9. $s_m = 1.86t^2 \Rightarrow v_m = 3.72t$ and solving $3.72t = 27.8 \Rightarrow t \approx 7.5$ sec on Mars; $s_j = 11.44t^2 \Rightarrow v_j = 22.88t$ and solving
   $22.88t = 27.8 \Rightarrow t \approx 1.2$ sec on Jupiter.

11. $s = 15t - \frac{1}{2}g_s t^2 \Rightarrow v = 15 - g_s t$ so that $v = 0 \Rightarrow 15 - g_s t = 0 \Rightarrow g_s = \frac{15}{t}$. Therefore $g_s = \frac{15}{20} = \frac{3}{4} = 0.75$ m/sec^2

13. (a) $s = 179 - 16t^2 \Rightarrow v = -32t \Rightarrow$ speed $= |v| = 32t$ ft/sec and $a = -32$ ft/sec^2
    (b) $s = 0 \Rightarrow 179 - 16t^2 = 0 \Rightarrow t = \sqrt{\frac{179}{16}} \approx 3.3$ sec
    (c) When $t = \sqrt{\frac{179}{16}}$, $v = -32\sqrt{\frac{179}{16}} = -8\sqrt{179} \approx -107.0$ ft/sec

15. (a) at 2 and 7 seconds   (b) between 3 and 6 seconds: $3 \leq t \leq 6$

(c)

(d)

17. (a) 190 ft/sec   (b) 2 sec
    (c) at 8 sec, 0 ft/sec   (d) 10.8 sec, 90 ft/sec
    (e) From t = 8 until t = 10.8 sec, a total of 2.8 sec
    (f) Greatest acceleration happens 2 sec after launch
    (g) From t = 2 to t = 10.8 sec; during this period, $a = \frac{v(10.8) - v(2)}{10.8 - 2} \approx -32$ ft/sec^2

19. $s = 490t^2 \Rightarrow v = 980t \Rightarrow a = 980$
    (a) Solving $160 = 490t^2 \Rightarrow t = \frac{4}{7}$ sec. The average velocity was $\frac{s(4/7) - s(0)}{4/7} = 280$ cm/sec.
    (b) At the 160 cm mark the balls are falling at $v(4/7) = 560$ cm/sec. The acceleration at the 160 cm mark
         was 980 cm/sec^2.
    (c) The light was flashing at a rate of $\frac{17}{4/7} = 29.75$ flashes per second.

21. C = position, A = velocity, and B = acceleration. Neither A nor C can be the derivative of B because B's
    derivative is constant. Graph C cannot be the derivative of A either, because A has some negative slopes while
    C has only positive values. So, C (being the derivative of neither A nor B) must be the graph of position.
    Curve C has both positive and negative slopes, so its derivative, the velocity, must be A and not B. That
    leaves B for acceleration.

23. (a) $c(100) = 11,000 \Rightarrow c_{av} = \frac{11,000}{100} = \$110$
    (b) $c(x) = 2000 + 100x - .1x^2 \Rightarrow c'(x) = 100 - .2x$. Marginal cost $= c'(x) \Rightarrow$ the marginal cost of producing 100
         machines is $c'(100) = \$80$
    (c) The cost of producing the 101st machine is $c(101) - c(100) = 100 - \frac{201}{10} = \$79.90$

25. $b(t) = 10^6 + 10^4 t - 10^3 t^2 \Rightarrow b'(t) = 10^4 - (2)(10^3 t) = 10^3(10 - 2t)$
    (a) $b'(0) = 10^4$ bacteria/hr   (b) $b'(5) = 0$ bacteria/hr
    (c) $b'(10) = -10^4$ bacteria/hr

27. (a) $y = 6\left(1 - \frac{t}{12}\right)^2 = 6\left(1 - \frac{t}{6} + \frac{t^2}{144}\right) \Rightarrow \frac{dy}{dt} = \frac{t}{12} - 1$
    (b) The largest value of $\frac{dy}{dt}$ is 0 m/h when t = 12 and the fluid level is falling the slowest at that time. The
         smallest value of $\frac{dy}{dt}$ is -1 m/h, when t = 0, and the fluid level is falling the fastest at that time.
    (c) In this situation, $\frac{dy}{dt} \leq 0 \Rightarrow$ the graph of y is
         always decreasing. As $\frac{dy}{dt}$ increases in value,
         the slope of the graph of y increases from -1
         to 0 over the interval $0 \leq t \leq 12$.

29. 200 km/hr $= 55\frac{5}{9}$ m/sec $= \frac{500}{9}$ m/sec, and D $= \frac{10}{9}t^2$ $\Rightarrow$ V $= \frac{20}{9}t$. Thus V $= \frac{500}{9}$ $\Rightarrow$ $\frac{20}{9}t = \frac{500}{9}$ $\Rightarrow$ t = 25 sec. When
t = 25, D $= \frac{10}{9}(25)^2 = \frac{6250}{9}$ m

31.

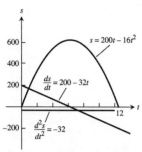

(a) v = 0 when t = 6.25 sec
(b) v > 0 when $0 \le t < 6.25$ $\Rightarrow$ body moves up; v < 0 when $6.25 < t \le 12.5$ $\Rightarrow$ body moves down
(c) body changes direction at t = 6.25 sec
(d) body speeds up on (6.25, 12.5] and slows down on [0, 6.25)
(e) The body is moving fastest at the endpoints t = 0 and t = 12.5 when it is traveling 200 ft/sec. It's moving slowest at t = 6.25 when the speed is 0.
(f) When t = 6.25 the body is s = 625 m from the origin and farthest away.

33.

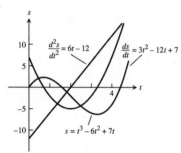

(a) v = 0 when t $= \frac{6 \pm \sqrt{15}}{3}$ sec
(b) v < 0 when $\frac{6 - \sqrt{15}}{3} < t < \frac{6 + \sqrt{15}}{3}$ $\Rightarrow$ body moves left; v > 0 when $0 \le t < \frac{6 - \sqrt{15}}{3}$ or $\frac{6 + \sqrt{15}}{3} < t \le 4$
$\Rightarrow$ body moves right
(c) body changes direction at t $= \frac{6 \pm \sqrt{15}}{3}$ sec
(d) body speeds up on $\left(\frac{6 - \sqrt{15}}{3}, 2\right) \cup \left(\frac{6 + \sqrt{15}}{3}, 4\right]$ and slows down on $\left[0, \frac{6 - \sqrt{15}}{3}\right) \cup \left(2, \frac{6 + \sqrt{15}}{3}\right)$.
(e) The body is moving fastest at t = 0 and t = 4 when it is moving 7 units/sec and slowest at t $= \frac{6 \pm \sqrt{15}}{3}$ sec
(f) When t $= \frac{6 + \sqrt{15}}{3}$ the body is at position s ≈ -6.303 units and farthest from the origin.

3.4 DERIVATIVES OF TRIGONOMETRIC FUNCTIONS

1. y $= -10x + 3\cos x$ $\Rightarrow$ $\frac{dy}{dx} = -10 + 3\frac{d}{dx}(\cos x) = -10 - 3\sin x$

3. y $= \csc x - 4\sqrt{x} + 7$ $\Rightarrow$ $\frac{dy}{dx} = -\csc x \cot x - \frac{4}{2\sqrt{x}} + 0 = -\csc x \cot x - \frac{2}{\sqrt{x}}$

5. $y = (\sec x + \tan x)(\sec x - \tan x) \Rightarrow \frac{dy}{dx} = (\sec x + \tan x)\frac{d}{dx}(\sec x - \tan x) + (\sec x - \tan x)\frac{d}{dx}(\sec x + \tan x)$

$= (\sec x + \tan x)(\sec x \tan x - \sec^2 x) + (\sec x - \tan x)(\sec x \tan x + \sec^2 x)$

$= (\sec^2 x \tan x + \sec x \tan^2 x - \sec^3 x - \sec^2 x \tan x) + (\sec^2 x \tan x - \sec x \tan^2 x + \sec^3 x - \tan x \sec^2 x) = 0.$

$\left(\text{Note also that } y = \sec^2 x - \tan^2 x = (\tan^2 x + 1) - \tan^2 x = 1 \Rightarrow \frac{dy}{dx} = 0.\right)$

7. $y = \frac{\cot x}{1 + \cot x} \Rightarrow \frac{dy}{dx} = \frac{(1 + \cot x)\frac{d}{dx}(\cot x) - (\cot x)\frac{d}{dx}(1 + \cot x)}{(1 + \cot x)^2} = \frac{(1 + \cot x)(-\csc^2 x) - (\cot x)(-\csc^2 x)}{(1 + \cot x)^2}$

$= \frac{-\csc^2 x - \csc^2 x \cot x + \csc^2 x \cot x}{(1 + \cot x)^2} = \frac{-\csc^2 x}{(1 + \cot x)^2}$

9. $y = \frac{4}{\cos x} + \frac{1}{\tan x} = 4 \sec x + \cot x \Rightarrow \frac{dy}{dx} = 4 \sec x \tan x - \csc^2 x$

11. $y = x^2 \sin x + 2x \cos x - 2 \sin x \Rightarrow \frac{dy}{dx} = (x^2 \cos x + (\sin x)(2x)) + ((2x)(-\sin x) + (\cos x)(2)) - 2 \cos x$

$= x^2 \cos x + 2x \sin x - 2x \sin x + 2 \cos x - 2 \cos x = x^2 \cos x$

13. $s = \tan t - e^{-t} \Rightarrow \frac{ds}{dt} = \sec^2 t + e^{-t}$

15. $s = \frac{1 + \csc t}{1 - \csc t} \Rightarrow \frac{ds}{dt} = \frac{(1 - \csc t)(-\csc t \cot t) - (1 + \csc t)(\csc t \cot t)}{(1 - \csc t)^2}$

$= \frac{-\csc t \cot t + \csc^2 t \cot t - \csc t \cot t - \csc^2 t \cot t}{(1 - \csc t)^2} = \frac{-2 \csc t \cot t}{(1 - \csc t)^2}$

17. $r = 4 - \theta^2 \sin \theta \Rightarrow \frac{dr}{d\theta} = -\left(\theta^2 \frac{d}{d\theta}(\sin \theta) + (\sin \theta)(2\theta)\right) = -(\theta^2 \cos \theta + 2\theta \sin \theta) = -\theta(\theta \cos \theta + 2 \sin \theta)$

19. $r = \sec \theta \csc \theta \Rightarrow \frac{dr}{d\theta} = (\sec \theta)(-\csc \theta \cot \theta) + (\csc \theta)(\sec \theta \tan \theta)$

$= \left(\frac{-1}{\cos \theta}\right)\left(\frac{1}{\sin \theta}\right)\left(\frac{\cos \theta}{\sin \theta}\right) + \left(\frac{1}{\sin \theta}\right)\left(\frac{1}{\cos \theta}\right)\left(\frac{\sin \theta}{\cos \theta}\right) = \frac{-1}{\sin^2 \theta} + \frac{1}{\cos^2 \theta} = \sec^2 \theta - \csc^2 \theta$

21. $p = 5 + \frac{1}{\cot q} = 5 + \tan q \Rightarrow \frac{dp}{dq} = \sec^2 q$

23. $p = \frac{\sin q + \cos q}{\cos q} \Rightarrow \frac{dp}{dq} = \frac{(\cos q)(\cos q - \sin q) - (\sin q + \cos q)(-\sin q)}{\cos^2 q}$

$= \frac{\cos^2 q - \cos q \sin q + \sin^2 q + \cos q \sin q}{\cos^2 q} = \frac{1}{\cos^2 q} = \sec^2 q$

25. (a) $y = \csc x \Rightarrow y' = -\csc x \cot x \Rightarrow y'' = -((\csc x)(-\csc^2 x) + (\cot x)(-\csc x \cot x)) = \csc^3 x + \csc x \cot^2 x$

$= (\csc x)(\csc^2 x + \cot^2 x) = (\csc x)(\csc^2 x + \csc^2 x - 1) = 2 \csc^3 x - \csc x$

(b) $y = \sec x \Rightarrow y' = \sec x \tan x \Rightarrow y'' = (\sec x)(\sec^2 x) + (\tan x)(\sec x \tan x) = \sec^3 x + \sec x \tan^2 x$

$= (\sec x)(\sec^2 x + \tan^2 x) = (\sec x)(\sec^2 x + \sec^2 x - 1) = 2 \sec^3 x - \sec x$

27. $y = \sin x \Rightarrow y' = \cos x \Rightarrow$ slope of tangent at
$x = -\pi$ is $y'(-\pi) = \cos(-\pi) = -1$; slope of
tangent at $x = 0$ is $y'(0) = \cos(0) = 1$; and
slope of tangent at $x = \frac{3\pi}{2}$ is $y'\left(\frac{3\pi}{2}\right) = \cos\frac{3\pi}{2}$
$= 0$. The tangent at $(-\pi, 0)$ is $y - 0 = -1(x + \pi)$,
or $y = -x - \pi$; the tangent at $(0, 0)$ is
$y - 0 = 1(x - 0)$, or $y = x$; and the tangent at
$\left(\frac{3\pi}{2}, -1\right)$ is $y = -1$.

29. $y = \sec x \Rightarrow y' = \sec x \tan x \Rightarrow$ slope of tangent at

$x = -\frac{\pi}{3}$ is $\sec\left(-\frac{\pi}{3}\right)\tan\left(-\frac{\pi}{3}\right) = -2\sqrt{3}$; slope of tangent

at $x = \frac{\pi}{4}$ is $\sec\left(\frac{\pi}{4}\right)\tan\left(\frac{\pi}{4}\right) = \sqrt{2}$. The tangent at the point

$\left(-\frac{\pi}{3}, \sec\left(-\frac{\pi}{3}\right)\right) = \left(-\frac{\pi}{3}, 2\right)$ is $y - 2 = -2\sqrt{3}\left(x + \frac{\pi}{3}\right)$;

the tangent at the point $\left(\frac{\pi}{4}, \sec\left(\frac{\pi}{4}\right)\right) = \left(\frac{\pi}{4}, \sqrt{2}\right)$ is $y - \sqrt{2}$

$= \sqrt{2}\left(x - \frac{\pi}{4}\right)$.

31. Yes, $y = x + \sin x \Rightarrow y' = 1 + \cos x$; horizontal tangent occurs where $1 + \cos x = 0 \Rightarrow \cos x = -1 \Rightarrow x = \pi$

33. No, $y = x - \cot x \Rightarrow y' = 1 + \csc^2 x$; horizontal tangent occurs where $1 + \csc^2 x = 0 \Rightarrow \csc^2 x = -1$. But there are no x-values for which $\csc^2 x = -1$.

35. We want all points on the curve where the tangent

line has slope 2. Thus, $y = \tan x \Rightarrow y' = \sec^2 x$ so

that $y' = 2 \Rightarrow \sec^2 x = 2 \Rightarrow \sec x = \pm\sqrt{2}$

$\Rightarrow x = \pm\frac{\pi}{4}$. Then the tangent line at $\left(\frac{\pi}{4}, 1\right)$ has

equation $y - 1 = 2\left(x - \frac{\pi}{4}\right)$; the tangent line at

$\left(-\frac{\pi}{4}, -1\right)$ has equation $y + 1 = 2\left(x + \frac{\pi}{4}\right)$.

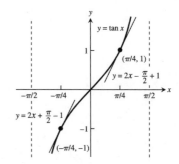

37. $y = 4 + \cot x - 2\csc x \Rightarrow y' = -\csc^2 x + 2\csc x \cot x = -\left(\frac{1}{\sin x}\right)\left(\frac{1 - 2\cos x}{\sin x}\right)$

(a) When $x = \frac{\pi}{2}$, then $y' = -1$; the tangent line is $y = -x + \frac{\pi}{2} + 2$.

(b) To find the location of the horizontal tangent set $y' = 0 \Rightarrow 1 - 2\cos x = 0 \Rightarrow x = \frac{\pi}{3}$ radians. When $x = \frac{\pi}{3}$, then $y = 4 - \sqrt{3}$ is the horizontal tangent.

39. $\lim\limits_{x \to 2} \sin\left(\frac{1}{x} - \frac{1}{2}\right) = \sin\left(\frac{1}{2} - \frac{1}{2}\right) = \sin 0 = 0$

41. $\lim\limits_{x \to 0} \sec\left[e^x + \pi\tan\left(\frac{\pi}{4\sec x}\right) - 1\right] = \sec\left[1 + \pi\tan\left(\frac{\pi}{4\sec 0}\right) - 1\right] = = \sec\left[\pi\tan\left(\frac{\pi}{4}\right)\right] = \sec\pi = -1$

43. $\lim\limits_{t \to 0} \tan\left(1 - \frac{\sin t}{t}\right) = \tan\left(1 - \lim\limits_{t \to 0}\frac{\sin t}{t}\right) = \tan(1 - 1) = 0$

45. $s = 2 - 2\sin t \Rightarrow v = \frac{ds}{dt} = -2\cos t \Rightarrow a = \frac{dv}{dt} = 2\sin t \Rightarrow j = \frac{da}{dt} = 2\cos t$. Therefore, velocity $= v\left(\frac{\pi}{4}\right)$

$= -\sqrt{2}$ m/sec; speed $= \left|v\left(\frac{\pi}{4}\right)\right| = \sqrt{2}$ m/sec; acceleration $= a\left(\frac{\pi}{4}\right) = \sqrt{2}$ m/sec^2; jerk $= j\left(\frac{\pi}{4}\right) = \sqrt{2}$ m/sec^3.

47. $\lim\limits_{x \to 0} f(x) = \lim\limits_{x \to 0}\frac{\sin^2 3x}{x^2} = \lim\limits_{x \to 0} 9\left(\frac{\sin 3x}{3x}\right)\left(\frac{\sin 3x}{3x}\right) = 9$ so that f is continuous at $x = 0 \Rightarrow \lim\limits_{x \to 0} f(x) = f(0)$

$\Rightarrow 9 = c$.

49. $\frac{d^{999}}{dx^{999}}(\cos x) = \sin x$ because $\frac{d^4}{dx^4}(\cos x) = \cos x \Rightarrow$ the derivative of cos x any number of times that is a

multiple of 4 is cos x. Thus, dividing 999 by 4 gives $999 = 249 \cdot 4 + 3 \Rightarrow \frac{d^{999}}{dx^{999}}(\cos x)$

$= \frac{d^3}{dx^3}\left[\frac{d^{249 \cdot 4}}{dx^{249 \cdot 4}}(\cos x)\right] = \frac{d^3}{dx^3}(\cos x) = \sin x$.

51.

As h takes on the values of 1, 0.5, 0.3 and 0.1 the corresponding dashed curves of $y = \frac{\sin(x+h) - \sin x}{h}$ get closer and closer to the black curve $y = \cos x$ because $\frac{d}{dx}(\sin x) = \lim_{h \to 0} \frac{\sin(x+h) - \sin x}{h} = \cos x$. The same is true as h takes on the values of $-1, -0.5, -0.3$ and -0.1.

53. (a)

The dashed curves of $y = \frac{\sin(x+h) - \sin(x-h)}{2h}$ are closer to the black curve $y = \cos x$ than the corresponding dashed curves in Exercise 51 illustrating that the centered difference quotient is a better approximation of the derivative of this function.

(b)

The dashed curves of $y = \frac{\cos(x+h) - \cos(x-h)}{2h}$ are closer to the black curve $y = -\sin x$ than the corresponding dashed curves in Exercise 52 illustrating that the centered difference quotient is a better approximation of the derivative of this function.

55. $y = \tan x \Rightarrow y' = \sec^2 x$, so the smallest value $y' = \sec^2 x$ takes on is $y' = 1$ when $x = 0$; y' has no maximum value since $\sec^2 x$ has no largest value on $\left(-\frac{\pi}{2}, \frac{\pi}{2}\right)$; y' is never negative since $\sec^2 x \geq 1$.

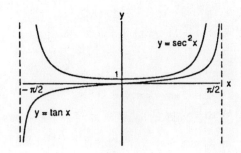

57. $y = \frac{\sin x}{x}$ appears to cross the y-axis at $y = 1$, since

$\lim\limits_{x \to 0} \frac{\sin x}{x} = 1$; $y = \frac{\sin 2x}{x}$ appears to cross the y-axis

at $y = 2$, since $\lim\limits_{x \to 0} \frac{\sin 2x}{x} = 2$; $y = \frac{\sin 4x}{x}$ appears to

cross the y-axis at $y = 4$, since $\lim\limits_{x \to 0} \frac{\sin 4x}{x} = 4$.

However, none of these graphs actually cross the y-axis

since $x = 0$ is not in the domain of the functions. Also,

$\lim\limits_{x \to 0} \frac{\sin 5x}{x} = 5$, $\lim\limits_{x \to 0} \frac{\sin (-3x)}{x} = -3$, and $\lim\limits_{x \to 0} \frac{\sin kx}{x}$

$= k \Rightarrow$ the graphs of $y = \frac{\sin 5x}{x}$, $y = \frac{\sin (-3x)}{x}$, and

$y = \frac{\sin kx}{x}$ approach 5, −3, and k, respectively, as

$x \to 0$. However, the graphs do not actually cross the

y-axis.

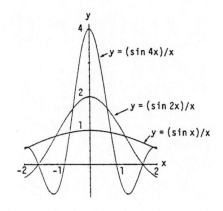

3.5 THE CHAIN RULE AND PARAMETRIC EQUATIONS

1. $f(u) = 6u - 9 \Rightarrow f'(u) = 6 \Rightarrow f'(g(x)) = 6$; $g(x) = \frac{1}{2}x^4 \Rightarrow g'(x) = 2x^3$; therefore $\frac{dy}{dx} = f'(g(x))g'(x)$
    $= 6 \cdot 2x^3 = 12x^3$

3. $f(u) = \sin u \Rightarrow f'(u) = \cos u \Rightarrow f'(g(x)) = \cos (3x + 1)$; $g(x) = 3x + 1 \Rightarrow g'(x) = 3$; therefore $\frac{dy}{dx} = f'(g(x))g'(x)$
    $= (\cos (3x + 1))(3) = 3 \cos (3x + 1)$

5. $f(u) = \cos u \Rightarrow f'(u) = -\sin u \Rightarrow f'(g(x)) = -\sin (\sin x)$; $g(x) = \sin x \Rightarrow g'(x) = \cos x$; therefore
    $\frac{dy}{dx} = f'(g(x))g'(x) = -(\sin (\sin x)) \cos x$

7. $f(u) = \tan u \Rightarrow f'(u) = \sec^2 u \Rightarrow f'(g(x)) = \sec^2 (10x - 5)$; $g(x) = 10x - 5 \Rightarrow g'(x) = 10$; therefore
    $\frac{dy}{dx} = f'(g(x))g'(x) = (\sec^2 (10x - 5)) (10) = 10 \sec^2 (10x - 5)$

9. With $u = (2x + 1)$, $y = u^5$: $\frac{dy}{dx} = \frac{dy}{du} \frac{du}{dx} = 5u^4 \cdot 2 = 10(2x + 1)^4$

11. With $u = \left(1 - \frac{x}{7}\right)$, $y = u^{-7}$: $\frac{dy}{dx} = \frac{dy}{du} \frac{du}{dx} = -7u^{-8} \cdot \left(-\frac{1}{7}\right) = \left(1 - \frac{x}{7}\right)^{-8}$

13. With $u = (1 - 6x)$, $y = u^{2/3} \Rightarrow \frac{dy}{dx} = \frac{dy}{du} \frac{du}{dx} = \frac{2}{3}u^{-1/3}(-6) = -4(1 - 6x)^{-1/3}$

15. With $u = \tan x$, $y = \sec u$: $\frac{dy}{dx} = \frac{dy}{du} \frac{du}{dx} = (\sec u \tan u) (\sec^2 x) = (\sec (\tan x) \tan (\tan x)) \sec^2 x$

17. With $u = \sin x$, $y = u^3$: $\frac{dy}{dx} = \frac{dy}{du} \frac{du}{dx} = 3u^2 \cos x = 3 (\sin^2 x) (\cos x)$

19. With $u = -5x$, $y = e^u$: $\frac{dy}{dx} = \frac{dy}{du} \frac{du}{dx} = e^u(-5) = -5e^{-5x}$

21. With $u = 5 - 7x$, $y = e^u$: $\frac{dy}{dx} = \frac{dy}{du} \frac{du}{dx} = e^u(-7) = -7e^{5-7x}$

23. $p = \sqrt{3 - t} = (3 - t)^{1/2} \Rightarrow \frac{dp}{dt} = \frac{1}{2} (3 - t)^{-1/2} \cdot \frac{d}{dt} (3 - t) = -\frac{1}{2} (3 - t)^{-1/2} = \frac{-1}{2\sqrt{3 - t}}$

25. $s = \frac{4}{3\pi} \sin 3t + \frac{4}{5\pi} \cos 5t \Rightarrow \frac{ds}{dt} = \frac{4}{3\pi} \cos 3t \cdot \frac{d}{dt} (3t) + \frac{4}{5\pi} (-\sin 5t) \cdot \frac{d}{dt} (5t) = \frac{4}{\pi} \cos 3t - \frac{4}{\pi} \sin 5t$
     $= \frac{4}{\pi} (\cos 3t - \sin 5t)$

27. $r = (\csc\theta + \cot\theta)^{-1} \Rightarrow \frac{dr}{d\theta} = -(\csc\theta + \cot\theta)^{-2}\frac{d}{d\theta}(\csc\theta + \cot\theta) = \frac{\csc\theta\cot\theta + \csc^2\theta}{(\csc\theta + \cot\theta)^2} = \frac{\csc\theta(\cot\theta + \csc\theta)}{(\csc\theta + \cot\theta)^2}$

$= \frac{\csc\theta}{\csc\theta + \cot\theta}$

29. $y = x^2\sin^4 x + x\cos^{-2} x \Rightarrow \frac{dy}{dx} = x^2\frac{d}{dx}(\sin^4 x) + \sin^4 x \cdot \frac{d}{dx}(x^2) + x\frac{d}{dx}(\cos^{-2} x) + \cos^{-2} x \cdot \frac{d}{dx}(x)$

$= x^2\left(4\sin^3 x\frac{d}{dx}(\sin x)\right) + 2x\sin^4 x + x\left(-2\cos^{-3} x \cdot \frac{d}{dx}(\cos x)\right) + \cos^{-2} x$

$= x^2\left(4\sin^3 x\cos x\right) + 2x\sin^4 x + x((-2\cos^{-3} x)(-\sin x)) + \cos^{-2} x$

$= 4x^2\sin^3 x\cos x + 2x\sin^4 x + 2x\sin x\cos^{-3} x + \cos^{-2} x$

31. $y = \frac{1}{21}(3x - 2)^7 + \left(4 - \frac{1}{2x^2}\right)^{-1} \Rightarrow \frac{dy}{dx} = \frac{7}{21}(3x - 2)^6 \cdot \frac{d}{dx}(3x - 2) + (-1)\left(4 - \frac{1}{2x^2}\right)^{-2} \cdot \frac{d}{dx}\left(4 - \frac{1}{2x^2}\right)$

$= \frac{7}{21}(3x - 2)^6 \cdot 3 + (-1)\left(4 - \frac{1}{2x^2}\right)^{-2}\left(\frac{1}{x^3}\right) = (3x - 2)^6 - \frac{1}{x^3\left(4 - \frac{1}{2x^2}\right)^2}$

33. $y = (4x + 3)^4(x + 1)^{-3} \Rightarrow \frac{dy}{dx} = (4x + 3)^4(-3)(x + 1)^{-4} \cdot \frac{d}{dx}(x + 1) + (x + 1)^{-3}(4)(4x + 3)^3 \cdot \frac{d}{dx}(4x + 3)$

$= (4x + 3)^4(-3)(x + 1)^{-4}(1) + (x + 1)^{-3}(4)(4x + 3)^3(4) = -3(4x + 3)^4(x + 1)^{-4} + 16(4x + 3)^3(x + 1)^{-3}$

$= \frac{(4x + 3)^3}{(x + 1)^4}\left[-3(4x + 3) + 16(x + 1)\right] = \frac{(4x + 3)^3(4x + 7)}{(x + 1)^4}$

35. $y = xe^{-x} + e^{3x} \Rightarrow y' = x \cdot e^{-x}(-1) + (1) \cdot e^{-x} + 3e^{3x} = (1 - x)e^{-x} + 3e^{3x}$

37. $y = (x^2 - 2x + 2)e^{5x/2} \Rightarrow y' = (x^2 - 2x + 2) \cdot e^{5x/2}\left(\frac{5}{2}\right) + (2x - 2) \cdot e^{5x/2} = \left(\frac{5}{2}x^2 - 3x + 3\right)e^{5x/2}$

39. $h(x) = x\tan\left(2\sqrt{x}\right) + 7 \Rightarrow h'(x) = x\frac{d}{dx}\left(\tan\left(2x^{1/2}\right)\right) + \tan\left(2x^{1/2}\right) \cdot \frac{d}{dx}(x) + 0$

$= x\sec^2\left(2x^{1/2}\right) \cdot \frac{d}{dx}\left(2x^{1/2}\right) + \tan\left(2x^{1/2}\right) = x\sec^2\left(2\sqrt{x}\right) \cdot \frac{1}{\sqrt{x}} + \tan\left(2\sqrt{x}\right) = \sqrt{x}\sec^2\left(2\sqrt{x}\right) + \tan\left(2\sqrt{x}\right)$

41. $f(\theta) = \left(\frac{\sin\theta}{1 + \cos\theta}\right)^2 \Rightarrow f'(\theta) = 2\left(\frac{\sin\theta}{1 + \cos\theta}\right) \cdot \frac{d}{d\theta}\left(\frac{\sin\theta}{1 + \cos\theta}\right) = \frac{2\sin\theta}{1 + \cos\theta} \cdot \frac{(1 + \cos\theta)(\cos\theta) - (\sin\theta)(-\sin\theta)}{(1 + \cos\theta)^2}$

$= \frac{(2\sin\theta)(\cos\theta + \cos^2\theta + \sin^2\theta)}{(1 + \cos\theta)^3} = \frac{(2\sin\theta)(\cos\theta + 1)}{(1 + \cos\theta)^3} = \frac{2\sin\theta}{(1 + \cos\theta)^2}$

43. $h(\theta) = \sqrt[3]{1 + \cos(2\theta)} = (1 + \cos 2\theta)^{1/3} \Rightarrow h'(\theta) = \frac{1}{3}(1 + \cos 2\theta)^{-2/3} \cdot (-\sin 2\theta) \cdot 2 = -\frac{2}{3}(\sin 2\theta)(1 + \cos 2\theta)^{-2/3}$

45. $q = \sin\left(\frac{t}{\sqrt{t + 1}}\right) \Rightarrow \frac{dq}{dt} = \cos\left(\frac{t}{\sqrt{t + 1}}\right) \cdot \frac{d}{dt}\left(\frac{t}{\sqrt{t + 1}}\right) = \cos\left(\frac{t}{\sqrt{t + 1}}\right) \cdot \frac{\sqrt{t + 1}(1) - t \cdot \frac{d}{dt}\left(\sqrt{t + 1}\right)}{\left(\sqrt{t + 1}\right)^2}$

$= \cos\left(\frac{t}{\sqrt{t + 1}}\right) \cdot \frac{\sqrt{t + 1} - \frac{t}{2\sqrt{t + 1}}}{t + 1} = \cos\left(\frac{t}{\sqrt{t + 1}}\right)\left(\frac{2(t + 1) - t}{2(t + 1)^{3/2}}\right) = \left(\frac{t + 2}{2(t + 1)^{3/2}}\right)\cos\left(\frac{t}{\sqrt{t + 1}}\right)$

47. $y = \cos\left(e^{-\theta^2}\right) \Rightarrow \frac{dy}{d\theta} = -\sin\left(e^{-\theta^2}\right)\frac{d}{d\theta}\left(e^{-\theta^2}\right) = \left(-\sin\left(e^{-\theta^2}\right)\right)\left(e^{-\theta^2}\right)\frac{d}{d\theta}(-\theta^2) = 2\theta e^{-\theta^2}\sin\left(e^{-\theta^2}\right)$

49. $y = \sin^2(\pi t - 2) \Rightarrow \frac{dy}{dt} = 2\sin(\pi t - 2) \cdot \frac{d}{dt}\sin(\pi t - 2) = 2\sin(\pi t - 2) \cdot \cos(\pi t - 2) \cdot \frac{d}{dt}(\pi t - 2)$

$= 2\pi\sin(\pi t - 2)\cos(\pi t - 2)$

51. $y = (1 + \cos 2t)^{-4} \Rightarrow \frac{dy}{dt} = -4(1 + \cos 2t)^{-5} \cdot \frac{d}{dt}(1 + \cos 2t) = -4(1 + \cos 2t)^{-5}(-\sin 2t) \cdot \frac{d}{dt}(2t) = \frac{8\sin 2t}{(1 + \cos 2t)^5}$

53. $y = e^{\cos^2(\pi t - 1)} \Rightarrow \frac{dy}{dt} = e^{\cos^2(\pi t - 1)} \cdot 2\cos(\pi t - 1) \cdot (-\sin(\pi t - 1)) \cdot \pi = -2\pi\sin(\pi t - 1)\cos(\pi t - 1)e^{\cos^2(\pi t - 1)}$

55. $y = \sin(\cos(2t - 5)) \Rightarrow \frac{dy}{dt} = \cos(\cos(2t - 5)) \cdot \frac{d}{dt}\cos(2t - 5) = \cos(\cos(2t - 5)) \cdot (-\sin(2t - 5)) \cdot \frac{d}{dt}(2t - 5)$

$= -2\cos(\cos(2t - 5))(\sin(2t - 5))$

57. $y = \left[1 + \tan^4\left(\frac{t}{12}\right)\right]^3 \Rightarrow \frac{dy}{dt} = 3\left[1 + \tan^4\left(\frac{t}{12}\right)\right]^2 \cdot \frac{d}{dt}\left[1 + \tan^4\left(\frac{t}{12}\right)\right] = 3\left[1 + \tan^4\left(\frac{t}{12}\right)\right]^2\left[4\tan^3\left(\frac{t}{12}\right) \cdot \frac{d}{dt}\tan\left(\frac{t}{12}\right)\right]$

$= 12\left[1 + \tan^4\left(\frac{t}{12}\right)\right]^2\left[\tan^3\left(\frac{t}{12}\right)\sec^2\left(\frac{t}{12}\right) \cdot \frac{1}{12}\right] = \left[1 + \tan^4\left(\frac{t}{12}\right)\right]^2\left[\tan^3\left(\frac{t}{12}\right)\sec^2\left(\frac{t}{12}\right)\right]$

59. $y = (1 + \cos(t^2))^{1/2} \Rightarrow \frac{dy}{dt} = \frac{1}{2}(1 + \cos(t^2))^{-1/2} \cdot \frac{d}{dt}(1 + \cos(t^2)) = \frac{1}{2}(1 + \cos(t^2))^{-1/2}\left(-\sin(t^2) \cdot \frac{d}{dt}(t^2)\right)$

$= -\frac{1}{2}(1 + \cos(t^2))^{-1/2}(\sin(t^2)) \cdot 2t = -\frac{t\sin(t^2)}{\sqrt{1 + \cos(t^2)}}$

61. $y = \left(1 + \frac{1}{x}\right)^3 \Rightarrow y' = 3\left(1 + \frac{1}{x}\right)^2\left(-\frac{1}{x^2}\right) = -\frac{3}{x^2}\left(1 + \frac{1}{x}\right)^2 \Rightarrow y'' = \left(-\frac{3}{x^2}\right) \cdot \frac{d}{dx}\left(1 + \frac{1}{x}\right)^2 - \left(1 + \frac{1}{x}\right)^2 \cdot \frac{d}{dx}\left(\frac{3}{x^2}\right)$

$= \left(-\frac{3}{x^2}\right)\left(2\left(1 + \frac{1}{x}\right)\left(-\frac{1}{x^2}\right)\right) + \left(\frac{6}{x^3}\right)\left(1 + \frac{1}{x}\right)^2 = \frac{6}{x^4}\left(1 + \frac{1}{x}\right) + \frac{6}{x^3}\left(1 + \frac{1}{x}\right)^2 = \frac{6}{x^3}\left(1 + \frac{1}{x}\right)\left(\frac{1}{x} + 1 + \frac{1}{x}\right)$

$= \frac{6}{x^3}\left(1 + \frac{1}{x}\right)\left(1 + \frac{2}{x}\right)$

63. $y = \frac{1}{9}\cot(3x - 1) \Rightarrow y' = -\frac{1}{9}\csc^2(3x - 1)(3) = -\frac{1}{3}\csc^2(3x - 1) \Rightarrow y'' = \left(-\frac{2}{3}\right)(\csc(3x - 1) \cdot \frac{d}{dx}\csc(3x - 1))$

$= -\frac{2}{3}\csc(3x - 1)(-\csc(3x - 1)\cot(3x - 1) \cdot \frac{d}{dx}(3x - 1)) = 2\csc^2(3x - 1)\cot(3x - 1)$

65. $y = e^{x^2} + 5x \Rightarrow y' = 2xe^{x^2} + 5 \Rightarrow y'' = 2x \cdot e^{x^2}(2x) + 2e^{x^2} = (4x^2 + 2)e^{x^2}$

67. $g(x) = \sqrt{x} \Rightarrow g'(x) = \frac{1}{2\sqrt{x}} \Rightarrow g(1) = 1$ and $g'(1) = \frac{1}{2}$; $f(u) = u^5 + 1 \Rightarrow f'(u) = 5u^4 \Rightarrow f'(g(1)) = f'(1) = 5$;

therefore, $(f \circ g)'(1) = f'(g(1)) \cdot g'(1) = 5 \cdot \frac{1}{2} = \frac{5}{2}$

69. $g(x) = 5\sqrt{x} \Rightarrow g'(x) = \frac{5}{2\sqrt{x}} \Rightarrow g(1) = 5$ and $g'(1) = \frac{5}{2}$; $f(u) = \cot\left(\frac{\pi u}{10}\right) \Rightarrow f'(u) = -\csc^2\left(\frac{\pi u}{10}\right)\left(\frac{\pi}{10}\right)$

$= \frac{-\pi}{10}\csc^2\left(\frac{\pi u}{10}\right) \Rightarrow f'(g(1)) = f'(5) = -\frac{\pi}{10}\csc^2\left(\frac{\pi}{2}\right) = -\frac{\pi}{10}$; therefore, $(f \circ g)'(1) = f'(g(1))g'(1) = -\frac{\pi}{10} \cdot \frac{5}{2}$

$= -\frac{\pi}{4}$

71. $g(x) = 10x^2 + x + 1 \Rightarrow g'(x) = 20x + 1 \Rightarrow g(0) = 1$ and $g'(0) = 1$; $f(u) = \frac{2u}{u^2+1} \Rightarrow f'(u) = \frac{(u^2+1)(2) - (2u)(2u)}{(u^2+1)^2}$

$= \frac{-2u^2 + 2}{(u^2+1)^2} \Rightarrow f'(g(0)) = f'(1) = 0$; therefore, $(f \circ g)'(0) = f'(g(0))g'(0) = 0 \cdot 1 = 0$

73. (a) $y = 2f(x) \Rightarrow \frac{dy}{dx} = 2f'(x) \Rightarrow \left.\frac{dy}{dx}\right|_{x=2} = 2f'(2) = 2\left(\frac{1}{3}\right) = \frac{2}{3}$

(b) $y = f(x) + g(x) \Rightarrow \frac{dy}{dx} = f'(x) + g'(x) \Rightarrow \left.\frac{dy}{dx}\right|_{x=3} = f'(3) + g'(3) = 2\pi + 5$

(c) $y = f(x) \cdot g(x) \Rightarrow \frac{dy}{dx} = f(x)g'(x) + g(x)f'(x) \Rightarrow \left.\frac{dy}{dx}\right|_{x=3} = f(3)g'(3) + g(3)f'(3) = 3 \cdot 5 + (-4)(2\pi) = 15 - 8\pi$

(d) $y = \frac{f(x)}{g(x)} \Rightarrow \frac{dy}{dx} = \frac{g(x)f'(x) - f(x)g'(x)}{[g(x)]^2} \Rightarrow \left.\frac{dy}{dx}\right|_{x=2} = \frac{g(2)f'(2) - f(2)g'(2)}{[g(2)]^2} = \frac{(2)\left(\frac{1}{3}\right) - (8)(-3)}{2^2} = \frac{37}{6}$

(e) $y = f(g(x)) \Rightarrow \frac{dy}{dx} = f'(g(x))g'(x) \Rightarrow \left.\frac{dy}{dx}\right|_{x=2} = f'(g(2))g'(2) = f'(2)(-3) = \frac{1}{3}(-3) = -1$

(f) $y = (f(x))^{1/2} \Rightarrow \frac{dy}{dx} = \frac{1}{2}(f(x))^{-1/2} \cdot f'(x) = \frac{f'(x)}{2\sqrt{f(x)}} \Rightarrow \left.\frac{dy}{dx}\right|_{x=2} = \frac{f'(2)}{2\sqrt{f(2)}} = \frac{\left(\frac{1}{3}\right)}{2\sqrt{8}} = \frac{1}{6\sqrt{8}} = \frac{1}{12\sqrt{2}} = \frac{\sqrt{2}}{24}$

(g) $y = (g(x))^{-2} \Rightarrow \frac{dy}{dx} = -2(g(x))^{-3} \cdot g'(x) \Rightarrow \left.\frac{dy}{dx}\right|_{x=3} = -2(g(3))^{-3}g'(3) = -2(-4)^{-3} \cdot 5 = \frac{5}{32}$

(h) $y = ((f(x))^2 + (g(x))^2)^{1/2} \Rightarrow \frac{dy}{dx} = \frac{1}{2}((f(x))^2 + (g(x))^2)^{-1/2}(2f(x) \cdot f'(x) + 2g(x) \cdot g'(x))$

$\Rightarrow \left.\frac{dy}{dx}\right|_{x=2} = \frac{1}{2}((f(2))^2 + (g(2))^2)^{-1/2}(2f(2)f'(2) + 2g(2)g'(2)) = \frac{1}{2}(8^2 + 2^2)^{-1/2}\left(2 \cdot 8 \cdot \frac{1}{3} + 2 \cdot 2 \cdot (-3)\right)$

$= -\frac{5}{3\sqrt{17}}$

75. $\frac{ds}{dt} = \frac{ds}{d\theta} \cdot \frac{d\theta}{dt}$: $s = \cos\theta \Rightarrow \frac{ds}{d\theta} = -\sin\theta \Rightarrow \left.\frac{ds}{d\theta}\right|_{\theta=\frac{3\pi}{2}} = -\sin\left(\frac{3\pi}{2}\right) = 1$ so that $\frac{ds}{dt} = \frac{ds}{d\theta} \cdot \frac{d\theta}{dt} = 1 \cdot 5 = 5$

77. With $y = x$, we should get $\frac{dy}{dx} = 1$ for both (a) and (b):

(a) $y = \frac{u}{5} + 7 \Rightarrow \frac{dy}{du} = \frac{1}{5}$; $u = 5x - 35 \Rightarrow \frac{du}{dx} = 5$; therefore, $\frac{dy}{dx} = \frac{dy}{du} \cdot \frac{du}{dx} = \frac{1}{5} \cdot 5 = 1$, as expected

(b) $y = 1 + \frac{1}{u} \Rightarrow \frac{dy}{du} = -\frac{1}{u^2}$; $u = (x-1)^{-1} \Rightarrow \frac{du}{dx} = -(x-1)^{-2}(1) = \frac{-1}{(x-1)^2}$; therefore $\frac{dy}{dx} = \frac{dy}{du} \cdot \frac{du}{dx}$

$= \frac{-1}{u^2} \cdot \frac{-1}{(x-1)^2} = \frac{-1}{((x-1)^{-1})^2} \cdot \frac{-1}{(x-1)^2} = (x-1)^2 \cdot \frac{1}{(x-1)^2} = 1$, again as expected

79. $y = 2\tan\left(\frac{\pi x}{4}\right) \Rightarrow \frac{dy}{dx} = \left(2\sec^2 \frac{\pi x}{4}\right)\left(\frac{\pi}{4}\right) = \frac{\pi}{2}\sec^2\frac{\pi x}{4}$

(a) $\frac{dy}{dx}\Big|_{x=1} = \frac{\pi}{2}\sec^2\left(\frac{\pi}{4}\right) = \pi \Rightarrow$ slope of tangent is 2; thus, $y(1) = 2\tan\left(\frac{\pi}{4}\right) = 2$ and $y'(1) = \pi \Rightarrow$ tangent line is

given by $y - 2 = \pi(x-1) \Rightarrow y = \pi x + 2 - \pi$

(b) $y' = \frac{\pi}{2}\sec^2\left(\frac{\pi x}{4}\right)$ and the smallest value the secant function can have in $-2 < x < 2$ is $1 \Rightarrow$ the minimum

value of y' is $\frac{\pi}{2}$ and that occurs when $\frac{\pi}{2} = \frac{\pi}{2}\sec^2\left(\frac{\pi x}{4}\right) \Rightarrow 1 = \sec^2\left(\frac{\pi x}{4}\right) \Rightarrow \pm 1 = \sec\left(\frac{\pi x}{4}\right) \Rightarrow x = 0.$

81. $x = \cos 2t$, $y = \sin 2t$, $0 \le t \le \pi$

$\Rightarrow \cos^2 2t + \sin^2 2t = 1 \Rightarrow x^2 + y^2 = 1$

83. $x = 4\cos t$, $y = 2\sin t$, $0 \le t \le 2\pi$

$\Rightarrow \frac{16\cos^2 t}{16} + \frac{4\sin^2 t}{4} = 1 \Rightarrow \frac{x^2}{16} + \frac{y^2}{4} = 1$

85. $x = 3t$, $y = 9t^2$, $-\infty < t < \infty \Rightarrow y = x^2$

87. $x = 2t - 5$, $y = 4t - 7$, $-\infty < t < \infty$

$\Rightarrow x + 5 = 2t \Rightarrow 2(x+5) = 4t$

$\Rightarrow y = 2(x+5) - 7 \Rightarrow y = 2x + 3$

89. $x = t$, $y = \sqrt{1 - t^2}$, $-1 \le t \le 0$
$\Rightarrow y = \sqrt{1 - x^2}$

91. $x = \sec^2 t - 1$, $y = \tan t$, $-\frac{\pi}{2} < t < \frac{\pi}{2}$

$\Rightarrow \sec^2 t - 1 = \tan^2 t \Rightarrow x = y^2$

93. (a) $x = a\cos t$, $y = -a\sin t$, $0 \le t \le 2\pi$ (b) $x = a\cos t$, $y = a\sin t$, $0 \le t \le 2\pi$

(c) $x = a\cos t$, $y = -a\sin t$, $0 \le t \le 4\pi$ (d) $x = a\cos t$, $y = a\sin t$, $0 \le t \le 4\pi$

95. Using $(-1, -3)$ we create the parametric equations $x = -1 + at$ and $y = -3 + bt$, representing a line which goes through $(-1, -3)$ at $t = 0$. We determine a and b so that the line goes through $(4, 1)$ when $t = 1$. Since $4 = -1 + a \Rightarrow a = 5$. Since $1 = -3 + b \Rightarrow b = 4$. Therefore, one possible parameterization is $x = -1 + 5t$, $y = -3 - 4t$, $0 \le t \le 1$.

97. The lower half of the parabola is given by $x = y^2 + 1$ for $y \le 0$. Substituting t for y, we obtain one possible parameterization $x = t^2 + 1$, $y = t$, $t \le 0$.

99. For simplicity, we assume that x and y are linear functions of t and that the point (x, y) starts at $(2, 3)$ for $t = 0$ and passes through $(-1, -1)$ at $t = 1$. Then $x = f(t)$, where $f(0) = 2$ and $f(1) = -1$.
    Since slope $= \frac{\Delta x}{\Delta t} = \frac{-1-2}{1-0} = -3$, $x = f(t) = -3t + 2 = 2 - 3t$. Also, $y = g(t)$, where $g(0) = 3$ and $g(1) = -1$.
    Since slope $= \frac{\Delta y}{\Delta t} = \frac{-1-3}{1-0} = -4$. $y = g(t) = -4t + 3 = 3 - 4t$.
    One possible parameterization is: $x = 2 - 3t$, $y = 3 - 4t$, $t \ge 0$.

101. $t = \frac{\pi}{4} \Rightarrow x = 2 \cos \frac{\pi}{4} = \sqrt{2}$, $y = 2 \sin \frac{\pi}{4} = \sqrt{2}$; $\frac{dx}{dt} = -2 \sin t$, $\frac{dy}{dt} = 2 \cos t \Rightarrow \frac{dy}{dx} = \frac{dy/dt}{dx/dt} = \frac{2 \cos t}{-2 \sin t} = -\cot t$
    $\Rightarrow \frac{dy}{dx}\Big|_{t=\frac{\pi}{4}} = -\cot \frac{\pi}{4} = -1$; tangent line is $y - \sqrt{2} = -1\left(x - \sqrt{2}\right)$ or $y = -x + 2\sqrt{2}$; $\frac{dy'}{dt} = \csc^2 t$
    $\Rightarrow \frac{d^2y}{dx^2} = \frac{dy'/dt}{dx/dt} = \frac{\csc^2 t}{-2 \sin t} = -\frac{1}{2 \sin^3 t} \Rightarrow \frac{d^2y}{dx^2}\Big|_{t=\frac{\pi}{4}} = -\sqrt{2}$

103. $t = \frac{1}{4} \Rightarrow x = \frac{1}{4}$, $y = \frac{1}{2}$; $\frac{dx}{dt} = 1$, $\frac{dy}{dt} = \frac{1}{2\sqrt{t}} \Rightarrow \frac{dy}{dx} = \frac{dy/dt}{dx/dt} = \frac{1}{2\sqrt{t}} \Rightarrow \frac{dy}{dx}\Big|_{t=\frac{1}{4}} = \frac{1}{2\sqrt{\frac{1}{4}}} = 1$; tangent line is
    $y - \frac{1}{2} = 1 \cdot \left(x - \frac{1}{4}\right)$ or $y = x + \frac{1}{4}$; $\frac{dy'}{dt} = -\frac{1}{4}t^{-3/2} \Rightarrow \frac{d^2y}{dx^2} = \frac{dy'/dt}{dx/dt} = -\frac{1}{4}t^{-3/2} \Rightarrow \frac{d^2y}{dx^2}\Big|_{t=\frac{1}{4}} = -2$

105. $t = -1 \Rightarrow x = 5$, $y = 1$; $\frac{dx}{dt} = 4t$, $\frac{dy}{dt} = 4t^3 \Rightarrow \frac{dy}{dx} = \frac{dy/dt}{dx/dt} = \frac{4t^3}{4t} = t^2 \Rightarrow \frac{dy}{dx}\Big|_{t=-1} = (-1)^2 = 1$; tangent line is
    $y - 1 = 1 \cdot (x - 5)$ or $y = x - 4$; $\frac{dy'}{dt} = 2t \Rightarrow \frac{d^2y}{dx^2} = \frac{dy'/dt}{dx/dt} = \frac{2t}{4t} = \frac{1}{2} \Rightarrow \frac{d^2y}{dx^2}\Big|_{t=-1} = \frac{1}{2}$

107. $t = \frac{\pi}{2} \Rightarrow x = \cos \frac{\pi}{2} = 0$, $y = 1 + \sin \frac{\pi}{2} = 2$; $\frac{dx}{dt} = -\sin t$, $\frac{dy}{dt} = \cos t \Rightarrow \frac{dy}{dx} = \frac{\cos t}{-\sin t} = -\cot t$
    $\Rightarrow \frac{dy}{dx}\Big|_{t=\frac{\pi}{2}} = -\cot \frac{\pi}{2} = 0$; tangent line is $y = 2$; $\frac{dy'}{dt} = \csc^2 t \Rightarrow \frac{d^2y}{dx^2} = \frac{\csc^2 t}{-\sin t} = -\csc^3 t \Rightarrow \frac{d^2y}{dx^2}\Big|_{t=\frac{\pi}{2}} = -1$

109. $s = A \cos (2\pi bt) \Rightarrow v = \frac{ds}{dt} = -A \sin (2\pi bt)(2\pi b) = -2\pi bA \sin (2\pi bt)$. If we replace b with 2b to double the frequency, the velocity formula gives $v = -4\pi bA \sin (4\pi bt) \Rightarrow$ doubling the frequency causes the velocity to double. Also $v = -2\pi bA \sin (2\pi bt) \Rightarrow a = \frac{dv}{dt} = -4\pi^2 b^2 A \cos (2\pi bt)$. If we replace b with 2b in the acceleration formula, we get $a = -16\pi^2 b^2 A \cos (4\pi bt) \Rightarrow$ doubling the frequency causes the acceleration to quadruple. Finally, $a = -4\pi^2 b^2 A \cos (2\pi bt) \Rightarrow j = \frac{da}{dt} = 8\pi^3 b^3 A \sin (2\pi bt)$. If we replace b with 2b in the jerk formula, we get $j = 64\pi^3 b^3 A \sin (4\pi bt) \Rightarrow$ doubling the frequency multiplies the jerk by a factor of 8.

111. $s = (1 + 4t)^{1/2} \Rightarrow v = \frac{ds}{dt} = \frac{1}{2}(1 + 4t)^{-1/2}(4) = 2(1 + 4t)^{-1/2} \Rightarrow v(6) = 2(1 + 4 \cdot 6)^{-1/2} = \frac{2}{5}$ m/sec;
    $v = 2(1 + 4t)^{-1/2} \Rightarrow a = \frac{dv}{dt} = -\frac{1}{2} \cdot 2(1 + 4t)^{-3/2}(4) = -4(1 + 4t)^{-3/2} \Rightarrow a(6) = -4(1 + 4 \cdot 6)^{-3/2} = -\frac{4}{125}$ m/sec^2

113. v proportional to $\frac{1}{\sqrt{s}} \Rightarrow v = \frac{k}{\sqrt{s}}$ for some constant k $\Rightarrow \frac{dv}{ds} = -\frac{k}{2s^{3/2}}$. Thus, $a = \frac{dv}{dt} = \frac{dv}{ds} \cdot \frac{ds}{dt} = \frac{dv}{ds} \cdot v$
    $= -\frac{k}{2s^{3/2}} \cdot \frac{k}{\sqrt{s}} = -\frac{k^2}{2}\left(\frac{1}{s^2}\right) \Rightarrow$ acceleration is a constant times $\frac{1}{s^2}$ so a is inversely proportional to s^2.

115. $T = 2\pi\sqrt{\frac{L}{g}} \Rightarrow \frac{dT}{dL} = 2\pi \cdot \frac{1}{2\sqrt{\frac{L}{g}}} \cdot \frac{1}{g} = \frac{\pi}{g\sqrt{\frac{L}{g}}} = \frac{\pi}{\sqrt{gL}}$. Therefore, $\frac{dT}{du} = \frac{dT}{dL} \cdot \frac{dL}{du} = \frac{\pi}{\sqrt{gL}} \cdot kL = \frac{\pi k\sqrt{L}}{\sqrt{g}} = \frac{1}{2} \cdot 2\pi k\sqrt{\frac{L}{g}} = \frac{kT}{2}$,
    as required.

117. The graph of $y = (f \circ g)(x)$ has a horizontal tangent at $x = 1$ provided that $(f \circ g)'(1) = 0 \Rightarrow f'(g(1))g'(1) = 0$
$\Rightarrow$ either $f'(g(1)) = 0$ or $g'(1) = 0$ (or both) $\Rightarrow$ either the graph of f has a horizontal tangent at $u = g(1)$, or the graph of g has a horizontal tangent at $x = 1$ (or both).

119. As $h \to 0$, the graph of $y = \frac{\sin 2(x+h) - \sin 2x}{h}$
approaches the graph of $y = 2 \cos 2x$ because
$\lim\limits_{h \to 0} \frac{\sin 2(x+h) - \sin 2x}{h} = \frac{d}{dx}(\sin 2x) = 2 \cos 2x.$

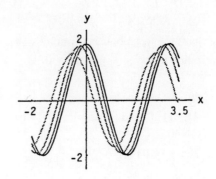

121. $\frac{dx}{dt} = \cos t$ and $\frac{dy}{dt} = 2 \cos 2t \Rightarrow \frac{dy}{dx} = \frac{dy/dt}{dx/dt} = \frac{2 \cos 2t}{\cos t} = \frac{2(2 \cos^2 t - 1)}{\cos t}$; then $\frac{dy}{dx} = 0 \Rightarrow \frac{2(2 \cos^2 t - 1)}{\cos t} = 0$
$\Rightarrow 2 \cos^2 t - 1 = 0 \Rightarrow \cos t = \pm \frac{1}{\sqrt{2}} \Rightarrow t = \frac{\pi}{4}, \frac{3\pi}{4}, \frac{5\pi}{4}, \frac{7\pi}{4}$. In the 1st quadrant: $t = \frac{\pi}{4} \Rightarrow x = \sin \frac{\pi}{4} = \frac{\sqrt{2}}{2}$ and
$y = \sin 2\left(\frac{\pi}{4}\right) = 1 \Rightarrow \left(\frac{\sqrt{2}}{2}, 1\right)$ is the point where the tangent line is horizontal. At the origin: $x = 0$ and $y = 0$
$\Rightarrow \sin t = 0 \Rightarrow t = 0$ or $t = \pi$ and $\sin 2t = 0 \Rightarrow t = 0, \frac{\pi}{2}, \pi, \frac{3\pi}{2}$; thus $t = 0$ and $t = \pi$ give the tangent lines at
the origin. Tangents at origin: $\frac{dy}{dx}\Big|_{t=0} = 2 \Rightarrow y = 2x$ and $\frac{dy}{dx}\Big|_{t=\pi} = -2 \Rightarrow y = -2x$

123. From the power rule, with $y = x^{1/4}$, we get $\frac{dy}{dx} = \frac{1}{4} x^{-3/4}$. From the chain rule, $y = \sqrt{\sqrt{\sqrt{x}}}$
$\Rightarrow \frac{dy}{dx} = \frac{1}{2\sqrt{\sqrt{x}}} \cdot \frac{d}{dx}\left(\sqrt{x}\right) = \frac{1}{2\sqrt{\sqrt{x}}} \cdot \frac{1}{2\sqrt{x}} = \frac{1}{4} x^{-3/4}$, in agreement.

125. (a)

(b) $\frac{df}{dt} = 1.27324 \sin 2t + 0.42444 \sin 6t + 0.2546 \sin 10t + 0.18186 \sin 14t$

(c) The curve of $y = \frac{df}{dt}$ approximates $y = \frac{dg}{dt}$
the best when t is not $-\pi, -\frac{\pi}{2}, 0, \frac{\pi}{2}$, nor π.

3.6 IMPLICIT DIFFERENTIATION

1. $x^2y + xy^2 = 6$:

   Step 1: $\left(x^2 \frac{dy}{dx} + y \cdot 2x\right) + \left(x \cdot 2y \frac{dy}{dx} + y^2 \cdot 1\right) = 0$

   Step 2: $x^2 \frac{dy}{dx} + 2xy \frac{dy}{dx} = -2xy - y^2$

   Step 3: $\frac{dy}{dx}(x^2 + 2xy) = -2xy - y^2$

   Step 4: $\frac{dy}{dx} = \frac{-2xy - y^2}{x^2 + 2xy}$

3. $2xy + y^2 = x + y$:

   Step 1: $\left(2x \frac{dy}{dx} + 2y\right) + 2y \frac{dy}{dx} = 1 + \frac{dy}{dx}$

   Step 2: $2x \frac{dy}{dx} + 2y \frac{dy}{dx} - \frac{dy}{dx} = 1 - 2y$

   Step 3: $\frac{dy}{dx}(2x + 2y - 1) = 1 - 2y$

   Step 4: $\frac{dy}{dx} = \frac{1 - 2y}{2x + 2y - 1}$

5. $x^2(x - y)^2 = x^2 - y^2$:

   Step 1: $x^2\left[2(x - y)\left(1 - \frac{dy}{dx}\right)\right] + (x - y)^2(2x) = 2x - 2y \frac{dy}{dx}$

   Step 2: $-2x^2(x - y) \frac{dy}{dx} + 2y \frac{dy}{dx} = 2x - 2x^2(x - y) - 2x(x - y)^2$

   Step 3: $\frac{dy}{dx}[-2x^2(x - y) + 2y] = 2x[1 - x(x - y) - (x - y)^2]$

   Step 4: $\frac{dy}{dx} = \frac{2x[1 - x(x - y) - (x - y)^2]}{-2x^2(x - y) + 2y} = \frac{x[1 - x(x - y) - (x - y)^2]}{y - x^2(x - y)} = \frac{x(1 - x^2 + xy - x^2 + 2xy - y^2)}{x^2y - x^3 + y}$

   $= \frac{x - 2x^3 + 3x^2y - xy^2}{x^2y - x^3 + y}$

7. $y^2 = \frac{x - 1}{x + 1} \Rightarrow 2y \frac{dy}{dx} = \frac{(x + 1) - (x - 1)}{(x + 1)^2} = \frac{2}{(x + 1)^2} \Rightarrow \frac{dy}{dx} = \frac{1}{y(x + 1)^2}$

9. $x = \tan y \Rightarrow 1 = (\sec^2 y) \frac{dy}{dx} \Rightarrow \frac{dy}{dx} = \frac{1}{\sec^2 y} = \cos^2 y$

11. $e^{2x} = \sin(x + 3y) \Rightarrow 2e^{2x} = (1 + 3y')\cos(x + 3y) \Rightarrow 1 + 3y' = \frac{2e^{2x}}{\cos(x + 3y)} \Rightarrow 3y' = \frac{2e^{2x}}{\cos(x + 3y)} - 1$

   $\Rightarrow y' = \frac{2e^{2x} - \cos(x + 3y)}{3\cos(x + 3y)}$

13. $y \sin\left(\frac{1}{y}\right) = 1 - xy \Rightarrow y\left[\cos\left(\frac{1}{y}\right) \cdot (-1) \frac{1}{y^2} \cdot \frac{dy}{dx}\right] + \sin\left(\frac{1}{y}\right) \cdot \frac{dy}{dx} = -x \frac{dy}{dx} - y \Rightarrow$

   $\frac{dy}{dx}\left[-\frac{1}{y}\cos\left(\frac{1}{y}\right) + \sin\left(\frac{1}{y}\right) + x\right] = -y \Rightarrow \frac{dy}{dx} = \frac{-y}{-\frac{1}{y}\cos\left(\frac{1}{y}\right) + \sin\left(\frac{1}{y}\right) + x} = \frac{-y^2}{y \sin\left(\frac{1}{y}\right) - \cos\left(\frac{1}{y}\right) + xy}$

15. $\theta^{1/2} + r^{1/2} = 1 \Rightarrow \frac{1}{2}\theta^{-1/2} + \frac{1}{2}r^{-1/2} \cdot \frac{dr}{d\theta} = 0 \Rightarrow \frac{dr}{d\theta}\left[\frac{1}{2\sqrt{r}}\right] = \frac{-1}{2\sqrt{\theta}} \Rightarrow \frac{dr}{d\theta} = -\frac{2\sqrt{r}}{2\sqrt{\theta}} = -\frac{\sqrt{r}}{\sqrt{\theta}}$

17. $\sin(r\theta) = \frac{1}{2} \Rightarrow [\cos(r\theta)]\left(r + \theta \frac{dr}{d\theta}\right) = 0 \Rightarrow \frac{dr}{d\theta}[\theta \cos(r\theta)] = -r \cos(r\theta) \Rightarrow \frac{dr}{d\theta} = \frac{-r \cos(r\theta)}{\theta \cos(r\theta)} = -\frac{r}{\theta}, \cos(r\theta) \neq 0$

19. $x^2 + y^2 = 1 \Rightarrow 2x + 2yy' = 0 \Rightarrow 2yy' = -2x \Rightarrow \frac{dy}{dx} = y' = -\frac{x}{y}$; now to find $\frac{d^2y}{dx^2}$, $\frac{d}{dx}(y') = \frac{d}{dx}\left(-\frac{x}{y}\right)$

   $\Rightarrow y'' = \frac{y(-1) + xy'}{y^2} = \frac{-y + x\left(-\frac{x}{y}\right)}{y^2}$ since $y' = -\frac{x}{y} \Rightarrow \frac{d^2y}{dx^2} = y'' = \frac{-y^2 - x^2}{y^3} = \frac{-y^2 - (1 - y^2)}{y^3} = \frac{-1}{y^3}$

21. $y^2 = e^{x^2} + 2x \Rightarrow 2yy' = 2x\,e^{x^2} + 2 \Rightarrow \frac{dy}{dx} = \frac{x\,e^{x^2}+1}{y} \Rightarrow \frac{d^2y}{dx^2} = \frac{y\left(2x^2 e^{x^2}+e^{x^2}\right)-\left(x\,e^{x^2}+1\right)y'}{y^2}$

$= \frac{y\left(2x^2 e^{x^2}+e^{x^2}\right)-\left(x\,e^{x^2}+1\right)\cdot\frac{x\,e^{x^2}+1}{y}}{y^2} = \frac{y^2\left(2x^2 e^{x^2}+e^{x^2}\right)-\left(x^2 e^{2x^2}+2x\,e^{x^2}+1\right)}{y^3} = \frac{\left(2x^2 y^2 + y^2 - 2x\right)e^{x^2} - x^2 e^{2x^2} - 1}{y^3}$

23. $2\sqrt{y} = x - y \Rightarrow y^{-1/2}y' = 1 - y' \Rightarrow y'\left(y^{-1/2}+1\right) = 1 \Rightarrow \frac{dy}{dx} = y' = \frac{1}{y^{-1/2}+1} = \frac{\sqrt{y}}{\sqrt{y}+1}$; we can

differentiate the equation $y'\left(y^{-1/2}+1\right) = 1$ again to find y'': $y'\left(-\frac{1}{2}y^{-3/2}y'\right) + \left(y^{-1/2}+1\right)y'' = 0$

$\Rightarrow \left(y^{-1/2}+1\right)y'' = \frac{1}{2}[y']^2 y^{-3/2} \Rightarrow \frac{d^2y}{dx^2} = y'' = \frac{\frac{1}{2}\left(\frac{1}{y^{-1/2}+1}\right)^2 y^{-3/2}}{\left(y^{-1/2}+1\right)} = \frac{1}{2y^{3/2}\left(y^{-1/2}+1\right)^3} = \frac{1}{2\left(1+\sqrt{y}\right)^3}$

25. $x^3 + y^3 = 16 \Rightarrow 3x^2 + 3y^2 y' = 0 \Rightarrow 3y^2 y' = -3x^2 \Rightarrow y' = -\frac{x^2}{y^2}$; we differentiate $y^2 y' = -x^2$ to find y'':

$y^2 y'' + y'[2y\cdot y'] = -2x \Rightarrow y^2 y'' = -2x - 2y[y']^2 \Rightarrow y'' = \frac{-2x-2y\left(-\frac{x^2}{y^2}\right)^2}{y^2} = \frac{-2x-\frac{2x^4}{y^3}}{y^2}$

$= \frac{-2xy^3-2x^4}{y^5} \Rightarrow \frac{d^2y}{dx^2}\Big|_{(2,2)} = \frac{-32-32}{32} = -2$

27. $y^2 + x^2 = y^4 - 2x$ at $(-2, 1)$ and $(-2, -1) \Rightarrow 2y\frac{dy}{dx} + 2x = 4y^3\frac{dy}{dx} - 2 \Rightarrow 2y\frac{dy}{dx} - 4y^3\frac{dy}{dx} = -2 - 2x$

$\Rightarrow \frac{dy}{dx}\left(2y - 4y^3\right) = -2 - 2x \Rightarrow \frac{dy}{dx} = \frac{x+1}{2y^3-y} \Rightarrow \frac{dy}{dx}\Big|_{(-2,1)} = -1$ and $\frac{dy}{dx}\Big|_{(-2,-1)} = 1$

29. $x^2 + xy - y^2 = 1 \Rightarrow 2x + y + xy' - 2yy' = 0 \Rightarrow (x - 2y)y' = -2x - y \Rightarrow y' = \frac{2x+y}{2y-x}$;
    (a) the slope of the tangent line $m = y'|_{(2,3)} = \frac{7}{4} \Rightarrow$ the tangent line is $y - 3 = \frac{7}{4}(x - 2) \Rightarrow y = \frac{7}{4}x - \frac{1}{2}$
    (b) the normal line is $y - 3 = -\frac{4}{7}(x - 2) \Rightarrow y = -\frac{4}{7}x + \frac{29}{7}$

31. $x^2 y^2 = 9 \Rightarrow 2xy^2 + 2x^2 yy' = 0 \Rightarrow x^2 yy' = -xy^2 \Rightarrow y' = -\frac{y}{x}$;
    (a) the slope of the tangent line $m = y'|_{(-1,3)} = -\frac{y}{x}\Big|_{(-1,3)} = 3 \Rightarrow$ the tangent line is $y - 3 = 3(x + 1)$
        $\Rightarrow y = 3x + 6$
    (b) the normal line is $y - 3 = -\frac{1}{3}(x + 1) \Rightarrow y = -\frac{1}{3}x + \frac{8}{3}$

33. $6x^2 + 3xy + 2y^2 + 17y - 6 = 0 \Rightarrow 12x + 3y + 3xy' + 4yy' + 17y' = 0 \Rightarrow y'(3x + 4y + 17) = -12x - 3y$
    $\Rightarrow y' = \frac{-12x-3y}{3x+4y+17}$;
    (a) the slope of the tangent line $m = y'|_{(-1,0)} = \frac{-12x-3y}{3x+4y+17}\Big|_{(-1,0)} = \frac{6}{7} \Rightarrow$ the tangent line is $y - 0 = \frac{6}{7}(x + 1)$
        $\Rightarrow y = \frac{6}{7}x + \frac{6}{7}$
    (b) the normal line is $y - 0 = -\frac{7}{6}(x + 1) \Rightarrow y = -\frac{7}{6}x - \frac{7}{6}$

35. $2xy + \pi \sin y = 2\pi \Rightarrow 2xy' + 2y + \pi(\cos y)y' = 0 \Rightarrow y'(2x + \pi \cos y) = -2y \Rightarrow y' = \frac{-2y}{2x+\pi\cos y}$;
    (a) the slope of the tangent line $m = y'|_{(1,\frac{\pi}{2})} = \frac{-2y}{2x+\pi\cos y}\Big|_{(1,\frac{\pi}{2})} = -\frac{\pi}{2} \Rightarrow$ the tangent line is
        $y - \frac{\pi}{2} = -\frac{\pi}{2}(x - 1) \Rightarrow y = -\frac{\pi}{2}x + \pi$
    (b) the normal line is $y - \frac{\pi}{2} = \frac{2}{\pi}(x - 1) \Rightarrow y = \frac{2}{\pi}x - \frac{2}{\pi} + \frac{\pi}{2}$

37. $y = 2\sin(\pi x - y) \Rightarrow y' = 2[\cos(\pi x - y)]\cdot(\pi - y') \Rightarrow y'[1 + 2\cos(\pi x - y)] = 2\pi \cos(\pi x - y)$
    $\Rightarrow y' = \frac{2\pi\cos(\pi x - y)}{1+2\cos(\pi x - y)}$;
    (a) the slope of the tangent line $m = y'|_{(1,0)} = \frac{2\pi\cos(\pi x - y)}{1+2\cos(\pi x - y)}\Big|_{(1,0)} = 2\pi \Rightarrow$ the tangent line is
        $y - 0 = 2\pi(x - 1) \Rightarrow y = 2\pi x - 2\pi$
    (b) the normal line is $y - 0 = -\frac{1}{2\pi}(x - 1) \Rightarrow y = -\frac{x}{2\pi} + \frac{1}{2\pi}$

39. Solving $x^2 + xy + y^2 = 7$ and $y = 0 \Rightarrow x^2 = 7 \Rightarrow x = \pm\sqrt{7} \Rightarrow \left(-\sqrt{7}, 0\right)$ and $\left(\sqrt{7}, 0\right)$ are the points where the curve crosses the x-axis. Now $x^2 + xy + y^2 = 7 \Rightarrow 2x + y + xy' + 2yy' = 0 \Rightarrow (x + 2y)y' = -2x - y$

$\Rightarrow y' = -\frac{2x+y}{x+2y} \Rightarrow m = -\frac{2x+y}{x+2y} \Rightarrow$ the slope at $\left(-\sqrt{7}, 0\right)$ is $m = -\frac{-2\sqrt{7}}{-\sqrt{7}} = -2$ and the slope at $\left(\sqrt{7}, 0\right)$ is

$m = -\frac{2\sqrt{7}}{\sqrt{7}} = -2$. Since the slope is -2 in each case, the corresponding tangents must be parallel.

41. $y^4 = y^2 - x^2 \Rightarrow 4y^3 y' = 2yy' - 2x \Rightarrow 2\left(2y^3 - y\right)y' = -2x \Rightarrow y' = \frac{x}{y - 2y^3}$; the slope of the tangent line at

$\left(\frac{\sqrt{3}}{4}, \frac{\sqrt{3}}{2}\right)$ is $\frac{x}{y-2y^3}\bigg|_{\left(\frac{\sqrt{3}}{4}, \frac{\sqrt{3}}{2}\right)} = \frac{\frac{\sqrt{3}}{4}}{\frac{\sqrt{3}}{2} - \frac{6\sqrt{3}}{8}} = \frac{\frac{1}{4}}{\frac{1}{2} - \frac{3}{4}} = \frac{1}{2-3} = -1$; the slope of the tangent line at $\left(\frac{\sqrt{3}}{4}, \frac{1}{2}\right)$

is $\frac{x}{y-2y^3}\bigg|_{\left(\frac{\sqrt{3}}{4}, \frac{1}{2}\right)} = \frac{\frac{\sqrt{3}}{4}}{\frac{1}{2} - \frac{2}{8}} = \frac{2\sqrt{3}}{4-2} = \sqrt{3}$

43. $y^4 - 4y^2 = x^4 - 9x^2 \Rightarrow 4y^3 y' - 8yy' = 4x^3 - 18x \Rightarrow y'\left(4y^3 - 8y\right) = 4x^3 - 18x \Rightarrow y' = \frac{4x^3 - 18x}{4y^3 - 8y} = \frac{2x^3 - 9x}{2y^3 - 4y}$

$= \frac{x\left(2x^2 - 9\right)}{y\left(2y^2 - 4\right)} = m$; $(-3, 2)$: $m = \frac{(-3)(18 - 9)}{2(8 - 4)} = -\frac{27}{8}$; $(-3, -2)$: $m = \frac{27}{8}$; $(3, 2)$: $m = \frac{27}{8}$; $(3, -2)$: $m = -\frac{27}{8}$

45. $x^2 - 2tx + 2t^2 = 4 \Rightarrow 2x \frac{dx}{dt} - 2x - 2t \frac{dx}{dt} + 4t = 0 \Rightarrow (2x - 2t) \frac{dx}{dt} = 2x - 4t \Rightarrow \frac{dx}{dt} = \frac{2x - 4t}{2x - 2t} = \frac{x - 2t}{x - t}$;

$2y^3 - 3t^2 = 4 \Rightarrow 6y^2 \frac{dy}{dt} - 6t = 0 \Rightarrow \frac{dy}{dt} = \frac{6t}{6y^2} = \frac{t}{y^2}$; thus $\frac{dy}{dx} = \frac{dy/dt}{dx/dt} = \frac{\left(\frac{t}{y^2}\right)}{\left(\frac{x-2t}{x-t}\right)} = \frac{t(x-t)}{y^2(x-2t)}$; $t = 2$

$\Rightarrow x^2 - 2(2)x + 2(2)^2 = 4 \Rightarrow x^2 - 4x + 4 = 0 \Rightarrow (x-2)^2 = 0 \Rightarrow x = 2$; $t = 2 \Rightarrow 2y^3 - 3(2)^2 = 4$

$\Rightarrow 2y^3 = 16 \Rightarrow y^3 = 8 \Rightarrow y = 2$; therefore $\frac{dy}{dx}\bigg|_{t=2} = \frac{2(2-2)}{(2)^2(2 - 2(2))} = 0$

47. $x + 2x^{3/2} = t^2 + t \Rightarrow \frac{dx}{dt} + 3x^{1/2} \frac{dx}{dt} = 2t + 1 \Rightarrow \left(1 + 3x^{1/2}\right) \frac{dx}{dt} = 2t + 1 \Rightarrow \frac{dx}{dt} = \frac{2t+1}{1+3x^{1/2}}$; $y\sqrt{t+1} + 2t\sqrt{y} = 4$

$\Rightarrow \frac{dy}{dt} \sqrt{t+1} + y\left(\frac{1}{2}\right)(t+1)^{-1/2} + 2\sqrt{y} + 2t\left(\frac{1}{2}y^{-1/2}\right)\frac{dy}{dt} = 0 \Rightarrow \frac{dy}{dt}\sqrt{t+1} + \frac{y}{2\sqrt{t+1}} + 2\sqrt{y} + \left(\frac{t}{\sqrt{y}}\right)\frac{dy}{dt} = 0$

$\Rightarrow \left(\sqrt{t+1} + \frac{t}{\sqrt{y}}\right)\frac{dy}{dt} = \frac{-y}{2\sqrt{t+1}} - 2\sqrt{y} \Rightarrow \frac{dy}{dt} = \frac{\left(\frac{-y}{2\sqrt{t+1}} - 2\sqrt{y}\right)}{\left(\sqrt{t+1} + \frac{t}{\sqrt{y}}\right)} = \frac{-y\sqrt{y} - 4y\sqrt{t+1}}{2\sqrt{y}(t+1) + 2t\sqrt{t+1}}$; thus

$\frac{dy}{dx} = \frac{dy/dt}{dx/dt} = \frac{\left(\frac{-y\sqrt{y} - 4y\sqrt{t+1}}{2\sqrt{y}(t+1) + 2t\sqrt{t+1}}\right)}{\left(\frac{2t+1}{1+3x^{1/2}}\right)}$; $t = 0 \Rightarrow x + 2x^{3/2} = 0 \Rightarrow x\left(1 + 2x^{1/2}\right) = 0 \Rightarrow x = 0$; $t = 0$

$\Rightarrow y\sqrt{0+1} + 2(0)\sqrt{y} = 4 \Rightarrow y = 4$; therefore $\frac{dy}{dx}\bigg|_{t=0} = \frac{\left(\frac{-4\sqrt{4} - 4(4)\sqrt{0+1}}{2\sqrt{4(0+1)} + 2(0)\sqrt{0+1}}\right)}{\left(\frac{2(0)+1}{1+3(0)^{1/2}}\right)} = -6$

49. $x^2 + 2xy - 3y^2 = 0 \Rightarrow 2x + 2xy' + 2y - 6yy' = 0 \Rightarrow y'(2x - 6y) = -2x - 2y \Rightarrow y' = \frac{x+y}{3y-x} \Rightarrow$ the slope of the

tangent line $m = y'|_{(1,1)} = \frac{x+y}{3y-x}\bigg|_{(1,1)} = 1 \Rightarrow$ the equation of the normal line at $(1, 1)$ is $y - 1 = -1(x - 1)$

$\Rightarrow y = -x + 2$. To find where the normal line intersects the curve we substitute into its equation:

$x^2 + 2x(2 - x) - 3(2 - x)^2 = 0 \Rightarrow x^2 + 4x - 2x^2 - 3\left(4 - 4x + x^2\right) = 0 \Rightarrow -4x^2 + 16x - 12 = 0$

$\Rightarrow x^2 - 4x + 3 = 0 \Rightarrow (x - 3)(x - 1) = 0 \Rightarrow x = 3$ and $y = -x + 2 = -1$. Therefore, the normal to the curve

at $(1, 1)$ intersects the curve at the point $(3, -1)$. Note that it also intersects the curve at $(1, 1)$.

51. $y^2 = x \Rightarrow \frac{dy}{dx} = \frac{1}{2y}$. If a normal is drawn from $(a, 0)$ to (x_1, y_1) on the curve its slope satisfies $\frac{y_1 - 0}{x_1 - a} = -2y_1$

$\Rightarrow y_1 = -2y_1(x_1 - a)$ or $a = x_1 + \frac{1}{2}$. Since $x_1 \geq 0$ on the curve, we must have that $a \geq \frac{1}{2}$. By symmetry, the

two points on the parabola are $\left(x_1, \sqrt{x_1}\right)$ and $\left(x_1, -\sqrt{x_1}\right)$. For the normal to be perpendicular,

$\left(\frac{\sqrt{x_1}}{x_1 - a}\right)\left(\frac{\sqrt{x_1}}{a - x_1}\right) = -1 \Rightarrow \frac{x_1}{(a - x_1)^2} = 1 \Rightarrow x_1 = (a - x_1)^2 \Rightarrow x_1 = \left(x_1 + \frac{1}{2} - x_1\right)^2 \Rightarrow x_1 = \frac{1}{4}$ and $y_1 = \pm\frac{1}{2}$.

Therefore, $\left(\frac{1}{4}, \pm\frac{1}{2}\right)$ and $a = \frac{3}{4}$.

53. $xy^3 + x^2y = 6 \Rightarrow x\left(3y^2 \frac{dy}{dx}\right) + y^3 + x^2 \frac{dy}{dx} + 2xy = 0 \Rightarrow \frac{dy}{dx}(3xy^2 + x^2) = -y^3 - 2xy \Rightarrow \frac{dy}{dx} = \frac{-y^3 - 2xy}{3xy^2 + x^2}$

$= -\frac{y^3 + 2xy}{3xy^2 + x^2}$; also, $xy^3 + x^2y = 6 \Rightarrow x(3y^2) + y^3 \frac{dx}{dy} + x^2 + y\left(2x\frac{dx}{dy}\right) = 0 \Rightarrow \frac{dx}{dy}(y^3 + 2xy) = -3xy^2 - x^2$

$\Rightarrow \frac{dx}{dy} = -\frac{3xy^2 + x^2}{y^3 + 2xy}$; thus $\frac{dx}{dy}$ appears to equal $\frac{1}{\frac{dy}{dx}}$. The two different treatments view the graphs as functions

symmetric across the line $y = x$, so their slopes are reciprocals of one another at the corresponding points
(a, b) and (b, a).

55. $x^4 + 4y^2 = 1$:

   (a) $y^2 = \frac{1 - x^4}{4} \Rightarrow y = \pm\frac{1}{2}\sqrt{1 - x^4} \Rightarrow \frac{dy}{dx} = \pm\frac{1}{4}(1 - x^4)^{-1/2}(-4x^3) = \frac{\pm x^3}{(1 - x^4)^{1/2}}$;

     differentiating implicitly, we find, $4x^3 + 8y\frac{dy}{dx} = 0 \Rightarrow \frac{dy}{dx} = \frac{-4x^3}{8y} = \frac{-4x^3}{8\left(\pm\frac{1}{2}\sqrt{1 - x^4}\right)} = \frac{\pm x^3}{(1 - x^4)^{1/2}}$.

   (b)

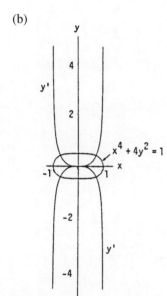

3.7 DERIVATIVES OF INVERSE FUNCTIONS AND LOGARITHMS

1. (a) $y = 2x + 3 \Rightarrow 2x = y - 3$ (b)
      $\Rightarrow x = \frac{y}{2} - \frac{3}{2} \Rightarrow f^{-1}(x) = \frac{x}{2} - \frac{3}{2}$

   (c) $\frac{df}{dx}\Big|_{x=-1} = 2, \frac{df^{-1}}{dx}\Big|_{x=1} = \frac{1}{2}$

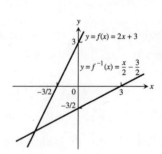

3. (a) $y = 5 - 4x \Rightarrow 4x = 5 - y$ (b)
      $\Rightarrow x = \frac{5}{4} - \frac{y}{4} \Rightarrow f^{-1}(x) = \frac{5}{4} - \frac{x}{4}$

   (c) $\frac{df}{dx}\Big|_{x=1/2} = -4, \frac{df^{-1}}{dx}\Big|_{x=3} = -\frac{1}{4}$

5. (a) $f(g(x)) = \left(\sqrt[3]{x}\right)^3 = x$, $g(f(x)) = \sqrt[3]{x^3} = x$

(b)

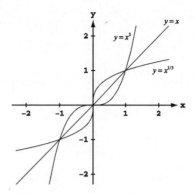

(c) $f'(x) = 3x^2 \Rightarrow f'(1) = 3, f'(-1) = 3$;

$g'(x) = \frac{1}{3}x^{-2/3} \Rightarrow g'(1) = \frac{1}{3}, g'(-1) = \frac{1}{3}$

(d) The line $y = 0$ is tangent to $f(x) = x^3$ at $(0, 0)$;

the line $x = 0$ is tangent to $g(x) = \sqrt[3]{x}$ at $(0, 0)$

7. $\frac{df}{dx} = 3x^2 - 6x \Rightarrow \frac{df^{-1}}{dx}\Big|_{x=f(3)} = \frac{1}{\frac{df}{dx}}\Big|_{x=3} = \frac{1}{9}$

9. $\frac{df^{-1}}{dx}\Big|_{x=4} = \frac{df^{-1}}{dx}\Big|_{x=f(2)} = \frac{1}{\frac{df}{dx}}\Big|_{x=2} = \frac{1}{\left(\frac{1}{3}\right)} = 3$

11. $y = \ln 3x \Rightarrow y' = \left(\frac{1}{3x}\right)(3) = \frac{1}{x}$

13. $y = \ln(t^2) \Rightarrow \frac{dy}{dt} = \left(\frac{1}{t^2}\right)(2t) = \frac{2}{t}$

15. $y = \ln\frac{3}{x} = \ln 3x^{-1} \Rightarrow \frac{dy}{dx} = \left(\frac{1}{3x^{-1}}\right)(-3x^{-2}) = -\frac{1}{x}$

17. $y = \ln(\theta + 1) \Rightarrow \frac{dy}{d\theta} = \left(\frac{1}{\theta+1}\right)(1) = \frac{1}{\theta+1}$

19. $y = \ln x^3 \Rightarrow \frac{dy}{dx} = \left(\frac{1}{x^3}\right)(3x^2) = \frac{3}{x}$

21. $y = t(\ln t)^2 \Rightarrow \frac{dy}{dt} = (\ln t)^2 + 2t(\ln t) \cdot \frac{d}{dt}(\ln t) = (\ln t)^2 + \frac{2t\ln t}{t} = (\ln t)^2 + 2\ln t$

23. $y = \frac{x^4}{4}\ln x - \frac{x^4}{16} \Rightarrow \frac{dy}{dx} = x^3 \ln x + \frac{x^4}{4} \cdot \frac{1}{x} - \frac{4x^3}{16} = x^3 \ln x$

$f = \ln t \qquad f' = t$

$g = t \qquad g' =$

25. $y = \frac{\ln t}{t} \Rightarrow \frac{dy}{dt} = \frac{t\left(\frac{1}{t}\right) - (\ln t)(1)}{t^2} = \frac{1 - \ln t}{t^2}$

27. $y = \frac{\ln x}{1 + \ln x} \Rightarrow y' = \frac{(1+\ln x)\left(\frac{1}{x}\right) - (\ln x)\left(\frac{1}{x}\right)}{(1 + \ln x)^2} = \frac{\frac{1}{x} + \frac{\ln x}{x} - \frac{\ln x}{x}}{(1 + \ln x)^2} = \frac{1}{x(1 + \ln x)^2}$

29. $y = \ln(\ln x) \Rightarrow y' = \left(\frac{1}{\ln x}\right)\left(\frac{1}{x}\right) = \frac{1}{x \ln x}$

31. $y = \theta[\sin(\ln\theta) + \cos(\ln\theta)] \Rightarrow \frac{dy}{d\theta} = [\sin(\ln\theta) + \cos(\ln\theta)] + \theta\left[\cos(\ln\theta) \cdot \frac{1}{\theta} - \sin(\ln\theta) \cdot \frac{1}{\theta}\right]$

$= \sin(\ln\theta) + \cos(\ln\theta) + \cos(\ln\theta) - \sin(\ln\theta) = 2\cos(\ln\theta)$

33. $y = \ln\frac{1}{x\sqrt{x+1}} = -\ln x - \frac{1}{2}\ln(x+1) \Rightarrow y' = -\frac{1}{x} - \frac{1}{2}\left(\frac{1}{x+1}\right) = -\frac{2(x+1)+x}{2x(x+1)} = -\frac{3x+2}{2x(x+1)}$

35. $y = \frac{1 + \ln t}{1 - \ln t} \Rightarrow \frac{dy}{dt} = \frac{(1-\ln t)\left(\frac{1}{t}\right) - (1+\ln t)\left(\frac{-1}{t}\right)}{(1-\ln t)^2} = \frac{\frac{1}{t} - \frac{\ln t}{t} + \frac{1}{t} + \frac{\ln t}{t}}{(1-\ln t)^2} = \frac{2}{t(1-\ln t)^2}$

37. $y = \ln(\sec(\ln\theta)) \Rightarrow \frac{dy}{d\theta} = \frac{1}{\sec(\ln\theta)} \cdot \frac{d}{d\theta}(\sec(\ln\theta)) = \frac{\sec(\ln\theta)\tan(\ln\theta)}{\sec(\ln\theta)} \cdot \frac{d}{d\theta}(\ln\theta) = \frac{\tan(\ln\theta)}{\theta}$

39. $y = \ln\left(\frac{(x^2+1)^5}{\sqrt{1-x}}\right) = 5\ln(x^2+1) - \frac{1}{2}\ln(1-x) \Rightarrow y' = \frac{5 \cdot 2x}{x^2+1} - \frac{1}{2}\left(\frac{1}{1-x}\right)(-1) = \frac{10x}{x^2+1} + \frac{1}{2(1-x)}$

41. $y = \sqrt{x(x+1)} = (x(x+1))^{1/2} \Rightarrow \ln y = \frac{1}{2}\ln(x(x+1)) \Rightarrow 2\ln y = \ln(x) + \ln(x+1) \Rightarrow \frac{2y'}{y} = \frac{1}{x} + \frac{1}{x+1}$

$\Rightarrow y' = \left(\frac{1}{2}\right)\sqrt{x(x+1)}\left(\frac{1}{x} + \frac{1}{x+1}\right) = \frac{\sqrt{x(x+1)}(2x+1)}{2x(x+1)} = \frac{2x+1}{2\sqrt{x(x+1)}}$

43. $y = \sqrt{\frac{t}{t+1}} = \left(\frac{t}{t+1}\right)^{1/2} \Rightarrow \ln y = \frac{1}{2}[\ln t - \ln(t+1)] \Rightarrow \frac{1}{y}\frac{dy}{dt} = \frac{1}{2}\left(\frac{1}{t} - \frac{1}{t+1}\right)$

$\Rightarrow \frac{dy}{dt} = \frac{1}{2}\sqrt{\frac{t}{t+1}}\left(\frac{1}{t} - \frac{1}{t+1}\right) = \frac{1}{2}\sqrt{\frac{t}{t+1}}\left[\frac{1}{t(t+1)}\right] = \frac{1}{2\sqrt{t}(t+1)^{3/2}}$

45. $y = \sqrt{\theta+3}(\sin\theta) = (\theta+3)^{1/2}\sin\theta \Rightarrow \ln y = \frac{1}{2}\ln(\theta+3) + \ln(\sin\theta) \Rightarrow \frac{1}{y}\frac{dy}{d\theta} = \frac{1}{2(\theta+3)} + \frac{\cos\theta}{\sin\theta}$

$\Rightarrow \frac{dy}{d\theta} = \sqrt{\theta+3}(\sin\theta)\left[\frac{1}{2(\theta+3)} + \cot\theta\right]$

47. $y = t(t+1)(t+2) \Rightarrow \ln y = \ln t + \ln(t+1) + \ln(t+2) \Rightarrow \frac{1}{y}\frac{dy}{dt} = \frac{1}{t} + \frac{1}{t+1} + \frac{1}{t+2}$

$\Rightarrow \frac{dy}{dt} = t(t+1)(t+2)\left(\frac{1}{t} + \frac{1}{t+1} + \frac{1}{t+2}\right) = t(t+1)(t+2)\left[\frac{(t+1)(t+2)+t(t+2)+t(t+1)}{t(t+1)(t+2)}\right] = 3t^2 + 6t + 2$

49. $y = \frac{\theta+5}{\theta\cos\theta} \Rightarrow \ln y = \ln(\theta+5) - \ln\theta - \ln(\cos\theta) \Rightarrow \frac{1}{y}\frac{dy}{d\theta} = \frac{1}{\theta+5} - \frac{1}{\theta} + \frac{\sin\theta}{\cos\theta}$

$\Rightarrow \frac{dy}{d\theta} = \left(\frac{\theta+5}{\theta\cos\theta}\right)\left(\frac{1}{\theta+5} - \frac{1}{\theta} + \tan\theta\right)$

51. $y = \frac{x\sqrt{x^2+1}}{(x+1)^{2/3}} \Rightarrow \ln y = \ln x + \frac{1}{2}\ln(x^2+1) - \frac{2}{3}\ln(x+1) \Rightarrow \frac{y'}{y} = \frac{1}{x} + \frac{x}{x^2+1} - \frac{2}{3(x+1)}$

$\Rightarrow y' = \frac{x\sqrt{x^2+1}}{(x+1)^{2/3}}\left[\frac{1}{x} + \frac{x}{x^2+1} - \frac{2}{3(x+1)}\right]$

53. $y = \sqrt[3]{\frac{x(x-2)}{x^2+1}} \Rightarrow \ln y = \frac{1}{3}\left[\ln x + \ln(x-2) - \ln(x^2+1)\right] \Rightarrow \frac{y'}{y} = \frac{1}{3}\left(\frac{1}{x} + \frac{1}{x-2} - \frac{2x}{x^2+1}\right)$

$\Rightarrow y' = \frac{1}{3}\sqrt[3]{\frac{x(x-2)}{x^2+1}}\left(\frac{1}{x} + \frac{1}{x-2} - \frac{2x}{x^2+1}\right)$

55. $y = \ln(\cos^2\theta) \Rightarrow \frac{dy}{d\theta} = \frac{1}{\cos^2\theta}\cdot 2\cos\theta\cdot(-\sin\theta) = -2\tan\theta$

57. $y = \ln(3te^{-t}) = \ln 3 + \ln t + \ln e^{-t} = \ln 3 + \ln t - t \Rightarrow \frac{dy}{dt} = \frac{1}{t} - 1 = \frac{1-t}{t}$

59. $y = \ln\frac{e^\theta}{1+e^\theta} = \ln e^\theta - \ln(1+e^\theta) = \theta - \ln(1+e^\theta) \Rightarrow \frac{dy}{d\theta} = 1 - \left(\frac{1}{1+e^\theta}\right)\frac{d}{d\theta}(1+e^\theta) = 1 - \frac{e^\theta}{1+e^\theta} = \frac{1}{1+e^\theta}$

61. $y = e^{(\cos t + \ln t)} = e^{\cos t}e^{\ln t} = te^{\cos t} \Rightarrow \frac{dy}{dt} = e^{\cos t} + te^{\cos t}\frac{d}{dt}(\cos t) = (1 - t\sin t)e^{\cos t}$

63. $\ln y = e^y \sin x \Rightarrow \left(\frac{1}{y}\right)y' = (y'e^y)(\sin x) + e^y\cos x \Rightarrow y'\left(\frac{1}{y} - e^y\sin x\right) = e^y\cos x$

$\Rightarrow y'\left(\frac{1-ye^y\sin x}{y}\right) = e^y\cos x \Rightarrow y' = \frac{ye^y\cos x}{1-ye^y\sin x}$

65. $x^y = y^x \Rightarrow \ln x^y = \ln y^x \Rightarrow y\ln x = x\ln y \Rightarrow y\cdot\frac{1}{x} + y'\cdot\ln x = x\cdot\frac{1}{y}\cdot y' + (1)\cdot\ln y \Rightarrow \ln x\cdot y' - \frac{x}{y}\cdot y' = \ln y - \frac{y}{x}$

$\Rightarrow y' = \frac{\ln y - \frac{y}{x}}{\ln x - \frac{x}{y}} = \frac{xy\ln y - y^2}{xy\ln x - x^2} = \frac{y}{x}\left(\frac{x\ln y - y}{y\ln x - x}\right)$

67. $y = 2^x \Rightarrow y' = 2^x\ln 2$

69. $y = 5^{\sqrt{s}} \Rightarrow \frac{dy}{ds} = 5^{\sqrt{s}}(\ln 5)\left(\frac{1}{2}s^{-1/2}\right) = \left(\frac{\ln 5}{2\sqrt{s}}\right)5^{\sqrt{s}}$

71. $y = x^\pi \Rightarrow y' = \pi x^{(\pi-1)}$

73. $y = \log_2 5\theta = \frac{\ln 5\theta}{\ln 2} \Rightarrow \frac{dy}{d\theta} = \left(\frac{1}{\ln 2}\right)\left(\frac{1}{5\theta}\right)(5) = \frac{1}{\theta\ln 2}$

75. $y = \log_4 x + \log_4 x^2 = \frac{\ln x}{\ln 4} + \frac{\ln x^2}{\ln 4} = \frac{\ln x}{\ln 4} + 2\frac{\ln x}{\ln 4} = 3\frac{\ln x}{\ln 4} \Rightarrow y' = \frac{3}{x\ln 4}$

77. $y = \log_2 r \cdot \log_4 r = \left(\frac{\ln r}{\ln 2}\right)\left(\frac{\ln r}{\ln 4}\right) = \frac{\ln^2 r}{(\ln 2)(\ln 4)} \Rightarrow \frac{dy}{dr} = \left[\frac{1}{(\ln 2)(\ln 4)}\right](2 \ln r)\left(\frac{1}{r}\right) = \frac{2 \ln r}{r(\ln 2)(\ln 4)}$

79. $y = \log_3\left(\left(\frac{x+1}{x-1}\right)^{\ln 3}\right) = \frac{\ln\left(\frac{x+1}{x-1}\right)^{\ln 3}}{\ln 3} = \frac{(\ln 3)\ln\left(\frac{x+1}{x-1}\right)}{\ln 3} = \ln\left(\frac{x+1}{x-1}\right) = \ln(x+1) - \ln(x-1)$

$\Rightarrow \frac{dy}{dx} = \frac{1}{x+1} - \frac{1}{x-1} = \frac{-2}{(x+1)(x-1)}$

81. $y = \theta \sin(\log_7 \theta) = \theta \sin\left(\frac{\ln \theta}{\ln 7}\right) \Rightarrow \frac{dy}{d\theta} = \sin\left(\frac{\ln \theta}{\ln 7}\right) + \theta\left[\cos\left(\frac{\ln \theta}{\ln 7}\right)\right]\left(\frac{1}{\theta \ln 7}\right) = \sin(\log_7 \theta) + \frac{1}{\ln 7}\cos(\log_7 \theta)$

83. $y = \log_5 e^x = \frac{\ln e^x}{\ln 5} = \frac{x}{\ln 5} \Rightarrow y' = \frac{1}{\ln 5}$

85. $y = 3^{\log_2 t} = 3^{(\ln t)/(\ln 2)} \Rightarrow \frac{dy}{dt} = \left[3^{(\ln t)/(\ln 2)}(\ln 3)\right]\left(\frac{1}{t \ln 2}\right) = \frac{1}{t}(\log_2 3)\, 3^{\log_2 t}$

87. $y = \log_2\left(8t^{\ln 2}\right) = \frac{\ln 8 + \ln\left(t^{\ln 2}\right)}{\ln 2} = \frac{3 \ln 2 + (\ln 2)(\ln t)}{\ln 2} = 3 + \ln t \Rightarrow \frac{dy}{dt} = \frac{1}{t}$

89. $y = (x+1)^x \Rightarrow \ln y = \ln(x+1)^x = x \ln(x+1) \Rightarrow \frac{y'}{y} = \ln(x+1) + x \cdot \frac{1}{(x+1)} \Rightarrow y' = (x+1)^x\left[\frac{x}{x+1} + \ln(x+1)\right]$

91. $y = \left(\sqrt{t}\right)^t = \left(t^{1/2}\right)^t = t^{t/2} \Rightarrow \ln y = \ln t^{t/2} = \left(\frac{t}{2}\right)\ln t \Rightarrow \frac{1}{y}\frac{dy}{dt} = \left(\frac{1}{2}\right)(\ln t) + \left(\frac{t}{2}\right)\left(\frac{1}{t}\right) = \frac{\ln t}{2} + \frac{1}{2}$

$\Rightarrow \frac{dy}{dt} = \left(\sqrt{t}\right)^t\left(\frac{\ln t}{2} + \frac{1}{2}\right)$

93. $y = (\sin x)^x \Rightarrow \ln y = \ln(\sin x)^x = x \ln(\sin x) \Rightarrow \frac{y'}{y} = \ln(\sin x) + x\left(\frac{\cos x}{\sin x}\right) \Rightarrow y' = (\sin x)^x\left[\ln(\sin x) + x \cot x\right]$

95. $y = x^{\ln x}, x > 0 \Rightarrow \ln y = (\ln x)^2 \Rightarrow \frac{y'}{y} = 2(\ln x)\left(\frac{1}{x}\right) \Rightarrow y' = \left(x^{\ln x}\right)\left(\frac{\ln x^2}{x}\right)$

97. $(g \circ f)(x) = x \Rightarrow g(f(x)) = x \Rightarrow g'(f(x))f'(x) = 1$

99. $y = A \sin(\ln x) + B \cos(\ln x) \Rightarrow y' = A \cos(\ln x) \cdot \frac{1}{x} - B \sin(\ln x) \cdot \frac{1}{x} = (A \cos(\ln x) - B \sin(\ln x)) \cdot \frac{1}{x}$

$\Rightarrow y'' = (A \cos(\ln x) - B \sin(\ln x)) \cdot \frac{-1}{x^2} + \left(-A \sin(\ln x) \cdot \frac{1}{x} - B \cos(\ln x) \cdot \frac{1}{x}\right) \cdot \frac{1}{x}$

$= (-A(\cos(\ln x) + \sin(\ln x)) + B(\sin(\ln x) - \cos(\ln x))) \cdot \frac{1}{x^2}$

$\Rightarrow x^2 y'' + x y' + y = (-A(\cos(\ln x) + \sin(\ln x)) + B(\sin(\ln x) - \cos(\ln x))) + (A \cos(\ln x) - B \sin(\ln x))$

$+ (A \sin(\ln x) + B \cos(\ln x)) = 0$

3.8 INVERSE TRIGONOMETRIC FUNCTIONS

1. (a) $\frac{\pi}{4}$ (b) $-\frac{\pi}{3}$ (c) $\frac{\pi}{6}$ 3. (a) $-\frac{\pi}{6}$ (b) $\frac{\pi}{4}$ (c) $-\frac{\pi}{3}$

5. (a) $\frac{\pi}{3}$ (b) $\frac{3\pi}{4}$ (c) $\frac{\pi}{6}$ 7. (a) $\frac{3\pi}{4}$ (b) $\frac{\pi}{6}$ (c) $\frac{2\pi}{3}$

9. (a) $\frac{\pi}{4}$ (b) $-\frac{\pi}{3}$ (c) $\frac{\pi}{6}$ 11. (a) $\frac{3\pi}{4}$ (b) $\frac{\pi}{6}$ (c) $\frac{2\pi}{3}$

13. $\lim_{x \to 1^-} \sin^{-1} x = \frac{\pi}{2}$ 15. $\lim_{x \to \infty} \tan^{-1} x = \frac{\pi}{2}$

17. $\lim_{x \to \infty} \sec^{-1} x = \frac{\pi}{2}$ 19. $\lim_{x \to \infty} \csc^{-1} x = \lim_{x \to \infty} \sin^{-1}\left(\frac{1}{x}\right) = 0$

21. $y = \cos^{-1}\left(x^2\right) \Rightarrow \frac{dy}{dx} = -\frac{2x}{\sqrt{1-(x^2)^2}} = \frac{-2x}{\sqrt{1-x^4}}$ 23. $y = \sin^{-1}\sqrt{2t} \Rightarrow \frac{dy}{dt} = \frac{\sqrt{2}}{\sqrt{1-\left(\sqrt{2t}\right)^2}} = \frac{\sqrt{2}}{\sqrt{1-2t^2}}$

25. $y = \sec^{-1}(2s+1) \Rightarrow \frac{dy}{ds} = \frac{2}{|2s+1|\sqrt{(2s+1)^2-1}} = \frac{2}{|2s+1|\sqrt{4s^2+4s}} = \frac{1}{|2s+1|\sqrt{s^2+s}}$

27. $y = \csc^{-1}(x^2+1) \Rightarrow \frac{dy}{dx} = -\frac{2x}{|x^2+1|\sqrt{(x^2+1)^2-1}} = \frac{-2x}{(x^2+1)\sqrt{x^4+2x^2}}$

29. $y = \sec^{-1}\left(\frac{1}{t}\right) = \cos^{-1}t \Rightarrow \frac{dy}{dt} = \frac{-1}{\sqrt{1-t^2}}$

31. $y = \cot^{-1}\sqrt{t} = \cot^{-1}t^{1/2} \Rightarrow \frac{dy}{dt} = -\frac{\left(\frac{1}{2}\right)t^{-1/2}}{1+(t^{1/2})^2} = \frac{-1}{2\sqrt{t}(1+t)}$

33. $y = \ln(\tan^{-1}x) \Rightarrow \frac{dy}{dx} = \frac{\left(\frac{1}{1+x^2}\right)}{\tan^{-1}x} = \frac{1}{(\tan^{-1}x)(1+x^2)}$

35. $y = \csc^{-1}(e^t) \Rightarrow \frac{dy}{dt} = -\frac{e^t}{|e^t|\sqrt{(e^t)^2-1}} = \frac{-1}{\sqrt{e^{2t}-1}}$

37. $y = s\sqrt{1-s^2} + \cos^{-1}s = s(1-s^2)^{1/2} + \cos^{-1}s \Rightarrow \frac{dy}{ds} = (1-s^2)^{1/2} + s\left(\frac{1}{2}\right)(1-s^2)^{-1/2}(-2s) - \frac{1}{\sqrt{1-s^2}}$

$= \sqrt{1-s^2} - \frac{s^2}{\sqrt{1-s^2}} - \frac{1}{\sqrt{1-s^2}} = \sqrt{1-s^2} - \frac{s^2+1}{\sqrt{1-s^2}} = \frac{1-s^2-s^2-1}{\sqrt{1-s^2}} = \frac{-2s^2}{\sqrt{1-s^2}}$

39. $y = \tan^{-1}\sqrt{x^2-1} + \csc^{-1}x = \tan^{-1}(x^2-1)^{1/2} + \csc^{-1}x \Rightarrow \frac{dy}{dx} = \frac{\left(\frac{1}{2}\right)(x^2-1)^{-1/2}(2x)}{1+\left[(x^2-1)^{1/2}\right]^2} - \frac{1}{|x|\sqrt{x^2-1}}$

$= \frac{1}{x\sqrt{x^2-1}} - \frac{1}{|x|\sqrt{x^2-1}} = 0$, for $x > 1$

41. $y = x\sin^{-1}x + \sqrt{1-x^2} = x\sin^{-1}x + (1-x^2)^{1/2} \Rightarrow \frac{dy}{dx} = \sin^{-1}x + x\left(\frac{1}{\sqrt{1-x^2}}\right) + \left(\frac{1}{2}\right)(1-x^2)^{-1/2}(-2x)$

$= \sin^{-1}x + \frac{x}{\sqrt{1-x^2}} - \frac{x}{\sqrt{1-x^2}} = \sin^{-1}x$

43. The angle α is the large angle between the wall and the right end of the blackboard minus the small angle between the left end of the blackboard and the wall $\Rightarrow \alpha = \cot^{-1}\left(\frac{x}{15}\right) - \cot^{-1}\left(\frac{x}{3}\right)$.

45. Take each square as a unit square. From the diagram we have the following: the smallest angle α has a tangent of $1 \Rightarrow \alpha = \tan^{-1}1$; the middle angle β has a tangent of $2 \Rightarrow \beta = \tan^{-1}2$; and the largest angle γ has a tangent of $3 \Rightarrow \gamma = \tan^{-1}3$. The sum of these three angles is $\pi \Rightarrow \alpha + \beta + \gamma = \pi$ $\Rightarrow \tan^{-1}1 + \tan^{-1}2 + \tan^{-1}3 = \pi$.

47. If $x = 1$: $\sin^{-1}(1) + \cos^{-1}(1) = \frac{\pi}{2} + 0 = \frac{\pi}{2}$.
   If $x = 0$: $\sin^{-1}(0) + \cos^{-1}(0) = 0 + \frac{\pi}{2} = \frac{\pi}{2}$.
   If $x = -1$: $\sin^{-1}(-1) + \cos^{-1}(-1) = -\frac{\pi}{2} + \pi = \frac{\pi}{2}$.
   The identity $\sin^{-1}(x) + \cos^{-1}(x) = \frac{\pi}{2}$ has been established for x in $(0, 1)$, by Figure 1.6.7. So now if x is in $(-1, 0)$, note that $-x$ is in $(0, 1)$, and we have that
   $\sin^{-1}(x) + \cos^{-1}(x) = -\sin^{-1}(-x) + \cos^{-1}(x)$ since $\sin^{-1}$ is odd
   $= -\sin^{-1}(-x) + \pi - \cos^{-1}(-x)$ by Eq. 3, Section 1.6
   $= -(\sin^{-1}(-x) + \cos^{-1}(-x)) + \pi$
   $= -\frac{\pi}{2} + \pi$
   $= \frac{\pi}{2}$

   This establishes the identity for all x in $[-1, 1]$.

49. (a) Defined; there is an angle whose tangent is 2.

    (b) Not defined; there is no angle whose cosine is 2.

51. (a) Not defined; there is no angle whose secant is 0.

    (b) Not defined; there is no angle whose sine is $\sqrt{2}$.

53. $\csc^{-1} u = \frac{\pi}{2} - \sec^{-1} u \Rightarrow \frac{d}{dx}\left(\csc^{-1} u\right) = \frac{d}{dx}\left(\frac{\pi}{2} - \sec^{-1} u\right) = 0 - \frac{\frac{du}{dx}}{|u|\sqrt{u^2-1}} = -\frac{\frac{du}{dx}}{|u|\sqrt{u^2-1}}, \; |u| > 1$

55. $f(x) = \sec x \Rightarrow f'(x) = \sec x \tan x \Rightarrow \left.\frac{df^{-1}}{dx}\right|_{x=b} = \frac{1}{\left.\frac{df}{dx}\right|_{x=f^{-1}(b)}} = \frac{1}{\sec(\sec^{-1} b)\tan(\sec^{-1} b)} = \frac{1}{b\left(\pm\sqrt{b^2-1}\right)}.$

    Since the slope of $\sec^{-1} x$ is always positive, we the right sign by writing $\frac{d}{dx}\sec^{-1} x = \frac{1}{|x|\sqrt{x^2-1}}.$

57. The functions f and g have the same derivative (for $x \geq 0$), namely $\frac{1}{\sqrt{x}(x+1)}$. The functions therefore differ

    by a constant. To identify the constant we can set x equal to 0 in the equation $f(x) = g(x) + C$, obtaining

    $\sin^{-1}(-1) = 2\tan^{-1}(0) + C \Rightarrow -\frac{\pi}{2} = 0 + C \Rightarrow C = -\frac{\pi}{2}$. For $x \geq 0$, we have $\sin^{-1}\left(\frac{x-1}{x+1}\right) = 2\tan^{-1}\sqrt{x} - \frac{\pi}{2}$.

59. (a) $\sec^{-1} 1.5 = \cos^{-1}\frac{1}{1.5} \approx 0.84107$ (b) $\csc^{-1}(-1.5) = \sin^{-1}\left(-\frac{1}{1.5}\right) \approx -0.72973$

    (c) $\cot^{-1} 2 = \frac{\pi}{2} - \tan^{-1} 2 \approx 0.46365$

61. (a) Domain: all real numbers except those having
the form $\frac{\pi}{2} + k\pi$ where k is an integer.
Range: $-\frac{\pi}{2} < y < \frac{\pi}{2}$

    (b) Domain: $-\infty < x < \infty$; Range: $-\infty < y < \infty$
The graph of $y = \tan^{-1}(\tan x)$ is periodic, the
graph of $y = \tan(\tan^{-1} x) = x$ for $-\infty \leq x < \infty$.

63. (a) Domain: $-\infty < x < \infty$; Range: $0 \leq y \leq \pi$

    (b) Domain: $-1 \leq x \leq 1$; Range: $-1 \leq y \leq 1$
The graph of $y = \cos^{-1}(\cos x)$ is periodic; the
graph of $y = \cos(\cos^{-1} x) = x$ for $-1 \leq x \leq 1$.

65. The graphs are identical for $y = 2 \sin(2 \tan^{-1} x)$

$= 4 \left[\sin(\tan^{-1} x) \right] \left[\cos(\tan^{-1} x) \right] = 4 \left(\frac{x}{\sqrt{x^2+1}} \right) \left(\frac{1}{\sqrt{x^2+1}} \right)$

$= \frac{4x}{x^2+1}$ from the triangle

67. The values of f increase over the interval $[-1, 1]$ because $f' > 0$, and the graph of f steepens as the values of f' increase towards the ends of the interval. The graph of f is concave down to the left of the origin where $f'' < 0$, and concave up to the right of the origin where $f'' > 0$. There is an inflection point at $x = 0$ where $f'' = 0$ and f' has a local minimum value.

3.9 RELATED RATES

1. $A = \pi r^2 \Rightarrow \frac{dA}{dt} = 2\pi r \frac{dr}{dt}$

3. (a) $V = \pi r^2 h \Rightarrow \frac{dV}{dt} = \pi r^2 \frac{dh}{dt}$ (b) $V = \pi r^2 h \Rightarrow \frac{dV}{dt} = 2\pi r h \frac{dr}{dt}$
   (c) $V = \pi r^2 h \Rightarrow \frac{dV}{dt} = \pi r^2 \frac{dh}{dt} + 2\pi r h \frac{dr}{dt}$

5. (a) $\frac{dV}{dt} = 1$ volt/sec   (b) $\frac{dI}{dt} = -\frac{1}{3}$ amp/sec
   (c) $\frac{dV}{dt} = R\left(\frac{dI}{dt}\right) + I\left(\frac{dR}{dt}\right) \Rightarrow \frac{dR}{dt} = \frac{1}{I}\left(\frac{dV}{dt} - R\frac{dI}{dt}\right) \Rightarrow \frac{dR}{dt} = \frac{1}{I}\left(\frac{dV}{dt} - \frac{V}{I}\frac{dI}{dt}\right)$
   (d) $\frac{dR}{dt} = \frac{1}{2}\left[1 - \frac{12}{2}\left(-\frac{1}{3}\right)\right] = \left(\frac{1}{2}\right)(3) = \frac{3}{2}$ ohms/sec, R is increasing

7. (a) $s = \sqrt{x^2+y^2} = (x^2+y^2)^{1/2} \Rightarrow \frac{ds}{dt} = \frac{x}{\sqrt{x^2+y^2}}\frac{dx}{dt}$
   (b) $s = \sqrt{x^2+y^2} = (x^2+y^2)^{1/2} \Rightarrow \frac{ds}{dt} = \frac{x}{\sqrt{x^2+y^2}}\frac{dx}{dt} + \frac{y}{\sqrt{x^2+y^2}}\frac{dy}{dt}$
   (c) $s = \sqrt{x^2+y^2} \Rightarrow s^2 = x^2+y^2 \Rightarrow 2s\frac{ds}{dt} = 2x\frac{dx}{dt} + 2y\frac{dy}{dt} \Rightarrow 2s \cdot 0 = 2x\frac{dx}{dt} + 2y\frac{dy}{dt} \Rightarrow \frac{dx}{dt} = -\frac{y}{x}\frac{dy}{dt}$

9. (a) $A = \frac{1}{2}ab\sin\theta \Rightarrow \frac{dA}{dt} = \frac{1}{2}ab\cos\theta\frac{d\theta}{dt}$ (b) $A = \frac{1}{2}ab\sin\theta \Rightarrow \frac{dA}{dt} = \frac{1}{2}ab\cos\theta\frac{d\theta}{dt} + \frac{1}{2}b\sin\theta\frac{da}{dt}$
   (c) $A = \frac{1}{2}ab\sin\theta \Rightarrow \frac{dA}{dt} = \frac{1}{2}ab\cos\theta\frac{d\theta}{dt} + \frac{1}{2}b\sin\theta\frac{da}{dt} + \frac{1}{2}a\sin\theta\frac{db}{dt}$

11. Given $\frac{d\ell}{dt} = -2$ cm/sec, $\frac{dw}{dt} = 2$ cm/sec, $\ell = 12$ cm and $w = 5$ cm.
    (a) $A = \ell w \Rightarrow \frac{dA}{dt} = \ell\frac{dw}{dt} + w\frac{d\ell}{dt} \Rightarrow \frac{dA}{dt} = 12(2) + 5(-2) = 14$ cm^2/sec, increasing
    (b) $P = 2\ell + 2w \Rightarrow \frac{dP}{dt} = 2\frac{d\ell}{dt} + 2\frac{dw}{dt} = 2(-2) + 2(2) = 0$ cm/sec, constant
    (c) $D = \sqrt{w^2 + \ell^2} = (w^2 + \ell^2)^{1/2} \Rightarrow \frac{dD}{dt} = \frac{1}{2}(w^2+\ell^2)^{-1/2}\left(2w\frac{dw}{dt} + 2\ell\frac{d\ell}{dt}\right) \Rightarrow \frac{dD}{dt} = \frac{w\frac{dw}{dt} + \ell\frac{d\ell}{dt}}{\sqrt{w^2+\ell^2}}$
    $= \frac{(5)(2) + (12)(-2)}{\sqrt{25+144}} = -\frac{14}{13}$ cm/sec, decreasing

13. Given: $\frac{dx}{dt} = 5$ ft/sec, the ladder is 13 ft long, and x = 12, y = 5 at the instant of time

    (a) Since $x^2 + y^2 = 169 \Rightarrow \frac{dy}{dt} = -\frac{x}{y} \frac{dx}{dt} = -\left(\frac{12}{5}\right)(5) = -12$ ft/sec, the ladder is sliding down the wall

    (b) The area of the triangle formed by the ladder and walls is $A = \frac{1}{2} xy \Rightarrow \frac{dA}{dt} = \left(\frac{1}{2}\right)\left(x \frac{dy}{dt} + y \frac{dx}{dt}\right)$. The area

        is changing at $\frac{1}{2}\left[12(-12) + 5(5)\right] = -\frac{119}{2} = -59.5$ ft^2/sec.

    (c) $\cos\theta = \frac{x}{13} \Rightarrow -\sin\theta \frac{d\theta}{dt} = \frac{1}{13} \cdot \frac{dx}{dt} \Rightarrow \frac{d\theta}{dt} = -\frac{1}{13\sin\theta} \cdot \frac{dx}{dt} = -\left(\frac{1}{5}\right)(5) = -1$ rad/sec

15. Let s represent the distance between the girl and the kite and x represents the horizontal distance between the

    girl and kite $\Rightarrow s^2 = (300)^2 + x^2 \Rightarrow \frac{ds}{dt} = \frac{x}{s} \frac{dx}{dt} = \frac{400(25)}{500} = 20$ ft/sec.

17. $V = \frac{1}{3}\pi r^2 h$, $h = \frac{3}{8}(2r) = \frac{3r}{4} \Rightarrow r = \frac{4h}{3} \Rightarrow V = \frac{1}{3}\pi \left(\frac{4h}{3}\right)^2 h = \frac{16\pi h^3}{27} \Rightarrow \frac{dV}{dt} = \frac{16\pi h^2}{9} \frac{dh}{dt}$

    (a) $\frac{dh}{dt}\big|_{h=4} = \left(\frac{9}{16\pi 4^2}\right)(10) = \frac{90}{256\pi} \approx 0.1119$ m/sec = 11.19 cm/sec

    (b) $r = \frac{4h}{3} \Rightarrow \frac{dr}{dt} = \frac{4}{3} \frac{dh}{dt} = \frac{4}{3}\left(\frac{90}{256\pi}\right) = \frac{15}{32\pi} \approx 0.1492$ m/sec = 14.92 cm/sec

19. (a) $V = \frac{\pi}{3} y^2 (3R - y) \Rightarrow \frac{dV}{dt} = \frac{\pi}{3}\left[2y(3R - y) + y^2(-1)\right] \frac{dy}{dt} \Rightarrow \frac{dy}{dt} = \left[\frac{\pi}{3}(6Ry - 3y^2)\right]^{-1} \frac{dV}{dt} \Rightarrow$ at R = 13 and

        y = 8 we have $\frac{dy}{dt} = \frac{1}{144\pi}(-6) = \frac{-1}{24\pi}$ m/min

    (b) The hemisphere is on the circle $r^2 + (13 - y)^2 = 169 \Rightarrow r = \sqrt{26y - y^2}$ m

    (c) $r = (26y - y^2)^{1/2} \Rightarrow \frac{dr}{dt} = \frac{1}{2}(26y - y^2)^{-1/2}(26 - 2y)\frac{dy}{dt} \Rightarrow \frac{dr}{dt} = \frac{13 - y}{\sqrt{26y - y^2}} \frac{dy}{dt} \Rightarrow \frac{dr}{dt}\big|_{y=8} = \frac{13 - 8}{\sqrt{26 \cdot 8 - 64}}\left(\frac{-1}{24\pi}\right)$

        $= \frac{-5}{288\pi}$ m/min

21. If $V = \frac{4}{3}\pi r^3$, r = 5, and $\frac{dV}{dt} = 100\pi$ ft^3/min, then $\frac{dV}{dt} = 4\pi r^2 \frac{dr}{dt} \Rightarrow \frac{dr}{dt} = 1$ ft/min. Then $S = 4\pi r^2 \Rightarrow \frac{dS}{dt}$

    $= 8\pi r \frac{dr}{dt} = 8\pi(5)(1) = 40\pi$ ft^2/min, the rate at which the surface area is increasing.

23. Let s represent the distance between the bicycle and balloon, h the height of the balloon and x the horizontal

    distance between the balloon and the bicycle. The relationship between the variables is $s^2 = h^2 + x^2$

    $\Rightarrow \frac{ds}{dt} = \frac{1}{s}\left(h \frac{dh}{dt} + x \frac{dx}{dt}\right) \Rightarrow \frac{ds}{dt} = \frac{1}{85}\left[68(1) + 51(17)\right] = 11$ ft/sec.

25. $y = QD^{-1} \Rightarrow \frac{dy}{dt} = D^{-1}\frac{dQ}{dt} - QD^{-2}\frac{dD}{dt} = \frac{1}{41}(0) - \frac{233}{(41)^2}(-2) = \frac{466}{1681}$ L/min $\Rightarrow$ increasing about 0.2772 L/min

27. Let P(x, y) represent a point on the curve $y = x^2$ and θ the angle of inclination of a line containing P and the

    origin. Consequently, $\tan\theta = \frac{y}{x} \Rightarrow \tan\theta = \frac{x^2}{x} = x \Rightarrow \sec^2\theta \frac{d\theta}{dt} = \frac{dx}{dt} \Rightarrow \frac{d\theta}{dt} = \cos^2\theta \frac{dx}{dt}$. Since $\frac{dx}{dt} = 10$ m/sec

    and $\cos^2\theta\big|_{x=3} = \frac{x^2}{y^2 + x^2} = \frac{3^2}{9^2 + 3^2} = \frac{1}{10}$, we have $\frac{d\theta}{dt}\big|_{x=3} = 1$ rad/sec.

29. The distance from the origin is $s = \sqrt{x^2 + y^2}$ and we wish to find $\frac{ds}{dt}\big|_{(5,12)}$

    $= \frac{1}{2}(x^2 + y^2)^{-1/2}\left(2x \frac{dx}{dt} + 2y \frac{dy}{dt}\right)\big|_{(5,12)} = \frac{(5)(-1) + (12)(-5)}{\sqrt{25 + 144}} = -5$ m/sec

31. Let $s = 16t^2$ represent the distance the ball has fallen,
    h the distance between the ball and the ground, and I
    the distance between the shadow and the point directly
    beneath the ball. Accordingly, s + h = 50 and since
    the triangle LOQ and triangle PRQ are similar we have

    $I = \frac{30h}{50 - h} \Rightarrow h = 50 - 16t^2$ and $I = \frac{30(50 - 16t^2)}{50 - (50 - 16t^2)}$

    $= \frac{1500}{16t^2} - 30 \Rightarrow \frac{dI}{dt} = -\frac{1500}{8t^3} \Rightarrow \frac{dI}{dt}\big|_{t=\frac{1}{2}} = -1500$ ft/sec.

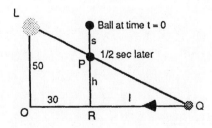

33. The volume of the ice is $V = \frac{4}{3}\pi r^3 - \frac{4}{3}\pi 4^3 \Rightarrow \frac{dV}{dt} = 4\pi r^2 \frac{dr}{dt} \Rightarrow \frac{dr}{dt}\big|_{r=6} = \frac{-5}{72\pi}$ in./min when $\frac{dV}{dt} = -10$ in^3/min, the thickness of the ice is decreasing at $\frac{5}{72\pi}$ in/min. The surface area is $S = 4\pi r^2 \Rightarrow \frac{dS}{dt} = 8\pi r \frac{dr}{dt} \Rightarrow \frac{dS}{dt}\big|_{r=6} = 48\pi\left(\frac{-5}{72\pi}\right)$ $= -\frac{10}{3}$ in^2/min, the outer surface area of the ice is decreasing at $\frac{10}{3}$ in^2/min.

35. When x represents the length of the shadow, then $\tan\theta = \frac{80}{x} \Rightarrow \sec^2\theta \frac{d\theta}{dt} = -\frac{80}{x^2}\frac{dx}{dt} \Rightarrow \frac{dx}{dt} = \frac{-x^2\sec^2\theta}{80}\frac{d\theta}{dt}$.

We are given that $\frac{d\theta}{dt} = 0.27° = \frac{3\pi}{2000}$ rad/min. At x = 60, $\cos\theta = \frac{3}{5} \Rightarrow$

$\left|\frac{dx}{dt}\right| = \left|\frac{-x^2\sec^2\theta}{80}\frac{d\theta}{dt}\right|\Big|_{\left(\frac{d\theta}{dt}=\frac{3\pi}{2000}\text{ and }\sec\theta=\frac{5}{3}\right)} = \frac{3\pi}{16}$ ft/min ≈ 0.589 ft/min ≈ 7.1 in./min.

37. Let x represent distance of the player from second base and s the distance to third base. Then $\frac{dx}{dt} = -16$ ft/sec

(a) $s^2 = x^2 + 8100 \Rightarrow 2s\frac{ds}{dt} = 2x\frac{dx}{dt} \Rightarrow \frac{ds}{dt} = \frac{x}{s}\frac{dx}{dt}$. When the player is 30 ft from first base, x = 60

$\Rightarrow s = 30\sqrt{13}$ and $\frac{ds}{dt} = \frac{60}{30\sqrt{13}}(-16) = \frac{-32}{\sqrt{13}} \approx -8.875$ ft/sec

(b) $\cos\theta_1 = \frac{90}{s} \Rightarrow -\sin\theta_1\frac{d\theta_1}{dt} = -\frac{90}{s^2}\frac{ds}{dt} \Rightarrow \frac{d\theta_1}{dt} = \frac{90}{s^2\sin\theta_1}\cdot\frac{ds}{dt} = \frac{90}{sx}\cdot\frac{ds}{dt}$. Therefore, x = 60 and s = $30\sqrt{13}$

$\Rightarrow \frac{d\theta_1}{dt} = \frac{90}{\left(30\sqrt{13}\right)(60)}\cdot\left(\frac{-32}{\sqrt{13}}\right) = \frac{-8}{65}$ rad/sec; $\sin\theta_2 = \frac{90}{s} \Rightarrow \cos\theta_2\frac{d\theta_2}{dt} = -\frac{90}{s^2}\cdot\frac{ds}{dt} \Rightarrow \frac{d\theta_2}{dt} = \frac{-90}{s^2\cos\theta_2}\cdot\frac{ds}{dt}$

$= \frac{-90}{sx}\cdot\frac{ds}{dt}$. Therefore, x = 60 and s = $30\sqrt{13} \Rightarrow \frac{d\theta_2}{dt} = \frac{8}{65}$ rad/sec.

(c) $\frac{d\theta_1}{dt} = \frac{90}{s^2\sin\theta_1}\cdot\frac{ds}{dt} = \frac{90}{\left(s^2\cdot\frac{x}{s}\right)}\cdot\left(\frac{x}{s}\right)\cdot\left(\frac{dx}{dt}\right) = \left(\frac{90}{s^2}\right)\left(\frac{dx}{dt}\right) = \left(\frac{90}{x^2+8100}\right)\frac{dx}{dt} \Rightarrow \lim_{x\to 0}\frac{d\theta_1}{dt}$

$= \lim_{x\to 0}\left(\frac{90}{x^2+8100}\right)(-15) = -\frac{1}{6}$ rad/sec; $\frac{d\theta_2}{dt} = \frac{-90}{s^2\cos\theta_2}\cdot\frac{ds}{dt} = \left(\frac{-90}{s^2\cdot\frac{x}{s}}\right)\left(\frac{x}{s}\right)\left(\frac{dx}{dt}\right) = \left(\frac{-90}{s^2}\right)\left(\frac{dx}{dt}\right)$

$= \left(\frac{-90}{x^2+8100}\right)\frac{dx}{dt} \Rightarrow \lim_{x\to 0}\frac{d\theta_2}{dt} = \frac{1}{6}$ rad/sec

3.10 LINEARIZATION AND DIFFERENTIALS

1. $f(x) = x^3 - 2x + 3 \Rightarrow f'(x) = 3x^2 - 2 \Rightarrow L(x) = f'(2)(x-2) + f(2) = 10(x-2) + 7 \Rightarrow L(x) = 10x - 13$ at x = 2

3. $f(x) = x + \frac{1}{x} \Rightarrow f'(x) = 1 - x^{-2} \Rightarrow L(x) = f(1) + f'(1)(x-1) = 2 + 0(x-1) = 2$

5. $f(x) = \tan x \Rightarrow f'(x) = \sec^2 x \Rightarrow L(x) = f(\pi) + f'(\pi)(x-\pi) = 0 + 1(x-\pi) = x - \pi$

7. $f(x) = x^2 + 2x \Rightarrow f'(x) = 2x + 2 \Rightarrow L(x) = f'(0)(x-0) + f(0) = 2(x-0) + 0 \Rightarrow L(x) = 2x$ at x = 0

9. $f(x) = 2x^2 + 4x - 3 \Rightarrow f'(x) = 4x + 4 \Rightarrow L(x) = f'(-1)(x+1) + f(-1) = 0(x+1) + (-5) \Rightarrow L(x) = -5$ at x = -1

11. $f(x) = \sqrt[3]{x} = x^{1/3} \Rightarrow f'(x) = \left(\frac{1}{3}\right)x^{-2/3} \Rightarrow L(x) = f'(8)(x-8) + f(8) = \frac{1}{12}(x-8) + 2 \Rightarrow L(x) = \frac{1}{12}x + \frac{4}{3}$ at x = 8

13. $f(x) = e^{-x} \Rightarrow f'(x) = -e^{-x} \Rightarrow L(x) = f(0) + f'(0)(x-0) = -x + 1$

15. $f'(x) = k(1+x)^{k-1}$. We have $f(0) = 1$ and $f'(0) = k$. $L(x) = f(0) + f'(0)(x-0) = 1 + k(x-0) = 1 + kx$

17. (a) $(1.0002)^{50} = (1+0.0002)^{50} \approx 1 + 50(0.0002) = 1 + .01 = 1.01$

(b) $\sqrt[3]{1.009} = (1+0.009)^{1/3} \approx 1 + \left(\frac{1}{3}\right)(0.009) = 1 + 0.003 = 1.003$

19. $y = x^3 - 3\sqrt{x} = x^3 - 3x^{1/2} \Rightarrow dy = \left(3x^2 - \frac{3}{2}x^{-1/2}\right)dx \Rightarrow dy = \left(3x^2 - \frac{3}{2\sqrt{x}}\right)dx$

21. $y = \frac{2x}{1+x^2} \Rightarrow dy = \left(\frac{(2)(1+x^2) - (2x)(2x)}{(1+x^2)^2}\right)dx = \frac{2-2x^2}{(1+x^2)^2}dx$

23. $2y^{3/2} + xy - x = 0 \Rightarrow 3y^{1/2}\,dy + y\,dx + x\,dy - dx = 0 \Rightarrow (3y^{1/2} + x)\,dy = (1 - y)\,dx \Rightarrow dy = \frac{1-y}{3\sqrt{y}+x}\,dx$

25. $y = \sin(5\sqrt{x}) = \sin(5x^{1/2}) \Rightarrow dy = (\cos(5x^{1/2}))(\frac{5}{2}x^{-1/2})\,dx \Rightarrow dy = \frac{5\cos(5\sqrt{x})}{2\sqrt{x}}\,dx$

27. $y = 4\tan\left(\frac{x^3}{3}\right) \Rightarrow dy = 4\left(\sec^2\left(\frac{x^3}{3}\right)\right)(x^2)\,dx \Rightarrow dy = 4x^2\sec^2\left(\frac{x^3}{3}\right)\,dx$

29. $y = 3\csc(1 - 2\sqrt{x}) = 3\csc(1 - 2x^{1/2}) \Rightarrow dy = 3(-\csc(1 - 2x^{1/2}))\cot(1 - 2x^{1/2})(-x^{-1/2})\,dx$
    $\Rightarrow dy = \frac{3}{\sqrt{x}}\csc(1 - 2\sqrt{x})\cot(1 - 2\sqrt{x})\,dx$

31. $y = e^{\sqrt{x}} \Rightarrow dy = \frac{e^{\sqrt{x}}}{2\sqrt{x}}\,dx$

33. $y = \ln(1 + x^2) \Rightarrow dy = \frac{2x}{1+x^2}\,dx$

35. $y = \tan^{-1}\left(e^{x^2}\right) \Rightarrow dy = \frac{1}{1 + \left(e^{x^2}\right)^2} \cdot e^{x^2} \cdot 2x\,dx = \frac{2x\,e^{x^2}}{1 + e^{2x^2}}\,dx$

37. $y = \sec^{-1}(e^{-x}) \Rightarrow dy = \frac{1}{e^{-x}\sqrt{(e^{-x})^2 - 1}} \cdot (-e^{-x})\,dx = \frac{-1}{\sqrt{\left(\frac{1}{e^x}\right)^2 - 1}}\,dx = \frac{-e^x}{\sqrt{1 - e^{2x}}}\,dx$

39. $f(x) = x^2 + 2x$, $x_0 = 1$, $dx = 0.1 \Rightarrow f'(x) = 2x + 2$
    (a) $\Delta f = f(x_0 + dx) - f(x_0) = f(1.1) - f(1) = 3.41 - 3 = 0.41$
    (b) $df = f'(x_0)\,dx = [2(1) + 2](0.1) = 0.4$
    (c) $|\Delta f - df| = |0.41 - 0.4| = 0.01$

41. $f(x) = x^3 - x$, $x_0 = 1$, $dx = 0.1 \Rightarrow f'(x) = 3x^2 - 1$
    (a) $\Delta f = f(x_0 + dx) - f(x_0) = f(1.1) - f(1) = .231$
    (b) $df = f'(x_0)\,dx = [3(1)^2 - 1](.1) = .2$
    (c) $|\Delta f - df| = |.231 - .2| = .031$

43. $f(x) = x^{-1}$, $x_0 = 0.5$, $dx = 0.1 \Rightarrow f'(x) = -x^{-2}$
    (a) $\Delta f = f(x_0 + dx) - f(x_0) = f(.6) - f(.5) = -\frac{1}{3}$
    (b) $df = f'(x_0)\,dx = (-4)\left(\frac{1}{10}\right) = -\frac{2}{5}$
    (c) $|\Delta f - df| = \left|-\frac{1}{3} + \frac{2}{5}\right| = \frac{1}{15}$

45. $V = \frac{4}{3}\pi r^3 \Rightarrow dV = 4\pi r_0^2\,dr$

47. $S = 6x^2 \Rightarrow dS = 12x_0\,dx$

49. $V = \pi r^2 h$, height constant $\Rightarrow dV = 2\pi r_0 h\,dr$

51. Given $r = 2$ m, $dr = .02$ m
    (a) $A = \pi r^2 \Rightarrow dA = 2\pi r\,dr = 2\pi(2)(.02) = .08\pi$ m^2
    (b) $\left(\frac{.08\pi}{4\pi}\right)(100\%) = 2\%$

53. The volume of a cylinder is $V = \pi r^2 h$. When h is held fixed, we have $\frac{dV}{dr} = 2\pi rh$, and so $dV = 2\pi rh\,dr$. For $h = 30$ in.,
    $r = 6$ in., and $dr = 0.5$ in., the volume of the material in the shell is approximately $dV = 2\pi rh\,dr = 2\pi(6)(30)(0.5)$
    $= 180\pi \approx 565.5$ in^3.

55. $V = \pi h^3 \Rightarrow dV = 3\pi h^2 \, dh$; recall that $\Delta V \approx dV$. Then $|\Delta V| \le (1\%)(V) = \frac{(1)(\pi h^3)}{100} \Rightarrow |dV| \le \frac{(1)(\pi h^3)}{100}$

$\Rightarrow |3\pi h^2 \, dh| \le \frac{(1)(\pi h^3)}{100} \Rightarrow |dh| \le \frac{1}{300} h = \left(\frac{1}{3}\%\right) h$. Therefore the greatest tolerated error in the measurement of h is $\frac{1}{3}\%$.

57. $V = \pi r^2 h$, h is constant $\Rightarrow dV = 2\pi rh \, dr$; recall that $\Delta V \approx dV$. We want $|\Delta V| \le \frac{1}{1000} V \Rightarrow |dV| \le \frac{\pi r^2 h}{1000}$

$\Rightarrow |2\pi rh \, dr| \le \frac{\pi r^2 h}{1000} \Rightarrow |dr| \le \frac{r}{2000} = (.05\%)r \Rightarrow$ a .05% variation in the radius can be tolerated.

59. $W = a + \frac{b}{g} = a + bg^{-1} \Rightarrow dW = -bg^{-2} \, dg = -\frac{b \, dg}{g^2} \Rightarrow \frac{dW_{moon}}{dW_{earth}} = \frac{\left(-\frac{b \, dg}{(5.2)^2}\right)}{\left(-\frac{b \, dg}{(32)^2}\right)} = \left(\frac{32}{5.2}\right)^2 = 37.87$, so a change of

gravity on the moon has about 38 times the effect that a change of the same magnitude has on Earth.

61. The error in measurement $dx = (1\%)(10) = 0.1$ cm; $V = x^3 \Rightarrow dV = 3x^2 \, dx = 3(10)^2(0.1) = 30$ cm$^3 \Rightarrow$ the percentage error in the volume calculation is $\left(\frac{30}{1000}\right)(100\%) = 3\%$

63. Given $D = 100$ cm, $dD = 1$ cm, $V = \frac{4}{3}\pi \left(\frac{D}{2}\right)^3 = \frac{\pi D^3}{6} \Rightarrow dV = \frac{\pi}{2} D^2 \, dD = \frac{\pi}{2} (100)^2(1) = \frac{10^4 \pi}{2}$. Then $\frac{dV}{V} (100\%)$

$= \left[\frac{\frac{10^4 \pi}{2}}{\frac{10^6 \pi}{6}}\right] (10^2\%) = \left[\frac{\frac{10^6 \pi}{2}}{\frac{10^6 \pi}{6}}\right] \% = 3\%$

65. $E(x) = f(x) - g(x) \Rightarrow E(x) = f(x) - m(x - a) - c$. Then $E(a) = 0 \Rightarrow f(a) - m(a - a) - c = 0 \Rightarrow c = f(a)$. Next

we calculate m: $\lim\limits_{x \to a} \frac{E(x)}{x - a} = 0 \Rightarrow \lim\limits_{x \to a} \frac{f(x) - m(x - a) - c}{x - a} = 0 \Rightarrow \lim\limits_{x \to a} \left[\frac{f(x) - f(a)}{x - a} - m\right] = 0$ (since $c = f(a)$)

$\Rightarrow f'(a) - m = 0 \Rightarrow m = f'(a)$. Therefore, $g(x) = m(x - a) + c = f'(a)(x - a) + f(a)$ is the linear approximation, as claimed.

67. (a) $f(x) = 2^x \Rightarrow f'(x) = 2^x \ln 2$; $L(x) = (2^0 \ln 2)x + 2^0 = x \ln 2 + 1 \approx 0.69x + 1$

(b)

3.11 HYPERBOLIC FUNCTIONS

1. $\sinh x = -\frac{3}{4} \Rightarrow \cosh x = \sqrt{1 + \sinh^2 x} = \sqrt{1 + \left(-\frac{3}{4}\right)^2} = \sqrt{1 + \frac{9}{16}} = \sqrt{\frac{25}{16}} = \frac{5}{4}$, $\tanh x = \frac{\sinh x}{\cosh x} = \frac{\left(-\frac{3}{4}\right)}{\left(\frac{5}{4}\right)} = -\frac{3}{5}$,

   $\coth x = \frac{1}{\tanh x} = -\frac{5}{3}$, $\operatorname{sech} x = \frac{1}{\cosh x} = \frac{4}{5}$, and $\operatorname{csch} x = \frac{1}{\sin x} = -\frac{4}{3}$

3. $\cosh x = \frac{17}{15}$, $x > 0 \Rightarrow \sinh x = \sqrt{\cosh^2 x - 1} = \sqrt{\left(\frac{17}{15}\right)^2 - 1} = \sqrt{\frac{289}{225} - 1} = \sqrt{\frac{64}{225}} = \frac{8}{15}$, $\tanh x = \frac{\sinh x}{\cosh x} = \frac{\left(\frac{8}{15}\right)}{\left(\frac{17}{15}\right)}$

   $= \frac{8}{17}$, $\coth x = \frac{1}{\tanh x} = \frac{17}{8}$, $\operatorname{sech} x = \frac{1}{\cosh x} = \frac{15}{17}$, and $\operatorname{csch} x = \frac{1}{\sinh x} = \frac{15}{8}$

5. $2 \cosh (\ln x) = 2 \left(\frac{e^{\ln x} + e^{-\ln x}}{2}\right) = e^{\ln x} + \frac{1}{e^{\ln x}} = x + \frac{1}{x}$

7. $\cosh 5x + \sinh 5x = \frac{e^{5x}+e^{-5x}}{2} + \frac{e^{5x}-e^{-5x}}{2} = e^{5x}$

9. $(\sinh x + \cosh x)^4 = \left(\frac{e^x-e^{-x}}{2} + \frac{e^x+e^{-x}}{2}\right)^4 = (e^x)^4 = e^{4x}$

11. (a) $\sinh 2x = \sinh(x+x) = \sinh x \cosh x + \cosh x \sinh x = 2 \sinh x \cosh x$

    (b) $\cosh 2x = \cosh(x+x) = \cosh x \cosh x + \sinh x \sin x = \cosh^2 x + \sinh^2 x$

13. $y = 6 \sinh \frac{x}{3} \Rightarrow \frac{dy}{dx} = 6 \left(\cosh \frac{x}{3}\right)\left(\frac{1}{3}\right) = 2 \cosh \frac{x}{3}$

15. $y = 2\sqrt{t} \tanh \sqrt{t} = 2t^{1/2} \tanh t^{1/2} \Rightarrow \frac{dy}{dt} = \left[\operatorname{sech}^2\left(t^{1/2}\right)\right]\left(\frac{1}{2}t^{-1/2}\right)\left(2t^{1/2}\right) + \left(\tanh t^{1/2}\right)\left(t^{-1/2}\right)$

    $= \operatorname{sech}^2 \sqrt{t} + \frac{\tanh \sqrt{t}}{\sqrt{t}}$

17. $y = \ln(\sinh z) \Rightarrow \frac{dy}{dz} = \frac{\cosh z}{\sinh z} = \coth z$

19. $y = (\operatorname{sech} \theta)(1 - \ln \operatorname{sech} \theta) \Rightarrow \frac{dy}{d\theta} = \left(-\frac{-\operatorname{sech} \theta \tanh \theta}{\operatorname{sech} \theta}\right)(\operatorname{sech} \theta) + (-\operatorname{sech} \theta \tanh \theta)(1 - \ln \operatorname{sech} \theta)$

    $= \operatorname{sech} \theta \tanh \theta - (\operatorname{sech} \theta \tanh \theta)(1 - \ln \operatorname{sech} \theta) = (\operatorname{sech} \theta \tanh \theta)[1 - (1 - \ln \operatorname{sech} \theta)]$

    $= (\operatorname{sech} \theta \tanh \theta)(\ln \operatorname{sech} \theta)$

21. $y = \ln \cosh v - \frac{1}{2} \tanh^2 v \Rightarrow \frac{dy}{dv} = \frac{\sinh v}{\cosh v} - \left(\frac{1}{2}\right)(2 \tanh v)(\operatorname{sech}^2 v) = \tanh v - (\tanh v)(\operatorname{sech}^2 v)$

    $= (\tanh v)\left(1 - \operatorname{sech}^2 v\right) = (\tanh v)(\tanh^2 v) = \tanh^3 v$

23. $y = (x^2 + 1) \operatorname{sech}(\ln x) = (x^2 + 1)\left(\frac{2}{e^{\ln x}+e^{-\ln x}}\right) = (x^2 + 1)\left(\frac{2}{x+x^{-1}}\right) = (x^2 + 1)\left(\frac{2x}{x^2+1}\right) = 2x \Rightarrow \frac{dy}{dx} = 2$

25. $y = \sinh^{-1} \sqrt{x} = \sinh^{-1}\left(x^{1/2}\right) \Rightarrow \frac{dy}{dx} = \frac{\left(\frac{1}{2}\right)x^{-1/2}}{\sqrt{1+(x^{1/2})^2}} = \frac{1}{2\sqrt{x}\sqrt{1+x}} = \frac{1}{2\sqrt{x(1+x)}}$

27. $y = (1 - \theta) \tanh^{-1} \theta \Rightarrow \frac{dy}{d\theta} = (1 - \theta)\left(\frac{1}{1-\theta^2}\right) + (-1) \tanh^{-1} \theta = \frac{1}{1+\theta} - \tanh^{-1} \theta$

29. $y = (1 - t) \coth^{-1} \sqrt{t} = (1 - t) \coth^{-1}\left(t^{1/2}\right) \Rightarrow \frac{dy}{dt} = (1 - t)\left[\frac{\left(\frac{1}{2}\right)t^{-1/2}}{1-(t^{1/2})^2}\right] + (-1) \coth^{-1}\left(t^{1/2}\right) = \frac{1}{2\sqrt{t}} - \coth^{-1} \sqrt{t}$

31. $y = \cos^{-1} x - x \operatorname{sech}^{-1} x \Rightarrow \frac{dy}{dx} = \frac{-1}{\sqrt{1-x^2}} - \left[x\left(\frac{-1}{x\sqrt{1-x^2}}\right) + (1) \operatorname{sech}^{-1} x\right] = \frac{-1}{\sqrt{1-x^2}} + \frac{1}{\sqrt{1-x^2}} - \operatorname{sech}^{-1} x$

    $= -\operatorname{sech}^{-1} x$

33. $y = \operatorname{csch}^{-1}\left(\frac{1}{2}\right)^{\theta} \Rightarrow \frac{dy}{d\theta} = -\frac{\left[\ln\left(\frac{1}{2}\right)\right]\left(\frac{1}{2}\right)^{\theta}}{\left(\frac{1}{2}\right)^{\theta}\sqrt{1+\left[\left(\frac{1}{2}\right)^{\theta}\right]^2}} = -\frac{\ln(1) - \ln(2)}{\sqrt{1+\left(\frac{1}{2}\right)^{2\theta}}} = \frac{\ln 2}{\sqrt{1+\left(\frac{1}{2}\right)^{2\theta}}}$

35. $y = \sinh^{-1}(\tan x) \Rightarrow \frac{dy}{dx} = \frac{\sec^2 x}{\sqrt{1+(\tan x)^2}} = \frac{\sec^2 x}{\sqrt{\sec^2 x}} = \frac{\sec^2 x}{|\sec x|} = \frac{|\sec x| \, |\sec x|}{|\sec x|} = |\sec x|$

37. $\sinh^{-1}\left(\frac{-5}{12}\right) = \ln\left(-\frac{5}{12} + \sqrt{\frac{25}{144}+1}\right) = \ln\left(\frac{2}{3}\right)$ 39. $\tanh^{-1}\left(-\frac{1}{2}\right) = \frac{1}{2}\ln\left(\frac{1-(1/2)}{1+(1/2)}\right) = -\frac{\ln 3}{2}$

41. $\operatorname{sech}^{-1}\left(\frac{3}{5}\right) = \ln\left(\frac{1+\sqrt{1-(9/25)}}{(3/5)}\right) = \ln 3$

43. Let $E(x) = \frac{f(x) + f(-x)}{2}$ and $O(x) = \frac{f(x) - f(-x)}{2}$. Then $E(x) + O(x) = \frac{f(x) + f(-x)}{2} + \frac{f(x) - f(-x)}{2}$

$= \frac{2f(x)}{2} = f(x)$. Also, $E(-x) = \frac{f(-x) + f(-(-x))}{2} = \frac{f(x) + f(-x)}{2} = E(x) \Rightarrow E(x)$ is even, and

$O(-x) = \frac{f(-x) - f(-(-x))}{2} = -\frac{f(x) - f(-x)}{2} = -O(x) \Rightarrow O(x)$ is odd. Consequently, $f(x)$ can be written as

a sum of an even and an odd function. If $f(x) = e^x$, then we can write $e^x = \frac{e^x + e^{-x}}{2} + \frac{e^x - e^{-x}}{2} = \sinh x + \cosh x$.

45. (a) $v = \sqrt{\frac{mg}{k}} \tanh\left(\sqrt{\frac{gk}{m}}\, t\right) \Rightarrow \frac{dv}{dt} = \sqrt{\frac{mg}{k}} \left[\text{sech}^2\left(\sqrt{\frac{gk}{m}}\, t\right)\right]\left(\sqrt{\frac{gk}{m}}\right) = g\,\text{sech}^2\left(\sqrt{\frac{gk}{m}}\, t\right)$.

Thus $m\frac{dv}{dt} = mg\,\text{sech}^2\left(\sqrt{\frac{gk}{m}}\, t\right) = mg\left(1 - \tanh^2\left(\sqrt{\frac{gk}{m}}\, t\right)\right) = mg - kv^2$. Also, since $\tanh x = 0$ when $x = 0$, $v = 0$

when $t = 0$.

(b) $\lim\limits_{t \to \infty} v = \lim\limits_{t \to \infty} \sqrt{\frac{mg}{k}} \tanh\left(\sqrt{\frac{kg}{m}}\, t\right) = \sqrt{\frac{mg}{k}} \lim\limits_{t \to \infty} \tanh\left(\sqrt{\frac{kg}{m}}\, t\right) = \sqrt{\frac{mg}{k}}\,(1) = \sqrt{\frac{mg}{k}}$

(c) $\sqrt{\frac{160}{0.005}} = \sqrt{\frac{160,000}{5}} = \frac{400}{\sqrt{5}} = 80\sqrt{5} \approx 178.89$ ft/sec

47. (a) $\lim\limits_{x \to \infty} \tanh x = \lim\limits_{x \to \infty} \frac{e^x - e^{-x}}{e^x + e^{-x}} = \lim\limits_{x \to \infty} \frac{e^x - \frac{1}{e^x}}{e^x + \frac{1}{e^x}} = \lim\limits_{x \to \infty} \frac{\left(e^x - \frac{1}{e^x}\right)}{\left(e^x + \frac{1}{e^x}\right)} \cdot \frac{\frac{1}{e^x}}{\frac{1}{e^x}} = \lim\limits_{x \to \infty} \frac{1 - \frac{1}{e^{2x}}}{1 + \frac{1}{e^{2x}}} = \frac{1-0}{1+0} = 1$

(b) $\lim\limits_{x \to -\infty} \tanh x = \lim\limits_{x \to -\infty} \frac{e^x - e^{-x}}{e^x + e^{-x}} = \lim\limits_{x \to -\infty} \frac{e^x - \frac{1}{e^x}}{e^x + \frac{1}{e^x}} = \lim\limits_{x \to -\infty} \frac{\left(e^x - \frac{1}{e^x}\right)}{\left(e^x + \frac{1}{e^x}\right)} \cdot \frac{e^x}{e^x} = \lim\limits_{x \to -\infty} \frac{e^{2x} - 1}{e^{2x} + 1} = \frac{0-1}{0+1} = -1$

(c) $\lim\limits_{x \to \infty} \sinh x = \lim\limits_{x \to \infty} \frac{e^x - e^{-x}}{2} = \lim\limits_{x \to \infty} \frac{e^x - \frac{1}{e^x}}{2} = \lim\limits_{x \to \infty} \left(\frac{e^x}{2} - \frac{1}{2e^x}\right) = \infty - 0 = \infty$

(d) $\lim\limits_{x \to -\infty} \sinh x = \lim\limits_{x \to -\infty} \frac{e^x - e^{-x}}{2} = \lim\limits_{x \to -\infty} \left(\frac{e^x}{2} - \frac{e^{-x}}{2}\right) = 0 - \infty = -\infty$

(e) $\lim\limits_{x \to \infty} \text{sech } x = \lim\limits_{x \to \infty} \frac{2}{e^x + e^{-x}} = \lim\limits_{x \to \infty} \frac{2}{e^x + \frac{1}{e^x}} \cdot \frac{\frac{1}{e^x}}{\frac{1}{e^x}} = \lim\limits_{x \to \infty} \frac{\frac{2}{e^x}}{1 + \frac{1}{e^{2x}}} = \frac{0}{1+0} = 0$

(f) $\lim\limits_{x \to \infty} \coth x = \lim\limits_{x \to \infty} \frac{e^x + e^{-x}}{e^x - e^{-x}} = \lim\limits_{x \to \infty} \frac{e^x + \frac{1}{e^x}}{e^x - \frac{1}{e^x}} = \lim\limits_{x \to \infty} \frac{\left(e^x + \frac{1}{e^x}\right)}{\left(e^x - \frac{1}{e^x}\right)} \cdot \frac{\frac{1}{e^x}}{\frac{1}{e^x}} = \lim\limits_{x \to \infty} \frac{1 + \frac{1}{e^{2x}}}{1 - \frac{1}{e^{2x}}} = \frac{1+0}{1-0} = 1$

(g) $\lim\limits_{x \to 0^+} \coth x = \lim\limits_{x \to 0^+} \frac{e^x + e^{-x}}{e^x - e^{-x}} = \lim\limits_{x \to 0^+} \frac{e^x + \frac{1}{e^x}}{e^x - \frac{1}{e^x}} \cdot \frac{e^x}{e^x} = \lim\limits_{x \to 0^+} \frac{e^{2x} + 1}{e^{2x} - 1} = +\infty$

(h) $\lim\limits_{x \to 0^-} \coth x = \lim\limits_{x \to 0^-} \frac{e^x + e^{-x}}{e^x - e^{-x}} = \lim\limits_{x \to 0^-} \frac{e^x + \frac{1}{e^x}}{e^x - \frac{1}{e^x}} \cdot \frac{e^x}{e^x} = \lim\limits_{x \to 0^-} \frac{e^{2x} + 1}{e^{2x} - 1} = -\infty$

(i) $\lim\limits_{x \to -\infty} \text{csch } x = \lim\limits_{x \to -\infty} \frac{2}{e^x - e^{-x}} = \lim\limits_{x \to -\infty} \frac{2}{e^x - \frac{1}{e^x}} \cdot \frac{e^x}{e^x} = \lim\limits_{x \to -\infty} \frac{2e^x}{e^{2x} - 1} = \frac{0}{0-1} = 0$

49. (a) $y = \frac{H}{w} \cosh\left(\frac{w}{H} x\right) \Rightarrow \tan\phi = \frac{dy}{dx} = \left(\frac{H}{w}\right)\left[\frac{w}{H} \sinh\left(\frac{w}{H} x\right)\right] = \sinh\left(\frac{w}{H} x\right)$

(b) The tension at P is given by $T\cos\phi = H \Rightarrow T = H\sec\phi = H\sqrt{1 + \tan^2\phi} = H\sqrt{1 + \left(\sinh\frac{w}{H} x\right)^2}$

$= H\cosh\left(\frac{w}{H} x\right) = w\left(\frac{H}{w}\right)\cosh\left(\frac{w}{H} x\right) = wy$

CHAPTER 3 PRACTICE EXERCISES

1. $y = x^5 - 0.125x^2 + 0.25x \Rightarrow \frac{dy}{dx} = 5x^4 - 0.25x + 0.25$

3. $y = x^3 - 3\left(x^2 + \pi^2\right) \Rightarrow \frac{dy}{dx} = 3x^2 - 3(2x + 0) = 3x^2 - 6x = 3x(x - 2)$

5. $y = (x + 1)^2\left(x^2 + 2x\right) \Rightarrow \frac{dy}{dx} = (x + 1)^2(2x + 2) + (x^2 + 2x)(2(x + 1)) = 2(x + 1)\left[(x + 1)^2 + x(x + 2)\right]$

$= 2(x + 1)\left(2x^2 + 4x + 1\right)$

7. $y = \left(\theta^2 + \sec\theta + 1\right)^3 \Rightarrow \frac{dy}{d\theta} = 3\left(\theta^2 + \sec\theta + 1\right)^2(2\theta + \sec\theta\tan\theta)$

9. $s = \frac{\sqrt{t}}{1 + \sqrt{t}} \Rightarrow \frac{ds}{dt} = \frac{(1 + \sqrt{t}) \cdot \frac{1}{2\sqrt{t}} - \sqrt{t}\left(\frac{1}{2\sqrt{t}}\right)}{(1 + \sqrt{t})^2} = \frac{(1 + \sqrt{t}) - \sqrt{t}}{2\sqrt{t}(1 + \sqrt{t})^2} = \frac{1}{2\sqrt{t}(1 + \sqrt{t})^2}$

11. $y = 2 \tan^2 x - \sec^2 x \Rightarrow \frac{dy}{dx} = (4 \tan x)(\sec^2 x) - (2 \sec x)(\sec x \tan x) = 2 \sec^2 x \tan x$

13. $s = \cos^4(1 - 2t) \Rightarrow \frac{ds}{dt} = 4 \cos^3(1 - 2t)(-\sin(1 - 2t))(-2) = 8 \cos^3(1 - 2t) \sin(1 - 2t)$

15. $s = (\sec t + \tan t)^5 \Rightarrow \frac{ds}{dt} = 5(\sec t + \tan t)^4(\sec t \tan t + \sec^2 t) = 5(\sec t)(\sec t + \tan t)^5$

17. $r = \sqrt{2\theta \sin\theta} = (2\theta \sin\theta)^{1/2} \Rightarrow \frac{dr}{d\theta} = \frac{1}{2}(2\theta \sin\theta)^{-1/2}(2\theta \cos\theta + 2 \sin\theta) = \frac{\theta \cos\theta + \sin\theta}{\sqrt{2\theta \sin\theta}}$

19. $r = \sin\sqrt{2\theta} = \sin(2\theta)^{1/2} \Rightarrow \frac{dr}{d\theta} = \cos(2\theta)^{1/2}\left(\frac{1}{2}(2\theta)^{-1/2}(2)\right) = \frac{\cos\sqrt{2\theta}}{\sqrt{2\theta}}$

21. $y = \frac{1}{2} x^2 \csc \frac{2}{x} \Rightarrow \frac{dy}{dx} = \frac{1}{2} x^2\left(-\csc \frac{2}{x} \cot \frac{2}{x}\right)\left(\frac{-2}{x^2}\right) + \left(\csc \frac{2}{x}\right)\left(\frac{1}{2} \cdot 2x\right) = \csc \frac{2}{x} \cot \frac{2}{x} + x \csc \frac{2}{x}$

23. $y = x^{-1/2} \sec(2x)^2 \Rightarrow \frac{dy}{dx} = x^{-1/2} \sec(2x)^2 \tan(2x)^2(2(2x) \cdot 2) + \sec(2x)^2\left(-\frac{1}{2} x^{-3/2}\right)$
$= 8x^{1/2} \sec(2x)^2 \tan(2x)^2 - \frac{1}{2} x^{-3/2} \sec(2x)^2 = \frac{1}{2} x^{1/2} \sec(2x)^2 \left[16 \tan(2x)^2 - x^{-2}\right]$ or $\frac{1}{2x^{3/2}} \sec(2x)^2 \left[16x^2 \tan(2x)^2 - 1\right]$

25. $y = 5 \cot x^2 \Rightarrow \frac{dy}{dx} = 5\left(-\csc^2 x^2\right)(2x) = -10x \csc^2(x^2)$

27. $y = x^2 \sin^2(2x^2) \Rightarrow \frac{dy}{dx} = x^2(2 \sin(2x^2))(\cos(2x^2))(4x) + \sin^2(2x^2)(2x) = 8x^3 \sin(2x^2) \cos(2x^2) + 2x \sin^2(2x^2)$

29. $s = \left(\frac{4t}{t+1}\right)^{-2} \Rightarrow \frac{ds}{dt} = -2\left(\frac{4t}{t+1}\right)^{-3}\left(\frac{(t+1)(4) - (4t)(1)}{(t+1)^2}\right) = -2\left(\frac{4t}{t+1}\right)^{-3}\frac{4}{(t+1)^2} = -\frac{(t+1)}{8t^3}$

31. $y = \left(\frac{\sqrt{x}}{x+1}\right)^2 \Rightarrow \frac{dy}{dx} = 2\left(\frac{\sqrt{x}}{x+1}\right) \cdot \frac{(x+1)\left(\frac{1}{2\sqrt{x}}\right) - (\sqrt{x})(1)}{(x+1)^2} = \frac{(x+1) - 2x}{(x+1)^3} = \frac{1-x}{(x+1)^3}$

33. $y = \sqrt{\frac{x^2 + x}{x^2}} = \left(1 + \frac{1}{x}\right)^{1/2} \Rightarrow \frac{dy}{dx} = \frac{1}{2}\left(1 + \frac{1}{x}\right)^{-1/2}\left(-\frac{1}{x^2}\right) = -\frac{1}{2x^2\sqrt{1 + \frac{1}{x}}}$

35. $r = \left(\frac{\sin\theta}{\cos\theta - 1}\right)^2 \Rightarrow \frac{dr}{d\theta} = 2\left(\frac{\sin\theta}{\cos\theta - 1}\right)\left[\frac{(\cos\theta - 1)(\cos\theta) - (\sin\theta)(-\sin\theta)}{(\cos\theta - 1)^2}\right]$
$= 2\left(\frac{\sin\theta}{\cos\theta - 1}\right)\left(\frac{\cos^2\theta - \cos\theta + \sin^2\theta}{(\cos\theta - 1)^2}\right) = \frac{(2 \sin\theta)(1 - \cos\theta)}{(\cos\theta - 1)^3} = \frac{-2 \sin\theta}{(\cos\theta - 1)^2}$

37. $y = (2x + 1)\sqrt{2x + 1} = (2x + 1)^{3/2} \Rightarrow \frac{dy}{dx} = \frac{3}{2}(2x + 1)^{1/2}(2) = 3\sqrt{2x + 1}$

39. $y = 3\left(5x^2 + \sin 2x\right)^{-3/2} \Rightarrow \frac{dy}{dx} = 3\left(-\frac{3}{2}\right)\left(5x^2 + \sin 2x\right)^{-5/2}[10x + (\cos 2x)(2)] = \frac{-9(5x + \cos 2x)}{(5x^2 + \sin 2x)^{5/2}}$

41. $y = 10e^{-x/5} \Rightarrow \frac{dy}{dx} = (10)\left(-\frac{1}{5}\right)e^{-x/5} = -2e^{-x/5}$

43. $y = \frac{1}{4} xe^{4x} - \frac{1}{16} e^{4x} \Rightarrow \frac{dy}{dx} = \frac{1}{4}\left[x(4e^{4x}) + e^{4x}(1)\right] - \frac{1}{16}(4e^{4x}) = xe^{4x} + \frac{1}{4} e^{4x} - \frac{1}{4} e^{4x} = xe^{4x}$

45. $y = \ln(\sin^2\theta) \Rightarrow \frac{dy}{d\theta} = \frac{2(\sin\theta)(\cos\theta)}{\sin^2\theta} = \frac{2 \cos\theta}{\sin\theta} = 2 \cot\theta$

47. $y = \log_2\left(\frac{x^2}{2}\right) = \frac{\ln\left(\frac{x^2}{2}\right)}{\ln 2} \Rightarrow \frac{dy}{dx} = \frac{1}{\ln 2}\left(\frac{x}{\left(\frac{x^2}{2}\right)}\right) = \frac{2}{(\ln 2)x}$

49. $y = 8^{-t} \Rightarrow \frac{dy}{dt} = 8^{-t}(\ln 8)(-1) = -8^{-t}(\ln 8)$ 51. $y = 5x^{3.6} \Rightarrow \frac{dy}{dx} = 5(3.6)x^{2.6} = 18x^{2.6}$

53. $y = (x+2)^{x+2} \Rightarrow \ln y = \ln(x+2)^{x+2} = (x+2)\ln(x+2) \Rightarrow \frac{y'}{y} = (x+2)\left(\frac{1}{x+2}\right) + (1)\ln(x+2)$

$\Rightarrow \frac{dy}{dx} = (x+2)^{x+2}[\ln(x+2)+1]$

55. $y = \sin^{-1}\sqrt{1-u^2} = \sin^{-1}(1-u^2)^{1/2} \Rightarrow \frac{dy}{du} = \frac{\frac{1}{2}(1-u^2)^{-1/2}(-2u)}{\sqrt{1-\left[(1-u^2)^{1/2}\right]^2}} = \frac{-u}{\sqrt{1-u^2}\sqrt{1-(1-u^2)}} = \frac{-u}{|u|\sqrt{1-u^2}}$

$= \frac{-u}{u\sqrt{1-u^2}} = \frac{-1}{\sqrt{1-u^2}}, 0 < u < 1$

57. $y = \ln(\cos^{-1}x) \Rightarrow y' = \frac{\left(\frac{-1}{\sqrt{1-x^2}}\right)}{\cos^{-1}x} = \frac{-1}{\sqrt{1-x^2}\cos^{-1}x}$

59. $y = t\tan^{-1}t - \left(\frac{1}{2}\right)\ln t \Rightarrow \frac{dy}{dt} = \tan^{-1}t + t\left(\frac{1}{1+t^2}\right) - \left(\frac{1}{2}\right)\left(\frac{1}{t}\right) = \tan^{-1}t + \frac{t}{1+t^2} - \frac{1}{2t}$

61. $y = z\sec^{-1}z - \sqrt{z^2-1} = z\sec^{-1}z - (z^2-1)^{1/2} \Rightarrow \frac{dy}{dz} = z\left(\frac{1}{|z|\sqrt{z^2-1}}\right) + (\sec^{-1}z)(1) - \frac{1}{2}(z^2-1)^{-1/2}(2z)$

$= \frac{z}{|z|\sqrt{z^2-1}} - \frac{z}{\sqrt{z^2-1}} + \sec^{-1}z = \frac{1-z}{\sqrt{z^2-1}} + \sec^{-1}z, z > 1$

63. $y = \csc^{-1}(\sec\theta) \Rightarrow \frac{dy}{d\theta} = \frac{-\sec\theta\tan\theta}{|\sec\theta|\sqrt{\sec^2\theta-1}} = -\frac{\tan\theta}{|\tan\theta|} = -1, 0 < \theta < \frac{\pi}{2}$

65. $xy + 2x + 3y = 1 \Rightarrow (xy' + y) + 2 + 3y' = 0 \Rightarrow xy' + 3y' = -2 - y \Rightarrow y'(x+3) = -2 - y \Rightarrow y' = -\frac{y+2}{x+3}$

67. $x^3 + 4xy - 3y^{4/3} = 2x \Rightarrow 3x^2 + \left(4x\frac{dy}{dx} + 4y\right) - 4y^{1/3}\frac{dy}{dx} = 2 \Rightarrow 4x\frac{dy}{dx} - 4y^{1/3}\frac{dy}{dx} = 2 - 3x^2 - 4y$

$\Rightarrow \frac{dy}{dx}\left(4x - 4y^{1/3}\right) = 2 - 3x^2 - 4y \Rightarrow \frac{dy}{dx} = \frac{2-3x^2-4y}{4x-4y^{1/3}}$

69. $(xy)^{1/2} = 1 \Rightarrow \frac{1}{2}(xy)^{-1/2}\left(x\frac{dy}{dx} + y\right) = 0 \Rightarrow x^{1/2}y^{-1/2}\frac{dy}{dx} = -x^{-1/2}y^{1/2} \Rightarrow \frac{dy}{dx} = -x^{-1}y \Rightarrow \frac{dy}{dx} = -\frac{y}{x}$

71. $y^2 = \frac{x}{x+1} \Rightarrow 2y\frac{dy}{dx} = \frac{(x+1)(1)-(x)(1)}{(x+1)^2} \Rightarrow \frac{dy}{dx} = \frac{1}{2y(x+1)^2}$

73. $e^{x+2y} = 1 \Rightarrow e^{x+2y}\left(1 + 2\frac{dy}{dx}\right) = 0 \Rightarrow \frac{dy}{dx} = -\frac{1}{2}$

75. $\ln\left(\frac{x}{y}\right) = 1 \Rightarrow \frac{1}{x/y}\frac{d}{dx}\left(\frac{x}{y}\right) = 0 \Rightarrow \frac{y(1)-x\frac{dy}{dx}}{y^2} = 0 \Rightarrow \frac{dy}{dx} = \frac{y}{x}$

77. $y e^{\tan^{-1}x} = 2 \Rightarrow y = 2e^{-\tan^{-1}x} \Rightarrow \frac{dy}{dx} = 2e^{-\tan^{-1}x}\frac{d}{dx}(-\tan^{-1}x) = -2e^{-\tan^{-1}x}\left(\frac{1}{1+x^2}\right) = -\frac{2e^{-\tan^{-1}x}}{1+x^2}$

79. $p^3 + 4pq - 3q^2 = 2 \Rightarrow 3p^2\frac{dp}{dq} + 4\left(p + q\frac{dp}{dq}\right) - 6q = 0 \Rightarrow 3p^2\frac{dp}{dq} + 4q\frac{dp}{dq} = 6q - 4p \Rightarrow \frac{dp}{dq}(3p^2 + 4q) = 6q - 4p$

$\Rightarrow \frac{dp}{dq} = \frac{6q-4p}{3p^2+4q}$

81. $r\cos 2s + \sin^2 s = \pi \Rightarrow r(-\sin 2s)(2) + (\cos 2s)\left(\frac{dr}{ds}\right) + 2\sin s\cos s = 0 \Rightarrow \frac{dr}{ds}(\cos 2s) = 2r\sin 2s - 2\sin s\cos s$

$\Rightarrow \frac{dr}{ds} = \frac{2r\sin 2s - \sin 2s}{\cos 2s} = \frac{(2r-1)(\sin 2s)}{\cos 2s} = (2r-1)(\tan 2s)$

83. (a) $x^3 + y^3 = 1 \Rightarrow 3x^2 + 3y^2 \frac{dy}{dx} = 0 \Rightarrow \frac{dy}{dx} = -\frac{x^2}{y^2} \Rightarrow \frac{d^2y}{dx^2} = \frac{y^2(-2x) - (-x^2)\left(2y\frac{dy}{dx}\right)}{y^4}$

$\Rightarrow \frac{d^2y}{dx^2} = \frac{-2xy^2 + (2yx^2)\left(-\frac{x^2}{y^2}\right)}{y^4} = \frac{-2xy^2 - \frac{2x^4}{y}}{y^4} = \frac{-2xy^3 - 2x^4}{y^5}$

(b) $y^2 = 1 - \frac{2}{x} \Rightarrow 2y\frac{dy}{dx} = \frac{2}{x^2} \Rightarrow \frac{dy}{dx} = \frac{1}{yx^2} \Rightarrow \frac{dy}{dx} = (yx^2)^{-1} \Rightarrow \frac{d^2y}{dx^2} = -(yx^2)^{-2}\left[y(2x) + x^2\frac{dy}{dx}\right]$

$\Rightarrow \frac{d^2y}{dx^2} = \frac{-2xy - x^2\left(\frac{1}{yx^2}\right)}{y^2x^4} = \frac{-2xy^2 - 1}{y^3x^4}$

85. (a) Let $h(x) = 6f(x) - g(x) \Rightarrow h'(x) = 6f'(x) - g'(x) \Rightarrow h'(1) = 6f'(1) - g'(1) = 6\left(\frac{1}{2}\right) - (-4) = 7$

(b) Let $h(x) = f(x)g^2(x) \Rightarrow h'(x) = f(x)(2g(x))g'(x) + g^2(x)f'(x) \Rightarrow h'(0) = 2f(0)g(0)g'(0) + g^2(0)f'(0)$
$= 2(1)(1)\left(\frac{1}{2}\right) + (1)^2(-3) = -2$

(c) Let $h(x) = \frac{f(x)}{g(x)+1} \Rightarrow h'(x) = \frac{(g(x)+1)f'(x) - f(x)g'(x)}{(g(x)+1)^2} \Rightarrow h'(1) = \frac{(g(1)+1)f'(1) - f(1)g'(1)}{(g(1)+1)^2}$
$= \frac{(5+1)\left(\frac{1}{2}\right) - 3(-4)}{(5+1)^2} = \frac{5}{12}$

(d) Let $h(x) = f(g(x)) \Rightarrow h'(x) = f'(g(x))g'(x) \Rightarrow h'(0) = f'(g(0))g'(0) = f'(1)\left(\frac{1}{2}\right) = \left(\frac{1}{2}\right)\left(\frac{1}{2}\right) = \frac{1}{4}$

(e) Let $h(x) = g(f(x)) \Rightarrow h'(x) = g'(f(x))f'(x) \Rightarrow h'(0) = g'(f(0))f'(0) = g'(1)f'(0) = (-4)(-3) = 12$

(f) Let $h(x) = (x + f(x))^{3/2} \Rightarrow h'(x) = \frac{3}{2}(x + f(x))^{1/2}(1 + f'(x)) \Rightarrow h'(1) = \frac{3}{2}(1 + f(1))^{1/2}(1 + f'(1))$
$= \frac{3}{2}(1 + 3)^{1/2}\left(1 + \frac{1}{2}\right) = \frac{9}{2}$

(g) Let $h(x) = f(x + g(x)) \Rightarrow h'(x) = f'(x + g(x))(1 + g'(x)) \Rightarrow h'(0) = f'(g(0))(1 + g'(0))$
$= f'(1)\left(1 + \frac{1}{2}\right) = \left(\frac{1}{2}\right)\left(\frac{3}{2}\right) = \frac{3}{4}$

87. $x = t^2 + \pi \Rightarrow \frac{dx}{dt} = 2t; \; y = 3\sin 2x \Rightarrow \frac{dy}{dx} = 3(\cos 2x)(2) = 6\cos 2x = 6\cos(2t^2 + 2\pi) = 6\cos(2t^2);$ thus,
$\frac{dy}{dt} = \frac{dy}{dx} \cdot \frac{dx}{dt} = 6\cos(2t^2) \cdot 2t \Rightarrow \left.\frac{dy}{dt}\right|_{t=0} = 6\cos(0) \cdot 0 = 0$

89. $\frac{dw}{ds} = \frac{dw}{dr} \cdot \frac{dr}{ds} = \left[\cos\left(e^{\sqrt{r}}\right)\left(e^{\sqrt{r}}\frac{1}{2\sqrt{r}}\right)\right]\left[3\cos\left(s + \frac{\pi}{6}\right)\right]$ at $x = 0, r = 3\sin\frac{\pi}{6} = \frac{3}{2}$
$\Rightarrow \frac{dw}{ds} = \cos\left(e^{\sqrt{3/2}}\right)\left(e^{\sqrt{3/2}}\frac{1}{2\sqrt{3/2}}\right)\left(3\cos\left(\frac{\pi}{6}\right)\right) = \frac{3\sqrt{3}e^{\sqrt{3/2}}}{4\sqrt{3/2}}\cos\left(e^{\sqrt{3/2}}\right) = \frac{3\sqrt{2}e^{\sqrt{3/2}}}{4}\cos\left(e^{\sqrt{3/2}}\right)$

91. $y^3 + y = 2\cos x \Rightarrow 3y^2\frac{dy}{dx} + \frac{dy}{dx} = -2\sin x \Rightarrow \frac{dy}{dx}(3y^2 + 1) = -2\sin x \Rightarrow \frac{dy}{dx} = \frac{-2\sin x}{3y^2 + 1} \Rightarrow \left.\frac{dy}{dx}\right|_{(0,1)}$
$= \frac{-2\sin(0)}{3+1} = 0; \; \frac{d^2y}{dx^2} = \frac{(3y^2+1)(-2\cos x) - (-2\sin x)\left(6y\frac{dy}{dx}\right)}{(3y^2+1)^2} \Rightarrow \left.\frac{d^2y}{dx^2}\right|_{(0,1)} = \frac{(3+1)(-2\cos 0) - (-2\sin 0)(6\cdot0)}{(3+1)^2} = -\frac{1}{2}$

93. $f(t) = \frac{1}{2t+1}$ and $f(t+h) = \frac{1}{2(t+h)+1} \Rightarrow \frac{f(t+h) - f(t)}{h} = \frac{\frac{1}{2(t+h)+1} - \frac{1}{2t+1}}{h} = \frac{2t+1 - (2t+2h+1)}{(2t+2h+1)(2t+1)h}$
$= \frac{-2h}{(2t+2h+1)(2t+1)h} = \frac{-2}{(2t+2h+1)(2t+1)} \Rightarrow f'(t) = \lim_{h \to 0}\frac{f(t+h)-f(t)}{h} = \lim_{h \to 0}\frac{-2}{(2t+2h+1)(2t+1)} = \frac{-2}{(2t+1)^2}$

95. (a)

$f(x) = \begin{cases} x^2, & -1 \le x < 0 \\ -x^2, & 0 \le x < 1 \end{cases}$

(b) $\lim_{x \to 0^-} f(x) = \lim_{x \to 0^-} x^2 = 0$ and $\lim_{x \to 0^+} f(x) = \lim_{x \to 0^+} -x^2 = 0 \Rightarrow \lim_{x \to 0} f(x) = 0$. Since $\lim_{x \to 0} f(x) = 0 = f(0)$ it
follows that f is continuous at $x = 0$.

(c) $\displaystyle\lim_{x \to 0^-} f'(x) = \lim_{x \to 0^-} (2x) = 0$ and $\displaystyle\lim_{x \to 0^+} f'(x) = \lim_{x \to 0^+} (-2x) = 0 \Rightarrow \lim_{x \to 0} f'(x) = 0$. Since this limit exists, it

follows that f is differentiable at x = 0.

97. (a)

$y = \begin{cases} x, & 0 \le x \le 1 \\ 2-x, & 1 < x \le 2 \end{cases}$

(b) $\displaystyle\lim_{x \to 1^-} f(x) = \lim_{x \to 1^-} x = 1$ and $\displaystyle\lim_{x \to 1^+} f(x) = \lim_{x \to 1^+} (2 - x) = 1 \Rightarrow \lim_{x \to 1} f(x) = 1$. Since $\displaystyle\lim_{x \to 1} f(x) = 1 = f(1)$, it

follows that f is continuous at x = 1.

(c) $\displaystyle\lim_{x \to 1^-} f'(x) = \lim_{x \to 1^-} 1 = 1$ and $\displaystyle\lim_{x \to 1^+} f'(x) = \lim_{x \to 1^+} -1 = -1 \Rightarrow \lim_{x \to 1^-} f'(x) \neq \lim_{x \to 1^+} f'(x)$, so $\displaystyle\lim_{x \to 1} f'(x)$ does

not exist $\Rightarrow$ f is not differentiable at x = 1.

99. $y = \frac{x}{2} + \frac{1}{2x-4} = \frac{1}{2}x + (2x - 4)^{-1} \Rightarrow \frac{dy}{dx} = \frac{1}{2} - 2(2x - 4)^{-2}$; the slope of the tangent is $-\frac{3}{2} \Rightarrow -\frac{3}{2}$

$= \frac{1}{2} - 2(2x - 4)^{-2} \Rightarrow -2 = -2(2x - 4)^{-2} \Rightarrow 1 = \frac{1}{(2x-4)^2} \Rightarrow (2x - 4)^2 = 1 \Rightarrow 4x^2 - 16x + 16 = 1$

$\Rightarrow 4x^2 - 16x + 15 = 0 \Rightarrow (2x - 5)(2x - 3) = 0 \Rightarrow x = \frac{5}{2}$ or $x = \frac{3}{2} \Rightarrow \left(\frac{5}{2}, \frac{9}{4}\right)$ and $\left(\frac{3}{2}, -\frac{1}{4}\right)$ are points on the

curve where the slope is $-\frac{3}{2}$.

101. $y = 2x^3 - 3x^2 - 12x + 20 \Rightarrow \frac{dy}{dx} = 6x^2 - 6x - 12$; the tangent is parallel to the x-axis when $\frac{dy}{dx} = 0$

$\Rightarrow 6x^2 - 6x - 12 = 0 \Rightarrow x^2 - x - 2 = 0 \Rightarrow (x - 2)(x + 1) = 0 \Rightarrow x = 2$ or $x = -1 \Rightarrow (2, 0)$ and $(-1, 27)$ are

points on the curve where the tangent is parallel to the x-axis.

103. $y = 2x^3 - 3x^2 - 12x + 20 \Rightarrow \frac{dy}{dx} = 6x^2 - 6x - 12$

(a) The tangent is perpendicular to the line $y = 1 - \frac{x}{24}$ when $\frac{dy}{dx} = -\left(\frac{1}{-\left(\frac{1}{24}\right)}\right) = 24$; $6x^2 - 6x - 12 = 24$

$\Rightarrow x^2 - x - 2 = 4 \Rightarrow x^2 - x - 6 = 0 \Rightarrow (x - 3)(x + 2) = 0 \Rightarrow x = -2$ or $x = 3 \Rightarrow (-2, 16)$ and $(3, 11)$ are

points where the tangent is perpendicular to $y = 1 - \frac{x}{24}$.

(b) The tangent is parallel to the line $y = \sqrt{2} - 12x$ when $\frac{dy}{dx} = -12 \Rightarrow 6x^2 - 6x - 12 = -12 \Rightarrow x^2 - x = 0$

$\Rightarrow x(x - 1) = 0 \Rightarrow x = 0$ or $x = 1 \Rightarrow (0, 20)$ and $(1, 7)$ are points where the tangent is parallel to $y = \sqrt{2} - 12x$.

105. $y = \tan x, -\frac{\pi}{2} < x < \frac{\pi}{2} \Rightarrow \frac{dy}{dx} = \sec^2 x$; now the slope

of $y = -\frac{x}{2}$ is $-\frac{1}{2} \Rightarrow$ the normal line is parallel to

$y = -\frac{x}{2}$ when $\frac{dy}{dx} = 2$. Thus, $\sec^2 x = 2 \Rightarrow \frac{1}{\cos^2 x} = 2$

$\Rightarrow \cos^2 x = \frac{1}{2} \Rightarrow \cos x = \frac{\pm 1}{\sqrt{2}} \Rightarrow x = -\frac{\pi}{4}$ and $x = \frac{\pi}{4}$

for $-\frac{\pi}{2} < x < \frac{\pi}{2} \Rightarrow \left(-\frac{\pi}{4}, -1\right)$ and $\left(\frac{\pi}{4}, 1\right)$ are points

where the normal is parallel to $y = -\frac{x}{2}$.

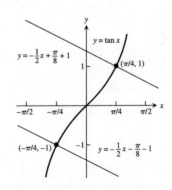

107. $y = x^2 + C \Rightarrow \frac{dy}{dx} = 2x$ and $y = x \Rightarrow \frac{dy}{dx} = 1$; the parabola is tangent to $y = x$ when $2x = 1 \Rightarrow x = \frac{1}{2} \Rightarrow y = \frac{1}{2}$;

thus, $\frac{1}{2} = \left(\frac{1}{2}\right)^2 + C \Rightarrow C = \frac{1}{4}$

109. The line through $(0, 3)$ and $(5, -2)$ has slope $m = \frac{3-(-2)}{0-5} = -1 \Rightarrow$ the line through $(0, 3)$ and $(5, -2)$ is

$y = -x + 3$; $y = \frac{c}{x+1} \Rightarrow \frac{dy}{dx} = \frac{-c}{(x+1)^2}$, so the curve is tangent to $y = -x + 3 \Rightarrow \frac{dy}{dx} = -1 = \frac{-c}{(x+1)^2}$

$\Rightarrow (x+1)^2 = c, x \neq -1$. Moreover, $y = \frac{c}{x+1}$ intersects $y = -x + 3 \Rightarrow \frac{c}{x+1} = -x + 3, x \neq -1$

$\Rightarrow c = (x+1)(-x+3), x \neq -1$. Thus $c = c \Rightarrow (x+1)^2 = (x+1)(-x+3) \Rightarrow (x+1)[x + 1 - (-x + 3)]$

$= 0, x \neq -1 \Rightarrow (x+1)(2x - 2) = 0 \Rightarrow x = 1$ (since $x \neq -1$) $\Rightarrow c = 4$.

111. $x^2 + 2y^2 = 9 \Rightarrow 2x + 4y \frac{dy}{dx} = 0 \Rightarrow \frac{dy}{dx} = -\frac{x}{2y} \Rightarrow \frac{dy}{dx}\Big|_{(1,2)} = -\frac{1}{4} \Rightarrow$ the tangent line is $y = 2 - \frac{1}{4}(x - 1)$

$= -\frac{1}{4}x + \frac{9}{4}$ and the normal line is $y = 2 + 4(x - 1) = 4x - 2$.

113. $xy + 2x - 5y = 2 \Rightarrow \left(x \frac{dy}{dx} + y\right) + 2 - 5\frac{dy}{dx} = 0 \Rightarrow \frac{dy}{dx}(x - 5) = -y - 2 \Rightarrow \frac{dy}{dx} = \frac{-y-2}{x-5} \Rightarrow \frac{dy}{dx}\Big|_{(3,2)} = 2$

$\Rightarrow$ the tangent line is $y = 2 + 2(x - 3) = 2x - 4$ and the normal line is $y = 2 + \frac{-1}{2}(x - 3) = -\frac{1}{2}x + \frac{7}{2}$.

115. $x + \sqrt{xy} = 6 \Rightarrow 1 + \frac{1}{2\sqrt{xy}}\left(x \frac{dy}{dx} + y\right) = 0 \Rightarrow x\frac{dy}{dx} + y = -2\sqrt{xy} \Rightarrow \frac{dy}{dx} = \frac{-2\sqrt{xy}-y}{x} \Rightarrow \frac{dy}{dx}\Big|_{(4,1)} = \frac{-5}{4}$

$\Rightarrow$ the tangent line is $y = 1 - \frac{5}{4}(x - 4) = -\frac{5}{4}x + 6$ and the normal line is $y = 1 + \frac{4}{5}(x - 4) = \frac{4}{5}x - \frac{11}{5}$.

117. $x^3y^3 + y^2 = x + y \Rightarrow \left[x^3\left(3y^2 \frac{dy}{dx}\right) + y^3(3x^2)\right] + 2y\frac{dy}{dx} = 1 + \frac{dy}{dx} \Rightarrow 3x^3y^2 \frac{dy}{dx} + 2y\frac{dy}{dx} - \frac{dy}{dx} = 1 - 3x^2y^3$

$\Rightarrow \frac{dy}{dx}(3x^3y^2 + 2y - 1) = 1 - 3x^2y^3 \Rightarrow \frac{dy}{dx} = \frac{1-3x^2y^3}{3x^3y^2+2y-1} \Rightarrow \frac{dy}{dx}\Big|_{(1,1)} = -\frac{2}{4}$, but $\frac{dy}{dx}\Big|_{(1,-1)}$ is undefined.

Therefore, the curve has slope $-\frac{1}{2}$ at $(1, 1)$ but the slope is undefined at $(1, -1)$.

119. $x = \frac{1}{2}\tan t$, $y = \frac{1}{2}\sec t \Rightarrow \frac{dy}{dx} = \frac{dy/dt}{dx/dt} = \frac{\frac{1}{2}\sec t \tan t}{\frac{1}{2}\sec^2 t} = \frac{\tan t}{\sec t} = \sin t \Rightarrow \frac{dy}{dx}\Big|_{t=\pi/3} = \sin \frac{\pi}{3} = \frac{\sqrt{3}}{2}$; $t = \frac{\pi}{3}$

$\Rightarrow x = \frac{1}{2}\tan \frac{\pi}{3} = \frac{\sqrt{3}}{2}$ and $y = \frac{1}{2}\sec \frac{\pi}{3} = 1 \Rightarrow y = \frac{\sqrt{3}}{2}x + \frac{1}{4}$; $\frac{d^2y}{dx^2} = \frac{dy'/dt}{dx/dt} = \frac{\cos t}{\frac{1}{2}\sec^2 t} = 2\cos^3 t \Rightarrow \frac{d^2y}{dx^2}\Big|_{t=\pi/3}$

$= 2\cos^3 \left(\frac{\pi}{3}\right) = \frac{1}{4}$

121. $B = $ graph of f, $A = $ graph of f'. Curve B cannot be the derivative of A because A has only negative slopes while some of B's values are positive.

123.

125. (a) $0, 0$ (b) largest 1700, smallest about 1400

127. $\lim\limits_{x \to 0} \frac{\sin x}{2x^2 - x} = \lim\limits_{x \to 0}\left[\left(\frac{\sin x}{x}\right) \cdot \frac{1}{(2x - 1)}\right] = (1)\left(\frac{1}{-1}\right) = -1$

129. $\lim\limits_{r \to 0} \frac{\sin r}{\tan 2r} = \lim\limits_{r \to 0}\left(\frac{\sin r}{r} \cdot \frac{2r}{\tan 2r} \cdot \frac{1}{2}\right) = \left(\frac{1}{2}\right)(1)\lim\limits_{r \to 0}\frac{\cos 2r}{\left(\frac{\sin 2r}{2r}\right)} = \left(\frac{1}{2}\right)(1)\left(\frac{1}{1}\right) = \frac{1}{2}$

131. $\lim\limits_{\theta \to \left(\frac{\pi}{2}\right)^-} \frac{4\tan^2 \theta + \tan \theta + 1}{\tan^2 \theta + 5} = \lim\limits_{\theta \to \left(\frac{\pi}{2}\right)^-} \frac{\left(4 + \frac{1}{\tan \theta} + \frac{1}{\tan^2 \theta}\right)}{\left(1 + \frac{5}{\tan^2 \theta}\right)} = \frac{(4+0+0)}{(1+0)} = 4$

133. $\lim\limits_{x \to 0} \frac{x \sin x}{2 - 2\cos x} = \lim\limits_{x \to 0} \frac{x \sin x}{2(1 - \cos x)} = \lim\limits_{x \to 0} \frac{x \sin x}{2\left(2 \sin^2\left(\frac{x}{2}\right)\right)} = \lim\limits_{x \to 0} \left[\frac{\frac{x}{2} \cdot \frac{x}{2}}{\sin^2\left(\frac{x}{2}\right)} \cdot \frac{\sin x}{x}\right]$

   $= \lim\limits_{x \to 0} \left[\frac{\left(\frac{x}{2}\right)}{\sin\left(\frac{x}{2}\right)} \cdot \frac{\left(\frac{x}{2}\right)}{\sin\left(\frac{x}{2}\right)} \cdot \frac{\sin x}{x}\right] = (1)(1)(1) = 1$

135. $\lim\limits_{x \to 0} \frac{\tan x}{x} = \lim\limits_{x \to 0} \left(\frac{1}{\cos x} \cdot \frac{\sin x}{x}\right) = 1$; let $\theta = \tan x \Rightarrow \theta \to 0$ as $x \to 0 \Rightarrow \lim\limits_{x \to 0} g(x) = \lim\limits_{x \to 0} \frac{\tan(\tan x)}{\tan x}$

   $= \lim\limits_{\theta \to 0} \frac{\tan \theta}{\theta} = 1$. Therefore, to make g continuous at the origin, define $g(0) = 1$.

137. $y = \frac{2(x^2 + 1)}{\sqrt{\cos 2x}} \Rightarrow \ln y = \ln\left(\frac{2(x^2 + 1)}{\sqrt{\cos 2x}}\right) = \ln(2) + \ln(x^2 + 1) - \frac{1}{2}\ln(\cos 2x) \Rightarrow \frac{y'}{y} = 0 + \frac{2x}{x^2 + 1} - \left(\frac{1}{2}\right)\frac{(-2\sin 2x)}{\cos 2x}$

   $\Rightarrow y' = \left(\frac{2x}{x^2 + 1} + \tan 2x\right) y = \frac{2(x^2 + 1)}{\sqrt{\cos 2x}} \left(\frac{2x}{x^2 + 1} + \tan 2x\right)$

139. $y = \left[\frac{(t + 1)(t - 1)}{(t - 2)(t + 3)}\right]^5 \Rightarrow \ln y = 5\left[\ln(t + 1) + \ln(t - 1) - \ln(t - 2) - \ln(t + 3)\right] \Rightarrow \left(\frac{1}{y}\right)\left(\frac{dy}{dt}\right)$

   $- 5\left(\frac{1}{t + 1} + \frac{1}{t - 1} - \frac{1}{t - 2} - \frac{1}{t + 3}\right) \Rightarrow \frac{dy}{dt} = 5\left[\frac{(t + 1)(t - 1)}{(t - 2)(t + 3)}\right]^5 \left(\frac{1}{t + 1} + \frac{1}{t - 1} - \frac{1}{t - 2} - \frac{1}{t + 3}\right)$

141. $y = (\sin \theta)^{\sqrt{\theta}} \Rightarrow \ln y = \sqrt{\theta}\ln(\sin \theta) \Rightarrow \left(\frac{1}{y}\right)\left(\frac{dy}{d\theta}\right) = \sqrt{\theta}\left(\frac{\cos \theta}{\sin \theta}\right) + \frac{1}{2}\theta^{-1/2}\ln(\sin \theta)$

   $\Rightarrow \frac{dy}{d\theta} = (\sin \theta)^{\sqrt{\theta}}\left(\sqrt{\theta}\cot \theta + \frac{\ln(\sin \theta)}{2\sqrt{\theta}}\right)$

143. $y = \sinh^2 3x \Rightarrow y' = 2\sinh 3x \cosh 3x \cdot 3 = 6\sinh 3x \cosh 3x \Rightarrow y' = 3\sinh 6x$

145. $\sin^{-1} x = \text{sech } y \Rightarrow \frac{1}{\sqrt{1 - x^2}} = -\text{sech } y \tanh y \frac{dy}{dx} \Rightarrow \frac{-1}{\text{sech } y \tanh y \sqrt{1 - x^2}} = \frac{dy}{dx} \Rightarrow \frac{dy}{dx} = -\frac{\cosh y \coth y}{\sqrt{1 - x^2}}$

147. $\tan^{-1} y = \tanh^{-1} x \Rightarrow \frac{1}{1 + y^2} \cdot \frac{dy}{dx} = \frac{1}{1 - x^2} \Rightarrow \frac{dy}{dx} = \frac{1 + y^2}{1 - x^2}$

149. $x = \cosh(\ln y) \Rightarrow 1 = \sinh(\ln y) \cdot \frac{1}{y} \cdot \frac{dy}{dx} \Rightarrow 1 = \frac{\sinh(\ln y)}{y} \cdot \frac{dy}{dx} \Rightarrow \frac{y}{\sinh(\ln y)} = \frac{dy}{dx} \Rightarrow \frac{dy}{dx} = y\,\text{csch}(\ln y)$

151. $y = \sinh^{-1}(\tan x) \Rightarrow \frac{dy}{dx} = \frac{1}{\sqrt{1 + (\tan x)^2}} \cdot \sec^2 x = \frac{\sec^2 x}{\sqrt{1 + \tan^2 x}} = \frac{\sec^2 x}{\sqrt{\sec^2 x}} = \sqrt{\sec^2 x} \Rightarrow \frac{dy}{dx} = |\sec x|$

153. (a) $S = 2\pi r^2 + 2\pi rh$ and h constant $\Rightarrow \frac{dS}{dt} = 4\pi r \frac{dr}{dt} + 2\pi h \frac{dr}{dt} = (4\pi r + 2\pi h)\frac{dr}{dt}$

   (b) $S = 2\pi r^2 + 2\pi rh$ and r constant $\Rightarrow \frac{dS}{dt} = 2\pi r \frac{dh}{dt}$

   (c) $S = 2\pi r^2 + 2\pi rh \Rightarrow \frac{dS}{dt} = 4\pi r \frac{dr}{dt} + 2\pi\left(r \frac{dh}{dt} + h \frac{dr}{dt}\right) = (4\pi r + 2\pi h)\frac{dr}{dt} + 2\pi r \frac{dh}{dt}$

   (d) S constant $\Rightarrow \frac{dS}{dt} = 0 \Rightarrow 0 = (4\pi r + 2\pi h)\frac{dr}{dt} + 2\pi r \frac{dh}{dt} \Rightarrow (2r + h)\frac{dr}{dt} = -r \frac{dh}{dt} \Rightarrow \frac{dr}{dt} = \frac{-r}{2r + h}\frac{dh}{dt}$

155. $A = \pi r^2 \Rightarrow \frac{dA}{dt} = 2\pi r \frac{dr}{dt}$; so r = 10 and $\frac{dr}{dt} = -\frac{2}{\pi}$ m/sec $\Rightarrow \frac{dA}{dt} = (2\pi)(10)\left(-\frac{2}{\pi}\right) = -40$ m^2/sec

157. $\frac{dR_1}{dt} = -1$ ohm/sec, $\frac{dR_2}{dt} = 0.5$ ohm/sec; and $\frac{1}{R} = \frac{1}{R_1} + \frac{1}{R_2} \Rightarrow \frac{-1}{R^2}\frac{dR}{dt} = \frac{-1}{R_1^2}\frac{dR_1}{dt} - \frac{1}{R_2^2}\frac{dR_2}{dt}$. Also,

   $R_1 = 75$ ohms and $R_2 = 50$ ohms $\Rightarrow \frac{1}{R} = \frac{1}{75} + \frac{1}{50} \Rightarrow R = 30$ ohms. Therefore, from the derivative equation,

   $\frac{-1}{(30)^2}\frac{dR}{dt} = \frac{-1}{(75)^2}(-1) - \frac{1}{(50)^2}(0.5) = \left(\frac{1}{5625} - \frac{1}{5000}\right) \Rightarrow \frac{dR}{dt} = (-900)\left(\frac{5000 - 5625}{5625 \cdot 5000}\right) = \frac{9(625)}{50(5625)} = \frac{1}{50} = 0.02$ ohm/sec.

159. Given $\frac{dx}{dt} = 10$ m/sec and $\frac{dy}{dt} = 5$ m/sec, let D be the distance from the origin $\Rightarrow D^2 = x^2 + y^2 \Rightarrow 2D \frac{dD}{dt}$

   $= 2x \frac{dx}{dt} + 2y \frac{dy}{dt} \Rightarrow D \frac{dD}{dt} = x \frac{dx}{dt} + y \frac{dy}{dt}$. When $(x, y) = (3, -4)$, $D = \sqrt{3^2 + (-4)^2} = 5$ and

   $5 \frac{dD}{dt} = (5)(10) + (12)(5) \Rightarrow \frac{dD}{dt} = \frac{110}{5} = 22$. Therefore, the particle is moving <u>away from</u> the origin at 22 m/sec (because the distance D is increasing).

161. (a) From the diagram we have $\frac{10}{h} = \frac{4}{r} \Rightarrow r = \frac{2}{5}$ h.

(b) $V = \frac{1}{3}\pi r^2 h = \frac{1}{3}\pi\left(\frac{2}{5}h\right)^2 h = \frac{4\pi h^3}{75} \Rightarrow \frac{dV}{dt} = \frac{4\pi h^2}{25}\frac{dh}{dt}$, so $\frac{dV}{dt} = -5$ and $h = 6 \Rightarrow \frac{dh}{dt} = -\frac{125}{144\pi}$ ft/min.

163. (a) From the sketch in the text, $\frac{d\theta}{dt} = -0.6$ rad/sec and $x = \tan\theta$. Also $x = \tan\theta \Rightarrow \frac{dx}{dt} = \sec^2\theta\,\frac{d\theta}{dt}$; at point A, $x = 0 \Rightarrow \theta = 0 \Rightarrow \frac{dx}{dt} = (\sec^2 0)(-0.6) = -0.6$. Therefore the speed of the light is $0.6 = \frac{3}{5}$ km/sec when it reaches point A.

(b) $\frac{(3/5)\text{ rad}}{\text{sec}} \cdot \frac{1\text{ rev}}{2\pi\text{ rad}} \cdot \frac{60\text{ sec}}{\text{min}} = \frac{18}{\pi}$ revs/min

165. (a) If $f(x) = \tan x$ and $x = -\frac{\pi}{4}$, then $f'(x) = \sec^2 x$, $f\left(-\frac{\pi}{4}\right) = -1$ and $f'\left(-\frac{\pi}{4}\right) = 2$. The linearization of $f(x)$ is $L(x) = 2\left(x + \frac{\pi}{4}\right) + (-1) = 2x + \frac{\pi-2}{2}$.

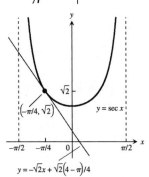

(b) If $f(x) = \sec x$ and $x = -\frac{\pi}{4}$, then $f'(x) = \sec x \tan x$, $f\left(-\frac{\pi}{4}\right) = \sqrt{2}$ and $f'\left(-\frac{\pi}{4}\right) = -\sqrt{2}$. The linearization of $f(x)$ is $L(x) = -\sqrt{2}\left(x + \frac{\pi}{4}\right) + \sqrt{2}$ $= -\sqrt{2}x + \frac{\sqrt{2}(4-\pi)}{4}$.

167. $f(x) = \sqrt{x+1} + \sin x - 0.5 = (x+1)^{1/2} + \sin x - 0.5 \Rightarrow f'(x) = \left(\frac{1}{2}\right)(x+1)^{-1/2} + \cos x$
$\Rightarrow L(x) = f'(0)(x-0) + f(0) = 1.5(x-0) + 0.5 \Rightarrow L(x) = 1.5x + 0.5$, the linearization of $f(x)$.

169. $S = \pi r\sqrt{r^2 + h^2}$, r constant $\Rightarrow dS = \pi r \cdot \frac{1}{2}(r^2 + h^2)^{-1/2} 2h\, dh = \frac{\pi r h}{\sqrt{r^2 + h^2}}dh$. Height changes from h_0 to $h_0 + dh$
$\Rightarrow dS = \frac{\pi r h_0(dh)}{\sqrt{r^2 + h_0^2}}$

171. $C = 2\pi r \Rightarrow r = \frac{C}{2\pi}$, $S = 4\pi r^2 = \frac{C^2}{\pi}$, and $V = \frac{4}{3}\pi r^3 = \frac{C^3}{6\pi^2}$. It also follows that $dr = \frac{1}{2\pi}\,dC$, $dS = \frac{2C}{\pi}\,dC$ and $dV = \frac{C^2}{2\pi^2}\,dC$. Recall that $C = 10$ cm and $dC = 0.4$ cm.

(a) $dr = \frac{0.4}{2\pi} = \frac{0.2}{\pi}$ cm $\Rightarrow \left(\frac{dr}{r}\right)(100\%) = \left(\frac{0.2}{\pi}\right)\left(\frac{2\pi}{10}\right)(100\%) = (.04)(100\%) = 4\%$

(b) $dS = \frac{20}{\pi}(0.4) = \frac{8}{\pi}$ cm $\Rightarrow \left(\frac{dS}{S}\right)(100\%) = \left(\frac{8}{\pi}\right)\left(\frac{\pi}{100}\right)(100\%) = 8\%$

(c) $dV = \frac{10^2}{2\pi^2}(0.4) = \frac{20}{\pi^2}$ cm $\Rightarrow \left(\frac{dV}{V}\right)(100\%) = \left(\frac{20}{\pi^2}\right)\left(\frac{6\pi^2}{1000}\right)(100\%) = 12\%$

CHAPTER 3 ADDITIONAL AND ADVANCED EXERCISES

1. (a) $\sin 2\theta = 2\sin\theta\cos\theta \Rightarrow \frac{d}{d\theta}(\sin 2\theta) = \frac{d}{d\theta}(2\sin\theta\cos\theta) \Rightarrow 2\cos 2\theta = 2[(\sin\theta)(-\sin\theta) + (\cos\theta)(\cos\theta)]$
$\Rightarrow \cos 2\theta = \cos^2\theta - \sin^2\theta$

(b) $\cos 2\theta = \cos^2\theta - \sin^2\theta \Rightarrow \frac{d}{d\theta}(\cos 2\theta) = \frac{d}{d\theta}(\cos^2\theta - \sin^2\theta) \Rightarrow -2\sin 2\theta = (2\cos\theta)(-\sin\theta) - (2\sin\theta)(\cos\theta)$
$\Rightarrow \sin 2\theta = \cos\theta\sin\theta + \sin\theta\cos\theta \Rightarrow \sin 2\theta = 2\sin\theta\cos\theta$

3. (a) $f(x) = \cos x \Rightarrow f'(x) = -\sin x \Rightarrow f''(x) = -\cos x$, and $g(x) = a + bx + cx^2 \Rightarrow g'(x) = b + 2cx \Rightarrow g''(x) = 2c$;
    also, $f(0) = g(0) \Rightarrow \cos(0) = a \Rightarrow a = 1; f'(0) = g'(0) \Rightarrow -\sin(0) = b \Rightarrow b = 0; f''(0) = g''(0)$
    $\Rightarrow -\cos(0) = 2c \Rightarrow c = -\frac{1}{2}$. Therefore, $g(x) = 1 - \frac{1}{2}x^2$.

    (b) $f(x) = \sin(x + a) \Rightarrow f'(x) = \cos(x + a)$, and $g(x) = b \sin x + c \cos x \Rightarrow g'(x) = b \cos x - c \sin x$; also,
    $f(0) = g(0) \Rightarrow \sin(a) = b \sin(0) + c \cos(0) \Rightarrow c = \sin a; f'(0) = g'(0) \Rightarrow \cos(a) = b \cos(0) - c \sin(0)$
    $\Rightarrow b = \cos a$. Therefore, $g(x) = \sin x \cos a + \cos x \sin a$.

    (c) When $f(x) = \cos x$, $f'''(x) = \sin x$ and $f^{(4)}(x) = \cos x$; when $g(x) = 1 - \frac{1}{2}x^2$, $g'''(x) = 0$ and $g^{(4)}(x) = 0$. Thus
    $f'''(0) = 0 = g'''(0)$ so the third derivatives agree at $x = 0$. However, the fourth derivatives do not agree since
    $f^{(4)}(0) = 1$ but $g^{(4)}(0) = 0$. In case (b), when $f(x) = \sin(x + a)$ and $g(x) = \sin x \cos a + \cos x \sin a$, notice that
    $f(x) = g(x)$ for all x, not just $x = 0$. Since this is an identity, we have $f^{(n)}(x) = g^{(n)}(x)$ for any x and any positive
    integer n.

5. If the circle $(x - h)^2 + (y - k)^2 = a^2$ and $y = x^2 + 1$ are tangent at $(1, 2)$, then the slope of this tangent is $m = 2x\big|_{(1,2)} = 2$
    and the tangent line is $y = 2x$. The line containing (h, k) and $(1, 2)$ is perpendicular to $y = 2x \Rightarrow \frac{k-2}{h-1} = -\frac{1}{2}$
    $\Rightarrow h = 5 - 2k \Rightarrow$ the location of the center is $(5 - 2k, k)$. Also, $(x - h)^2 + (y - k)^2 = a^2 \Rightarrow x - h + (y - k)y' = 0$
    $\Rightarrow 1 + (y')^2 + (y - k)y'' = 0 \Rightarrow y'' = \frac{1 + (y')^2}{k - y}$. At the point $(1, 2)$ we know $y' = 2$ from the tangent line and that $y'' = 2$
    from the parabola. Since the second derivatives are equal at $(1, 2)$ we obtain $2 = \frac{1 + (2)^2}{k - 2} \Rightarrow k = \frac{9}{2}$. Then
    $h = 5 - 2k = -4 \Rightarrow$ the circle is $(x + 4)^2 + \left(y - \frac{9}{2}\right)^2 = a^2$. Since $(1, 2)$ lies on the circle we have that $a = \frac{5\sqrt{5}}{2}$.

7. (a) $y = uv \Rightarrow \frac{dy}{dt} = \frac{du}{dt}v + u\frac{dv}{dt} = (0.04u)v + u(0.05v) = 0.09uv = 0.09y \Rightarrow$ the rate of growth of the total production is
    9% per year.

    (b) If $\frac{du}{dt} = -0.02u$ and $\frac{dv}{dt} = 0.03v$, then $\frac{dy}{dt} = (-0.02u)v + (0.03v)u = 0.01uv = 0.01y$, increasing at 1% per year.

9. Answers will vary. Here is one possibility.

11. (a) $s(t) = 64t - 16t^2 \Rightarrow v(t) = \frac{ds}{dt} = 64 - 32t = 32(2 - t)$. The maximum height is reached when $v(t) = 0$
    $\Rightarrow t = 2$ sec. The velocity when it leaves the hand is $v(0) = 64$ ft/sec.

    (b) $s(t) = 64t - 2.6t^2 \Rightarrow v(t) = \frac{ds}{dt} = 64 - 5.2t$. The maximum height is reached when $v(t) = 0 \Rightarrow t \approx 12.31$ sec.
    The maximum height is about $s(12.31) = 393.85$ ft.

13. $m(v^2 - v_0^2) = k(x_0^2 - x^2) \Rightarrow m\left(2v\frac{dv}{dt}\right) = k\left(-2x\frac{dx}{dt}\right) \Rightarrow m\frac{dv}{dt} = k\left(-\frac{2x}{2v}\right)\frac{dx}{dt} \Rightarrow m\frac{dv}{dt} = -kx\left(\frac{1}{v}\right)\frac{dx}{dt}$. Then
    substituting $\frac{dx}{dt} = v \Rightarrow m\frac{dv}{dt} = -kx$, as claimed.

15. (a) To be continuous at $x = \pi$ requires that $\lim_{x \to \pi^-} \sin x = \lim_{x \to \pi^+} (mx + b) \Rightarrow 0 = m\pi + b \Rightarrow m = -\frac{b}{\pi}$;

    (b) If $y' = \begin{cases} \cos x, & x < \pi \\ m, & x \geq \pi \end{cases}$ is differentiable at $x = \pi$, then $\lim_{x \to \pi^-} \cos x = m \Rightarrow m = -1$ and $b = \pi$.

17. (a) For all a, b and for all $x \neq 2$, f is differentiable at x. Next, f differentiable at $x = 2 \Rightarrow$ f continuous at $x = 2$
    $\Rightarrow \lim_{x \to 2^-} f(x) = f(2) \Rightarrow 2a = 4a - 2b + 3 \Rightarrow 2a - 2b + 3 = 0$. Also, f differentiable at $x \neq 2$
    $\Rightarrow f'(x) = \begin{cases} a, & x < 2 \\ 2ax - b, & x > 2 \end{cases}$. In order that $f'(2)$ exist we must have $a = 2a(2) - b \Rightarrow a = 4a - b \Rightarrow 3a = b$.
    Then $2a - 2b + 3 = 0$ and $3a = b \Rightarrow a = \frac{3}{4}$ and $b = \frac{9}{4}$.

    (b) For $x < 2$, the graph of f is a straight line having a slope of $\frac{3}{4}$ and passing through the origin; for $x \geq 2$, the graph of f is a parabola. At $x = 2$, the value of the y-coordinate on the parabola is $\frac{3}{2}$ which matches the y-coordinate of the point on the straight line at $x = 2$. In addition, the slope of the parabola at the match up point is $\frac{3}{4}$ which is equal to the slope of the straight line. Therefore, since the graph is differentiable at the match up point, the graph is smooth there.

19. f odd $\Rightarrow f(-x) = -f(x) \Rightarrow \frac{d}{dx}(f(-x)) = \frac{d}{dx}(-f(x)) \Rightarrow f'(-x)(-1) = -f'(x) \Rightarrow f'(-x) = f'(x) \Rightarrow f'$ is even.

21. Let $h(x) = (fg)(x) = f(x)\,g(x) \Rightarrow h'(x) = \lim_{x \to x_0} \frac{h(x) - h(x_0)}{x - x_0} = \lim_{x \to x_0} \frac{f(x)\,g(x) - f(x_0)\,g(x_0)}{x - x_0}$

$= \lim_{x \to x_0} \frac{f(x)\,g(x) - f(x)\,g(x_0) + f(x)\,g(x_0) - f(x_0)\,g(x_0)}{x - x_0} = \lim_{x \to x_0} \left[f(x) \left[\frac{g(x) - g(x_0)}{x - x_0} \right] \right] + \lim_{x \to x_0} \left[g(x_0) \left[\frac{f(x) - f(x_0)}{x - x_0} \right] \right]$

$= f(x_0) \lim_{x \to x_0} \left[\frac{g(x) - g(x_0)}{x - x_0} \right] + g(x_0)\,f'(x_0) = 0 \cdot \lim_{x \to x_0} \left[\frac{g(x) - g(x_0)}{x - x_0} \right] + g(x_0)\,f'(x_0) = g(x_0)\,f'(x_0)$, if g is

continuous at x_0. Therefore $(fg)(x)$ is differentiable at x_0 if $f(x_0) = 0$, and $(fg)'(x_0) = g(x_0)\,f'(x_0)$.

23. If $f(x) = x$ and $g(x) = x \sin\left(\frac{1}{x}\right)$, then $x^2 \sin\left(\frac{1}{x}\right)$ is differentiable at $x = 0$ because $f'(0) = 1$, $f(0) = 0$ and

$\lim_{x \to 0} x \sin\left(\frac{1}{x}\right) = \lim_{x \to 0} \frac{\sin\left(\frac{1}{x}\right)}{\frac{1}{x}} = \lim_{t \to \infty} \frac{\sin t}{t} = 0$ (so g is continuous at $x = 0$). In fact, from Exercise 21,

$h'(0) = g(0)\,f'(0) = 0$. However, for $x \neq 0$, $h'(x) = \left[x^2 \cos\left(\frac{1}{x}\right) \right]\left(-\frac{1}{x^2}\right) + 2x \sin\left(\frac{1}{x}\right)$. But

$\lim_{x \to 0} h'(x) = \lim_{x \to 0} \left[-\cos\left(\frac{1}{x}\right) + 2x \sin\left(\frac{1}{x}\right) \right]$ does not exist because $\cos\left(\frac{1}{x}\right)$ has no limit as $x \to 0$. Therefore, the derivative is not continuous at $x = 0$ because it has no limit there.

25. Step 1:   The formula holds for $n = 2$ (a single product) since $y = u_1 u_2 \Rightarrow \frac{dy}{dx} = \frac{du_1}{dx} u_2 + u_1 \frac{du_2}{dx}$.

Step 2:   Assume the formula holds for $n = k$:

$y = u_1 u_2 \cdots u_k \Rightarrow \frac{dy}{dx} = \frac{du_1}{dx} u_2 u_3 \cdots u_k + u_1 \frac{du_2}{dx} u_3 \cdots u_k + \ldots + u_1 u_2 \cdots u_{k-1} \frac{du_k}{dx}$.

If $y = u_1 u_2 \cdots u_k u_{k+1} = (u_1 u_2 \cdots u_k)\,u_{k+1}$, then $\frac{dy}{dx} = \frac{d(u_1 u_2 \cdots u_k)}{dx} u_{k+1} + u_1 u_2 \cdots u_k \frac{du_{k+1}}{dx}$

$= \left(\frac{du_1}{dx} u_2 u_3 \cdots u_k + u_1 \frac{du_2}{dx} u_3 \cdots u_k + \cdots + u_1 u_2 \cdots u_{k-1} \frac{du_k}{dx} \right) u_{k+1} + u_1 u_2 \cdots u_k \frac{du_{k+1}}{dx}$

$= \frac{du_1}{dx} u_2 u_3 \cdots u_{k+1} + u_1 \frac{du_2}{dx} u_3 \cdots u_{k+1} + \cdots + u_1 u_2 \cdots u_{k-1} \frac{du_k}{dx} u_{k+1} + u_1 u_2 \cdots u_k \frac{du_{k+1}}{dx}$.

Thus the original formula holds for $n = (k+1)$ whenever it holds for $n = k$.

27. (a) $T^2 = \frac{4\pi^2 L}{g} \Rightarrow L = \frac{T^2 g}{4\pi^2} \Rightarrow L = \frac{(1\ \sec^2)(32.2\ ft/sec^2)}{4\pi^2} \Rightarrow L \approx 0.8156$ ft

(b) $T^2 = \frac{4\pi^2 L}{g} \Rightarrow T = \frac{2\pi}{\sqrt{g}} \sqrt{L}$; $dT = \frac{2\pi}{\sqrt{g}} \cdot \frac{1}{2\sqrt{L}} dL = \frac{\pi}{\sqrt{Lg}} dL$; $dT = \frac{\pi}{\sqrt{(0.8156\ ft)(32.2\ ft/sec^2)}} (0.01\ ft) \approx 0.00613$ sec.

(c)   Since there are 86,400 sec in a day, we have $(0.00613\ \sec)(86{,}400\ \sec/day) \approx 529.6$ sec/day, or 8.83 min/day; the clock will lose about 8.83 min/day.

NOTES:

CHAPTER 4 APPLICATIONS OF DERIVATIVES

4.1 EXTREME VALUES OF FUNCTIONS

1. An absolute minimum at $x = c_2$, an absolute maximum at $x = b$. Theorem 1 guarantees the existence of such extreme values because h is continuous on $[a, b]$.

3. No absolute minimum. An absolute maximum at $x = c$. Since the function's domain is an open interval, the function does not satisfy the hypotheses of Theorem 1 and need not have absolute extreme values.

5. An absolute minimum at $x = a$ and an absolute maximum at $x = c$. Note that $y = g(x)$ is not continuous but still has extrema. When the hypothesis of Theorem 1 is satisfied then extrema are guaranteed, but when the hypothesis is not satisfied, absolute extrema may or may not occur.

7. Local minimum at $(-1, 0)$, local maximum at $(1, 0)$

9. Maximum at $(0, 5)$. Note that there is no minimum since the endpoint $(2, 0)$ is excluded from the graph.

11. Graph (c), since this the only graph that has positive slope at c.

13. Graph (d), since this is the only graph representing a funtion that is differentiable at b but not at a.

15. $f(x) = \frac{2}{3}x - 5 \Rightarrow f'(x) = \frac{2}{3} \Rightarrow$ no critical points;
    $f(-2) = -\frac{19}{3}, f(3) = -3 \Rightarrow$ the absolute maximum
    is -3 at $x = 3$ and the absolute minimum is $-\frac{19}{3}$ at
    $x = -2$

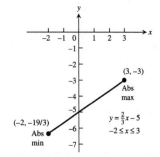

17. $f(x) = x^2 - 1 \Rightarrow f'(x) = 2x \Rightarrow$ a critical point at
    $x = 0$; $f(-1) = 0, f(0) = -1, f(2) = 3 \Rightarrow$ the absolute
    maximum is 3 at $x = 2$ and the absolute minimum is -1
    at $x = 0$

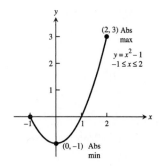

19. $F(x) = -\frac{1}{x^2} = -x^{-2} \Rightarrow F'(x) = 2x^{-3} = \frac{2}{x^3}$, however
$x = 0$ is not a critical point since 0 is not in the domain;
$F(0.5) = -4$, $F(2) = -0.25 \Rightarrow$ the absolute maximum is
-0.25 at $x = 2$ and the absolute minimum is -4 at
$x = 0.5$

21. $h(x) = \sqrt[3]{x} = x^{1/3} \Rightarrow h'(x) = \frac{1}{3} x^{-2/3} \Rightarrow$ a critical point
at $x = 0$; $h(-1) = -1$, $h(0) = 0$, $h(8) = 2 \Rightarrow$ the absolute
maximum is 2 at $x = 8$ and the absolute minimum is -1
at $x = -1$

23. $g(x) = \sqrt{4 - x^2} = (4 - x^2)^{1/2}$
$\Rightarrow g'(x) = \frac{1}{2}(4 - x^2)^{-1/2}(-2x) = \frac{-x}{\sqrt{4 - x^2}}$
$\Rightarrow$ critical points at $x = -2$ and $x = 0$, but not at $x = 2$
because 2 is not in the domain; $g(-2) = 0$, $g(0) = 2$,
$g(1) = \sqrt{3} \Rightarrow$ the absolute maximum is 2 at $x = 0$ and the
absolute minimum is 0 at $x = -2$

25. $f(\theta) = \sin \theta \Rightarrow f'(\theta) = \cos \theta \Rightarrow \theta = \frac{\pi}{2}$ is a critical point,
but $\theta = \frac{-\pi}{2}$ is not a critical point because $\frac{-\pi}{2}$ is not interior to
the domain; $f\left(\frac{-\pi}{2}\right) = -1$, $f\left(\frac{\pi}{2}\right) = 1$, $f\left(\frac{5\pi}{6}\right) = \frac{1}{2}$
$\Rightarrow$ the absolute maximum is 1 at $\theta = \frac{\pi}{2}$ and the absolute
minimum is -1 at $\theta = \frac{-\pi}{2}$

27. $g(x) = \csc x \Rightarrow g'(x) = -(\csc x)(\cot x) \Rightarrow$ a critical point
at $x = \frac{\pi}{2}$; $g\left(\frac{\pi}{3}\right) = \frac{2}{\sqrt{3}}$, $g\left(\frac{\pi}{2}\right) = 1$, $g\left(\frac{2\pi}{3}\right) = \frac{2}{\sqrt{3}} \Rightarrow$ the
absolute maximum is $\frac{2}{\sqrt{3}}$ at $x = \frac{\pi}{3}$ and $x = \frac{2\pi}{3}$, and the
absolute minimum is 1 at $x = \frac{\pi}{2}$

29. $f(t) = 2 - |t| = 2 - \sqrt{t^2} = 2 - (t^2)^{1/2}$
$\Rightarrow f'(t) = -\frac{1}{2}(t^2)^{-1/2}(2t) = -\frac{t}{\sqrt{t^2}} = -\frac{t}{|t|}$
$\Rightarrow$ a critical point at $t = 0$; $f(-1) = 1$,
$f(0) = 2$, $f(3) = -1 \Rightarrow$ the absolute maximum is 2 at $t = 0$
and the absolute minimum is -1 at $t = 3$

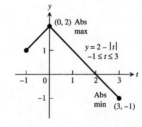

31. $g(x) = xe^{-x} \Rightarrow g'(x) = e^{-x} - xe^{-x}$
    $\Rightarrow$ a critical point at $x = 1$; $g(-1) = -e$, and $g(1) = \frac{1}{e}$,
    $\Rightarrow$ the absolute maximum is $\frac{1}{e}$ at $x = 1$ and the absolute
    minimum is $-e$ at $x = -1$

33. The first derivative $f'(x) = -\frac{1}{x^2} + \frac{1}{x}$ has a zero at $x = 1$.
    Critical point value: $f(1) = 1 + \ln 1 = 1$
    Endpoint values: $f(0.5) = 2 + \ln 0.5 \approx 1.307$;
    $\qquad\qquad f(4) = \frac{1}{4} + \ln 4 \approx 1.636$;
    Absolute maximum value is $\frac{1}{4} + \ln 4$ at $x = 4$;
    Absolute Minimum value is 1 at $x = 1$;
    Local maximum at $\left(\frac{1}{2}, 2 - \ln 2\right)$

35. $f(x) = x^{4/3} \Rightarrow f'(x) = \frac{4}{3}x^{1/3} \Rightarrow$ a critical point at $x = 0$; $f(-1) = 1$, $f(0) = 0$, $f(8) = 16 \Rightarrow$ the absolute
    maximum is 16 at $x = 8$ and the absolute minimum is 0 at $x = 0$

37. $g(\theta) = \theta^{3/5} \Rightarrow g'(\theta) = \frac{3}{5}\theta^{-2/5} \Rightarrow$ a critical point at $\theta = 0$; $g(-32) = -8$, $g(0) = 0$, $g(1) = 1 \Rightarrow$ the absolute
    maximum is 1 at $\theta = 1$ and the absolute minimum is -8 at $\theta = -32$

39. Minimum value is 1 at $x = 2$.

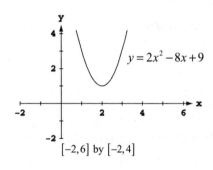

41. To find the exact values, note that that $y' = 3x^2 + 2x - 8$
    $= (3x - 4)(x + 2)$, which is zero when $x = -2$ or $x = \frac{4}{3}$.
    Local maximum at $(-1, 17)$; local minimum at $\left(\frac{4}{3}, -\frac{41}{27}\right)$

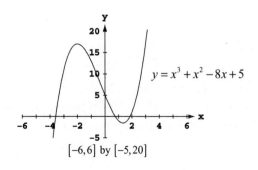

43. Minimum value is 0 when x = −1 or x = 1.

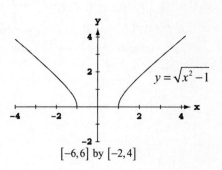

$y = \sqrt{x^2 - 1}$

$[-6, 6]$ by $[-2, 4]$

45. The actual graph of the function has asymptotes at $x = \pm 1$, so there are no extrema near these values. (This is an example of grapher failure.) There is a local minimum at (0, 1).

$y = \dfrac{1}{\sqrt[3]{1 - x^2}}$

$[-4.7, 4.7]$ by $[-3.1, 3.1]$

47. Maximum value is $\frac{1}{2}$ at x = 1; minimum value is $-\frac{1}{2}$ as x = −1.

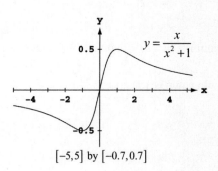

$y = \dfrac{x}{x^2 + 1}$

$[-5, 5]$ by $[-0.7, 0.7]$

49. $y = e^x + e^{-x} \Rightarrow y' = e^x - e^{-x} = e^x - \frac{1}{e^x} = \frac{e^{2x} - 1}{e^x}$, which is 0 at x = 0; an absolute minimum value is 2 at x = 0.

$y = e^x + e^{-x}$

51. $y = x \ln x \Rightarrow y' = x \cdot \frac{1}{x} + (1) \cdot \ln x = 1 + \ln x$, which is zero at $x = e^{-1}$; an absolute minimum value is $-\frac{1}{e}$ at $x = \frac{1}{e}$

$y = x \ln x$

53. $y = \cos^{-1}(x^2) \Rightarrow y' = \frac{-1}{\sqrt{1-(x^2)^2}} \cdot (2x) = \frac{-2x}{\sqrt{1-x^4}}$,

which is zero at $x = 0$; an absolute maximum value is $\frac{\pi}{2}$ at $x = 0$; an absolute minimum value is 0 at $x = 1$ and $x = -1$.

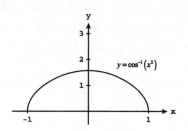

55. $y' = x^{2/3}(1) + \frac{2}{3}x^{-1/3}(x + 2) = \frac{5x+4}{3\sqrt[3]{x}}$

crit. pt.	derivative	extremum	value
$x = -\frac{4}{5}$	0	local max	$\frac{12}{25}10^{1/3} = 1.034$
$x = 0$	undefined	local min	0

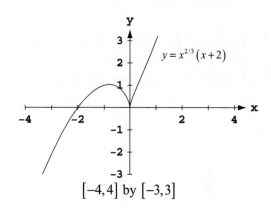

$[-4,4]$ by $[-3,3]$

57. $y' = x\frac{1}{2\sqrt{4-x^2}}(-2x) + (1)\sqrt{4-x^2}$

$= \frac{-x^2+(4-x^2)}{\sqrt{4-x^2}} = \frac{4-2x^2}{\sqrt{4-x^2}}$

crit. pt.	derivative	extremum	value
$x = -2$	undefined	local max	0
$x = -\sqrt{2}$	0	minimum	-2
$x = \sqrt{2}$	0	maximum	2
$x = 2$	undefined	local min	0

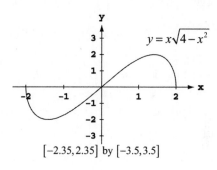

$[-2.35, 2.35]$ by $[-3.5, 3.5]$

59. $y' = \begin{cases} -2, & x < 1 \\ 1, & x > 1 \end{cases}$

crit. pt.	derivative	extremum	value
$x = 1$	undefined	minimum	2

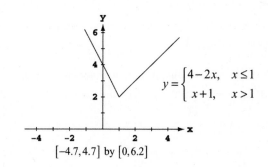

$[-4.7, 4.7]$ by $[0, 6.2]$

61. $y' = \begin{cases} -2x - 2, & x < 1 \\ -2x + 6, & x > 1 \end{cases}$

crit. pt.	derivative	extremum	value
$x = -1$	0	maximum	5
$x = 1$	undefined	local min	1
$x = 3$	0	maximum	5

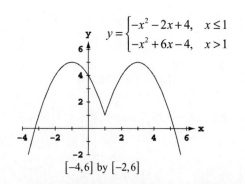

$[-4, 6]$ by $[-2, 6]$

63. (a) No, since $f'(x) = \frac{2}{3}(x-2)^{-1/3}$, which is undefined at $x = 2$.

    (b) The derivative is defined and nonzero for all $x \neq 2$. Also, $f(2) = 0$ and $f(x) > 0$ for all $x \neq 2$.

    (c) No, $f(x)$ need not have a global maximum because its domain is all real numbers. Any restriction of f to a closed interval of the form $[a, b]$ would have both a maximum value and minimum value on the interval.

    (d) The answers are the same as (a) and (b) with 2 replaced by a.

65. Yes, since $f(x) = |x| = \sqrt{x^2} = (x^2)^{1/2} \Rightarrow f'(x) = \frac{1}{2}(x^2)^{-1/2}(2x) = \frac{x}{(x^2)^{1/2}} = \frac{x}{|x|}$ is not defined at $x = 0$. Thus it is not required that f' be zero at a local extreme point since f' may be undefined there.

67. If $g(c)$ is a local minimum value of g, then $g(x) \geq g(c)$ for all x in some open interval (a, b) containing c. Since g is odd, $g(-x) = -g(x) \leq -g(c) = g(-c)$ for all $-x$ in the open interval $(-b, -a)$ containing $-c$. That is, g assumes a local maximum at the point $-c$. This is also clear from the graph of g because the graph of an odd function is symmetric about the origin.

69. (a) $V(x) = 160x - 52x^2 + 4x^3$

    $V'(x) = 160 - 104x + 12x^2 = 4(x-2)(3x-20)$

    The only critical point in the interval $(0, 5)$ is at $x = 2$. The maximum value of $V(x)$ is 144 at $x = 2$.

    (b) The largest possible volume of the box is 144 cubic units, and it occurs when $x = 2$ units.

71. $s = -\frac{1}{2}gt^2 + v_0t + s_0 \Rightarrow \frac{ds}{dt} = -gt + v_0 = 0 \Rightarrow t = \frac{v_0}{g}$. Now $s(t) = s_0 \Leftrightarrow t\left(-\frac{gt}{2} + v_0\right) = 0 \Leftrightarrow t = 0$ or $t = \frac{2v_0}{g}$.

    Thus $s\left(\frac{v_0}{g}\right) = -\frac{1}{2}g\left(\frac{v_0}{g}\right)^2 + v_0\left(\frac{v_0}{g}\right) + s_0 = \frac{v_0^2}{2g} + s_0 > s_0$ is the <u>maximum</u> height over the interval $0 \leq t \leq \frac{2v_0}{g}$.

73. Maximum value is 11 at $x = 5$;

    minimum value is 5 on the interval $[-3, 2]$;

    local maximum at $(-5, 9)$

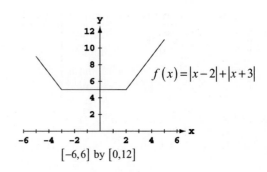

$f(x) = |x-2| + |x+3|$

$[-6, 6]$ by $[0, 12]$

75. Maximum value is 5 on the interval $[3, \infty)$;

    minimum value is -5 on the interval $(-\infty, -2]$.

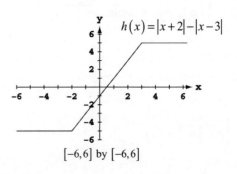

$h(x) = |x+2| - |x-3|$

$[-6, 6]$ by $[-6, 6]$

4.2 THE MEAN VALUE THEOREM

1. When $f(x) = x^2 + 2x - 1$ for $0 \leq x \leq 1$, then $\frac{f(1) - f(0)}{1 - 0} = f'(c) \Rightarrow 3 = 2c + 2 \Rightarrow c = \frac{1}{2}$.

3. When $f(x) = \sin^{-1}(x)$ for $-1 \le x \le 1$, then $f'(c) = \frac{f(1)-f(-1)}{1-(-1)} \Rightarrow \frac{1}{\sqrt{1-c^2}} = \frac{\frac{\pi}{2}-(-\frac{\pi}{2})}{2} \Rightarrow \sqrt{1-c^2} = \frac{2}{\pi} \Rightarrow 1-c^2 = \frac{4}{\pi^2}$

   $\Rightarrow c^2 = 1 - \frac{4}{\pi^2} \Rightarrow c = \pm\sqrt{1-\frac{4}{\pi^2}} \approx \pm 0.771$

5. Does not; $f(x)$ is not differentiable at $x = 0$ in $(-1, 8)$.

7. Does; $f(x)$ is continuous for every point of $[0, 1]$ and differentiable for every point in $(0, 1)$.

9. Since $f(x)$ is not continuous on $0 \le x \le 1$, Rolle's Theorem does not apply: $\lim\limits_{x \to 1^-} f(x) = \lim\limits_{x \to 1^-} x = 1$

   $\ne 0 = f(1)$.

11. (a) i

   ii

   iii

   iv

   (b) Let r_1 and r_2 be zeros of the polynomial $P(x) = x^n + a_{n-1}x^{n-1} + \ldots + a_1 x + a_0$, then $P(r_1) = P(r_2) = 0$. Since polynomials are everywhere continuous and differentiable, by Rolle's Theorem $P'(r) = 0$ for some r between r_1 and r_2, where $P'(x) = nx^{n-1} + (n-1)a_{n-1}x^{n-2} + \ldots + a_1$.

13. Since f'' exists throughout $[a, b]$ the derivative function f' is continuous there. If f' has more than one zero in $[a, b]$, say $f'(r_1) = f'(r_2) = 0$ for $r_1 \ne r_2$, then by Rolle's Theorem there is a c between r_1 and r_2 such that $f''(c) = 0$, contrary to $f'' > 0$ throughout $[a, b]$. Therefore f' has at most one zero in $[a, b]$. The same argument holds if $f'' < 0$ throughout $[a, b]$.

15. With $f(-2) = 11 > 0$ and $f(-1) = -1 < 0$ we conclude from the Intermediate Value Theorem that $f(x) = x^4 + 3x + 1$ has at least one zero between -2 and -1. Then $-2 < x < -1 \Rightarrow -8 < x^3 < -1$
   $\Rightarrow -32 < 4x^3 < -4 \Rightarrow -29 < 4x^3 + 3 < -1 \Rightarrow f'(x) < 0$ for $-2 < x < -1 \Rightarrow f(x)$ is decreasing on $[-2, -1]$
   $\Rightarrow f(x) = 0$ has exactly one solution in the interval $(-2, -1)$.

17. $g(t) = \sqrt{t} + \sqrt{t+1} - 4 \Rightarrow g'(t) = \frac{1}{2\sqrt{t}} + \frac{1}{2\sqrt{t+1}} > 0 \Rightarrow g(t)$ is increasing for t in $(0, \infty)$; $g(3) = \sqrt{3} - 2 < 0$
   and $g(15) = \sqrt{15} > 0 \Rightarrow g(t)$ has exactly one zero in $(0, \infty)$.

19. $r(\theta) = \theta + \sin^2\left(\frac{\theta}{3}\right) - 8 \Rightarrow r'(\theta) = 1 + \frac{2}{3}\sin\left(\frac{\theta}{3}\right)\cos\left(\frac{\theta}{3}\right) = 1 + \frac{1}{3}\sin\left(\frac{2\theta}{3}\right) > 0$ on $(-\infty, \infty) \Rightarrow r(\theta)$ is increasing on $(-\infty, \infty)$; $r(0) = -8$ and $r(8) = \sin^2\left(\frac{8}{3}\right) > 0 \Rightarrow r(\theta)$ has exactly one zero in $(-\infty, \infty)$.

21. $r(\theta) = \sec\theta - \frac{1}{\theta^3} + 5 \Rightarrow r'(\theta) = (\sec\theta)(\tan\theta) + \frac{3}{\theta^4} > 0$ on $\left(0, \frac{\pi}{2}\right) \Rightarrow r(\theta)$ is increasing on $\left(0, \frac{\pi}{2}\right)$;
   $r(0.1) \approx -994$ and $r(1.57) \approx 1260.5 \Rightarrow r(\theta)$ has exactly one zero in $\left(0, \frac{\pi}{2}\right)$.

23. By Corollary 1, $f'(x) = 0$ for all $x \Rightarrow f(x) = C$, where C is a constant. Since $f(-1) = 3$ we have $C = 3$
   $\Rightarrow f(x) = 3$ for all x.

25. $g(x) = x^2 \Rightarrow g'(x) = 2x = f'(x)$ for all x. By Corollary 2, $f(x) = g(x) + C$.
   (a) $f(0) = 0 \Rightarrow 0 = g(0) + C = 0 + C \Rightarrow C = 0 \Rightarrow f(x) = x^2 \Rightarrow f(2) = 4$
   (b) $f(1) = 0 \Rightarrow 0 = g(1) + C = 1 + C \Rightarrow C = -1 \Rightarrow f(x) = x^2 - 1 \Rightarrow f(2) = 3$
   (c) $f(-2) = 3 \Rightarrow 3 = g(-2) + C \Rightarrow 3 = 4 + C \Rightarrow C = -1 \Rightarrow f(x) = x^2 - 1 \Rightarrow f(2) = 3$

27. (a) $y = \frac{x^2}{2} + C$ (b) $y = \frac{x^3}{3} + C$ (c) $y = \frac{x^4}{4} + C$

29. (a) $y = \ln x + C$ if $x > 0$ and $y = \ln(-x) + C$ if $x < 0$, where C is a constant. (These functions can be combined as $y = \ln |x| + C$.)

   (b) $y = x - \ln x + C$ if $x > 0$ and $y = x - \ln(-x) + C$ if $x < 0$, where C is a constant. (These functions can be combined as $y = x - \ln |x| + C$.)

   (c) $y = 5x + \ln x + C$ if $x > 0$ and $y = 5x + \ln(-x) + C$ if $x < 0$, where C is a constant. (These functions can be combined as $y = 5x + \ln |x| + C$.)

31. (a) $y = -\frac{1}{2} \cos 2t + C$ (b) $y = 2 \sin \frac{t}{2} + C$

   (c) $y = -\frac{1}{2} \cos 2t + 2 \sin \frac{t}{2} + C$

33. $f(x) = x^2 - x + C; 0 = f(0) = 0^2 - 0 + C \Rightarrow C = 0 \Rightarrow f(x) = x^2 - x$

35. $f(x) = \frac{e^{2x}}{2} + C; f(0) = \frac{3}{2} \Rightarrow \frac{e^{2(0)}}{2} + C = \frac{3}{2} \Rightarrow C = 1 \Rightarrow f(x) = 1 + \frac{e^{2x}}{2}$

37. $v = \frac{ds}{dt} = 9.8t + 5 \Rightarrow s = 4.9t^2 + 5t + C$; at $s = 10$ and $t = 0$ we have $C = 10 \Rightarrow s = 4.9t^2 + 5t + 10$

39. $v = \frac{ds}{dt} = \sin(\pi t) \Rightarrow s = -\frac{1}{\pi} \cos(\pi t) + C$; at $s = 0$ and $t = 0$ we have $C = \frac{1}{\pi} \Rightarrow s = \frac{1 - \cos(\pi t)}{\pi}$

41. $a = \frac{dv}{dt} = e^t \Rightarrow v = e^t + C$; at $v = 20$ and $t = 0$ we have $C = 19 \Rightarrow v = e^t + 19$

   $v = \frac{ds}{dt} = e^t + 19 \Rightarrow s = e^t + 19t + C$; at $s = 5$ and $t = 0$ we have $C = 4 \Rightarrow s = e^t + 19t + 4$

43. $a = -4\sin(2t) \Rightarrow v = 2\cos(2t) + C_1$; at $v = 2$ and $t = 0$ we have $C_1 = 0 \Rightarrow v = 2\cos(2t) \Rightarrow s = \sin(2t) + C_2$; at $s = -3$ and $t = 0$ we have $C_2 = -3 \Rightarrow s = \sin(2t) - 3$

45. If T(t) is the temperature of the thermometer at time t, then $T(0) = -19°$ C and $T(14) = 100°$ C. From the Mean Value Theorem there exists a $0 < t_0 < 14$ such that $\frac{T(14) - T(0)}{14 - 0} = 8.5°$ C/sec $= T'(t_0)$, the rate at which the temperature was changing at $t = t_0$ as measured by the rising mercury on the thermometer.

47. Because its average speed was approximately 7.667 knots, and by the Mean Value Theorem, it must have been going that speed at least once during the trip.

49. Let d(t) represent the distance the automobile traveled in time t. The average speed over $0 \leq t \leq 2$ is $\frac{d(2) - d(0)}{2 - 0}$. The Mean Value Theorem says that for some $0 < t_0 < 2$, $d'(t_0) = \frac{d(2) - d(0)}{2 - 0}$. The value $d'(t_0)$ is the speed of the automobile at time t_0 (which is read on the speedometer).

51. The conclusion of the Mean Value Theorem yields $\frac{\frac{1}{b} - \frac{1}{a}}{b - a} = -\frac{1}{c^2} \Rightarrow c^2 \left(\frac{a - b}{ab} \right) = a - b \Rightarrow c = \sqrt{ab}$.

53. $f'(x) = [\cos x \sin (x + 2) + \sin x \cos (x + 2)] - 2 \sin (x + 1) \cos (x + 1) = \sin (x + x + 2) - \sin 2(x + 1)$
   $= \sin (2x + 2) - \sin (2x + 2) = 0$. Therefore, the function has the constant value $f(0) = -\sin^2 1 \approx -0.7081$ which explains why the graph is a horizontal line.

55. f(x) must be zero at least once between a and b by the Intermediate Value Theorem. Now suppose that f(x) is zero twice between a and b. Then by the Mean Value Theorem, $f'(x)$ would have to be zero at least once between the two zeros of f(x), but this can't be true since we are given that $f'(x) \neq 0$ on this interval. Therefore, f(x) is zero once and only once between a and b.

57. Yes. By Corollary 2 we have $f(x) = g(x) + c$ since $f'(x) = g'(x)$. If the graphs start at the same point $x = a$, then $f(a) = g(a) \Rightarrow c = 0 \Rightarrow f(x) = g(x)$.

59. By the Mean Value Theorem we have $\frac{f(b) - f(a)}{b - a} = f'(c)$ for some point c between a and b. Since $b - a > 0$ and $f(b) < f(a)$, we have $f(b) - f(a) < 0 \Rightarrow f'(c) < 0$.

61. $f'(x) = (1 + x^4 \cos x)^{-1} \Rightarrow f''(x) = -(1 + x^4 \cos x)^{-2}(4x^3 \cos x - x^4 \sin x)$
    $= -x^3(1 + x^4 \cos x)^{-2}(4 \cos x - x \sin x) < 0$ for $0 \le x \le 0.1 \Rightarrow f'(x)$ is decreasing when $0 \le x \le 0.1$
    $\Rightarrow \min f' \approx 0.9999$ and $\max f' = 1$. Now we have $0.9999 \le \frac{f(0.1) - 1}{0.1} \le 1 \Rightarrow 0.09999 \le f(0.1) - 1 \le 0.1$
    $\Rightarrow 1.09999 \le f(0.1) \le 1.1$.

63. (a) Suppose $x < 1$, then by the Mean Value Theorem $\frac{f(x) - f(1)}{x - 1} < 0 \Rightarrow f(x) > f(1)$. Suppose $x > 1$, then by the
    Mean Value Theorem $\frac{f(x) - f(1)}{x - 1} > 0 \Rightarrow f(x) > f(1)$. Therefore $f(x) \ge 1$ for all x since $f(1) = 1$.
    (b) Yes. From part (a), $\lim_{x \to 1^-} \frac{f(x) - f(1)}{x - 1} \le 0$ and $\lim_{x \to 1^+} \frac{f(x) - f(1)}{x - 1} \ge 0$. Since $f'(1)$ exists, these two one-sided
    limits are equal and have the value $f'(1) \Rightarrow f'(1) \le 0$ and $f'(1) \ge 0 \Rightarrow f'(1) = 0$.

65. Proof that $\ln\left(\frac{b}{x}\right) = \ln b - \ln x$: Note that $\frac{d}{dx}\ln\left(\frac{b}{x}\right) = \frac{1}{b/x} \cdot \frac{-b}{x^2} = \frac{-1}{x}$ and $\frac{d}{dx}(\ln b - \ln x) = \frac{-1}{x}$; by Corollary 2 of the Mean
    Value Theorem there is a constant C so that $\ln\left(\frac{b}{x}\right) = \ln b - \ln x + C$; if $x = b$, then $\ln 1 = \ln b - \ln b + C \Rightarrow C = 0$
    $\Rightarrow \ln\left(\frac{b}{x}\right) = \ln b - \ln x$.

67. $e^x e^{-x} = e^{(x-x)} = e^0 = 1 \Rightarrow e^{-x} = \frac{1}{e^x}$ for all x; $\frac{e^{x_1}}{e^{x_2}} = e^{x_1}\left(\frac{1}{e^{x_2}}\right) = e^{x_1}e^{-x_2} = e^{x_1 - x_2}$

4.3 MONOTONIC FUNCTIONS AND THE FIRST DERIVATIVE TEST

1. (a) $f'(x) = x(x - 1) \Rightarrow$ critical points at 0 and 1
   (b) $f' = +++ \mid --- \mid +++ \Rightarrow$ increasing on $(-\infty, 0)$ and $(1, \infty)$, decreasing on $(0, 1)$
       $ 0 1$
   (c) Local maximum at $x = 0$ and a local minimum at $x = 1$

3. (a) $f'(x) = (x - 1)^2(x + 2) \Rightarrow$ critical points at -2 and 1
   (b) $f' = --- \mid +++ \mid +++ \Rightarrow$ increasing on $(-2, 1)$ and $(1, \infty)$, decreasing on $(-\infty, -2)$
       $ -2 1$
   (c) No local maximum and a local minimum at $x = -2$

5. (a) $f'(x) = (x - 1)e^{-x} \Rightarrow$ critical point at $x = 1$
   (b) $f' = ---- \mid ++++ \Rightarrow$ decreasing on $(-\infty, 1]$, increasing on $[1, \infty)$
       $ 1$
   (c) Local (and absolute) minimum at $x = 1$

7. (a) $f'(x) = x^{-1/3}(x + 2) \Rightarrow$ critical points at -2 and 0
   (b) $f' = +++ \mid ---)(+++ \Rightarrow$ increasing on $(-\infty, -2)$ and $(0, \infty)$, decreasing on $(-2, 0)$
       $ -2 0$
   (c) Local maximum at $x = -2$, local minimum at $x = 0$

9. (a) $g(t) = -t^2 - 3t + 3 \Rightarrow g'(t) = -2t - 3 \Rightarrow$ a critical point at $t = -\frac{3}{2}$; $g' = +++ \mid ---$, increasing on
       $ -3/2$
       $\left(-\infty, -\frac{3}{2}\right)$, decreasing on $\left(-\frac{3}{2}, \infty\right)$
   (b) local maximum value of $g\left(-\frac{3}{2}\right) = \frac{21}{4}$ at $t = -\frac{3}{2}$
   (c) absolute maximum is $\frac{21}{4}$ at $t = -\frac{3}{2}$

(d)

11. (a) $h(x) = -x^3 + 2x^2 \Rightarrow h'(x) = -3x^2 + 4x = x(4 - 3x) \Rightarrow$ critical points at $x = 0, \frac{4}{3}$

$\Rightarrow h' = --- \mid +++ \mid ---$, increasing on $\left(0, \frac{4}{3}\right)$, decreasing on $(-\infty, 0)$ and $\left(\frac{4}{3}, \infty\right)$
$\quad\quad\quad\quad\; 0 \quad\; 4/3$

(b) local maximum value of $h\left(\frac{4}{3}\right) = \frac{32}{27}$ at $x = \frac{4}{3}$; local minimum value of $h(0) = 0$ at $x = 0$

(c) no absolute extrema

(d)

13. (a) $f(\theta) = 3\theta^2 - 4\theta^3 \Rightarrow f'(\theta) = 6\theta - 12\theta^2 = 6\theta(1 - 2\theta) \Rightarrow$ critical points at $\theta = 0, \frac{1}{2} \Rightarrow f' = --- \mid +++ \mid ---$,
$\quad\quad\quad\quad\quad\quad\quad\quad\quad\quad\quad\quad\quad\quad\quad\quad\quad\quad\quad\quad\quad\quad\quad\quad\quad\quad\quad\quad\quad\quad\quad\quad\quad\quad\quad\quad\quad\quad 0 \quad\; 1/2$

increasing on $\left(0, \frac{1}{2}\right)$, decreasing on $(-\infty, 0)$ and $\left(\frac{1}{2}, \infty\right)$

(b) a local maximum is $f\left(\frac{1}{2}\right) = \frac{1}{4}$ at $\theta = \frac{1}{2}$, a local minimum is $f(0) = 0$ at $\theta = 0$

(c) no absolute extrema

(d)

15. (a) $f(r) = 3r^3 + 16r \Rightarrow f'(r) = 9r^2 + 16 \Rightarrow$ no critical points $\Rightarrow f' = +++++$, increasing on $(-\infty, \infty)$, never decreasing

(b) no local extrema

(c) no absolute extrema

(d)

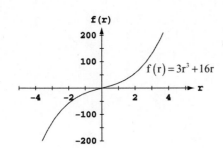

17. (a) $f(x) = x^4 - 8x^2 + 16 \Rightarrow f'(x) = 4x^3 - 16x = 4x(x+2)(x-2) \Rightarrow$ critical points at $x = 0$ and $x = \pm 2$
    $\Rightarrow f' = --- \mid \underset{-2}{+++} \mid \underset{0}{---} \mid \underset{2}{+++}$, increasing on $(-2, 0)$ and $(2, \infty)$, decreasing on $(-\infty, -2)$ and $(0, 2)$

    (b) a local maximum is $f(0) = 16$ at $x = 0$, local minima are $f(\pm 2) = 0$ at $x = \pm 2$

    (c) no absolute maximum; absolute minimum is 0 at $x = \pm 2$

    (d)

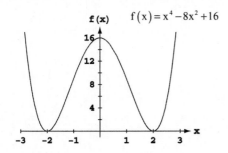

19. (a) $H(t) = \frac{3}{2}t^4 - t^6 \Rightarrow H'(t) = 6t^3 - 6t^5 = 6t^3(1+t)(1-t) \Rightarrow$ critical points at $t = 0, \pm 1$
    $\Rightarrow H' = +++ \mid \underset{-1}{---} \mid \underset{0}{+++} \mid \underset{1}{---}$, increasing on $(-\infty, -1)$ and $(0, 1)$, decreasing on $(-1, 0)$ and $(1, \infty)$

    (b) the local maxima are $H(-1) = \frac{1}{2}$ at $t = -1$ and $H(1) = \frac{1}{2}$ at $t = 1$, the local minimum is $H(0) = 0$ at $t = 0$

    (c) absolute maximum is $\frac{1}{2}$ at $t = \pm 1$; no absolute minimum

    (d)

21. (a) $g(x) = x\sqrt{8 - x^2} = x(8 - x^2)^{1/2} \Rightarrow g'(x) = (8 - x^2)^{1/2} + x\left(\frac{1}{2}\right)(8 - x^2)^{-1/2}(-2x) = \dfrac{2(2-x)(2+x)}{\sqrt{\left(2\sqrt{2}-x\right)\left(2\sqrt{2}+x\right)}}$

    $\Rightarrow$ critical points at $x = \pm 2, \pm 2\sqrt{2} \Rightarrow g' = (\underset{-2\sqrt{2}}{\quad} --- \mid \underset{-2}{+++} \mid \underset{2}{---}\underset{2\sqrt{2}}{)}$, increasing on $(-2, 2)$, decreasing on

    $\left(-2\sqrt{2}, -2\right)$ and $\left(2, 2\sqrt{2}\right)$

    (b) local maxima are $g(2) = 4$ at $x = 2$ and $g\left(-2\sqrt{2}\right) = 0$ at $x = -2\sqrt{2}$, local minima are $g(-2) = -4$ at

    $x = -2$ and $g\left(2\sqrt{2}\right) = 0$ at $x = 2\sqrt{2}$

    (c) absolute maximum is 4 at $x = 2$; absolute minimum is -4 at $x = -2$

(d)

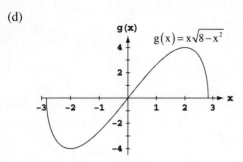

23. (a) $f(x) = \frac{x^2-3}{x-2} \Rightarrow f'(x) = \frac{2x(x-2)-(x^2-3)\,(1)}{(x-2)^2} = \frac{(x-3)(x-1)}{(x-2)^2} \Rightarrow$ critical points at $x = 1, 3$

$\Rightarrow f' = +++ \mid {-}{-}{-})({-}{-}{-} \mid +++$, increasing on $(-\infty, 1)$ and $(3, \infty)$, decreasing on $(1, 2)$ and $(2, 3)$,
$\qquad\qquad\quad 1 \qquad 2 \qquad 3$

discontinuous at $x = 2$

(b) a local maximum is $f(1) = 2$ at $x = 1$, a local minimum is $f(3) = 6$ at $x = 3$

(c) no absolute extrema

(d)

25. (a) $f(x) = x^{1/3}(x+8) = x^{4/3} + 8x^{1/3} \Rightarrow f'(x) = \frac{4}{3}x^{1/3} + \frac{8}{3}x^{-2/3} = \frac{4(x+2)}{3x^{2/3}} \Rightarrow$ critical points at $x = 0, -2$

$\Rightarrow f' = {-}{-}{-} \mid +++)(+++$, increasing on $(-2, 0) \cup (0, \infty)$, decreasing on $(-\infty, -2)$
$\qquad\qquad\quad -2 \qquad 0$

(b) no local maximum, a local minimum is $f(-2) = -6\sqrt[3]{2} \approx -7.56$ at $x = -2$

(c) no absolute maximum; absolute minimum is $-6\sqrt[3]{2}$ at $x = -2$

(d)

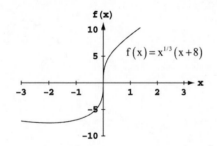

27. (a) $h(x) = x^{1/3}(x^2 - 4) = x^{7/3} - 4x^{1/3} \Rightarrow h'(x) = \frac{7}{3}x^{4/3} - \frac{4}{3}x^{-2/3} = \frac{\left(\sqrt{7}x+2\right)\left(\sqrt{7}x-2\right)}{3\sqrt[3]{x^2}} \Rightarrow$ critical points at

$x = 0, \frac{\pm 2}{\sqrt{7}} \Rightarrow h' = +++ \mid {-}{-}{-})({-}{-}{-} \mid +++$, increasing on $\left(-\infty, \frac{-2}{\sqrt{7}}\right)$ and $\left(\frac{2}{\sqrt{7}}, \infty\right)$, decreasing on
$\qquad\qquad\qquad\qquad -2/\sqrt{7} \qquad 0 \qquad 2/\sqrt{7}$

$\left(\frac{-2}{\sqrt{7}}, 0\right)$ and $\left(0, \frac{2}{\sqrt{7}}\right)$

(b) local maximum is $h\left(\frac{-2}{\sqrt{7}}\right) = \frac{24\sqrt[3]{2}}{7^{7/6}} \approx 3.12$ at $x = \frac{-2}{\sqrt{7}}$, the local minimum is $h\left(\frac{2}{\sqrt{7}}\right) = -\frac{24\sqrt[3]{2}}{7^{7/6}} \approx -3.12$

(c) no absolute extrema

(d)

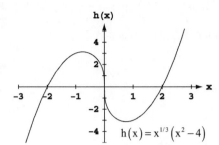

$h(x) = x^{1/3}(x^2 - 4)$

29. (a) $f(x) = e^{2x} + e^{-x} \Rightarrow f'(x) = 2e^{2x} - e^{-x} = 0 \Rightarrow e^{3x} = \frac{1}{2} \Rightarrow$ a critical point at $x = \frac{1}{3}\ln\left(\frac{1}{2}\right)$

$\Rightarrow f' = ----\left|++++\right.$, increasing on $\left(\frac{1}{3}\ln\left(\frac{1}{2}\right), \infty\right)$, decreasing on $\left(-\infty, \frac{1}{3}\ln\left(\frac{1}{2}\right)\right)$
$\qquad\qquad\quad \frac{1}{3}\ln\left(\frac{1}{2}\right)$

(b) a local minimum is $\frac{3}{2^{2/3}}$ at $x = \frac{1}{3}\ln\left(\frac{1}{2}\right)$; no local maximum

(c) an absolute minimum $\frac{3}{2^{2/3}}$ at $x = \frac{1}{3}\ln\left(\frac{1}{2}\right)$; no absolute maximum

(d)

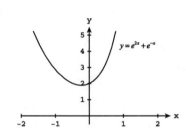

$y = e^{2x} + e^{-x}$

31. (a) $f(x) = x\ln x \Rightarrow f'(x) = 1 + \ln x \Rightarrow$ a critical point at $x = e^{-1} \Rightarrow f' = [\ ---\ |+++$, increasing on (e^{-1}, ∞),
$\qquad\qquad\qquad\qquad\qquad\qquad\qquad\qquad\qquad\qquad\qquad\qquad\qquad 0 \qquad e^{-1}$

decreasing on $(0, e^{-1})$

(b) A local minimum is $-e^{-1}$ at $x = e^{-1}$, no local maximum

(c) An absolute minimum is $-e^{-1}$ at $x = e^{-1}$, no absolute maximum

(d)

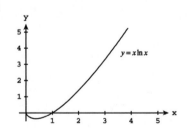

$y = x\ln x$

33. (a) $f(x) = 2x - x^2 \Rightarrow f'(x) = 2 - 2x \Rightarrow$ a critical point at $x = 1 \Rightarrow f' = +++|---]$ and $f(1) = 1$ and $f(2) = 0$
$\qquad\qquad\qquad\qquad\qquad\qquad\qquad\qquad\qquad\qquad\qquad\qquad\qquad\quad 1 \qquad 2$

a local maximum is 1 at $x = 1$, a local minimum is 0 at $x = 2$.

(b) There is an absolute maximum of 1 at $x = 1$; no absolute minimum.

(c)

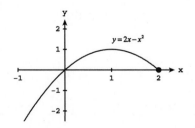

$y = 2x - x^2$

35. (a) $g(x) = x^2 - 4x + 4 \Rightarrow g'(x) = 2x - 4 = 2(x - 2) \Rightarrow$ a critical point at $x = 2 \Rightarrow g' = [\ {---}\ |\ {+++}$ and

$\qquad\qquad\qquad\qquad\qquad\qquad\qquad\qquad\qquad\qquad\qquad\qquad\qquad\quad 1 \qquad 2$

$g(1) = 1, g(2) = 0 \Rightarrow$ a local maximum is 1 at $x = 1$, a local minimum is $g(2) = 0$ at $x = 2$

(b) no absolute maximum; absolute minimum is 0 at $x = 2$

(c)

37. (a) $f(t) = 12t - t^3 \Rightarrow f'(t) = 12 - 3t^2 = 3(2 + t)(2 - t) \Rightarrow$ critical points at $t = \pm 2 \Rightarrow f' = [\ {---}\ |\ {+++}\ |\ {---}$

$\qquad\qquad\qquad\qquad\qquad\qquad\qquad\qquad\qquad\qquad\qquad\qquad\qquad\qquad\qquad\qquad -3 \quad -2 \quad 2$

and $f(-3) = -9$, $f(-2) = -16$, $f(2) = 16 \Rightarrow$ local maxima are -9 at $t = -3$ and 16 at $t = -2$, a local

minimum is -16 at $t = -2$

(b) absolute maximum is 16 at $t = 2$; no absolute minimum

(c)

39. (a) $h(x) = \frac{x^3}{3} - 2x^2 + 4x \Rightarrow h'(x) = x^2 - 4x + 4 = (x - 2)^2 \Rightarrow$ a critical point at $x = 2 \Rightarrow h' = [\ {+++}\ |\ {+++}$ and

$\qquad\qquad\qquad\qquad\qquad\qquad\qquad\qquad\qquad\qquad\qquad\qquad\qquad\qquad\qquad\qquad\qquad\qquad\qquad 0 \qquad 2$

$h(0) = 0 \Rightarrow$ no local maximum, a local minimum is 0 at $x = 0$

(b) no absolute maximum; absolute minimum is 0 at $x = 0$

(c)

41. (a) $f(x) = \frac{x}{2} - 2\sin\left(\frac{x}{2}\right) \Rightarrow f'(x) = \frac{1}{2} - \cos\left(\frac{x}{2}\right)$, $f'(x) = 0 \Rightarrow \cos\left(\frac{x}{2}\right) = \frac{1}{2} \Rightarrow$ a critical point at $x = \frac{2\pi}{3}$

$\Rightarrow f' = [\ {---}\ |\ {+++}]$ and $f(0) = 0$, $f\left(\frac{2\pi}{3}\right) = \frac{\pi}{3} - \sqrt{3}$, $f(2\pi) = \pi \Rightarrow$ local maxima are 0 at $x = 0$ and π

$\qquad\qquad 0 \quad 2\pi/3 \quad 2\pi$

at $x = 2\pi$, a local minimum is $\frac{\pi}{3} - \sqrt{3}$ at $x = \frac{2\pi}{3}$

(b) The graph of f rises when $f' > 0$, falls when $f' < 0$, and has a local minimum value at the point where f' changes from negative to positive.

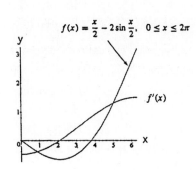

$f(x) = \frac{x}{2} - 2\sin\frac{x}{2}, \quad 0 \le x \le 2\pi$

43. (a) $f(x) = \csc^2 x - 2\cot x \Rightarrow f'(x) = 2(\csc x)(-\csc x)(\cot x) - 2(-\csc^2 x) = -2(\csc^2 x)(\cot x - 1) \Rightarrow$ a critical

point at $x = \frac{\pi}{4} \Rightarrow f' = (\; ---\; \underset{\underset{\pi/4}{|}}{} \; +++) \text{ and } f\left(\frac{\pi}{4}\right) = 0 \Rightarrow$ no local maximum, a local minimum is 0 at $x = \frac{\pi}{4}$

(b) The graph of f rises when $f' > 0$, falls when $f' < 0$, and has a local minimum value at the point where $f' = 0$ and the values of f' change from negative to positive. The graph of f steepens as $f'(x) \to \pm\infty$.

$f(x) = \csc^2 x - 2\cot x, \quad 0 < x < \pi$

45. $h(\theta) = 3\cos\left(\frac{\theta}{2}\right) \Rightarrow h'(\theta) = -\frac{3}{2}\sin\left(\frac{\theta}{2}\right) \Rightarrow h' = [\; ---\; \underset{\underset{2\pi}{]}}{\underset{0}{}} \;, (0,3) \text{ and } (2\pi, -3) \Rightarrow$ a local maximum is 3 at $\theta = 0$,

a local minimum is -3 at $\theta = 2\pi$

47. (a)

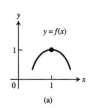

$y = f(x)$

(a)

(b)

$y = f(x)$

(b)

(c)

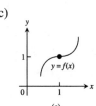

$y = f(x)$

(c)

(d)

$y = f(x)$

(d)

49. (a)

$y = g(x)$

(b)

$y = g(x)$

51. (a) $f(x) = \ln(\cos x) \Rightarrow f'(x) = -\frac{\sin x}{\cos x} = -\tan x = 0 \Rightarrow x = 0; f'(x) > 0 \text{ for } -\frac{\pi}{4} \le x < 0 \text{ and } f'(x) < 0 \text{ for }$

$0 < x \le \frac{\pi}{3} \Rightarrow$ there is a relative maximum at $x = 0$ with $f(0) = \ln(\cos 0) = \ln 1 = 0; f\left(-\frac{\pi}{4}\right) = \ln\left(\cos\left(-\frac{\pi}{4}\right)\right)$

$= \ln\left(\frac{1}{\sqrt{2}}\right) = -\frac{1}{2}\ln 2 \text{ and } f\left(\frac{\pi}{3}\right) = \ln\left(\cos\left(\frac{\pi}{3}\right)\right) = \ln\frac{1}{2} = -\ln 2$. Therefore, the absolute minimum occurs at

$x = \frac{\pi}{3}$ with $f\left(\frac{\pi}{3}\right) = -\ln 2$ and the absolute maximum occurs at $x = 0$ with $f(0) = 0$.

(b) $f(x) = \cos(\ln x) \Rightarrow f'(x) = \frac{-\sin(\ln x)}{x} = 0 \Rightarrow x = 1; f'(x) > 0$ for $\frac{1}{2} \le x < 1$ and $f'(x) < 0$ for $1 < x \le 2$

$\Rightarrow$ there is a relative maximum at $x = 1$ with $f(1) = \cos(\ln 1) = \cos 0 = 1; f\left(\frac{1}{2}\right) = \cos\left(\ln\left(\frac{1}{2}\right)\right) = \cos(-\ln 2)$

$= \cos(\ln 2)$ and $f(2) = \cos(\ln 2)$. Therefore, the absolute minimum occurs at $x = \frac{1}{2}$ and $x = 2$ with

$f\left(\frac{1}{2}\right) = f(2) = \cos(\ln 2)$, and the absolute maximum occurs at $x = 1$ with $f(1) = 1$.

53. $f(x) = e^x - 2x \Rightarrow f'(x) = e^x - 2; f'(x) = 0 \Rightarrow e^x = 2 \Rightarrow x = \ln 2; f(0) = 1$, the absolute maximum;

$f(\ln 2) = 2 - 2\ln 2 \approx 0.613706$, the absolute minimum; $f(1) = e - 2 \approx 0.71828$, a relative or local maximum since

$f''(x) = e^x$ is always positive.

55. $f(x) = x^2 \ln \frac{1}{x} \Rightarrow f'(x) = 2x \ln \frac{1}{x} + x^2\left(\frac{1}{\frac{1}{x}}\right)(-x^{-2}) = 2x \ln \frac{1}{x} - x = -x(2 \ln x + 1); f'(x) = 0 \Rightarrow x = 0$ or $\ln x = -\frac{1}{2}$.

Since $x = 0$ is not in the domain of f, $x = e^{-1/2} = \frac{1}{\sqrt{e}}$. Also, $f'(x) > 0$ for $0 < x < \frac{1}{\sqrt{e}}$ and $f'(x) < 0$ for $x > \frac{1}{\sqrt{e}}$.

Therefore, $f\left(\frac{1}{\sqrt{e}}\right) = \frac{1}{e} \ln \sqrt{e} = \frac{1}{e} \ln e^{1/2} = \frac{1}{2e} \ln e = \frac{1}{2e}$ is the absolute maximum value of f assumed at $x = \frac{1}{\sqrt{e}}$.

57. $f(x) = x^3 - 3x + 2 \Rightarrow f'(x) = 3x^2 - 3 = 3(x-1)(x+1) \Rightarrow f' = +++ | \underset{-1}{} --- | \underset{1}{} +++ \Rightarrow$ rising for $x = c = 2$ since

$f'(x) > 0$ for $x = c = 2$.

59. Let $x_1 \ne x_2$ be two numbers in the domain of an increasing function f. Then, either $x_1 < x_2$ or $x_1 > x_2$ which implies

$f(x_1) < f(x_2)$ or $f(x_1) > f(x_2)$, since $f(x)$ is increasing. In either case, $f(x_1) \ne f(x_2)$ and f is one-to-one. Similar arguments

hold if f is decreasing.

61. $f(x)$ is increasing since $x_2 > x_1 \Rightarrow 27x_2^3 > 27x_1^3; y = 27x^3 \Rightarrow x = \frac{1}{3} y^{1/3} \Rightarrow f^{-1}(x) = \frac{1}{3} x^{1/3}; \frac{df}{dx} = 81x^2$

$\Rightarrow \frac{df^{-1}}{dx} = \frac{1}{81x^2}\Big|_{\frac{1}{3}x^{1/3}} = \frac{1}{9x^{2/3}} = \frac{1}{9} x^{-2/3}$

63. $f(x)$ is decreasing since $x_2 > x_1 \Rightarrow (1-x_2)^3 < (1-x_1)^3; y = (1-x)^3 \Rightarrow x = 1 - y^{1/3} \Rightarrow f^{-1}(x) = 1 - x^{1/3};$

$\frac{df}{dx} = -3(1-x)^2 \Rightarrow \frac{df^{-1}}{dx} = \frac{1}{-3(1-x)^2}\Big|_{1-x^{1/3}} = \frac{-1}{3x^{2/3}} = -\frac{1}{3} x^{-2/3}$

4.4 CONCAVITY AND CURVE SKETCHING

1. $y = \frac{x^3}{3} - \frac{x^2}{2} - 2x + \frac{1}{3} \Rightarrow y' = x^2 - x - 2 = (x-2)(x+1) \Rightarrow y'' = 2x - 1 = 2\left(x - \frac{1}{2}\right)$. The graph is rising on

$(-\infty, -1)$ and $(2, \infty)$, falling on $(-1, 2)$, concave up on $\left(\frac{1}{2}, \infty\right)$ and concave down on $\left(-\infty, \frac{1}{2}\right)$. Consequently,

a local maximum is $\frac{3}{2}$ at $x = -1$, a local minimum is -3 at $x = 2$, and $\left(\frac{1}{2}, -\frac{3}{4}\right)$ is a point of inflection.

3. $y = \frac{3}{4}(x^2-1)^{2/3} \Rightarrow y' = \left(\frac{3}{4}\right)\left(\frac{2}{3}\right)(x^2-1)^{-1/3}(2x) = x(x^2-1)^{-1/3}, y' = ---) \underset{-1}{} (+++ | \underset{0}{} ---)(+++ \underset{1}{}$

$\Rightarrow$ the graph is rising on $(-1, 0)$ and $(1, \infty)$, falling on $(-\infty, -1)$ and $(0, 1) \Rightarrow$ a local maximum is $\frac{3}{4}$ at $x = 0$, local

minima are 0 at $x = \pm 1; y'' = (x^2-1)^{-1/3} + (x)\left(-\frac{1}{3}\right)(x^2-1)^{-4/3}(2x) = \frac{x^2-3}{3\sqrt[3]{(x^2-1)^4}}$,

$y'' = +++ | \underset{-\sqrt{3}}{} ---) \underset{-1}{} (---)(--- | \underset{\sqrt{3}}{} +++ \Rightarrow$ the graph is concave up on $\left(-\infty, -\sqrt{3}\right)$ and $\left(\sqrt{3}, \infty\right)$, concave

down on $\left(-\sqrt{3}, \sqrt{3}\right) \Rightarrow$ points of inflection at $\left(\pm\sqrt{3}, \frac{3\sqrt[3]{4}}{4}\right)$

5. $y = x + \sin 2x \Rightarrow y' = 1 + 2\cos 2x, y' = [--- | \underset{-2\pi/3 - \pi/3}{} +++ | \underset{\pi/3}{} ---] \underset{2\pi/3}{} \Rightarrow$ the graph is rising on $\left(-\frac{\pi}{3}, \frac{\pi}{3}\right)$, falling

on $\left(-\frac{2\pi}{3}, -\frac{\pi}{3}\right)$ and $\left(\frac{\pi}{3}, \frac{2\pi}{3}\right) \Rightarrow$ local maxima are $-\frac{2\pi}{3} + \frac{\sqrt{3}}{2}$ at $x = -\frac{2\pi}{3}$ and $\frac{\pi}{3} + \frac{\sqrt{3}}{2}$ at $x = \frac{\pi}{3}$, local minima are

$-\frac{\pi}{3} - \frac{\sqrt{3}}{2}$ at $x = -\frac{\pi}{3}$ and $\frac{2\pi}{3} - \frac{\sqrt{3}}{2}$ at $x = \frac{2\pi}{3}$; $y'' = -4\sin 2x$, $y'' = \begin{bmatrix} & --- & | & +++ & | & --- & | & +++ & \end{bmatrix}$ $\Rightarrow$ the
$\ -2\pi/3 \quad -\pi/2 \quad\ 0 \quad\ \pi/2 \quad 2\pi/3$

graph is concave up on $\left(-\frac{\pi}{2}, 0\right)$ and $\left(\frac{\pi}{2}, \frac{2\pi}{3}\right)$, concave down on $\left(-\frac{2\pi}{3}, -\frac{\pi}{2}\right)$ and $\left(0, \frac{\pi}{2}\right)$ $\Rightarrow$ points of inflection at
$\left(-\frac{\pi}{2}, -\frac{\pi}{2}\right)$, $(0, 0)$, and $\left(\frac{\pi}{2}, \frac{\pi}{2}\right)$

7. If $x \geq 0$, $\sin |x| = \sin x$ and if $x < 0$, $\sin |x| = \sin(-x)$
$= -\sin x$. From the sketch the graph is rising on
$\left(-\frac{3\pi}{2}, -\frac{\pi}{2}\right)$, $\left(0, \frac{\pi}{2}\right)$ and $\left(\frac{3\pi}{2}, 2\pi\right)$, falling on $\left(-2\pi, -\frac{3\pi}{2}\right)$,
$\left(-\frac{\pi}{2}, 0\right)$ and $\left(\frac{\pi}{2}, \frac{3\pi}{2}\right)$; local minima are -1 at $x = \pm\frac{3\pi}{2}$
and 0 at $x = 0$; local maxima are 1 at $x = \pm\frac{\pi}{2}$ and 0 at
$x = \pm 2\pi$; concave up on $(-2\pi, -\pi)$ and $(\pi, 2\pi)$, and
concave down on $(-\pi, 0)$ and $(0, \pi)$ $\Rightarrow$ points of inflection
are $(-\pi, 0)$ and $(\pi, 0)$

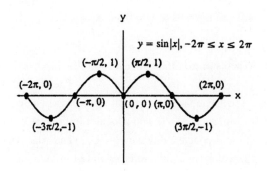

9. When $y = x^2 - 4x + 3$, then $y' = 2x - 4 = 2(x - 2)$ and
$y'' = 2$. The curve rises on $(2, \infty)$ and falls on $(-\infty, 2)$.
At $x = 2$ there is a minimum. Since $y'' > 0$, the curve is
concave up for all x.

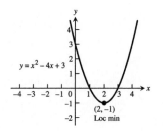

11. When $y = x^3 - 3x + 3$, then $y' = 3x^2 - 3 = 3(x - 1)(x + 1)$
and $y'' = 6x$. The curve rises on $(-\infty, -1) \cup (1, \infty)$ and
falls on $(-1, 1)$. At $x = -1$ there is a local maximum and at
$x = 1$ a local minimum. The curve is concave down on
$(-\infty, 0)$ and concave up on $(0, \infty)$. There is a point of
inflection at $x = 0$.

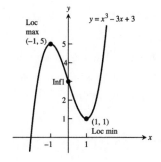

13. When $y = -2x^3 + 6x^2 - 3$, then $y' = -6x^2 + 12x$
$= -6x(x - 2)$ and $y'' = -12x + 12 = -12(x - 1)$. The
curve rises on $(0, 2)$ and falls on $(-\infty, 0)$ and $(2, \infty)$.
At $x = 0$ there is a local minimum and at $x = 2$ a local
maximum. The curve is concave up on $(-\infty, 1)$ and
concave down on $(1, \infty)$. At $x = 1$ there is a point of
inflection.

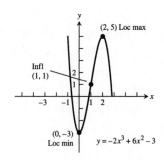

15. When $y = x^4 - 2x^2$, then $y' = 4x^3 - 4x = 4x(x + 1)(x - 1)$
and $y'' = 12x^2 - 4 = 12\left(x + \frac{1}{\sqrt{3}}\right)\left(x - \frac{1}{\sqrt{3}}\right)$. The curve
rises on $(-1, 0)$ and $(1, \infty)$ and falls on $(-\infty, -1)$ and $(0, 1)$.
At $x = \pm 1$ there are local minima and at $x = 0$ a local
maximum. The curve is concave up on $\left(-\infty, -\frac{1}{\sqrt{3}}\right)$ and
$\left(\frac{1}{\sqrt{3}}, \infty\right)$ and concave down on $\left(-\frac{1}{\sqrt{3}}, \frac{1}{\sqrt{3}}\right)$. At $x = \frac{\pm 1}{\sqrt{3}}$
there are points of inflection.

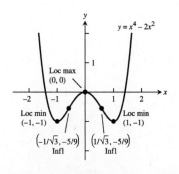

17. When $y = 4x^3 - x^4$, then $y' = 12x^2 - 4x^3 = 4x^2(3 - x)$ and $y'' = 24x - 12x^2 = 12x(2 - x)$. The curve rises on $(-\infty, 3)$ and falls on $(3, \infty)$. At $x = 3$ there is a local maximum, but there is no local minimum. The graph is concave up on $(0, 2)$ and concave down on $(-\infty, 0)$ and $(2, \infty)$. There are inflection points at $x = 0$ and $x = 2$.

19. When $y = x^5 - 5x^4$, then $y' = 5x^4 - 20x^3 = 5x^3(x - 4)$ and $y'' = 20x^3 - 60x^2 = 20x^2(x - 3)$. The curve rises on $(-\infty, 0)$ and $(4, \infty)$, and falls on $(0, 4)$. There is a local maximum at $x = 0$, and a local minimum at $x = 4$. The curve is concave down on $(-\infty, 3)$ and concave up on $(3, \infty)$. At $x = 3$ there is a point of inflection.

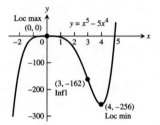

21. When $y = x + \sin x$, then $y' = 1 + \cos x$ and $y'' = -\sin x$. The curve rises on $(0, 2\pi)$. At $x = 0$ there is a local and absolute minimum and at $x = 2\pi$ there is a local and absolute maximum. The curve is concave down on $(0, \pi)$ and concave up on $(\pi, 2\pi)$. At $x = \pi$ there is a point of inflection.

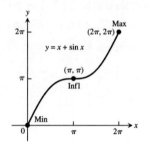

23. When $y = x^{1/5}$, then $y' = \frac{1}{5} x^{-4/5}$ and $y'' = -\frac{4}{25} x^{-9/5}$. The curve rises on $(-\infty, \infty)$ and there are no extrema. The curve is concave up on $(-\infty, 0)$ and concave down on $(0, \infty)$. At $x = 0$ there is a point of inflection.

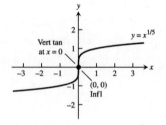

25. When $y = 2x - 3x^{2/3}$, then $y' = 2 - 2x^{-1/3}$ and $y'' = \frac{2}{3} x^{-4/3}$. The curve is rising on $(-\infty, 0)$ and $(1, \infty)$, and falling on $(0, 1)$. There is a local maximum at $x = 0$ and a local minimum at $x = 1$. The curve is concave up on $(-\infty, 0)$ and $(0, \infty)$. There are no points of inflection, but a cusp exists at $x = 0$.

27. When $y = x\sqrt{8 - x^2} = x\left(8 - x^2\right)^{1/2}$, then

$y' = \left(8 - x^2\right)^{1/2} + (x)\left(\frac{1}{2}\right)\left(8 - x^2\right)^{-1/2}(-2x)$

$= \left(8 - x^2\right)^{-1/2}\left(8 - 2x^2\right) = \dfrac{2(2 - x)(2 + x)}{\sqrt{\left(2\sqrt{2} + x\right)\left(2\sqrt{2} - x\right)}}$ and

$y'' = \left(-\frac{1}{2}\right)\left(8 - x^2\right)^{-\frac{3}{2}}(-2x)\left(8 - 2x^2\right) + \left(8 - x^2\right)^{-\frac{1}{2}}(-4x)$

$= \dfrac{2x\left(x^2 - 12\right)}{\sqrt{\left(8 - x^2\right)^3}}$. The curve is rising on $(-2, 2)$, and falling

on $\left(-2\sqrt{2}, -2\right)$ and $\left(2, 2\sqrt{2}\right)$. There are local minima

$x = -2$ and $x = 2\sqrt{2}$, and local maxima at $x = -2\sqrt{2}$ and

$x = 2$. The curve is concave up on $\left(-2\sqrt{2}, 0\right)$ and

concave down on $\left(0, 2\sqrt{2}\right)$. There is a point of inflection

at $x = 0$.

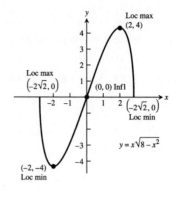

29. When $y = \frac{x^2 - 3}{x - 2}$, then $y' = \frac{2x(x - 2) - (x^2 - 3)(1)}{(x - 2)^2}$

$= \frac{(x - 3)(x - 1)}{(x - 2)^2}$ and

$y'' = \frac{(2x - 4)(x - 2)^2 - (x^2 - 4x + 3)2(x - 2)}{(x - 2)^4} = \frac{2}{(x - 2)^3}$.

The curve is rising on $(-\infty, 1)$ and $(3, \infty)$, and falling on
$(1, 2)$ and $(2, 3)$. There is a local maximum at $x = 1$ and a
local minimum at $x = 3$. The curve is concave down on
$(-\infty, 2)$ and concave up on $(2, \infty)$. There are no points
of inflection because $x = 2$ is not in the domain.

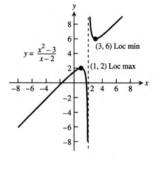

31. When $y = |x^2 - 1| = \begin{cases} x^2 - 1, & |x| \geq 1 \\ 1 - x^2, & |x| < 1 \end{cases}$, then

$y' = \begin{cases} 2x, & |x| > 1 \\ -2x, & |x| < 1 \end{cases}$ and $y'' = \begin{cases} 2, & |x| > 1 \\ -2, & |x| < 1 \end{cases}$. The

curve rises on $(-1, 0)$ and $(1, \infty)$ and falls on $(-\infty, -1)$
and $(0, 1)$. There is a local maximum at $x = 0$ and local
minima at $x = \pm 1$. The curve is concave up on $(-\infty, -1)$
and $(1, \infty)$, and concave down on $(-1, 1)$. There are no
points of inflection because y is not differentiable at $x = \pm 1$
(so there is no tangent line at those points).

33. When $y = x\,e^{1/x}$, then $y' = -\frac{x\,e^{1/x}}{x^2} + e^{1/x} = e^{1/x}\left(1 - \frac{1}{x}\right)$

and $y'' = e^{1/x}\left(\frac{1}{x^2}\right) + \left(1 - \frac{1}{x}\right)\left(-\frac{e^{1/x}}{x^2}\right) = \frac{e^{1/x}}{x^2}\left(\frac{1}{x}\right) = \frac{e^{1/x}}{x^3}$.

The curve is rising on $(1, \infty)$ and $(-\infty, 0)$ and falling on
$(0, 1)$. The curve is concave down on $(-\infty, 0)$ and concave
up on $(0, \infty)$. There is a local minimum of e at $x = 1$, but
there are no inflection points.

35. $y = \ln(3 - x^2)$, $\Rightarrow y' = \frac{-2x}{3-x^2} = \frac{2x}{x^2-3}$

$\Rightarrow y' = (\underset{-\sqrt{3}}{\ +++\ } | \underset{0}{\ ---\ }) \underset{\sqrt{3}}{} \Rightarrow$ the graph is rising on

$\left(-\sqrt{3}, 0\right)$, falling on $\left(0, \sqrt{3}\right)$; a local minimum is $\ln 3$ at

$x = 0$; $y'' = \frac{(x^2-3)(2) - (2x)(2x)}{(x^2-3)^2} = \frac{-2(x^2+3)}{(x^2-3)^2}$

$\Rightarrow y'' = (\underset{-\sqrt{3}}{\ ---\ })\underset{\sqrt{3}}{} \Rightarrow$ the graph is concave down on

$\left(-\sqrt{3}, \sqrt{3}\right)$.

37. $y = e^x - 2e^{-x} - 3x \Rightarrow y' = e^x + 2e^{-x} - 3 = \frac{(e^x)^2 - 3e^x + 2}{e^x}$

$= \frac{(e^x - 2)(e^x - 1)}{e^x} \Rightarrow y' = \underset{0}{\ +++\ } | \underset{\ln 2}{\ ---\ } | \ +++\ \Rightarrow$

the graph is incrreasing on $(-\infty, 0)$ and $(\ln 2, \infty)$,

decreasing on $(0, \ln 2)$; a local maximum is -1 at $x = 0$ and

a local minimum is $1 - 3\ln 2$ at $x = \ln 2$; $y'' = e^x - 2e^{-x}$

$= \frac{(e^x)^2 - 2}{e^x} \Rightarrow y'' = \underset{\frac{1}{2}\ln 2}{\ --- \ } | \ +++\ \Rightarrow$ the graph is

concave up on $\left(\frac{1}{2}\ln 2, \infty\right)$, concave down on $\left(-\infty, \frac{1}{2}\ln 2\right)$

$\Rightarrow$ point of inflection at $\left(\frac{1}{2}\ln 2, -\frac{3}{2}\ln 2\right)$.

39. $y = \ln(\cos x) \Rightarrow y' = \frac{-\sin x}{\cos x} = -\tan x$

$\Rightarrow y' = \underset{-\frac{7\pi}{2}}{\ ....\ }) \text{ none} \underset{-\frac{5\pi}{2}}{(} \underset{-2\pi}{\ +++\ } | \underset{-\frac{3\pi}{2}}{\ ---\ }) \text{ none} \underset{-\frac{\pi}{2}}{(} \underset{0}{\ +++\ } | \underset{\frac{\pi}{2}}{\ ---\ }) \text{ none} \underset{\frac{3\pi}{2}}{(} \underset{2\pi}{\ +++\ } | \underset{\frac{5\pi}{2}}{\ ---\ }) \text{ none} \underset{\frac{7\pi}{2}}{(\ ...}$

$\Rightarrow$ the graph is incrreasing on $\ldots, \left(-\frac{5\pi}{2}, -2\pi\right), \left(-\frac{\pi}{2}, 0\right)$,

$\left(\frac{3\pi}{2}, 2\pi\right), \ldots$, decreasing on $\left(-2\pi, -\frac{3\pi}{2}\right), \left(0, \frac{\pi}{2}\right), \left(2\pi, \frac{5\pi}{2}\right)$;

local maxima are 0 at $x = 0, \pm 2\pi, \pm 4\pi, \ldots$; $y'' = -\sec^2 x$

$\frac{-1}{\cos^2 x} \Rightarrow$ the graph is concave down on $\left(-\frac{5\pi}{2}, -\frac{3\pi}{2}\right), \left(-\frac{\pi}{2}, \frac{\pi}{2}\right)$,

$\left(\frac{3\pi}{2}, \frac{5\pi}{2}\right), \ldots$

41. $y = \frac{1}{1+e^{-x}} = \frac{e^x}{e^x+1} \Rightarrow y' = \frac{(e^x+1)e^x - e^x \cdot e^x}{(e^x+1)^2} = \frac{e^x}{(e^x+1)^2}$

$\Rightarrow y' = \ +++\ \Rightarrow$ the graph is incrreasing on $(-\infty, \infty)$;

$y'' = \frac{(e^x+1)^2 \cdot e^x - e^x \cdot 2(e^x+1)e^x}{(e^x+1)^4} = \frac{e^x(1-e^x)}{(e^x+1)^3}$

$\Rightarrow y'' = \underset{0}{\ +++\ } | \ --- \ \Rightarrow$ the graph is concave up on

$(-\infty, 0)$, concave down on $(0, \infty) \Rightarrow$ point of inflection

is $\left(0, \frac{1}{2}\right)$.

43. $y' = 2 + x - x^2 = (1 + x)(2 - x)$, $y' = --- \mid +++ \mid ---$
    ${-1}{2}$

$\Rightarrow$ rising on $(-1, 2)$, falling on $(-\infty, -1)$ and $(2, \infty)$
$\Rightarrow$ there is a local maximum at $x = 2$ and a local minimum
at $x = -1$; $y'' = 1 - 2x$, $y'' = +++ \mid ---$
${1/2}$

$\Rightarrow$ concave up on $\left(-\infty, \frac{1}{2}\right)$, concave down on $\left(\frac{1}{2}, \infty\right)$
$\Rightarrow$ a point of inflection at $x = \frac{1}{2}$

45. $y' = x(x - 3)^2$, $y' = --- \mid +++ \mid +++ \Rightarrow$ rising on
    ${0}{3}$

$(0, \infty)$, falling on $(-\infty, 0) \Rightarrow$ no local maximum, but there
is a local minimum at $x = 0$; $y'' = (x - 3)^2 + x(2)(x - 3)$
$= 3(x - 3)(x - 1)$, $y'' = +++ \mid --- \mid +++ \Rightarrow$ concave
${1}{3}$

up on $(-\infty, 1)$ and $(3, \infty)$, concave down on $(1, 3) \Rightarrow$
points of inflection at $x = 1$ and $x = 3$

47. $y' = x(x^2 - 12) = x\left(x - 2\sqrt{3}\right)\left(x + 2\sqrt{3}\right)$,

$y' = --- \mid +++ \mid --- \mid +++ \Rightarrow$ rising on
${-2\sqrt{3}}{0}{2\sqrt{3}}$

$\left(-2\sqrt{3}, 0\right)$ and $\left(2\sqrt{3}, \infty\right)$, falling on $\left(-\infty, -2\sqrt{3}\right)$

and $\left(0, 2\sqrt{3}\right) \Rightarrow$ a local maximum at $x = 0$, local minima

at $x = \pm 2\sqrt{3}$; $y'' = (1)(x^2 - 12) + (x)(2x)$
$= 3(x - 2)(x + 2)$, $y'' = +++ \mid --- \mid +++$
${-2}{2}$

$\Rightarrow$ concave up on $(-\infty, -2)$ and $(2, \infty)$, concave down on
$(-2, 2) \Rightarrow$ points of inflection at $x = \pm 2$

49. $y' = \sec^2 x$, $y' = (+++) \Rightarrow$ rising on $\left(-\frac{\pi}{2}, \frac{\pi}{2}\right)$,
    ${-\pi/2}{\pi/2}$

never falling $\Rightarrow$ no local extrema;
$y'' = 2(\sec x)(\sec x)(\tan x) = 2(\sec^2 x)(\tan x)$,
$y'' = (--- \mid +++) \Rightarrow$ concave up on $\left(0, \frac{\pi}{2}\right)$,
${-\pi/2}{0}{\pi/2}$
concave down on $\left(-\frac{\pi}{2}, 0\right)$, 0 is a opoint of inflection.

51. $y' = \cot \frac{\theta}{2}$, $y' = (+++ \mid ---) \Rightarrow$ rising on $(0, \pi)$,
    ${0}{\pi}{2\pi}$

falling on $(\pi, 2\pi) \Rightarrow$ a local maximum at $\theta = \pi$, no local
minimum; $y'' = -\frac{1}{2}\csc^2 \frac{\theta}{2}$, $y'' = (---) \Rightarrow$ never
${0}{2\pi}$
concave up, concave down on $(0, 2\pi) \Rightarrow$ no points of
inflection

53. $y' = \tan^2 \theta - 1 = (\tan \theta - 1)(\tan \theta + 1)$,

$y' = ($ $\underset{-\pi/2}{} +++ \,|\, \underset{-\pi/4}{} --- \,|\, \underset{\pi/4}{} +++ \underset{\pi/2}{})$ $\Rightarrow$ rising on

$\left(-\frac{\pi}{2}, -\frac{\pi}{4}\right)$ and $\left(\frac{\pi}{4}, \frac{\pi}{2}\right)$, falling on $\left(-\frac{\pi}{4}, \frac{\pi}{4}\right)$

$\Rightarrow$ a local maximum at $\theta = -\frac{\pi}{4}$, a local minimum at $\theta = \frac{\pi}{4}$;

$y'' = 2 \tan \theta \sec^2 \theta$, $y'' = ($ $\underset{-\pi/2}{} --- \,|\, \underset{0}{} +++ \underset{\pi/2}{})$

$\Rightarrow$ concave up on $\left(0, \frac{\pi}{2}\right)$, concave down on $\left(-\frac{\pi}{2}, 0\right)$

$\Rightarrow$ a point of inflection at $\theta = 0$

55. $y' = \cos t$, $y' = [$ $\underset{0}{} +++ \,|\, \underset{\pi/2}{} --- \,|\, \underset{3\pi/2}{} +++ \underset{2\pi}{}]$ $\Rightarrow$ rising on

$\left(0, \frac{\pi}{2}\right)$ and $\left(\frac{3\pi}{2}, 2\pi\right)$, falling on $\left(\frac{\pi}{2}, \frac{3\pi}{2}\right)$ $\Rightarrow$ local maxima at

$t = \frac{\pi}{2}$ and $t = 2\pi$, local minima at $t = 0$ and $t = \frac{3\pi}{2}$;

$y'' = -\sin t$, $y'' = [$ $\underset{0}{} --- \,|\, \underset{\pi}{} +++ \underset{2\pi}{}]$

$\Rightarrow$ concave up on $(\pi, 2\pi)$, concave down

on $(0, \pi)$ $\Rightarrow$ a point of inflection at $t = \pi$

57. $y' = (x + 1)^{-2/3}$, $y' = +++ \underset{-1}{)} (+++$ $\Rightarrow$ rising on

$(-\infty, \infty)$, never falling $\Rightarrow$ no local extrema;

$y'' = -\frac{2}{3}(x + 1)^{-5/3}$, $y'' = +++ \underset{-1}{)} (---$

$\Rightarrow$ concave up on $(-\infty, -1)$, concave down on $(-1, \infty)$

$\Rightarrow$ a point of inflection and vertical tangent at $x = -1$

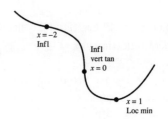

59. $y' = x^{-2/3}(x - 1)$, $y' = --- \underset{0}{)} (--- \,|\, \underset{1}{} +++$ $\Rightarrow$ rising on

$(1, \infty)$, falling on $(-\infty, 1)$ $\Rightarrow$ no local maximum, but a

local minimum at $x = 1$; $y'' = \frac{1}{3}x^{-2/3} + \frac{2}{3}x^{-5/3}$

$= \frac{1}{3}x^{-5/3}(x + 2)$, $y'' = +++ \,|\, \underset{-2}{} --- \underset{0}{)} (+++$

$\Rightarrow$ concave up on $(-\infty, -2)$ and $(0, \infty)$, concave down on

$(-2, 0)$ $\Rightarrow$ points of inflection at $x = -2$ and $x = 0$, and a

vertical tangent at $x = 0$

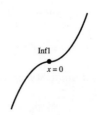

61. $y' = \begin{cases} -2x, & x \le 0 \\ 2x, & x > 0 \end{cases}$, $y' = +++ \underset{0}{|} +++$ $\Rightarrow$ rising on

$(-\infty, \infty)$ $\Rightarrow$ no local extrema; $y'' = \begin{cases} -2, & x < 0 \\ 2, & x > 0 \end{cases}$,

$y'' = --- \underset{0}{)} (+++$ $\Rightarrow$ concave up on $(0, \infty)$, concave

down on $(-\infty, 0)$ $\Rightarrow$ a point of inflection at $x = 0$

63. The graph of $y = f''(x)$ $\Rightarrow$ the graph of $y = f(x)$ is concave up on $(0, \infty)$, concave down on $(-\infty, 0)$ $\Rightarrow$ a point of inflection at $x = 0$; the graph of $y = f'(x)$

$\Rightarrow$ $y' = +++ \mid --- \mid +++$ $\Rightarrow$ the graph $y = f(x)$ has both a local maximum and a local minimum

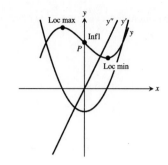

65. The graph of $y = f''(x)$ $\Rightarrow$ $y'' = --- \mid +++ \mid ---$

$\Rightarrow$ the graph of $y = f(x)$ has two points of inflection, the graph of $y = f'(x)$ $\Rightarrow$ $y' = --- \mid +++$ $\Rightarrow$ the graph of $y = f(x)$ has a local minimum

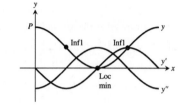

67.

Point	y'	y''
P	$-$	$+$
Q	$+$	0
R	$+$	$-$
S	0	$-$
T	$-$	$-$

69.

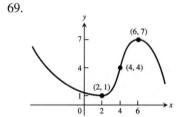

71. Graphs printed in color can shift during a press run, so your values may differ somewhat from those given here.

(a) The body is moving away from the origin when |displacement| is increasing as t increases, $0 < t < 2$ and $6 < t < 9.5$; the body is moving toward the origin when |displacement| is decreasing as t increases, $2 < t < 6$ and $9.5 < t < 15$.

(b) The velocity will be zero when the slope of the tangent line for $y = s(t)$ is horizontal. The velocity is zero when t is approximately 2, 6, or 9.5 sec.

(c) The acceleration will be zero at those values of t where the curve $y = s(t)$ has points of inflection. The acceleration is zero when t is approximately 4, 7.5, or 12.5 sec.

(d) The acceleration is positive when the concavity is up, $4 < t < 7.5$ and $12.5 < t < 15$; the acceleration is negative when the concavity is down, $0 < t < 4$ and $7.5 < t < 12.5$

73. The marginal cost is $\frac{dc}{dx}$ which changes from decreasing to increasing when its derivative $\frac{d^2c}{dx^2}$ is zero. This is a point of inflection of the cost curve and occurs when the production level x is approximately 60 thousand units.

75. When $y' = (x - 1)^2(x - 2)$, then $y'' = 2(x - 1)(x - 2) + (x - 1)^2$. The curve falls on $(-\infty, 2)$ and rises on $(2, \infty)$. At $x = 2$ there is a local minimum. There is no local maximum. The curve is concave upward on $(-\infty, 1)$ and $\left(\frac{5}{3}, \infty\right)$, and concave downward on $\left(1, \frac{5}{3}\right)$. At $x = 1$ or $x = \frac{5}{3}$ there are inflection points.

77. The graph must be concave down for x > 0 because
$f''(x) = -\frac{1}{x^2} < 0$.

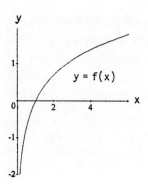

79. The curve will have a point of inflection at x = 1 if 1 is a solution of $y'' = 0$; $y = x^3 + bx^2 + cx + d$
$\Rightarrow y' = 3x^2 + 2bx + c \Rightarrow y'' = 6x + 2b$ and $6(1) + 2b = 0 \Rightarrow b = -3$.

81. (a) $f(x) = ax^2 + bx + c = a\left(x^2 + \frac{b}{a}x\right) + c = a\left(x^2 + \frac{b}{a}x + \frac{b^2}{4a^2}\right) - \frac{b^2}{4a} + c = a\left(x + \frac{b}{2a}\right)^2 - \frac{b^2 - 4ac}{4a}$ a parabola

whose vertex is at $x = -\frac{b}{2a} \Rightarrow$ the coordinates of the vertex are $\left(-\frac{b}{2a}, -\frac{b^2 - 4ac}{4a}\right)$

(b) The second derivative, $f''(x) = 2a$, describes concavity $\Rightarrow$ when a > 0 the parabola is concave up and
when a < 0 the parabola is concave down.

83. A quadratic curve never has an inflection point. If $y = ax^2 + bx + c$ where $a \neq 0$, then $y' = 2ax + b$ and
$y'' = 2a$. Since 2a is a constant, it is not possible for y'' to change signs.

85. If $y = x^5 - 5x^4 - 240$, then $y' = 5x^3(x - 4)$ and
$y'' = 20x^2(x - 3)$. The zeros of y' are extrema, and
there is a point of inflection at x = 3.

87. If $y = \frac{4}{5}x^5 + 16x^2 - 25$, then $y' = 4x(x^3 + 8)$ and
$y'' = 16(x^3 + 2)$. The zeros of y' and y'' are
extrema and points of inflection, respectively.

89. The graph of f falls where $f' < 0$, rises where $f' > 0$,
and has horizontal tangents where $f' = 0$. It has local
minima at points where f' changes from negative to
positive and local maxima where f' changes from
positive to negative. The graph of f is concave down
where $f'' < 0$ and concave up where $f'' > 0$. It has an
inflection point each time f'' changes sign, provided a tangent
line exists there.

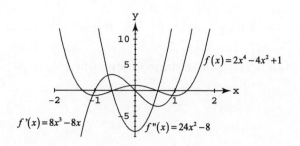

91. (a) It appears to control the number and magnitude of the local extrema. If $k < 0$, there is a local maximum to the left of the origin and a local minimum to the right. The larger the magnitude of k ($k < 0$), the greater the magnitude of the extrema. If $k > 0$, the graph has only positive slopes and lies entirely in the first and third quadrants with no local extrema. The graph becomes increasingly steep and straight as $k \to \infty$.

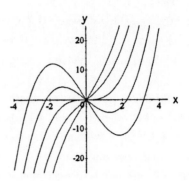

(b) $f'(x) = 3x^2 + k \Rightarrow$ the discriminant $0^2 - 4(3)(k) = -12k$ is positive for $k < 0$, zero for $k = 0$, and negative for $k > 0$; f' has two zeros $x = \pm\sqrt{-\frac{k}{3}}$ when $k < 0$, one zero $x = 0$ when $k = 0$ and no real zeros when $k > 0$; the sign of k controls the number of local extrema.

(c) As $k \to \infty$, $f'(x) \to \infty$ and the graph becomes increasingly steep and straight. As $k \to -\infty$, the crest of the graph (local maximum) in the second quadrant becomes increasingly high and the trough (local minimum) in the fourth quadrant becomes increasingly deep.

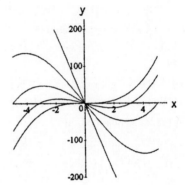

93. (a) If $y = x^{2/3}(x^2 - 2)$, then $y' = \frac{4}{3}x^{-1/3}(2x^2 - 1)$ and $y'' = \frac{4}{9}x^{-4/3}(10x^2 + 1)$. The curve rises on $\left(-\frac{1}{\sqrt{2}}, 0\right)$ and $\left(\frac{1}{\sqrt{2}}, \infty\right)$ and falls on $\left(-\infty, -\frac{1}{\sqrt{2}}\right)$ and $\left(0, \frac{1}{\sqrt{2}}\right)$. The curve is concave up on $(-\infty, 0)$ and $(0, \infty)$.

(b) A cusp since $\lim_{x \to 0^-} y' = \infty$ and $\lim_{x \to 0^+} y' = -\infty$.

$y = x^{2/3}(x^2 - 2)$

95. Yes: $y = x^2 + 3\sin 2x \Rightarrow y' = 2x + 6\cos 2x$. The graph of y' is zero near -3 and this indicates a horizontal tangent near $x = -3$.

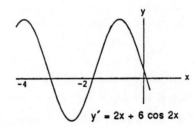

$y' = 2x + 6\cos 2x$

4.5 APPLIED OPTIMIZATION

1. Let ℓ and w represent the length and width of the rectangle, respectively. With an area of 16 in.², we have that $(\ell)(w) = 16 \Rightarrow w = 16\ell^{-1} \Rightarrow$ the perimeter is $P = 2\ell + 2w = 2\ell + 32\ell^{-1}$ and $P'(\ell) = 2 - \frac{32}{\ell^2} = \frac{2(\ell^2 - 16)}{\ell^2}$. Solving $P'(\ell) = 0 \Rightarrow \frac{2(\ell+4)(\ell-4)}{\ell^2} = 0 \Rightarrow \ell = -4, 4$. Since $\ell > 0$ for the length of a rectangle, ℓ must be 4 and $w = 4 \Rightarrow$ the perimeter is 16 in., a minimum since $P''(\ell) = \frac{16}{\ell^3} > 0$.

3. (a) The line containing point P also contains the points $(0, 1)$ and $(1, 0) \Rightarrow$ the line containing P is $y = 1 - x$ $\Rightarrow$ a general point on that line is $(x, 1 - x)$.

(b) The area $A(x) = 2x(1 - x)$, where $0 \le x \le 1$.

(c) When $A(x) = 2x - 2x^2$, then $A'(x) = 0 \Rightarrow 2 - 4x = 0 \Rightarrow x = \frac{1}{2}$. Since $A(0) = 0$ and $A(1) = 0$, we conclude that $A\left(\frac{1}{2}\right) = \frac{1}{2}$ sq units is the largest area. The dimensions are 1 unit by $\frac{1}{2}$ unit.

5. The volume of the box is $V(x) = x(15 - 2x)(8 - 2x)$
    $= 120x - 46x^2 + 4x^3$, where $0 \le x \le 4$. Solving $V'(x) = 0$
    $\Rightarrow 120 - 92x + 12x^2 = 4(6 - x)(5 - 3x) = 0 \Rightarrow x = \frac{5}{3}$
    or 6, but 6 is not in the domain. Since $V(0) = V(4) = 0$,
    $V\left(\frac{5}{3}\right) = \frac{2450}{27} \approx 91$ in^3 must be the maximum volume of
    the box with dimensions $\frac{14}{3} \times \frac{35}{3} \times \frac{5}{3}$ inches.

7. The area is $A(x) = x(800 - 2x)$, where $0 \le x \le 400$.
    Solving $A'(x) = 800 - 4x = 0 \Rightarrow x = 200$. With
    $A(0) = A(400) = 0$, the maximum area is
    $A(200) = 80,000$ m^2. The dimensions are 200 m by 400 m.

9. (a) We minimize the weight $= tS$ where S is the surface area, and t is the thickness of the steel walls of the tank. The surace area is $S = x^2 + 4xy$ where x is the length of a side of the square base of the tank, and y is its depth. The volume of the tank must be 500ft$^3 \Rightarrow y = \frac{500}{x^2}$. Therefore, the weight of the tank is $w(x) = t\left(x^2 + \frac{2000}{x}\right)$. Treating the thickness as a constant gives $w'(x) = t\left(2x - \frac{2000}{x^2}\right)$ for x.0. The critical value is at $x = 10$. Since $w''(10) = t\left(2 + \frac{4000}{10^3}\right) > 0$, there is a minimum at $x = 10$. Therefore, the optimum dimensions of the tank are 10 ft on the base edges and 5 ft deep.

(b) Minimizing the surface area of the tank minimizes its weight for a given wall thickness. The thickness of the steel walls would likely be determined by other considerations such as structural requirements.

11. The area of the printing is $(y - 4)(x - 8) = 50$.
     Consequently, $y = \left(\frac{50}{x - 8}\right) + 4$. The area of the paper is
     $A(x) = x\left(\frac{50}{x - 8} + 4\right)$, where $8 < x$. Then
     $A'(x) = \left(\frac{50}{x - 8} + 4\right) - x\left(\frac{50}{(x - 8)^2}\right) = \frac{4(x - 8)^2 - 400}{(x - 8)^2} = 0$
     $\Rightarrow$ the critical points are -2 and 18, but -2 is not in the
     domain. Thus $A''(18) > 0 \Rightarrow$ at $x = 18$ we have
     a minimum. Therefore the dimensions 18 by 9 inches
     minimize the amount of paper.

13. The area of the triangle is $A(\theta) = \frac{ab \sin \theta}{2}$, where $0 < \theta < \pi$.
     Solving $A'(\theta) = 0 \Rightarrow \frac{ab \cos \theta}{2} = 0 \Rightarrow \theta = \frac{\pi}{2}$. Since $A''(\theta)$
     $= -\frac{ab \sin \theta}{2} \Rightarrow A''\left(\frac{\pi}{2}\right) < 0$, there is a maximum at $\theta = \frac{\pi}{2}$.

15. With a volume of 1000 cm and $V = \pi r^2 h$, then $h = \frac{1000}{\pi r^2}$. The amount of aluminum used per can is
     $A = 8r^2 + 2\pi rh = 8r^2 + \frac{2000}{r}$. Then $A'(r) = 16r - \frac{2000}{r^2} = 0 \Rightarrow \frac{8r^3 - 1000}{r^2} = 0 \Rightarrow$ the critical points are 0 and 5,
     but $r = 0$ results in no can. Since $A''(r) = 16 + \frac{1000}{r^3} > 0$ we have a minimum at $r = 5 \Rightarrow h = \frac{40}{\pi}$ and $h{:}r = 8{:}\pi$.

17. (a) The" sides" of the suitcase will measure $24 - 2x$ in. by $18 - 2x$ in. and will be $2x$ in. apart, so the volume formula is
         $V(x) = 2x(24 - 2x)(18 - 2x) = 8x^3 - 168x^2 + 862x$.

(b) We require x > 0, 2x < 18, and 2x < 24. Combining these requirements, the domain is the interval (0, 9).

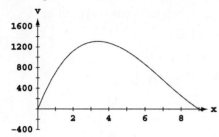

(c) The maximum volume is approximately 1309.95 in.^{3} when x ≈ 3.39 in.

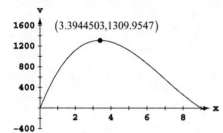

(3.3944503, 1309.9547)

(d) $V'(x) = 24x^2 - 336x + 864 = 24(x^2 - 14x + 36)$. The critical point is at $x = \frac{14 \pm \sqrt{(-14)^2 - 4(1)(36)}}{2(1)} = \frac{14 \pm \sqrt{52}}{2}$

$= 7 \pm \sqrt{13}$, that is, x ≈ 3.39 or x ≈ 10.61. We discard the larger value because it is not in the domain. Since $V''(x) = 24(2x - 14)$ which is negative when x ≈ 3.39, the critical point corresponds to the maximum volume. The maximum value occurs at $x = 7 - \sqrt{13} ≈ 3.39$, which confirms the results in (c).

(e) $8x^3 - 168x^2 + 862x = 1120 \Rightarrow 8(x^3 - 21x^2 + 108x - 140) = 0 \Rightarrow 8(x - 2)(x - 5)(x - 14) = 0$. Since 14 is not in the fomain, the possible values of x are x = 2 in. or x = 5 in.

(f) The dimensions of the resulting box are 2x in., (24 − 2x) in., and (18 − 2x). Each of these measurements must be positive, so that gives the domain of (0, 9).

19. Let the radius of the cylinder be r cm, 0 < r < 10. Then the height is $2\sqrt{100 - r^2}$ and the volume is

$V(r) = 2\pi r^2 \sqrt{100 - r^2}$ cm^3. Then, $V'(r) = 2\pi r^2 \left(\frac{1}{\sqrt{100 - r^2}}\right)(-2r) + \left(2\pi \sqrt{100 - r^2}\right)(2r)$

$= \frac{-2\pi r^3 + 4\pi r(100 - r^2)}{\sqrt{100 - r^2}} = \frac{2\pi r(200 - 3r^2)}{\sqrt{100 - r^2}}$. The critical point for 0 < r < 10 occurs at $r = \sqrt{\frac{200}{3}} = 10\sqrt{\frac{2}{3}}$. Since $V'(r) > 0$ for

$0 < r < 10\sqrt{\frac{2}{3}}$ and $V'(r) < 0$ for $10\sqrt{\frac{2}{3}} < r < 10$, the critical point corresponds to the maximum volume. The

dimensions are $r = 10\sqrt{\frac{2}{3}} ≈ 8.16$ cm and $h = \frac{20}{\sqrt{3}} ≈ 11.55$ cm, and the volume is $\frac{4000\pi}{3\sqrt{3}} ≈ 2418.40$ cm^3.

21. (a) From the diagram we have 3h + 2w = 108 and

$V = h^2 w \Rightarrow V(h) = h^2 \left(54 - \frac{3}{2}h\right) = 54h^2 - \frac{3}{2}h^3$.

Then $V'(h) = 108h - \frac{9}{2}h^2 = \frac{9}{2}h(24 - h) = 0$

$\Rightarrow$ h = 0 or h = 24, but h = 0 results in no box. Since

$V''(h) = 108 - 9h < 0$ at h = 24, we have a maximum

volume at h = 24 and $w = 54 - \frac{3}{2}h = 18$.

(b)

10000 ┤   (24, 10368)
      │   Abs max
8000 ┤
6000 ┤
4000 ┤ $\left(v = 54h^2 - \frac{3}{2}h^3\right)$
2000 ┤
   0 ┼──┬──┬──┬──┬──┬──┬──┬──→ h
      5 10 15 20 25 30 35

23. The fixed volume is $V = \pi r^2 h + \frac{2}{3}\pi r^3 \Rightarrow h = \frac{V}{\pi r^2} - \frac{2r}{3}$, where h is the height of the cylinder and r is the radius of the hemisphere. To minimize the cost we must minimize surface area of the cylinder added to twice the surface area of the hemisphere. Thus, we minimize $C = 2\pi rh + 4\pi r^2 = 2\pi r\left(\frac{V}{\pi r^2} - \frac{2r}{3}\right) + 4\pi r^2 = \frac{2V}{r} + \frac{8}{3}\pi r^2$.

Then $\frac{dC}{dr} = -\frac{2V}{r^2} + \frac{16}{3}\pi r = 0 \Rightarrow V = \frac{8}{3}\pi r^3 \Rightarrow r = \left(\frac{3V}{8\pi}\right)^{1/3}$. From the volume equation, $h = \frac{V}{\pi r^2} - \frac{2r}{3}$

$= \frac{4V^{1/3}}{\pi^{1/3}\cdot 3^{2/3}} - \frac{2\cdot 3^{1/3}\cdot V^{1/3}}{3\cdot 2\cdot \pi^{1/3}} = \frac{3^{1/3}\cdot 2\cdot 4\cdot V^{1/3} - 2\cdot 3^{1/3}\cdot V^{1/3}}{3\cdot 2\cdot \pi^{1/3}} = \left(\frac{3V}{\pi}\right)^{1/3}$. Since $\frac{d^2C}{dr^2} = \frac{4V}{r^3} + \frac{16}{3}\pi > 0$, these dimensions do minimize the cost.

25. (a) From the diagram we have: $\overline{AP} = x$, $\overline{RA} = \sqrt{L - x^2}$,

$\overline{PB} = 8.5 - x$, $\overline{CH} = \overline{DR} = 11 - \overline{RA} = 11 - \sqrt{L - x^2}$,

$\overline{QB} = \sqrt{x^2 - (8.5 - x)^2}$, $\overline{HQ} = 11 - \overline{CH} - \overline{QB}$

$= 11 - \left[11 - \sqrt{L - x^2} + \sqrt{x^2 - (8.5 - x)^2}\right]$

$= \sqrt{L - x^2} - \sqrt{x^2 - (8.5 - x)^2}$, $\overline{RQ}^2 = \overline{RH}^2 + \overline{HQ}^2$

$= (8.5)^2 + \left(\sqrt{L - x^2} - \sqrt{x^2 - (8.5 - x)^2}\right)^2$. It

follows that $\overline{RP}^2 = \overline{PQ}^2 + \overline{RQ}^2 \Rightarrow L^2 = x^2 + \left(\sqrt{L^2 - x^2} - \sqrt{x^2 - (x - 8.5)^2}\right)^2 + (8.5)^2$

$\Rightarrow L^2 = x^2 + L^2 - x^2 - 2\sqrt{L^2 - x^2}\sqrt{17x - (8.5)^2} + 17x - (8.5)^2 + (8.5)^2$

$\Rightarrow 17^2 x^2 = 4(L^2 - x^2)(17x - (8.5)^2) \Rightarrow L^2 = x^2 + \frac{17^2 x^2}{4[17x - (8.5)^2]} = \frac{17x^3}{17x - (8.5)^2}$

$= \frac{17x^3}{17x - \left(\frac{17}{2}\right)^2} = \frac{4x^3}{4x - 17} = \frac{2x^3}{2x - 8.5}$.

(b) If $f(x) = \frac{4x^3}{4x - 17}$ is minimized, then L^2 is minimized. Now $f'(x) = \frac{4x^2(8x - 51)}{(4x - 17)^2} \Rightarrow f'(x) < 0$ when $x < \frac{51}{8}$ and $f'(x) > 0$ when $x > \frac{51}{8}$. Thus L^2 is minimized when $x = \frac{51}{8}$.

(c) When $x = \frac{51}{8}$, then $L \approx 11.0$ in.

27. Note that $h^2 + r^2 = 3$ and so $r = \sqrt{3 - h^2}$. Then the volume is given by $V = \frac{\pi}{3}r^2 h = \frac{\pi}{3}(3 - h^2)h = \pi h - \frac{\pi}{3}h^3$ for $0 < h < \sqrt{3}$, and so $\frac{dV}{dh} = \pi - \pi r^2 = \pi(1 - r^2)$. The critical point (for $h > 0$) occurs at $h = 1$. Since $\frac{dV}{dh} > 0$ for $0 < h < 1$, and $\frac{dV}{dh} < 0$ for $1 < h < \sqrt{3}$, the critical point corresponds to the maximum volume. The cone of greatest volume has radius $\sqrt{2}$ m, height 1m, and volume $\frac{2\pi}{3}$ m³.

29. If $f(x) = x^2 + \frac{a}{x}$, then $f'(x) = 2x - ax^{-2}$ and $f''(x) = 2 + 2ax^{-3}$. The critical points are 0 and $\sqrt[3]{\frac{a}{2}}$, but $x \neq 0$. Now $f''\left(\sqrt[3]{\frac{a}{2}}\right) = 6 > 0 \Rightarrow$ at $x = \sqrt[3]{\frac{a}{2}}$ there is a local minimum. However, no local maximum exists for any a.

31. (a) $s(t) = -16t^2 + 96t + 112 \Rightarrow v(t) = s'(t) = -32t + 96$. At $t = 0$, the velocity is $v(0) = 96$ ft/sec.

(b) The maximum height ocurs when $v(t) = 0$, when $t = 3$. The maximum height is $s(3) = 256$ ft and it occurs at $t = 3$ sec.

(c) Note that $s(t) = -16t^2 + 96t + 112 = -16(t + 1)(t - 7)$, so $s = 0$ at $t = -1$ or $t = 7$. Choosing the positive value of t, the velocity when $s = 0$ is $v(7) = -128$ ft/sec.

33. $\frac{8}{x} = \frac{h}{x+27} \Rightarrow h = 8 + \frac{216}{x}$ and $L(x) = \sqrt{h^2 + (x+27)^2}$

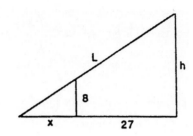

$= \sqrt{\left(8 + \frac{216}{x}\right)^2 + (x+27)^2}$ when $x \geq 0$. Note that $L(x)$ is

minimized when $f(x) = \left(8 + \frac{216}{x}\right)^2 + (x+27)^2$ is

minimized. If $f'(x) = 0$, then

$2\left(8 + \frac{216}{x}\right)\left(-\frac{216}{x^2}\right) + 2(x+27) = 0$

$\Rightarrow (x+27)\left(1 - \frac{1728}{x^3}\right) = 0 \Rightarrow x = -27$ (not acceptable

since distance is never negative or $x = 12$. Then $L(12) = \sqrt{2197} \approx 46.87$ ft.

35. (a) From the situation we have $w^2 = 144 - d^2$. The stiffness of the beam is $S = kwd^3 = kd^3(144 - d^2)^{1/2}$,

where $0 \leq d \leq 12$. Also, $S'(d) = \frac{4kd^2(108 - d^2)}{\sqrt{144 - d^2}} \Rightarrow$ critical points at $0, 12,$ and $6\sqrt{3}$. Both $d = 0$ and

$d = 12$ cause $S = 0$. The maximum occurs at $d = 6\sqrt{3}$. The dimensions are 6 by $6\sqrt{3}$ inches.

(b)

(c)

Both graphs indicate the same maximum value and are consistent with each other. The changing of k has
no effect.

37. (a) $s = 10\cos(\pi t) \Rightarrow v = -10\pi\sin(\pi t) \Rightarrow$ speed $= |10\pi\sin(\pi t)| = 10\pi|\sin(\pi t)| \Rightarrow$ the maximum speed is
$10\pi \approx 31.42$ cm/sec since the maximum value of $|\sin(\pi t)|$ is 1; the cart is moving the fastest at $t = 0.5$ sec,
1.5 sec, 2.5 sec and 3.5 sec when $|\sin(\pi t)|$ is 1. At these times the distance is $s = 10\cos\left(\frac{\pi}{2}\right) = 0$ cm and
$a = -10\pi^2\cos(\pi t) \Rightarrow |a| = 10\pi^2|\cos(\pi t)| \Rightarrow |a| = 0$ cm/sec^2

(b) $|a| = 10\pi^2|\cos(\pi t)|$ is greatest at $t = 0.0$ sec, 1.0 sec, 2.0 sec, 3.0 sec and 4.0 sec, and at these times the
magnitude of the cart's position is $|s| = 10$ cm from the rest position and the speed is 0 cm/sec.

39.

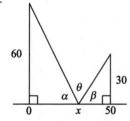

From the diagram above we have $\alpha + \beta + \theta = \pi \Rightarrow \theta = \pi - \alpha - \beta$. From the left triangle we have

$\cot\alpha = \frac{x}{60} \Rightarrow \cot^{-1}\left(\frac{x}{60}\right) = \alpha$, and from the triangle on the right side we have $\cot\beta = \frac{50 - x}{60} \Rightarrow \cot^{-1}\left(\frac{50 - x}{30}\right) = \beta$

$\Rightarrow \theta = \pi - \cot^{-1}\left(\frac{x}{60}\right) - \cot^{-1}\left(\frac{50 - x}{30}\right) \Rightarrow \frac{d\theta}{dx} = 0 - \left(-\frac{1}{1 + \left(\frac{x}{60}\right)^2} \cdot \frac{1}{60}\right) - \left(-\frac{1}{1 + \left(\frac{50 - x}{30}\right)^2} \cdot \frac{-1}{30}\right)$

$\frac{d\theta}{dx} = \frac{1}{60\left(1 + \left(\frac{x}{60}\right)^2\right)} - \frac{1}{30\left(1 + \left(\frac{50 - x}{30}\right)^2\right)} = \frac{1}{60 + \frac{x^2}{60}} - \frac{1}{30 + \frac{(50 - x)^2}{30}} = \frac{60}{60^2 + x^2} - \frac{30}{30^2 + (50 - x)^2}$

$\frac{60}{60^2 + x^2} - \frac{30}{30^2 + (50 - x)^2} = 0 \Rightarrow 60\left(30^2 + (50 - x)^2\right) - 30(60^2 + x^2) \Rightarrow 30x^2 - 6000x + 96000 = 0$

$\Rightarrow x = 20\left(5 \pm \sqrt{17}\right)$. Since $0 \leq x \leq 50 \Rightarrow x = 20\left(5 - \sqrt{17}\right) \approx 17.54$. $\theta'(10) = \frac{60}{60^2 + 10^2} - \frac{30}{30^2 + (50 - 10)^2}$

$= \frac{60}{3700} - \frac{30}{2500} = \frac{39}{9250} > 0$ and $\theta'(20) = \frac{60}{60^2 + 20^2} - \frac{30}{30^2 + (50 - 20)^2} = \frac{60}{4000} - \frac{30}{1800} = -\frac{1}{600}$.

$\Rightarrow$ the maximum angle is $\theta = \pi - \cot^{-1}\left(\frac{20\left(5-\sqrt{17}\right)}{60}\right) - \cot^{-1}\left(\frac{50-20\left(5-\sqrt{17}\right)}{30}\right) \approx 1.10917$ rad or

$63.55°$ when the solar station is ≈ 17.54 m west of the left (60 m) building.

41. If $v = kax - kx^2$, then $v' = ka - 2kx$ and $v'' = -2k$, so $v' = 0 \Rightarrow x = \frac{a}{2}$. At $x = \frac{a}{2}$ there is a maximum since $v''\left(\frac{a}{2}\right) = -2k < 0$. The maximum value of v is $\frac{ka^2}{4}$.

43. The profit is $p = nx - nc = n(x - c) = [a(x - c)^{-1} + b(100 - x)](x - c) = a + b(100 - x)(x - c)$
    $= a + (bc + 100b)x - 100bc - bx^2$. Then $p'(x) = bc + 100b - 2bx$ and $p''(x) = -2b$. Solving $p'(x) = 0 \Rightarrow x = \frac{c}{2} + 50$.
    At $x = \frac{c}{2} + 50$ there is a maximum profit since $p''(x) = -2b < 0$ for all x.

45. (a) $A(q) = kmq^{-1} + cm + \frac{h}{2}q$, where $q > 0 \Rightarrow A'(q) = -kmq^{-2} + \frac{h}{2} = \frac{hq^2 - 2km}{2q^2}$ and $A''(q) = 2kmq^{-3}$. The
    critical points are $-\sqrt{\frac{2km}{h}}$, 0, and $\sqrt{\frac{2km}{h}}$, but only $\sqrt{\frac{2km}{h}}$ is in the domain. Then $A''\left(\sqrt{\frac{2km}{h}}\right) > 0 \Rightarrow$ at
    $q = \sqrt{\frac{2km}{h}}$ there is a minimum average weekly cost.

    (b) $A(q) = \frac{(k+bq)m}{q} + cm + \frac{h}{2}q = kmq^{-1} + bm + cm + \frac{h}{2}q$, where $q > 0 \Rightarrow A'(q) = 0$ at $q = \sqrt{\frac{2km}{h}}$ as in (a).
    Also $A''(q) = 2kmq^{-3} > 0$ so the most economical quantity to order is still $q = \sqrt{\frac{2km}{h}}$ which minimizes the
    average weekly cost.

47. The profit $p(x) = r(x) - c(x) = 6x - (x^3 - 6x^2 + 15x) = -x^3 + 6x^2 - 9x$, where $x \geq 0$. Then $p'(x) = -3x^2 + 12x - 9$
    $= -3(x - 3)(x - 1)$ and $p''(x) = -6x + 12$. The critical points are 1 and 3. Thus $p''(1) = 6 > 0 \Rightarrow$ at $x = 1$ there is a
    local minimum, and $p''(3) = -6 < 0 \Rightarrow$ at $x = 3$ there is a local maximum. But $p(3) = 0 \Rightarrow$ the best you can do is
    break even.

49. We have $\frac{dR}{dM} = CM - M^2$. Solving $\frac{d^2R}{dM^2} = C - 2M = 0 \Rightarrow M = \frac{C}{2}$. Also, $\frac{d^3R}{dM^3} = -2 < 0 \Rightarrow$ at $M = \frac{C}{2}$ there is a
    maximum.

51. If $x > 0$, then $(x - 1)^2 \geq 0 \Rightarrow x^2 + 1 \geq 2x \Rightarrow \frac{x^2+1}{x} \geq 2$. In particular if a, b, c and d are positive integers,
    then $\left(\frac{a^2+1}{a}\right)\left(\frac{b^2+1}{b}\right)\left(\frac{c^2+1}{c}\right)\left(\frac{d^2+1}{d}\right) \geq 16$.

53. At $x = c$, the tangents to the curves are parallel. Justification: The vertical distance between the curves is
    $D(x) = f(x) - g(x)$, so $D'(x) = f'(x) - g'(x)$. The maximum value of D will occur at a point c where $D' = 0$. At
    such a point, $f'(c) - g'(c) = 0$, or $f'(c) = g'(c)$.

55. (a) If $y = \cot x - \sqrt{2} \csc x$ where $0 < x < \pi$, then $y' = (\csc x)\left(\sqrt{2} \cot x - \csc x\right)$. Solving $y' = 0 \Rightarrow \cos x = \frac{1}{\sqrt{2}}$
    $\Rightarrow x = \frac{\pi}{4}$. For $0 < x < \frac{\pi}{4}$ we have $y' > 0$, and $y' < 0$ when $\frac{\pi}{4} < x < \pi$. Therefore, at $x = \frac{\pi}{4}$ there is a maximum
    value of $y = -1$.

(b)

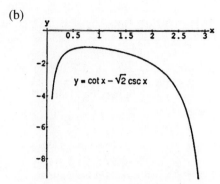

The graph confirms the findings in (a).

57. (a) The square of the distance is $D(x) = \left(x - \frac{3}{2}\right)^2 + \left(\sqrt{x} + 0\right)^2 = x^2 - 2x + \frac{9}{4}$, so $D'(x) = 2x - 2$ and the critical point occurs at $x = 1$. Since $D'(x) < 0$ for $x < 1$ and $D'(x) > 0$ for $x > 1$, the critical point corresponds to the minimum distance. The minimum distance is $\sqrt{D(1)} = \frac{\sqrt{5}}{2}$.

(b)

The minimum distance is from the point $\left(\frac{3}{2}, 0\right)$ to the point $(1, 1)$ on the graph of $y = \sqrt{x}$, and this occurs at the value $x = 1$ where $D(x)$, the distance squared, has its minimum value.

4.6 INDETERMINATE FORMS AND L'HÔPITAL'S RULE

1. l'Hôpital: $\lim\limits_{x \to 2} \frac{x-2}{x^2-4} = \frac{1}{2x}\Big|_{x=2} = \frac{1}{4}$ or $\lim\limits_{x \to 2} \frac{x-2}{x^2-4} = \lim\limits_{x \to 2} \frac{x-2}{(x-2)(x+2)} = \lim\limits_{x \to 2} \frac{1}{x+2} = \frac{1}{4}$

3. l'Hôpital: $\lim\limits_{x \to \infty} \frac{5x^2 - 3x}{7x^2 + 1} = \lim\limits_{x \to \infty} \frac{10x - 3}{14x} = \lim\limits_{x \to \infty} \frac{10}{14} = \frac{5}{7}$ or $\lim\limits_{x \to \infty} \frac{5x^2 - 3x}{7x^2 + 1} = \lim\limits_{x \to \infty} \frac{5 - \frac{3}{x}}{7 + \frac{1}{x}} = \frac{5}{7}$

5. l'Hôpital: $\lim\limits_{x \to 0} \frac{1 - \cos x}{x^2} = \lim\limits_{x \to 0} \frac{\sin x}{2x} = \lim\limits_{x \to 0} \frac{\cos x}{2} = \frac{1}{2}$ or $\lim\limits_{x \to 0} \frac{1 - \cos x}{x^2} = \lim\limits_{x \to 0} \left[\frac{(1 - \cos x)}{x^2}\left(\frac{1 + \cos x}{1 + \cos x}\right)\right]$

$= \lim\limits_{x \to 0} \frac{\sin^2 x}{x^2(1 + \cos x)} = \lim\limits_{x \to 0} \left[\left(\frac{\sin x}{x}\right)\left(\frac{\sin x}{x}\right)\left(\frac{1}{1 + \cos x}\right)\right] = \frac{1}{2}$

7. $\lim\limits_{x \to 2} \frac{x-2}{x^2-4} = \lim\limits_{x \to 2} \frac{1}{2x} = \frac{1}{4}$

9. $\lim\limits_{t \to -3} \frac{t^3 - 4t + 15}{t^2 - t - 12} = \lim\limits_{t \to -3} \frac{3t^2 - 4}{2t - 1} = \frac{3(-3)^2 - 4}{2(-3) - 1} = -\frac{23}{7}$

11. $\lim\limits_{x \to \infty} \frac{5x^3 - 2x}{7x^3 + 3} = \lim\limits_{x \to \infty} \frac{15x^2 - 2}{21x^2} = \lim\limits_{x \to \infty} \frac{30x}{42x} = \lim\limits_{x \to \infty} \frac{30}{42} = \frac{5}{7}$

13. $\lim\limits_{t \to 0} \frac{\sin t^2}{t} = \lim\limits_{t \to 0} \frac{(\cos t^2)(2t)}{1} = 0$

15. $\lim\limits_{x \to 0} \dfrac{8x^2}{\cos x - 1} = \lim\limits_{x \to 0} \dfrac{16x}{-\sin x} = \lim\limits_{x \to 0} \dfrac{16}{-\cos x} = \dfrac{16}{-1} = -16$

17. $\lim\limits_{\theta \to \pi/2} \dfrac{2\theta - \pi}{\cos(2\pi - \theta)} = \lim\limits_{\theta \to \pi/2} \dfrac{2}{\sin(2\pi - \theta)} = \dfrac{2}{\sin\left(\frac{3\pi}{2}\right)} = -2$

19. $\lim\limits_{\theta \to \pi/2} \dfrac{1 - \sin\theta}{1 + \cos 2\theta} = \lim\limits_{\theta \to \pi/2} \dfrac{-\cos\theta}{-2\sin 2\theta} = \lim\limits_{\theta \to \pi/2} \dfrac{\sin\theta}{-4\cos 2\theta} = \dfrac{1}{(-4)(-1)} = \dfrac{1}{4}$

21. $\lim\limits_{x \to 0} \dfrac{x^2}{\ln(\sec x)} = \lim\limits_{x \to 0} \dfrac{2x}{\left(\frac{\sec x \tan x}{\sec x}\right)} = \lim\limits_{x \to 0} \dfrac{2x}{\tan x} = \lim\limits_{x \to 0} \dfrac{2}{\sec^2 x} = \dfrac{2}{1^2} = 2$

23. $\lim\limits_{t \to 0} \dfrac{t(1 - \cos t)}{t - \sin t} = \lim\limits_{t \to 0} \dfrac{(1 - \cos t) + t(\sin t)}{1 - \cos t} = \lim\limits_{t \to 0} \dfrac{\sin t + (\sin t + t \cos t)}{\sin t} = \lim\limits_{t \to 0} \dfrac{\cos t + \cos t + \cos t - t \sin t}{\cos t} = \dfrac{1 + 1 + 1 - 0}{1} = 3$

25. $\lim\limits_{x \to (\pi/2)^-} \left(x - \dfrac{\pi}{2}\right)\sec x = \lim\limits_{x \to (\pi/2)^-} \dfrac{\left(x - \frac{\pi}{2}\right)}{\cos x} = \lim\limits_{x \to (\pi/2)^-} \left(\dfrac{1}{-\sin x}\right) = \dfrac{1}{-1} = -1$

27. $\lim\limits_{\theta \to 0} \dfrac{3^{\sin\theta} - 1}{\theta} = \lim\limits_{\theta \to 0} \dfrac{3^{\sin\theta}(\ln 3)(\cos\theta)}{1} = \dfrac{(3^0)(\ln 3)(1)}{1} = \ln 3$

29. $\lim\limits_{x \to 0} \dfrac{x \, 2^x}{2^x - 1} = \lim\limits_{x \to 0} \dfrac{(1)(2^x) + (x)(\ln 2)(2^x)}{(\ln 2)(2^x)} = \dfrac{1 \cdot 2^0 + 0}{(\ln 2) \cdot 2^0} = \dfrac{1}{\ln 2}$

31. $\lim\limits_{x \to \infty} \dfrac{\ln(x+1)}{\log_2 x} = \lim\limits_{x \to \infty} \dfrac{\ln(x+1)}{\left(\frac{\ln x}{\ln 2}\right)} = (\ln 2)\lim\limits_{x \to \infty} \dfrac{\left(\frac{1}{x+1}\right)}{\left(\frac{1}{x}\right)} = (\ln 2)\lim\limits_{x \to \infty} \dfrac{x}{x+1} = (\ln 2)\lim\limits_{x \to \infty} \dfrac{1}{1} = \ln 2$

33. $\lim\limits_{x \to 0^+} \dfrac{\ln(x^2 + 2x)}{\ln x} = \lim\limits_{x \to 0^+} \dfrac{\left(\frac{2x+2}{x^2+2x}\right)}{\left(\frac{1}{x}\right)} = \lim\limits_{x \to 0^+} \dfrac{2x^2 + 2x}{x^2 + 2x} = \lim\limits_{x \to 0^+} \dfrac{4x+2}{2x+2} = \lim\limits_{x \to 0^+} \dfrac{2}{2} = 1$

35. $\lim\limits_{y \to 0} \dfrac{\sqrt{5y + 25} - 5}{y} = \lim\limits_{y \to 0} \dfrac{(5y + 25)^{1/2} - 5}{y} = \lim\limits_{y \to 0} \dfrac{\left(\frac{1}{2}\right)(5y + 25)^{-1/2}(5)}{1} = \lim\limits_{y \to 0} \dfrac{5}{2\sqrt{5y + 25}} = \dfrac{1}{2}$

37. $\lim\limits_{x \to \infty}[\ln 2x - \ln(x + 1)] = \lim\limits_{x \to \infty} \ln\left(\dfrac{2x}{x+1}\right) = \ln\left(\lim\limits_{x \to \infty} \dfrac{2x}{x+1}\right) = \ln\left(\lim\limits_{x \to \infty} \dfrac{2}{1}\right) = \ln 2$

39. $\lim\limits_{h \to 0} \dfrac{\sin(a + h) - \sin a}{h} = \lim\limits_{h \to 0} \dfrac{\cos(a + h) - \cos a}{1} = 0$

41. $\lim\limits_{x \to 1^+}\left(\dfrac{1}{x-1} - \dfrac{1}{\ln x}\right) = \lim\limits_{x \to 1^+}\left(\dfrac{\ln x - (x-1)}{(x-1)(\ln x)}\right) = \lim\limits_{x \to 1^+}\left(\dfrac{\frac{1}{x} - 1}{(\ln x) + (x-1)\left(\frac{1}{x}\right)}\right) = \lim\limits_{x \to 1^+}\left(\dfrac{1 - x}{(x \ln x) + x - 1}\right)$

$= \lim\limits_{x \to 1^+}\left(\dfrac{-1}{(\ln x + 1) + 1}\right) = \dfrac{-1}{(0 + 1) + 1} = -\dfrac{1}{2}$

43. $\lim\limits_{\theta \to 0} \dfrac{\cos\theta - 1}{e^\theta - \theta - 1} = \lim\limits_{\theta \to 0} \dfrac{-\sin\theta}{e^\theta - 1} = \lim\limits_{\theta \to 0} \dfrac{-\cos\theta}{e^\theta} = -1$

45. $\lim\limits_{t \to \infty} \dfrac{e^t + t^2}{e^t - 1} = \lim\limits_{t \to \infty} \dfrac{e^t + 2t}{e^t} = \lim\limits_{t \to \infty} \dfrac{e^t + 2}{e^t} = \lim\limits_{t \to \infty} \dfrac{e^t}{e^t} = 1$

47. The limit leads to the indeterminate form 1^∞. Let $f(x) = x^{1/(1-x)} \Rightarrow \ln f(x) = \ln\left(x^{1/(1-x)}\right) = \dfrac{\ln x}{1 - x}$. Now

$\lim\limits_{x \to 1^+} \ln f(x) = \lim\limits_{x \to 1^+} \dfrac{\ln x}{1 - x} = \lim\limits_{x \to 1^+} \dfrac{\left(\frac{1}{x}\right)}{-1} = -1$. Therefore $\lim\limits_{x \to 1^+} x^{1/(1-x)} = \lim\limits_{x \to 1^+} f(x) = \lim\limits_{x \to 1^+} e^{\ln f(x)} = e^{-1} = \dfrac{1}{e}$

49. The limit leads to the indeterminate form ∞^0. Let $f(x) = (\ln x)^{1/x} \Rightarrow \ln f(x) = \ln(\ln x)^{1/x} = \dfrac{\ln(\ln x)}{x}$. Now

$\lim\limits_{x \to \infty} \ln f(x) = \lim\limits_{x \to \infty} \dfrac{\ln(\ln x)}{x} = \lim\limits_{x \to \infty} \dfrac{\left(\frac{1}{x \ln x}\right)}{1} = 0$. Therefore $\lim\limits_{x \to \infty} (\ln x)^{1/x} = \lim\limits_{x \to \infty} f(x)$

$= \lim\limits_{x \to \infty} e^{\ln f(x)} = e^0 = 1$

51. The limit leads to the indeterminate form 0^0. Let $f(x) = x^{-1/\ln x} \Rightarrow \ln f(x) = -\frac{\ln x}{\ln x} = -1$. Therefore

$$\lim_{x \to 0^+} x^{-1/\ln x} = \lim_{x \to 0^+} f(x) = \lim_{x \to 0^+} e^{\ln f(x)} = e^{-1} = \frac{1}{e}$$

53. The limit leads to the indeterminate form ∞^0. Let $f(x) = (1 + 2x)^{1/(2\ln x)} \Rightarrow \ln f(x) = \frac{\ln(1+2x)}{2\ln x}$

$$\Rightarrow \lim_{x \to \infty} \ln f(x) = \lim_{x \to \infty} \frac{\ln(1+2x)}{2\ln x} = \lim_{x \to \infty} \frac{x}{1+2x} = \lim_{x \to \infty} \frac{1}{2} = \frac{1}{2}. \text{ Therefore } \lim_{x \to \infty} (1+2x)^{1/(2\ln x)}$$
$$= \lim_{x \to \infty} f(x) = \lim_{x \to \infty} e^{\ln f(x)} = e^{1/2}$$

55. The limit leads to the indeterminate form 0^0. Let $f(x) = x^x \Rightarrow \ln f(x) = x \ln x \Rightarrow \ln f(x) = \frac{\ln x}{\left(\frac{1}{x}\right)}$

$$= \lim_{x \to 0^+} \ln f(x) = \lim_{x \to 0^+} \frac{\ln x}{\left(\frac{1}{x}\right)} = \lim_{x \to 0^+} \frac{\left(\frac{1}{x}\right)}{\left(-\frac{1}{x^2}\right)} = \lim_{x \to 0^+} (-x) = 0. \text{ Therefore } \lim_{x \to 0^+} x^x = \lim_{x \to 0^+} f(x)$$
$$= \lim_{x \to 0^+} e^{\ln f(x)} = e^0 = 1$$

57. $\lim_{x \to \infty} \frac{\sqrt{9x+1}}{\sqrt{x+1}} = \sqrt{\lim_{x \to \infty} \frac{9x+1}{x+1}} = \sqrt{\lim_{x \to \infty} \frac{9}{1}} = \sqrt{9} = 3$

59. $\lim_{x \to \pi/2^-} \frac{\sec x}{\tan x} = \lim_{x \to \pi/2^-} \left(\frac{1}{\cos x}\right)\left(\frac{\cos x}{\sin x}\right) = \lim_{x \to \pi/2^-} \frac{1}{\sin x} = 1$

61. Part (b) is correct because part (a) is neither in the $\frac{0}{0}$ nor $\frac{\infty}{\infty}$ form and so l'Hôpital's rule may not be used.

63. Part (d) is correct, the other parts are indeterminate forms and cannot be calculated by the incorrect arithmetic

65. If $f(x)$ is to be continuous at $x = 0$, then $\lim_{x \to 0} f(x) = f(0) \Rightarrow c = f(0) = \lim_{x \to 0} \frac{9x - 3\sin 3x}{5x^3} = \lim_{x \to 0} \frac{9 - 9\cos 3x}{15x^2}$

$$= \lim_{x \to 0} \frac{27 \sin 3x}{30x} = \lim_{x \to 0} \frac{81 \cos 3x}{30} = \frac{27}{10}.$$

67. The graph indicates a limit near -1. The limit leads to the

indeterminate form $\frac{0}{0}$: $\lim_{x \to 1} \frac{2x^2 - (3x+1)\sqrt{x+2}}{x-1}$

$$= \lim_{x \to 1} \frac{2x^2 - 3x^{3/2} - x^{1/2} + 2}{x-1} = \lim_{x \to 1} \frac{4x - \frac{9}{2}x^{1/2} - \frac{1}{2}x^{-1/2}}{1}$$

$$= \frac{4 - \frac{9}{2} - \frac{1}{2}}{1} = \frac{4-5}{1} = -1$$

69. Let $f(k) = \left(1 + \frac{r}{k}\right)^k \Rightarrow \ln f(k) = \frac{\ln(1 + rk^{-1})}{k^{-1}} \Rightarrow \lim_{k \to \infty} \frac{\ln(1 + rk^{-1})}{k^{-1}} = \lim_{k \to \infty} \frac{\left(\frac{-rk^{-2}}{1+rk^{-1}}\right)}{-k^{-2}} = \lim_{k \to \infty} \frac{r}{1+rk^{-1}}$

$$= \lim_{k \to \infty} \frac{rk}{k+r} = \lim_{k \to \infty} \frac{r}{1} = r. \text{ Therefore } \lim_{k \to \infty} \left(1 + \frac{r}{k}\right)^k = \lim_{k \to \infty} f(k) = \lim_{k \to \infty} e^{\ln f(k)} = e^r.$$

71. (a) We should assign the value 1 to $f(x) = (\sin x)^x$ to
     make it continuous at $x = 0$.

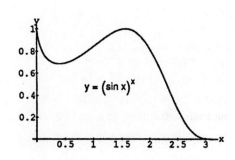

(b) $\ln f(x) = x \ln (\sin x) = \frac{\ln (\sin x)}{\left(\frac{1}{x}\right)} \Rightarrow \lim_{x \to 0^+} \ln f(x) = \lim_{x \to 0^+} \frac{\ln (\sin x)}{\left(\frac{1}{x}\right)} = \lim_{x \to 0^+} \frac{\left(\frac{1}{\sin x}\right)(\cos x)}{\left(-\frac{1}{x^2}\right)} = \lim_{x \to 0} \frac{-x^2}{\tan x}$

$= \lim_{x \to 0} \frac{-2x}{\sec^2 x} = 0 \Rightarrow \lim_{x \to 0} f(x) = e^0 = 1$

(c) The maximum value of $f(x)$ is close to 1 near the point $x \approx 1.55$ (see the graph in part (a)).

(d) The root in question is near 1.57.

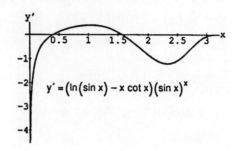

$y' = \left(\ln (\sin x) - x \cot x\right)\left(\sin x\right)^x$

4.7 NEWTON'S METHOD

1. $y = x^2 + x - 1 \Rightarrow y' = 2x + 1 \Rightarrow x_{n+1} = x_n - \frac{x_n^2 + x_n - 1}{2x_n + 1}$; $x_0 = 1 \Rightarrow x_1 = 1 - \frac{1+1-1}{2+1} = \frac{2}{3}$

$\Rightarrow x_2 = \frac{2}{3} - \frac{\frac{4}{9}+\frac{2}{3}-1}{\frac{4}{3}+1} \Rightarrow x_2 = \frac{2}{3} - \frac{4+6-9}{12+9} = \frac{2}{3} - \frac{1}{21} = \frac{13}{21} \approx .61905$; $x_0 = -1 \Rightarrow x_1 = 1 - \frac{1-1-1}{-2+1} = -2$

$\Rightarrow x_2 = -2 - \frac{4-2-1}{-4+1} = -\frac{5}{3} \approx -1.66667$

3. $y = x^4 + x - 3 \Rightarrow y' = 4x^3 + 1 \Rightarrow x_{n+1} = x_n - \frac{x_n^4 + x_n - 3}{4x_n^3 + 1}$; $x_0 = 1 \Rightarrow x_1 = 1 - \frac{1+1-3}{4+1} = \frac{6}{5}$

$\Rightarrow x_2 = \frac{6}{5} - \frac{\frac{1296}{625}+\frac{6}{5}-3}{\frac{864}{125}+1} = \frac{6}{5} - \frac{1296+750-1875}{4320+625} = \frac{6}{5} - \frac{171}{4945} = \frac{5763}{4945} \approx 1.16542$; $x_0 = -1 \Rightarrow x_1 = -1 - \frac{1-1-3}{-4+1}$

$= -2 \Rightarrow x_2 = -2 - \frac{16-2-3}{-32+1} = -2 + \frac{11}{31} = -\frac{51}{31} \approx -1.64516$

5. One obvious root is $x = 0$. Graphing e^{-x} and $2x + 1$ shows that $x = 0$ is the only root. Taking a naive approach we can use Newton's Method to estimate the root as follows: Let $f(x) = e^{-x} - 2x - 1$, $x_0 = 1$, and $x_{n+1} = x_n - \frac{f(x_n)}{f'(x_n)}$

$= x_n + \frac{e^{-x_n} - 2x_n - 1}{e^{-x_n} + 2}$. Performing iterations on a calculator, spreadsheet, or CAS gives $x_1 = -0.111594$,

$x_2 = -0.00215192$, $x_3 = -0.000000773248$. You may get different results depending upon what you select for $f(x)$ and x_0, and what calculator or computer you may use.

7. $f(x_0) = 0$ and $f'(x_0) \neq 0 \Rightarrow x_{n+1} = x_n - \frac{f(x_n)}{f'(x_n)}$ gives $x_1 = x_0 \Rightarrow x_2 = x_0 \Rightarrow x_n = x_0$ for all $n \geq 0$. That is, all of the approximations in Newton's method will be the root of $f(x) = 0$.

9. If $x_0 = h > 0 \Rightarrow x_1 = x_0 - \frac{f(x_0)}{f'(x_0)} = h - \frac{f(h)}{f'(h)}$

$= h - \frac{\sqrt{h}}{\left(\frac{1}{2\sqrt{h}}\right)} = h - \left(\sqrt{h}\right)\left(2\sqrt{h}\right) = -h$;

if $x_0 = -h < 0 \Rightarrow x_1 = x_0 - \frac{f(x_0)}{f'(x_0)} = -h - \frac{f(-h)}{f'(-h)}$

$= -h - \frac{\sqrt{h}}{\left(\frac{-1}{2\sqrt{h}}\right)} = -h + \left(\sqrt{h}\right)\left(2\sqrt{h}\right) = h.$

$y = \begin{cases} \sqrt{x}, & x \geq 0 \\ \sqrt{-x}, & x < 0 \end{cases}$

11. i) is equivalent to solving $x^3 - 3x - 1 = 0$.

ii) is equivalent to solving $x^3 - 3x - 1 = 0$.

iii) is equivalent to solving $x^3 - 3x - 1 = 0$.

iv) is equivalent to solving $x^3 - 3x - 1 = 0$.

All four equations are equivalent.

13. $f(x) = \tan x - 2x \Rightarrow f'(x) = \sec^2 x - 2 \Rightarrow x_{n+1} = x_n - \frac{\tan(x_n) - 2x_n}{\sec^2(x_n)}$; $x_0 = 1 \Rightarrow x_1 = 12920445 \Rightarrow x_2 = 1.155327774$

$\Rightarrow x_{16} = x_{17} = 1.165561185$

15. (a) The graph of $f(x) = \sin 3x - 0.99 + x^2$ in the window
$-2 \le x \le 2, -2 \le y \le 3$ suggests three roots.
However, when you zoom in on the x-axis near $x = 1.2$,
you can see that the graph lies above the axis there.
There are only two roots, one near $x = -1$, the other
near $x = 0.4$.

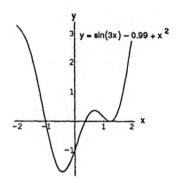

(b) $f(x) = \sin 3x - 0.99 + x^2 \Rightarrow f'(x) = 3 \cos 3x + 2x$

$\Rightarrow x_{n+1} = x_n - \frac{\sin(3x_n) - 0.99 + x_n^2}{3 \cos(3x_n) + 2x_n}$ and the solutions
are approximately 0.35003501505249 and
-1.0261731615301

17. $f(x) = 2x^4 - 4x^2 + 1 \Rightarrow f'(x) = 8x^3 - 8x \Rightarrow x_{n+1} = x_n - \frac{2x_n^4 - 4x_n^2 + 1}{8x_n^3 - 8x_n}$; if $x_0 = -2$, then $x_6 = -1.30656296$; if
$x_0 = -0.5$, then $x_3 = -0.5411961$; the roots are approximately ± 0.5411961 and ± 1.30656296 because $f(x)$ is an even
function.

19. Graphing e^{-x^2} and $x^2 - x + 1$ shows that there are two places where the curves intersect, one at $x = 0$ and the other
between $x = 0.5$ and $x = 0.6$. Let $f(x) = e^{-x^2} - x^2 + x - 1$, $x_0 = 0.5$, and $x_{n+1} = x_n - \frac{f(x_n)}{f'(x_n)} = x_n - \frac{e^{-x_n^2} - x_n^2 + x_n - 1}{1 - 2x_n - 2x_n e^{-x_n^2}}$.
Performing iterations on a calculator, spreadsheet, or CAS gives $x_1 = 0.536981$, $x_2 = 0.534856$, $x_3 = 0.53485$,
$x_4 = 0.53485$. (You may get different results depending upon what you select for $f(x)$ and x_0, and what calculator or
computer you may use.) Therefore, the two curves intersect at $x = 0$ and $x = 0.53485$.

21. If $f(x) = x^3 + 2x - 4$, then $f(1) = -1 < 0$ and $f(2) = 8 > 0 \Rightarrow$ by the Intermediate Value Theorem the equation
$x^3 + 2x - 4 = 0$ has a solution between 1 and 2. Consequently, $f'(x) = 3x^2 + 2$ and $x_{n+1} = x_n - \frac{x_n^3 + 2x_n - 4}{3x_n^2 + 2}$.
Then $x_0 = 1 \Rightarrow x_1 = 1.2 \Rightarrow x_2 = 1.17975 \Rightarrow x_3 = 1.179509 \Rightarrow x_4 = 1.1795090 \Rightarrow$ the root is approximately
1.17951.

23. $f(x) = 4x^4 - 4x^2 \Rightarrow f'(x) = 16x^3 - 8x \Rightarrow x_{i+1} = x_i - \frac{f(x_i)}{f'(x_i)} = x_i - \frac{x_i^3 - x_i}{4x_i^2 - 2}$. Iterations are performed using the
procedure in problem 13 in this section.

(a) For $x_0 = -2$ or $x_0 = -0.8$, $x_i \to -1$ as i gets large.

(b) For $x_0 = -0.5$ or $x_0 = 0.25$, $x_i \to 0$ as i gets large.

(c) For $x_0 = 0.8$ or $x_0 = 2$, $x_i \to 1$ as i gets large.

(d) (If your calculator has a CAS, put it in exact mode, otherwise approximate the radicals with a decimal value.)
For $x_0 = -\frac{\sqrt{21}}{7}$ or $x_0 = -\frac{\sqrt{21}}{7}$, Newton's method does not converge. The values of x_i alternate between
$x_0 = -\frac{\sqrt{21}}{7}$ or $x_0 = -\frac{\sqrt{21}}{7}$ as i increases.

25. $f(x) = (x - 1)^{40} \Rightarrow f'(x) = 40(x - 1)^{39} \Rightarrow x_{n+1} = x_n - \frac{(x_n - 1)^{40}}{40(x_n - 1)^{39}} = \frac{39x_n + 1}{40}$. With $x_0 = 2$, our computer
gave $x_{87} = x_{88} = x_{89} = \cdots = x_{200} = 1.11051$, coming within 0.11051 of the root $x = 1$.

4.8 ANTIDERIVATIVES

1. (a) x^2 (b) $\frac{x^3}{3}$ (c) $\frac{x^3}{3} - x^2 + x$

3. (a) x^{-3} (b) $-\frac{x^{-3}}{3}$ (c) $-\frac{x^{-3}}{3} + x^2 + 3x$

5. (a) $\frac{-1}{x}$ (b) $\frac{-5}{x}$ (c) $2x + \frac{5}{x}$

7. (a) $\sqrt{x^3}$ (b) $\sqrt{x}$ (c) $\frac{2}{3}\sqrt{x^3} + 2\sqrt{x}$

9. (a) $x^{2/3}$ (b) $x^{1/3}$ (c) $x^{-1/3}$

11. (a) $\ln|x|$ (b) $7\ln|x|$ (c) $x - 5\ln|x|$

13. (a) $\cos(\pi x)$ (b) $-3\cos x$ (c) $\frac{-\cos(\pi x)}{\pi} + \cos(3x)$

15. (a) $\tan x$ (b) $2\tan\left(\frac{x}{3}\right)$ (c) $-\frac{2}{3}\tan\left(\frac{3x}{2}\right)$

17. (a) $-\csc x$ (b) $\frac{1}{5}\csc(5x)$ (c) $2\csc\left(\frac{\pi x}{2}\right)$

19. (a) $\frac{1}{3}e^{3x}$ (b) $-e^{-x}$ (c) $2e^{x/2}$

21. (a) $\frac{1}{\ln 3} \cdot 3^x$ (b) $\frac{-1}{\ln 2} \cdot 2^{-x}$ (c) $\frac{1}{\ln(5/3)} \cdot \left(\frac{5}{3}\right)^x$

23. (a) $2\sin^{-1}x$ (b) $\frac{1}{2}\tan^{-1}x$ (c) $\frac{1}{2}\tan^{-1}(2x)$

25. $\int (x+1)\,dx = \frac{x^2}{2} + x + C$ 27. $\int \left(3t^2 + \frac{t}{2}\right) dt = t^3 + \frac{t^2}{4} + C$

29. $\int (2x^3 - 5x + 7)\,dx = \frac{1}{2}x^4 - \frac{5}{2}x^2 + 7x + C$

31. $\int \left(\frac{1}{x^2} - x^2 - \frac{1}{3}\right) dx = \int \left(x^{-2} - x^2 - \frac{1}{3}\right) dx = \frac{x^{-1}}{-1} - \frac{x^3}{3} - \frac{1}{3}x + C = -\frac{1}{x} - \frac{x^3}{3} - \frac{x}{3} + C$

33. $\int x^{-1/3}\,dx = \frac{x^{2/3}}{\frac{2}{3}} + C = \frac{3}{2}x^{2/3} + C$

35. $\int \left(\sqrt{x} + \sqrt[3]{x}\right) dx = \int \left(x^{1/2} + x^{1/3}\right) dx = \frac{x^{3/2}}{\frac{3}{2}} + \frac{x^{4/3}}{\frac{4}{3}} + C = \frac{2}{3}x^{3/2} + \frac{3}{4}x^{4/3} + C$

37. $\int \left(8y - \frac{2}{y^{1/4}}\right) dy = \int \left(8y - 2y^{-1/4}\right) dy = \frac{8y^2}{2} - 2\left(\frac{y^{3/4}}{\frac{3}{4}}\right) + C = 4y^2 - \frac{8}{3}y^{3/4} + C$

39. $\int 2x\left(1 - x^{-3}\right) dx = \int \left(2x - 2x^{-2}\right) dx = \frac{2x^2}{2} - 2\left(\frac{x^{-1}}{-1}\right) + C = x^2 + \frac{2}{x} + C$

41. $\int \frac{t\sqrt{t}+\sqrt{t}}{t^2}\,dt = \int \left(\frac{t^{3/2}}{t^2} + \frac{t^{1/2}}{t^2}\right) dt = \int \left(t^{-1/2} + t^{-3/2}\right) dt = \frac{t^{1/2}}{\frac{1}{2}} + \left(\frac{t^{-1/2}}{-\frac{1}{2}}\right) + C = 2\sqrt{t} - \frac{2}{\sqrt{t}} + C$

43. $\int -2\cos t\,dt = -2\sin t + C$ 45. $\int 7\sin\frac{\theta}{3}\,d\theta = -21\cos\frac{\theta}{3} + C$

47. $\int -3 \csc^2 x \, dx = 3 \cot x + C$

49. $\int \frac{\csc \theta \cot \theta}{2} \, d\theta = -\frac{1}{2} \csc \theta + C$

51. $\int (e^{3x} + 5e^{-x}) \, dx = \frac{e^{3x}}{3} - 5e^{-x} + C$

53. $\int (e^{-x} + 4^x) \, dx = -e^{-x} + \frac{4^x}{\ln 4} + C$

55. $\int (4 \sec x \tan x - 2 \sec^2 x) \, dx = 4 \sec x - 2 \tan x + C$

57. $\int (\sin 2x - \csc^2 x) \, dx = -\frac{1}{2} \cos 2x + \cot x + C$

59. $\int \frac{1 + \cos 4t}{2} \, dt = \int \left(\frac{1}{2} + \frac{1}{2} \cos 4t \right) dt = \frac{1}{2} t + \frac{1}{2} \left(\frac{\sin 4t}{4} \right) + C = \frac{t}{2} + \frac{\sin 4t}{8} + C$

61. $\int \left(\frac{1}{x} - \frac{5}{x^2 + 1} \right) dx = \ln |x| - 5 \tan^{-1} x + C$

63. $\int 3x^{\sqrt{3}} \, dx = \frac{3x^{(\sqrt{3} + 1)}}{\sqrt{3} + 1} + C$

65. $\int (1 + \tan^2 \theta) \, d\theta = \int \sec^2 \theta \, d\theta = \tan \theta + C$

67. $\int \cot^2 x \, dx = \int (\csc^2 x - 1) \, dx = -\cot x - x + C$

69. $\int \cos \theta (\tan \theta + \sec \theta) \, d\theta = \int (\sin \theta + 1) \, d\theta = -\cos \theta + \theta + C$

71. $\frac{d}{dx} \left(\frac{(7x - 2)^4}{28} + C \right) = \frac{4(7x - 2)^3 (7)}{28} = (7x - 2)^3$

73. $\frac{d}{dx} \left(\frac{1}{5} \tan (5x - 1) + C \right) = \frac{1}{5} (\sec^2 (5x - 1)) (5) = \sec^2 (5x - 1)$

75. $\frac{d}{dx} \left(\frac{-1}{x + 1} + C \right) = (-1)(-1)(x + 1)^{-2} = \frac{1}{(x + 1)^2}$

77. $\frac{d}{dx} (\ln(x + 1) + C) = \frac{1}{x + 1}$

79. $\frac{d}{dx} \left(\frac{1}{a} \tan^{-1} \left(\frac{x}{a} \right) + C \right) = \frac{1}{a} \cdot \frac{1}{1 + \left(\frac{x}{a} \right)^2} \cdot \frac{d}{dx} \left(\frac{x}{a} \right) = \frac{1}{a^2 \left(1 + \frac{x^2}{a^2} \right)} = \frac{1}{a^2 + x^2}$

81. If $y = \ln x - \frac{1}{2} \ln (1 + x^2) - \frac{\tan^{-1} x}{x} + C$, then $dy = \left[\frac{1}{x} - \frac{x}{1 + x^2} - \frac{\left(\frac{x}{1 + x^2} \right) - \tan^{-1} x}{x^2} \right] dx$

$= \left(\frac{1}{x} - \frac{x}{1 + x^2} - \frac{1}{x(1 + x^2)} + \frac{\tan^{-1} x}{x^2} \right) dx = \frac{x(1 + x^2) - x^3 - x + (\tan^{-1} x)(1 + x^2)}{x^2 (1 + x^2)} \, dx = \frac{\tan^{-1} x}{x^2} \, dx$,

which verifies the formula

83. (a) Wrong: $\frac{d}{dx} \left(\frac{x^2}{2} \sin x + C \right) = \frac{2x}{2} \sin x + \frac{x^2}{2} \cos x = x \sin x + \frac{x^2}{2} \cos x \neq x \sin x$

   (b) Wrong: $\frac{d}{dx} (-x \cos x + C) = -\cos x + x \sin x \neq x \sin x$

   (c) Right: $\frac{d}{dx} (-x \cos x + \sin x + C) = -\cos x + x \sin x + \cos x = x \sin x$

85. (a) Wrong: $\frac{d}{dx} \left(\frac{(2x + 1)^3}{3} + C \right) = \frac{3(2x + 1)^2 (2)}{3} = 2(2x + 1)^2 \neq (2x + 1)^2$

   (b) Wrong: $\frac{d}{dx} ((2x + 1)^3 + C) = 3(2x + 1)^2 (2) = 6(2x + 1)^2 \neq 3(2x + 1)^2$

   (c) Right: $\frac{d}{dx} ((2x + 1)^3 + C) = 6(2x + 1)^2$

87. Graph (b), because $\frac{dy}{dx} = 2x \Rightarrow y = x^2 + C$. Then $y(1) = 4 \Rightarrow C = 3$.

89. $\frac{dy}{dx} = 2x - 7 \Rightarrow y = x^2 - 7x + C$; at $x = 2$ and $y = 0$ we have $0 = 2^2 - 7(2) + C \Rightarrow C = 10 \Rightarrow y = x^2 - 7x + 10$

91. $\frac{dy}{dx} = \frac{1}{x^2} + x = x^{-2} + x \Rightarrow y = -x^{-1} + \frac{x^2}{2} + C$; at $x = 2$ and $y = 1$ we have $1 = -2^{-1} + \frac{2^2}{2} + C \Rightarrow C = -\frac{1}{2}$

$\Rightarrow y = -x^{-1} + \frac{x^2}{2} - \frac{1}{2}$ or $y = -\frac{1}{x} + \frac{x^2}{2} - \frac{1}{2}$

93. $\frac{dy}{dx} = 3x^{-2/3} \Rightarrow y = \frac{3x^{1/3}}{\frac{1}{3}} + C = 9$; at $x = 9x^{1/3} + C$; at $x = -1$ and $y = -5$ we have $-5 = 9(-1)^{1/3} + C \Rightarrow C = 4$

$\Rightarrow y = 9x^{1/3} + 4$

95. $\frac{ds}{dt} = 1 + \cos t \Rightarrow s = t + \sin t + C$; at $t = 0$ and $s = 4$ we have $4 = 0 + \sin 0 + C \Rightarrow C = 4 \Rightarrow s = t + \sin t + 4$

97. $\frac{dr}{d\theta} = -\pi \sin \pi\theta \Rightarrow r = \cos(\pi\theta) + C$; at $r = 0$ and $\theta = 0$ we have $0 = \cos(\pi 0) + C \Rightarrow C = -1 \Rightarrow r = \cos(\pi\theta) - 1$

99. $\frac{dv}{dt} = \frac{1}{2} \sec t \tan t \Rightarrow v = \frac{1}{2} \sec t + C$; at $v = 1$ and $t = 0$ we have $1 = \frac{1}{2} \sec(0) + C \Rightarrow C = \frac{1}{2} \Rightarrow v = \frac{1}{2} \sec t + \frac{1}{2}$

101. $\frac{dv}{dt} = \frac{3}{t\sqrt{t^2-1}}, t > 1 \Rightarrow v = 3 \sec^{-1} t + C$; at $t = 2$ and $v = 0$ we have $0 = 3 \sec^{-1} 2 + C \Rightarrow C = -\pi \Rightarrow v = 3 \sec^{-1} t - \pi$

103. $\frac{d^2y}{dx^2} = 2 - 6x \Rightarrow \frac{dy}{dx} = 2x - 3x^2 + C_1$; at $\frac{dy}{dx} = 4$ and $x = 0$ we have $4 = 2(0) - 3(0)^2 + C_1 \Rightarrow C_1 = 4$

$\Rightarrow \frac{dy}{dx} = 2x - 3x^2 + 4 \Rightarrow y = x^2 - x^3 + 4x + C_2$; at $y = 1$ and $x = 0$ we have $1 = 0^2 - 0^3 + 4(0) + C_2 \Rightarrow C_2 = 1$

$\Rightarrow y = x^2 - x^3 + 4x + 1$

105. $\frac{d^2r}{dt^2} = \frac{2}{t^3} = 2t^{-3} \Rightarrow \frac{dr}{dt} = -t^{-2} + C_1$; at $\frac{dr}{dt} = 1$ and $t = 1$ we have $1 = -(1)^{-2} + C_1 \Rightarrow C_1 = 2 \Rightarrow \frac{dr}{dt} = -t^{-2} + 2$

$\Rightarrow r = t^{-1} + 2t + C_2$; at $r = 1$ and $t = 1$ we have $1 = 1^{-1} + 2(1) + C_2 \Rightarrow C_2 = -2 \Rightarrow r = t^{-1} + 2t - 2$ or

$r = \frac{1}{t} + 2t - 2$

107. $\frac{d^3y}{dx^3} = 6 \Rightarrow \frac{d^2y}{dx^2} = 6x + C_1$; at $\frac{d^2y}{dx^2} = -8$ and $x = 0$ we have $-8 = 6(0) + C_1 \Rightarrow C_1 = -8 \Rightarrow \frac{d^2y}{dx^2} = 6x - 8$

$\Rightarrow \frac{dy}{dx} = 3x^2 - 8x + C_2$; at $\frac{dy}{dx} = 0$ and $x = 0$ we have $0 = 3(0)^2 - 8(0) + C_2 \Rightarrow C_2 = 0 \Rightarrow \frac{dy}{dx} = 3x^2 - 8x$

$\Rightarrow y = x^3 - 4x^2 + C_3$; at $y = 5$ and $x = 0$ we have $5 = 0^3 - 4(0)^2 + C_3 \Rightarrow C_3 = 5 \Rightarrow y = x^3 - 4x^2 + 5$

109. $y^{(4)} = -\sin t + \cos t \Rightarrow y''' = \cos t + \sin t + C_1$; at $y''' = 7$ and $t = 0$ we have $7 = \cos(0) + \sin(0) + C_1$

$\Rightarrow C_1 = 6 \Rightarrow y''' = \cos t + \sin t + 6 \Rightarrow y'' = \sin t - \cos t + 6t + C_2$; at $y'' = -1$ and $t = 0$ we have

$-1 = \sin(0) - \cos(0) + 6(0) + C_2 \Rightarrow C_2 = 0 \Rightarrow y'' = \sin t - \cos t + 6t \Rightarrow y' = -\cos t - \sin t + 3t^2 + C_3$;

at $y' = -1$ and $t = 0$ we have $-1 = -\cos(0) - \sin(0) + 3(0)^2 + C_3 \Rightarrow C_3 = 0 \Rightarrow y' = -\cos t - \sin t + 3t^2$

$\Rightarrow y = -\sin t + \cos t + t^3 + C_4$; at $y = 0$ and $t = 0$ we have $0 = -\sin(0) + \cos(0) + 0^3 + C_4 \Rightarrow C_4 = -1$

$\Rightarrow y = -\sin t + \cos t + t^3 - 1$

111. $m = y' = 3\sqrt{x} = 3x^{1/2} \Rightarrow y = 2x^{3/2} + C$; at $(9, 4)$ we have $4 = 2(9)^{3/2} + C \Rightarrow C = -50 \Rightarrow y = 2x^{3/2} - 50$

113. $\frac{dy}{dx} = 1 - \frac{4}{3}x^{1/3} \Rightarrow y = \int \left(1 - \frac{4}{3}x^{1/3}\right) dx = x - x^{4/3} + C$; at $(1, 0.5)$ on the curve we have $0.5 = 1 - 1^{4/3} + C$

$\Rightarrow C = 0.5 \Rightarrow y = x - x^{4/3} + \frac{1}{2}$

115. $\frac{dy}{dx} = \sin x - \cos x \Rightarrow y = \int (\sin x - \cos x) dx = -\cos x - \sin x + C$; at $(-\pi, -1)$ on the curve we have

$-1 = -\cos(-\pi) - \sin(-\pi) + C \Rightarrow C = -2 \Rightarrow y = -\cos x - \sin x - 2$

117. (a) $\frac{ds}{dt} = 9.8t - 3 \Rightarrow s = 4.9t^2 - 3t + C$; (i) at $s = 5$ and $t = 0$ we have $C = 5 \Rightarrow s = 4.9t^2 - 3t + 5$;

displacement $= s(3) - s(1) = ((4.9)(9) - 9 + 5) - (4.9 - 3 + 5) = 33.2$ units; (ii) at $s = -2$ and $t = 0$ we have

$C = -2 \Rightarrow s = 4.9t^2 - 3t - 2$; displacement $= s(3) - s(1) = ((4.9)(9) - 9 - 2) - (4.9 - 3 - 2) = 33.2$ units;

(iii) at $s = s_0$ and $t = 0$ we have $C = s_0 \Rightarrow s = 4.9t^2 - 3t + s_0$; displacement $= s(3) - s(1)$

$= ((4.9)(9) - 9 + s_0) - (4.9 - 3 + s_0) = 33.2$ units

(b) True. Given an antiderivative f(t) of the velocity function, we know that the body's position function is $s = f(t) + C$ for some constant C. Therefore, the displacement from $t = a$ to $t = b$ is $(f(b) + C) - (f(a) + C)$ $= f(b) - f(a)$. Thus we can find the displacement from any antiderivative f as the numerical difference $f(b) - f(a)$ without knowing the exact values of C and s.

119. Step 1: $\frac{d^2s}{dt^2} = -k \Rightarrow \frac{ds}{dt} = -kt + C_1$; at $\frac{ds}{dt} = 88$ and $t = 0$ we have $C_1 = 88 \Rightarrow \frac{ds}{dt} = -kt + 88 \Rightarrow$

   $s = -k\left(\frac{t^2}{2}\right) + 88t + C_2$; at $s = 0$ and $t = 0$ we have $C_2 = 0 \Rightarrow s = -\frac{kt^2}{2} + 88t$

   Step 2: $\frac{ds}{dt} = 0 \Rightarrow 0 = -kt + 88 \Rightarrow t = \frac{88}{k}$

   Step 3: $242 = \frac{-k\left(\frac{88}{k}\right)^2}{2} + 88\left(\frac{88}{k}\right) \Rightarrow 242 = -\frac{(88)^2}{2k} + \frac{(88)^2}{k} \Rightarrow 242 = \frac{(88)^2}{2k} \Rightarrow k = 16$

121. (a) $v = \int a\,dt = \int \left(15t^{1/2} - 3t^{-1/2}\right)dt = 10t^{3/2} - 6t^{1/2} + C$; $\frac{ds}{dt}(1) = 4 \Rightarrow 4 = 10(1)^{3/2} - 6(1)^{1/2} + C \Rightarrow C = 0$
   $\Rightarrow v = 10t^{3/2} - 6t^{1/2}$

   (b) $s = \int v\,dt = \int \left(10t^{3/2} - 6t^{1/2}\right)dt = 4t^{5/2} - 4t^{3/2} + C$; $s(1) = 0 \Rightarrow 0 = 4(1)^{5/2} - 4(1)^{3/2} + C \Rightarrow C = 0$
   $\Rightarrow s = 4t^{5/2} - 4t^{3/2}$

123. $\frac{d^2s}{dt^2} = a \Rightarrow \frac{ds}{dt} = \int a\,dt = at + C$; $\frac{ds}{dt} = v_0$ when $t = 0 \Rightarrow C = v_0 \Rightarrow \frac{ds}{dt} = at + v_0 \Rightarrow s = \frac{at^2}{2} + v_0 t + C_1$; $s = s_0$
   when $t = 0 \Rightarrow s_0 = \frac{a(0)^2}{2} + v_0(0) + C_1 \Rightarrow C_1 = s_0 \Rightarrow s = \frac{at^2}{2} + v_0 t + s_0$

125. (a) $\int f(x)\,dx = 1 - \sqrt{x} + C_1 = -\sqrt{x} + C$ (b) $\int g(x)\,dx = x + 2 + C_1 = x + C$

   (c) $\int -f(x)\,dx = -\left(1 - \sqrt{x}\right) + C_1 = \sqrt{x} + C$ (d) $\int -g(x)\,dx = -(x + 2) + C_1 = -x + C$

   (e) $\int [f(x) + g(x)]\,dx = \left(1 - \sqrt{x}\right) + (x + 2) + C_1 = x - \sqrt{x} + C$

   (f) $\int [f(x) - g(x)]\,dx = \left(1 - \sqrt{x}\right) - (x + 2) + C_1 = -x - \sqrt{x} + C$

CHAPTER 4 PRACTICE EXERCISES

1. No, since $f(x) = x^3 + 2x + \tan x \Rightarrow f'(x) = 3x^2 + 2 + \sec^2 x > 0 \Rightarrow f(x)$ is always increasing on its domain

3. No absolute minimum because $\lim_{x \to \infty} (7 + x)(11 - 3x)^{1/3} = -\infty$. Next $f'(x) =$
   $(11 - 3x)^{1/3} - (7 + x)(11 - 3x)^{-2/3} = \frac{(11-3x)-(7+x)}{(11-3x)^{2/3}} = \frac{4(1-x)}{(11-3x)^{2/3}} \Rightarrow x = 1$ and $x = \frac{11}{3}$ are critical points.
   Since $f' > 0$ if $x < 1$ and $f' < 0$ if $x > 1$, $f(1) = 16$ is the absolute maximum.

5. $g(x) = e^x - x \Rightarrow g'(x) = e^x - 1 \Rightarrow g' = \underset{0}{--- \mid +++} \Rightarrow$ the graph is decreasing on $(-\infty, 0)$, increasing on $(0, \infty)$;
   an absolute minimum value is 1 at $x = 0$; $x = 0$ is the only critical point of g; there is no absolute maximum value

7. $f(x) = x - 2\ln x$ on $1 \le x \le 3 \Rightarrow f'(x) = 1 - \frac{2}{x} \Rightarrow f' = \underset{1}{\mid} \underset{2}{--- \mid} \underset{3}{+++ \mid} \Rightarrow$ the graph is decreasing on $(1, 2)$,
   increasing on $(2, 3)$; an absolute minimum value is $2 - 2\ln 2$ at $x = 2$; an absolute maximum value is 1 at $x = 1$.

9. Yes, because at each point of $[0, 1)$ except $x = 0$, the function's value is a local minimum value as well as a local maximum value. At $x = 0$ the function's value, 0, is not a local minimum value because each open interval around $x = 0$ on the x-axis contains points to the left of 0 where f equals -1.

11. No, because the interval $0 < x < 1$ fails to be closed. The Extreme Value Theorem says that if the function is continuous throughout a finite closed interval $a \le x \le b$ then the existence of absolute extrema is guaranteed on that interval.

13. (a) There appear to be local minima at $x = -1.75$ and 1.8. Points of inflection are indicated at approximately $x = 0$ and $x = \pm 1$.

(b) $f'(x) = x^7 - 3x^5 - 5x^4 + 15x^2 = x^2(x^2 - 3)(x^3 - 5)$. The pattern $y' = $ --- | +++ | +++ | --- | +++

$\qquad\qquad\qquad\qquad\qquad\qquad\qquad\qquad -\sqrt{3} \quad 0 \quad \sqrt[3]{5} \quad \sqrt{3}$

indicates a local maximum at $x = \sqrt[3]{5}$ and local minima at $x = \pm\sqrt{3}$.

(c)

15. (a) $g(t) = \sin^2 t - 3t \Rightarrow g'(t) = 2 \sin t \cos t - 3 = \sin(2t) - 3 \Rightarrow g' < 0 \Rightarrow g(t)$ is always falling and hence must decrease on every interval in its domain.

(b) One, since $\sin^2 t - 3t - 5 = 0$ and $\sin^2 t - 3t = 5$ have the same solutions: $f(t) = \sin^2 t - 3t - 5$ has the same derivative as $g(t)$ in part (a) and is always decreasing with $f(-3) > 0$ and $f(0) < 0$. The Intermediate Value Theorem guarantees the continuous function f has a root in $[-3, 0]$.

17. (a) $f(x) = x^4 + 2x^2 - 2 \Rightarrow f'(x) = 4x^3 + 4x$. Since $f(0) = -2 < 0$, $f(1) = 1 > 0$ and $f'(x) \geq 0$ for $0 \leq x \leq 1$, we may conclude from the Intermediate Value Theorem that $f(x)$ has exactly one solution when $0 \leq x \leq 1$.

(b) $x^2 = \frac{-2 \pm \sqrt{4+8}}{2} > 0 \Rightarrow x^2 = \sqrt{3} - 1$ and $x \geq 0 \Rightarrow x \approx \sqrt{.7320508076} \approx .8555996772$

19. Let $V(t)$ represent the volume of the water in the reservoir at time t, in minutes, let $V(0) = a_0$ be the initial amount and $V(1440) = a_0 + (1400)(43,560)(7.48)$ gallons be the amount of water contained in the reservoir after the rain, where $24 \text{ hr} = 1440 \text{ min}$. Assume that $V(t)$ is continuous on $[0, 1440]$ and differentiable on $(0, 1440)$. The Mean Value Theorem says that for some t_0 in $(0, 1440)$ we have $V'(t_0) = \frac{V(1440) - V(0)}{1440 - 0}$

$= \frac{a_0 + (1400)(43,560)(7.48) - a_0}{1440} = \frac{456,160,320 \text{ gal}}{1440 \text{ min}} = 316,778$ gal/min. Therefore at t_0 the reservoir's volume was increasing at a rate in excess of 225,000 gal/min.

21. No, $\frac{x}{x+1} = 1 + \frac{-1}{x+1} \Rightarrow \frac{x}{x+1}$ differs from $\frac{-1}{x+1}$ by the constant 1. Both functions have the same derivative $\frac{d}{dx}\left(\frac{x}{x+1}\right) = \frac{(x+1) - x(1)}{(x+1)^2} = \frac{1}{(x+1)^2} = \frac{d}{dx}\left(\frac{-1}{x+1}\right)$.

23. The global minimum value of $\frac{1}{2}$ occurs at $x = 2$.

25. (a) $t = 0, 6, 12$ (b) $t = 3, 9$ (c) $6 < t < 12$ (d) $0 < t < 6, 12 < t < 14$

27.

29.

31.

33.

35.

37.

39.

41.

43. (a) $y' = 16 - x^2 \Rightarrow y' = \begin{array}{c} {} \\ --- \end{array} \underset{-4}{\mid} \begin{array}{c} {} \\ +++ \end{array} \underset{4}{\mid} \begin{array}{c} {} \\ --- \end{array} \Rightarrow$ the curve is rising on $(-4, 4)$, falling on $(-\infty, -4)$ and $(4, \infty)$

$\Rightarrow$ a local maximum at $x = 4$ and a local minimum at $x = -4$; $y'' = -2x \Rightarrow y'' = +++ \underset{0}{\mid} \ ---$ $\Rightarrow$ the curve

is concave up on $(-\infty, 0)$, concave down on $(0, \infty)$ $\Rightarrow$ a point of inflection at $x = 0$

(b)

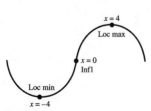

45. (a) $y' = 6x(x+1)(x-2) = 6x^3 - 6x^2 - 12x \Rightarrow y' = ---\underset{-1}{|} +++\underset{0}{|} ---\underset{2}{|} +++ \Rightarrow$ the graph is rising on $(-1, 0)$

and $(2, \infty)$, falling on $(-\infty, -1)$ and $(0, 2) \Rightarrow$ a local maximum at $x = 0$, local minima at $x = -1$ and

$x = 2$; $y'' = 18x^2 - 12x - 12 = 6(3x^2 - 2x - 2) = 6\left(x - \frac{1-\sqrt{7}}{3}\right)\left(x - \frac{1+\sqrt{7}}{3}\right) \Rightarrow$

$y'' = +++\underset{\frac{1-\sqrt{7}}{3}}{|} ---\underset{\frac{1+\sqrt{7}}{3}}{|} +++ \Rightarrow$ the curve is concave up on $\left(-\infty, \frac{1-\sqrt{7}}{3}\right)$ and $\left(\frac{1+\sqrt{7}}{3}, \infty\right)$, concave down

on $\left(\frac{1-\sqrt{7}}{3}, \frac{1+\sqrt{7}}{3}\right) \Rightarrow$ points of inflection at $x = \frac{1 \pm \sqrt{7}}{3}$

(b)

47. (a) $y' = x^4 - 2x^2 = x^2(x^2 - 2) \Rightarrow y' = +++\underset{-\sqrt{2}}{|} ---\underset{0}{|} ---\underset{\sqrt{2}}{|} +++ \Rightarrow$ the curve is rising on $\left(-\infty, -\sqrt{2}\right)$ and

$\left(\sqrt{2}, \infty\right)$, falling on $\left(-\sqrt{2}, \sqrt{2}\right) \Rightarrow$ a local maximum at $x = -\sqrt{2}$ and a local minimum at $x = \sqrt{2}$;

$y'' = 4x^3 - 4x = 4x(x-1)(x+1) \Rightarrow y'' = ---\underset{-1}{|} +++\underset{0}{|} ---\underset{1}{|} +++ \Rightarrow$ concave up on $(-1, 0)$ and $(1, \infty)$,

concave down on $(-\infty, -1)$ and $(0, 1) \Rightarrow$ points of inflection at $x = 0$ and $x = \pm 1$

(b)

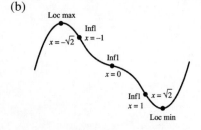

49. The values of the first derivative indicate that the curve is rising on $(0, \infty)$ and falling on $(-\infty, 0)$. The slope
of the curve approaches $-\infty$ as $x \to 0^-$, and approaches ∞ as $x \to 0^+$ and $x \to 1$. The curve should therefore
have a cusp and local minimum at $x = 0$, and a vertical tangent at $x = 1$.

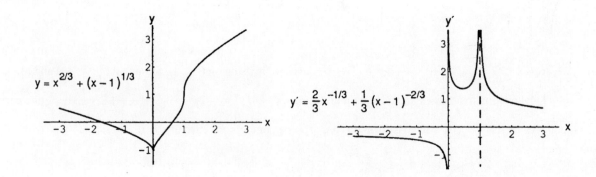

51. The values of the first derivative indicate that the curve is always rising. The slope of the curve approaches ∞ as $x \to 0$ and as $x \to 1$, indicating vertical tangents at both $x = 0$ and $x = 1$.

53. $y = \frac{x+1}{x-3} = 1 + \frac{4}{x-3}$

55. $y = \frac{x^2+1}{x} = x + \frac{1}{x}$

 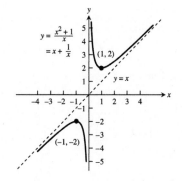

57. $y = \frac{x^3+2}{2x} = \frac{x^2}{2} + \frac{1}{x}$

59. $y = \frac{x^2-4}{x^2-3} = 1 - \frac{1}{x^2-3}$

61. $\displaystyle\lim_{x \to 1} \frac{x^2+3x-4}{x-1} = \lim_{x \to 1} \frac{2x+3}{1} = 5$

63. $\displaystyle\lim_{x \to \pi} \frac{\tan x}{x} = \frac{\tan \pi}{\pi} = 0$

65. $\displaystyle\lim_{x \to 0} \frac{\sin^2 x}{\tan(x^2)} = \lim_{x \to 0} \frac{2\sin x \cdot \cos x}{2x\sec^2(x^2)} = \lim_{x \to 0} \frac{\sin(2x)}{2x\sec^2(x^2)} = \lim_{x \to 0} \frac{2\cos(2x)}{2x\left(2\sec^2(x^2)\tan(x^2)\cdot 2x\right) + 2\sec^2(x^2)} = \frac{2}{0+2\cdot 1} = 1$

67. $\displaystyle\lim_{x \to \pi/2^-} \sec(7x)\cos(3x) = \lim_{x \to \pi/2^-} \frac{\cos(3x)}{\cos(7x)} = \lim_{x \to \pi/2^-} \frac{-3\sin(3x)}{-7\sin(7x)} = \frac{3}{7}$

69. $\displaystyle\lim_{x \to 0} (\csc x - \cot x) = \lim_{x \to 0} \frac{1-\cos x}{\sin x} = \lim_{x \to 0} \frac{\sin x}{\cos x} = \frac{0}{1} = 0$

71. $\lim\limits_{x \to \infty} \left(\sqrt{x^2 + x + 1} - \sqrt{x^2 - x} \right) = \lim\limits_{x \to \infty} \left(\sqrt{x^2 + x + 1} - \sqrt{x^2 - x} \right) \cdot \dfrac{\sqrt{x^2 + x + 1} + \sqrt{x^2 - x}}{\sqrt{x^2 + x + 1} + \sqrt{x^2 - x}}$

$= \lim\limits_{x \to \infty} \dfrac{2x + 1}{\sqrt{x^2 + x + 1} + \sqrt{x^2 - x}}$

Notice that $x = \sqrt{x^2}$ for $x > 0$ so this is equivalent to

$= \lim\limits_{x \to \infty} \dfrac{\frac{2x + 1}{x}}{\sqrt{\frac{x^2 + x + 1}{x^2}} + \sqrt{\frac{x^2 - x}{x^2}}} = \lim\limits_{x \to \infty} \dfrac{2 + \frac{1}{x}}{\sqrt{1 + \frac{1}{x} + \frac{1}{x^2}} + \sqrt{1 - \frac{1}{x}}} = \dfrac{2}{\sqrt{1} + \sqrt{1}} = 1$

73. The limit leads to the indeterminate form $\frac{0}{0}$: $\lim\limits_{x \to 0} \dfrac{10^x - 1}{x} = \lim\limits_{x \to 0} \dfrac{(\ln 10)10^x}{1} = \ln 10$

75. The limit leads to the indeterminate form $\frac{0}{0}$: $\lim\limits_{x \to 0} \dfrac{2^{\sin x} - 1}{e^x - 1} = \lim\limits_{x \to 0} \dfrac{2^{\sin x}(\ln 2)(\cos x)}{e^x} = \ln 2$

77. The limit leads to the indeterminate form $\frac{0}{0}$: $\lim\limits_{x \to 0} \dfrac{5 - 5\cos x}{e^x - x - 1} = \lim\limits_{x \to 0} \dfrac{5 \sin x}{e^x - 1} = \lim\limits_{x \to 0} \dfrac{5 \cos x}{e^x} = 5$

79. The limit leads to the indeterminate form $\frac{0}{0}$: $\lim\limits_{t \to 0^+} \dfrac{t - \ln(1 + 2t)}{t^2} = \lim\limits_{t \to 0^+} \dfrac{\left(1 - \frac{2}{1 + 2t}\right)}{2t} = -\infty$

81. The limit leads to the indeterminate form $\frac{0}{0}$: $\lim\limits_{t \to 0^+} \left(\dfrac{e^t}{t} - \dfrac{1}{t} \right) = \lim\limits_{t \to 0^+} \left(\dfrac{e^t - 1}{t} \right) = \lim\limits_{t \to 0^+} \dfrac{e^t}{1} = 1$

83. $\lim\limits_{x \to \infty} \left(1 + \dfrac{b}{x} \right)^{kx} = \lim\limits_{x \to \infty} \left[\left(1 + \dfrac{1}{x/b} \right)^{x/b} \right]^{bk} = e^{bk}$

85. (a) Maximize $f(x) = \sqrt{x} - \sqrt{36 - x} = x^{1/2} - (36 - x)^{1/2}$ where $0 \le x \le 36$

$\Rightarrow f'(x) = \frac{1}{2} x^{-1/2} - \frac{1}{2}(36 - x)^{-1/2}(-1) = \dfrac{\sqrt{36 - x} + \sqrt{x}}{2\sqrt{x}\sqrt{36 - x}} \Rightarrow$ derivative fails to exist at 0 and 36; $f(0) = -6$,

and $f(36) = 6 \Rightarrow$ the numbers are 0 and 36

(b) Maximize $g(x) = \sqrt{x} + \sqrt{36 - x} = x^{1/2} + (36 - x)^{1/2}$ where $0 \le x \le 36$

$\Rightarrow g'(x) = \frac{1}{2} x^{-1/2} + \frac{1}{2}(36 - x)^{-1/2}(-1) = \dfrac{\sqrt{36 - x} - \sqrt{x}}{2\sqrt{x}\sqrt{36 - x}} \Rightarrow$ critical points at 0, 18 and 36; $g(0) = 6$,

$g(18) = 2\sqrt{18} = 6\sqrt{2}$ and $g(36) = 6 \Rightarrow$ the numbers are 18 and 18

87. $A(x) = \frac{1}{2} (2x) (27 - x^2)$ for $0 \le x \le \sqrt{27}$

$\Rightarrow A'(x) = 3(3 + x)(3 - x)$ and $A''(x) = -6x.$

The critical points are -3 and 3, but -3 is not in the

domain. Since $A''(3) = -18 < 0$ and $A\left(\sqrt{27}\right) = 0,$

the maximum occurs at $x = 3 \Rightarrow$ the largest area is

$A(3) = 54$ sq units.

89. From the diagram we have $\left(\frac{h}{2}\right)^2 + r^2 = \left(\sqrt{3}\right)^2$

$\Rightarrow r^2 = \frac{12 - h^2}{4}.$ The volume of the cylinder is

$V = \pi r^2 h = \pi \left(\frac{12 - h^2}{4} \right) h = \frac{\pi}{4} (12h - h^3),$ where

$0 \le h \le 2\sqrt{3}.$ Then $V'(h) = \frac{3\pi}{4} (2 + h)(2 - h)$

$\Rightarrow$ the critical points are -2 and 2, but -2 is not in

the domain. At $h = 2$ there is a maximum since

$V''(2) = -3\pi < 0.$ The dimensions of the largest

cylinder are radius $= \sqrt{2}$ and height $= 2.$

91. The profit $P = 2px + py = 2px + p\left(\frac{40 - 10x}{5 - x}\right)$, where p is the profit on grade B tires and $0 \le x \le 4$. Thus

$P'(x) = \frac{2p}{(5 - x)^2}(x^2 - 10x + 20) \Rightarrow$ the critical points are $\left(5 - \sqrt{5}\right)$, 5, and $\left(5 + \sqrt{5}\right)$, but only $\left(5 - \sqrt{5}\right)$ is in

the domain. Now $P'(x) > 0$ for $0 < x < \left(5 - \sqrt{5}\right)$ and $P'(x) < 0$ for $\left(5 - \sqrt{5}\right) < x < 4 \Rightarrow$ at $x = \left(5 - \sqrt{5}\right)$ there

is a local maximum. Also $P(0) = 8p$, $P\left(5 - \sqrt{5}\right) = 4p\left(5 - \sqrt{5}\right) \approx 11p$, and $P(4) = 8p \Rightarrow$ at $x = \left(5 - \sqrt{5}\right)$ there

is an absolute maximum. The maximum occurs when $x = \left(5 - \sqrt{5}\right)$ and $y = 2\left(5 - \sqrt{5}\right)$, the units are

hundreds of tires, i.e., $x \approx 276$ tires and $y \approx 553$ tires.

93. The dimensions will be x in. by $10 - 2x$ in. by $16 - 2x$ in., so $V(x) = x(10 - 2x)(16 - 2x) = 4x^3 - 52x^2 + 160x$ for
$0 < x < 5$. Then $V'(x) = 12x^2 - 104x + 160 = 4(x - 2)(3x - 20)$, so the critical point in the correct domain is $x = 2$.
This critical point corresponds to the maximum possible volume because $V'(x) > 0$ for $0 < x < 2$ and $V'(x) < 0$ for
$2 < x < 5$. The box of largest volume has a height of 2 in. and a base measuring 6 in. by 12 in., and its volume is 144 in.[3]
Graphical support:

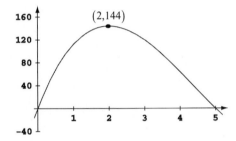

95. $g(x) = 3x - x^3 + 4 \Rightarrow g(2) = 2 > 0$ and $g(3) = -14 < 0 \Rightarrow g(x) = 0$ in the interval $[2, 3]$ by the Intermediate

Value Theorem. Then $g'(x) = 3 - 3x^2 \Rightarrow x_{n+1} = x_n - \frac{3x_n - x_n^3 + 4}{3 - 3x_n^2}$; $x_0 = 2 \Rightarrow x_1 = 2.\overline{22} \Rightarrow x_2 = 2.196215$, and

so forth to $x_5 = 2.195823345$.

97. $\int (x^3 + 5x - 7)\, dx = \frac{x^4}{4} + \frac{5x^2}{2} - 7x + C$

99. $\int \left(3\sqrt{t} + \frac{4}{t^2}\right) dt = \int \left(3t^{1/2} + 4t^{-2}\right) dt = \frac{3t^{3/2}}{\left(\frac{3}{2}\right)} + \frac{4t^{-1}}{-1} + C = 2t^{3/2} - \frac{4}{t} + C$

101. Let $u = r + 5 \Rightarrow du = dr$

$\int \frac{dr}{(r + 5)^2} = \int \frac{du}{u^2} = \int u^{-2}\, du = \frac{u^{-1}}{-1} + C = -u^{-1} + C = -\frac{1}{(r + 5)} + C$

103. Let $u = \theta^2 + 1 \Rightarrow du = 2\theta\, d\theta \Rightarrow \frac{1}{2}\, du = \theta\, d\theta$

$\int 3\theta\sqrt{\theta^2 + 1}\, d\theta = \int \sqrt{u}\left(\frac{3}{2}\, du\right) = \frac{3}{2}\int u^{1/2}\, du = \frac{3}{2}\left(\frac{u^{3/2}}{\frac{3}{2}}\right) + C = u^{3/2} + C = \left(\theta^2 + 1\right)^{3/2} + C$

105. Let $u = 1 + x^4 \Rightarrow du = 4x^3\, dx \Rightarrow \frac{1}{4}\, du = x^3\, dx$

$\int x^3 \left(1 + x^4\right)^{-1/4} dx = \int u^{-1/4}\left(\frac{1}{4}\, du\right) = \frac{1}{4}\int u^{-1/4}\, du = \frac{1}{4}\left(\frac{u^{3/4}}{\frac{3}{4}}\right) + C = \frac{1}{3}u^{3/4} + C = \frac{1}{3}\left(1 + x^4\right)^{3/4} + C$

107. Let $u = \frac{s}{10} \Rightarrow du = \frac{1}{10}\, ds \Rightarrow 10\, du = ds$

$\int \sec^2 \frac{s}{10}\, ds = \int \left(\sec^2 u\right)(10\, du) = 10\int \sec^2 u\, du = 10\tan u + C = 10\tan\frac{s}{10} + C$

109. Let $u = \sqrt{2}\,\theta \Rightarrow du = \sqrt{2}\,d\theta \Rightarrow \frac{1}{\sqrt{2}}\,du = d\theta$

$\int \csc \sqrt{2}\theta \cot \sqrt{2}\theta\, d\theta = \int (\csc u \cot u)\left(\frac{1}{\sqrt{2}}\,du\right) = \frac{1}{\sqrt{2}}(-\csc u) + C = -\frac{1}{\sqrt{2}} \csc \sqrt{2}\theta + C$

111. Let $u = \frac{x}{4} \Rightarrow du = \frac{1}{4}\,dx \Rightarrow 4\,du = dx$

$\int \sin^2 \frac{x}{4}\,dx = \int (\sin^2 u)\,(4\,du) = \int 4\left(\frac{1 - \cos 2u}{2}\right)du = 2\int (1 - \cos 2u)\,du = 2\left(u - \frac{\sin 2u}{2}\right) + C$
$= 2u - \sin 2u + C = 2\left(\frac{x}{4}\right) - \sin 2\left(\frac{x}{4}\right) + C = \frac{x}{2} - \sin \frac{x}{2} + C$

113. $\int \left(\frac{3}{x} - x\right)dx = 3\ln|x| - \frac{x^2}{2} + C$

115. $\int \left(\frac{1}{2}e^t - e^{-t}\right)dt = \frac{1}{2}e^t - \frac{e^{-t}}{-1} + C = \frac{1}{2}e^t + e^{-t} + C$

117. $\int (\theta^{1-\pi})d\theta = \frac{\theta^{2-\pi}}{2-\pi} + C$

119. $\int \frac{3}{2x\sqrt{x^2-1}}\,dx = \frac{3}{2}\int \frac{1}{x\sqrt{x^2-1}}\,dx = \frac{3}{2}\sec^{-1}x + C$

121. $y = \int \frac{x^2+1}{x^2}\,dx = \int (1 + x^{-2})\,dx = x - x^{-1} + C = x - \frac{1}{x} + C;\, y = -1$ when $x = 1 \Rightarrow 1 - \frac{1}{1} + C = -1$
$\Rightarrow C = -1 \Rightarrow y = x - \frac{1}{x} - 1$

123. $\frac{dr}{dt} = \int \left(15\sqrt{t} + \frac{3}{\sqrt{t}}\right)dt = \int (15t^{1/2} + 3t^{-1/2})\,dt = 10t^{3/2} + 6t^{1/2} + C;\, \frac{dr}{dt} = 8$ when $t = 1$

$\Rightarrow 10(1)^{3/2} + 6(1)^{1/2} + C = 8 \Rightarrow C = -8.$ Thus $\frac{dr}{dt} = 10t^{3/2} + 6t^{1/2} - 8 \Rightarrow r = \int (10t^{3/2} + 6t^{1/2} - 8)\,dt$

$= 4t^{5/2} + 4t^{3/2} - 8t + C;\, r = 0$ when $t = 1 \Rightarrow 4(1)^{5/2} + 4(1)^{3/2} - 8(1) + C_1 = 0 \Rightarrow C_1 = 0.$ Therefore,
$r = 4t^{5/2} + 4t^{3/2} - 8t$

125. Yes, $\sin^{-1} x$ and $-\cos^{-1} x$ differ by the constant $\frac{\pi}{2}$

127. $A = xy = xe^{-x^2} \Rightarrow \frac{dA}{dx} = e^{-x^2} + (x)(-2x)\,e^{-x^2} = e^{-x^2}(1 - 2x^2).$ Solving $\frac{dA}{dx} = 0 \Rightarrow 1 - 2x^2 = 0$

$\Rightarrow x = \frac{1}{\sqrt{2}};\, \frac{dA}{dx} < 0$ for $x > \frac{1}{\sqrt{2}}$ and $\frac{dA}{dx} > 0$ for $0 < x < \frac{1}{\sqrt{2}} \Rightarrow$ absolute maximum of $\frac{1}{\sqrt{2}}\,e^{-1/2} = \frac{1}{\sqrt{2e}}$ at

$x = \frac{1}{\sqrt{2}}$ units long by $y = e^{-1/2} = \frac{1}{\sqrt{e}}$ units high.

129. $y = x \ln 2x - x \Rightarrow y' = x\left(\frac{2}{2x}\right) + \ln(2x) - 1 = \ln 2x;$
solving $y' = 0 \Rightarrow x = \frac{1}{2};\, y' > 0$ for $x > \frac{1}{2}$ and $y' < 0$ for
$x < \frac{1}{2} \Rightarrow$ relative minimum of $-\frac{1}{2}$ at $x = \frac{1}{2};\, f\left(\frac{1}{2e}\right) = -\frac{1}{e}$
and $f\left(\frac{e}{2}\right) = 0 \Rightarrow$ absolute minimum is $-\frac{1}{2}$ at $x = \frac{1}{2}$ and
the absolute maximum is 0 at $x = \frac{e}{2}$

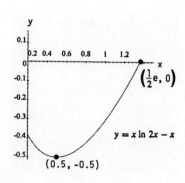

131. $f(x) = e^{x/\sqrt{x^4+1}}$ for all x in $(-\infty, \infty)$; $f'(x) = \left[\dfrac{\left(\sqrt{x^4+1}\right)\cdot 1 - x\left(\frac{2x^3}{\sqrt{x^4+1}}\right)}{\left(\sqrt{x^4+1}\right)^2} \right] e^{x/\sqrt{x^4+1}} = \dfrac{1-x^4}{\left(\sqrt{x^4+1}\right)^3} e^{x/\sqrt{x^4+1}}$

$= \dfrac{(1-x^2)(1+x^2)}{(x^4+1)^{3/2}} e^{x/\sqrt{x^4+1}} = 0 \Rightarrow 1 - x^2 = 0 \Rightarrow x = \pm 1$ are critical points. Consider the behavior of f as $x \to \pm\infty$;

$\lim\limits_{x \to \infty} e^{x/\sqrt{x^4+1}} = \lim\limits_{x \to -\infty} e^{x/\sqrt{x^4+1}} = 1$ as suggested by the following table (14 digit precision, 12 digits displayed):

x	$x/\sqrt{x^4+1}$	$e^{x/\sqrt{x^4+1}}$
$-\infty$	0	1
⋮	⋮	⋮
-100000	$-0.0000\ 10000\ 0000\ 00000$	$0.9999\ 9000\ 0050$
-10000	$-0.0001\ 0000\ 0000\ 000$	$0.9999\ 0000\ 5000$
-1000	$-0.0010\ 0000\ 0000\ 00$	$0.9990\ 0049\ 9833$
-100	$-0.0099\ 9999\ 9950\ 00$	$0.9900\ 4983\ 3799$
-10	$-0.0999\ 9500\ 0375\ 0$	$0.9048\ 4194\ 1895$
0	0	1
10	$0.0999\ 9500\ 0375\ 0$	$1.1051\ 6539\ 265$
100	$0.0099\ 9999\ 9950\ 00$	$1.0100\ 5016\ 703$
1000	$0.0010\ 0000\ 0000\ 00$	$1.0010\ 0050\ 017$
10000	$0.0001\ 0000\ 0000\ 000$	$1.0001\ 0000\ 500$
100000	$0.0000\ 10000\ 0000\ 00000$	$1.0000\ 1000\ 005$
⋮	⋮	⋮
∞	0	1

Therefore, $y = 1$ is a horizontal asymptote in both directions. Check the critical points for absolute extreme values: $f(-1) = e^{-\sqrt{2}/2} \approx 0.4931$, $f(1) = e^{\sqrt{2}/2} \approx 2.0281 \Rightarrow$ the absolute minimum value of the function is $e^{-\sqrt{2}/2}$ at $x = -1$, and the absolute maximum value is $e^{\sqrt{2}/2}$ at $x = 1$.

133. (a) $y = \dfrac{\ln x}{\sqrt{x}} \Rightarrow y' = \dfrac{1}{x\sqrt{x}} - \dfrac{\ln x}{2x^{3/2}} = \dfrac{2 - \ln x}{2x\sqrt{x}}$

$\Rightarrow y'' = -\dfrac{3}{4} x^{-5/2}(2 - \ln x) - \dfrac{1}{2} x^{-5/2} = x^{-5/2}\left(\dfrac{3}{4}\ln x - 2\right)$;

solving $y' = 0 \Rightarrow \ln x = 2 \Rightarrow x = e^2$; $y' < 0$ for $x > e^2$ and and $y' > 0$ for $x < e^2 \Rightarrow$ a maximum of $\dfrac{2}{e}$; $y'' = 0$

$\Rightarrow \ln x = \dfrac{8}{3} \Rightarrow x = e^{8/3}$; the curve is concave down on $\left(0, e^{8/3}\right)$ and concave up on $\left(e^{8/3}, \infty\right)$; so there is an inflection point at $\left(e^{8/3}, \dfrac{8}{3e^{4/3}}\right)$.

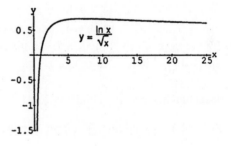

(b) $y = e^{-x^2} \Rightarrow y' = -2xe^{-x^2} \Rightarrow y'' = -2e^{-x^2} + 4x^2 e^{-x^2}$

$= (4x^2 - 2)e^{-x^2}$; solving $y' = 0 \Rightarrow x = 0$; $y' < 0$ for $x > 0$ and $y' > 0$ for $x < 0 \Rightarrow$ a maximum at $x = 0$ of $e^0 = 1$; there are points of inflection at $x = \pm\dfrac{1}{\sqrt{2}}$; the curve is concave down for $-\dfrac{1}{\sqrt{2}} < x < \dfrac{1}{\sqrt{2}}$ and concave up otherwise.

(c) $y = (1+x)e^{-x} \Rightarrow y' = e^{-x} - (1+x)e^{-x} = -xe^{-x}$

$\Rightarrow y'' = -e^{-x} + xe^{-x} = (x-1)e^{-x}$; solving $y' = 0$ $\Rightarrow -xe^{-x} = 0 \Rightarrow x = 0$; $y' < 0$ for $x > 0$ and $y' > 0$ for $x < 0 \Rightarrow$ a maximum at $x = 0$ of $(1+0)e^0 = 1$; there is a point of inflection at $x = 1$ and the curve is concave up for $x > 1$ and concave down for $x < 1$.

135   In the interval $\pi < x < 2\pi$ the function $\sin x < 0$
      $\Rightarrow (\sin x)^{\sin x}$ is not defined for all values in that
      interval or its translation by 2π.

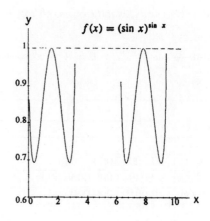

$f(x) = (\sin x)^{\sin x}$

CHAPTER 4   ADDITIONAL AND ADVANCED EXERCISES

1.   If M and m are the maximum and minimum values, respectively, then $m \le f(x) \le M$ for all $x \in I$. If $m = M$
     then f is constant on I.

3.   On an open interval the extreme values of a continuous function (if any) must occur at an interior critical
     point. On a half-open interval the extreme values of a continuous function may be at a critical point or at the
     closed endpoint. Extreme values occur only where $f' = 0$, f' does not exist, or at the endpoints of the interval.
     Thus the extreme points will not be at the ends of an open interval.

5.   (a)   If $y' = 6(x+1)(x-2)^2$, then $y' < 0$ for $x < -1$ and $y' > 0$ for $x > -1$. The sign pattern is
           $f' = --- \mid +++ \mid +++ \Rightarrow$ f has a local minimum at $x = -1$. Also $y'' = 6(x-2)^2 + 12(x+1)(x-2)$
           $\qquad\qquad\;\; -1 \qquad 2$
           $= 6(x-2)(3x) \Rightarrow y'' > 0$ for $x < 0$ or $x > 2$, while $y'' < 0$ for $0 < x < 2$. Therefore f has points of inflection
           at $x = 0$ and $x = 2$. There is no local maximum.
     (b)   If $y' = 6x(x+1)(x-2)$, then $y' < 0$ for $x < -1$ and $0 < x < 2$; $y' > 0$ for $-1 < x < 0$ and $x > 2$. The sign
           sign pattern is $y' = --- \mid +++ \mid --- \mid +++$. Therefore f has a local maximum at $x = 0$ and
           $\qquad\qquad\qquad\qquad\quad -1 \quad\;\; 0 \quad\;\; 2$
           local minima at $x = -1$ and $x = 2$. Also, $y'' = 18\left[x - \left(\frac{1-\sqrt{7}}{3}\right)\right]\left[x - \left(\frac{1+\sqrt{7}}{3}\right)\right]$, so $y'' < 0$ for
           $\frac{1-\sqrt{7}}{3} < x < \frac{1+\sqrt{7}}{3}$ and $y'' > 0$ for all other $x \Rightarrow$ f has points of inflection at $x = \frac{1\pm\sqrt{7}}{3}$.

7.   If f is continuous on $[a, c)$ and $f'(x) \le 0$ on $[a, c)$, then by the Mean Value Theorem for all $x \in [a, c)$ we have
     $\frac{f(c) - f(x)}{c - x} \le 0 \Rightarrow f(c) - f(x) \le 0 \Rightarrow f(x) \ge f(c)$. Also if f is continuous on $(c, b]$ and $f'(x) \ge 0$ on $(c, b]$, then for
     all $x \in (c, b]$ we have $\frac{f(x) - f(c)}{x - c} \ge 0 \Rightarrow f(x) - f(c) \ge 0 \Rightarrow f(x) \ge f(c)$. Therefore $f(x) \ge f(c)$ for all $x \in [a, b]$.

9.   No. Corollary 1 requires that $f'(x) = 0$ for <u>all</u> x in some interval I, not $f'(x) = 0$ at a single point in I.

11. From (ii), $f(-1) = \frac{-1+a}{b-c+2} = 0 \Rightarrow a = 1$; from (iii), either $1 = \lim_{x \to \infty} f(x)$ or $1 = \lim_{x \to -\infty} f(x)$. In either case,
     $\lim_{x \to \pm\infty} f(x) = \lim_{x \to \pm\infty} \frac{x+1}{bx^2 + cx + 2} = \lim_{x \to \pm\infty} \frac{1 + \frac{1}{x}}{bx + c + \frac{2}{x}} = 1 \Rightarrow b = 0$ and $c = 1$. For if $b = 1$, then
     $\lim_{x \to \pm\infty} \frac{1 + \frac{1}{x}}{x + c + \frac{2}{x}} = 0$ and if $c = 0$, then $\lim_{x \to \pm\infty} \frac{1 + \frac{1}{x}}{bx + \frac{2}{x}} = \lim_{x \to \pm\infty} \frac{1 + \frac{1}{x}}{\frac{2}{x}} = \pm\infty$. Thus $a = 1$, $b = 0$, and $c = 1$.

13. The area of the $\triangle ABC$ is $A(x) = \frac{1}{2}(2)\sqrt{1-x^2} = (1-x^2)^{1/2}$,

where $0 \le x \le 1$. Thus $A'(x) = \frac{-x}{\sqrt{1-x^2}} \Rightarrow 0$ and ± 1 are

critical points. Also $A(\pm 1) = 0$ so $A(0) = 1$ is the

maximum. When $x = 0$ the $\triangle ABC$ is isosceles since

$AC = BC = \sqrt{2}$.

15. The time it would take the water to hit the ground from height y is $\sqrt{\frac{2y}{g}}$, where g is the acceleration of

gravity. The product of time and exit velocity (rate) yields the distance the water travels:

$D(y) = \sqrt{\frac{2y}{g}}\sqrt{64(h-y)} = 8\sqrt{\frac{2}{g}}(hy - y^2)^{1/2}, 0 \le y \le h \Rightarrow D'(y) = -4\sqrt{\frac{2}{g}}(hy - y^2)^{-1/2}(h - 2y) \Rightarrow 0, \frac{h}{2}$ and h

are critical points. Now $D(0) = 0$, $D\left(\frac{h}{2}\right) = \frac{8h}{\sqrt{g}}$ and $D(h) = 0 \Rightarrow$ the best place to drill the hole is at $y = \frac{h}{2}$.

17. The surface area of the cylinder is $S = 2\pi r^2 + 2\pi rh$. From

the diagram we have $\frac{r}{R} = \frac{H-h}{H} \Rightarrow h = \frac{RH - rH}{R}$ and

$S(r) = 2\pi r(r + h) = 2\pi r\left(r + H - r\frac{H}{R}\right)$

$= 2\pi\left(1 - \frac{H}{R}\right)r^2 + 2\pi Hr$, where $0 \le r \le R$.

Case 1: $H < R \Rightarrow S(r)$ is a quadratic equation containing

the origin and concave upward $\Rightarrow S(r)$ is maximum at

$r = R$.

Case 2: $H = R \Rightarrow S(r)$ is a linear equation containing the

origin with a positive slope $\Rightarrow S(r)$ is maximum at

$r = R$.

Case 3: $H > R \Rightarrow S(r)$ is a quadratic equation containing the origin and concave downward. Then

$\frac{dS}{dr} = 4\pi\left(1 - \frac{H}{R}\right)r + 2\pi H$ and $\frac{dS}{dr} = 0 \Rightarrow 4\pi\left(1 - \frac{H}{R}\right)r + 2\pi H = 0 \Rightarrow r = \frac{RH}{2(H-R)}$. For simplification

we let $r^* = \frac{RH}{2(H-R)}$.

(a) If $R < H < 2R$, then $0 > H - 2R \Rightarrow H > 2(H - R) \Rightarrow \frac{RH}{2(H-R)} > R$ which is impossible.

(b) If $H = 2R$, then $r^* = \frac{2R^2}{2R} = R \Rightarrow S(r)$ is maximum at $r = R$.

(c) If $H > 2R$, then $2R + H < 2H \Rightarrow H < 2(H - R) \Rightarrow \frac{H}{2(H-R)} < 1 \Rightarrow \frac{RH}{2(H-R)} < R \Rightarrow r^* < R$. Therefore,

$S(r)$ is a maximum at $r = r^* = \frac{RH}{2(H-R)}$.

Conclusion: If $H \in (0, R]$ or $H = 2R$, then the maximum surface area is at $r = R$. If $H \in (R, 2R)$, then $r > R$

which is not possible. If $H \in (2R, \infty)$, then the maximum is at $r = r^* = \frac{RH}{2(H-R)}$.

19. (a) $\lim\limits_{x \to 0} \frac{2\sin(5x)}{3x} = \lim\limits_{x \to 0} \frac{2\sin(5x)}{\frac{3}{5}(5x)} = \lim\limits_{x \to 0} \frac{10}{3}\frac{\sin(5x)}{(5x)} = \frac{10}{3} \cdot 1 = \frac{10}{3}$

(b) $\lim\limits_{x \to 0} \sin(5x)\cot(3x) = \lim\limits_{x \to 0} \frac{\sin(5x)\cos(3x)}{\sin(3x)} = \lim\limits_{x \to 0} \frac{-3\sin(5x)\sin(3x) + 5\cos(5x)\cos(3x)}{3\cos(3x)} = \frac{5}{3}$

(c) $\lim\limits_{x \to 0} x\csc^2\sqrt{2x} = \lim\limits_{x \to 0} \frac{x}{\sin^2\sqrt{2x}} = \lim\limits_{x \to 0} \frac{1}{\frac{2\sin\sqrt{2x}\cos\sqrt{2x}}{\sqrt{2x}}} = \lim\limits_{x \to 0} \frac{\sqrt{2x}}{\sin(2\sqrt{2x})} = \lim\limits_{x \to 0} \frac{\frac{1}{\sqrt{2x}}}{\cos(2\sqrt{2x})\frac{2}{\sqrt{2x}}}$

$= \lim\limits_{x \to 0} \frac{1}{\cos(2\sqrt{2x}) \cdot 2} = \frac{1}{2}$

(d) $\lim\limits_{x \to \pi/2} (\sec x - \tan x) = \lim\limits_{x \to \pi/2} \frac{1 - \sin x}{\cos x} = \lim\limits_{x \to \pi/2} \frac{-\cos x}{-\sin x} = 0$.

(e) $\lim\limits_{x \to 0} \frac{x - \sin x}{x - \tan x} = \lim\limits_{x \to 0} \frac{1 - \cos x}{1 - \sec^2 x} = \lim\limits_{x \to 0} \frac{1 - \cos x}{-\tan^2 x} = \lim\limits_{x \to 0} \frac{\cos x - 1}{\tan^2 x} = \lim\limits_{x \to 0} \frac{-\sin x}{2\tan x\sec^2 x} = \lim\limits_{x \to 0} \frac{-\sin x}{\frac{2\sin x}{\cos^3 x}} =$

$= \lim\limits_{x \to 0} \frac{\cos^3 x}{-2} = -\frac{1}{2}$

(f) $\lim\limits_{x \to 0} \frac{\sin(x^2)}{x\sin x} = \lim\limits_{x \to 0} \frac{2x\cos(x^2)}{x\cos x + \sin x} = \lim\limits_{x \to 0} \frac{-(2x^2)\sin(x^2) + 2\cos(x^2)}{-x\sin x + 2\cos x} = \frac{2}{2} = 1$

(g) $\lim\limits_{x \to 0} \frac{\sec x - 1}{x^2} = \lim\limits_{x \to 0} \frac{\sec x \tan x}{2x} = \lim\limits_{x \to 0} \frac{\sec^3 x + \tan^2 x \sec x}{2} = \frac{1+0}{2} = \frac{1}{2}$

(h) $\lim\limits_{x \to 2} \frac{x^3 - 8}{x^2 - 4} = \lim\limits_{x \to 2} \frac{(x-2)(x^2+2x+4)}{(x-2)(x+2)} = \lim\limits_{x \to 2} \frac{x^2+2x+4}{x+2} = \frac{4+4+4}{4} = 3$

21. (a) The profit function is $P(x) = (c - ex)x - (a + bx) = -ex^2 + (c - b)x - a$. $P'(x) = -2ex + c - b = 0$
$\Rightarrow x = \frac{c-b}{2e}$. $P''(x) = -2e < 0$ if $e > 0$ so that the profit function is maximized at $x = \frac{c-b}{2e}$.

(b) The price therefore that corresponds to a production level yeilding a maximum profit is
$p\Big|_{x=\frac{c-b}{2e}} = c - e\left(\frac{c-b}{2e}\right) = \frac{c+b}{2}$ dollars.

(c) The weekly profit at this production level is $P(x) = -e\left(\frac{c-b}{2e}\right)^2 + (c - b)\left(\frac{c-b}{2e}\right) - a = \frac{(c-b)^2}{4e} - a$.

(d) The tax increases cost to the new profit function is $F(x) = (c - ex)x - (a + bx + tx) = -ex^2 + (c - b - t)x - a$.
Now $F'(x) = -2ex + c - b - t = 0$ when $x = \frac{t+b-c}{-2e} = \frac{c-b-t}{2e}$. Since $F''(x) = -2e < 0$ if $e > 0$, F is maximized
when $x = \frac{c-b-t}{2e}$ units per week. Thus the price per unit is $p = c - e\left(\frac{c-b-t}{2e}\right) = \frac{c+b+t}{2}$ dollars. Thus, such a tax
increases the cost per unit by $\frac{c+b+t}{2} - \frac{c+b}{2} = \frac{t}{2}$ dollars if units are priced to maximize profit.

23. $x_1 = x_0 - \frac{f(x_0)}{f'(x_0)} = x_0 - \frac{x_0^q - a}{qx_0^{q-1}} = x_0 - \frac{qx_0^q - x_0^q - a}{qx_0^{q-1}} = \frac{x_0^q(q-1) - a}{qx_0^{q-1}} = x_0\left(\frac{q-1}{q}\right) + \frac{a}{x_0^{q-1}}\left(\frac{1}{q}\right)$ so that x_1 is a weighted average of x_0
and $\frac{a}{x_0^{q-1}}$ with weights $m_0 = \frac{q-1}{q}$ and $m_1 = \frac{1}{q}$.

In the case where $x_0 = \frac{a}{x_0^{q-1}}$ we have $x_0^q = a$ and $x_1 = \frac{a}{x_0^{q-1}}\left(\frac{q-1}{q}\right) + \frac{a}{x_0^{q-1}}\left(\frac{1}{q}\right) = \frac{a}{x_0^{q-1}}\left(\frac{q-1}{q} + \frac{1}{q}\right) = \frac{a}{x_0^{q-1}}$.

25. $\frac{ds}{dt} = ks \Rightarrow \frac{ds}{s} = k\,dt \Rightarrow \ln s = kt + C \Rightarrow s = s_0 e^{kt}$
$\Rightarrow$ the 14th century model of free fall was exponential;
note that the motion starts too slowly at first and then
becomes too fast after about 7 seconds

27. (a) $a(t) = s''(t) = -k \ (k > 0) \Rightarrow s'(t) = -kt + C_1$, where $s'(0) = 88 \Rightarrow C_1 = 88 \Rightarrow s'(t) = -kt + 88$. So
$s(t) = \frac{-kt^2}{2} + 88t + C_2$ where $s(0) = 0 \Rightarrow C_2 = 0$ so $s(t) = \frac{-kt^2}{2} + 88t$. Now $s(t) = 100$ when
$\frac{-kt^2}{2} + 88t = 100$. Solving for t we obtain $t = \frac{88 \pm \sqrt{88^2 - 200k}}{k}$. At such t we want $s'(t) = 0$, thus
$-k\left(\frac{88 + \sqrt{88^2 - 200k}}{k}\right) + 88 = 0$ or $-k\left(\frac{88 - \sqrt{88^2 - 200k}}{k}\right) + 88 = 0$. In either case we obtain $88^2 - 200k = 0$
so that $k = \frac{88^2}{200} \approx 38.72$ ft/sec^2.

(b) The initial condition that $s'(0) = 44$ ft/sec implies that $s'(t) = -kt + 44$ and $s(t) = \frac{-kt^2}{2} + 44t$ where k is as above.
The car is stopped at a time t such that $s'(t) = -kt + 44 = 0 \Rightarrow t = \frac{44}{k}$. At this time the car has traveled a distance
$s\left(\frac{44}{k}\right) = \frac{-k}{2}\left(\frac{44}{k}\right)^2 + 44\left(\frac{44}{k}\right) = \frac{44^2}{2k} = \frac{968}{k} = 968\left(\frac{200}{88^2}\right) = 25$ feet. Thus halving the initial velocity quarters
stopping distance.

29. Yes. The curve $y = x$ satisfies all three conditions since $\frac{dy}{dx} = 1$ everywhere, when $x = 0$, $y = 0$, and $\frac{d^2y}{dx^2} = 0$ everywhere.

31. $s''(t) = a = -t^2 \Rightarrow v = s'(t) = \frac{-t^3}{3} + C$. We seek $v_0 = s'(0) = C$. We know that $s(t^*) = b$ for some t^* and s is at a
maximum for this t^*. Since $s(t) = \frac{-t^4}{12} + Ct + k$ and $s(0) = 0$ we have that $s(t) = \frac{-t^4}{12} + Ct$ and also $s'(t^*) = 0$ so that
$t^* = (3C)^{1/3}$. So $\frac{[-(3C)^{1/3}]^4}{12} + C(3C)^{1/3} = b \Rightarrow (3C)^{1/3}\left(C - \frac{3C}{12}\right) = b \Rightarrow (3C)^{1/3}\left(\frac{3C}{4}\right) = b \Rightarrow 3^{1/3}C^{4/3} = \frac{4b}{3}$
$\Rightarrow C = \frac{(4b)^{3/4}}{3}$. Thus $v_0 = s'(0) = \frac{(4b)^{3/4}}{3} = \frac{2\sqrt{2}}{3}b^{3/4}$.

33. The graph of $f(x) = ax^2 + bx + c$ with $a > 0$ is a parabola opening upwards. Thus $f(x) \geq 0$ for all x if $f(x) = 0$ for at most one real value of x. The solutions to $f(x) = 0$ are, by the quadratic equation $\dfrac{-2b \pm \sqrt{(2b)^2 - 4ac}}{2a}$. Thus we require $(2b)^2 - 4ac \leq 0 \Rightarrow b^2 - ac \leq 0$.

35. (a) $L = k\left(\dfrac{a - b\cot\theta}{R^4} + \dfrac{b\csc\theta}{r^4}\right) \Rightarrow \dfrac{dL}{d\theta} = k\left(\dfrac{b\csc^2\theta}{R^4} - \dfrac{b\csc\theta\cot\theta}{r^4}\right)$; solving $\dfrac{dL}{d\theta} = 0$

$\Rightarrow r^4 b\csc^2\theta - bR^4\csc\theta\cot\theta = 0 \Rightarrow (b\csc\theta)(r^4\csc\theta - R^4\cot\theta) = 0$; but $b\csc\theta \neq 0$ since

$\theta \neq \dfrac{\pi}{2} \Rightarrow r^4\csc\theta - R^4\cot\theta = 0 \Rightarrow \cos\theta = \dfrac{r^4}{R^4} \Rightarrow \theta = \cos^{-1}\left(\dfrac{r^4}{R^4}\right)$, the critical value of θ

(b) $\theta = \cos^{-1}\left(\dfrac{5}{6}\right)^4 \approx \cos^{-1}(0.48225) \approx 61°$

NOTES:

CHAPTER 5 INTEGRATION

5.1 ESTIMATING WITH FINITE SUMS

1. $f(x) = x^2$

Since f is increasing on $[0, 1]$, we use left endpoints to obtain lower sums and right endpoints to obtain upper sums.

(a) $\triangle x = \frac{1-0}{2} = \frac{1}{2}$ and $x_i = i\triangle x = \frac{i}{2} \Rightarrow$ a lower sum is $\sum_{i=0}^{1} \left(\frac{i}{2}\right)^2 \cdot \frac{1}{2} = \frac{1}{2}\left(0^2 + \left(\frac{1}{2}\right)^2\right) = \frac{1}{8}$

(b) $\triangle x = \frac{1-0}{4} = \frac{1}{4}$ and $x_i = i\triangle x = \frac{i}{4} \Rightarrow$ a lower sum is $\sum_{i=0}^{3} \left(\frac{i}{4}\right)^2 \cdot \frac{1}{4} = \frac{1}{4}\left(0^2 + \left(\frac{1}{4}\right)^2 + \left(\frac{1}{2}\right)^2 + \left(\frac{3}{4}\right)^2\right) = \frac{1}{4} \cdot \frac{7}{8} = \frac{7}{32}$

(c) $\triangle x = \frac{1-0}{2} = \frac{1}{2}$ and $x_i = i\triangle x = \frac{i}{2} \Rightarrow$ an upper sum is $\sum_{i=1}^{2} \left(\frac{i}{2}\right)^2 \cdot \frac{1}{2} = \frac{1}{2}\left(\left(\frac{1}{2}\right)^2 + 1^2\right) = \frac{5}{8}$

(d) $\triangle x = \frac{1-0}{4} = \frac{1}{4}$ and $x_i = i\triangle x = \frac{i}{4} \Rightarrow$ an upper sum is $\sum_{i=1}^{4} \left(\frac{i}{4}\right)^2 \cdot \frac{1}{4} = \frac{1}{4}\left(\left(\frac{1}{4}\right)^2 + \left(\frac{1}{2}\right)^2 + \left(\frac{3}{4}\right)^2 + 1^2\right) = \frac{1}{4} \cdot \left(\frac{30}{16}\right) = \frac{15}{32}$

3. $f(x) = \frac{1}{x}$

Since f is decreasing on $[1, 5]$, we use left endpoints to obtain upper sums and right endpoints to obtain lower sums.

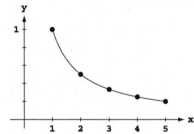

(a) $\triangle x = \frac{5-1}{2} = 2$ and $x_i = 1 + i\triangle x = 1 + 2i \Rightarrow$ a lower sum is $\sum_{i=1}^{2} \frac{1}{x_i} \cdot 2 = 2\left(\frac{1}{3} + \frac{1}{5}\right) = \frac{16}{15}$

(b) $\triangle x = \frac{5-1}{4} = 1$ and $x_i = 1 + i\triangle x = 1 + i \Rightarrow$ a lower sum is $\sum_{i=1}^{4} \frac{1}{x_i} \cdot 1 = 1\left(\frac{1}{2} + \frac{1}{3} + \frac{1}{4} + \frac{1}{5}\right) = \frac{77}{60}$

(c) $\triangle x = \frac{5-1}{2} = 2$ and $x_i = 1 + i\triangle x = 1 + 2i \Rightarrow$ an upper sum is $\sum_{i=0}^{1} \frac{1}{x_i} \cdot 2 = 2\left(1 + \frac{1}{3}\right) = \frac{8}{3}$

(d) $\triangle x = \frac{5-1}{4} = 1$ and $x_i = 1 + i\triangle x = 1 + i \Rightarrow$ an upper sum is $\sum_{i=0}^{3} \frac{1}{x_i} \cdot 1 = 1\left(1 + \frac{1}{2} + \frac{1}{3} + \frac{1}{4}\right) = \frac{25}{12}$

5. $f(x) = x^2$

Using 2 rectangles $\Rightarrow \triangle x = \frac{1-0}{2} = \frac{1}{2} \Rightarrow \frac{1}{2}\left(f\left(\frac{1}{4}\right) + f\left(\frac{3}{4}\right)\right)$

$= \frac{1}{2}\left(\left(\frac{1}{4}\right)^2 + \left(\frac{3}{4}\right)^2\right) = \frac{10}{32} = \frac{5}{16}$

Using 4 rectangles $\Rightarrow \triangle x = \frac{1-0}{4} = \frac{1}{4}$

$\Rightarrow \frac{1}{4}\left(f\left(\frac{1}{8}\right) + f\left(\frac{3}{8}\right) + f\left(\frac{5}{8}\right) + f\left(\frac{7}{8}\right)\right)$

$= \frac{1}{4}\left(\left(\frac{1}{8}\right)^2 + \left(\frac{3}{8}\right)^2 + \left(\frac{5}{8}\right)^2 + \left(\frac{7}{8}\right)^2\right) = \frac{21}{64}$

7. $f(x) = \frac{1}{x}$

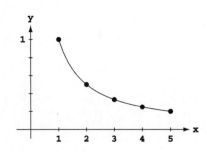

Using 2 rectangles $\Rightarrow \triangle x = \frac{5-1}{2} = 2 \Rightarrow 2(f(2) + f(4))$

$= 2\left(\frac{1}{2} + \frac{1}{4}\right) = \frac{3}{2}$

Using 4 rectangles $\Rightarrow \triangle x = \frac{5-1}{4} = 1$

$\Rightarrow 1\left(f\left(\frac{3}{2}\right) + f\left(\frac{5}{2}\right) + f\left(\frac{7}{2}\right) + f\left(\frac{9}{2}\right)\right)$

$= 1\left(\frac{2}{3} + \frac{2}{5} + \frac{2}{7} + \frac{2}{9}\right) = \frac{1488}{3\cdot5\cdot7\cdot9} = \frac{496}{5\cdot7\cdot9} = \frac{496}{315}$

9. (a) $D \approx (0)(1) + (12)(1) + (22)(1) + (10)(1) + (5)(1) + (13)(1) + (11)(1) + (6)(1) + (2)(1) + (6)(1) = 87$ inches
    (b) $D \approx (12)(1) + (22)(1) + (10)(1) + (5)(1) + (13)(1) + (11)(1) + (6)(1) + (2)(1) + (6)(1) + (0)(1) = 87$ inches

11. (a) $D \approx (0)(10) + (44)(10) + (15)(10) + (35)(10) + (30)(10) + (44)(10) + (35)(10) + (15)(10) + (22)(10)$
        $+ (35)(10) + (44)(10) + (30)(10) = 3490$ feet ≈ 0.66 miles
    (b) $D \approx (44)(10) + (15)(10) + (35)(10) + (30)(10) + (44)(10) + (35)(10) + (15)(10) + (22)(10) + (35)(10)$
        $+ (44)(10) + (30)(10) + (35)(10) = 3840$ feet ≈ 0.73 miles

13. (a) Because the acceleration is decreasing, an upper estimate is obtained using left end-points in summing
        acceleration $\cdot \Delta t$. Thus, $\Delta t = 1$ and speed $\approx [32.00 + 19.41 + 11.77 + 7.14 + 4.33](1) = 74.65$ ft/sec
    (b) Using right end-points we obtain a lower estimate: speed $\approx [19.41 + 11.77 + 7.14 + 4.33 + 2.63](1) = 45.28$ ft/sec
    (c) Upper estimates for the speed at each second are:

t	0	1	2	3	4	5
v	0	32.00	51.41	63.18	70.32	74.65

Thus, the distance fallen when $t = 3$ seconds is $s \approx [32.00 + 51.41 + 63.18](1) = 146.59$ ft.

15. Partition $[0, 2]$ into the four subintervals $[0, 0.5]$, $[0.5, 1]$, $[1, 1.5]$, and $[1.5, 2]$. The midpoints of these subintervals are
    $m_1 = 0.25$, $m_2 = 0.75$, $m_3 = 1.25$, and $m_4 = 1.75$. The heights of the four approximating rectangles are
    $f(m_1) = (0.25)^3 = \frac{1}{64}$, $f(m_2) = (0.75)^3 = \frac{27}{64}$, $f(m_3) = (1.25)^3 = \frac{125}{64}$, and $f(m_4) = (1.75)^3 = \frac{343}{64}$

    Notice that the average value is approximated by $\frac{1}{2}\left[\left(\frac{1}{4}\right)^3\left(\frac{1}{2}\right) + \left(\frac{3}{4}\right)^3\left(\frac{1}{2}\right) + \left(\frac{5}{4}\right)^3\left(\frac{1}{2}\right) + \left(\frac{7}{4}\right)^3\left(\frac{1}{2}\right)\right] = \frac{31}{16}$

    $= \frac{1}{\text{length of } [0,2]} \cdot \left[\begin{array}{c}\text{approximate area under}\\ \text{curve } f(x) = x^3\end{array}\right]$. We use this observation in solving the next several exercises.

17. Partition $[0, 2]$ into the four subintervals $[0, 0.5]$, $[0.5, 1]$, $[1, 1.5]$, and $[1.5, 2]$. The midpoints of the subintervals
    are $m_1 = 0.25$, $m_2 = 0.75$, $m_3 = 1.25$, and $m_4 = 1.75$. The heights of the four approximating rectangles are

    $f(m_1) = \frac{1}{2} + \sin^2\frac{\pi}{4} = \frac{1}{2} + \frac{1}{2} = 1$, $f(m_2) = \frac{1}{2} + \sin^2\frac{3\pi}{4} = \frac{1}{2} + \frac{1}{2} = 1$, $f(m_3) = \frac{1}{2} + \sin^2\frac{5\pi}{4} = \frac{1}{2} + \left(-\frac{1}{\sqrt{2}}\right)^2$

    $= \frac{1}{2} + \frac{1}{2} = 1$, and $f(m_4) = \frac{1}{2} + \sin^2\frac{7\pi}{4} = \frac{1}{2} + \left(-\frac{1}{\sqrt{2}}\right)^2 = 1$. The width of each rectangle is $\Delta x = \frac{1}{2}$. Thus,

Area $\approx (1 + 1 + 1 + 1)\left(\frac{1}{2}\right) = 2 \Rightarrow$ average value $\approx \frac{\text{area}}{\text{length of } [0,2]} = \frac{2}{2} = 1$.

19. Since the leakage is increasing, an upper estimate uses right endpoints and a lower estimate uses left endpoints:

(a) upper estimate $= (70)(1) + (97)(1) + (136)(1) + (190)(1) + (265)(1) = 758$ gal,
    lower estimate $= (50)(1) + (70)(1) + (97)(1) + (136)(1) + (190)(1) = 543$ gal.

(b) upper estimate $= (70 + 97 + 136 + 190 + 265 + 369 + 516 + 720) = 2363$ gal,
    lower estimate $= (50 + 70 + 97 + 136 + 190 + 265 + 369 + 516) = 1693$ gal.

(c) worst case: $2363 + 720t = 25{,}000 \Rightarrow t \approx 31.4$ hrs;
    best case: $1693 + 720t = 25{,}000 \Rightarrow t \approx 32.4$ hrs

21. (a) The diagonal of the square has length 2, so the side length is $\sqrt{2}$. Area $= \left(\sqrt{2}\right)^2 = 2$

(b) Think of the octagon as a collection of 16 right triangles with a hypotenuse of length 1 and an acute angle measuring $\frac{2\pi}{16} = \frac{\pi}{8}$.
    Area $= 16\left(\frac{1}{2}\right)\left(\sin \frac{\pi}{8}\right)\left(\cos \frac{\pi}{8}\right) = 4 \sin \frac{\pi}{4} = 2\sqrt{2} \approx 2.828$

(c) Think of the 16-gon as a collection of 32 right triangles with a hypotenuse of length 1 and an acute angle measuring $\frac{2\pi}{32} = \frac{\pi}{16}$.
    Area $= 32\left(\frac{1}{2}\right)\left(\sin \frac{\pi}{16}\right)\left(\cos \frac{\pi}{16}\right) = 8 \sin \frac{\pi}{8} = 2\sqrt{2} \approx 3.061$

(d) Each area is less than the area of the circle, π. As n increases, the area approaches π.

5.2 SIGMA NOTATION AND LIMITS OF FINITE SUMS

1. $\displaystyle\sum_{k=1}^{2} \frac{6k}{k+1} = \frac{6(1)}{1+1} + \frac{6(2)}{2+1} = \frac{6}{2} + \frac{12}{3} = 7$

3. $\displaystyle\sum_{k=1}^{4} \cos k\pi = \cos(1\pi) + \cos(2\pi) + \cos(3\pi) + \cos(4\pi) = -1 + 1 - 1 + 1 = 0$

5. $\displaystyle\sum_{k=1}^{3} (-1)^{k+1} \sin \frac{\pi}{k} = (-1)^{1+1} \sin \frac{\pi}{1} + (-1)^{2+1} \sin \frac{\pi}{2} + (-1)^{3+1} \sin \frac{\pi}{3} = 0 - 1 + \frac{\sqrt{3}}{2} = \frac{\sqrt{3}-2}{2}$

7. (a) $\displaystyle\sum_{k=1}^{6} 2^{k-1} = 2^{1-1} + 2^{2-1} + 2^{3-1} + 2^{4-1} + 2^{5-1} + 2^{6-1} = 1 + 2 + 4 + 8 + 16 + 32$

(b) $\displaystyle\sum_{k=0}^{5} 2^k = 2^0 + 2^1 + 2^2 + 2^3 + 2^4 + 2^5 = 1 + 2 + 4 + 8 + 16 + 32$

(c) $\displaystyle\sum_{k=-1}^{4} 2^{k+1} = 2^{-1+1} + 2^{0+1} + 2^{1+1} + 2^{2+1} + 2^{3+1} + 2^{4+1} = 1 + 2 + 4 + 8 + 16 + 32$

All of them represent $1 + 2 + 4 + 8 + 16 + 32$

9. (a) $\displaystyle\sum_{k=2}^{4} \frac{(-1)^{k-1}}{k-1} = \frac{(-1)^{2-1}}{2-1} + \frac{(-1)^{3-1}}{3-1} + \frac{(-1)^{4-1}}{4-1} = -1 + \frac{1}{2} - \frac{1}{3}$

(b) $\displaystyle\sum_{k=0}^{2} \frac{(-1)^k}{k+1} = \frac{(-1)^0}{0+1} + \frac{(-1)^1}{1+1} + \frac{(-1)^2}{2+1} = 1 - \frac{1}{2} + \frac{1}{3}$

(c) $\displaystyle\sum_{k=-1}^{1} \frac{(-1)^k}{k+2} = \frac{(-1)^{-1}}{-1+2} + \frac{(-1)^0}{0+2} + \frac{(-1)^1}{1+2} = -1 + \frac{1}{2} - \frac{1}{3}$

(a) and (c) are equivalent; (b) is not equivalent to the other two.

11. $\displaystyle\sum_{k=1}^{6} k$ 13. $\displaystyle\sum_{k=1}^{4} \frac{1}{2^k}$ 15. $\displaystyle\sum_{k=1}^{5} (-1)^{k+1} \frac{1}{k}$

17. (a) $\sum\limits_{k=1}^{n} 3a_k = 3 \sum\limits_{k=1}^{n} a_k = 3(-5) = -15$

(b) $\sum\limits_{k=1}^{n} \frac{b_k}{6} = \frac{1}{6} \sum\limits_{k=1}^{n} b_k = \frac{1}{6}(6) = 1$

(c) $\sum\limits_{k=1}^{n} (a_k + b_k) = \sum\limits_{k=1}^{n} a_k + \sum\limits_{k=1}^{n} b_k = -5 + 6 = 1$

(d) $\sum\limits_{k=1}^{n} (a_k - b_k) = \sum\limits_{k=1}^{n} a_k - \sum\limits_{k=1}^{n} b_k = -5 - 6 = -11$

(e) $\sum\limits_{k=1}^{n} (b_k - 2a_k) = \sum\limits_{k=1}^{n} b_k - 2 \sum\limits_{k=1}^{n} a_k = 6 - 2(-5) = 16$

19. (a) $\sum\limits_{k=1}^{10} k = \frac{10(10+1)}{2} = 55$ (b) $\sum\limits_{k=1}^{10} k^2 = \frac{10(10+1)(2(10)+1)}{6} = 385$

(c) $\sum\limits_{k=1}^{10} k^3 = \left[\frac{10(10+1)}{2} \right]^2 = 55^2 = 3025$

21. $\sum\limits_{k=1}^{7} -2k = -2 \sum\limits_{k=1}^{7} k = -2 \left(\frac{7(7+1)}{2} \right) = -56$

23. $\sum\limits_{k=1}^{6} (3 - k^2) = \sum\limits_{k=1}^{6} 3 - \sum\limits_{k=1}^{6} k^2 = 3(6) - \frac{6(6+1)(2(6)+1)}{6} = -73$

25. $\sum\limits_{k=1}^{5} k(3k + 5) = \sum\limits_{k=1}^{5} (3k^2 + 5k) = 3 \sum\limits_{k=1}^{5} k^2 + 5 \sum\limits_{k=1}^{5} k = 3 \left(\frac{5(5+1)(2(5)+1)}{6} \right) + 5 \left(\frac{5(5+1)}{2} \right) = 240$

27. $\sum\limits_{k=1}^{5} \frac{k^3}{225} + \left(\sum\limits_{k=1}^{5} k \right)^3 = \frac{1}{225} \sum\limits_{k=1}^{5} k^3 + \left(\sum\limits_{k=1}^{5} k \right)^3 = \frac{1}{225} \left(\frac{5(5+1)}{2} \right)^2 + \left(\frac{5(5+1)}{2} \right)^3 = 3376$

29. (a)   (b)   (c)

31. (a)   (b)   (c)

33. $|x_1 - x_0| = |1.2 - 0| = 1.2$, $|x_2 - x_1| = |1.5 - 1.2| = 0.3$, $|x_3 - x_2| = |2.3 - 1.5| = 0.8$, $|x_4 - x_3| = |2.6 - 2.3| = 0.3$, and $|x_5 - x_4| = |3 - 2.6| = 0.4$; the largest is $\|P\| = 1.2$.

35. $f(x) = 1 - x^2$

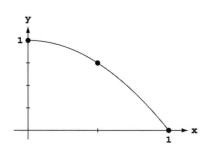

Since f is decreasing on $[0, 1]$ we use left endpoints to obtain upper sums. $\triangle x = \frac{1-0}{n} = \frac{1}{n}$ and $x_i = i\triangle x = \frac{i}{n}$. So an upper sum

is $\sum\limits_{i=0}^{n-1} (1 - x_i^2)\frac{1}{n} = \frac{1}{n} \sum\limits_{i=0}^{n-1} \left(1 - \left(\frac{i}{n}\right)^2\right) = \frac{1}{n^3} \sum\limits_{i=0}^{n-1} (n^2 - i^2)$

$= \frac{n^3}{n^3} - \frac{1}{n^3}\sum\limits_{i=0}^{n} i^2 = 1 - \frac{(n-1)n(2(n-1)+1)}{6n^3} = 1 - \frac{2n^3 - 3n^2 + n}{6n^3}$

$= 1 - \frac{2 - \frac{3}{n} + \frac{1}{n^2}}{6}$. Thus,

$\lim\limits_{n\to\infty} \sum\limits_{i=0}^{n-1}(1 - x_i^2)\frac{1}{n} = \lim\limits_{n\to\infty}\left(1 - \frac{2 - \frac{3}{n} + \frac{1}{n^2}}{6}\right) = 1 - \frac{1}{3} = \frac{2}{3}$

37. $f(x) = x^2 + 1$

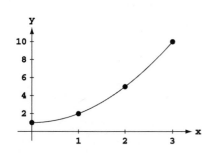

Since f is increasing on $[0, 3]$ we use right endpoints to obtain upper sums. $\triangle x = \frac{3-0}{n} = \frac{3}{n}$ and $x_i = i\triangle x = \frac{3i}{n}$. So an upper

sum is $\sum\limits_{i=1}^{n}(x_i^2 + 1)\frac{3}{n} = \sum\limits_{i=1}^{n}\left(\left(\frac{3i}{n}\right)^2 + 1\right)\frac{3}{n} = \frac{3}{n}\sum\limits_{i=1}^{n}\left(\frac{9i^2}{n^2} + 1\right)$

$= \frac{27}{n}\sum\limits_{i=1}^{n} i^2 + \frac{3}{n} \cdot n = \frac{27}{n^3}\left(\frac{n(n+1)(2n+1)}{6}\right) + 3$

$= \frac{9(2n^3 + 3n^2 + n)}{2n^3} + 3 = \frac{18 + \frac{27}{n} + \frac{9}{n^2}}{2} + 3$. Thus,

$\lim\limits_{n\to\infty} \sum\limits_{i=1}^{n}(x_i^2 + 1)\frac{3}{n} = \lim\limits_{n\to\infty}\left(\frac{18 + \frac{27}{n} + \frac{9}{n^2}}{2} + 3\right) = 9 + 3 = 12$.

39. $f(x) = x + x^2 = x(1 + x)$

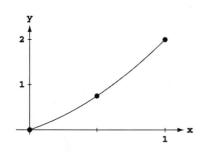

Since f is increasing on $[0, 1]$ we use right endpoints to obtain upper sums. $\triangle x = \frac{1-0}{n} = \frac{1}{n}$ and $x_i = i\triangle x = \frac{i}{n}$. So an upper sum

is $\sum\limits_{i=1}^{n}(x_i + x_i^2)\frac{1}{n} = \sum\limits_{i=1}^{n}\left(\frac{i}{n} + \left(\frac{i}{n}\right)^2\right)\frac{1}{n} = \frac{1}{n^2}\sum\limits_{i=1}^{n} i + \frac{1}{n^3}\sum\limits_{i=1}^{n} i^2$

$= \frac{1}{n^2}\left(\frac{n(n+1)}{2}\right) + \frac{1}{n^3}\left(\frac{n(n+1)(2n+1)}{6}\right) = \frac{n^2 + n}{2n^2} + \frac{2n^3 + 3n^2 + n}{6n^3}$

$= \frac{1 + \frac{1}{n}}{2} + \frac{2 + \frac{3}{n} + \frac{1}{n^2}}{6}$. Thus, $\lim\limits_{n\to\infty} \sum\limits_{i=1}^{n}(x_i + x_i^2)\frac{1}{n}$

$= \lim\limits_{n\to\infty}\left[\left(\frac{1 + \frac{1}{n}}{2}\right) + \left(\frac{2 + \frac{3}{n} + \frac{1}{n^2}}{6}\right)\right] = \frac{1}{2} + \frac{2}{6} = \frac{5}{6}$.

5.3 THE DEFINITE INTEGRAL

1. $\displaystyle\int_0^2 x^2 \, dx$

3. $\displaystyle\int_{-7}^5 (x^2 - 3x) \, dx$

5. $\displaystyle\int_2^3 \frac{1}{1-x} \, dx$

7. $\displaystyle\int_{-\pi/4}^0 (\sec x) \, dx$

9. (a) $\displaystyle\int_2^2 g(x) \, dx = 0$

   (b) $\displaystyle\int_5^1 g(x) \, dx = -\int_1^5 g(x) \, dx = -8$

   (c) $\displaystyle\int_1^2 3f(x) \, dx = 3\int_1^2 f(x) \, dx = 3(-4) = -12$

   (d) $\displaystyle\int_2^5 f(x) \, dx = \int_1^5 f(x) \, dx - \int_1^2 f(x) \, dx = 6 - (-4) = 10$

   (e) $\displaystyle\int_1^5 [f(x) - g(x)] \, dx = \int_1^5 f(x) \, dx - \int_1^5 g(x) \, dx = 6 - 8 = -2$

   (f) $\displaystyle\int_1^5 [4f(x) - g(x)] \, dx = 4\int_1^5 f(x) \, dx - \int_1^5 g(x) \, dx = 4(6) - 8 = 16$

11. (a) $\int_1^2 f(u)\, du = \int_1^2 f(x)\, dx = 5$

  (b) $\int_1^2 \sqrt{3}\, f(z)\, dz = \sqrt{3} \int_1^2 f(z)\, dz = 5\sqrt{3}$

  (c) $\int_2^1 f(t)\, dt = -\int_1^2 f(t)\, dt = -5$

  (d) $\int_1^2 [-f(x)]\, dx = -\int_1^2 f(x)\, dx = -5$

13. (a) $\int_3^4 f(z)\, dz = \int_0^4 f(z)\, dz - \int_0^3 f(z)\, dz = 7 - 3 = 4$

  (b) $\int_4^3 f(t)\, dt = -\int_3^4 f(t)\, dt = -4$

15. The area of the trapezoid is $A = \frac{1}{2}(B + b)h$

  $= \frac{1}{2}(5 + 2)(6) = 21 \Rightarrow \int_{-2}^4 \left(\frac{x}{2} + 3\right) dx$

  $= 21$ square units

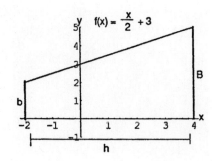

17. The area of the semicircle is $A = \frac{1}{2}\pi r^2 = \frac{1}{2}\pi(3)^2$

  $= \frac{9}{2}\pi \Rightarrow \int_{-3}^3 \sqrt{9 - x^2}\, dx = \frac{9}{2}\pi$ square units

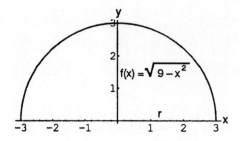

19. The area of the triangle on the left is $A = \frac{1}{2}bh = \frac{1}{2}(2)(2)$

  $= 2$. The area of the triangle on the right is $A = \frac{1}{2}bh$

  $= \frac{1}{2}(1)(1) = \frac{1}{2}$. Then, the total area is 2.5

  $\Rightarrow \int_{-2}^1 |x|\, dx = 2.5$ square units

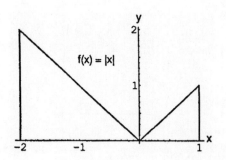

21. The area of the triangular peak is $A = \frac{1}{2}bh = \frac{1}{2}(2)(1) = 1$.

  The area of the rectangular base is $S = \ell w = (2)(1) = 2$.

  Then the total area is $3 \Rightarrow \int_{-1}^1 (2 - |x|)\, dx = 3$ square units

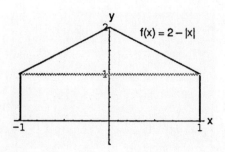

23. $\int_0^b \frac{x}{2}\, dx = \frac{1}{2}\,(b)\left(\frac{b}{2}\right) = \frac{b^2}{4}$

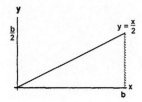

25. $\int_a^b 2s\, ds = \frac{1}{2}\,b(2b) - \frac{1}{2}\,a(2a) = b^2 - a^2$

27. $\int_1^{\sqrt{2}} x\, dx = \frac{\left(\sqrt{2}\right)^2}{2} - \frac{(1)^2}{2} = \frac{1}{2}$

29. $\int_\pi^{2\pi} \theta\, d\theta = \frac{(2\pi)^2}{2} - \frac{\pi^2}{2} = \frac{3\pi^2}{2}$

31. $\int_0^{\sqrt[3]{7}} x^2\, dx = \frac{\left(\sqrt[3]{7}\right)^3}{3} = \frac{7}{3}$

33. $\int_0^{1/2} t^2\, dt = \frac{\left(\frac{1}{2}\right)^3}{3} = \frac{1}{24}$

35. $\int_a^{2a} x\, dx = \frac{(2a)^2}{2} - \frac{a^2}{2} = \frac{3a^2}{2}$

37. $\int_0^{\sqrt[3]{b}} x^2\, dx = \frac{\left(\sqrt[3]{b}\right)^3}{3} = \frac{b}{3}$

39. $\int_3^1 7\, dx = 7(1-3) = -14$

41. $\int_0^2 5x\, dx = 5\int_0^2 x\, dx = 5\left[\frac{2^2}{2} - \frac{0^2}{2}\right] = 10$

43. $\int_0^2 (2t-3)\, dt = 2\int_1^1 t\, dt - \int_0^2 3\, dt = 2\left[\frac{2^2}{2} - \frac{0^2}{2}\right] - 3(2-0) = 4 - 6 = -2$

45. $\int_2^1 \left(1 + \frac{z}{2}\right) dz = \int_2^1 1\, dz + \int_2^1 \frac{z}{2}\, dz = \int_2^1 1\, dz - \frac{1}{2}\int_1^2 z\, dz = 1[1-2] - \frac{1}{2}\left[\frac{2^2}{2} - \frac{1^2}{2}\right] = -1 - \frac{1}{2}\left(\frac{3}{2}\right) = -\frac{7}{4}$

47. $\int_1^2 3u^2\, du = 3\int_1^2 u^2\, du = 3\left[\int_0^2 u^2\, du - \int_0^1 u^2\, du\right] = 3\left(\left[\frac{2^3}{3} - \frac{0^3}{3}\right] - \left[\frac{1^3}{3} - \frac{0^3}{3}\right]\right) = 3\left[\frac{2^3}{3} - \frac{1^3}{3}\right] = 3\left(\frac{7}{3}\right) = 7$

49. $\int_0^2 (3x^2 + x - 5)\, dx = 3\int_0^2 x^2\, dx + \int_0^2 x\, dx - \int_0^2 5\, dx = 3\left[\frac{2^3}{3} - \frac{0^3}{3}\right] + \left[\frac{2^2}{2} - \frac{0^2}{2}\right] - 5[2-0] = (8+2) - 10 = 0$

51. Let $\Delta x = \frac{b-0}{n} = \frac{b}{n}$ and let $x_0 = 0$, $x_1 = \Delta x$,
$x_2 = 2\Delta x, \dots, x_{n-1} = (n-1)\Delta x$, $x_n = n\Delta x = b$.
Let the c_k's be the right end-points of the subintervals
$\Rightarrow c_1 = x_1$, $c_2 = x_2$, and so on. The rectangles
defined have areas:

$\quad f(c_1)\,\Delta x = f(\Delta x)\,\Delta x = 3(\Delta x)^2\,\Delta x = 3(\Delta x)^3$
$\quad f(c_2)\,\Delta x = f(2\Delta x)\,\Delta x = 3(2\Delta x)^2\,\Delta x = 3(2)^2(\Delta x)^3$
$\quad f(c_3)\,\Delta x = f(3\Delta x)\,\Delta x = 3(3\Delta x)^2\,\Delta x = 3(3)^2(\Delta x)^3$
$\quad \vdots$
$\quad f(c_n)\,\Delta x = f(n\Delta x)\,\Delta x = 3(n\Delta x)^2\,\Delta x = 3(n)^2(\Delta x)^3$

Then $S_n = \sum_{k=1}^n f(c_k)\,\Delta x = \sum_{k=1}^n 3k^2(\Delta x)^3$

$= 3(\Delta x)^3 \sum_{k=1}^n k^2 = 3\left(\frac{b^3}{n^3}\right)\left(\frac{n(n+1)(2n+1)}{6}\right)$

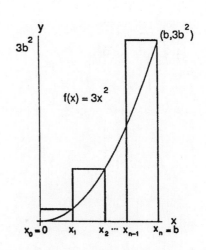

$$= \frac{b^3}{2}\left(2 + \frac{3}{n} + \frac{1}{n^2}\right) \Rightarrow \int_0^b 3x^2 \, dx = \lim_{n \to \infty} \frac{b^3}{2}\left(2 + \frac{3}{n} + \frac{1}{n^2}\right) = b^3.$$

53. Let $\Delta x = \frac{b-0}{n} = \frac{b}{n}$ and let $x_0 = 0$, $x_1 = \Delta x$,
    $x_2 = 2\Delta x, \ldots , x_{n-1} = (n-1)\Delta x$, $x_n = n\Delta x = b$.
    Let the c_k's be the right end-points of the subintervals
    $\Rightarrow c_1 = x_1$, $c_2 = x_2$, and so on. The rectangles
    defined have areas:

    $\qquad f(c_1)\,\Delta x = f(\Delta x)\,\Delta x = 2(\Delta x)(\Delta x) = 2(\Delta x)^2$
    $\qquad f(c_2)\,\Delta x = f(2\Delta x)\,\Delta x = 2(2\Delta x)(\Delta x) = 2(2)(\Delta x)^2$
    $\qquad f(c_3)\,\Delta x = f(3\Delta x)\,\Delta x = 2(3\Delta x)(\Delta x) = 2(3)(\Delta x)^2$
    $\qquad \vdots$
    $\qquad f(c_n)\,\Delta x = f(n\Delta x)\,\Delta x = 2(n\Delta x)(\Delta x) = 2(n)(\Delta x)^2$

    Then $S_n = \displaystyle\sum_{k=1}^{n} f(c_k)\,\Delta x = \sum_{k=1}^{n} 2k(\Delta x)^2$

    $$= 2(\Delta x)^2 \sum_{k=1}^{n} k = 2\left(\frac{b^2}{n^2}\right)\left(\frac{n(n+1)}{2}\right)$$

    $$= b^2\left(1 + \frac{1}{n}\right) \Rightarrow \int_0^b 2x \, dx = \lim_{n \to \infty} b^2\left(1 + \frac{1}{n}\right) = b^2.$$

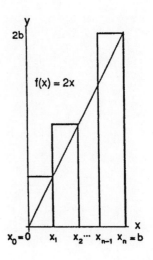

55. $\mathrm{av}(f) = \left(\frac{1}{\sqrt{3}-0}\right)\displaystyle\int_0^{\sqrt{3}} (x^2 - 1)\,dx$

    $$= \frac{1}{\sqrt{3}}\int_0^{\sqrt{3}} x^2 \, dx - \frac{1}{\sqrt{3}}\int_0^{\sqrt{3}} 1 \, dx$$

    $$= \frac{1}{\sqrt{3}}\left(\frac{\left(\sqrt{3}\right)^3}{3}\right) - \frac{1}{\sqrt{3}}\left(\sqrt{3} - 0\right) = 1 - 1 = 0.$$

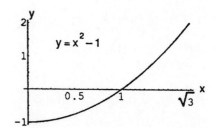

57. $\mathrm{av}(f) = \left(\frac{1}{1-0}\right)\displaystyle\int_0^1 (-3x^2 - 1)\,dx =$

    $$= -3\int_0^1 x^2 \, dx - \int_0^1 1 \, dx = -3\left(\frac{1^3}{3}\right) - (1 - 0)$$

    $$= -2.$$

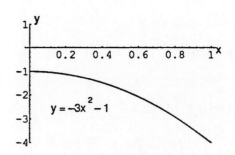

59. $\mathrm{av}(f) = \left(\frac{1}{3-0}\right)\displaystyle\int_0^3 (t-1)^2 \, dt$

    $$= \frac{1}{3}\int_0^3 t^2 \, dt - \frac{2}{3}\int_0^3 t \, dt + \frac{1}{3}\int_0^3 1 \, dt$$

    $$= \frac{1}{3}\left(\frac{3^3}{3}\right) - \frac{2}{3}\left(\frac{3^2}{2} - \frac{0^2}{2}\right) + \frac{1}{3}(3 - 0) = 1.$$

61. (a) $\text{av}(g) = \left(\frac{1}{1-(-1)}\right) \int_{-1}^{1} (|x| - 1)\, dx$

$= \frac{1}{2} \int_{-1}^{0} (-x - 1)\, dx + \frac{1}{2} \int_{0}^{1} (x - 1)\, dx$

$= -\frac{1}{2} \int_{-1}^{0} x\, dx - \frac{1}{2} \int_{-1}^{0} 1\, dx + \frac{1}{2} \int_{0}^{1} x\, dx - \frac{1}{2} \int_{0}^{1} 1\, dx$

$= -\frac{1}{2}\left(\frac{0^2}{2} - \frac{(-1)^2}{2}\right) - \frac{1}{2}(0 - (-1)) + \frac{1}{2}\left(\frac{1^2}{2} - \frac{0^2}{2}\right) - \frac{1}{2}(1 - 0)$

$= -\frac{1}{2}.$

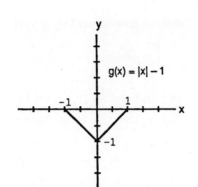

(b) $\text{av}(g) = \left(\frac{1}{3-1}\right) \int_{1}^{3} (|x| - 1)\, dx = \frac{1}{2} \int_{1}^{3} (x - 1)\, dx$

$= \frac{1}{2} \int_{1}^{3} x\, dx - \frac{1}{2} \int_{1}^{3} 1\, dx = \frac{1}{2}\left(\frac{3^2}{2} - \frac{1^2}{2}\right) - \frac{1}{2}(3 - 1)$

$= 1.$

(c) $\text{av}(g) = \left(\frac{1}{3-(-1)}\right) \int_{-1}^{3} (|x| - 1)\, dx$

$= \frac{1}{4} \int_{-1}^{1} (|x| - 1)\, dx + \frac{1}{4} \int_{1}^{3} (|x| - 1)\, dx$

$= \frac{1}{4}(-1 + 2) = \frac{1}{4}$ (see parts (a) and (b) above).

63. To find where $x - x^2 \geq 0$, let $x - x^2 = 0 \Rightarrow x(1 - x) = 0 \Rightarrow x = 0$ or $x = 1$. If $0 < x < 1$, then $0 < x - x^2 \Rightarrow a = 0$ and $b = 1$ maximize the integral.

65. $f(x) = \frac{1}{1+x^2}$ is decreasing on $[0, 1] \Rightarrow$ maximum value of f occurs at $0 \Rightarrow \max f = f(0) = 1$; minimum value of f occurs at $1 \Rightarrow \min f = f(1) = \frac{1}{1+1^2} = \frac{1}{2}$. Therefore, $(1 - 0)\min f \leq \int_{0}^{1} \frac{1}{1+x^2}\, dx \leq (1 - 0)\max f$

$\Rightarrow \frac{1}{2} \leq \int_{0}^{1} \frac{1}{1+x^2}\, dx \leq 1.$ That is, an upper bound $= 1$ and a lower bound $= \frac{1}{2}$.

67. $-1 \leq \sin(x^2) \leq 1$ for all $x \Rightarrow (1 - 0)(-1) \leq \int_{0}^{1} \sin(x^2)\, dx \leq (1 - 0)(1)$ or $\int_{0}^{1} \sin x^2\, dx \leq 1 \Rightarrow \int_{0}^{1} \sin x^2\, dx$ cannot equal 2.

69. If $f(x) \geq 0$ on $[a, b]$, then $\min f \geq 0$ and $\max f \geq 0$ on $[a, b]$. Now, $(b - a)\min f \leq \int_{a}^{b} f(x)\, dx \leq (b - a)\max f$. Then $b \geq a \Rightarrow b - a \geq 0 \Rightarrow (b - a)\min f \geq 0 \Rightarrow \int_{a}^{b} f(x)\, dx \geq 0.$

71. $\sin x \leq x$ for $x \geq 0 \Rightarrow \sin x - x \leq 0$ for $x \geq 0 \Rightarrow \int_{0}^{1} (\sin x - x)\, dx \leq 0$ (see Exercise 70) $\Rightarrow \int_{0}^{1} \sin x\, dx - \int_{0}^{1} x\, dx$

$\leq 0 \Rightarrow \int_{0}^{1} \sin x\, dx \leq \int_{0}^{1} x\, dx \Rightarrow \int_{0}^{1} \sin x\, dx \leq \left(\frac{1^2}{2} - \frac{0^2}{2}\right) \Rightarrow \int_{0}^{1} \sin x\, dx \leq \frac{1}{2}.$ Thus an upper bound is $\frac{1}{2}$.

73. Yes, for the following reasons: $\text{av}(f) = \frac{1}{b-a}\int_a^b f(x)\,dx$ is a constant K. Thus $\int_a^b \text{av}(f)\,dx = \int_a^b K\,dx$

$= K(b-a) \Rightarrow \int_a^b \text{av}(f)\,dx = (b-a)K = (b-a)\cdot\frac{1}{b-a}\int_a^b f(x)\,dx = \int_a^b f(x)\,dx.$

75. Consider the partition P that subdivides the interval $[a, b]$ into n subintervals of width $\triangle x = \frac{b-a}{n}$ and let c_k be the right endpoint of each subinterval. So the partition is $P = \{a, a+\frac{b-a}{n}, a+\frac{2(b-a)}{n}, \ldots, a+\frac{n(b-a)}{n}\}$ and $c_k = a+\frac{k(b-a)}{n}$.

We get the Riemann sum $\sum_{k=1}^n f(c_k)\triangle x = \sum_{k=1}^n c\cdot\frac{b-a}{n} = \frac{c(b-a)}{n}\sum_{k=1}^n 1 = \frac{c(b-a)}{n}\cdot n = c(b-a)$. As $n\to\infty$ and $\|P\|\to 0$

this expression remains $c(b-a)$. Thus, $\int_a^b c\,dx = c(b-a)$.

77. (a) $U = \max_1 \triangle x + \max_2 \triangle x + \ldots + \max_n \triangle x$ where $\max_1 = f(x_1), \max_2 = f(x_2), \ldots, \max_n = f(x_n)$ since f is increasing on $[a, b]$; $L = \min_1 \triangle x + \min_2 \triangle x + \ldots + \min_n \triangle x$ where $\min_1 = f(x_0), \min_2 = f(x_1), \ldots$, $\min_n = f(x_{n-1})$ since f is increasing on $[a, b]$. Therefore
$U - L = (\max_1 - \min_1)\triangle x + (\max_2 - \min_2)\triangle x + \ldots + (\max_n - \min_n)\triangle x$
$= (f(x_1) - f(x_0))\triangle x + (f(x_2) - f(x_1))\triangle x + \ldots + (f(x_n) - f(x_{n-1}))\triangle x = (f(x_n) - f(x_0))\triangle x = (f(b) - f(a))\triangle x.$

(b) $U = \max_1 \triangle x_1 + \max_2 \triangle x_2 + \ldots + \max_n \triangle x_n$ where $\max_1 = f(x_1), \max_2 = f(x_2), \ldots, \max_n = f(x_n)$ since f is increasing on$[a, b]$; $L = \min_1 \triangle x_1 + \min_2 \triangle x_2 + \ldots + \min_n \triangle x_n$ where
$\min_1 = f(x_0), \min_2 = f(x_1), \ldots, \min_n = f(x_{n-1})$ since f is increasing on $[a, b]$. Therefore
$U - L = (\max_1 - \min_1)\triangle x_1 + (\max_2 - \min_2)\triangle x_2 + \ldots + (\max_n - \min_n)\triangle x_n$
$= (f(x_1) - f(x_0))\triangle x_1 + (f(x_2) - f(x_1))\triangle x_2 + \ldots + (f(x_n) - f(x_{n-1}))\triangle x_n$
$\leq (f(x_1) - f(x_0))\triangle x_{max} + (f(x_2) - f(x_1))\triangle x_{max} + \ldots + (f(x_n) - f(x_{n-1}))\triangle x_{max}$. Then
$U - L \leq (f(x_n) - f(x_0))\triangle x_{max} = (f(b) - f(a))\triangle x_{max} = |f(b) - f(a)|\triangle x_{max}$ since $f(b) \geq f(a)$. Thus
$\lim_{\|P\|\to 0}(U - L) = \lim_{\|P\|\to 0}(f(b) - f(a))\triangle x_{max} = 0$, since $\triangle x_{max} = \|P\|$.

79. (a) Partition $\left[0, \frac{\pi}{2}\right]$ into n subintervals, each of length $\triangle x = \frac{\pi}{2n}$ with points $x_0 = 0, x_1 = \triangle x$,
$x_2 = 2\triangle x, \ldots, x_n = n\triangle x = \frac{\pi}{2}$. Since $\sin x$ is increasing on $\left[0, \frac{\pi}{2}\right]$, the upper sum U is the sum of the areas of the circumscribed rectangles of areas $f(x_1)\triangle x = (\sin \triangle x)\triangle x, f(x_2)\triangle x = (\sin 2\triangle x)\triangle x, \ldots, f(x_n)\triangle x$
$= (\sin n\triangle x)\triangle x$. Then $U = (\sin \triangle x + \sin 2\triangle x + \ldots + \sin n\triangle x)\triangle x = \left[\dfrac{\cos\frac{\triangle x}{2} - \cos\left(\left(n+\frac{1}{2}\right)\triangle x\right)}{2\sin\frac{\triangle x}{2}}\right]\triangle x$

$= \left[\dfrac{\cos\frac{\pi}{4n} - \cos\left(\left(n+\frac{1}{2}\right)\frac{\pi}{2n}\right)}{2\sin\frac{\pi}{4n}}\right]\left(\frac{\pi}{2n}\right) = \dfrac{\pi\left(\cos\frac{\pi}{4n} - \cos\left(\frac{\pi}{2}+\frac{\pi}{4n}\right)\right)}{4n\sin\frac{\pi}{4n}} = \dfrac{\cos\frac{\pi}{4n} - \cos\left(\frac{\pi}{2}+\frac{\pi}{4n}\right)}{\left(\frac{\sin\frac{\pi}{4n}}{\frac{\pi}{4n}}\right)}$

(b) The area is $\int_0^{\pi/2} \sin x\,dx = \lim_{n\to\infty} \dfrac{\cos\frac{\pi}{4n} - \cos\left(\frac{\pi}{2}+\frac{\pi}{4n}\right)}{\left(\frac{\sin\frac{\pi}{4n}}{\frac{\pi}{4n}}\right)} = \dfrac{1 - \cos\frac{\pi}{2}}{1} = 1.$

81. By Exercise 80, $U - L = \sum_{i=1}^n \triangle x_i\cdot M_i - \sum_{i=1}^n \triangle x_i\cdot m_i$ where $M_i = \max\{f(x)$ on the ith subinterval$\}$ and

$m_i = \min\{f(x)$ on the ith subinterval$\}$. Thus $U - L = \sum_{i=1}^n (M_i - m_i)\triangle x_i < \sum_{i=1}^n \epsilon\cdot\triangle x_i$ provided $\triangle x_i < \delta$ for each

$i = 1, \ldots, n$. Since $\sum_{i=1}^n \epsilon\cdot\triangle x_i = \epsilon\sum_{i=1}^n \triangle x_i = \epsilon(b-a)$ the result, $U - L < \epsilon(b-a)$ follows.

5.4 THE FUNDAMENTAL THEOREM OF CALCULUS

1. $\int_{-2}^0 (2x+5)\,dx = [x^2 + 5x]_{-2}^0 = (0^2 + 5(0)) - ((-2)^2 + 5(-2)) = 6$

3. $\int_0^4 \left(3x - \frac{x^3}{4}\right)dx = \left[\frac{3x^2}{2} - \frac{x^4}{16}\right]_0^4 = \left(\frac{3(4)^2}{2} - \frac{4^4}{16}\right) - \left(\frac{3(0)^2}{2} - \frac{(0)^4}{16}\right) = 8$

5. $\int_0^1 \left(x^2 + \sqrt{x}\right) dx = \left[\frac{x^3}{3} + \frac{2}{3} x^{3/2}\right]_0^1 = \left(\frac{1}{3} + \frac{2}{3}\right) - 0 = 1$

7. $\int_1^{32} x^{-6/5} dx = \left[-5x^{-1/5}\right]_1^{32} = \left(-\frac{5}{2}\right) - (-5) = \frac{5}{2}$

9. $\int_0^\pi \sin x \, dx = [-\cos x]_0^\pi = (-\cos \pi) - (-\cos 0) = -(-1) - (-1) = 2$

11. $\int_0^{\pi/3} 2 \sec^2 x \, dx = [2 \tan x]_0^{\pi/3} = \left(2 \tan \left(\frac{\pi}{3}\right)\right) - (2 \tan 0) = 2\sqrt{3} - 0 = 2\sqrt{3}$

13. $\int_{\pi/4}^{3\pi/4} \csc \theta \cot \theta \, d\theta = [-\csc \theta]_{\pi/4}^{3\pi/4} = \left(-\csc \left(\frac{3\pi}{4}\right)\right) - \left(-\csc \left(\frac{\pi}{4}\right)\right) = -\sqrt{2} - \left(-\sqrt{2}\right) = 0$

15. $\int_{\pi/2}^0 \frac{1 + \cos 2t}{2} dt = \int_{\pi/2}^0 \left(\frac{1}{2} + \frac{1}{2} \cos 2t\right) dt = \left[\frac{1}{2} t + \frac{1}{4} \sin 2t\right]_{\pi/2}^0 = \left(\frac{1}{2}(0) + \frac{1}{4} \sin 2(0)\right) - \left(\frac{1}{2}\left(\frac{\pi}{2}\right) + \frac{1}{4} \sin 2\left(\frac{\pi}{2}\right)\right)$
$= -\frac{\pi}{4}$

17. $\int_{-\pi/2}^{\pi/2} (8y^2 + \sin y) \, dy = \left[\frac{8y^3}{3} - \cos y\right]_{-\pi/2}^{\pi/2} = \left(\frac{8\left(\frac{\pi}{2}\right)^3}{3} - \cos \frac{\pi}{2}\right) - \left(\frac{8\left(-\frac{\pi}{2}\right)^3}{3} - \cos \left(-\frac{\pi}{2}\right)\right) = \frac{2\pi^3}{3}$

19. $\int_1^{-1} (r + 1)^2 \, dr = \int_1^{-1} (r^2 + 2r + 1) \, dr = \left[\frac{r^3}{3} + r^2 + r\right]_1^{-1} = \left(\frac{(-1)^3}{3} + (-1)^2 + (-1)\right) - \left(\frac{1^3}{3} + 1^2 + 1\right) = -\frac{8}{3}$

21. $\int_{\sqrt{2}}^1 \left(\frac{u^7}{2} - \frac{1}{u^5}\right) du = \int_{\sqrt{2}}^1 \left(\frac{u^7}{2} - u^{-5}\right) du = \left[\frac{u^8}{16} + \frac{1}{4u^4}\right]_{\sqrt{2}}^1 = \left(\frac{1^8}{16} + \frac{1}{4(1)^4}\right) - \left(\frac{\left(\sqrt{2}\right)^8}{16} + \frac{1}{4\left(\sqrt{2}\right)^4}\right) = -\frac{3}{4}$

23. $\int_1^{\sqrt{2}} \frac{s^2 + \sqrt{s}}{s^2} \, ds = \int_1^{\sqrt{2}} \left(1 + s^{-3/2}\right) ds = \left[s - \frac{2}{\sqrt{s}}\right]_1^{\sqrt{2}} = \left(\sqrt{2} - \frac{2}{\sqrt{\sqrt{2}}}\right) - \left(1 - \frac{2}{\sqrt{1}}\right) = \sqrt{2} - 2^{3/4} + 1$
$= \sqrt{2} - \sqrt[4]{8} + 1$

25. $\int_{-4}^4 |x| \, dx = \int_{-4}^0 |x| \, dx + \int_0^4 |x| \, dx = -\int_{-4}^0 x \, dx + \int_0^4 x \, dx = \left[-\frac{x^2}{2}\right]_{-4}^0 + \left[\frac{x^2}{2}\right]_0^4 = \left(-\frac{0^2}{2} + \frac{(-4)^2}{2}\right) + \left(\frac{4^2}{2} - \frac{0^2}{2}\right)$
$= 16$

27. $\int_0^{\ln 2} e^{3x} dx = \frac{1}{3} e^{3x} \Big|_0^{\ln 2} = \frac{1}{3} e^{3 \ln 2} - \frac{1}{3} e^0 = \frac{1}{3} e^{\ln 8} - \frac{1}{3} = \frac{8}{3} - \frac{1}{3} = \frac{7}{3}$

29. $\int_0^1 \frac{4}{1 + x^2} dx = 4 \tan^{-1} x \Big|_0^1 = 4 \tan^{-1} 1 - 4 \tan^{-1} 0 = 4\left(\frac{\pi}{4}\right) - 4(0) = \pi$

31. $\int_2^4 x^{\pi - 1} dx = \frac{x^\pi}{\pi} \Big|_2^4 = \frac{1}{\pi}(4^\pi - 2^\pi)$

33. $\int_0^1 x e^{x^2} dx = \frac{1}{2} e^{x^2} \Big|_0^1 = \frac{1}{2} e^1 - \frac{1}{2} e^0 = \frac{1}{2}(e - 1)$

35. (a) $\int_0^{\sqrt{x}} \cos t \, dt = [\sin t]_0^{\sqrt{x}} = \sin \sqrt{x} - \sin 0 = \sin \sqrt{x} \Rightarrow \frac{d}{dx}\left(\int_0^{\sqrt{x}} \cos t \, dt\right) = \frac{d}{dx}\left(\sin \sqrt{x}\right) = \cos \sqrt{x} \left(\frac{1}{2} x^{-1/2}\right)$
$= \frac{\cos \sqrt{x}}{2\sqrt{x}}$

(b) $\frac{d}{dx}\left(\int_0^{\sqrt{x}} \cos t\, dt\right) = (\cos \sqrt{x})\left(\frac{d}{dx}(\sqrt{x})\right) = (\cos \sqrt{x})\left(\frac{1}{2}x^{-1/2}\right) = \frac{\cos \sqrt{x}}{2\sqrt{x}}$

37. (a) $\int_0^{t^4} \sqrt{u}\, du = \int_0^{t^4} u^{1/2}\, du = \left[\frac{2}{3}u^{3/2}\right]_0^{t^4} = \frac{2}{3}(t^4)^{3/2} - 0 = \frac{2}{3}t^6 \Rightarrow \frac{d}{dt}\left(\int_0^{t^4}\sqrt{u}\,du\right) = \frac{d}{dt}\left(\frac{2}{3}t^6\right) = 4t^5$

(b) $\frac{d}{dt}\left(\int_0^{t^4}\sqrt{u}\,du\right) = \sqrt{t^4}\left(\frac{d}{dt}(t^4)\right) = t^2(4t^3) = 4t^5$

39. (a) $\int_0^{x^3} e^{-t}dt = -e^{-t}\Big|_0^{x^3} = -e^{-x^3} + 1 \Rightarrow \frac{d}{dx}\left(\int_0^{x^3}e^{-t}dt\right) = \frac{d}{dx}\left(-e^{-x^3}+1\right) = 3x^2e^{-x^3}$

(b) $\frac{d}{dx}\left(\int_0^{x^3}e^{-t}dt\right) = e^{-x^3}\cdot\frac{d}{dx}(x^3) = 3x^2e^{-x^3}$

41. $y = \int_0^x \sqrt{1+t^2}\, dt \Rightarrow \frac{dy}{dx} = \sqrt{1+x^2}$

43. $y = \int_{\sqrt{x}}^0 \sin t^2\, dt = -\int_0^{\sqrt{x}} \sin t^2\, dt \Rightarrow \frac{dy}{dx} = -\left(\sin(\sqrt{x})^2\right)\left(\frac{d}{dx}(\sqrt{x})\right) = -(\sin x)\left(\frac{1}{2}x^{-1/2}\right) = -\frac{\sin x}{2\sqrt{x}}$

45. $y = \int_0^{\sin x} \frac{dt}{\sqrt{1-t^2}},\ |x| < \frac{\pi}{2} \Rightarrow \frac{dy}{dx} = \frac{1}{\sqrt{1-\sin^2 x}}\left(\frac{d}{dx}(\sin x)\right) = \frac{1}{\sqrt{\cos^2 x}}(\cos x) = \frac{\cos x}{|\cos x|} = \frac{\cos x}{\cos x} = 1$ since $|x| < \frac{\pi}{2}$

47. $y = \int_0^{e^{x^2}} \frac{1}{\sqrt{t}}dt \Rightarrow \frac{dy}{dx} = \frac{1}{\sqrt{e^{x^2}}}\cdot\frac{d}{dx}\left(e^{x^2}\right) = \frac{1}{e^{\frac{1}{2}x^2}}\cdot 2xe^{x^2} = 2xe^{\frac{1}{2}x^2}$

49. $y = \int_0^{\sin^{-1}t} \cos t\, dt \Rightarrow \frac{dy}{dx} = \cos(\sin^{-1}x)\cdot\frac{d}{dx}(\sin^{-1}x) = \sqrt{1-x^2}\cdot\frac{1}{\sqrt{1-x^2}} = 1$

51. $-x^2 - 2x = 0 \Rightarrow -x(x+2) = 0 \Rightarrow x = 0$ or $x = -2$; Area

$= -\int_{-3}^{-2}(-x^2-2x)dx + \int_{-2}^{0}(-x^2-2x)dx - \int_0^2(-x^2-2x)dx$

$= -\left[-\frac{x^3}{3}-x^2\right]_{-3}^{-2} + \left[-\frac{x^3}{3}-x^2\right]_{-2}^{0} - \left[-\frac{x^3}{3}-x^2\right]_0^2$

$= -\left(\left(-\frac{(-2)^3}{3}-(-2)^2\right) - \left(-\frac{(-3)^3}{3}-(-3)^2\right)\right)$

$+ \left(\left(-\frac{0^3}{3}-0^2\right) - \left(-\frac{(-2)^3}{3}-(-2)^2\right)\right)$

$- \left(\left(-\frac{2^3}{3}-2^2\right) - \left(-\frac{0^3}{3}-0^2\right)\right) = \frac{28}{3}$

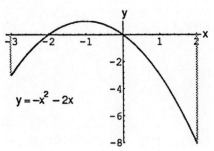

$y = -x^2 - 2x$

53. $x^3 - 3x^2 + 2x = 0 \Rightarrow x(x^2-3x+2) = 0$

$\Rightarrow x(x-2)(x-1) = 0 \Rightarrow x = 0, 1,$ or 2;

Area $= \int_0^1 (x^3-3x^2+2x)dx - \int_1^2 (x^3-3x^2+2x)dx$

$= \left[\frac{x^4}{4}-x^3+x^2\right]_0^1 - \left[\frac{x^4}{4}-x^3+x^2\right]_1^2$

$= \left(\frac{1^4}{4}-1^3+1^2\right) - \left(\frac{0^4}{4}-0^3+0^2\right)$

$- \left[\left(\frac{2^4}{4}-2^3+2^2\right) - \left(\frac{1^4}{4}-1^3+1^2\right)\right] = \frac{1}{2}$

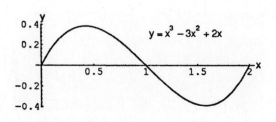

$y = x^3 - 3x^2 + 2x$

55. $x^{1/3} = 0 \Rightarrow x = 0$; Area $= -\int_{-1}^{0} x^{1/3} \, dx + \int_{0}^{8} x^{1/3} \, dx$

$= \left[-\frac{3}{4} x^{4/3} \right]_{-1}^{0} + \left[\frac{3}{4} x^{4/3} \right]_{0}^{8}$

$= \left(-\frac{3}{4} (0)^{4/3} \right) - \left(-\frac{3}{4} (-1)^{4/3} \right) + \left(\frac{3}{4} (8)^{4/3} \right) - \left(\frac{3}{4} (0)^{4/3} \right)$

$= \frac{51}{4}$

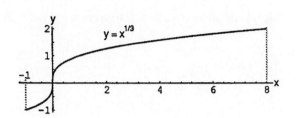

57. The area of the rectangle bounded by the lines $y = 2$, $y = 0$, $x = \pi$, and $x = 0$ is 2π. The area under the curve $y = 1 + \cos x$ on $[0, \pi]$ is $\int_{0}^{\pi} (1 + \cos x) \, dx = [x + \sin x]_{0}^{\pi} = (\pi + \sin \pi) - (0 + \sin 0) = \pi$. Therefore the area of the shaded region is $2\pi - \pi = \pi$.

59. On $\left[-\frac{\pi}{4}, 0 \right]$: The area of the rectangle bounded by the lines $y = \sqrt{2}$, $y = 0$, $\theta = 0$, and $\theta = -\frac{\pi}{4}$ is $\sqrt{2} \left(\frac{\pi}{4} \right)$

$= \frac{\pi\sqrt{2}}{4}$. The area between the curve $y = \sec \theta \tan \theta$ and $y = 0$ is $-\int_{-\pi/4}^{0} \sec \theta \tan \theta \, d\theta = [-\sec \theta]_{-\pi/4}^{0}$

$= (-\sec 0) - \left(-\sec \left(-\frac{\pi}{4} \right) \right) = \sqrt{2} - 1$. Therefore the area of the shaded region on $\left[-\frac{\pi}{4}, 0 \right]$ is $\frac{\pi\sqrt{2}}{4} + \left(\sqrt{2} - 1 \right)$.

On $\left[0, \frac{\pi}{4} \right]$: The area of the rectangle bounded by $\theta = \frac{\pi}{4}$, $\theta = 0$, $y = \sqrt{2}$, and $y = 0$ is $\sqrt{2} \left(\frac{\pi}{4} \right) = \frac{\pi\sqrt{2}}{4}$. The area

under the curve $y = \sec \theta \tan \theta$ is $\int_{0}^{\pi/4} \sec \theta \tan \theta \, d\theta = [\sec \theta]_{0}^{\pi/4} = \sec \frac{\pi}{4} - \sec 0 = \sqrt{2} - 1$. Therefore the area

of the shaded region on $\left[0, \frac{\pi}{4} \right]$ is $\frac{\pi\sqrt{2}}{4} - \left(\sqrt{2} - 1 \right)$. Thus, the area of the total shaded region is

$\left(\frac{\pi\sqrt{2}}{4} + \sqrt{2} - 1 \right) + \left(\frac{\pi\sqrt{2}}{4} - \sqrt{2} + 1 \right) = \frac{\pi\sqrt{2}}{2}$.

61. $y = \int_{\pi}^{x} \frac{1}{t} \, dt - 3 \Rightarrow \frac{dy}{dx} = \frac{1}{x}$ and $y(\pi) = \int_{\pi}^{\pi} \frac{1}{t} \, dt - 3 = 0 - 3 = -3 \Rightarrow$ (d) is a solution to this problem.

63. $y = \int_{0}^{x} \sec t \, dt + 4 \Rightarrow \frac{dy}{dx} = \sec x$ and $y(0) = \int_{0}^{0} \sec t \, dt + 4 = 0 + 4 = 4 \Rightarrow$ (b) is a solution to this problem.

65. $y = \int_{2}^{x} \sec t \, dt + 3$ 67. $s = \int_{t_0}^{t} f(x) \, dx + s_0$

69. Area $= \int_{-b/2}^{b/2} \left(h - \left(\frac{4h}{b^2} \right) x^2 \right) \, dx = \left[hx - \frac{4hx^3}{3b^2} \right]_{-b/2}^{b/2}$

$= \left(h \left(\frac{b}{2} \right) - \frac{4h \left(\frac{b}{2} \right)^3}{3b^2} \right) - \left(h \left(-\frac{b}{2} \right) - \frac{4h \left(-\frac{b}{2} \right)^3}{3b^2} \right)$

$= \left(\frac{bh}{2} - \frac{bh}{6} \right) - \left(-\frac{bh}{2} + \frac{bh}{6} \right) = bh - \frac{bh}{3} = \frac{2}{3} bh$

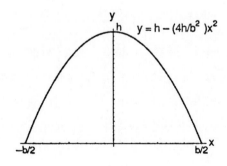

71. $\frac{dc}{dx} = \frac{1}{2\sqrt{x}} = \frac{1}{2} x^{-1/2} \Rightarrow c = \int_{0}^{x} \frac{1}{2} t^{-1/2} dt = \left[t^{1/2} \right]_{0}^{x} = \sqrt{x}$

$c(100) - c(1) = \sqrt{100} - \sqrt{1} = \$9.00$

73. $k > 0 \Rightarrow$ one arch of $y = \sin kx$ will occur over the interval $\left[0, \frac{\pi}{k} \right] \Rightarrow$ the area $= \int_{0}^{\pi/k} \sin kx \, dx = \left[-\frac{1}{k} \cos kx \right]_{0}^{\pi/k}$

$= -\frac{1}{k} \cos \left(k \left(\frac{\pi}{k} \right) \right) - \left(-\frac{1}{k} \cos (0) \right) = \frac{2}{k}$

75. $\int_1^x f(t)\,dt = x^2 - 2x + 1 \Rightarrow f(x) = \frac{d}{dx}\int_1^x f(t)\,dt = \frac{d}{dx}(x^2 - 2x + 1) = 2x - 2$

77. $f(x) = 2 - \int_2^{x+1} \frac{9}{1+t}\,dt \Rightarrow f'(x) = -\frac{9}{1+(x+1)} = \frac{-9}{x+2} \Rightarrow f'(1) = -3; f(1) = 2 - \int_2^{1+1} \frac{9}{1+t}\,dt = 2 - 0 = 2;$
$L(x) = -3(x-1) + f(1) = -3(x-1) + 2 = -3x + 5$

79. (a) True: since f is continuous, g is differentiable by Part 1 of the Fundamental Theorem of Calculus.
    (b) True: g is continuous because it is differentiable.
    (c) True, since $g'(1) = f(1) = 0$.
    (d) False, since $g''(1) = f'(1) > 0$.
    (e) True, since $g'(1) = 0$ and $g''(1) = f'(1) > 0$.
    (f) False: $g''(x) = f'(x) > 0$, so g'' never changes sign.
    (g) True, since $g'(1) = f(1) = 0$ and $g'(x) = f(x)$ is an increasing function of x (because $f'(x) > 0$).

5.5 INDEFINTE INTEGRALS AND THE SUBSTITUTION RULE

1. Let $u = 3x \Rightarrow du = 3\,dx \Rightarrow \frac{1}{3}\,du = dx$
   $\int \sin 3x\,dx = \int \frac{1}{3}\sin u\,du = -\frac{1}{3}\cos u + C = -\frac{1}{3}\cos 3x + C$

3. Let $u = 2t \Rightarrow du = 2\,dt \Rightarrow \frac{1}{2}\,du = dt$
   $\int \sec 2t \tan 2t\,dt = \int \frac{1}{2}\sec u \tan u\,du = \frac{1}{2}\sec u + C = \frac{1}{2}\sec 2t + C$

5. Let $u = 7x - 2 \Rightarrow du = 7\,dx \Rightarrow \frac{1}{7}\,du = dx$
   $\int 28(7x-2)^{-5}\,dx = \int \frac{1}{7}(28)u^{-5}\,du = \int 4u^{-5}\,du = -u^{-4} + C = -(7x-2)^{-4} + C$

7. Let $u = 1 - r^3 \Rightarrow du = -3r^2\,dr \Rightarrow -3\,du = 9r^2\,dr$
   $\int \frac{9r^2\,dr}{\sqrt{1-r^3}} = \int -3u^{-1/2}\,du = -3(2)u^{1/2} + C = -6(1-r^3)^{1/2} + C$

9. Let $u = x^{3/2} - 1 \Rightarrow du = \frac{3}{2}x^{1/2}\,dx \Rightarrow \frac{2}{3}\,du = \sqrt{x}\,dx$
   $\int \sqrt{x}\sin^2\left(x^{3/2} - 1\right)\,dx = \int \frac{2}{3}\sin^2 u\,du = \frac{2}{3}\left(\frac{u}{2} - \frac{1}{4}\sin 2u\right) + C = \frac{1}{3}\left(x^{3/2} - 1\right) - \frac{1}{6}\sin\left(2x^{3/2} - 2\right) + C$

11. (a) Let $u = \cot 2\theta \Rightarrow du = -2\csc^2 2\theta\,d\theta \Rightarrow -\frac{1}{2}\,du = \csc^2 2\theta\,d\theta$
    $\int \csc^2 2\theta \cot 2\theta\,d\theta = -\int \frac{1}{2}u\,du = -\frac{1}{2}\left(\frac{u^2}{2}\right) + C = -\frac{u^2}{4} + C = -\frac{1}{4}\cot^2 2\theta + C$
    (b) Let $u = \csc 2\theta \Rightarrow du = -2\csc 2\theta \cot 2\theta\,d\theta \Rightarrow -\frac{1}{2}\,du = \csc 2\theta \cot 2\theta\,d\theta$
    $\int \csc^2 2\theta \cot 2\theta\,d\theta = \int -\frac{1}{2}u\,du = -\frac{1}{2}\left(\frac{u^2}{2}\right) + C = -\frac{u^2}{4} + C = -\frac{1}{4}\csc^2 2\theta + C$

13. Let $u = 3 - 2s \Rightarrow du = -2\,ds \Rightarrow -\frac{1}{2}\,du = ds$
    $\int \sqrt{3-2s}\,ds = \int \sqrt{u}\left(-\frac{1}{2}\,du\right) = -\frac{1}{2}\int u^{1/2}\,du = \left(-\frac{1}{2}\right)\left(\frac{2}{3}u^{3/2}\right) + C = -\frac{1}{3}(3-2s)^{3/2} + C$

15. Let $u = 5s + 4 \Rightarrow du = 5\,ds \Rightarrow \frac{1}{5}\,du = ds$
    $\int \frac{1}{\sqrt{5s+4}}\,ds = \int \frac{1}{\sqrt{u}}\left(\frac{1}{5}\,du\right) = \frac{1}{5}\int u^{-1/2}\,du = \left(\frac{1}{5}\right)\left(2u^{1/2}\right) + C = \frac{2}{5}\sqrt{5s+4} + C$

17. Let $u = 1 - \theta^2 \Rightarrow du = -2\theta\,d\theta \Rightarrow -\frac{1}{2}\,du = \theta\,d\theta$
    $\int \theta\sqrt[4]{1-\theta^2}\,d\theta = \int \sqrt[4]{u}\left(-\frac{1}{2}\,du\right) = -\frac{1}{2}\int u^{1/4}\,du = \left(-\frac{1}{2}\right)\left(\frac{4}{5}u^{5/4}\right) + C = -\frac{2}{5}\left(1-\theta^2\right)^{5/4} + C$

19. Let $u = 1 + \sqrt{x} \Rightarrow du = \frac{1}{2\sqrt{x}} dx \Rightarrow 2\,du = \frac{1}{\sqrt{x}} dx$

$$\int \frac{1}{\sqrt{x}\,(1+\sqrt{x})^2}\,dx = \int \frac{2\,du}{u^2} = -\frac{2}{u} + C = \frac{-2}{1+\sqrt{x}} + C$$

21. Let $u = 3z + 4 \Rightarrow du = 3\,dz \Rightarrow \frac{1}{3}\,du = dz$

$$\int \cos(3z+4)\,dz = \int (\cos u)\left(\frac{1}{3}\,du\right) = \frac{1}{3}\int \cos u\,du = \frac{1}{3}\sin u + C = \frac{1}{3}\sin(3z+4) + C$$

23. Let $u = \cos x \Rightarrow du = -\sin x\,dx \Rightarrow -du = \sin x\,dx$

$$\int \tan x\,dx = \int \frac{\sin x}{\cos x}\,dx = -\int \frac{1}{u}\,du = -\ln|u| + C = -\ln|\cos x| + C = \ln|\cos x|^{-1} + C = \ln|\sec x| + C$$

25. Let $u = \frac{r^3}{18} - 1 \Rightarrow du = \frac{r^2}{6}\,dr \Rightarrow 6\,du = r^2\,dr$

$$\int r^2 \left(\frac{r^3}{18} - 1\right)^5 dr = \int u^5\,(6\,du) = 6\int u^5\,du = 6\left(\frac{u^6}{6}\right) + C = \left(\frac{r^3}{18} - 1\right)^6 + C$$

27. Let $u = x^{3/2} + 1 \Rightarrow du = \frac{3}{2}x^{1/2}\,dx \Rightarrow \frac{2}{3}\,du = x^{1/2}\,dx$

$$\int x^{1/2} \sin\left(x^{3/2} + 1\right)\,dx = \int (\sin u)\left(\frac{2}{3}\,du\right) = \frac{2}{3}\int \sin u\,du = \frac{2}{3}(-\cos u) + C = -\frac{2}{3}\cos\left(x^{3/2} + 1\right) + C$$

29. Let $u = \cos(2t+1) \Rightarrow du = -2\sin(2t+1)\,dt \Rightarrow -\frac{1}{2}\,du = \sin(2t+1)\,dt$

$$\int \frac{\sin(2t+1)}{\cos^2(2t+1)}\,dt = \int -\frac{1}{2}\frac{du}{u^2} = \frac{1}{2u} + C = \frac{1}{2\cos(2t+1)} + C$$

31. Let $u = \sin\frac{1}{\theta} \Rightarrow du = \left(\cos\frac{1}{\theta}\right)\left(-\frac{1}{\theta^2}\right)d\theta \Rightarrow -du = \frac{1}{\theta^2}\cos\frac{1}{\theta}\,d\theta$

$$\int \frac{1}{\theta^2}\sin\frac{1}{\theta}\cos\frac{1}{\theta}\,d\theta = \int -u\,du = -\frac{1}{2}u^2 + C = -\frac{1}{2}\sin^2\frac{1}{\theta} + C$$

33. Let $u = \frac{1}{t} - 1 = t^{-1} - 1 \Rightarrow du = -t^{-2}\,dt \Rightarrow -du = \frac{1}{t^2}\,dt$

$$\int \frac{1}{t^2}\cos\left(\frac{1}{t} - 1\right)dt = \int (\cos u)(-du) = -\int \cos u\,du = -\sin u + C = -\sin\left(\frac{1}{t} - 1\right) + C$$

35. Let $u = s^3 + 2s^2 - 5s + 5 \Rightarrow du = (3s^2 + 4s - 5)\,ds$

$$\int \left(s^3 + 2s^2 - 5s + 5\right)(3s^2 + 4s - 5)\,ds = \int u\,du = \frac{u^2}{2} + C = \frac{\left(s^3 + 2s^2 - 5s + 5\right)^2}{2} + C$$

37. Let $u = 1 - \frac{1}{x} \Rightarrow du = \frac{1}{x^2}\,dx$

$$\int \sqrt{\frac{x-1}{x^5}}\,dx = \int \frac{1}{x^2}\sqrt{\frac{x-1}{x}}\,dx = \int \frac{1}{x^2}\sqrt{1 - \frac{1}{x}}\,dx = \int \sqrt{u}\,du = \int u^{1/2}\,du = \frac{2}{3}u^{3/2} + C = \frac{2}{3}\left(1 - \frac{1}{x}\right)^{3/2} + C$$

39. Let $u = \sin x \Rightarrow du = \cos x\,dx$

$$\int (\cos x)\,e^{\sin x}\,dx = \int e^u\,du = e^u + C = e^{\sin x} + C$$

41. Let $u = e^{\sqrt{x}} + 1 \Rightarrow du = \frac{1}{2\sqrt{x}}e^{\sqrt{x}}dx \Rightarrow 2\,du = \frac{1}{\sqrt{x}}e^{\sqrt{x}}dx = \frac{1}{\sqrt{x}\,e^{-\sqrt{x}}}\,dx$

$$\int \frac{1}{\sqrt{x}\,e^{-\sqrt{x}}}\sec^2\left(e^{\sqrt{x}} + 1\right)dx = 2\int \sec^2 u\,du = 2\tan u + C = 2\tan\left(e^{\sqrt{x}} + 1\right) + C$$

43. Let $u = \ln x \Rightarrow du = \frac{1}{x}dx$

$$\int \frac{1}{x\ln x}\,dx = \int \frac{1}{u}\,du = \ln|u| + C = \ln|\ln x| + C$$

45. Let $u = e^{-z} + 1 \Rightarrow du = -e^{-z} dz \Rightarrow -du = e^{-z} dz$

$\int \frac{dz}{1 + e^z} = \int \frac{1}{1 + e^z} \cdot \frac{e^{-z}}{e^{-z}} dz = \int \frac{e^{-z}}{e^{-z} + 1} dz = -\int \frac{1}{u} du = -\ln|u| + C = -\ln(e^{-z} + 1) + C = -\ln\left(\frac{1 + e^z}{e^z}\right) + C$

$= -(\ln(1 + e^z) - \ln e^z) + C = z - \ln(1 + e^z) + C$

47. Let $u = \frac{2}{3}r \Rightarrow du = \frac{2}{3}dr \Rightarrow \frac{3}{2}du = dr$

$\int \frac{5}{9 + 4r^2} dr = \frac{5}{9} \int \frac{1}{1 + \left(\frac{2}{3}r\right)^2} dr = \frac{5}{9} \int \frac{\frac{3}{2}}{1 + u^2} du = \frac{5}{6}\tan^{-1}u + C = \frac{5}{6}\tan^{-1}\left(\frac{2}{3}r\right) + C$

49. $\int \frac{e^{\sin^{-1}x}}{\sqrt{1 - x^2}} dx = \int e^u du$, where $u = \sin^{-1}x$ and $du = \frac{dx}{\sqrt{1 - x^2}}$

$= e^u + C = e^{\sin^{-1}x} + C$

51. $\int \frac{(\sin^{-1}x)^2}{\sqrt{1 - x^2}} dx = \int u^2 du$, where $u = \sin^{-1}x$ and $du = \frac{dx}{\sqrt{1 - x^2}}$

$= \frac{u^3}{3} + C = \frac{(\sin^{-1}x)^3}{3} + C$

53. $\int \frac{1}{(\tan^{-1}y)(1 + y^2)} dy = \int \frac{\left(\frac{1}{1 + y^2}\right)}{\tan^{-1}y} dy = \int \frac{1}{u} du$, where $u = \tan^{-1}y$ and $du = \frac{dy}{1 + y^2}$

$= \ln|u| + C = \ln|\tan^{-1}y| + C$

55. (a) Let $u = \tan x \Rightarrow du = \sec^2 x \, dx$; $v = u^3 \Rightarrow dv = 3u^2 du \Rightarrow 6 \, dv = 18u^2 du$; $w = 2 + v \Rightarrow dw = dv$

$\int \frac{18 \tan^2 x \sec^2 x}{(2 + \tan^3 x)^2} dx = \int \frac{18u^2}{(2 + u^3)^2} du = \int \frac{6 \, dv}{(2 + v)^2} = \int \frac{6 \, dw}{w^2} = 6 \int w^{-2} dw = -6w^{-1} + C = -\frac{6}{2 + v} + C$

$= -\frac{6}{2 + u^3} + C = -\frac{6}{2 + \tan^3 x} + C$

(b) Let $u = \tan^3 x \Rightarrow du = 3 \tan^2 x \sec^2 x \, dx \Rightarrow 6 \, du = 18 \tan^2 x \sec^2 x \, dx$; $v = 2 + u \Rightarrow dv = du$

$\int \frac{18 \tan^2 x \sec^2 x}{(2 + \tan^3 x)^2} dx = \int \frac{6 \, du}{(2 + u)^2} = \int \frac{6 \, dv}{v^2} = -\frac{6}{v} + C = -\frac{6}{2 + u} + C = -\frac{6}{2 + \tan^3 x} + C$

(c) Let $u = 2 + \tan^3 x \Rightarrow du = 3 \tan^2 x \sec^2 x \, dx \Rightarrow 6 \, du = 18 \tan^2 x \sec^2 x \, dx$

$\int \frac{18 \tan^2 x \sec^2 x}{(2 + \tan^3 x)^2} dx = \int \frac{6 \, du}{u^2} = -\frac{6}{u} + C = -\frac{6}{2 + \tan^3 x} + C$

57. Let $u = 3(2r - 1)^2 + 6 \Rightarrow du = 6(2r - 1)(2) \, dr \Rightarrow \frac{1}{12} du = (2r - 1) \, dr$; $v = \sqrt{u} \Rightarrow dv = \frac{1}{2\sqrt{u}} du \Rightarrow \frac{1}{6} dv = \frac{1}{12\sqrt{u}} du$

$\int \frac{(2r - 1)\cos\sqrt{3(2r - 1)^2 + 6}}{\sqrt{3(2r - 1)^2 + 6}} dr = \int \left(\frac{\cos\sqrt{u}}{\sqrt{u}}\right)\left(\frac{1}{12} du\right) = \int (\cos v)\left(\frac{1}{6} dv\right) = \frac{1}{6}\sin v + C = \frac{1}{6}\sin\sqrt{u} + C$

$= \frac{1}{6}\sin\sqrt{3(2r - 1)^2 + 6} + C$

59. Let $u = 3t^2 - 1 \Rightarrow du = 6t \, dt \Rightarrow 2 \, du = 12t \, dt$

$s = \int 12t(3t^2 - 1)^3 dt = \int u^3(2 \, du) = 2\left(\frac{1}{4}u^4\right) + C = \frac{1}{2}u^4 + C = \frac{1}{2}(3t^2 - 1)^4 + C$;

$s = 3$ when $t = 1 \Rightarrow 3 = \frac{1}{2}(3 - 1)^4 + C \Rightarrow 3 = 8 + C \Rightarrow C = -5 \Rightarrow s = \frac{1}{2}(3t^2 - 1)^4 - 5$

61. Let $u = t + \frac{\pi}{12} \Rightarrow du = dt$

$s = \int 8 \sin^2\left(t + \frac{\pi}{12}\right) dt = \int 8 \sin^2 u \, du = 8\left(\frac{u}{2} - \frac{1}{4}\sin 2u\right) + C = 4\left(t + \frac{\pi}{12}\right) - 2\sin\left(2t + \frac{\pi}{6}\right) + C$;

$s = 8$ when $t = 0 \Rightarrow 8 = 4\left(\frac{\pi}{12}\right) - 2\sin\left(\frac{\pi}{6}\right) + C \Rightarrow C = 8 - \frac{\pi}{3} + 1 = 9 - \frac{\pi}{3}$

$\Rightarrow s = 4\left(t + \frac{\pi}{12}\right) - 2\sin\left(2t + \frac{\pi}{6}\right) + 9 - \frac{\pi}{3} = 4t - 2\sin\left(2t + \frac{\pi}{6}\right) + 9$

63. Let $u = 2t - \frac{\pi}{2} \Rightarrow du = 2 \, dt \Rightarrow -2 \, du = -4 \, dt$

$\frac{ds}{dt} = \int -4 \sin\left(2t - \frac{\pi}{2}\right) dt = \int (\sin u)(-2 \, du) = 2 \cos u + C_1 = 2 \cos\left(2t - \frac{\pi}{2}\right) + C_1$;

at $t = 0$ and $\frac{ds}{dt} = 100$ we have $100 = 2 \cos\left(-\frac{\pi}{2}\right) + C_1 \Rightarrow C_1 = 100 \Rightarrow \frac{ds}{dt} = 2 \cos\left(2t - \frac{\pi}{2}\right) + 100$

$\Rightarrow s = \int \left(2 \cos \left(2t - \frac{\pi}{2}\right) + 100\right) dt = \int (\cos u + 50) \, du = \sin u + 50u + C_2 = \sin \left(2t - \frac{\pi}{2}\right) + 50 \left(2t - \frac{\pi}{2}\right) + C_2;$

at $t = 0$ and $s = 0$ we have $0 = \sin \left(-\frac{\pi}{2}\right) + 50 \left(-\frac{\pi}{2}\right) + C_2 \Rightarrow C_2 = 1 + 25\pi$

$\Rightarrow s = \sin \left(2t - \frac{\pi}{2}\right) + 100t - 25\pi + (1 + 25\pi) \Rightarrow s = \sin \left(2t - \frac{\pi}{2}\right) + 100t + 1$

65. Let $u = 2t \Rightarrow du = 2 \, dt \Rightarrow 3 \, du = 6 \, dt$

$s = \int 6 \sin 2t \, dt = \int (\sin u)(3 \, du) = -3 \cos u + C = -3 \cos 2t + C;$

at $t = 0$ and $s = 0$ we have $0 = -3 \cos 0 + C \Rightarrow C = 3 \Rightarrow s = 3 - 3 \cos 2t \Rightarrow s\left(\frac{\pi}{2}\right) = 3 - 3 \cos (\pi) = 6 \text{ m}$

67. All three integrations are correct. In each case, the derivative of the function on the right is the integrand on the left, and each formula has an arbitrary constant for generating the remaining antiderivatives. Moreover, $\sin^2 x + C_1 = 1 - \cos^2 x + C_1 \Rightarrow C_2 = 1 + C_1;$ also $-\cos^2 x + C_2 = -\frac{\cos 2x}{2} - \frac{1}{2} + C_2 \Rightarrow C_3 = C_2 - \frac{1}{2} = C_1 + \frac{1}{2}.$

5.6 SUBSTITUTION AND AREA BETWEEN CURVES

1. (a) Let $u = y + 1 \Rightarrow du = dy; y = 0 \Rightarrow u = 1, y = 3 \Rightarrow u = 4$

$\int_0^3 \sqrt{y+1} \, dy = \int_1^4 u^{1/2} \, du = \left[\frac{2}{3} u^{3/2}\right]_1^4 = \left(\frac{2}{3}\right)(4)^{3/2} - \left(\frac{2}{3}\right)(1)^{3/2} = \left(\frac{2}{3}\right)(8) - \left(\frac{2}{3}\right)(1) = \frac{14}{3}$

   (b) Use the same substitution for u as in part (a); $y = -1 \Rightarrow u = 0, y = 0 \Rightarrow u = 1$

$\int_{-1}^0 \sqrt{y+1} \, dy = \int_0^1 u^{1/2} \, du = \left[\frac{2}{3} u^{3/2}\right]_0^1 = \left(\frac{2}{3}\right)(1)^{3/2} - 0 = \frac{2}{3}$

3. (a) Let $u = \tan x \Rightarrow du = \sec^2 x \, dx; x = 0 \Rightarrow u = 0, x = \frac{\pi}{4} \Rightarrow u = 1$

$\int_0^{\pi/4} \tan x \sec^2 x \, dx = \int_0^1 u \, du = \left[\frac{u^2}{2}\right]_0^1 = \frac{1^2}{2} - 0 = \frac{1}{2}$

   (b) Use the same substitution as in part (a); $x = -\frac{\pi}{4} \Rightarrow u = -1, x = 0 \Rightarrow u = 0$

$\int_{-\pi/4}^0 \tan x \sec^2 x \, dx = \int_{-1}^0 u \, du = \left[\frac{u^2}{2}\right]_{-1}^0 = 0 - \frac{1}{2} = -\frac{1}{2}$

5. (a) $u = 1 + t^4 \Rightarrow du = 4t^3 \, dt \Rightarrow \frac{1}{4} \, du = t^3 \, dt; t = 0 \Rightarrow u = 1, t = 1 \Rightarrow u = 2$

$\int_0^1 t^3 \left(1 + t^4\right)^3 \, dt = \int_1^2 \frac{1}{4} u^3 \, du = \left[\frac{u^4}{16}\right]_1^2 = \frac{2^4}{16} - \frac{1^4}{16} = \frac{15}{16}$

   (b) Use the same substitution as in part (a); $t = -1 \Rightarrow u = 2, t = 1 \Rightarrow u = 2$

$\int_{-1}^1 t^3 \left(1 + t^4\right)^3 \, dt = \int_2^2 \frac{1}{4} u^3 \, du = 0$

7. (a) Let $u = 4 + r^2 \Rightarrow du = 2r \, dr \Rightarrow \frac{1}{2} \, du = r \, dr; r = -1 \Rightarrow u = 5, r = 1 \Rightarrow u = 5$

$\int_{-1}^1 \frac{5r}{(4+r^2)^2} \, dr = 5 \int_5^5 \frac{1}{2} u^{-2} \, du = 0$

   (b) Use the same substitution as in part (a); $r = 0 \Rightarrow u = 4, r = 1 \Rightarrow u = 5$

$\int_0^1 \frac{5r}{(4+r^2)^2} \, dr = 5 \int_4^5 \frac{1}{2} u^{-2} \, du = 5 \left[-\frac{1}{2} u^{-1}\right]_4^5 = 5 \left(-\frac{1}{2} (5)^{-1}\right) - 5 \left(-\frac{1}{2} (4)^{-1}\right) = \frac{1}{8}$

9. (a) Let $u = x^2 + 1 \Rightarrow du = 2x \, dx \Rightarrow 2 \, du = 4x \, dx; x = 0 \Rightarrow u = 1, x = \sqrt{3} \Rightarrow u = 4$

$\int_0^{\sqrt{3}} \frac{4x}{\sqrt{x^2+1}} \, dx = \int_1^4 \frac{2}{\sqrt{u}} \, du = \int_1^4 2u^{-1/2} \, du = \left[4u^{1/2}\right]_1^4 = 4(4)^{1/2} - 4(1)^{1/2} = 4$

   (b) Use the same substitution as in part (a); $x = -\sqrt{3} \Rightarrow u = 4, x = \sqrt{3} \Rightarrow u = 4$

$\int_{-\sqrt{3}}^{\sqrt{3}} \frac{4x}{\sqrt{x^2+1}} \, dx = \int_4^4 \frac{2}{\sqrt{u}} \, du = 0$

11. (a) Let $u = 1 - \cos 3t \;\Rightarrow\; du = 3 \sin 3t \, dt \;\Rightarrow\; \frac{1}{3} du = \sin 3t \, dt; \, t = 0 \;\Rightarrow\; u = 0, \, t = \frac{\pi}{6} \;\Rightarrow\; u = 1 - \cos \frac{\pi}{2} = 1$

$$\int_0^{\pi/6} (1 - \cos 3t) \sin 3t \, dt = \int_0^1 \frac{1}{3} u \, du = \left[\frac{1}{3} \left(\frac{u^2}{2} \right) \right]_0^1 = \frac{1}{6} (1)^2 - \frac{1}{6} (0)^2 = \frac{1}{6}$$

(b) Use the same substitution as in part (a); $t = \frac{\pi}{6} \;\Rightarrow\; u = 1, \, t = \frac{\pi}{3} \;\Rightarrow\; u = 1 - \cos \pi = 2$

$$\int_{\pi/6}^{\pi/3} (1 - \cos 3t) \sin 3t \, dt = \int_1^2 \frac{1}{3} u \, du = \left[\frac{1}{3} \left(\frac{u^2}{2} \right) \right]_1^2 = \frac{1}{6} (2)^2 - \frac{1}{6} (1)^2 = \frac{1}{2}$$

13. (a) Let $u = 4 + 3 \sin z \;\Rightarrow\; du = 3 \cos z \, dz \;\Rightarrow\; \frac{1}{3} du = \cos z \, dz; \, z = 0 \;\Rightarrow\; u = 4, \, z = 2\pi \;\Rightarrow\; u = 4$

$$\int_0^{2\pi} \frac{\cos z}{\sqrt{4 + 3 \sin z}} \, dz = \int_4^4 \frac{1}{\sqrt{u}} \left(\frac{1}{3} du \right) = 0$$

(b) Use the same substitution as in part (a); $z = -\pi \;\Rightarrow\; u = 4 + 3 \sin (-\pi) = 4, \, z = \pi \;\Rightarrow\; u = 4$

$$\int_{-\pi}^{\pi} \frac{\cos z}{\sqrt{4 + 3 \sin z}} \, dz = \int_4^4 \frac{1}{\sqrt{u}} \left(\frac{1}{3} du \right) = 0$$

15. Let $u = t^5 + 2t \;\Rightarrow\; du = (5t^4 + 2) \, dt; \, t = 0 \;\Rightarrow\; u = 0, \, t = 1 \;\Rightarrow\; u = 3$

$$\int_0^1 \sqrt{t^5 + 2t} \, (5t^4 + 2) \, dt = \int_0^3 u^{1/2} \, du = \left[\frac{2}{3} u^{3/2} \right]_0^3 = \frac{2}{3} (3)^{3/2} - \frac{2}{3} (0)^{3/2} = 2\sqrt{3}$$

17. Let $u = \cos 2\theta \;\Rightarrow\; du = -2 \sin 2\theta \, d\theta \;\Rightarrow\; -\frac{1}{2} du = \sin 2\theta \, d\theta; \, \theta = 0 \;\Rightarrow\; u = 1, \, \theta = \frac{\pi}{6} \;\Rightarrow\; u = \cos 2 \left(\frac{\pi}{6} \right) = \frac{1}{2}$

$$\int_0^{\pi/6} \cos^{-3} 2\theta \sin 2\theta \, d\theta = \int_1^{1/2} u^{-3} \left(-\frac{1}{2} du \right) = -\frac{1}{2} \int_1^{1/2} u^{-3} \, du = \left[-\frac{1}{2} \left(\frac{u^{-2}}{-2} \right) \right]_1^{1/2} = \frac{1}{4 \left(\frac{1}{2} \right)^2} - \frac{1}{4(1)^2} = \frac{3}{4}$$

19. Let $u = 5 - 4 \cos t \;\Rightarrow\; du = 4 \sin t \, dt \;\Rightarrow\; \frac{1}{4} du = \sin t \, dt; \, t = 0 \;\Rightarrow\; u = 5 - 4 \cos 0 = 1, \, t = \pi \;\Rightarrow$
$u = 5 - 4 \cos \pi = 9$

$$\int_0^{\pi} 5 (5 - 4 \cos t)^{1/4} \sin t \, dt = \int_1^9 5u^{1/4} \left(\frac{1}{4} du \right) = \frac{5}{4} \int_1^9 u^{1/4} \, du = \left[\frac{5}{4} \left(\frac{4}{5} u^{5/4} \right) \right]_1^9 = 9^{5/4} - 1 = 3^{5/2} - 1$$

21. Let $u = 4y - y^2 + 4y^3 + 1 \;\Rightarrow\; du = (4 - 2y + 12y^2) \, dy; \, y = 0 \;\Rightarrow\; u = 1, \, y = 1 \;\Rightarrow\; u = 4(1) - (1)^2 + 4(1)^3 + 1 = 8$

$$\int_0^1 (4y - y^2 + 4y^3 + 1)^{-2/3} (12y^2 - 2y + 4) \, dy = \int_1^8 u^{-2/3} \, du = \left[3u^{1/3} \right]_1^8 = 3(8)^{1/3} - 3(1)^{1/3} = 3$$

23. Let $u = \theta^{3/2} \;\Rightarrow\; du = \frac{3}{2} \theta^{1/2} \, d\theta \;\Rightarrow\; \frac{2}{3} du = \sqrt{\theta} \, d\theta; \, \theta = 0 \;\Rightarrow\; u = 0, \, \theta = \sqrt[3]{\pi^2} \;\Rightarrow\; u = \pi$

$$\int_0^{\sqrt[3]{\pi^2}} \sqrt{\theta} \cos^2 (\theta^{3/2}) \, d\theta = \int_0^{\pi} \cos^2 u \left(\frac{2}{3} du \right) = \left[\frac{2}{3} \left(\frac{u}{2} + \frac{1}{4} \sin 2u \right) \right]_0^{\pi} = \frac{2}{3} \left(\frac{\pi}{2} + \frac{1}{4} \sin 2\pi \right) - \frac{2}{3} (0) = \frac{\pi}{3}$$

25. Let $u = \tan \theta \;\Rightarrow\; du = \sec^2 \theta \, d\theta; \, \theta = 0 \;\Rightarrow\; u = 0, \, \theta = \frac{\pi}{4} \;\Rightarrow\; u = 1;$

$$\int_0^{\pi/4} \left(1 + e^{\tan \theta} \right) \sec^2 \theta \, d\theta = \int_0^{\pi/4} \sec^2 \theta \, d\theta + \int_0^1 e^u \, du = \left[\tan \theta \right]_0^{\pi/4} + \left[e^u \right]_0^1 = \left[\tan \left(\frac{\pi}{4} \right) - \tan (0) \right] + (e^1 - e^0)$$
$$= (1 - 0) + (e - 1) = e$$

27. $\int_0^{\pi} \frac{\sin t}{2 - \cos t} \, dt = \left[\ln |2 - \cos t| \right]_0^{\pi} = \ln 3 - \ln 1 = \ln 3;$ or let $u = 2 - \cos t \;\Rightarrow\; du = \sin t \, dt$ with $t = 0$

$\;\Rightarrow\; u = 1$ and $t = \pi \;\Rightarrow\; u = 3 \;\Rightarrow\; \int_0^{\pi} \frac{\sin t}{2 - \cos t} \, dt = \int_1^3 \frac{1}{u} \, du = \left[\ln |u| \right]_1^3 = \ln 3 - \ln 1 = \ln 3$

29. Let $u = \ln x \;\Rightarrow\; du = \frac{1}{x} \, dx; \, x = 1 \;\Rightarrow\; u = 0$ and $x = 2 \;\Rightarrow\; u = \ln 2;$

$$\int_1^2 \frac{2 \ln x}{x} \, dx = \int_0^{\ln 2} 2u \, du = \left[u^2 \right]_0^{\ln 2} = (\ln 2)^2$$

31. Let $u = \ln x \;\Rightarrow\; du = \frac{1}{x} \, dx; \, x = 2 \;\Rightarrow\; u = \ln 2$ and $x = 4 \;\Rightarrow\; u = \ln 4;$

$$\int_2^4 \frac{dx}{x(\ln x)^2} = \int_{\ln 2}^{\ln 4} u^{-2} \, du = \left[-\frac{1}{u} \right]_{\ln 2}^{\ln 4} = -\frac{1}{\ln 4} + \frac{1}{\ln 2} = -\frac{1}{\ln 2^2} + \frac{1}{\ln 2} = -\frac{1}{2 \ln 2} + \frac{1}{\ln 2} = \frac{1}{2 \ln 2} = \frac{1}{\ln 4}$$

33. Let $u = \cos \frac{x}{2} \Rightarrow du = -\frac{1}{2} \sin \frac{x}{2}\, dx \Rightarrow -2\, du = \sin \frac{x}{2}\, dx;\ x = 0 \Rightarrow u = 1$ and $x = \frac{\pi}{2} \Rightarrow u = \frac{1}{\sqrt{2}}$;

$$\int_0^{\pi/2} \tan \frac{x}{2}\, dx = \int_0^{\pi/2} \frac{\sin \frac{x}{2}}{\cos \frac{x}{2}}\, dx = -2 \int_1^{1/\sqrt{2}} \frac{du}{u} = \left[-2 \ln |u|\right]_1^{1/\sqrt{2}} = -2 \ln \frac{1}{\sqrt{2}} = 2 \ln \sqrt{2} = \ln 2$$

35. Let $u = \sin \frac{\theta}{3} \Rightarrow du = \frac{1}{3} \cos \frac{\theta}{3}\, d\theta \Rightarrow 6\, du = 2 \cos \frac{\theta}{3}\, d\theta;\ \theta = \frac{\pi}{2} \Rightarrow u = \frac{1}{2}$ and $\theta = \pi \Rightarrow u = \frac{\sqrt{3}}{2}$;

$$\int_{\pi/2}^{\pi} 2 \cot \frac{\theta}{3}\, d\theta = \int_{\pi/2}^{\pi} \frac{2 \cos \frac{\theta}{3}}{\sin \frac{\theta}{3}}\, d\theta = 6 \int_{1/2}^{\sqrt{3}/2} \frac{du}{u} = 6\left[\ln |u|\right]_{1/2}^{\sqrt{3}/2} = 6\left(\ln \frac{\sqrt{3}}{2} - \ln \frac{1}{2}\right) = 6 \ln \sqrt{3} = \ln 27$$

37. $\int_{-\pi/2}^{\pi/2} \frac{2 \cos \theta\, d\theta}{1 + (\sin \theta)^2} = 2 \int_{-1}^{1} \frac{du}{1 + u^2}$, where $u = \sin \theta$ and $du = \cos \theta\, d\theta;\ \theta = -\frac{\pi}{2} \Rightarrow u = -1,\ \theta = \frac{\pi}{2} \Rightarrow u = 1$

$$= \left[2 \tan^{-1} u\right]_{-1}^{1} = 2\left(\tan^{-1} 1 - \tan^{-1}(-1)\right) = 2\left[\frac{\pi}{4} - \left(-\frac{\pi}{4}\right)\right] = \pi$$

39. $\int_0^{\ln \sqrt{3}} \frac{e^x\, dx}{1 + e^{2x}} = \int_1^{\sqrt{3}} \frac{du}{1 + u^2}$, where $u = e^x$ and $du = e^x\, dx;\ x = 0 \Rightarrow u = 1,\ x = \ln \sqrt{3} \Rightarrow u = \sqrt{3}$

$$= \left[\tan^{-1} u\right]_1^{\sqrt{3}} = \tan^{-1} \sqrt{3} - \tan^{-1} 1 = \frac{\pi}{3} - \frac{\pi}{4} = \frac{\pi}{12}$$

41. $\int_0^1 \frac{4\, ds}{\sqrt{4 - s^2}} = \left[4 \sin^{-1} \frac{s}{2}\right]_0^1 = 4\left(\sin^{-1} \frac{1}{2} - \sin^{-1} 0\right) = 4\left(\frac{\pi}{6} - 0\right) = \frac{2\pi}{3}$

43. $\int_{\sqrt{2}}^{2} \frac{\sec^2 (\sec^{-1} x)}{x\sqrt{x^2 - 1}}\, dx = \int_{\pi/4}^{\pi/3} \sec^2 u\, du$, where $u = \sec^{-1} x$ and $du = \frac{dx}{x\sqrt{x^2 - 1}};\ x = \sqrt{2} \Rightarrow u = \frac{\pi}{4},\ x = 2 \Rightarrow u = \frac{\pi}{3}$

$$= \left[\tan u\right]_{\pi/4}^{\pi/3} = \tan \frac{\pi}{3} - \tan \frac{\pi}{4} = \sqrt{3} - 1$$

45. $\int_{-1}^{-\sqrt{2}/2} \frac{dy}{y\sqrt{4y^2 - 1}} = \int_{-2}^{-\sqrt{2}} \frac{du}{u\sqrt{u^2 - 1}}$, where $u = 2y$ and $du = 2\, dy;\ y = -1 \Rightarrow u = -2,\ y = -\frac{\sqrt{2}}{2} \Rightarrow u = -\sqrt{2}$

$$= \left[\sec^{-1} |u|\right]_{-2}^{-\sqrt{2}} = \sec^{-1} \left|-\sqrt{2}\right| - \sec^{-1} |-2| = \frac{\pi}{4} - \frac{\pi}{3} = -\frac{\pi}{12}$$

47. Let $u = 4 - x^2 \Rightarrow du = -2x\, dx \Rightarrow -\frac{1}{2}\, du = x\, dx;\ x = -2 \Rightarrow u = 0,\ x = 0 \Rightarrow u = 4,\ x = 2 \Rightarrow u = 0$

$$A = -\int_{-2}^{0} x\sqrt{4 - x^2}\, dx + \int_0^2 x\sqrt{4 - x^2}\, dx = -\int_0^4 -\frac{1}{2} u^{1/2}\, du + \int_4^0 -\frac{1}{2} u^{1/2}\, du = 2 \int_0^4 \frac{1}{2} u^{1/2}\, du = \int_0^4 u^{1/2}\, du$$

$$= \left[\frac{2}{3} u^{3/2}\right]_0^4 = \frac{2}{3}(4)^{3/2} - \frac{2}{3}(0)^{3/2} = \frac{16}{3}$$

49. Let $u = 1 + \cos x \Rightarrow du = -\sin x\, dx \Rightarrow -du = \sin x\, dx;\ x = -\pi \Rightarrow u = 1 + \cos(-\pi) = 0,\ x = 0$
$\Rightarrow u = 1 + \cos 0 = 2$

$$A = -\int_{-\pi}^{0} 3(\sin x)\sqrt{1 + \cos x}\, dx = -\int_0^2 3u^{1/2}(-du) = 3 \int_0^2 u^{1/2}\, du = \left[2u^{3/2}\right]_0^2 = 2(2)^{3/2} - 2(0)^{3/2} = 2^{5/2}$$

51. For the sketch given, $a = 0,\ b = \pi;\ f(x) - g(x) = 1 - \cos^2 x = \sin^2 x = \frac{1 - \cos 2x}{2}$;

$$A = \int_0^{\pi} \frac{(1 - \cos 2x)}{2}\, dx = \frac{1}{2} \int_0^{\pi} (1 - \cos 2x)\, dx = \frac{1}{2}\left[x - \frac{\sin 2x}{2}\right]_0^{\pi} = \frac{1}{2}\left[(\pi - 0) - (0 - 0)\right] = \frac{\pi}{2}$$

53. For the sketch given, $a = -2,\ b = 2;\ f(x) - g(x) = 2x^2 - (x^4 - 2x^2) = 4x^2 - x^4$;

$$A = \int_{-2}^{2} (4x^2 - x^4)\, dx = \left[\frac{4x^3}{3} - \frac{x^5}{5}\right]_{-2}^{2} = \left(\frac{32}{3} - \frac{32}{5}\right) - \left[-\frac{32}{3} - \left(-\frac{32}{5}\right)\right] = \frac{64}{3} - \frac{64}{5} = \frac{320 - 192}{15} = \frac{128}{15}$$

55. For the sketch given, $c = 0,\ d = 1;\ f(y) - g(y) = (12y^2 - 12y^3) - (2y^2 - 2y) = 10y^2 - 12y^3 + 2y$;

$$A = \int_0^1 (10y^2 - 12y^3 + 2y)\, dy = \int_0^1 10y^2\, dy - \int_0^1 12y^3\, dy + \int_0^1 2y\, dy = \left[\frac{10}{3} y^3\right]_0^1 - \left[\frac{12}{4} y^4\right]_0^1 + \left[\frac{2}{2} y^2\right]_0^1$$

$$= \left(\frac{10}{3} - 0\right) - (3 - 0) + (1 - 0) = \frac{4}{3}$$

57. We want the area between the line $y = 1$, $0 \le x \le 2$, and the curve $y = \frac{x^2}{4}$, *minus* the area of a triangle

(formed by $y = x$ and $y = 1$) with base 1 and height 1. Thus, $A = \int_0^2 \left(1 - \frac{x^2}{4}\right) dx - \frac{1}{2}(1)(1) = \left[x - \frac{x^3}{12}\right]_0^2 - \frac{1}{2}$

$= \left(2 - \frac{8}{12}\right) - \frac{1}{2} = 2 - \frac{2}{3} - \frac{1}{2} = \frac{5}{6}$

59. AREA $=$ A1 $+$ A2

   A1:   For the sketch given, $a = -3$ and we find b by solving the equations $y = x^2 - 4$ and $y = -x^2 - 2x$

   simultaneously for x: $x^2 - 4 = -x^2 - 2x \Rightarrow 2x^2 + 2x - 4 = 0 \Rightarrow 2(x + 2)(x - 1) \Rightarrow x = -2$ or $x = 1$ so

   $b = -2$: $f(x) - g(x) = (x^2 - 4) - (-x^2 - 2x) = 2x^2 + 2x - 4 \Rightarrow$ A1 $= \int_{-3}^{-2} (2x^2 + 2x - 4)\, dx$

   $= \left[\frac{2x^3}{3} + \frac{2x^2}{2} - 4x\right]_{-3}^{-2} = \left(-\frac{16}{3} + 4 + 8\right) - (-18 + 9 + 12) = 9 - \frac{16}{3} = \frac{11}{3}$;

   A2:   For the sketch given, $a = -2$ and $b = 1$: $f(x) - g(x) = (-x^2 - 2x) - (x^2 - 4) = -2x^2 - 2x + 4$

   $\Rightarrow$ A2 $= -\int_{-2}^{1} (2x^2 + 2x - 4)\, dx = -\left[\frac{2x^3}{3} + x^2 - 4x\right]_{-2}^{1} = -\left(\frac{2}{3} + 1 - 4\right) + \left(-\frac{16}{3} + 4 + 8\right)$

   $= -\frac{2}{3} - 1 + 4 - \frac{16}{3} + 4 + 8 = 9$;

   Therefore, AREA $=$ A1 $+$ A2 $= \frac{11}{3} + 9 = \frac{38}{3}$

61. AREA $=$ A1 $+$ A2 $+$ A3

   A1:   For the sketch given, $a = -2$ and $b = -1$: $f(x) - g(x) = (-x + 2) - (4 - x^2) = x^2 - x - 2$

   $\Rightarrow$ A1 $= \int_{-2}^{-1} (x^2 - x - 2)\, dx = \left[\frac{x^3}{3} - \frac{x^2}{2} - 2x\right]_{-2}^{-1} = \left(-\frac{1}{3} - \frac{1}{2} + 2\right) - \left(-\frac{8}{3} - \frac{4}{2} + 4\right) = \frac{7}{3} - \frac{1}{2} = \frac{14 - 3}{6} = \frac{11}{6}$;

   A2:   For the sketch given, $a = -1$ and $b = 2$: $f(x) - g(x) = (4 - x^2) - (-x + 2) = -(x^2 - x - 2)$

   $\Rightarrow$ A2 $= -\int_{-1}^{2} (x^2 - x - 2)\, dx = -\left[\frac{x^3}{3} - \frac{x^2}{2} - 2x\right]_{-1}^{2} = -\left(\frac{8}{3} - \frac{4}{2} - 4\right) + \left(-\frac{1}{3} - \frac{1}{2} + 2\right) = -3 + 8 - \frac{1}{2} = \frac{9}{2}$;

   A3:   For the sketch given, $a = 2$ and $b = 3$: $f(x) - g(x) = (-x + 2) - (4 - x^2) = x^2 - x - 2$

   $\Rightarrow$ A3 $= \int_{2}^{3} (x^2 - x - 2)\, dx = \left[\frac{x^3}{3} - \frac{x^2}{2} - 2x\right]_{2}^{3} = \left(\frac{27}{3} - \frac{9}{2} - 6\right) - \left(\frac{8}{3} - \frac{4}{2} - 4\right) = 9 - \frac{9}{2} - \frac{8}{3}$;

   Therefore, AREA $=$ A1 $+$ A2 $+$ A3 $= \frac{11}{6} + \frac{9}{2} + \left(9 - \frac{9}{2} - \frac{8}{3}\right) = 9 - \frac{5}{6} = \frac{49}{6}$

63. $a = -2$, $b = 2$;

   $f(x) - g(x) = 2 - (x^2 - 2) = 4 - x^2$

   $\Rightarrow A = \int_{-2}^{2} (4 - x^2)dx = \left[4x - \frac{x^3}{3}\right]_{-2}^{2} = \left(8 - \frac{8}{3}\right) - \left(-8 + \frac{8}{3}\right)$

   $= 2 \cdot \left(\frac{24}{3} - \frac{8}{3}\right) = \frac{32}{3}$

65. $a = 0$, $b = 2$;

   $f(x) - g(x) = 8x - x^4 \Rightarrow A = \int_0^2 (8x - x^4)\, dx$

   $= \left[\frac{8x^2}{2} - \frac{x^5}{5}\right]_0^2 = 16 - \frac{32}{5} = \frac{80 - 32}{5} = \frac{48}{5}$

67. Limits of integration: $x^2 = -x^2 + 4x \Rightarrow 2x^2 - 4x = 0$

$\Rightarrow 2x(x - 2) = 0 \Rightarrow a = 0 \text{ and } b = 2;$

$f(x) - g(x) = (-x^2 + 4x) - x^2 = -2x^2 + 4x$

$\Rightarrow A = \int_0^2 (-2x^2 + 4x) \, dx = \left[\frac{-2x^3}{3} + \frac{4x^2}{2}\right]_0^2$

$= -\frac{16}{3} + \frac{16}{2} = \frac{-32 + 48}{6} = \frac{8}{3}$

69. Limits of integration: $x^4 - 4x^2 + 4 = x^2$

$\Rightarrow x^4 - 5x^2 + 4 = 0 \Rightarrow (x^2 - 4)(x^2 - 1) = 0$

$\Rightarrow (x + 2)(x - 2)(x + 1)(x - 1) = 0 \Rightarrow x = -2, -1, 1, 2;$

$f(x) - g(x) = (x^4 - 4x^2 + 4) - x^2 = x^4 - 5x^2 + 4$ and

$g(x) - f(x) = x^2 - (x^4 - 4x^2 + 4) = -x^4 + 5x^2 - 4$

$\Rightarrow A = \int_{-2}^{-1} (-x^4 + 5x^2 - 4)dx + \int_{-1}^{1} (x^4 - 5x^2 + 4)dx$

$+ \int_1^2 (-x^4 + 5x^2 - 4)dx$

$= \left[-\frac{x^5}{5} + \frac{5x^3}{3} - 4x\right]_{-2}^{-1} + \left[\frac{x^5}{5} - \frac{5x^3}{3} + 4x\right]_{-1}^{1} + \left[\frac{-x^5}{5} + \frac{5x^3}{3} - 4x\right]_1^2$

$= \left(\frac{1}{5} - \frac{5}{3} + 4\right) - \left(\frac{32}{5} - \frac{40}{3} + 8\right) + \left(\frac{1}{5} - \frac{5}{3} + 4\right) - \left(-\frac{1}{5} + \frac{5}{3} - 4\right) + \left(-\frac{32}{5} + \frac{40}{3} - 8\right) - \left(-\frac{1}{5} + \frac{5}{3} - 4\right)$

$= -\frac{60}{5} + \frac{60}{3} = \frac{300 - 180}{15} = 8$

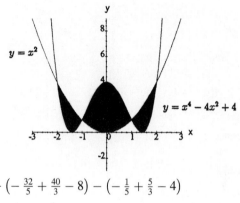

71. Limits of integration: $y = \sqrt{|x|} = \begin{cases} \sqrt{-x}, & x \le 0 \\ \sqrt{x}, & x \ge 0 \end{cases}$ and

$5y = x + 6 \text{ or } y = \frac{x}{5} + \frac{6}{5};$ for $x \le 0$: $\sqrt{-x} = \frac{x}{5} + \frac{6}{5}$

$\Rightarrow 5\sqrt{-x} = x + 6 \Rightarrow 25(-x) = x^2 + 12x + 36$

$\Rightarrow x^2 + 37x + 36 = 0 \Rightarrow (x + 1)(x + 36) = 0$

$\Rightarrow x = -1, -36 \text{ (but } x = -36 \text{ is not a solution)};$

for $x \ge 0$: $5\sqrt{x} = x + 6 \Rightarrow 25x = x^2 + 12x + 36$

$\Rightarrow x^2 - 13x + 36 = 0 \Rightarrow (x - 4)(x - 9) = 0$

$\Rightarrow x = 4, 9;$ there are three intersection points and

$A = \int_{-1}^{0} \left(\frac{x+6}{5} - \sqrt{-x}\right)dx + \int_0^4 \left(\frac{x+6}{5} - \sqrt{x}\right)dx + \int_4^9 \left(\sqrt{x} - \frac{x+6}{5}\right) dx$

$= \left[\frac{(x+6)^2}{10} + \frac{2}{3}(-x)^{3/2}\right]_{-1}^{0} + \left[\frac{(x+6)^2}{10} - \frac{2}{3}x^{3/2}\right]_0^4 + \left[\frac{2}{3}x^{3/2} - \frac{(x+6)^2}{10}\right]_4^9$

$= \left(\frac{36}{10} - \frac{25}{10} - \frac{2}{3}\right) + \left(\frac{100}{10} - \frac{2}{3} \cdot 4^{3/2} - \frac{36}{10} + 0\right) + \left(\frac{2}{3} \cdot 9^{3/2} - \frac{225}{10} - \frac{2}{3} \cdot 4^{3/2} + \frac{100}{10}\right) = -\frac{50}{10} + \frac{20}{3} = \frac{5}{3}$

73. Limits of integration: $c = 0$ and $d = 3;$

$f(y) - g(y) = 2y^2 - 0 = 2y^2$

$\Rightarrow A = \int_0^3 2y^2 \, dy = \left[\frac{2y^3}{3}\right]_0^3 = 2 \cdot 9 = 18$

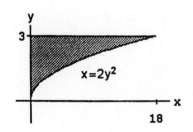

75. Limits of integration: $4x = y^2 - 4$ and $4x = 16 + y$

$\Rightarrow y^2 - 4 = 16 + y \Rightarrow y^2 - y - 20 = 0 \Rightarrow$

$(y - 5)(y + 4) = 0 \Rightarrow c = -4$ and $d = 5$;

$f(y) - g(y) = \left(\frac{16+y}{4}\right) - \left(\frac{y^2-4}{4}\right) = \frac{-y^2+y+20}{4}$

$\Rightarrow A = \frac{1}{4} \int_{-4}^{5} (-y^2 + y + 20)\, dy$

$= \frac{1}{4} \left[-\frac{y^3}{3} + \frac{y^2}{2} + 20y\right]_{-4}^{5}$

$= \frac{1}{4}\left(-\frac{125}{3} + \frac{25}{2} + 100\right) - \frac{1}{4}\left(\frac{64}{3} + \frac{16}{2} - 80\right)$

$= \frac{1}{4}\left(-\frac{189}{3} + \frac{9}{2} + 180\right) = \frac{243}{8}$

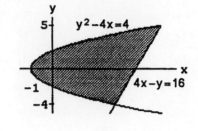

77. Limits of integration: $x = -y^2$ and $x = 2 - 3y^2$

$\Rightarrow -y^2 = 2 - 3y^2 \Rightarrow 2y^2 - 2 = 0$

$\Rightarrow 2(y - 1)(y + 1) = 0 \Rightarrow c = -1$ and $d = 1$;

$f(y) - g(y) = (2 - 3y^2) - (-y^2) = 2 - 2y^2 = 2(1 - y^2)$

$\Rightarrow A = 2 \int_{-1}^{1} (1 - y^2)\, dy = 2\left[y - \frac{y^3}{3}\right]_{-1}^{1}$

$= 2\left(1 - \frac{1}{3}\right) - 2\left(-1 + \frac{1}{3}\right) = 4\left(\frac{2}{3}\right) = \frac{8}{3}$

79. Limits of integration: $x = y^2 - 1$ and $x = |y|\sqrt{1 - y^2}$

$\Rightarrow y^2 - 1 = |y|\sqrt{1 - y^2} \Rightarrow y^4 - 2y^2 + 1 = y^2(1 - y^2)$

$\Rightarrow y^4 - 2y^2 + 1 = y^2 - y^4 \Rightarrow 2y^4 - 3y^2 + 1 = 0$

$\Rightarrow (2y^2 - 1)(y^2 - 1) = 0 \Rightarrow 2y^2 - 1 = 0$ or $y^2 - 1 = 0$

$\Rightarrow y^2 = \frac{1}{2}$ or $y^2 = 1 \Rightarrow y = \pm\frac{\sqrt{2}}{2}$ or $y = \pm 1$.

Substitution shows that $\frac{\pm\sqrt{2}}{2}$ are not solutions $\Rightarrow y = \pm 1$;

for $-1 \le y \le 0$, $f(x) - g(x) = -y\sqrt{1 - y^2} - (y^2 - 1)$

$= 1 - y^2 - y(1 - y^2)^{1/2}$, and by symmetry of the graph,

$A = 2 \int_{-1}^{0} \left[1 - y^2 - y(1 - y^2)^{1/2}\right] dy$

$= 2 \int_{-1}^{0} (1 - y^2)\, dy - 2\int_{-1}^{0} y(1 - y^2)^{1/2}\, dy$

$= 2\left[y - \frac{y^3}{3}\right]_{-1}^{0} + 2\left(\frac{1}{2}\right)\left[\frac{2(1 - y^2)^{3/2}}{3}\right]_{-1}^{0} = 2\left[(0 - 0) - \left(-1 + \frac{1}{3}\right)\right] + \left(\frac{2}{3} - 0\right) = 2$

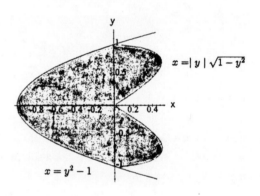

81. Limits of integration: $y = -4x^2 + 4$ and $y = x^4 - 1$

$\Rightarrow x^4 - 1 = -4x^2 + 4 \Rightarrow x^4 + 4x^2 - 5 = 0$

$\Rightarrow (x^2 + 5)(x - 1)(x + 1) = 0 \Rightarrow a = -1$ and $b = 1$;

$f(x) - g(x) = -4x^2 + 4 - x^4 + 1 = -4x^2 - x^4 + 5$

$\Rightarrow A = \int_{-1}^{1} (-4x^2 - x^4 + 5)\, dx = \left[-\frac{4x^3}{3} - \frac{x^5}{5} + 5x\right]_{-1}^{1}$

$= \left(-\frac{4}{3} - \frac{1}{5} + 5\right) - \left(\frac{4}{3} + \frac{1}{5} - 5\right) = 2\left(-\frac{4}{3} - \frac{1}{5} + 5\right) = \frac{104}{15}$

83. Limits of integration: $x = 4 - 4y^2$ and $x = 1 - y^4$

$\Rightarrow 4 - 4y^2 = 1 - y^4 \Rightarrow y^4 - 4y^2 + 3 = 0$

$\Rightarrow \left(y - \sqrt{3}\right)\left(y + \sqrt{3}\right)(y - 1)(y + 1) = 0 \Rightarrow c = -1$

and $d = 1$ since $x \geq 0$; $f(y) - g(y) = (4 - 4y^2) - (1 - y^4)$

$= 3 - 4y^2 + y^4 \Rightarrow A = \int_{-1}^{1}(3 - 4y^2 + y^4)\,dy$

$= \left[3y - \frac{4y^3}{3} + \frac{y^5}{5}\right]_{-1}^{1} = 2\left(3 - \frac{4}{3} + \frac{1}{5}\right) = \frac{56}{15}$

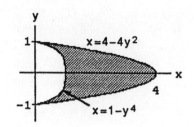

85. $a = 0, b = \pi$; $f(x) - g(x) = 2 \sin x - \sin 2x$

$\Rightarrow A = \int_{0}^{\pi}(2 \sin x - \sin 2x)\,dx = \left[-2 \cos x + \frac{\cos 2x}{2}\right]_{0}^{\pi}$

$= \left[-2(-1) + \frac{1}{2}\right] - \left(-2 \cdot 1 + \frac{1}{2}\right) = 4$

87. $a = -1, b = 1$; $f(x) - g(x) = (1 - x^2) - \cos\left(\frac{\pi x}{2}\right)$

$\Rightarrow A = \int_{-1}^{1}\left[1 - x^2 - \cos\left(\frac{\pi x}{2}\right)\right]dx = \left[x - \frac{x^3}{3} - \frac{2}{\pi}\sin\left(\frac{\pi x}{2}\right)\right]_{-1}^{1}$

$= \left(1 - \frac{1}{3} - \frac{2}{\pi}\right) - \left(-1 + \frac{1}{3} + \frac{2}{\pi}\right) = 2\left(\frac{2}{3} - \frac{2}{\pi}\right) = \frac{4}{3} - \frac{4}{\pi}$

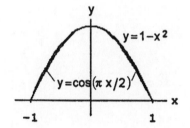

89. $a = -\frac{\pi}{4}, b = \frac{\pi}{4}$; $f(x) - g(x) = \sec^2 x - \tan^2 x$

$\Rightarrow A = \int_{-\pi/4}^{\pi/4}(\sec^2 x - \tan^2 x)\,dx$

$= \int_{-\pi/4}^{\pi/4}[\sec^2 x - (\sec^2 x - 1)]\,dx$

$= \int_{-\pi/4}^{\pi/4}1 \cdot dx = [x]_{-\pi/4}^{\pi/4} = \frac{\pi}{4} - \left(-\frac{\pi}{4}\right) = \frac{\pi}{2}$

91. $c = 0, d = \frac{\pi}{2}$; $f(y) - g(y) = 3 \sin y \sqrt{\cos y} - 0 = 3 \sin y \sqrt{\cos y}$

$\Rightarrow A = 3\int_{0}^{\pi/2} \sin y \sqrt{\cos y}\,dy = -3\left[\frac{2}{3}(\cos y)^{3/2}\right]_{0}^{\pi/2}$

$= -2(0 - 1) = 2$

93. $A = A_1 + A_2$

Limits of integration: $x = y^3$ and $x = y \Rightarrow y = y^3$

$\Rightarrow y^3 - y = 0 \Rightarrow y(y-1)(y+1) = 0 \Rightarrow c_1 = -1, d_1 = 0$

and $c_2 = 0, d_2 = 1; f_1(y) - g_1(y) = y^3 - y$ and

$f_2(y) - g_2(y) = y - y^3 \Rightarrow$ by symmetry about the origin,

$A_1 + A_2 = 2A_2 \Rightarrow A = 2\int_0^1 (y - y^3)\, dy = 2\left[\frac{y^2}{2} - \frac{y^4}{4}\right]_0^1$

$= 2\left(\frac{1}{2} - \frac{1}{4}\right) = \frac{1}{2}$

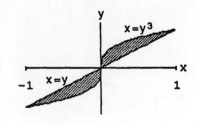

95. $A = A_1 + A_2$

Limits of integration: $y = x$ and $y = \frac{1}{x^2} \Rightarrow x = \frac{1}{x^2}, x \neq 0$

$\Rightarrow x^3 = 1 \Rightarrow x = 1, f_1(x) - g_1(x) = x - 0 = x$

$\Rightarrow A_1 = \int_0^1 x\, dx = \left[\frac{x^2}{2}\right]_0^1 = \frac{1}{2}; f_2(x) - g_2(x) = \frac{1}{x^2} - 0$

$= x^{-2} \Rightarrow A_2 = \int_1^2 x^{-2}\, dx = \left[\frac{-1}{x}\right]_1^2 = -\frac{1}{2} + 1 = \frac{1}{2};$

$A = A_1 + A_2 = \frac{1}{2} + \frac{1}{2} = 1$

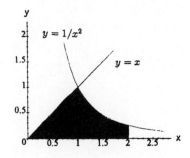

97. $\int_1^5 (\ln 2x - \ln x)\, dx = \int_1^5 (-\ln x + \ln 2 + \ln x)\, dx = (\ln 2)\int_1^5 dx = (\ln 2)(5-1) = \ln 2^4 = \ln 16$

99. $\int_0^{\ln 3} (e^{2x} - e^x)\, dx = \left[\frac{e^{2x}}{2} - e^x\right]_0^{\ln 3} = \left(\frac{e^{2\ln 3}}{2} - e^{\ln 3}\right) - \left(\frac{e^0}{2} - e^0\right) = \left(\frac{9}{2} - 3\right) - \left(\frac{1}{2} - 1\right) = \frac{8}{2} - 2 = 2$

101. $A = \int_{-2}^2 \frac{2x}{1+x^2}\, dx = 2\int_0^2 \frac{2x}{1+x^2}\, dx; \left[u = 1 + x^2 \Rightarrow du = 2x\, dx; x = 0 \Rightarrow u = 1, x = 2 \Rightarrow u = 5\right]$

$\rightarrow A = 2\int_1^5 \frac{1}{u}\, du = 2\left[\ln |u|\right]_1^5 = 2(\ln 5 - \ln 1) = 2\ln 5$

103. (a) The coordinates of the points of intersection of the
line and parabola are $c = x^2 \Rightarrow x = \pm\sqrt{c}$ and $y = c$

(b) $f(y) - g(y) = \sqrt{y} - (-\sqrt{y}) = 2\sqrt{y} \Rightarrow$ the area of

the lower section is, $A_L = \int_0^c [f(y) - g(y)]\, dy$

$= 2\int_0^c \sqrt{y}\, dy = 2\left[\frac{2}{3} y^{3/2}\right]_0^c = \frac{4}{3} c^{3/2}$. The area of

the entire shaded region can be found by setting $c = 4$: $A = \left(\frac{4}{3}\right) 4^{3/2} = \frac{4 \cdot 8}{3} = \frac{32}{3}$. Since we want c to divide the

region into subsections of equal area we have $A = 2A_L \Rightarrow \frac{32}{3} = 2\left(\frac{4}{3} c^{3/2}\right) \Rightarrow c = 4^{2/3}$

(c) $f(x) - g(x) = c - x^2 \Rightarrow A_L = \int_{-\sqrt{c}}^{\sqrt{c}} [f(x) - g(x)]\, dx = \int_{-\sqrt{c}}^{\sqrt{c}} (c - x^2)\, dx = \left[cx - \frac{x^3}{3}\right]_{-\sqrt{c}}^{\sqrt{c}} = 2\left[c^{3/2} - \frac{c^{3/2}}{3}\right]$

$= \frac{4}{3} c^{3/2}$. Again, the area of the whole shaded region can be found by setting $c = 4 \Rightarrow A = \frac{32}{3}$. From the

condition $A = 2A_L$, we get $\frac{4}{3} c^{3/2} = \frac{32}{3} \Rightarrow c = 4^{2/3}$ as in part (b).

105. Limits of integration: $y = 1 + \sqrt{x}$ and $y = \frac{2}{\sqrt{x}}$

$\Rightarrow 1 + \sqrt{x} = \frac{2}{\sqrt{x}}, x \neq 0 \Rightarrow \sqrt{x} + x = 2 \Rightarrow x = (2 - x)^2$

$\Rightarrow x = 4 - 4x + x^2 \Rightarrow x^2 - 5x + 4 = 0$

$\Rightarrow (x - 4)(x - 1) = 0 \Rightarrow x = 1, 4$ (but $x = 4$ does not

satisfy the equation); $y = \frac{2}{\sqrt{x}}$ and $y = \frac{x}{4} \Rightarrow \frac{2}{\sqrt{x}} = \frac{x}{4}$

$\Rightarrow 8 = x\sqrt{x} \Rightarrow 64 = x^3 \Rightarrow x = 4$.

Therefore, $\text{AREA} = A_1 + A_2$: $f_1(x) - g_1(x)$

$= (1 + x^{1/2}) - \frac{x}{4} \Rightarrow A_1 = \int_0^1 (1 + x^{1/2} - \frac{x}{4}) \, dx$

$= [x + \frac{2}{3} x^{3/2} - \frac{x^2}{8}]_0^1$

$= (1 + \frac{2}{3} - \frac{1}{8}) - 0 = \frac{37}{24}; f_2(x) - g_2(x) = 2x^{-1/2} - \frac{x}{4} \Rightarrow A_2 = \int_1^4 (2x^{-1/2} - \frac{x}{4}) \, dx = [4x^{1/2} - \frac{x^2}{8}]_1^4$

$= (4 \cdot 2 - \frac{16}{8}) - (4 - \frac{1}{8}) = 4 - \frac{15}{8} = \frac{17}{8}$; Therefore, $\text{AREA} = A_1 + A_2 = \frac{37}{24} + \frac{17}{8} = \frac{37+51}{24} = \frac{88}{24} = \frac{11}{3}$

107. Area between parabola and $y = a^2$: $A = 2\int_0^a (a^2 - x^2) \, dx = 2[a^2 x - \frac{1}{3} x^3]_0^a = 2(a^3 - \frac{a^3}{3}) - 0 = \frac{4a^3}{3}$;

Area of triangle AOC: $\frac{1}{2}(2a)(a^2) = a^3$; limit of ratio $= \lim\limits_{a \to 0^+} \frac{a^3}{(\frac{4a^3}{3})} = \frac{3}{4}$ which is independent of a.

109. Neither one; they are both zero. Neither integral takes into account the changes in the formulas for the region's upper and lower bounding curves at $x = 0$. The area of the shaded region is actually

$A = \int_{-1}^0 [-x - (x)] \, dx + \int_0^1 [x - (-x)] \, dx = \int_{-1}^0 -2x \, dx + \int_0^1 2x \, dx = 2.$

111. Let $u = 2x \Rightarrow du = 2 \, dx \Rightarrow \frac{1}{2} du = dx; x = 1 \Rightarrow u = 2, x = 3 \Rightarrow u = 6$

$\int_1^3 \frac{\sin 2x}{x} \, dx = \int_2^6 \frac{\sin u}{(\frac{u}{2})} (\frac{1}{2} du) = \int_2^6 \frac{\sin u}{u} \, du = [F(u)]_2^6 = F(6) - F(2)$

113. (a) Let $u = -x \Rightarrow du = -dx; x = -1 \Rightarrow u = 1, x = 0 \Rightarrow u = 0$

f odd $\Rightarrow f(-x) = -f(x)$. Then $\int_{-1}^0 f(x) \, dx = \int_1^0 f(-u)(-du) = \int_1^0 -f(u)(-du) = \int_1^0 f(u) \, du = -\int_0^1 f(u) \, du$

$= -3$

(b) Let $u = -x \Rightarrow du = -dx; x = -1 \Rightarrow u = 1, x = 0 \Rightarrow u = 0$

f even $\Rightarrow f(-x) = f(x)$. Then $\int_{-1}^0 f(x) \, dx = \int_1^0 f(-u)(-du) = -\int_1^0 f(u) \, du = \int_0^1 f(u) \, du = 3$

115. Let $u = a - x \Rightarrow du = -dx; x = 0 \Rightarrow u = a, x = a \Rightarrow u = 0$

$I = \int_0^a \frac{f(x) \, dx}{f(x) + f(a-x)} = \int_a^0 \frac{f(a-u)}{f(a-u) + f(u)} (-du) = \int_0^a \frac{f(a-u) \, du}{f(u) + f(a-u)} = \int_0^a \frac{f(a-x) \, dx}{f(x) + f(a-x)}$

$\Rightarrow I + I = \int_0^a \frac{f(x) \, dx}{f(x) + f(a-x)} + \int_0^a \frac{f(a-x) \, dx}{f(x) + f(a-x)} = \int_0^a \frac{f(x) + f(a-x)}{f(x) + f(a-x)} \, dx = \int_0^a dx = [x]_0^a = a - 0 = a.$

Therefore, $2I = a \Rightarrow I = \frac{a}{2}$.

117. Let $u = x + c \Rightarrow du = dx; x = a - c \Rightarrow u = a, x = b - c \Rightarrow u = b$

$\int_{a-c}^{b-c} f(x + c) \, dx = \int_a^b f(u) \, du = \int_a^b f(x) \, dx$

5.7 THE LOGARITHM DEFINED AS AN INTEGRAL

1. $\int_{-3}^{-2} \frac{1}{x} \, dx = [\ln |x|]_{-3}^{-2} = \ln 2 - \ln 3 = \ln \frac{2}{3}$

3. $\int \frac{2y}{y^2 - 25} \, dy = \ln |y^2 - 25| + C$

5. Let $u = 6 + 3\tan t \Rightarrow du = 3\sec^2 t\, dt$;

$$\int \frac{3\sec^2 t}{6 + 3\tan t}\, dt = \int \frac{du}{u} = \ln|u| + C = \ln|6 + 3\tan t| + C$$

7. $\int \frac{dx}{2\sqrt{x} + 2x} = \int \frac{dx}{2\sqrt{x}\,(1 + \sqrt{x})}$; let $u = 1 + \sqrt{x} \Rightarrow du = \frac{1}{2\sqrt{x}}\, dx$; $\int \frac{dx}{2\sqrt{x}\,(1 + \sqrt{x})} = \int \frac{du}{u} = \ln|u| + C$

$\quad = \ln\left|1 + \sqrt{x}\right| + C = \ln\left(1 + \sqrt{x}\right) + C$

9. $\int_{\ln 2}^{\ln 3} e^x\, dx = [e^x]_{\ln 2}^{\ln 3} = e^{\ln 3} - e^{\ln 2} = 3 - 2 = 1$ 11. $\int 8e^{(x+1)}\, dx = 8e^{(x+1)} + C$

13. $\int_{\ln 4}^{\ln 9} e^{x/2}\, dx = \left[2e^{x/2}\right]_{\ln 4}^{\ln 9} = 2\left[e^{(\ln 9)/2} - e^{(\ln 4)/2}\right] = 2\left(e^{\ln 3} - e^{\ln 2}\right) = 2(3 - 2) = 2$

15. Let $u = r^{1/2} \Rightarrow du = \frac{1}{2}r^{-1/2}\, dr \Rightarrow 2\, du = r^{-1/2}\, dr$;

$$\int \frac{e^{\sqrt{r}}}{\sqrt{r}}\, dr = \int e^{r^{1/2}} \cdot r^{-1/2}\, dr = 2\int e^u\, du = 2e^u + C = 2e^{r^{1/2}} + C = 2e^{\sqrt{r}} + C$$

17. Let $u = -t^2 \Rightarrow du = -2t\, dt \Rightarrow -du = 2t\, dt$;

$$\int 2te^{-t^2}\, dt = -\int e^u\, du = -e^u + C = -e^{-t^2} + C$$

19. Let $u = \frac{1}{x} \Rightarrow du = -\frac{1}{x^2}\, dx \Rightarrow -du = \frac{1}{x^2}\, dx$;

$$\int \frac{e^{1/x}}{x^2}\, dx = \int -e^u\, du = -e^u + C = -e^{1/x} + C$$

21. Let $u = \sec \pi t \Rightarrow du = \pi \sec \pi t \tan \pi t\, dt \Rightarrow \frac{du}{\pi} = \sec \pi t \tan \pi t\, dt$;

$$\int e^{\sec(\pi t)}\sec(\pi t)\tan(\pi t)\, dt = \frac{1}{\pi}\int e^u\, du = \frac{e^u}{\pi} + C = \frac{e^{\sec(\pi t)}}{\pi} + C$$

23. Let $u = e^v \Rightarrow du = e^v\, dv \Rightarrow 2\, du = 2e^v\, dv$; $v = \ln\frac{\pi}{6} \Rightarrow u = \frac{\pi}{6}$, $v = \ln\frac{\pi}{2} \Rightarrow u = \frac{\pi}{2}$;

$$\int_{\ln(\pi/6)}^{\ln(\pi/2)} 2e^v \cos e^v\, dv = 2\int_{\pi/6}^{\pi/2} \cos u\, du = [2\sin u]_{\pi/6}^{\pi/2} = 2\left[\sin\left(\frac{\pi}{2}\right) - \sin\left(\frac{\pi}{6}\right)\right] = 2\left(1 - \frac{1}{2}\right) = 1$$

25. Let $u = 1 + e^r \Rightarrow du = e^r\, dr$;

$$\int \frac{e^r}{1 + e^r}\, dr = \int \frac{1}{u}\, du = \ln|u| + C = \ln(1 + e^r) + C$$

27. $\int_0^1 2^{-\theta}\, d\theta = \int_0^1 \left(\frac{1}{2}\right)^\theta\, d\theta = \left[\frac{\left(\frac{1}{2}\right)^\theta}{\ln\left(\frac{1}{2}\right)}\right]_0^1 = \frac{\frac{1}{2}}{\ln\left(\frac{1}{2}\right)} - \frac{1}{\ln\left(\frac{1}{2}\right)} = -\frac{\frac{1}{2}}{\ln\left(\frac{1}{2}\right)} = \frac{-1}{2(\ln 1 - \ln 2)} = \frac{1}{2\ln 2}$

29. Let $u = x^2 \Rightarrow du = 2x\, dx \Rightarrow \frac{1}{2}\, du = x\, dx$; $x = 1 \Rightarrow u = 1$, $x = \sqrt{2} \Rightarrow u = 2$;

$$\int_1^{\sqrt{2}} x2^{(x^2)}\, dx = \int_1^2 \left(\frac{1}{2}\right)2^u\, du = \frac{1}{2}\left[\frac{2^u}{\ln 2}\right]_1^2 = \left(\frac{1}{2\ln 2}\right)(2^2 - 2^1) = \frac{1}{\ln 2}$$

31. Let $u = \cos t \Rightarrow du = -\sin t\, dt \Rightarrow -du = \sin t\, dt$; $t = 0 \Rightarrow u = 1$, $t = \frac{\pi}{2} \Rightarrow u = 0$;

$$\int_0^{\pi/2} 7^{\cos t}\sin t\, dt = -\int_1^0 7^u\, du = \left[-\frac{7^u}{\ln 7}\right]_1^0 = \left(\frac{-1}{\ln 7}\right)(7^0 - 7) = \frac{6}{\ln 7}$$

33. Let $u = x^{2x} \Rightarrow \ln u = 2x \ln x \Rightarrow \frac{1}{u}\frac{du}{dx} = 2\ln x + (2x)\left(\frac{1}{x}\right) \Rightarrow \frac{du}{dx} = 2u(\ln x + 1) \Rightarrow \frac{1}{2}\, du = x^{2x}(1 + \ln x)\, dx$;

$\quad x = 2 \Rightarrow u = 2^4 = 16$, $x = 4 \Rightarrow u = 4^8 = 65{,}536$;

$$\int_2^4 x^{2x}(1 + \ln x)\, dx = \frac{1}{2}\int_{16}^{65{,}536} du = \frac{1}{2}[u]_{16}^{65{,}536} = \frac{1}{2}(65{,}536 - 16) = \frac{65{,}520}{2} = 32{,}760$$

35. $\int_0^3 \left(\sqrt{2}+1\right) x^{\sqrt{2}}\, dx = \left[x^{\left(\sqrt{2}+1\right)}\right]_0^3 = 3^{\left(\sqrt{2}+1\right)}$

37. $\int \frac{\log_{10} x}{x}\, dx = \int \left(\frac{\ln x}{\ln 10}\right)\left(\frac{1}{x}\right) dx;\ \left[u = \ln x\ \Rightarrow\ du = \frac{1}{x}\, dx\right]$

$\rightarrow \int \left(\frac{\ln x}{\ln 10}\right)\left(\frac{1}{x}\right) dx = \frac{1}{\ln 10}\int u\, du = \left(\frac{1}{\ln 10}\right)\left(\frac{1}{2} u^2\right) + C = \frac{(\ln x)^2}{2 \ln 10} + C$

39. $\int_1^4 \frac{\ln 2 \log_2 x}{x}\, dx = \int_1^4 \left(\frac{\ln 2}{x}\right)\left(\frac{\ln x}{\ln 2}\right) dx = \int_1^4 \frac{\ln x}{x}\, dx = \left[\frac{1}{2}(\ln x)^2\right]_1^4 = \frac{1}{2}\left[(\ln 4)^2 - (\ln 1)^2\right] = \frac{1}{2}(\ln 4)^2 = \frac{1}{2}(2 \ln 2)^2 = 2(\ln 2)^2$

41. $\int_0^2 \frac{\log_2(x+2)}{x+2}\, dx = \frac{1}{\ln 2}\int_0^2 \left[\ln(x+2)\right]\left(\frac{1}{x+2}\right) dx = \left(\frac{1}{\ln 2}\right)\left[\frac{(\ln(x+2))^2}{2}\right]_0^2 = \left(\frac{1}{\ln 2}\right)\left[\frac{(\ln 4)^2}{2} - \frac{(\ln 2)^2}{2}\right]$

$= \left(\frac{1}{\ln 2}\right)\left[\frac{4(\ln 2)^2}{2} - \frac{(\ln 2)^2}{2}\right] = \frac{3}{2}\ln 2$

43. $\int_0^9 \frac{2 \log_{10}(x+1)}{x+1}\, dx = \frac{2}{\ln 10}\int_0^9 \ln(x+1)\left(\frac{1}{x+1}\right) dx = \left(\frac{2}{\ln 10}\right)\left[\frac{(\ln(x+1))^2}{2}\right]_0^9 = \left(\frac{2}{\ln 10}\right)\left[\frac{(\ln 10)^2}{2} - \frac{(\ln 1)^2}{2}\right] = \ln 10$

45. $\int \frac{dx}{x \log_{10} x} = \int \left(\frac{\ln 10}{\ln x}\right)\left(\frac{1}{x}\right) dx = (\ln 10)\int \left(\frac{1}{\ln x}\right)\left(\frac{1}{x}\right) dx;\ \left[u = \ln x\ \Rightarrow\ du = \frac{1}{x}\, dx\right]$

$\rightarrow (\ln 10)\int \left(\frac{1}{\ln x}\right)\left(\frac{1}{x}\right) dx = (\ln 10)\int \frac{1}{u}\, du = (\ln 10)\ln|u| + C = (\ln 10)\ln|\ln x| + C$

47. $\frac{dy}{dt} = e^t \sin(e^t - 2)\ \Rightarrow\ y = \int e^t \sin(e^t - 2)\, dt;$

let $u = e^t - 2\ \Rightarrow\ du = e^t\, dt\ \Rightarrow\ y = \int \sin u\, du = -\cos u + C = -\cos(e^t - 2) + C;\ y(\ln 2) = 0$

$\Rightarrow\ -\cos\left(e^{\ln 2} - 2\right) + C = 0\ \Rightarrow\ -\cos(2 - 2) + C = 0\ \Rightarrow\ C = \cos 0 = 1;$ thus, $y = 1 - \cos(e^t - 2)$

49. $\frac{d^2 y}{dx^2} = 2e^{-x}\ \Rightarrow\ \frac{dy}{dx} = -2e^{-x} + C;\ x = 0$ and $\frac{dy}{dx} = 0\ \Rightarrow\ 0 = -2e^0 + C\ \Rightarrow\ C = 2;$ thus $\frac{dy}{dx} = -2e^{-x} + 2$

$\Rightarrow\ y = 2e^{-x} + 2x + C_1;\ x = 0$ and $y = 1\ \Rightarrow\ 1 = 2e^0 + C_1\ \Rightarrow\ C_1 = -1\ \Rightarrow\ y = 2e^{-x} + 2x - 1 = 2\left(e^{-x} + x\right) - 1$

51. $\frac{dy}{dx} = 1 + \frac{1}{x}$ at $(1, 3)\ \Rightarrow\ y = x + \ln|x| + C;\ y = 3$ at $x = 1\ \Rightarrow\ C = 2\ \Rightarrow\ y = x + \ln|x| + 2$

53. (a) $L(x) = f(0) + f'(0) \cdot x,$ and $f(x) = \ln(1 + x)\ \Rightarrow\ f'(x)\big|_{x=0} = \frac{1}{1+x}\big|_{x=0} = 1\ \Rightarrow\ L(x) = \ln 1 + 1 \cdot x\ \Rightarrow\ L(x) = x$

(b) Let $f(x) = \ln(x + 1).$ Since $f''(x) = -\frac{1}{(x+1)^2} < 0$ on $[0, 0.1],$ the graph of f is concave down on this interval and the largest error in the linear approximation will occur when $x = 0.1.$ This error is $0.1 - \ln(1.1) \approx 0.00469$ to five decimal places.

(c) The approximation $y = x$ for $\ln(1 + x)$ is best for smaller positive values of x; in particular for $0 \le x \le 0.1$ in the graph. As x increases, so does the error $x - \ln(1 + x).$ From the graph an upper bound for the error is $0.5 - \ln(1 + 0.5) \approx 0.095;$ i.e., $|E(x)| \le 0.095$ for $0 \le x \le 0.5.$ Note from the graph that $0.1 - \ln(1 + 0.1) \approx 0.00469$ estimates the error in replacing $\ln(1 + x)$ by x over $0 \le x \le 0.1.$ This is consistent with the estimate given in part (b) above.

55. Note that $y = \ln x$ and $e^y = x$ are the same curve; $\int_1^a \ln x\, dx = $ area under the curve between 1 and a;

$\int_0^{\ln a} e^y\, dy = $ area to the left of the curve between 0 and $\ln a.$ The sum of these areas is equal to the area of the rectangle

$\Rightarrow \int_1^a \ln x\, dx + \int_0^{\ln a} e^y\, dy = a \ln a.$

57. (a) The point of tangency is (p, ln p) and $m_{tangent} = \frac{1}{p}$ since $\frac{dy}{dx} = \frac{1}{x}$. The tangent line passes through $(0, 0) \Rightarrow$ the

equation of the tangent line is $y = \frac{1}{p}x$. The tangent line also passes through$(p, ln p) \Rightarrow ln\ p = \frac{1}{p}p = 1 \Rightarrow p = e$, and

the tangent line equation is $y = \frac{1}{e}x$.

(b) $\frac{d^2y}{dx^2} = -\frac{1}{x^2}$ for $x \neq 0 \Rightarrow y = ln\ x$ is concave downward over its domain. Therefore, $y = ln\ x$ lies below the graph of

$y = \frac{1}{e}x$ for all $x > 0$, $x \neq e$, and $ln\ x < \frac{x}{e}$ for $x > 0$, $x \neq e$.

(c) Multiplying by e, $e\ ln\ x < x$ or $ln\ x^e < x$.

(d) Exponentiating both sides of $ln\ x^e < x$, we have $e^{ln\ x^e} < e^x$, or $x^e < e^x$ for all positive $x \neq e$.

(e) Let $x = \pi$ to see that $\pi^e < e^\pi$. Therefore, e^π is bigger.

CHAPTER 5 PRACTICE EXERCISES

1. (a) Each time subinterval is of length $\Delta t = 0.4$ sec. The distance traveled over each subinterval, using the

midpoint rule, is $\Delta h = \frac{1}{2}(v_i + v_{i+1})\Delta t$, where v_i is the velocity at the left endpoint and v_{i+1} the velocity at

the right endpoint of the subinterval. We then add Δh to the height attained so far at the left endpoint v_i to

arrive at the height associated with velocity v_{i+1} at the right endpoint. Using this methodology we build

the following table based on the figure in the text:

t (sec)	0	0.4	0.8	1.2	1.6	2.0	2.4	2.8	3.2	3.6	4.0	4.4	4.8	5.2	5.6	6.0
v (fps)	0	10	25	55	100	190	180	165	150	140	130	115	105	90	76	65
h (ft)	0	2	9	25	56	114	188	257	320	378	432	481	525	564	592	620.2

t (sec)	6.4	6.8	7.2	7.6	8.0
v (fps)	50	37	25	12	0
h (ft)	643.2	660.6	672	679.4	681.8

NOTE: Your table values may vary slightly from ours depending on the v-values you read from the graph.
Remember that some shifting of the graph occurs in the printing process.
The total height attained is about 680 ft.

(b) The graph is based on the table in part (a).

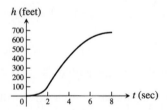

3. (a) $\sum\limits_{k=1}^{10} \frac{a_k}{4} = \frac{1}{4}\sum\limits_{k=1}^{10} a_k = \frac{1}{4}(-2) = -\frac{1}{2}$ (b) $\sum\limits_{k=1}^{10}(b_k - 3a_k) = \sum\limits_{k=1}^{10} b_k - 3\sum\limits_{k=1}^{10} a_k = 25 - 3(-2) = 31$

(c) $\sum\limits_{k=1}^{10}(a_k + b_k - 1) = \sum\limits_{k=1}^{10} a_k + \sum\limits_{k=1}^{10} b_k - \sum\limits_{k=1}^{10} 1 = -2 + 25 - (1)(10) = 13$

(d) $\sum\limits_{k=1}^{10}\left(\frac{5}{2} - b_k\right) = \sum\limits_{k=1}^{10}\frac{5}{2} - \sum\limits_{k=1}^{10} b_k = \frac{5}{2}(10) - 25 = 0$

5. Let $u = 2x - 1 \Rightarrow du = 2\ dx \Rightarrow \frac{1}{2}du = dx; x = 1 \Rightarrow u = 1, x = 5 \Rightarrow u = 9$

$\int_1^5 (2x - 1)^{-1/2}\ dx = \int_1^9 u^{-1/2}\left(\frac{1}{2}du\right) = \left[u^{1/2}\right]_1^9 = 3 - 1 = 2$

7. Let $u = \frac{x}{2} \Rightarrow 2\ du = dx; x = -\pi \Rightarrow u = -\frac{\pi}{2}, x = 0 \Rightarrow u = 0$

$\int_{-\pi}^0 \cos\left(\frac{x}{2}\right)\ dx = \int_{-\pi/2}^0 (\cos u)(2\ du) = [2\sin u]_{-\pi/2}^0 = 2\sin 0 - 2\sin\left(-\frac{\pi}{2}\right) = 2(0 - (-1)) = 2$

9. (a) $\int_{-2}^{2} f(x)\, dx = \frac{1}{3}\int_{-2}^{2} 3\, f(x)\, dx = \frac{1}{3}(12) = 4$

   (b) $\int_{2}^{5} f(x)\, dx = \int_{-2}^{5} f(x)\, dx - \int_{-2}^{2} f(x)\, dx = 6 - 4 = 2$

   (c) $\int_{5}^{-2} g(x)\, dx = -\int_{-2}^{5} g(x)\, dx = -2$

   (d) $\int_{-2}^{5} (-\pi\, g(x))\, dx = -\pi \int_{-2}^{5} g(x)\, dx = -\pi(2) = -2\pi$

   (e) $\int_{-2}^{5} \left(\frac{f(x) + g(x)}{5} \right) dx = \frac{1}{5}\int_{-2}^{5} f(x)\, dx + \frac{1}{5}\int_{-2}^{5} g(x)\, dx = \frac{1}{5}(6) + \frac{1}{5}(2) = \frac{8}{5}$

11. $x^2 - 4x + 3 = 0 \Rightarrow (x - 3)(x - 1) = 0 \Rightarrow x = 3 \text{ or } x = 1;$

   Area $= \int_{0}^{1} (x^2 - 4x + 3)\, dx - \int_{1}^{3} (x^2 - 4x + 3)\, dx$

   $= \left[\frac{x^3}{3} - 2x^2 + 3x \right]_{0}^{1} - \left[\frac{x^3}{3} - 2x^2 + 3x \right]_{1}^{3}$

   $= \left[\left(\frac{1^3}{3} - 2(1)^2 + 3(1) \right) - 0 \right]$

   $\quad - \left[\left(\frac{3^3}{3} - 2(3)^2 + 3(3) \right) - \left(\frac{1^3}{3} - 2(1)^2 + 3(1) \right) \right]$

   $= \left(\frac{1}{3} + 1 \right) - \left[0 - \left(\frac{1}{3} + 1 \right) \right] = \frac{8}{3}$

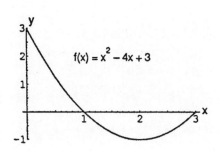

13. $5 - 5x^{2/3} = 0 \Rightarrow 1 - x^{2/3} = 0 \Rightarrow x = \pm 1;$

   Area $= \int_{-1}^{1} \left(5 - 5x^{2/3} \right) dx - \int_{1}^{8} \left(5 - 5x^{2/3} \right) dx$

   $= \left[5x - 3x^{5/3} \right]_{-1}^{1} - \left[5x - 3x^{5/3} \right]_{1}^{8}$

   $= \left[\left(5(1) - 3(1)^{5/3} \right) - \left(5(-1) - 3(-1)^{5/3} \right) \right]$

   $\quad - \left[\left(5(8) - 3(8)^{5/3} \right) - \left(5(1) - 3(1)^{5/3} \right) \right]$

   $= [2 - (-2)] - [(40 - 96) - 2] = 62$

15. $f(x) = x,\ g(x) = \frac{1}{x^2},\ a = 1,\ b = 2 \Rightarrow A = \int_{a}^{b} [f(x) - g(x)]\, dx$

   $= \int_{1}^{2} \left(x - \frac{1}{x^2} \right) dx = \left[\frac{x^2}{2} + \frac{1}{x} \right]_{1}^{2} = \left(\frac{4}{2} + \frac{1}{2} \right) - \left(\frac{1}{2} + 1 \right) = 1$

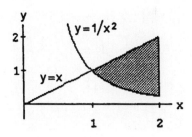

17. $f(x) = \left(1 - \sqrt{x} \right)^2,\ g(x) = 0,\ a = 0,\ b = 1 \Rightarrow A = \int_{a}^{b} [f(x) - g(x)]\, dx = \int_{0}^{1} \left(1 - \sqrt{x} \right)^2 dx = \int_{0}^{1} \left(1 - 2\sqrt{x} + x \right) dx$

   $= \int_{0}^{1} \left(1 - 2x^{1/2} + x \right) dx = \left[x - \frac{4}{3}x^{3/2} + \frac{x^2}{2} \right]_{0}^{1} = 1 - \frac{4}{3} + \frac{1}{2} = \frac{1}{6}(6 - 8 + 3) = \frac{1}{6}$

19. $f(y) = 2y^2,\ g(y) = 0,\ c = 0,\ d = 3$

   $\Rightarrow A = \int_{c}^{d} [f(y) - g(y)]\, dy = \int_{0}^{3} (2y^2 - 0)\, dy$

   $= 2\int_{0}^{3} y^2\, dy = \frac{2}{3}\left[y^3 \right]_{0}^{3} = 18$

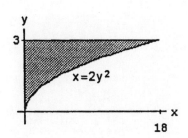

21. Let us find the intersection points: $\frac{y^2}{4} = \frac{y+2}{4}$

$\Rightarrow y^2 - y - 2 = 0 \Rightarrow (y-2)(y+1) = 0 \Rightarrow y = -1$

or $y = 2 \Rightarrow c = -1, d = 2$; $f(y) = \frac{y+2}{4}$, $g(y) = \frac{y^2}{4}$

$\Rightarrow A = \int_c^d [f(y) - g(y)]\, dy = \int_{-1}^2 \left(\frac{y+2}{4} - \frac{y^2}{4}\right) dy$

$= \frac{1}{4}\int_{-1}^2 (y + 2 - y^2)\, dy = \frac{1}{4}\left[\frac{y^2}{2} + 2y - \frac{y^3}{3}\right]_{-1}^2$

$= \frac{1}{4}\left[\left(\frac{4}{2} + 4 - \frac{8}{3}\right) - \left(\frac{1}{2} - 2 + \frac{1}{3}\right)\right] = \frac{9}{8}$

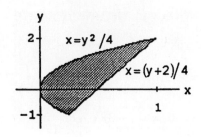

23. $f(x) = x$, $g(x) = \sin x$, $a = 0$, $b = \frac{\pi}{4}$

$\Rightarrow A = \int_a^b [f(x) - g(x)]\, dx = \int_0^{\pi/4} (x - \sin x)\, dx$

$= \left[\frac{x^2}{2} + \cos x\right]_0^{\pi/4} = \left(\frac{\pi^2}{32} + \frac{\sqrt{2}}{2}\right) - 1$

25. $a = 0$, $b = \pi$, $f(x) - g(x) = 2\sin x - \sin 2x$

$\Rightarrow A = \int_0^\pi (2\sin x - \sin 2x)\, dx = \left[-2\cos x + \frac{\cos 2x}{2}\right]_0^\pi$

$= \left[-2\cdot(-1) + \frac{1}{2}\right] - \left(-2\cdot 1 + \frac{1}{2}\right) = 4$

27. $f(y) = \sqrt{y}$, $g(y) = 2 - y$, $c = 1$, $d = 2$

$\Rightarrow A = \int_c^d [f(y) - g(y)]\, dy = \int_1^2 \left[\sqrt{y} - (2 - y)\right] dy$

$= \int_1^2 (\sqrt{y} - 2 + y)\, dy = \left[\frac{2}{3}y^{3/2} - 2y + \frac{y^2}{2}\right]_1^2$

$= \left(\frac{4}{3}\sqrt{2} - 4 + 2\right) - \left(\frac{2}{3} - 2 + \frac{1}{2}\right) = \frac{4}{3}\sqrt{2} - \frac{7}{6} = \frac{8\sqrt{2}-7}{6}$

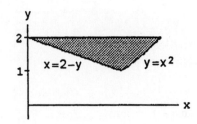

29. $f(x) = x^3 - 3x^2 = x^2(x - 3) \Rightarrow f'(x) = 3x^2 - 6x = 3x(x - 2) \Rightarrow f' = +++ \mid ---- \mid +++$
                                                                                   $\quad\quad 0 \quad\quad 2$

$\Rightarrow f(0) = 0$ is a maximum and $f(2) = -4$ is a minimum. $A = -\int_0^3 (x^3 - 3x^2)\, dx = -\left[\frac{x^4}{4} - x^3\right]_0^3$

$= -\left(\frac{81}{4} - 27\right) = \frac{27}{4}$

31. The area above the x-axis is $A_1 = \int_0^1 \left(y^{2/3} - y\right) dy$

$= \left[\frac{3y^{5/3}}{5} - \frac{y^2}{2}\right]_0^1 = \frac{1}{10}$; the area below the x-axis is

$A_2 = \int_{-1}^0 \left(y^{2/3} - y\right) dy = \left[\frac{3y^{5/3}}{5} - \frac{y^2}{2}\right]_{-1}^0 = \frac{11}{10}$

$\Rightarrow$ the total area is $A_1 + A_2 = \frac{6}{5}$

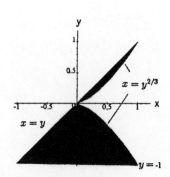

33. $A = \int_1^e \frac{2 \ln x}{x} \, dx = \int_0^1 2u \, du = [u^2]_0^1 = 1$, where

$u = \ln x$ and $du = \frac{1}{x} \, dx$; $x = 1 \Rightarrow u = 0$, $x = e \Rightarrow u = 1$

35. $y = x^2 + \int_1^x \frac{1}{t} \, dt \Rightarrow \frac{dy}{dx} = 2x + \frac{1}{x} \Rightarrow \frac{d^2y}{dx^2} = 2 - \frac{1}{x^2}$; $y(1) = 1 + \int_1^1 \frac{1}{t} \, dt = 1$ and $y'(1) = 2 + 1 = 3$

37. $y = \int_5^x \frac{\sin t}{t} \, dt - 3 \Rightarrow \frac{dy}{dx} = \frac{\sin x}{x}$; $x = 5 \Rightarrow y = \int_5^5 \frac{\sin t}{t} \, dt - 3 = -3$

39. $\frac{dy}{dx} = \frac{1}{\sqrt{1-x^2}} \Rightarrow dy = \frac{dx}{\sqrt{1-x^2}} \Rightarrow y = \sin^{-1} x + C$; $x = 0$ and $y = 0 \Rightarrow 0 = \sin^{-1} 0 + C \Rightarrow C = 0 \Rightarrow y = \sin^{-1} x$

41. $\frac{dy}{dx} = \frac{1}{x\sqrt{x^2-1}} \Rightarrow dy = \frac{dx}{x\sqrt{x^2-1}} \Rightarrow y = \sec^{-1} |x| + C$; $x = 2$ and $y = \pi \Rightarrow \pi = \sec^{-1} 2 + C \Rightarrow C = \pi - \sec^{-1} 2$

$= \pi - \frac{\pi}{3} = \frac{2\pi}{3} \Rightarrow y = \sec^{-1}(x) + \frac{2\pi}{3}$, $x > 1$

43. Let $u = \cos x \Rightarrow du = -\sin x \, dx \Rightarrow -du = \sin x \, dx$

$\int 2(\cos x)^{-1/2} \sin x \, dx = \int 2u^{-1/2}(-du) = -2 \int u^{-1/2} \, du = -2\left(\frac{u^{1/2}}{\frac{1}{2}}\right) + C = -4u^{1/2} + C = -4(\cos x)^{1/2} + C$

45. Let $u = 2\theta + 1 \Rightarrow du = 2 \, d\theta \Rightarrow \frac{1}{2} \, du = d\theta$

$\int [2\theta + 1 + 2\cos(2\theta + 1)] \, d\theta = \int (u + 2\cos u)\left(\frac{1}{2} \, du\right) = \frac{u^2}{4} + \sin u + C_1 = \frac{(2\theta+1)^2}{4} + \sin(2\theta + 1) + C_1$

$= \theta^2 + \theta + \sin(2\theta + 1) + C$, where $C = C_1 + \frac{1}{4}$ is still an arbitrary constant

47. $\int \left(t - \frac{2}{t}\right)\left(t + \frac{2}{t}\right) dt = \int \left(t^2 - \frac{4}{t^2}\right) dt = \int (t^2 - 4t^{-2}) \, dt = \frac{t^3}{3} - 4\left(\frac{t^{-1}}{-1}\right) + C = \frac{t^3}{3} + \frac{4}{t} + C$

49. Let $u = 2t^{3/2} \Rightarrow du = 3\sqrt{t} \, dt \Rightarrow \frac{1}{3} du = \sqrt{t} \, dt$

$\int \sqrt{t} \sin(2t^{3/2}) \, dt = \frac{1}{3} \int \sin u \, du = -\frac{1}{3}\cos u + C = -\frac{1}{3}\cos(2t^{3/2}) + C$

51. $\int e^x \sec^2(e^x - 7) \, dx = \int \sec^2 u \, du$, where $u = e^x - 7$ and $du = e^x \, dx$

$= \tan u + C = \tan(e^x - 7) + C$

53. $\int (\sec^2 x) e^{\tan x} \, dx = \int e^u \, du$, where $u = \tan x$ and $du = \sec^2 x \, dx$

$= e^u + C = e^{\tan x} + C$

55. $\int_{-1}^1 \frac{1}{3x-4} \, dx = \frac{1}{3} \int_{-7}^{-1} \frac{1}{u} \, du$, where $u = 3x - 4$, $du = 3 \, dx$; $x = -1 \Rightarrow u = -7$, $x = 1 \Rightarrow u = -1$

$= \frac{1}{3} [\ln |u|]_{-7}^{-1} = \frac{1}{3} [\ln |-1| - \ln |-7|] = \frac{1}{3} [0 - \ln 7] = -\frac{\ln 7}{3}$

57. $\int_0^4 \frac{2t}{t^2-25} \, dt = \int_{-25}^{-9} \frac{1}{u} \, du$, where $u = t^2 - 25$, $du = 2t \, dt$; $t = 0 \Rightarrow u = -25$, $t = 4 \Rightarrow u = -9$

$= [\ln |u|]_{-25}^{-9} = \ln |-9| - \ln |-25| = \ln 9 - \ln 25 = \ln \frac{9}{25}$

59. $\int \frac{(\ln x)^{-3}}{x} \, dx = \int u^{-3} \, du$, where $u = \ln x$ and $du = \frac{1}{x} \, dx$

$= \frac{u^{-2}}{-2} + C = -\frac{1}{2}(\ln x)^{-2} + C$

61. $\int x3^{x^2}\, dx = \frac{1}{2}\int 3^u\, du$, where $u = x^2$ and $du = 2x\, dx$

$$= \frac{1}{2\ln 3}(3^u) + C = \frac{1}{2\ln 3}(3^{x^2}) + C$$

63. $\int \frac{3\, dr}{\sqrt{1 - 4(r-1)^2}} = \frac{3}{2}\int \frac{du}{\sqrt{1 - u^2}}$, where $u = 2(r - 1)$ and $du = 2\, dr$

$$= \frac{3}{2}\sin^{-1} u + C = \frac{3}{2}\sin^{-1} 2(r - 1) + C$$

65. $\int \frac{dx}{2 + (x-1)^2} = \int \frac{du}{2 + u^2}$, where $u = x - 1$ and $du = dx$

$$= \frac{1}{\sqrt{2}}\tan^{-1}\frac{u}{\sqrt{2}} + C = \frac{1}{\sqrt{2}}\tan^{-1}\left(\frac{x-1}{\sqrt{2}}\right) + C$$

67. $\int \frac{dx}{(2x-1)\sqrt{(2x-1)^2 - 4}} = \frac{1}{2}\int \frac{du}{u\sqrt{u^2 - 4}}$, where $u = 2x - 1$ and $du = 2\, dx$

$$= \frac{1}{2}\cdot\frac{1}{2}\sec^{-1}\left|\frac{u}{2}\right| + C = \frac{1}{4}\sec^{-1}\left|\frac{2x-1}{2}\right| + C$$

69. $\int \frac{e^{\sin^{-1}\sqrt{x}}}{2\sqrt{x - x^2}}\, dx = \int e^u\, du$, where $u = \sin^{-1}\sqrt{x}$ and $du = \frac{dx}{2\sqrt{x - x^2}}$

$$= e^u + C = e^{\sin^{-1}\sqrt{x}} + C$$

71. $\int \frac{1}{\sqrt{\tan^{-1} y}\,(1 + y^2)}\, dy = \int \frac{\left(\frac{1}{1+y^2}\right)}{\sqrt{\tan^{-1} y}}\, dy = \int u^{-1/2}\, du$, where $u = \tan^{-1} y$ and $du = \frac{dy}{1 + y^2}$

$$= 2u^{1/2} + C = 2\sqrt{\tan^{-1} y} + C$$

73. $\int_{-1}^{1}(3x^2 - 4x + 7)\, dx = [x^3 - 2x^2 + 7x]_{-1}^{1} = [1^3 - 2(1)^2 + 7(1)] - [(-1)^3 - 2(-1)^2 + 7(-1)] = 6 - (-10) = 16$

75. $\int_{1}^{2} \frac{4}{v^2}\, dv = \int_{1}^{2} 4v^{-2}\, dv = [-4v^{-1}]_{1}^{2} = \left(\frac{-4}{2}\right) - \left(\frac{-4}{1}\right) = 2$

77. $\int_{1}^{4} \frac{dt}{t\sqrt{t}} = \int_{1}^{4} \frac{dt}{t^{3/2}} = \int_{1}^{4} t^{-3/2}\, dt = [-2t^{-1/2}]_{1}^{4} = \frac{-2}{\sqrt{4}} - \frac{(-2)}{\sqrt{1}} = 1$

79. Let $u = 2x + 1 \Rightarrow du = 2\, dx \Rightarrow 18\, du = 36\, dx$; $x = 0 \Rightarrow u = 1, x = 1 \Rightarrow u = 3$

$$\int_{0}^{1} \frac{36\, dx}{(2x+1)^3} = \int_{1}^{3} 18u^{-3}\, du = \left[\frac{18u^{-2}}{-2}\right]_{1}^{3} = \left[\frac{-9}{u^2}\right]_{1}^{3} = \left(\frac{-9}{3^2}\right) - \left(\frac{-9}{1^2}\right) = 8$$

81. Let $u = 1 - x^{2/3} \Rightarrow du = -\frac{2}{3}x^{-1/3}\, dx \Rightarrow -\frac{3}{2}\, du = x^{-1/3}\, dx$; $x = \frac{1}{8} \Rightarrow u = 1 - \left(\frac{1}{8}\right)^{2/3} = \frac{3}{4}$,

$x = 1 \Rightarrow u = 1 - 1^{2/3} = 0$

$$\int_{1/8}^{1} x^{-1/3}\left(1 - x^{2/3}\right)^{3/2}\, dx = \int_{3/4}^{0} u^{3/2}\left(-\frac{3}{2}\, du\right) = \left[\left(-\frac{3}{2}\right)\left(\frac{u^{5/2}}{\frac{5}{2}}\right)\right]_{3/4}^{0} = \left[-\frac{3}{5}u^{5/2}\right]_{3/4}^{0} = -\frac{3}{5}(0)^{5/2} - \left(-\frac{3}{5}\right)\left(\frac{3}{4}\right)^{5/2}$$

$$= \frac{27\sqrt{3}}{160}$$

83. Let $u = 5r \Rightarrow du = 5\, dr \Rightarrow \frac{1}{5}\, du = dr$; $r = 0 \Rightarrow u = 0, r = \pi \Rightarrow u = 5\pi$

$$\int_{0}^{\pi} \sin^2 5r\, dr = \int_{0}^{5\pi}(\sin^2 u)\left(\frac{1}{5}\, du\right) = \frac{1}{5}\left[\frac{u}{2} - \frac{\sin 2u}{4}\right]_{0}^{5\pi} = \left(\frac{\pi}{2} - \frac{\sin 10\pi}{20}\right) - \left(0 - \frac{\sin 0}{20}\right) = \frac{\pi}{2}$$

85. $\int_{0}^{\pi/3} \sec^2 \theta\, d\theta = [\tan \theta]_{0}^{\pi/3} = \tan\frac{\pi}{3} - \tan 0 = \sqrt{3}$

87. Let $u = \frac{x}{6} \Rightarrow du = \frac{1}{6} dx \Rightarrow 6\,du = dx; x = \pi \Rightarrow u = \frac{\pi}{6}, x = 3\pi \Rightarrow u = \frac{\pi}{2}$

$$\int_{\pi}^{3\pi} \cot^2 \frac{x}{6}\, dx = \int_{\pi/6}^{\pi/2} 6 \cot^2 u\, du = 6\int_{\pi/6}^{\pi/2} (\csc^2 u - 1)\, du = [6(-\cot u - u)]_{\pi/6}^{\pi/2} = 6\left(-\cot \frac{\pi}{2} - \frac{\pi}{2}\right) - 6\left(-\cot \frac{\pi}{6} - \frac{\pi}{6}\right)$$
$$= 6\sqrt{3} - 2\pi$$

89. $\displaystyle\int_{-\pi/3}^{0} \sec x \tan x\, dx = [\sec x]_{-\pi/3}^{0} = \sec 0 - \sec\left(-\frac{\pi}{3}\right) = 1 - 2 = -1$

91. Let $u = \sin x \Rightarrow du = \cos x\, dx; x = 0 \Rightarrow u = 0, x = \frac{\pi}{2} \Rightarrow u = 1$

$$\int_{0}^{\pi/2} 5(\sin x)^{3/2} \cos x\, dx = \int_{0}^{1} 5u^{3/2}\, du = \left[5\left(\tfrac{2}{5}\right) u^{5/2}\right]_{0}^{1} = \left[2u^{5/2}\right]_{0}^{1} = 2(1)^{5/2} - 2(0)^{5/2} = 2$$

93. Let $u = 1 + 3\sin^2 x \Rightarrow du = 6\sin x \cos x\, dx \Rightarrow \frac{1}{2} du = 3\sin x \cos x\, dx; x = 0 \Rightarrow u = 1, x = \frac{\pi}{2}$
$\Rightarrow u = 1 + 3\sin^2 \frac{\pi}{2} = 4$

$$\int_{0}^{\pi/2} \frac{3\sin x \cos x}{\sqrt{1+3\sin^2 x}}\, dx = \int_{1}^{4} \frac{1}{\sqrt{u}} \left(\tfrac{1}{2} du\right) = \int_{1}^{4} \tfrac{1}{2} u^{-1/2}\, du = \left[\tfrac{1}{2}\left(\tfrac{u^{1/2}}{\frac{1}{2}}\right)\right]_{1}^{4} = \left[u^{1/2}\right]_{1}^{4} = 4^{1/2} - 1^{1/2} = 1$$

95. $\displaystyle\int_{1}^{4} \left(\frac{x}{8} + \frac{1}{2x}\right) dx = \frac{1}{2}\int_{1}^{4} \left(\frac{1}{4} x + \frac{1}{x}\right) dx = \frac{1}{2}\left[\frac{1}{8} x^2 + \ln|x|\right]_{1}^{4} = \frac{1}{2}\left[\left(\frac{16}{8} + \ln 4\right) - \left(\frac{1}{8} + \ln 1\right)\right] = \frac{15}{16} + \frac{1}{2}\ln 4$
$= \frac{15}{16} + \ln\sqrt{4} = \frac{15}{16} + \ln 2$

97. $\displaystyle\int_{-2}^{-1} e^{-(x+1)}\, dx = -\int_{1}^{0} e^u\, du$, where $u = -(x+1), du = -dx; x = -2 \Rightarrow u = 1, x = -1 \Rightarrow u = 0$
$= -[e^u]_{1}^{0} = -(e^0 - e^1) = e - 1$

99. $\displaystyle\int_{1}^{\ln 5} e^r (3e^r + 1)^{-3/2}\, dr = \frac{1}{3}\int_{4}^{16} u^{-3/2}\, du$, where $u = 3e^r + 1, du = 3e^r dr; r = 0 \Rightarrow u = 4, r = \ln 5 \Rightarrow u = 16$
$= -\frac{2}{3}\left[u^{-1/2}\right]_{4}^{16} = -\frac{2}{3}\left(16^{-1/2} - 4^{-1/2}\right) = \left(-\frac{2}{3}\right)\left(\frac{1}{4} - \frac{1}{2}\right) = \left(-\frac{2}{3}\right)\left(-\frac{1}{4}\right) = \frac{1}{6}$

101. $\displaystyle\int_{1}^{e} \frac{1}{x}(1 + 7\ln x)^{-1/3}\, dx = \frac{1}{7}\int_{1}^{8} u^{-1/3}\, du$, where $u = 1 + 7\ln x, du = \frac{7}{x} dx, x = 1 \Rightarrow u = 1, x = e \Rightarrow u = 8$
$= \frac{3}{14}\left[u^{2/3}\right]_{1}^{8} = \frac{3}{14}\left(8^{2/3} - 1^{2/3}\right) = \left(\frac{3}{14}\right)(4 - 1) = \frac{9}{14}$

103. $\displaystyle\int_{1}^{8} \frac{\log_4 \theta}{\theta}\, d\theta = \frac{1}{\ln 4}\int_{1}^{8} (\ln \theta)\left(\frac{1}{\theta}\right) d\theta = \frac{1}{\ln 4}\int_{0}^{\ln 8} u\, du$, where $u = \ln \theta, du = \frac{1}{\theta} d\theta, \theta = 1 \Rightarrow u = 0, \theta = 8 \Rightarrow u = \ln 8$
$= \frac{1}{2\ln 4}\left[u^2\right]_{0}^{\ln 8} = \frac{1}{\ln 16}\left[(\ln 8)^2 - 0^2\right] = \frac{(3\ln 2)^2}{4\ln 2} = \frac{9\ln 2}{4}$

105. $\displaystyle\int_{-3/4}^{3/4} \frac{6}{\sqrt{9-4x^2}}\, dx = 3\int_{-3/4}^{3/4} \frac{2}{\sqrt{3^2 - (2x)^2}}\, dx = 3\int_{-3/2}^{3/2} \frac{1}{\sqrt{3^2 - u^2}}\, du$, where $u = 2x, du = 2\, dx$;
$$x = -\frac{3}{4} \Rightarrow u = -\frac{3}{2}, x = \frac{3}{4} \Rightarrow u = \frac{3}{2}$$
$= 3\left[\sin^{-1}\left(\frac{u}{3}\right)\right]_{-3/2}^{3/2} = 3\left[\sin^{-1}\left(\frac{1}{2}\right) - \sin^{-1}\left(-\frac{1}{2}\right)\right] = 3\left[\frac{\pi}{6} - \left(-\frac{\pi}{6}\right)\right] = 3\left(\frac{\pi}{3}\right) = \pi$

107. $\displaystyle\int_{-2}^{2} \frac{3}{4+3t^2}\, dt = \sqrt{3}\int_{-2}^{2} \frac{\sqrt{3}}{2^2 + \left(\sqrt{3}t\right)^2}\, dt = \sqrt{3}\int_{-2\sqrt{3}}^{2\sqrt{3}} \frac{1}{2^2 + u^2}\, du$, where $u = \sqrt{3}t, du = \sqrt{3}\, dt$;
$$t = -2 \Rightarrow u = -2\sqrt{3}, t = 2 \Rightarrow u = 2\sqrt{3}$$
$= \sqrt{3}\left[\frac{1}{2}\tan^{-1}\left(\frac{u}{2}\right)\right]_{-2\sqrt{3}}^{2\sqrt{3}} = \frac{\sqrt{3}}{2}\left[\tan^{-1}\left(\sqrt{3}\right) - \tan^{-1}\left(-\sqrt{3}\right)\right] = \frac{\sqrt{3}}{2}\left[\frac{\pi}{3} - \left(-\frac{\pi}{3}\right)\right] = \frac{\pi}{\sqrt{3}}$

109. $\int \frac{1}{y\sqrt{4y^2-1}}\,dy = \int \frac{2}{(2y)\sqrt{(2y)^2-1}}\,dy = \int \frac{1}{u\sqrt{u^2-1}}\,du$, where $u = 2y$ and $du = 2\,dy$

$\quad = \sec^{-1}|u| + C = \sec^{-1}|2y| + C$

111. $\int_{\sqrt{2/3}}^{2/3} \frac{1}{|y|\sqrt{9y^2-1}}\,dy = \int_{\sqrt{2/3}}^{2/3} \frac{3}{|3y|\sqrt{(3y)^2-1}}\,dy = \int_{\sqrt{2}}^{2} \frac{1}{|u|\sqrt{u^2-1}}\,du$, where $u = 3y$, $du = 3\,dy$;

$\qquad\qquad\qquad\qquad\qquad\qquad\qquad\qquad\qquad\qquad\qquad y = \frac{\sqrt{2}}{3} \Rightarrow u = \sqrt{2},\, y = \frac{2}{3} \Rightarrow u = 2$

$\quad = \left[\sec^{-1} u\right]_{\sqrt{2}}^{2} = \left[\sec^{-1} 2 - \sec^{-1}\sqrt{2}\right] = \frac{\pi}{3} - \frac{\pi}{4} = \frac{\pi}{12}$

113. (a) $\text{av}(f) = \frac{1}{1-(-1)} \int_{-1}^{1} (mx + b)\,dx = \frac{1}{2}\left[\frac{mx^2}{2} + bx\right]_{-1}^{1} = \frac{1}{2}\left[\left(\frac{m(1)^2}{2} + b(1)\right) - \left(\frac{m(-1)^2}{2} + b(-1)\right)\right] = \frac{1}{2}(2b) = b$

(b) $\text{av}(f) = \frac{1}{k-(-k)} \int_{-k}^{k} (mx + b)\,dx = \frac{1}{2k}\left[\frac{mx^2}{2} + bx\right]_{-k}^{k} = \frac{1}{2k}\left[\left(\frac{m(k)^2}{2} + b(k)\right) - \left(\frac{m(-k)^2}{2} + b(-k)\right)\right]$

$\quad = \frac{1}{2k}(2bk) = b$

115. $f'_{\text{av}} = \frac{1}{b-a}\int_{a}^{b} \sqrt{ax} f'(x)\,dx = \frac{1}{b-a}[f(x)]_{a}^{b} = \frac{1}{b-a}[f(b) - f(a)] = \frac{f(b)-f(a)}{b-a}$ so the average value of f' over $[a, b]$ is the slope of the secant line joining the points $(a, f(a))$ and $(b, f(b))$, which is the average rate of change of f over $[a, b]$.

117. (a) $\frac{d}{dx}(x \ln x - x + C) = x \cdot \frac{1}{x} + \ln x - 1 + 0 = \ln x$

(b) average value $= \frac{1}{e-1}\int_{1}^{e} \ln x\,dx = \frac{1}{e-1}[x \ln x - x]_{1}^{e} = \frac{1}{e-1}[(e \ln e - e) - (1 \ln 1 - 1)]$

$\quad = \frac{1}{e-1}(e - e + 1) = \frac{1}{e-1}$

119. We want to evaluate

$\frac{1}{365-0}\int_{0}^{365} f(x)\,dx = \frac{1}{365}\int_{0}^{365}\left(37\sin\left[\frac{2\pi}{365}(x - 101)\right] + 25\right)dx = \frac{37}{365}\int_{0}^{365} \sin\left[\frac{2\pi}{365}(x - 101)\right]dx + \frac{25}{365}\int_{0}^{365} dx$

Notice that the period of $y = \sin\left[\frac{2\pi}{365}(x - 101)\right]$ is $\frac{2\pi}{\frac{2\pi}{365}} = 365$ and that we are integrating this function over an iterval of

length 365. Thus the value of $\frac{37}{365}\int_{0}^{365} \sin\left[\frac{2\pi}{365}(x - 101)\right]dx + \frac{25}{365}\int_{0}^{365} dx$ is $\frac{37}{365} \cdot 0 + \frac{25}{365} \cdot 365 = 25$.

121. $\frac{dy}{dx} = \sqrt{2 + \cos^3 x}$

123. $\frac{dy}{dx} = \frac{d}{dx}\left(-\int_{1}^{x} \frac{6}{3+t^4}dt\right) = -\frac{6}{3+x^4}$

125. $y = \int_{\ln x^2}^{0} e^{\cos t}\,dt = -\int_{0}^{\ln x^2} e^{\cos t}\,dt \Rightarrow \frac{dy}{dx} = -e^{\cos(\ln x^2)} \cdot \frac{d}{dx}(\ln x^2) = -\frac{2}{x} e^{\cos(\ln x^2)}$

127. $y = \int_{0}^{\sin^{-1}x} \frac{dt}{\sqrt{1-2t^2}} \Rightarrow \frac{dy}{dx} = \frac{1}{\sqrt{1-2(\sin^{-1}x)^2}} \cdot \frac{d}{dx}(\sin^{-1}x) = \frac{1}{\sqrt{1-2(\sin^{-1}x)^2}} \cdot \frac{1}{\sqrt{1-x^2}}$

129. Yes. The function f, being differentiable on $[a, b]$, is then continuous on $[a, b]$. The Fundamental Theorem of Calculus says that every continuous function on $[a, b]$ is the derivative of a function on $[a, b]$.

131. $y = \int_{1}^{x} \sqrt{1+t^2}\,dt = -\int_{1}^{x} \sqrt{1+t^2}\,dt \Rightarrow \frac{dy}{dx} = \frac{d}{dx}\left[-\int_{1}^{x} \sqrt{1+t^2}\,dt\right] = -\frac{d}{dx}\left[\int_{1}^{x} \sqrt{1+t^2}\,dt\right] = -\sqrt{1+x^2}$

133. We estimate the area A using midpoints of the vertical intervals, and we will estimate the width of the parking lot on each interval by averaging the widths at top and bottom. This gives the estimate

$$A \approx 15 \cdot \left(\frac{0+36}{2} + \frac{36+54}{2} + \frac{54+51}{2} + \frac{51+49.5}{2} + \frac{49.5+54}{2} + \frac{54+64.4}{2} + \frac{64.4+67.5}{2} + \frac{67.5+42}{2}\right) \approx 5961 \text{ ft}^2$$

The cost is Area · ($2.10/\text{ft}^2$) $\approx$ (5961 ft^2) ($2.10/\text{ft}^2$) = \$12,518.10 $\Rightarrow$ the job cannot be done for \$11,000.

135. $y = \ln x \Rightarrow \frac{dy}{dx} = \frac{1}{x} \, ; \frac{dy}{dt} = \frac{dy}{dx} \frac{dx}{dt} \Rightarrow \frac{dy}{dt} = \left(\frac{1}{x}\right) \sqrt{x} = \frac{1}{\sqrt{x}} \Rightarrow \frac{dy}{dt}\Big|_{e^2} = \frac{1}{e}$ m/sec

CHAPTER 5 ADDITIONAL AND ADVANCED EXERCISES

1. (a) Yes, because $\int_0^1 f(x) \, dx = \frac{1}{7} \int_0^1 7f(x) \, dx = \frac{1}{7}(7) = 1$

   (b) No. For example, $\int_0^1 8x \, dx = [4x^2]_0^1 = 4$, but $\int_0^1 \sqrt{8x} \, dx = \left[2\sqrt{2}\left(\frac{x^{3/2}}{\frac{3}{2}}\right)\right]_0^1 = \frac{4\sqrt{2}}{3}\left(1^{3/2} - 0^{3/2}\right)$

   $= \frac{4\sqrt{2}}{3} \neq \sqrt{4}$

3. $y = \frac{1}{a}\int_0^x f(t) \sin a(x - t) \, dt = \frac{1}{a}\int_0^x f(t) \sin ax \cos at \, dt - \frac{1}{a}\int_0^x f(t) \cos ax \sin at \, dt$

   $= \frac{\sin ax}{a}\int_0^x f(t) \cos at \, dt - \frac{\cos ax}{a}\int_0^x f(t) \sin at \, dt \Rightarrow \frac{dy}{dx} = \cos ax \left(\int_0^x f(t) \cos at \, dt\right)$

   $+ \frac{\sin ax}{a}\left(\frac{d}{dx}\int_0^x f(t) \cos at \, dt\right) + \sin ax \int_0^x f(t) \sin at \, dt - \frac{\cos ax}{a}\left(\frac{d}{dx}\int_0^x f(t) \sin at \, dt\right)$

   $= \cos ax \int_0^x f(t) \cos at \, dt + \frac{\sin ax}{a}(f(x) \cos ax) + \sin ax \int_0^x f(t) \sin at \, dt - \frac{\cos ax}{a}(f(x) \sin ax)$

   $\Rightarrow \frac{dy}{dx} = \cos ax \int_0^x f(t) \cos at \, dt + \sin ax \int_0^x f(t) \sin at \, dt$. Next,

   $\frac{d^2y}{dx^2} = -a \sin ax \int_0^x f(t) \cos at \, dt + (\cos ax)\left(\frac{d}{dx}\int_0^x f(t) \cos at \, dt\right) + a \cos ax \int_0^x f(t) \sin at \, dt$

   $+ (\sin ax)\left(\frac{d}{dx}\int_0^x f(t) \sin at \, dt\right) = -a \sin ax \int_0^x f(t) \cos at \, dt + (\cos ax)f(x) \cos ax$

   $+ a \cos ax \int_0^x f(t) \sin at \, dt + (\sin ax)f(x) \sin ax = -a \sin ax \int_0^x f(t) \cos at \, dt + a \cos ax \int_0^x f(t) \sin at \, dt + f(x)$.

   Therefore, $y'' + a^2 y = a \cos ax \int_0^x f(t) \sin at \, dt - a \sin ax \int_0^x f(t) \cos at \, dt + f(x)$

   $+ a^2\left(\frac{\sin ax}{a}\int_0^x f(t) \cos at \, dt - \frac{\cos ax}{a}\int_0^x f(t) \sin at \, dt\right) = f(x)$. Note also that $y'(0) = y(0) = 0$.

5. (a) $\int_0^{x^2} f(t) \, dt = x \cos \pi x \Rightarrow \frac{d}{dx}\int_0^{x^2} f(t) \, dt = \cos \pi x - \pi x \sin \pi x \Rightarrow f(x^2)(2x) = \cos \pi x - \pi x \sin \pi x$

   $\Rightarrow f(x^2) = \frac{\cos \pi x - \pi x \sin \pi x}{2x}$. Thus, $x = 2 \Rightarrow f(4) = \frac{\cos 2\pi - 2\pi \sin 2\pi}{4} = \frac{1}{4}$

   (b) $\int_0^{f(x)} t^2 \, dt = \left[\frac{t^3}{3}\right]_0^{f(x)} = \frac{1}{3}(f(x))^3 \Rightarrow \frac{1}{3}(f(x))^3 = x \cos \pi x \Rightarrow (f(x))^3 = 3x \cos \pi x \Rightarrow f(x) = \sqrt[3]{3x \cos \pi x}$

   $\Rightarrow f(4) = \sqrt[3]{3(4) \cos 4\pi} = \sqrt[3]{12}$

7. $\int_1^b f(x) \, dx = \sqrt{b^2 + 1} - \sqrt{2} \Rightarrow f(b) = \frac{d}{db}\int_1^b f(x) \, dx = \frac{1}{2}(b^2 + 1)^{-1/2}(2b) = \frac{b}{\sqrt{b^2 + 1}} \Rightarrow f(x) = \frac{x}{\sqrt{x^2 + 1}}$

9. $\frac{dy}{dx} = 3x^2 + 2 \Rightarrow y = \int (3x^2 + 2) \, dx = x^3 + 2x + C$. Then $(1, -1)$ on the curve $\Rightarrow 1^3 + 2(1) + C = -1 \Rightarrow C = -4$

   $\Rightarrow y = x^3 + 2x - 4$

11. $\int_{-8}^{3} f(x)\,dx = \int_{-8}^{0} x^{2/3}\,dx + \int_{0}^{3} -4\,dx$

$= \left[\frac{3}{5} x^{5/3}\right]_{-8}^{0} + [-4x]_{0}^{3}$

$= \left(0 - \frac{3}{5}(-8)^{5/3}\right) + (-4(3) - 0) = \frac{96}{5} - 12$

$= \frac{36}{5}$

13. $\int_{0}^{2} g(t)\,dt = \int_{0}^{1} t\,dt + \int_{1}^{2} \sin \pi t\,dt$

$= \left[\frac{t^2}{2}\right]_{0}^{1} + \left[-\frac{1}{\pi} \cos \pi t\right]_{1}^{2}$

$= \left(\frac{1}{2} - 0\right) + \left[-\frac{1}{\pi} \cos 2\pi - \left(-\frac{1}{\pi} \cos \pi\right)\right]$

$= \frac{1}{2} - \frac{2}{\pi}$

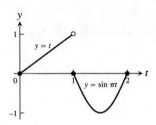

15. $\int_{-2}^{2} f(x)\,dx = \int_{-2}^{-1} dx + \int_{-1}^{1} (1 - x^2)\,dx + \int_{1}^{2} 2\,dx$

$= [x]_{-2}^{-1} + \left[x - \frac{x^3}{3}\right]_{-1}^{1} + [2x]_{1}^{2}$

$= (-1 - (-2)) + \left[\left(1 - \frac{1^3}{3}\right) - \left(-1 - \frac{(-1)^3}{3}\right)\right] + \left[2(2) - 2(1)\right]$

$= 1 + \frac{2}{3} - \left(-\frac{2}{3}\right) + 4 - 2 = \frac{13}{3}$

17. Ave. value $= \frac{1}{b-a}\int_{a}^{b} f(x)\,dx = \frac{1}{2-0}\int_{0}^{2} f(x)\,dx = \frac{1}{2}\left[\int_{0}^{1} x\,dx + \int_{1}^{2} (x - 1)\,dx\right] = \frac{1}{2}\left[\frac{x^2}{2}\right]_{0}^{1} + \frac{1}{2}\left[\frac{x^2}{2} - x\right]_{1}^{2}$

$= \frac{1}{2}\left[\left(\frac{1^2}{2} - 0\right) + \left(\frac{2^2}{2} - 2\right) - \left(\frac{1^2}{2} - 1\right)\right] = \frac{1}{2}$

19. $\int \sinh u\,du = \int \frac{e^u - e^{-u}}{2}\,du = \int \left(\frac{e^u}{2} - \frac{e^{-u}}{2}\right)du = \frac{1}{2}\int e^u\,du - \frac{1}{2}\int e^{-u}\,du = \frac{1}{2}e^u + \frac{1}{2}e^{-u} + C = \frac{e^u + e^{-u}}{2} + C = \cosh u + C$

$\int \cosh u\,du = \int \frac{e^u + e^{-u}}{2}\,du = \int \left(\frac{e^u}{2} + \frac{e^{-u}}{2}\right)du = \frac{1}{2}\int e^u\,du + \frac{1}{2}\int e^{-u}\,du = \frac{1}{2}e^u - \frac{1}{2}e^{-u} + C = \frac{e^u - e^{-u}}{2} + C = \sinh u + C$

$\int \text{sech}^2 u\,du = \int \left(\frac{2}{e^u + e^{-u}}\right)^2 du = \int \frac{4}{(e^u + e^{-u})^2}\,du = \int \frac{2+2}{(e^u + e^{-u})^2}\,du = \int \left(\frac{2}{(e^u + e^{-u})^2} + \frac{2}{(e^u + e^{-u})^2}\right) du$

$= \int \left(\frac{2}{(e^u + e^{-u})^2} \cdot \frac{e^{-2u}}{e^{-2u}} + \frac{2}{(e^u + e^{-u})^2} \cdot \frac{e^{2u}}{e^{2u}}\right) du = \int \left(\frac{2e^{-2u}}{(e^u + e^{-u})^2(e^{-2u})} + \frac{2e^{2u}}{(e^u + e^{-u})^2(e^{2u})}\right) du$

$= \int \left(\frac{2e^{-2u}}{(1 + e^{-2u})^2} + \frac{2e^{2u}}{(e^{2u} + 1)^2}\right) du = \int (1 + e^{-2u})^{-2}(2e^{-2u})\,du + \int (e^{2u} + 1)^{-2}(2e^{2u})\,du$

$= -\frac{(1 + e^{-2u})^{-1}}{-1} + \frac{(e^{2u} + 1)^{-1}}{-1} + C = \frac{1}{1 + e^{-2u}} - \frac{1}{e^{2u} + 1} + C = \frac{1}{1 + e^{-2u}} \cdot \frac{e^u}{e^u} - \frac{1}{e^{2u} + 1} \cdot \frac{e^{-u}}{e^{-u}} + C$

$= \frac{e^u}{e^u + e^{-u}} - \frac{e^{-u}}{e^u + e^{-u}} + C = \frac{e^u - e^{-u}}{e^u + e^{-u}} + C = \tanh u + C$

$\int \text{csch}^2 u\,du = \int \left(\frac{2}{e^u - e^{-u}}\right)^2 du = \int \frac{4}{(e^u - e^{-u})^2}\,du = \int \frac{2+2}{(e^u - e^{-u})^2}\,du = \int \left(\frac{2}{(e^u - e^{-u})^2} + \frac{2}{(e^u - e^{-u})^2}\right) du$

$= \int \left(\frac{2}{(e^u - e^{-u})^2} \cdot \frac{e^{-2u}}{e^{-2u}} + \frac{2}{(e^u - e^{-u})^2} \cdot \frac{e^{2u}}{e^{2u}}\right) du = \int \left(\frac{2e^{-2u}}{(e^u - e^{-u})^2(e^{-2u})} + \frac{2e^{2u}}{(e^u - e^{-u})^2(e^{2u})}\right) du$

$= \int \left(\frac{2e^{-2u}}{(1 - e^{-2u})^2} + \frac{2e^{2u}}{(e^{2u} - 1)^2}\right) du = \int (1 - e^{-2u})^{-2}(2e^{-2u})\,du + \int (e^{2u} - 1)^{-2}(2e^{2u})\,du$

$= \frac{(1 - e^{-2u})^{-1}}{-1} + \frac{(e^{2u} - 1)^{-1}}{-1} + C = -\frac{1}{1 - e^{-2u}} - \frac{1}{e^{2u} - 1} + C = -\frac{1}{1 - e^{-2u}} \cdot \frac{e^u}{e^u} - \frac{1}{e^{2u} - 1} \cdot \frac{e^{-u}}{e^{-u}} + C$

$= -\frac{e^u}{e^u - e^{-u}} - \frac{e^{-u}}{e^u - e^{-u}} + C = \frac{-e^u - e^{-u}}{e^u - e^{-u}} + C = -\frac{e^u + e^{-u}}{e^u - e^{-u}} + C = -\coth u + C$

$\int \text{sech}\,u \tanh u\,du = \int \frac{2}{e^u + e^{-u}} \cdot \frac{e^u - e^{-u}}{e^u + e^{-u}}\,du = 2\int \frac{e^u - e^{-u}}{(e^u + e^{-u})^2}\,du \quad [w = e^u + e^{-u}, dw = (e^u - e^{-u})du]$

$= 2\int \frac{1}{w^2}\,dw = -2w^{-1} + C = -\frac{2}{e^u + e^{-u}} + C = -\text{sech}\,u + C$

$\int \text{csch } u \coth u \, du = \int \frac{2}{e^u - e^{-u}} \cdot \frac{e^u + e^{-u}}{e^u - e^{-u}} \, du = 2 \int \frac{e^u + e^{-u}}{(e^u - e^{-u})^2} \, du \; [w = e^u - e^{-u}, \, dw = (e^u + e^{-u}) du]$

$$= 2 \int \frac{1}{w^2} \, dw = -2w^{-1} + C = -\frac{2}{e^u - e^{-u}} + C = -\text{csch } u + C$$

21. (a) $\int_0^{2\sqrt{3}} \frac{dx}{\sqrt{4+x^2}} = \left[\sinh^{-1} \frac{x}{2} \right]_0^{2\sqrt{3}} = \sinh^{-1} \sqrt{3} - \sinh 0 = \sinh^{-1} \sqrt{3}$

   (b) $\sinh^{-1} \sqrt{3} = \ln \left(\sqrt{3} + \sqrt{3+1} \right) = \ln \left(\sqrt{3} + 2 \right)$

23. (a) $\int_{5/4}^2 \frac{1}{1-x^2} \, dx = \left[\coth^{-1} x \right]_{5/4}^2 = \coth^{-1} 2 - \coth^{-1} \frac{5}{4}$

   (b) $\coth^{-1} 2 - \coth^{-1} \frac{5}{4} = \frac{1}{2} \left[\ln 3 - \ln \left(\frac{9/4}{1/4} \right) \right] = \frac{1}{2} \ln \frac{1}{3}$

25. (a) $\int_{1/5}^{3/13} \frac{dx}{x\sqrt{1-16x^2}} = \int_{4/5}^{12/13} \frac{du}{u\sqrt{a^2-u^2}}$, where $u = 4x$, $du = 4 \, dx$, $a = 1$

   $= \left[-\text{sech}^{-1} u \right]_{4/5}^{12/13} = -\text{sech}^{-1} \frac{12}{13} + \text{sech}^{-1} \frac{4}{5}$

   (b) $-\text{sech}^{-1} \frac{12}{13} + \text{sech}^{-1} \frac{4}{5} = -\ln \left(\frac{1+\sqrt{1-(12/13)^2}}{(12/13)} \right) + \ln \left(\frac{1+\sqrt{1-(4/5)^2}}{(4/5)} \right)$

   $= -\ln \left(\frac{13+\sqrt{169-144}}{12} \right) + \ln \left(\frac{5+\sqrt{25-16}}{4} \right) = \ln \left(\frac{5+3}{4} \right) - \ln \left(\frac{13+5}{12} \right) = \ln 2 - \ln \frac{3}{2}$

   $= \ln \left(2 \cdot \frac{2}{3} \right) = \ln \frac{4}{3}$

27. (a) $\int_0^\pi \frac{\cos x}{\sqrt{1+\sin^2 x}} \, dx = \int_0^0 \frac{1}{\sqrt{1+u^2}} \, du = \left[\sinh^{-1} u \right]_0^0 = \sinh^{-1} 0 - \sinh^{-1} 0 = 0$, where $u = \sin x$, $du = \cos x \, dx$

   (b) $\sinh^{-1} 0 - \sinh^{-1} 0 = \ln \left(0 + \sqrt{0+1} \right) - \ln \left(0 + \sqrt{0+1} \right) = 0$

29. $\lim\limits_{b \to 1^-} \int_0^b \frac{1}{\sqrt{1-x^2}} \, dx = \lim\limits_{b \to 1^-} \left[\sin^{-1} x \right]_0^b = \lim\limits_{b \to 1^-} (\sin^{-1} b - \sin^{-1} 0) = \lim\limits_{b \to 1^-} (\sin^{-1} b - 0) = \lim\limits_{b \to 1^-} \sin^{-1} b = \frac{\pi}{2}$

31. $\lim\limits_{x \to \infty} \left(\frac{1}{n+1} + \frac{1}{n+2} + \dots + \frac{1}{2n} \right) = \lim\limits_{x \to \infty} \left(\left(\frac{1}{n} \right) \left[\frac{1}{1+\left(\frac{1}{n}\right)} \right] + \left(\frac{1}{n} \right) \left[\frac{1}{1+2\left(\frac{1}{n}\right)} \right] + \dots + \left(\frac{1}{n} \right) \left[\frac{1}{1+n\left(\frac{1}{n}\right)} \right] \right)$

   which can be interpreted as a Riemann sum with partitioning $\Delta x = \frac{1}{n} \Rightarrow \lim\limits_{x \to \infty} \left(\frac{1}{n+1} + \frac{1}{n+2} + \dots + \frac{1}{2n} \right)$

   $= \int_0^1 \frac{1}{1+x} \, dx = \left[\ln(1+x) \right]_0^1 = \ln 2$

33. Let $f(x) = x^5$ on $[0, 1]$. Partition $[0, 1]$ into n subintervals with $\Delta x = \frac{1-0}{n} = \frac{1}{n}$. Then $\frac{1}{n}, \frac{2}{n}, \dots, \frac{n}{n}$ are the

   right-hand endpoints of the subintervals. Since f is increasing on $[0, 1]$, $U = \sum\limits_{j=1}^\infty \left(\frac{i}{n} \right)^5 \left(\frac{1}{n} \right)$ is the upper sum for

   $f(x) = x^5$ on $[0, 1] \Rightarrow \lim\limits_{n \to \infty} \sum\limits_{j=1}^\infty \left(\frac{i}{n} \right)^5 \left(\frac{1}{n} \right) = \lim\limits_{n \to \infty} \frac{1}{n} \left[\left(\frac{1}{n} \right)^5 + \left(\frac{2}{n} \right)^5 + \dots + \left(\frac{n}{n} \right)^5 \right] = \lim\limits_{n \to \infty} \left[\frac{1^5 + 2^5 + \dots + n^5}{n^6} \right]$

   $= \int_0^1 x^5 \, dx = \left[\frac{x^6}{6} \right]_0^1 = \frac{1}{6}$

35. Let $y = f(x)$ on $[0, 1]$. Partition $[0, 1]$ into n subintervals with $\Delta x = \frac{1-0}{n} = \frac{1}{n}$. Then $\frac{1}{n}, \frac{2}{n}, \dots, \frac{n}{n}$ are the

   right-hand endpoints of the subintervals. Since f is continuous on $[0, 1]$, $\sum\limits_{j=1}^\infty f \left(\frac{i}{n} \right) \left(\frac{1}{n} \right)$ is a Riemann sum of

   $y = f(x)$ on $[0, 1] \Rightarrow \lim\limits_{n \to \infty} \sum\limits_{j=1}^\infty f \left(\frac{i}{n} \right) \left(\frac{1}{n} \right) = \lim\limits_{n \to \infty} \frac{1}{n} \left[f \left(\frac{1}{n} \right) + f \left(\frac{2}{n} \right) + \dots + f \left(\frac{n}{n} \right) \right] = \int_0^1 f(x) \, dx$

37. (a) Let the polygon be inscribed in a circle of radius r. If we draw a radius from the center of the circle (and the polygon) to each vertex of the polygon, we have n isosceles triangles formed (the equal sides are equal to r, the radius of the circle) and a vertex angle of θ_n where $\theta_n = \frac{2\pi}{n}$. The area of each triangle is $A_n = \frac{1}{2} r^2 \sin \theta_n \Rightarrow$ the area of the polygon is $A = nA_n = \frac{nr^2}{2} \sin \theta_n = \frac{nr^2}{2} \sin \frac{2\pi}{n}$.

(b) $\lim\limits_{n \to \infty} A = \lim\limits_{n \to \infty} \frac{nr^2}{2} \sin \frac{2\pi}{n} = \lim\limits_{n \to \infty} \frac{n\pi r^2}{2\pi} \sin \frac{2\pi}{n} = \lim\limits_{n \to \infty} (\pi r^2) \frac{\sin\left(\frac{2\pi}{n}\right)}{\left(\frac{2\pi}{n}\right)} = (\pi r^2) \lim\limits_{2\pi/n \to 0} \frac{\sin\left(\frac{2\pi}{n}\right)}{\left(\frac{2\pi}{n}\right)} = \pi r^2$

39. $A_1 = \int_1^e \frac{2\log_2 x}{x} dx = \frac{2}{\ln 2} \int_1^e \frac{\ln x}{x} dx = \left[\frac{(\ln x)^2}{\ln 2} \right]_1^e = \frac{1}{\ln 2}$; $A_2 = \int_1^e \frac{2\log_4 x}{4} dx = \frac{2}{\ln 4} \int_1^e \frac{\ln x}{x} dx$

$= \left[\frac{(\ln x)^2}{2\ln 2} \right]_1^e = \frac{1}{2\ln 2} \Rightarrow A_1 : A_2 = 2 : 1$

41. $f(x) = e^{g(x)} \Rightarrow f'(x) = e^{g(x)} g'(x)$, where $g'(x) = \frac{x}{1+x^4} \Rightarrow f'(2) = e^0 \left(\frac{2}{1+16} \right) = \frac{2}{17}$

43. (a) $g(1) = \int_1^1 f(t) \, dt = 0$

(b) $g(3) = \int_1^3 f(t) \, dt = -\frac{1}{2}(2)(1) = -1$

(c) $g(-1) = \int_1^{-1} f(t) \, dt = -\int_{-1}^1 f(t) \, dt = -\frac{1}{4}(\pi \, 2^2) = -\pi$

(d) $g'(x) = f(x) = 0 \Rightarrow x = -3, 1, 3$ and the sign chart for $g'(x) = f(x)$ is $\underset{-3}{|} +++ \underset{1}{|} --- \underset{3}{|} +++$. So g has a relative maximum at $x = 1$.

(e) $g'(-1) = f(-1) = 2$ is the slope and $g(-1) = \int_1^{-1} f(t) \, dt = -\pi$, by (c). Thus the equation is $y + \pi = 2(x + 1)$ $y = 2x + 2 - \pi$.

(f) $g''(x) = f'(x) = 0$ at $x = -1$ and $g''(x) = f'(x)$ is negative on $(-3, -1)$ and positive on $(-1, 1)$ so there is an inflection point for g at $x = -1$. We notice that $g''(x) = f'(x) < 0$ for x on $(-1, 2)$ and $g''(x) = f'(x) > 0$ for x on $(2, 4)$, even though $g''(2)$ does not exist, g has a tangent line at $x = 2$, so there is an inflection point at $x = 2$.

(g) g is continuous on $[-3, 4]$ and so it attains its absolute maximum and minimum values on this interval. We saw in (d) that $g'(x) = 0 \Rightarrow x = -3, 1, 3$. We have that

$g(-3) = \int_1^{-3} f(t) \, dt = -\int_{-3}^1 f(t) \, dt = -\frac{\pi \, 2^2}{2} = -2\pi$

$g(1) = \int_1^1 f(t) \, dt = 0$

$g(3) = \int_1^3 f(t) \, dt = -1$

$g(4) = \int_1^4 f(t) \, dt = -1 + \frac{1}{2} \cdot 1 \cdot 1 = -\frac{1}{2}$

Thus, the absolute minimum is -2π and the absolute maximum is 0. Thus, the range is $[-2\pi, 0]$.

45. The area of the shaded region is $\int_0^1 \sin^{-1} x \, dx = \int_0^1 \sin^{-1} y \, dy$, which is the same as the area of the region to the left of the curve $y = \sin x$ (and part of the rectangle formed by the coordinate axes and dashed lines $y = 1$, $x = \frac{\pi}{2}$). The area of the rectangle is $\frac{\pi}{2} = \int_0^1 \sin^{-1} y \, dy + \int_0^{\pi/2} \sin x \, dx$, so we have

$\frac{\pi}{2} = \int_0^1 \sin^{-1} x \, dx + \int_0^{\pi/2} \sin x \, dx \Rightarrow \int_0^{\pi/2} \sin x \, dx = \frac{\pi}{2} - \int_0^1 \sin^{-1} x \, dx$.

47. $f(x) = \int_{1/x}^x \frac{1}{t} \, dt \Rightarrow f'(x) = \frac{1}{x}\left(\frac{dx}{dx}\right) - \left(\frac{1}{\frac{1}{x}}\right)\left(\frac{d}{dx}\left(\frac{1}{x}\right)\right) = \frac{1}{x} - x\left(-\frac{1}{x^2}\right) = \frac{1}{x} + \frac{1}{x} = \frac{2}{x}$

49. $g(y) = \int_{\sqrt{y}}^{2\sqrt{y}} \sin t^2 \, dt \Rightarrow g'(y) = \left(\sin \left(2\sqrt{y}\right)^2\right)\left(\frac{d}{dy}\left(2\sqrt{y}\right)\right) - \left(\sin \left(\sqrt{y}\right)^2\right)\left(\frac{d}{dy}\left(\sqrt{y}\right)\right) = \frac{\sin 4y}{\sqrt{y}} - \frac{\sin y}{2\sqrt{y}}$

51. $y = \int_{x^2/2}^{x^2} \ln \sqrt{t}\, dt \Rightarrow \frac{dy}{dx} = \left(\ln \sqrt{x^2}\right) \cdot \frac{d}{dx}\left(x^2\right) - \left(\ln \sqrt{\frac{x^2}{2}}\right) \cdot \frac{d}{dx}\left(\frac{x^2}{2}\right) = 2x \ln|x| - x \ln \frac{|x|}{\sqrt{2}}$

53. $\int_0^{\ln x} \sin e^t\, dt \Rightarrow y' = \left(\sin e^{\ln x}\right) \cdot \frac{d}{dx}(\ln x) = \frac{\sin x}{x}$

55. $f(x) = \int_x^{x+3} t(5 - t)\, dt \Rightarrow f'(x) = (x + 3)(5 - (x + 3))\left(\frac{d}{dx}(x + 3)\right) - x(5 - x)\left(\frac{dx}{dx}\right) = (x + 3)(2 - x) - x(5 - x)$

$= 6 - x - x^2 - 5x + x^2 = 6 - 6x.$ Thus $f'(x) = 0 \Rightarrow 6 - 6x = 0 \Rightarrow x = 1.$ Also, $f''(x) = -6 < 0 \Rightarrow x = 1$ gives a maximum.

NOTES:

CHAPTER 6 APPLICATIONS OF DEFINITE INTEGRALS

6.1 VOLUMES BY SLICING AND ROTATION ABOUT AN AXIS

1. $A(x) = \frac{(\text{diagonal})^2}{2} = \frac{\left(\sqrt{x} - (-\sqrt{x})\right)^2}{2} = 2x$ (see Exercise 1c); $a = 0, b = 4$;

   $V = \int_a^b A(x)\, dx = \int_0^4 2x\, dx = [x^2]_0^4 = 16$

3. $A(x) = (\text{edge})^2 = \left[\sqrt{1 - x^2} - \left(-\sqrt{1 - x^2}\right)\right]^2 = \left(2\sqrt{1 - x^2}\right)^2 = 4(1 - x^2)$; $a = -1, b = 1$;

   $V = \int_a^b A(x)\, dx = \int_{-1}^1 4(1 - x^2)\, dx = 4\left[x - \frac{x^3}{3}\right]_{-1}^1 = 8\left(1 - \frac{1}{3}\right) = \frac{16}{3}$

5. (a) STEP 1) $A(x) = \frac{1}{2}(\text{side}) \cdot (\text{side}) \cdot \left(\sin \frac{\pi}{3}\right) = \frac{1}{2} \cdot \left(2\sqrt{\sin x}\right) \cdot \left(2\sqrt{\sin x}\right)\left(\sin \frac{\pi}{3}\right) = \sqrt{3}\sin x$

   STEP 2) $a = 0, b = \pi$

   STEP 3) $V = \int_a^b A(x)\, dx = \sqrt{3}\int_0^\pi \sin x\, dx = \left[-\sqrt{3}\cos x\right]_0^\pi = \sqrt{3}(1 + 1) = 2\sqrt{3}$

   (b) STEP 1) $A(x) = (\text{side})^2 = \left(2\sqrt{\sin x}\right)\left(2\sqrt{\sin x}\right) = 4\sin x$

   STEP 2) $a = 0, b = \pi$

   STEP 3) $V = \int_a^b A(x)\, dx = \int_0^\pi 4\sin x\, dx = [-4\cos x]_0^\pi = 8$

7. $A(y) = \frac{\pi}{4}(\text{diameter})^2 = \frac{\pi}{4}\left(\sqrt{5}y^2 - 0\right)^2 = \frac{5\pi}{4}y^4$;

   $c = 0, d = 2; V = \int_c^d A(y)\, dy = \int_0^2 \frac{5\pi}{4}y^4\, dy$

   $= \left[\left(\frac{5\pi}{4}\right)\left(\frac{y^5}{5}\right)\right]_0^2 = \frac{\pi}{4}(2^5 - 0) = 8\pi$

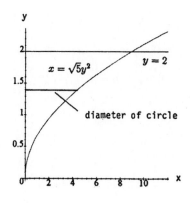

9. (a) $A(x) = \frac{\pi}{4}(\text{diameter})^2 = \frac{\pi}{4}\left[\frac{1}{\sqrt{1 + x^2}} - \left(-\frac{1}{\sqrt{1 + x^2}}\right)\right]^2 = \frac{\pi}{1 + x^2} \Rightarrow V = \int_a^b A(x)\, dx = \int_{-1}^1 \frac{\pi\, dx}{1 + x^2}$

   $= \pi\left[\tan^{-1} x\right]_{-1}^1 = (\pi)(2)\left(\frac{\pi}{4}\right) = \frac{\pi^2}{2}$

   (b) $A(x) = (\text{edge})^2 = \left[\frac{1}{\sqrt{1 + x^2}} - \left(-\frac{1}{\sqrt{1 + x^2}}\right)\right]^2 = \frac{4}{1 + x^2} \Rightarrow V = \int_a^b A(x)\, dx = \int_{-1}^1 \frac{4\, dx}{1 + x^2}$

   $= 4\left[\tan^{-1} x\right]_{-1}^1 = 4\left[\tan^{-1}(1) - \tan^{-1}(-1)\right] = 4\left[\frac{\pi}{4} - \left(-\frac{\pi}{4}\right)\right] = 2\pi$

11. (a) It follows from Cavalieri's Principle that the volume of a column is the same as the volume of a right prism with a square base of side length s and altitude h. Thus, STEP 1) $A(x) = (\text{side length})^2 = s^2$;

    STEP 2) $a = 0, b = h$; STEP 3) $V = \int_a^b A(x)\, dx = \int_0^h s^2\, dx = s^2 h$

    (b) From Cavalieri's Principle we conclude that the volume of the column is the same as the volume of the prism described above, regardless of the number of turns $\Rightarrow V = s^2 h$

13. $R(x) = y = 1 - \frac{x}{2}$ $\Rightarrow$ $V = \int_0^2 \pi[R(x)]^2 \, dx = \pi \int_0^2 \left(1 - \frac{x}{2}\right)^2 dx = \pi \int_0^2 \left(1 - x + \frac{x^2}{4}\right) dx = \pi \left[x - \frac{x^2}{2} + \frac{x^3}{12}\right]_0^2$

$= \pi \left(2 - \frac{4}{2} + \frac{8}{12}\right) = \frac{2\pi}{3}$

15. $R(x) = \tan\left(\frac{\pi}{4} y\right)$; $u = \frac{\pi}{4} y$ $\Rightarrow$ $du = \frac{\pi}{4} dy$ $\Rightarrow$ $4 \, du = \pi \, dy$; $y = 0$ $\Rightarrow$ $u = 0$, $y = 1$ $\Rightarrow$ $u = \frac{\pi}{4}$;

$V = \int_0^1 \pi[R(y)]^2 \, dy = \pi \int_0^1 \left[\tan\left(\frac{\pi}{4} y\right)\right]^2 dy = 4 \int_0^{\pi/4} \tan^2 u \, du = 4 \int_0^{\pi/4} (-1 + \sec^2 u) \, du = 4[-u + \tan u]_0^{\pi/4}$

$= 4\left(-\frac{\pi}{4} + 1 - 0\right) = 4 - \pi$

17. $R(x) = x^2$ $\Rightarrow$ $V = \int_0^2 \pi[R(x)]^2 \, dx = \pi \int_0^2 (x^2)^2 \, dx$

$= \pi \int_0^2 x^4 \, dx = \pi \left[\frac{x^5}{5}\right]_0^2 = \frac{32\pi}{5}$

19. $R(x) = \sqrt{9 - x^2}$ $\Rightarrow$ $V = \int_{-3}^3 \pi[R(x)]^2 \, dx = \pi \int_{-3}^3 (9 - x^2) \, dx$

$= \pi \left[9x - \frac{x^3}{3}\right]_{-3}^3 = 2\pi \left[9(3) - \frac{27}{3}\right] = 2 \cdot \pi \cdot 18 = 36\pi$

21. $R(x) = \sqrt{\cos x}$ $\Rightarrow$ $V = \int_0^{\pi/2} \pi[R(x)]^2 \, dx = \pi \int_0^{\pi/2} \cos x \, dx$

$= \pi [\sin x]_0^{\pi/2} = \pi(1 - 0) = \pi$

23. $R(x) = e^{-x}$ $\Rightarrow$ $V = \int_0^1 \pi[R(x)]^2 \, dx = \pi \int_0^1 (e^{-x})^2 \, dx$

$= \pi \int_0^1 e^{-2x} \, dx = -\frac{\pi}{2} e^{-2x} \Big|_0^1 = -\frac{\pi}{2}(e^{-2} - 1)$

$= \frac{\pi}{2}\left(1 - \frac{1}{e^2}\right) = \frac{\pi(e^2 - 1)}{2e^2}$

25. $V = \frac{\pi}{4} \int_{1/4}^4 \frac{1}{x} \, dx = \frac{\pi}{4} [\ln x]_{1/4}^4 = \frac{\pi}{4}\left(\ln 4 - \ln \frac{1}{4}\right) = \frac{\pi}{2} \ln 4$

27. $R(x) = \sqrt{2} - \sec x \tan x \Rightarrow V = \int_0^{\pi/4} \pi [R(x)]^2 \, dx$

$= \pi \int_0^{\pi/4} \left(\sqrt{2} - \sec x \tan x\right)^2 dx$

$= \pi \int_0^{\pi/4} \left(2 - 2\sqrt{2} \sec x \tan x + \sec^2 x \tan^2 x\right) dx$

$= \pi \left(\int_0^{\pi/4} 2 \, dx - 2\sqrt{2} \int_0^{\pi/4} \sec x \tan x \, dx + \int_0^{\pi/4} (\tan x)^2 \sec^2 x \, dx\right)$

$= \pi \left([2x]_0^{\pi/4} - 2\sqrt{2} [\sec x]_0^{\pi/4} + \left[\frac{\tan^3 x}{3}\right]_0^{\pi/4}\right)$

$= \pi \left[\left(\frac{\pi}{2} - 0\right) - 2\sqrt{2} \left(\sqrt{2} - 1\right) + \frac{1}{3}(1^3 - 0)\right] = \pi \left(\frac{\pi}{2} + 2\sqrt{2} - \frac{11}{3}\right)$

29. $R(y) = \sqrt{5} \cdot y^2 \Rightarrow V = \int_{-1}^1 \pi [R(y)]^2 \, dy = \pi \int_{-1}^1 5y^4 \, dy$

$= \pi [y^5]_{-1}^1 = \pi[1 - (-1)] = 2\pi$

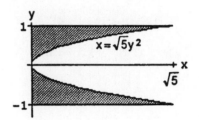

31. $R(y) = \sqrt{2 \sin 2y} \Rightarrow V = \int_0^{\pi/2} \pi [R(y)]^2 \, dy$

$= \pi \int_0^{\pi/2} 2 \sin 2y \, dy = \pi [-\cos 2y]_0^{\pi/2}$

$= \pi[1 - (-1)] = 2\pi$

33. $R(y) = \frac{2}{\sqrt{y+1}} \Rightarrow V = \int_0^3 \pi [R(y)]^2 \, dy = 4\pi \int_0^3 \frac{1}{y+1} \, dy$

$= 4\pi [\ln|y + 1|]_0^3 = 4\pi[\ln 4 - \ln 1] = 4\pi \ln 4$

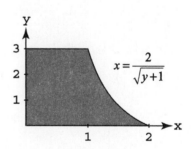

35. For the sketch given, $a = -\frac{\pi}{2}$, $b = \frac{\pi}{2}$; $R(x) = 1$, $r(x) = \sqrt{\cos x}$; $V = \int_a^b \pi \left([R(x)]^2 - [r(x)]^2\right) dx$

$= \int_{-\pi/2}^{\pi/2} \pi(1 - \cos x) \, dx = 2\pi \int_0^{\pi/2} (1 - \cos x) \, dx = 2\pi[x - \sin x]_0^{\pi/2} = 2\pi \left(\frac{\pi}{2} - 1\right) = \pi^2 - 2\pi$

37. $r(x) = x$ and $R(x) = 1 \Rightarrow V = \int_0^1 \pi \left([R(x)]^2 - [r(x)]^2\right) dx$

$= \int_0^1 \pi (1 - x^2) \, dx = \pi \left[x - \frac{x^3}{3}\right]_0^1 = \pi \left[\left(1 - \frac{1}{3}\right) - 0\right] = \frac{2\pi}{3}$

39. $r(x) = x^2 + 1$ and $R(x) = x + 3$

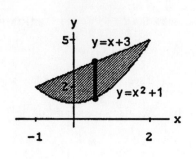

$\Rightarrow V = \int_{-1}^{2} \pi \left([R(x)]^2 - [r(x)]^2\right) dx$

$= \pi \int_{-1}^{2} \left[(x+3)^2 - \left(x^2 + 1\right)^2\right] dx$

$= \pi \int_{-1}^{2} \left[(x^2 + 6x + 9) - (x^4 + 2x^2 + 1)\right] dx$

$= \pi \int_{-1}^{2} \left(-x^4 - x^2 + 6x + 8\right) dx$

$= \pi \left[-\frac{x^5}{5} - \frac{x^3}{3} + \frac{6x^2}{2} + 8x\right]_{-1}^{2}$

$= \pi \left[\left(-\frac{32}{5} - \frac{8}{3} + \frac{24}{2} + 16\right) - \left(\frac{1}{5} + \frac{1}{3} + \frac{6}{2} - 8\right)\right] = \pi \left(-\frac{33}{5} - 3 + 28 - 3 + 8\right) = \pi \left(\frac{5 \cdot 30 - 33}{5}\right) = \frac{117\pi}{5}$

41. $r(x) = \sec x$ and $R(x) = \sqrt{2}$

$\Rightarrow V = \int_{-\pi/4}^{\pi/4} \pi \left([R(x)]^2 - [r(x)]^2\right) dx$

$= \pi \int_{-\pi/4}^{\pi/4} (2 - \sec^2 x) \, dx = \pi [2x - \tan x]_{-\pi/4}^{\pi/4}$

$= \pi \left[\left(\frac{\pi}{2} - 1\right) - \left(-\frac{\pi}{2} + 1\right)\right] = \pi(\pi - 2)$

43. $r(y) = 1$ and $R(y) = 1 + y$

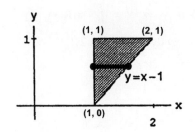

$\Rightarrow V = \int_{0}^{1} \pi \left([R(y)]^2 - [r(y)]^2\right) dy$

$= \pi \int_{0}^{1} \left[(1+y)^2 - 1\right] dy = \pi \int_{0}^{1} \left(1 + 2y + y^2 - 1\right) dy$

$= \pi \int_{0}^{1} \left(2y + y^2\right) dy = \pi \left[y^2 + \frac{y^3}{3}\right]_{0}^{1} = \pi \left(1 + \frac{1}{3}\right) = \frac{4\pi}{3}$

45. $R(y) = 2$ and $r(y) = \sqrt{y}$

$\Rightarrow V = \int_{0}^{4} \pi \left([R(y)]^2 - [r(y)]^2\right) dy$

$= \pi \int_{0}^{4} (4 - y) \, dy = \pi \left[4y - \frac{y^2}{2}\right]_{0}^{4} = \pi(16 - 8) = 8\pi$

47. $R(y) = 2$ and $r(y) = 1 + \sqrt{y}$

$\Rightarrow V = \int_{0}^{1} \pi \left([R(y)]^2 - [r(y)]^2\right) dy$

$= \pi \int_{0}^{1} \left[4 - \left(1 + \sqrt{y}\right)^2\right] dy$

$= \pi \int_{0}^{1} \left(4 - 1 - 2\sqrt{y} - y\right) dy$

$= \pi \int_{0}^{1} \left(3 - 2\sqrt{y} - y\right) dy$

$= \pi \left[3y - \frac{4}{3} y^{3/2} - \frac{y^2}{2}\right]_{0}^{1}$

$= \pi \left(3 - \frac{4}{3} - \frac{1}{2}\right) = \pi \left(\frac{18 - 8 - 3}{6}\right) = \frac{7\pi}{6}$

49. (a) $r(x) = \sqrt{x}$ and $R(x) = 2$

$\Rightarrow V = \int_0^4 \pi \left([R(x)]^2 - [r(x)]^2\right) dx$

$= \pi \int_0^4 (4 - x) \, dx = \pi \left[4x - \frac{x^2}{2}\right]_0^4 = \pi(16 - 8) = 8\pi$

(b) $r(y) = 0$ and $R(y) = y^2$

$\Rightarrow V = \int_0^2 \pi \left([R(y)]^2 - [r(y)]^2\right) dy$

$= \pi \int_0^2 y^4 \, dy = \pi \left[\frac{y^5}{5}\right]_0^2 = \frac{32\pi}{5}$

(c) $r(x) = 0$ and $R(x) = 2 - \sqrt{x} \Rightarrow V = \int_0^4 \pi \left([R(x)]^2 - [r(x)]^2\right) dx = \pi \int_0^4 \left(2 - \sqrt{x}\right)^2 dx$

$= \pi \int_0^4 \left(4 - 4\sqrt{x} + x\right) dx = \pi \left[4x - \frac{8x^{3/2}}{3} + \frac{x^2}{2}\right]_0^4 = \pi \left(16 - \frac{64}{3} + \frac{16}{2}\right) = \frac{8\pi}{3}$

(d) $r(y) = 4 - y^2$ and $R(y) = 4 \Rightarrow V = \int_0^2 \pi \left([R(y)]^2 - [r(y)]^2\right) dy = \pi \int_0^2 \left[16 - (4 - y^2)^2\right] dy$

$= \pi \int_0^2 \left(16 - 16 + 8y^2 - y^4\right) dy = \pi \int_0^2 \left(8y^2 - y^4\right) dy = \pi \left[\frac{8}{3} y^3 - \frac{y^5}{5}\right]_0^2 = \pi \left(\frac{64}{3} - \frac{32}{5}\right) = \frac{224\pi}{15}$

51. (a) $r(x) = 0$ and $R(x) = 1 - x^2$

$\Rightarrow V = \int_{-1}^1 \pi \left([R(x)]^2 - [r(x)]^2\right) dx$

$= \pi \int_{-1}^1 \left(1 - x^2\right)^2 dx = \pi \int_{-1}^1 \left(1 - 2x^2 + x^4\right) dx$

$= \pi \left[x - \frac{2x^3}{3} + \frac{x^5}{5}\right]_{-1}^1 = 2\pi \left(1 - \frac{2}{3} + \frac{1}{5}\right)$

$= 2\pi \left(\frac{15 - 10 + 3}{15}\right) = \frac{16\pi}{15}$

(b) $r(x) = 1$ and $R(x) = 2 - x^2 \Rightarrow V = \int_{-1}^1 \pi \left([R(x)]^2 - [r(x)]^2\right) dx = \pi \int_{-1}^1 \left[(2 - x^2)^2 - 1\right] dx$

$= \pi \int_{-1}^1 \left(4 - 4x^2 + x^4 - 1\right) dx = \pi \int_{-1}^1 \left(3 - 4x^2 + x^4\right) dx = \pi \left[3x - \frac{4}{3} x^3 + \frac{x^5}{5}\right]_{-1}^1 = 2\pi \left(3 - \frac{4}{3} + \frac{1}{5}\right)$

$= \frac{2\pi}{15} (45 - 20 + 3) = \frac{56\pi}{15}$

(c) $r(x) = 1 + x^2$ and $R(x) = 2 \Rightarrow V = \int_{-1}^1 \pi \left([R(x)]^2 - [r(x)]^2\right) dx = \pi \int_{-1}^1 \left[4 - (1 + x^2)^2\right] dx$

$= \pi \int_{-1}^1 \left(4 - 1 - 2x^2 - x^4\right) dx = \pi \int_{-1}^1 \left(3 - 2x^2 - x^4\right) dx = \pi \left[3x - \frac{2}{3} x^3 - \frac{x^5}{5}\right]_{-1}^1 = 2\pi \left(3 - \frac{2}{3} - \frac{1}{5}\right)$

$= \frac{2\pi}{15} (45 - 10 - 3) = \frac{64\pi}{15}$

6.2 VOLUME BY CYLINDRICAL SHELLS

1. For the sketch given, $a = 0$, $b = 2$;

$V = \int_a^b 2\pi \left(\begin{smallmatrix}\text{shell}\\\text{radius}\end{smallmatrix}\right) \left(\begin{smallmatrix}\text{shell}\\\text{height}\end{smallmatrix}\right) dx = \int_0^2 2\pi x \left(1 + \frac{x^2}{4}\right) dx = 2\pi \int_0^2 \left(x + \frac{x^3}{4}\right) dx = 2\pi \left[\frac{x^2}{2} + \frac{x^4}{16}\right]_0^2 = 2\pi \left(\frac{4}{2} + \frac{16}{16}\right)$

$= 2\pi \cdot 3 = 6\pi$

3. For the sketch given, $c = 0$, $d = \sqrt{2}$;

$V = \int_c^d 2\pi \left(\begin{smallmatrix}\text{shell}\\\text{radius}\end{smallmatrix}\right) \left(\begin{smallmatrix}\text{shell}\\\text{height}\end{smallmatrix}\right) dy = \int_0^{\sqrt{2}} 2\pi y \cdot (y^2) \, dy = 2\pi \int_0^{\sqrt{2}} y^3 \, dy = 2\pi \left[\frac{y^4}{4}\right]_0^{\sqrt{2}} = 2\pi$

5. For the sketch given, $a = 0$, $b = \sqrt{3}$;

$V = \int_a^b 2\pi \left(\begin{smallmatrix}\text{shell}\\\text{radius}\end{smallmatrix}\right) \left(\begin{smallmatrix}\text{shell}\\\text{height}\end{smallmatrix}\right) dx = \int_0^{\sqrt{3}} 2\pi x \cdot \left(\sqrt{x^2 + 1}\right) dx;$

$\left[u = x^2 + 1 \Rightarrow du = 2x \, dx; \, x = 0 \Rightarrow u = 1, \, x = \sqrt{3} \Rightarrow u = 4\right]$

$\rightarrow V = \pi \int_1^4 u^{1/2} \, du = \pi \left[\frac{2}{3} u^{3/2}\right]_1^4 = \frac{2\pi}{3} \left(4^{3/2} - 1\right) = \left(\frac{2\pi}{3}\right)(8 - 1) = \frac{14\pi}{3}$

7. $a = 0, b = 2$;

$$V = \int_a^b 2\pi \left(\begin{smallmatrix} \text{shell} \\ \text{radius} \end{smallmatrix}\right) \left(\begin{smallmatrix} \text{shell} \\ \text{height} \end{smallmatrix}\right) dx = \int_0^2 2\pi x \left[x - \left(-\frac{x}{2}\right)\right] dx$$

$$= \int_0^2 2\pi x^2 \cdot \frac{3}{2} \, dx = \pi \int_0^2 3x^2 \, dx = \pi \left[x^3\right]_0^2 = 8\pi$$

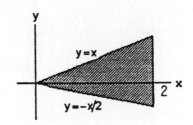

9. $a = 0, b = 1$;

$$V = \int_a^b 2\pi \left(\begin{smallmatrix} \text{shell} \\ \text{radius} \end{smallmatrix}\right) \left(\begin{smallmatrix} \text{shell} \\ \text{height} \end{smallmatrix}\right) dx = \int_0^1 2\pi x \left[(2 - x) - x^2\right] dx$$

$$= 2\pi \int_0^1 (2x - x^2 - x^3) \, dx = 2\pi \left[x^2 - \frac{x^3}{3} - \frac{x^4}{4}\right]_0^1$$

$$= 2\pi \left(1 - \frac{1}{3} - \frac{1}{4}\right) = 2\pi \left(\frac{12 - 4 - 3}{12}\right) = \frac{10\pi}{12} = \frac{5\pi}{6}$$

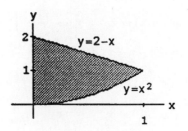

11. $a = 0, b = 1$;

$$V = \int_a^b 2\pi \left(\begin{smallmatrix} \text{shell} \\ \text{radius} \end{smallmatrix}\right) \left(\begin{smallmatrix} \text{shell} \\ \text{height} \end{smallmatrix}\right) dx = \int_0^1 2\pi x \left[\sqrt{x} - (2x - 1)\right] dx$$

$$= 2\pi \int_0^1 \left(x^{3/2} - 2x^2 + x\right) dx = 2\pi \left[\frac{2}{5} x^{5/2} - \frac{2}{3} x^3 + \frac{1}{2} x^2\right]_0^1$$

$$= 2\pi \left(\frac{2}{5} - \frac{2}{3} + \frac{1}{2}\right) = 2\pi \left(\frac{12 - 20 + 15}{30}\right) = \frac{7\pi}{15}$$

13. (a) $xf(x) = \begin{cases} x \cdot \frac{\sin x}{x}, & 0 < x \le \pi \\ x, & x = 0 \end{cases} \Rightarrow xf(x) = \begin{cases} \sin x, & 0 < x \le \pi \\ 0, & x = 0 \end{cases}$; since $\sin 0 = 0$ we have

$$xf(x) = \begin{cases} \sin x, & 0 < x \le \pi \\ \sin x, & x = 0 \end{cases} \Rightarrow xf(x) = \sin x, 0 \le x \le \pi$$

(b) $V = \int_a^b 2\pi \left(\begin{smallmatrix} \text{shell} \\ \text{radius} \end{smallmatrix}\right) \left(\begin{smallmatrix} \text{shell} \\ \text{height} \end{smallmatrix}\right) dx = \int_0^\pi 2\pi x \cdot f(x) \, dx$ and $x \cdot f(x) = \sin x, 0 \le x \le \pi$ by part (a)

$$\Rightarrow V = 2\pi \int_0^\pi \sin x \, dx = 2\pi [-\cos x]_0^\pi = 2\pi(-\cos \pi + \cos 0) = 4\pi$$

15. $c = 0, d = 2$;

$$V = \int_c^d 2\pi \left(\begin{smallmatrix} \text{shell} \\ \text{radius} \end{smallmatrix}\right) \left(\begin{smallmatrix} \text{shell} \\ \text{height} \end{smallmatrix}\right) dy = \int_0^2 2\pi y \left[\sqrt{y} - (-y)\right] dy$$

$$= 2\pi \int_0^2 \left(y^{3/2} + y^2\right) dy = 2\pi \left[\frac{2y^{5/2}}{5} + \frac{y^3}{3}\right]_0^2$$

$$= 2\pi \left[\frac{2}{5} \left(\sqrt{2}\right)^5 + \frac{2^3}{3}\right] = 2\pi \left(\frac{8\sqrt{2}}{5} + \frac{8}{3}\right) = 16\pi \left(\frac{\sqrt{2}}{5} + \frac{1}{3}\right)$$

$$= \frac{16\pi}{15} \left(3\sqrt{2} + 5\right)$$

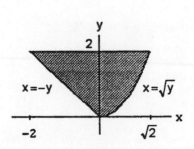

17. $c = 0, d = 2$;

$V = \int_c^d 2\pi \left(\begin{smallmatrix}\text{shell}\\\text{radius}\end{smallmatrix}\right) \left(\begin{smallmatrix}\text{shell}\\\text{height}\end{smallmatrix}\right) dy = \int_0^2 2\pi y \, (2y - y^2) dy$

$= 2\pi \int_0^2 (2y^2 - y^3) \, dy = 2\pi \left[\frac{2y^3}{3} - \frac{y^4}{4} \right]_0^2 = 2\pi \left(\frac{16}{3} - \frac{16}{4} \right)$

$= 32\pi \left(\frac{1}{3} - \frac{1}{4} \right) = \frac{32\pi}{12} = \frac{8\pi}{3}$

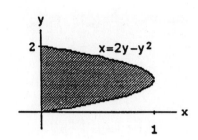

19. $c = 0, d = 1$;

$V = \int_c^d 2\pi \left(\begin{smallmatrix}\text{shell}\\\text{radius}\end{smallmatrix}\right) \left(\begin{smallmatrix}\text{shell}\\\text{height}\end{smallmatrix}\right) dy = 2\pi \int_0^1 y[y - (-y)] dy$

$= 2\pi \int_0^1 2y^2 \, dy = \frac{4\pi}{3} [y^3]_0^1 = \frac{4\pi}{3}$

21. $c = 0, d = 2$;

$V = \int_c^d 2\pi \left(\begin{smallmatrix}\text{shell}\\\text{radius}\end{smallmatrix}\right) \left(\begin{smallmatrix}\text{shell}\\\text{height}\end{smallmatrix}\right) dy = \int_0^2 2\pi y \left[(2 + y) - y^2 \right] dy$

$= 2\pi \int_0^2 (2y + y^2 - y^3) \, dy = 2\pi \left[y^2 + \frac{y^3}{3} - \frac{y^4}{4} \right]_0^2$

$= 2\pi \left(4 + \frac{8}{3} - \frac{16}{4} \right) = \frac{\pi}{6} (48 + 32 - 48) = \frac{16\pi}{3}$

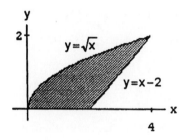

23. (a) $V = \int_c^d 2\pi \left(\begin{smallmatrix}\text{shell}\\\text{radius}\end{smallmatrix}\right) \left(\begin{smallmatrix}\text{shell}\\\text{height}\end{smallmatrix}\right) dy = \int_0^1 2\pi y \cdot 12 \, (y^2 - y^3) \, dy = 24\pi \int_0^1 (y^3 - y^4) \, dy = 24\pi \left[\frac{y^4}{4} - \frac{y^5}{5} \right]_0^1$

$= 24\pi \left(\frac{1}{4} - \frac{1}{5} \right) = \frac{24\pi}{20} = \frac{6\pi}{5}$

(b) $V = \int_c^d 2\pi \left(\begin{smallmatrix}\text{shell}\\\text{radius}\end{smallmatrix}\right) \left(\begin{smallmatrix}\text{shell}\\\text{height}\end{smallmatrix}\right) dy = \int_0^1 2\pi(1 - y) \left[12 \, (y^2 - y^3) \right] dy = 24\pi \int_0^1 (1 - y) (y^2 - y^3) \, dy$

$= 24\pi \int_0^1 (y^2 - 2y^3 + y^4) \, dy = 24\pi \left[\frac{y^3}{3} - \frac{y^4}{2} + \frac{y^5}{5} \right]_0^1 = 24\pi \left(\frac{1}{3} - \frac{1}{2} + \frac{1}{5} \right) = 24\pi \left(\frac{1}{30} \right) = \frac{4\pi}{5}$

(c) $V = \int_c^d 2\pi \left(\begin{smallmatrix}\text{shell}\\\text{radius}\end{smallmatrix}\right) \left(\begin{smallmatrix}\text{shell}\\\text{height}\end{smallmatrix}\right) dy = \int_0^1 2\pi \left(\frac{8}{5} - y \right) \left[12 \, (y^2 - y^3) \right] dy = 24\pi \int_0^1 \left(\frac{8}{5} - y \right) (y^2 - y^3) \, dy$

$= 24\pi \int_0^1 \left(\frac{8}{5} \, y^2 - \frac{13}{5} \, y^3 + y^4 \right) dy = 24\pi \left[\frac{8}{15} \, y^3 - \frac{13}{20} \, y^4 + \frac{y^5}{5} \right]_0^1 = 24\pi \left(\frac{8}{15} - \frac{13}{20} + \frac{1}{5} \right) = \frac{24\pi}{60} (32 - 39 + 12)$

$= \frac{24\pi}{12} = 2\pi$

(d) $V = \int_c^d 2\pi \left(\begin{smallmatrix}\text{shell}\\\text{radius}\end{smallmatrix}\right) \left(\begin{smallmatrix}\text{shell}\\\text{height}\end{smallmatrix}\right) dy = \int_0^1 2\pi \left(y + \frac{2}{5} \right) \left[12 \, (y^2 - y^3) \right] dy = 24\pi \int_0^1 \left(y + \frac{2}{5} \right) (y^2 - y^3) \, dy$

$= 24\pi \int_0^1 \left(y^3 - y^4 + \frac{2}{5} \, y^2 - \frac{2}{5} \, y^3 \right) dy = 24\pi \int_0^1 \left(\frac{2}{5} \, y^2 + \frac{3}{5} \, y^3 - y^4 \right) dy = 24\pi \left[\frac{2}{15} \, y^3 + \frac{3}{20} \, y^4 - \frac{y^5}{5} \right]_0^1$

$= 24\pi \left(\frac{2}{15} + \frac{3}{20} - \frac{1}{5} \right) = \frac{24\pi}{60} (8 + 9 - 12) = \frac{24\pi}{12} = 2\pi$

25. (a) About x-axis: $V = \int_c^d 2\pi \left(\begin{smallmatrix} \text{shell} \\ \text{radius} \end{smallmatrix}\right) \left(\begin{smallmatrix} \text{shell} \\ \text{height} \end{smallmatrix}\right) dy$

$= \int_0^1 2\pi y(\sqrt{y} - y) dy = 2\pi \int_0^1 (y^{3/2} - y^2) dy$

$= 2\pi \left[\frac{2}{5}y^{5/2} - \frac{1}{3}y^3\right]_0^1 = 2\pi \left(\frac{2}{5} - \frac{1}{3}\right) = \frac{2\pi}{15}$

About y-axis: $V = \int_a^b 2\pi \left(\begin{smallmatrix} \text{shell} \\ \text{radius} \end{smallmatrix}\right) \left(\begin{smallmatrix} \text{shell} \\ \text{height} \end{smallmatrix}\right) dx$

$= \int_0^1 2\pi x(x - x^2) dx = 2\pi \int_0^1 (x^2 - x^3) dx$

$= 2\pi \left[\frac{x^3}{3} - \frac{x^4}{4}\right]_0^1 = 2\pi \left(\frac{1}{3} - \frac{1}{4}\right) = \frac{\pi}{6}$

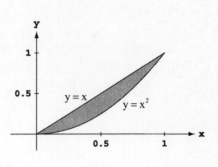

(b) About x-axis: $R(x) = x$ and $r(x) = x^2 \Rightarrow V = \int_a^b \pi\left[R(x)^2 - r(x)^2\right]dx = \int_0^1 \pi[x^2 - x^4]dx$

$= \pi\left[\frac{x^3}{3} - \frac{x^5}{5}\right]_0^1 = \pi\left(\frac{1}{3} - \frac{1}{5}\right) = \frac{2\pi}{15}$

About y-axis: $R(y) = \sqrt{y}$ and $r(y) = y \Rightarrow V = \int_c^d \pi\left[R(y)^2 - r(y)^2\right]dy = \int_0^1 \pi[y - y^2]dy$

$= \pi\left[\frac{y^2}{2} - \frac{y^3}{3}\right]_0^1 = \pi\left(\frac{1}{2} - \frac{1}{3}\right) = \frac{\pi}{6}$

27. (a) $V = \int_c^d 2\pi \left(\begin{smallmatrix} \text{shell} \\ \text{radius} \end{smallmatrix}\right) \left(\begin{smallmatrix} \text{shell} \\ \text{height} \end{smallmatrix}\right) dy = \int_1^2 2\pi y(y - 1) \, dy$

$= 2\pi \int_1^2 (y^2 - y) \, dy = 2\pi \left[\frac{y^3}{3} - \frac{y^2}{2}\right]_1^2$

$= 2\pi \left[\left(\frac{8}{3} - \frac{4}{2}\right) - \left(\frac{1}{3} - \frac{1}{2}\right)\right]$

$= 2\pi \left(\frac{7}{3} - 2 + \frac{1}{2}\right) = \frac{\pi}{3}(14 - 12 + 3) = \frac{5\pi}{3}$

(b) $V = \int_a^b 2\pi \left(\begin{smallmatrix} \text{shell} \\ \text{radius} \end{smallmatrix}\right) \left(\begin{smallmatrix} \text{shell} \\ \text{height} \end{smallmatrix}\right) dx = \int_1^2 2\pi x(2 - x) \, dx = 2\pi \int_1^2 (2x - x^2) \, dx = 2\pi \left[x^2 - \frac{x^3}{3}\right]_1^2$

$= 2\pi \left[\left(4 - \frac{8}{3}\right) - \left(1 - \frac{1}{3}\right)\right] = 2\pi \left[\left(\frac{12-8}{3}\right) - \left(\frac{3-1}{3}\right)\right] = 2\pi \left(\frac{4}{3} - \frac{2}{3}\right) = \frac{4\pi}{3}$

(c) $V = \int_a^b 2\pi \left(\begin{smallmatrix} \text{shell} \\ \text{radius} \end{smallmatrix}\right) \left(\begin{smallmatrix} \text{shell} \\ \text{height} \end{smallmatrix}\right) dx = \int_1^2 2\pi \left(\frac{10}{3} - x\right)(2 - x) \, dx = 2\pi \int_1^2 \left(\frac{20}{3} - \frac{16}{3}x + x^2\right) dx$

$= 2\pi \left[\frac{20}{3}x - \frac{8}{3}x^2 + \frac{1}{3}x^3\right]_1^2 = 2\pi \left[\left(\frac{40}{3} - \frac{32}{3} + \frac{8}{3}\right) - \left(\frac{20}{3} - \frac{8}{3} + \frac{1}{3}\right)\right] = 2\pi \left(\frac{3}{3}\right) = 2\pi$

(d) $V = \int_c^d 2\pi \left(\begin{smallmatrix} \text{shell} \\ \text{radius} \end{smallmatrix}\right) \left(\begin{smallmatrix} \text{shell} \\ \text{height} \end{smallmatrix}\right) dy = \int_1^2 2\pi(y - 1)(y - 1) \, dy = 2\pi\int_1^2 (y - 1)^2 = 2\pi \left[\frac{(y-1)^3}{3}\right]_1^2 = \frac{2\pi}{3}$

29. (a) $V = \int_c^d 2\pi \left(\begin{smallmatrix} \text{shell} \\ \text{radius} \end{smallmatrix}\right) \left(\begin{smallmatrix} \text{shell} \\ \text{height} \end{smallmatrix}\right) dy = \int_0^1 2\pi y(y - y^3) \, dy$

$= \int_0^1 2\pi (y^2 - y^4) \, dy = 2\pi \left[\frac{y^3}{3} - \frac{y^5}{5}\right]_0^1 = 2\pi \left(\frac{1}{3} - \frac{1}{5}\right)$

$= \frac{4\pi}{15}$

(b) $V = \int_c^d 2\pi \left(\begin{smallmatrix} \text{shell} \\ \text{radius} \end{smallmatrix}\right) \left(\begin{smallmatrix} \text{shell} \\ \text{height} \end{smallmatrix}\right) dy$

$= \int_0^1 2\pi(1 - y)(y - y^3) \, dy$

$= 2\pi \int_0^1 (y - y^2 - y^3 + y^4) \, dy = 2\pi \left[\frac{y^2}{2} - \frac{y^3}{3} - \frac{y^4}{4} + \frac{y^5}{5}\right]_0^1 = 2\pi \left(\frac{1}{2} - \frac{1}{3} - \frac{1}{4} + \frac{1}{5}\right) = \frac{2\pi}{60}(30 - 20 - 15 + 12) = \frac{7\pi}{30}$

31. (a) $V = \int_c^d 2\pi \left(\begin{smallmatrix}\text{shell}\\\text{radius}\end{smallmatrix}\right)\left(\begin{smallmatrix}\text{shell}\\\text{height}\end{smallmatrix}\right) dy = \int_0^2 2\pi y \left(\sqrt{8y} - y^2\right) dy$

$\qquad = 2\pi \int_0^2 \left(2\sqrt{2}\, y^{3/2} - y^3\right) dy = 2\pi \left[\frac{4\sqrt{2}}{5} y^{5/2} - \frac{y^4}{4}\right]_0^2$

$\qquad = 2\pi \left(\frac{4\sqrt{2}\cdot(\sqrt{2})^5}{5} - \frac{2^4}{4}\right) = 2\pi\left(\frac{4\cdot 2^3}{5} - \frac{4\cdot 4}{4}\right)$

$\qquad = 2\pi \cdot 4\left(\frac{8}{5} - 1\right) = \frac{8\pi}{5}(8 - 5) = \frac{24\pi}{5}$

(b) $V = \int_a^b 2\pi \left(\begin{smallmatrix}\text{shell}\\\text{radius}\end{smallmatrix}\right)\left(\begin{smallmatrix}\text{shell}\\\text{height}\end{smallmatrix}\right) dx = \int_0^4 2\pi x \left(\sqrt{x} - \frac{x^2}{8}\right) dx = 2\pi \int_0^4 \left(x^{3/2} - \frac{x^3}{8}\right) dx = 2\pi\left[\frac{2}{5}x^{5/2} - \frac{x^4}{32}\right]_0^4$

$\qquad = 2\pi\left(\frac{2\cdot 2^5}{5} - \frac{4^4}{32}\right) = 2\pi\left(\frac{2^6}{5} - \frac{2^8}{32}\right) = \frac{\pi\cdot 2^7}{160}(32 - 20) = \frac{\pi\cdot 2^9\cdot 3}{160} = \frac{\pi\cdot 2^4\cdot 3}{5} = \frac{48\pi}{5}$

33. (a) $V = \int_a^b \pi\left[R^2(x) - r^2(x)\right] dx = \pi \int_{1/16}^1 \left(x^{-1/2} - 1\right) dx$

$\qquad = \pi\left[2x^{1/2} - x\right]_{1/16}^1 = \pi\left[(2 - 1) - \left(2\cdot\frac{1}{4} - \frac{1}{16}\right)\right]$

$\qquad = \pi\left(1 - \frac{7}{16}\right) = \frac{9\pi}{16}$

(b) $V = \int_a^b 2\pi \left(\begin{smallmatrix}\text{shell}\\\text{radius}\end{smallmatrix}\right)\left(\begin{smallmatrix}\text{shell}\\\text{height}\end{smallmatrix}\right) dy = \int_1^2 2\pi y\left(\frac{1}{y^4} - \frac{1}{16}\right) dy$

$\qquad = 2\pi\int_1^2 \left(y^{-3} - \frac{y}{16}\right) dy = 2\pi\left[-\frac{1}{2}y^{-2} - \frac{y^2}{32}\right]_1^2$

$\qquad = 2\pi\left[\left(-\frac{1}{8} - \frac{1}{8}\right) - \left(-\frac{1}{2} - \frac{1}{32}\right)\right] = 2\pi\left(\frac{1}{4} + \frac{1}{32}\right)$

$\qquad = \frac{2\pi}{32}(8 + 1) = \frac{9\pi}{16}$

35. (a) *Disk*: $V = V_1 - V_2$

$\qquad V_1 = \int_{a_1}^{b_1} \pi[R_1(x)]^2\, dx$ and $V_2 = \int_{a_2}^{b_2} \pi[R_2(x)]^2$ with $R_1(x) = \sqrt{\frac{x+2}{3}}$ and $R_2(x) = \sqrt{x}$,

$\qquad a_1 = -2, b_1 = 1; a_2 = 0, b_2 = 1 \Rightarrow$ two integrals are required

(b) *Washer*: $V = V_1 - V_2$

$\qquad V_1 = \int_{a_1}^{b_1} \pi\left([R_1(x)]^2 - [r_1(x)]^2\right) dx$ with $R_1(x) = \sqrt{\frac{x+2}{3}}$ and $r_1(x) = 0$; $a_1 = -2$ and $b_1 = 0$;

$\qquad V_2 = \int_{a_2}^{b_2} \pi\left([R_2(x)]^2 - [r_2(x)]^2\right) dx$ with $R_2(x) = \sqrt{\frac{x+2}{3}}$ and $r_2(x) = \sqrt{x}$; $a_2 = 0$ and $b_2 = 1$

$\qquad \Rightarrow$ two integrals are required

(c) *Shell*: $V = \int_c^d 2\pi \left(\begin{smallmatrix}\text{shell}\\\text{radius}\end{smallmatrix}\right)\left(\begin{smallmatrix}\text{shell}\\\text{height}\end{smallmatrix}\right) dy = \int_c^d 2\pi y\left(\begin{smallmatrix}\text{shell}\\\text{height}\end{smallmatrix}\right) dy$ where shell height $= y^2 - (3y^2 - 2) = 2 - 2y^2$;

$\qquad c = 0$ and $d = 1$. Only *one* integral is required. It is, therefore preferable to use the *shell* method.
$\qquad$ However, whichever method you use, you will get $V = \pi$.

37. $W(a) = \int_{f(a)}^{f(a)} \pi\left[\left(f^{-1}(y)\right)^2 - a^2\right] dy = 0 = \int_a^a 2\pi x[f(a) - f(x)]\, dx = S(a)$; $W'(t) = \pi\left[\left(f^{-1}(f(t))\right)^2 - a^2\right] f'(t)$

$\qquad = \pi\left(t^2 - a^2\right) f'(t)$; also $S(t) = 2\pi f(t)\int_a^t x\, dx - 2\pi\int_a^t xf(x)\, dx = \left[\pi f(t)t^2 - \pi f(t)a^2\right] - 2\pi\int_a^t xf(x)\, dx$

$\qquad \Rightarrow S'(t) = \pi t^2 f'(t) + 2\pi t f(t) - \pi a^2 f'(t) - 2\pi t f(t) = \pi\left(t^2 - a^2\right) f'(t) \Rightarrow W'(t) = S'(t)$. Therefore, $W(t) = S(t)$

$\qquad$ for all $t \in [a, b]$.

39. $V = \int_a^b 2\pi \left(\begin{smallmatrix}\text{shell}\\\text{radius}\end{smallmatrix}\right)\left(\begin{smallmatrix}\text{shell}\\\text{height}\end{smallmatrix}\right) dx = \int_0^1 2\pi x\, e^{-x^2} dx = -\pi e^{-x^2}\Big|_0^1 = -\pi(e^{-1} - e^0) = \pi\left(1 - \frac{1}{e}\right)$

6.3 LENGTHS OF PLANE CURVES

1. $\frac{dx}{dt} = -1$ and $\frac{dy}{dt} = 3 \Rightarrow \sqrt{\left(\frac{dx}{dt}\right)^2 + \left(\frac{dy}{dt}\right)^2} = \sqrt{(-1)^2 + (3)^2} = \sqrt{10}$

$\Rightarrow$ Length $= \int_{-2/3}^{1} \sqrt{10}\, dt = \sqrt{10}\, [t]_{-2/3}^{1} = \sqrt{10} - \left(-\frac{2}{3}\sqrt{10}\right) = \frac{5\sqrt{10}}{3}$

3. $\frac{dx}{dt} = 3t^2$ and $\frac{dy}{dt} = 3t \Rightarrow \sqrt{\left(\frac{dx}{dt}\right)^2 + \left(\frac{dy}{dt}\right)^2} = \sqrt{(3t^2)^2 + (3t)^2} = \sqrt{9t^4 + 9t^2} = 3t\sqrt{t^2 + 1}$ $\left(\text{since } t \geq 0 \text{ on } \left[0, \sqrt{3}\right]\right)$

$\Rightarrow$ Length $= \int_{0}^{\sqrt{3}} 3t\sqrt{t^2 + 1}\, dt;\ \left[u = t^2 + 1 \Rightarrow \frac{3}{2}\, du = 3t\, dt;\ t = 0 \Rightarrow u = 1, t = \sqrt{3} \Rightarrow u = 4\right]$

$\to \int_{1}^{4} \frac{3}{2} u^{1/2}\, du = \left[u^{3/2}\right]_{1}^{4} = (8 - 1) = 7$

5. $\frac{dx}{dt} = (2t + 3)^{1/2}$ and $\frac{dy}{dt} = 1 + t \Rightarrow \sqrt{\left(\frac{dx}{dt}\right)^2 + \left(\frac{dy}{dt}\right)^2} = \sqrt{(2t + 3) + (1 + t)^2} = \sqrt{t^2 + 4t + 4} = |t + 2| = t + 2$

since $0 \leq t \leq 3 \Rightarrow$ Length $= \int_{0}^{3} (t + 2)\, dt = \left[\frac{t^2}{2} + 2t\right]_{0}^{3} = \frac{21}{2}$

7. $\frac{dx}{dt} = e^t - 1$ and $\frac{dy}{dt} = 2e^{t/2} \Rightarrow \sqrt{\left(\frac{dx}{dt}\right)^2 + \left(\frac{dy}{dt}\right)^2} = \sqrt{(e^t - 1)^2 + (2e^{t/2})^2} = \sqrt{e^{2t} + 2e^t + 1} = \sqrt{(e^t + 1)^2} = |e^t + 1|$

$= e^t + 1$ (since $e^t + 1 > 0$ for all t) $\Rightarrow L = \int_{0}^{3} (e^t + 1)\, dt = [e^t + t]_{0}^{3} = e^3 + 2$

9. $\frac{dy}{dx} = \frac{1}{3} \cdot \frac{3}{2} (x^2 + 2)^{1/2} \cdot 2x = \sqrt{(x^2 + 2)} \cdot x$

$\Rightarrow L = \int_{0}^{3} \sqrt{1 + (x^2 + 2) x^2}\, dx = \int_{0}^{3} \sqrt{1 + 2x^2 + x^4}\, dx$

$= \int_{0}^{3} \sqrt{(1 + x^2)^2}\, dx = \int_{0}^{3} (1 + x^2)\, dx = \left[x + \frac{x^3}{3}\right]_{0}^{3}$

$= 3 + \frac{27}{3} = 12$

11. $\frac{dx}{dy} = y^2 - \frac{1}{4y^2} \Rightarrow \left(\frac{dx}{dy}\right)^2 = y^4 - \frac{1}{2} + \frac{1}{16y^4}$

$\Rightarrow L = \int_{1}^{3} \sqrt{1 + y^4 - \frac{1}{2} + \frac{1}{16y^4}}\, dy$

$= \int_{1}^{3} \sqrt{y^4 + \frac{1}{2} + \frac{1}{16y^4}}\, dy$

$= \int_{1}^{3} \sqrt{\left(y^2 + \frac{1}{4y^2}\right)^2}\, dy = \int_{1}^{3} \left(y^2 + \frac{1}{4y^2}\right) dy$

$= \left[\frac{y^3}{3} - \frac{y^{-1}}{4}\right]_{1}^{3} = \left(\frac{27}{3} - \frac{1}{12}\right) - \left(\frac{1}{3} - \frac{1}{4}\right) = 9 - \frac{1}{12} - \frac{1}{3} + \frac{1}{4} = 9 + \frac{(-1 - 4 + 3)}{12} = 9 + \frac{(-2)}{12} = \frac{53}{6}$

13. $\frac{dx}{dy} = y^3 - \frac{1}{4y^3} \Rightarrow \left(\frac{dx}{dy}\right)^2 = y^6 - \frac{1}{2} + \frac{1}{16y^6}$

$\Rightarrow L = \int_{1}^{2} \sqrt{1 + y^6 - \frac{1}{2} + \frac{1}{16y^6}}\, dy$

$= \int_{1}^{2} \sqrt{y^6 + \frac{1}{2} + \frac{1}{16y^6}}\, dy = \int_{1}^{2} \sqrt{\left(y^3 + \frac{y^{-3}}{4}\right)^2}\, dy$

$= \int_{1}^{2} \left(y^3 + \frac{y^{-3}}{4}\right) dy = \left[\frac{y^4}{4} - \frac{y^{-2}}{8}\right]_{1}^{2}$

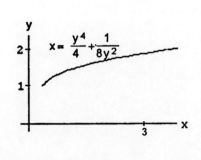

$= \left(\frac{16}{4} - \frac{1}{(16)(2)}\right) - \left(\frac{1}{4} - \frac{1}{8}\right) = 4 - \frac{1}{32} - \frac{1}{4} + \frac{1}{8} = \frac{128 - 1 - 8 + 4}{32} = \frac{123}{32}$

15. $\frac{dy}{dx} = x^{1/3} - \frac{1}{4}x^{-1/3} \Rightarrow \left(\frac{dy}{dx}\right)^2 = x^{2/3} - \frac{1}{2} + \frac{x^{-2/3}}{16}$

$\Rightarrow L = \int_1^8 \sqrt{1 + x^{2/3} - \frac{1}{2} + \frac{x^{-2/3}}{16}}\, dx$

$= \int_1^8 \sqrt{x^{2/3} + \frac{1}{2} + \frac{x^{-2/3}}{16}}\, dx$

$= \int_1^8 \sqrt{\left(x^{1/3} + \frac{1}{4}x^{-1/3}\right)^2}\, dx = \int_1^8 \left(x^{1/3} + \frac{1}{4}x^{-1/3}\right) dx$

$= \left[\frac{3}{4}x^{4/3} + \frac{3}{8}x^{2/3}\right]_1^8 = \frac{3}{8}\left[2x^{4/3} + x^{2/3}\right]_1^8$

$= \frac{3}{8}\left[(2 \cdot 2^4 + 2^2) - (2 + 1)\right] = \frac{3}{8}(32 + 4 - 3) = \frac{99}{8}$

17. $y = \sqrt{1 - x^2} = (1 - x^2)^{1/2} \Rightarrow y' = \left(\frac{1}{2}\right)(1 - x^2)^{-1/2}(-2x) \Rightarrow 1 + (y')^2 = \frac{1}{1 - x^2}$; $L = \int_{-1/2}^{1/2} \sqrt{1 + (y')^2}\, dx$

$= 2\int_0^{1/2} \frac{1}{\sqrt{1 - x^2}}\, dx = 2\left[\sin^{-1} x\right]_0^{1/2} = 2\left(\frac{\pi}{6} - 0\right) = \frac{\pi}{3}$

19. (a) $\frac{dy}{dx} = 2x \Rightarrow \left(\frac{dy}{dx}\right)^2 = 4x^2$

$\Rightarrow L = \int_{-1}^2 \sqrt{1 + \left(\frac{dy}{dx}\right)^2}\, dx$

$= \int_{-1}^2 \sqrt{1 + 4x^2}\, dx$

(c) $L \approx 6.13$

(b)

21. (a) $\frac{dx}{dy} = \cos y \Rightarrow \left(\frac{dx}{dy}\right)^2 = \cos^2 y$

$\Rightarrow L = \int_0^\pi \sqrt{1 + \cos^2 y}\, dy$

(c) $L \approx 3.82$

(b)

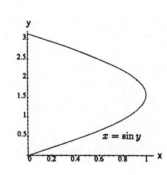

23. (a) $2y + 2 = 2\frac{dx}{dy} \Rightarrow \left(\frac{dx}{dy}\right)^2 = (y + 1)^2$

$\Rightarrow L = \int_{-1}^3 \sqrt{1 + (y + 1)^2}\, dy$

(c) $L \approx 9.29$

(b)

25. (a) $\frac{dy}{dx} = \tan x \Rightarrow \left(\frac{dy}{dx}\right)^2 = \tan^2 x$

       $\Rightarrow L = \int_0^{\pi/6} \sqrt{1 + \tan^2 x}\, dx = \int_0^{\pi/6} \sqrt{\frac{\sin^2 x + \cos^2 x}{\cos^2 x}}\, dx$

       $= \int_0^{\pi/6} \frac{dx}{\cos x} = \int_0^{\pi/6} \sec x\, dx$

  (c) $L \approx 0.55$

(b)

27. $\sqrt{2}\, x = \int_0^x \sqrt{1 + \left(\frac{dy}{dt}\right)^2}\, dt,\ x \geq 0 \Rightarrow \sqrt{2} = \sqrt{1 + \left(\frac{dy}{dx}\right)^2} \Rightarrow \frac{dy}{dx} = \pm 1 \Rightarrow y = f(x) = \pm x + C$ where C is any

real number.

29. (a) $\left(\frac{dy}{dx}\right)^2$ corresponds to $\frac{1}{4x}$ here, so take $\frac{dy}{dx}$ as $\frac{1}{2\sqrt{x}}$. Then $y = \sqrt{x} + C$ and since $(1, 1)$ lies on the curve, $C = 0$.

      So $y = \sqrt{x}$ from $(1, 1)$ to $(4, 2)$.

  (b) Only one. We know the derivative of the function and the value of the function at one value of x.

31. $L = \int_0^1 \sqrt{1 + \frac{e^x}{4}}\, dx \Rightarrow \frac{dy}{dx} = \frac{e^{x/2}}{2} \Rightarrow y = e^{x/2} + C;\ y(0) = 0 \Rightarrow 0 = e^0 + C \Rightarrow C = -1 \Rightarrow y = e^{x/2} - 1$

33. $\frac{dx}{dt} = \left(\frac{1}{\sec t + \tan t}\right)(\sec t \tan t + \sec^2 t) - \cos t = \sec t - \cos t$ and $\frac{dy}{dt} = -\sin t \Rightarrow \sqrt{\left(\frac{dx}{dt}\right)^2 + \left(\frac{dy}{dt}\right)^2}$

    $= \sqrt{(\sec t - \cos t)^2 + (-\sin t)^2} = \sqrt{\sec^2 t - 1} = \sqrt{\tan^2 t} = |\tan t| = \tan t$ since $0 \leq t \leq \frac{\pi}{3}$

    $\Rightarrow$ Length $= \int_0^{\pi/3} \tan t\, dt = \int_0^{\pi/3} \frac{\sin t}{\cos t}\, dt = [-\ln|\cos t|]\,_0^{\pi/3} = -\ln \frac{1}{2} + \ln 1 = \ln 2$

35. $\frac{dx}{dt} = e^t - e^{-t}$ and $\frac{dy}{dt} = -2 \Rightarrow \sqrt{\left(\frac{dx}{dt}\right)^2 + \left(\frac{dy}{dt}\right)^2} = \sqrt{(e^t - e^{-t})^2 + (-2)^2} = \sqrt{e^{2t} + 2 + e^{-2t}} = \sqrt{\frac{e^{4t} + 2e^{2t} + 1}{e^{2t}}}$

    $= \sqrt{\frac{(e^{2t} + 1)^2}{(e^t)^2}} = \frac{e^{2t} + 1}{e^t} = e^t + e^{-t} \Rightarrow$ Length $= \int_0^3 (e^t + e^{-t})dt = (e^t - e^{-t})\Big|_0^3 = (e^3 - e^{-3}) - (e^0 - e^0) = e^3 - \frac{1}{e^3}$

6.4 AREAS OF SURFACES OF REVOLUTION

1. (a) $\frac{dy}{dx} = \sec^2 x \Rightarrow \left(\frac{dy}{dx}\right)^2 = \sec^4 x$

      $\Rightarrow S = 2\pi \int_0^{\pi/4} (\tan x) \sqrt{1 + \sec^4 x}\, dx$

  (c) $S \approx 3.84$

(b)

3. (a) $xy = 1 \Rightarrow x = \frac{1}{y} \Rightarrow \frac{dx}{dy} = -\frac{1}{y^2} \Rightarrow \left(\frac{dx}{dy}\right)^2 = \frac{1}{y^4}$

   $\Rightarrow S = 2\pi \int_1^2 \frac{1}{y} \sqrt{1 + y^{-4}} \, dy$

   (c) $S \approx 5.02$

(b)

5. (a) $x^{1/2} + y^{1/2} = 3 \Rightarrow y = \left(3 - x^{1/2}\right)^2$

   $\Rightarrow \frac{dy}{dx} = 2\left(3 - x^{1/2}\right)\left(-\frac{1}{2}x^{-1/2}\right)$

   $\Rightarrow \left(\frac{dy}{dx}\right)^2 = \left(1 - 3x^{-1/2}\right)^2$

   $\Rightarrow S = 2\pi \int_1^4 \left(3 - x^{1/2}\right)^2 \sqrt{1 + \left(1 - 3x^{-1/2}\right)^2} \, dx$

   (c) $S \approx 63.37$

(b)

7. (a) $\frac{dx}{dy} = \tan y \Rightarrow \left(\frac{dx}{dy}\right)^2 = \tan^2 y$

   $\Rightarrow S = 2\pi \int_0^{\pi/3} \left(\int_0^y \tan t \, dt\right) \sqrt{1 + \tan^2 y} \, dy$

   $= 2\pi \int_0^{\pi/3} \left(\int_0^y \tan t \, dt\right) \sec y \, dy$

   (c) $S \approx 2.08$

(b)

9. $y = \frac{x}{2} \Rightarrow \frac{dy}{dx} = \frac{1}{2}; S = \int_a^b 2\pi y \sqrt{1 + \left(\frac{dy}{dx}\right)^2} \, dx \Rightarrow S = \int_0^4 2\pi \left(\frac{x}{2}\right) \sqrt{1 + \frac{1}{4}} \, dx = \frac{\pi\sqrt{5}}{2} \int_0^4 x \, dx$

   $= \frac{\pi\sqrt{5}}{2} \left[\frac{x^2}{2}\right]_0^4 = 4\pi\sqrt{5}$; Geometry formula: base circumference $= 2\pi(2)$, slant height $= \sqrt{4^2 + 2^2} = 2\sqrt{5}$

   $\Rightarrow$ Lateral surface area $= \frac{1}{2}(4\pi)\left(2\sqrt{5}\right) = 4\pi\sqrt{5}$ in agreement with the integral value

11. $\frac{dy}{dx} = \frac{1}{2}; S = \int_a^b 2\pi y \sqrt{1 + \left(\frac{dy}{dx}\right)^2} \, dx = \int_1^3 2\pi \frac{(x+1)}{2} \sqrt{1 + \left(\frac{1}{2}\right)^2} \, dx = \frac{\pi\sqrt{5}}{2} \int_1^3 (x + 1) \, dx = \frac{\pi\sqrt{5}}{2} \left[\frac{x^2}{2} + x\right]_1^3$

   $= \frac{\pi\sqrt{5}}{2} \left[\left(\frac{9}{2} + 3\right) - \left(\frac{1}{2} + 1\right)\right] = \frac{\pi\sqrt{5}}{2}(4 + 2) = 3\pi\sqrt{5}$; Geometry formula: $r_1 = \frac{1}{2} + \frac{1}{2} = 1, r_2 = \frac{3}{2} + \frac{1}{2} = 2$,

   slant height $= \sqrt{(2 - 1)^2 + (3 - 1)^2} = \sqrt{5} \Rightarrow$ Frustum surface area $= \pi(r_1 + r_2) \times$ slant height $= \pi(1 + 2)\sqrt{5}$

   $= 3\pi\sqrt{5}$ in agreement with the integral value

13. $\frac{dy}{dx} = \frac{x^2}{3} \Rightarrow \left(\frac{dy}{dx}\right)^2 = \frac{x^4}{9} \Rightarrow S = \int_0^2 \frac{2\pi x^3}{9} \sqrt{1 + \frac{x^4}{9}}\, dx;$

$\left[u = 1 + \frac{x^4}{9} \Rightarrow du = \frac{4}{9} x^3\, dx \Rightarrow \frac{1}{4}\, du = \frac{x^3}{9}\, dx; \right.$

$x = 0 \Rightarrow u = 1,\ x = 2 \Rightarrow u = \frac{25}{9} \Big]$

$\rightarrow S = 2\pi \int_1^{25/9} u^{1/2} \cdot \frac{1}{4}\, du = \frac{\pi}{2} \left[\frac{2}{3} u^{3/2} \right]_1^{25/9}$

$= \frac{\pi}{3} \left(\frac{125}{27} - 1 \right) = \frac{\pi}{3} \left(\frac{125-27}{27} \right) = \frac{98\pi}{81}$

15. $\frac{dy}{dx} = \frac{1}{2} \frac{(2-2x)}{\sqrt{2x-x^2}} = \frac{1-x}{\sqrt{2x-x^2}} \Rightarrow \left(\frac{dy}{dx}\right)^2 = \frac{(1-x)^2}{2x-x^2}$

$\Rightarrow S = \int_{0.5}^{1.5} 2\pi \sqrt{2x-x^2} \sqrt{1 + \frac{(1-x)^2}{2x-x^2}}\, dx$

$= 2\pi \int_{0.5}^{1.5} \sqrt{2x-x^2}\ \frac{\sqrt{2x-x^2+1-2x+x^2}}{\sqrt{2x-x^2}}\, dx$

$= 2\pi \int_{0.5}^{1.5} dx = 2\pi [x]_{0.5}^{1.5} = 2\pi$

17. $\frac{dx}{dy} = y^2 \Rightarrow \left(\frac{dx}{dy}\right)^2 = y^4 \Rightarrow S = \int_0^1 \frac{2\pi y^3}{3} \sqrt{1 + y^4}\, dy;$

$\left[u = 1 + y^4 \Rightarrow du = 4y^3\, dy \Rightarrow \frac{1}{4}\, du = y^3\, dy;\ y = 0 \right.$

$\Rightarrow u = 1,\ y = 1 \Rightarrow u = 2 \Big] \rightarrow S = \int_1^2 2\pi \left(\frac{1}{3}\right) u^{1/2} \left(\frac{1}{4}\, du\right)$

$= \frac{\pi}{6} \int_1^2 u^{1/2}\, du = \frac{\pi}{6} \left[\frac{2}{3} u^{3/2} \right]_1^2 = \frac{\pi}{9} \left(\sqrt{8} - 1 \right)$

19. $\frac{dx}{dy} = \frac{-1}{\sqrt{4-y}} \Rightarrow \left(\frac{dx}{dy}\right)^2 = \frac{1}{4-y} \Rightarrow S = \int_0^{15/4} 2\pi \cdot 2\sqrt{4-y} \sqrt{1 + \frac{1}{4-y}}\, dy = 4\pi \int_0^{15/4} \sqrt{(4-y)+1}\, dy$

$= 4\pi \int_0^{15/4} \sqrt{5-y}\, dy = -4\pi \left[\frac{2}{3}(5-y)^{3/2} \right]_0^{15/4} = -\frac{8\pi}{3} \left[\left(5 - \frac{15}{4}\right)^{3/2} - 5^{3/2} \right] = -\frac{8\pi}{3} \left[\left(\frac{5}{4}\right)^{3/2} - 5^{3/2} \right]$

$= \frac{8\pi}{3} \left(5\sqrt{5} - \frac{5\sqrt{5}}{8} \right) = \frac{8\pi}{3} \left(\frac{40\sqrt{5}-5\sqrt{5}}{8} \right) = \frac{35\pi\sqrt{5}}{3}$

21. $S = 2\pi \int_0^{\ln 2} \left(\frac{e^y + e^{-y}}{2} \right) \sqrt{1 + \left(\frac{e^y - e^{-y}}{2} \right)^2}\, dy = 2\pi \int_0^{\ln 2} \left(\frac{e^y + e^{-y}}{2} \right) \sqrt{1 + \frac{1}{4}(e^{2y} - 2 + e^{-2y})}\, dy$

$= 2\pi \int_0^{\ln 2} \left(\frac{e^y + e^{-y}}{2} \right) \sqrt{\left(\frac{e^y + e^{-y}}{2} \right)^2}\, dy = 2\pi \int_0^{\ln 2} \left(\frac{e^y + e^{-y}}{2} \right)^2 dy = \frac{\pi}{2} \int_0^{\ln 2} (e^{2y} + 2 + e^{-2y})\, dy$

$= \frac{\pi}{2} \left[\frac{1}{2} e^{2y} + 2y - \frac{1}{2} e^{-2y} \right]_0^{\ln 2} = \frac{\pi}{2} \left[\left(\frac{1}{2} e^{2\ln 2} + 2\ln 2 - \frac{1}{2} e^{-2\ln 2} \right) - \left(\frac{1}{2} + 0 - \frac{1}{2} \right) \right]$

$= \frac{\pi}{2} \left(\frac{1}{2} \cdot 4 + 2\ln 2 - \frac{1}{2} \cdot \frac{1}{4} \right) = \frac{\pi}{2} \left(2 - \frac{1}{8} + 2\ln 2 \right) = \pi \left(\frac{15}{16} + \ln 2 \right)$

23. $ds = \sqrt{dx^2 + dy^2} = \sqrt{\left(y^3 - \frac{1}{4y^3} \right)^2 + 1}\, dy = \sqrt{\left(y^6 - \frac{1}{2} + \frac{1}{16y^6} \right) + 1}\, dy = \sqrt{\left(y^6 + \frac{1}{2} + \frac{1}{16y^6} \right)}\, dy$

$= \sqrt{\left(y^3 + \frac{1}{4y^3} \right)^2}\, dy = \left(y^3 + \frac{1}{4y^3} \right) dy;\ S = \int_1^2 2\pi y\, ds = 2\pi \int_1^2 y \left(y^3 + \frac{1}{4y^3} \right) dy = 2\pi \int_1^2 \left(y^4 + \frac{1}{4} y^{-2} \right) dy$

$= 2\pi \left[\frac{y^5}{5} - \frac{1}{4} y^{-1} \right]_1^2 = 2\pi \left[\left(\frac{32}{5} - \frac{1}{8} \right) - \left(\frac{1}{5} - \frac{1}{4} \right) \right] = 2\pi \left(\frac{31}{5} + \frac{1}{8} \right) = \frac{2\pi}{40} (8 \cdot 31 + 5) = \frac{253\pi}{20}$

25. $y = \sqrt{a^2 - x^2} \Rightarrow \frac{dy}{dx} = \frac{1}{2} (a^2 - x^2)^{-1/2}(-2x) = \frac{-x}{\sqrt{a^2-x^2}} \Rightarrow \left(\frac{dy}{dx}\right)^2 = \frac{x^2}{(a^2-x^2)}$

$\Rightarrow S = 2\pi \int_{-a}^a \sqrt{a^2-x^2} \sqrt{1 + \frac{x^2}{(a^2-x^2)}}\, dx = 2\pi \int_{-a}^a \sqrt{(a^2-x^2)+x^2}\, dx = 2\pi \int_{-a}^a a\, dx = 2\pi a[x]_{-a}^a$

$= 2\pi a[a - (-a)] = (2\pi a)(2a) = 4\pi a^2$

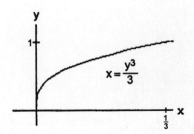

27. The area of the surface of one wok is $S = \int_c^d 2\pi x \sqrt{1 + \left(\frac{dx}{dy}\right)^2}\, dy$. Now, $x^2 + y^2 = 16^2 \Rightarrow x = \sqrt{16^2 - y^2}$

$\Rightarrow \frac{dx}{dy} = \frac{-y}{\sqrt{16^2 - y^2}} \Rightarrow \left(\frac{dx}{dy}\right)^2 = \frac{y^2}{16^2 - y^2}$; $S = \int_{-16}^{-7} 2\pi \sqrt{16^2 - y^2} \sqrt{1 + \frac{y^2}{16^2 - y^2}}\, dy = 2\pi \int_{-16}^{-7} \sqrt{(16^2 - y^2) + y^2}\, dy$

$= 2\pi \int_{-16}^{-7} 16\, dy = 32\pi \cdot 9 = 288\pi \approx 904.78$ cm^2. The enamel needed to cover one surface of one wok is

$V = S \cdot 0.5$ mm $= S \cdot 0.05$ cm $= (904.78)(0.05)$ cm$^3 = 45.24$ cm^3. For 5000 woks, we need

$5000 \cdot V = 5000 \cdot 45.24$ cm$^3 = (5)(45.24)L = 226.2L \Rightarrow 226.2$ liters of each color are needed.

29. $y = \sqrt{R^2 - x^2} \Rightarrow \frac{dy}{dx} = -\frac{1}{2} \frac{2x}{\sqrt{R^2 - x^2}} = \frac{-x}{\sqrt{R^2 - x^2}} \Rightarrow \left(\frac{dx}{dy}\right)^2 = \frac{x^2}{R^2 - x^2}$; $S = 2\pi \int_a^{a+h} \sqrt{R^2 - x^2} \sqrt{1 + \frac{x^2}{R^2 - x^2}}\, dx$

$= 2\pi \int_a^{a+h} \sqrt{(R^2 - x^2) + x^2}\, dx = 2\pi R \int_a^{a+h} dx = 2\pi Rh$

31. (a) An equation of the tangent line segment is
(see figure) $y = f(m_k) + f'(m_k)(x - m_k)$.
When $x = x_{k-1}$ we have
$r_1 = f(m_k) + f'(m_k)(x_{k-1} - m_k)$
$= f(m_k) + f'(m_k)\left(-\frac{\Delta x_k}{2}\right) = f(m_k) - f'(m_k)\frac{\Delta x_k}{2}$;
when $x = x_k$ we have
$r_2 = f(m_k) + f'(m_k)(x_k - m_k)$
$= f(m_k) + f'(m_k)\frac{\Delta x_k}{2}$;

(b) $L_k^2 = (\Delta x_k)^2 + (r_2 - r_1)^2$
$= (\Delta x_k)^2 + \left[f'(m_k)\frac{\Delta x_k}{2} - \left(-f'(m_k)\frac{\Delta x_k}{2}\right)\right]^2$
$= (\Delta x_k)^2 + [f'(m_k)\Delta x_k]^2 \Rightarrow L_k = \sqrt{(\Delta x_k)^2 + [f'(m_k)\Delta x_k]^2}$, as claimed

(c) From geometry it is a fact that the lateral surface area of the frustum obtained by revolving the tangent

line segment about the x-axis is given by $\Delta S_k = \pi(r_1 + r_2)L_k = \pi[2f(m_k)] \sqrt{(\Delta x_k)^2 + [f'(m_k)\Delta x_k]^2}$

using parts (a) and (b) above. Thus, $\Delta S_k = 2\pi f(m_k) \sqrt{1 + [f'(m_k)]^2}\, \Delta x_k$.

(d) $S = \lim_{n \to \infty} \sum_{k=1}^{n} \Delta S_k = \lim_{n \to \infty} \sum_{k=1}^{n} 2\pi f(m_k) \sqrt{1 + [f'(m_k)]^2}\, \Delta x_k = \int_a^b 2\pi f(x) \sqrt{1 + [f'(x)]^2}\, dx$

33. $\frac{dx}{dt} = -\sin t$ and $\frac{dy}{dt} = \cos t \Rightarrow \sqrt{\left(\frac{dx}{dt}\right)^2 + \left(\frac{dy}{dt}\right)^2} = \sqrt{(-\sin t)^2 + (\cos t)^2} = 1 \Rightarrow S = \int 2\pi y\, ds$

$= \int_0^{2\pi} 2\pi(2 + \sin t)(1)\, dt = 2\pi\left[2t - \cos t\right]_0^{2\pi} = 2\pi[(4\pi - 1) - (0 - 1)] = 8\pi^2$

35. $\frac{dx}{dt} = 1$ and $\frac{dy}{dt} = t + \sqrt{2} \Rightarrow \sqrt{\left(\frac{dx}{dt}\right)^2 + \left(\frac{dy}{dt}\right)^2} = \sqrt{1^2 + \left(t + \sqrt{2}\right)^2} = \sqrt{t^2 + 2\sqrt{2}t + 3} \Rightarrow S = \int 2\pi x\, ds$

$= \int_{-\sqrt{2}}^{\sqrt{2}} 2\pi\left(t + \sqrt{2}\right)\sqrt{t^2 + 2\sqrt{2}t + 3}\, dt$; $\left[u = t^2 + 2\sqrt{2}t + 3 \Rightarrow du = \left(2t + 2\sqrt{2}\right) dt; t = -\sqrt{2} \Rightarrow u = 1,\right.$

$\left. t = \sqrt{2} \Rightarrow u = 9\right] \to \int_1^9 \pi \sqrt{u}\, du = \left[\frac{2}{3}\pi u^{3/2}\right]_1^9 = \frac{2\pi}{3}(27 - 1) = \frac{52\pi}{3}$

37. $S = 2\pi \int_0^1 y \sqrt{\left(\frac{dx}{dt}\right)^2 + \left(\frac{dy}{dt}\right)^2}\, dt = 2\pi \int_0^1 4e^{t/2} \sqrt{(e^t - 1)^2 + (2e^{t/2})^2}\, dt = 8\pi \int_0^1 e^{t/2} \sqrt{e^{2t} + 2e^t + 1}\, dt$

$= 8\pi \int_0^1 e^{t/2} \sqrt{(e^t + 1)^2}\, dt = 8\pi \int_0^1 e^{t/2}(e^t + 1)dt = 8\pi \int_0^1 \left(e^{3t/2} + e^{t/2}\right)dt = 8\pi\left(\frac{2}{3}e^{3t/2} + 2e^{t/2}\right)\Big|_0^1$

$= 8\pi\left(\left(\frac{2}{3}e^{3/2} + 2e^{1/2}\right) - \frac{8}{3}\right) = \frac{16\pi}{3}\left(e^{3/2} + 3e^{1/2} - 4\right)$

39. $\frac{dx}{dt} = 2$ and $\frac{dy}{dt} = 1 \Rightarrow \sqrt{\left(\frac{dx}{dt}\right)^2 + \left(\frac{dy}{dt}\right)^2} = \sqrt{2^2 + 1^2} = \sqrt{5} \Rightarrow S = \int 2\pi y \, ds = \int_0^1 2\pi(t+1)\sqrt{5} \, dt$

$= 2\pi\sqrt{5} \left[\frac{t^2}{2} + t\right]_0^1 = 3\pi\sqrt{5}$. Check: slant height is $\sqrt{5} \Rightarrow$ Area is $\pi(1+2)\sqrt{5} = 3\pi\sqrt{5}$.

6.5 EXPONENTIAL CHANGE AND SEPARABLE DIFFERENTIAL EQUATIONS

1. (a) $y = e^{-x} \Rightarrow y' = -e^{-x} \Rightarrow 2y' + 3y = 2\left(-e^{-x}\right) + 3e^{-x} = e^{-x}$
   (b) $y = e^{-x} + e^{-3x/2} \Rightarrow y' = -e^{-x} - \frac{3}{2}e^{-3x/2} \Rightarrow 2y' + 3y = 2\left(-e^{-x} - \frac{3}{2}e^{-3x/2}\right) + 3\left(e^{-x} + e^{-3x/2}\right) = e^{-x}$
   (c) $y = e^{-x} + Ce^{-3x/2} \Rightarrow y' = -e^{-x} - \frac{3}{2}Ce^{-3x/2} \Rightarrow 2y' + 3y = 2\left(-e^{-x} - \frac{3}{2}Ce^{-3x/2}\right) + 3\left(e^{-x} + Ce^{-3x/2}\right) = e^{-x}$

3. $y = \frac{1}{x}\int_1^x \frac{e^t}{t} \, dt \Rightarrow y' = -\frac{1}{x^2}\int_1^x \frac{e^t}{t} \, dt + \left(\frac{1}{x}\right)\left(\frac{e^x}{x}\right) \Rightarrow x^2 y' = -\int_1^x \frac{e^t}{t} \, dt + e^x = -x\left(\frac{1}{x}\int_1^x \frac{e^t}{t} \, dt\right) + e^x = -xy + e^x$
   $\Rightarrow x^2 y' + xy = e^x$

5. $y = e^{-x}\tan^{-1}(2e^x) \Rightarrow y' = -e^{-x}\tan^{-1}(2e^x) + e^{-x}\left[\frac{1}{1+(2e^x)^2}\right](2e^x) = -e^{-x}\tan^{-1}(2e^x) + \frac{2}{1+4e^{2x}}$
   $\Rightarrow y' = -y + \frac{2}{1+4e^{2x}} \Rightarrow y' + y = \frac{2}{1+4e^{2x}}$; $y(-\ln 2) = e^{-(-\ln 2)}\tan^{-1}(2e^{-\ln 2}) = 2\tan^{-1} 1 = 2\left(\frac{\pi}{4}\right) = \frac{\pi}{2}$

7. $y = \frac{\cos x}{x} \Rightarrow y' = \frac{-x\sin x - \cos x}{x^2} \Rightarrow y' = -\frac{\sin x}{x} - \frac{1}{x}\left(\frac{\cos x}{x}\right) \Rightarrow y' = -\frac{\sin x}{x} - \frac{y}{x} \Rightarrow xy' = -\sin x - y$
   $\Rightarrow xy' + y = -\sin x$; $y\left(\frac{\pi}{2}\right) = \frac{\cos(\pi/2)}{(\pi/2)} = 0$

9. $2\sqrt{xy}\frac{dy}{dx} = 1 \Rightarrow 2x^{1/2}y^{1/2} \, dy = dx \Rightarrow 2y^{1/2} \, dy = x^{-1/2} \, dx \Rightarrow \int 2y^{1/2} \, dy = \int x^{-1/2} \, dx \Rightarrow 2\left(\frac{2}{3}y^{3/2}\right)$
   $= 2x^{1/2} + C_1 \Rightarrow \frac{2}{3}y^{3/2} - x^{1/2} = C$, where $C = \frac{1}{2}C_1$

11. $\frac{dy}{dx} = e^{x-y} \Rightarrow dy = e^x e^{-y} \, dx \Rightarrow e^y \, dy = e^x \, dx \Rightarrow \int e^y \, dy = \int e^x \, dx \Rightarrow e^y = e^x + C \Rightarrow e^y - e^x = C$

13. $\frac{dy}{dx} = \sqrt{y}\cos^2\sqrt{y} \Rightarrow dy = \left(\sqrt{y}\cos^2\sqrt{y}\right) dx \Rightarrow \frac{\sec^2\sqrt{y}}{\sqrt{y}} \, dy = dx \Rightarrow \int \frac{\sec^2\sqrt{y}}{\sqrt{y}} \, dy = \int dx$. In the integral on the left-hand
    side, substitute $u = \sqrt{y} \Rightarrow du = \frac{1}{2\sqrt{y}} \, dy \Rightarrow 2 \, du = \frac{1}{\sqrt{y}} \, dy$, and we have $\int \sec^2 u \, du = \int dx \Rightarrow 2\tan u = x + C$
    $\Rightarrow -x + 2\tan\sqrt{y} = C$

15. $\sqrt{x}\frac{dy}{dx} = e^{y+\sqrt{x}} \Rightarrow \frac{dy}{dx} = \frac{e^y e^{\sqrt{x}}}{\sqrt{x}} \Rightarrow dy = \frac{e^y e^{\sqrt{x}}}{\sqrt{x}} \, dx \Rightarrow e^{-y} \, dy = \frac{e^{\sqrt{x}}}{\sqrt{x}} \, dx \Rightarrow \int e^{-y} \, dy = \int \frac{e^{\sqrt{x}}}{\sqrt{x}} \, dx$. In the integral on the
    right-hand side, substitute $u = \sqrt{x} \Rightarrow du = \frac{1}{2\sqrt{x}} \, dx \Rightarrow 2 \, du = \frac{1}{\sqrt{x}} \, dx$, and we have $\int e^{-y} \, dy = 2\int e^u \, du$
    $\Rightarrow -e^{-y} = 2e^u + C_1 \Rightarrow -e^{-y} = 2e^{\sqrt{x}} + C$, where $C = -C_1$

17. $\frac{dy}{dx} = 2x\sqrt{1-y^2} \Rightarrow dy = 2x\sqrt{1-y^2} \, dx \Rightarrow \frac{dy}{\sqrt{1-y^2}} = 2x \, dx \Rightarrow \int \frac{dy}{\sqrt{1-y^2}} = \int 2x \, dx \Rightarrow \sin^{-1} y = x^2 + C$ since $|y| < 1$
    $\Rightarrow y = \sin(x^2 + C)$

19. (a) $y = y_0 e^{kt} \Rightarrow 0.99 y_0 = y_0 e^{1000k} \Rightarrow k = \frac{\ln 0.99}{1000} \approx -0.00001$
    (b) $0.9 = e^{(-0.00001)t} \Rightarrow (-0.00001)t = \ln(0.9) \Rightarrow t = \frac{\ln(0.9)}{-0.00001} \approx 10{,}536$ years
    (c) $y = y_0 e^{(20{,}000)k} \approx y_0 e^{-0.2} = y_0(0.82) \Rightarrow 82\%$

21. $\frac{dy}{dt} = -0.6y \Rightarrow y = y_0 e^{-0.6t}$; $y_0 = 100 \Rightarrow y = 100e^{-0.6t} \Rightarrow y = 100e^{-0.6} \approx 54.88$ grams when $t = 1$ hr

23. $L(x) = L_0 e^{-kx} \Rightarrow \frac{L_0}{2} = L_0 e^{-18k} \Rightarrow \ln \frac{1}{2} = -18k \Rightarrow k = \frac{\ln 2}{18} \approx 0.0385 \Rightarrow L(x) = L_0 e^{-0.0385x}$; when the intensity is one-tenth of the surface value, $\frac{L_0}{10} = L_0 e^{-0.0385x} \Rightarrow \ln 10 = 0.0385x \Rightarrow x \approx 59.8$ ft

25. $y = y_0 e^{kt}$ and $y_0 = 1 \Rightarrow y = e^{kt}$ at $y = 2$ and $t = 0.5$ we have $2 = e^{0.5k} \Rightarrow \ln 2 = 0.5k \Rightarrow k = \frac{\ln 2}{0.5} = \ln 4$. Therefore, $y = e^{(\ln 4)t} \Rightarrow y = e^{24\ln 4} = 4^{24} = 2.81474978 \times 10^{14}$ at the end of 24 hrs

27. (a) $10{,}000 e^{k(1)} = 7500 \Rightarrow e^k = 0.75 \Rightarrow k = \ln 0.75$ and $y = 10{,}000 e^{(\ln 0.75)t}$. Now $1000 = 10{,}000 e^{(\ln 0.75)t}$
    $\Rightarrow \ln 0.1 = (\ln 0.75)t \Rightarrow t = \frac{\ln 0.1}{\ln 0.75} \approx 8.00$ years (to the nearest hundredth of a year)

    (b) $1 = 10{,}000 e^{(\ln 0.75)t} \Rightarrow \ln 0.0001 = (\ln 0.75)t \Rightarrow t = \frac{\ln 0.0001}{\ln 0.75} \approx 32.02$ years (to the nearest hundredth of a year)

29. $0.9 P_0 = P_0 e^k \Rightarrow k = \ln 0.9$; when the well's output falls to one-fifth of its present value $P = 0.2 P_0$
    $\Rightarrow 0.2 P_0 = P_0 e^{(\ln 0.9)t} \Rightarrow 0.2 = e^{(\ln 0.9)t} \Rightarrow \ln(0.2) = (\ln 0.9)t \Rightarrow t = \frac{\ln 0.2}{\ln 0.9} \approx 15.28$ yr

31. $A = A_0 e^{kt}$ and $A_0 = 10 \Rightarrow A = 10 e^{kt}, 5 = 10 e^{k(24360)} \Rightarrow k = \frac{\ln(0.5)}{24360} \approx -0.000028454 \Rightarrow A = 10 e^{-0.000028454t}$, then $0.2(10) = 10 e^{-0.000028254t} \Rightarrow t = \frac{\ln 0.2}{-0.000028454} \approx 56563$ years

33. $y = y_0 e^{-kt} = y_0 e^{-(k)(3/k)} = y_0 e^{-3} = \frac{y_0}{e^3} < \frac{y_0}{20} = (0.05)(y_0) \Rightarrow$ after three mean lifetimes less than 5% remains

35. $T - T_s = (T_0 - T_s) e^{-kt}, T_0 = 90°C, T_s = 20°C, T = 60°C \Rightarrow 60 - 20 = 70 e^{-10k} \Rightarrow \frac{4}{7} = e^{-10k}$
    $\Rightarrow k = \frac{\ln\left(\frac{7}{4}\right)}{10} \approx 0.05596$

    (a) $35 - 20 = 70 e^{-0.05596t} \Rightarrow t \approx 27.5$ min is the total time $\Rightarrow$ it will take $27.5 - 10 = 17.5$ minutes longer to reach $35°C$

    (b) $T - T_s = (T_0 - T_s) e^{-kt}, T_0 = 90°C, T_s = -15°C \Rightarrow 35 + 15 = 105 e^{-0.05596t} \Rightarrow t \approx 13.26$ min

37. $T - T_s = (T_0 - T_s) e^{-kt} \Rightarrow 39 - T_s = (46 - T_s) e^{-10k}$ and $33 - T_s = (46 - T_s) e^{-20k} \Rightarrow \frac{39 - T_s}{46 - T_s} = e^{-10k}$ and
    $\frac{33 - T_s}{46 - T_s} = e^{-20k} = (e^{-10k})^2 \Rightarrow \frac{33 - T_s}{46 - T_s} = \left(\frac{39 - T_s}{46 - T_s}\right)^2 \Rightarrow (33 - T_s)(46 - T_s) = (39 - T_s)^2 \Rightarrow 1518 - 79 T_s + T_s^2$
    $= 1521 - 78 T_s + T_s^2 \Rightarrow -T_s = 3 \Rightarrow T_s = -3°C$

39. From Example 5, the half-life of carbon-14 is 5700 yr $\Rightarrow \frac{1}{2} c_0 = c_0 e^{-k(5700)} \Rightarrow k = \frac{\ln 2}{5700} \approx 0.0001216$
    $\Rightarrow c = c_0 e^{-0.0001216t} \Rightarrow (0.445) c_0 = c_0 e^{-0.0001216t} \Rightarrow t = \frac{\ln(0.445)}{-0.0001216} \approx 6659$ years

41. From Exercise 25, $k \approx 0.0001216$ for carbon-14. Thus, $c = c_0 e^{-0.0001216t} \Rightarrow (0.995) c_0 = c_0 e^{-0.0001216t}$
    $\Rightarrow t = \frac{\ln(0.995)}{-0.0001216} \approx 41$ years old

6.6 WORK

1. The force required to stretch the spring from its natural length of 2 m to a length of 5 m is $F(x) = kx$. The work done by F is $W = \int_0^3 F(x)\, dx = k \int_0^3 x\, dx = \frac{k}{2} [x^2]_0^3 = \frac{9k}{2}$. This work is equal to 1800 J $\Rightarrow \frac{9}{2} k = 1800 \Rightarrow k = 400$ N/m

3. We find the force constant from Hooke's law: $F = kx$. A force of 2 N stretches the spring to 0.02 m
    $\Rightarrow 2 = k \cdot (0.02) \Rightarrow k = 100 \frac{N}{m}$. The force of 4 N will stretch the rubber band y m, where $F = ky \Rightarrow y = \frac{F}{k}$
    $\Rightarrow y = \frac{4N}{100 \frac{N}{m}} \Rightarrow y = 0.04$ m $= 4$ cm. The work done to stretch the rubber band 0.04 m is $W = \int_0^{0.04} kx\, dx$
    $= 100 \int_0^{0.04} x\, dx = 100 \left[\frac{x^2}{2}\right]_0^{0.04} = \frac{(100)(0.04)^2}{2} = 0.08$ J

5. (a) We find the spring's constant from Hooke's law: $F = kx \Rightarrow k = \frac{F}{x} = \frac{21{,}714}{8-5} = \frac{21{,}714}{3} \Rightarrow k = 7238 \frac{\text{lb}}{\text{in}}$

    (b) The work done to compress the assembly the first half inch is $W = \int_0^{0.5} kx\,dx = 7238 \int_0^{0.5} x\,dx = 7238 \left[\frac{x^2}{2}\right]_0^{0.5}$

    $= (7238)\frac{(0.5)^2}{2} = \frac{(7238)(0.25)}{2} \approx 905$ in $\cdot$ lb. The work done to compress the assembly the second half inch is:

    $W = \int_{0.5}^{1.0} kx\,dx = 7238 \int_{0.5}^{1.0} x\,dx = 7238 \left[\frac{x^2}{2}\right]_{0.5}^{1.0} = \frac{7238}{2}\left[1 - (0.5)^2\right] = \frac{(7238)(0.75)}{2} \approx 2714$ in $\cdot$ lb

7. The force required to haul up the rope is equal to the rope's weight, which varies steadily and is proportional to x, the length of the rope still hanging: $F(x) = 0.624x$. The work done is: $W = \int_0^{50} F(x)\,dx = \int_0^{50} 0.624x\,dx$

    $= 0.624 \left[\frac{x^2}{2}\right]_0^{50} = 780$ J

9. The force required to lift the cable is equal to the weight of the cable paid out: $F(x) = (4.5)(180 - x)$ where x is the position of the car off the first floor. The work done is: $W = \int_0^{180} F(x)\,dx = 4.5 \int_0^{180} (180 - x)\,dx$

    $= 4.5 \left[180x - \frac{x^2}{2}\right]_0^{180} = 4.5 \left(180^2 - \frac{180^2}{2}\right) = \frac{4.5 \cdot 180^2}{2} = 72{,}900$ ft $\cdot$ lb

11. The force against the piston is $F = pA$. If $V = Ax$, where x is the height of the cylinder, then $dV = A\,dx$

    $\Rightarrow$ Work $= \int F\,dx = \int pA\,dx = \int_{(p_1, V_1)}^{(p_2, V_2)} p\,dV$.

13. Let $r =$ the constant rate of leakage. Since the bucket is leaking at a constant rate and the bucket is rising at a constant rate, the amount of water in the bucket is proportional to $(20 - x)$, the distance the bucket is being raised. The leakage rate of the water is 0.8 lb/ft raised and the weight of the water in the bucket is $F = 0.8(20 - x)$. So:

    $W = \int_0^{20} 0.8(20 - x)\,dx = 0.8 \left[20x - \frac{x^2}{2}\right]_0^{20} = 160$ ft $\cdot$ lb.

15. We will use the coordinate system given.

    (a) The typical slab between the planes at y and $y + \Delta y$ has a volume of $\Delta V = (10)(12)\,\Delta y = 120\,\Delta y$ ft^3. The force F required to lift the slab is equal to its weight: $F = 62.4\,\Delta V = 62.4 \cdot 120\,\Delta y$ lb. The distance through which F must act is about y ft, so the work done lifting the slab is about $\Delta W =$ force $\times$ distance $= 62.4 \cdot 120 \cdot y \cdot \Delta y$ ft $\cdot$ lb. The work it takes to lift all the water is approximately $W \approx \sum_0^{20} \Delta W$

    $= \sum_0^{20} 62.4 \cdot 120y \cdot \Delta y$ ft $\cdot$ lb. This is a Riemann sum for

    the function $62.4 \cdot 120y$ over the interval $0 \le y \le 20$. The work of pumping the tank empty is the limit of these sums:

    $W = \int_0^{20} 62.4 \cdot 120y\,dy = (62.4)(120) \left[\frac{y^2}{2}\right]_0^{20} = (62.4)(120)\left(\frac{400}{2}\right) = (62.4)(120)(200) = 1{,}497{,}600$ ft $\cdot$ lb

    (b) The time t it takes to empty the full tank with $\left(\frac{5}{11}\right)$–hp motor is t $= \frac{W}{250\,\frac{\text{ft-lb}}{\text{sec}}} = \frac{1{,}497{,}600\ \text{ft-lb}}{250\,\frac{\text{ft-lb}}{\text{sec}}} = 5990.4$ sec

    $= 1.664$ hr $\Rightarrow$ t ≈ 1 hr and 40 min

    (c) Following all the steps of part (a), we find that the work it takes to lower the water level 10 ft is

    $W = \int_0^{10} 62.4 \cdot 120y\,dy = (62.4)(120) \left[\frac{y^2}{2}\right]_0^{10} = (62.4)(120)\left(\frac{100}{2}\right) = 374{,}400$ ft $\cdot$ lb and the time is t $= \frac{W}{250\,\frac{\text{ft-lb}}{\text{sec}}}$

    $= 1497.6$ sec $= 0.416$ hr ≈ 25 min

(d) In a location where water weighs $62.26 \frac{lb}{ft^3}$:

    a) $W = (62.26)(24,000) = 1,494,240$ ft · lb.

    b) $t = \frac{1,494,240}{250} = 5976.96$ sec ≈ 1.660 hr $\Rightarrow$ t $\approx$ 1 hr and 40 min

    In a location where water weighs $62.59 \frac{lb}{ft^3}$

    a) $W = (62.59)(24,000) = 1,502,160$ ft · lb

    b) $t = \frac{1,502,160}{250} = 6008.64$ sec ≈ 1.669 hr $\Rightarrow$ t $\approx$ 1 hr and 40.1 min

17. The slab is a disk of area $\pi x^2 = \pi \left(\frac{y}{2}\right)^2$, thickness $\triangle y$, and height below the top of the tank $(10 - y)$. So the work to pump the oil in this slab, $\triangle W$, is $57(10 - y)\pi \left(\frac{y}{2}\right)^2$. The work to pump all the oil to the top of the tank is

$$W = \int_0^{10} \frac{57\pi}{4}(10y^2 - y^3)dy = \frac{57\pi}{4}\left[\frac{10y^3}{3} - \frac{y^4}{4}\right]_0^{10} = 11,875\pi \text{ ft · lb} \approx 37,306 \text{ ft · lb.}$$

19. The typical slab between the planes at y and and $y + \Delta y$ has a volume of $\Delta V = \pi(\text{radius})^2(\text{thickness}) = \pi \left(\frac{20}{2}\right)^2 \Delta y$

$= \pi \cdot 100 \, \Delta y$ ft³. The force F required to lift the slab is equal to its weight: $F = 51.2 \, \Delta V = 51.2 \cdot 100\pi \, \Delta y$ lb $\Rightarrow$ $F = 5120\pi \, \Delta y$ lb. The distance through which F must act is about $(30 - y)$ ft. The work it takes to lift all the

kerosene is approximately $W \approx \sum_0^{30} \Delta W = \sum_0^{30} 5120\pi(30 - y) \, \Delta y$ ft · lb which is a Riemann sum. The work to pump the

tank dry is the limit of these sums: $W = \int_0^{30} 5120\pi(30 - y) \, dy = 5120\pi \left[30y - \frac{y^2}{2}\right]_0^{30} = 5120\pi \left(\frac{900}{2}\right) = (5120)(450\pi)$

$\approx 7,238,229.48$ ft · lb

21. (a) Follow all the steps of Example 5 but make the substitution of $64.5 \frac{lb}{ft^3}$ for $57 \frac{lb}{ft^3}$. Then,

$$W = \int_0^8 \frac{64.5\pi}{4}(10 - y)y^2 \, dy = \frac{64.5\pi}{4}\left[\frac{10y^3}{3} - \frac{y^4}{4}\right]_0^8 = \frac{64.5\pi}{4}\left(\frac{10 \cdot 8^3}{3} - \frac{8^4}{4}\right) = \left(\frac{64.5\pi}{4}\right)(8^3)\left(\frac{10}{3} - 2\right)$$

$$= \frac{64.5\pi \cdot 8^3}{3} = 21.5\pi \cdot 8^3 \approx 34,582.65 \text{ ft · lb}$$

    (b) Exactly as done in Example 5 but change the distance through which F acts to distance $\approx (13 - y)$ ft. Then

$$W = \int_0^8 \frac{57\pi}{4}(13 - y)y^2 \, dy = \frac{57\pi}{4}\left[\frac{13y^3}{3} - \frac{y^4}{4}\right]_0^8 = \frac{57\pi}{4}\left(\frac{13 \cdot 8^3}{3} - \frac{8^4}{4}\right) = \left(\frac{57\pi}{4}\right)(8^3)\left(\frac{13}{3} - 2\right) = \frac{57\pi \cdot 8^3 \cdot 7}{3 \cdot 4}$$

$$= (19\pi)\left(8^2\right)(7)(2) \approx 53.482.5 \text{ ft · lb}$$

23. The typical slab between the planes at y and $y + \Delta y$ has a volume of about $\Delta V = \pi(\text{radius})^2(\text{thickness})$

$= \pi \left(\sqrt{25 - y^2}\right)^2 \Delta y$ m³. The force F(y) required to lift this slab is equal to its weight: $F(y) = 9800 \cdot \Delta V$

$= 9800\pi \left(\sqrt{25 - y^2}\right)^2 \Delta y = 9800\pi \left(25 - y^2\right) \Delta y$ N. The distance through which F(y) must act to lift the

slab to the level of 4 m above the top of the reservoir is about $(4 - y)$ m, so the work done is approximately

$\Delta W \approx 9800\pi \left(25 - y^2\right)(4 - y) \Delta y$ N · m. The work done lifting all the slabs from $y = -5$ m to $y = 0$ m is

approximately $W \approx \sum_{-5}^0 9800\pi \left(25 - y^2\right)(4 - y) \Delta y$ N · m. Taking the limit of these Riemann sums, we get

$$W = \int_{-5}^0 9800\pi \left(25 - y^2\right)(4 - y) \, dy = 9800\pi \int_{-5}^0 (100 - 25y - 4y^2 + y^3) \, dy = 9800\pi \left[100y - \frac{25}{2}y^2 - \frac{4}{3}y^3 + \frac{y^4}{4}\right]_{-5}^0$$

$$= -9800\pi \left(-500 - \frac{25 \cdot 25}{2} + \frac{4}{3} \cdot 125 + \frac{625}{4}\right) \approx 15,073,099.75 \text{ J}$$

25. $F = m\frac{dv}{dt} = mv\frac{dv}{dx}$ by the chain rule $\Rightarrow$ $W = \int_{x_1}^{x_2} mv \frac{dv}{dx} \, dx = m\int_{x_1}^{x_2} \left(v \frac{dv}{dx}\right) dx = m\left[\frac{1}{2}v^2(x)\right]_{x_1}^{x_2}$

$= \frac{1}{2}m\left[v^2(x_2) - v^2(x_1)\right] = \frac{1}{2}mv_2^2 - \frac{1}{2}mv_1^2$, as claimed.

27. 90 mph $= \frac{90 \text{ mi}}{1 \text{ hr}} \cdot \frac{1 \text{ hr}}{60 \text{ min}} \cdot \frac{1 \text{ min}}{60 \text{ sec}} \cdot \frac{5280 \text{ ft}}{1 \text{ mi}} = 132$ ft/sec; $m = \frac{0.3125 \text{ lb}}{32 \text{ ft/sec}^2} = \frac{0.3125}{32}$ slugs;

$W = \left(\frac{1}{2}\right)\left(\frac{0.3125 \text{ lb}}{32 \text{ ft/sec}^2}\right)(132 \text{ ft/sec})^2 \approx 85.1$ ft · lb

29. weight $= 2$ oz $= \frac{1}{8}$ lb $\Rightarrow$ m $= \frac{\frac{1}{8}}{32}$ slugs $= \frac{1}{256}$ slugs; 124 mph $= \frac{(124)(5280)}{(60)(60)} \approx 181.87$ ft/sec;

   W $= \left(\frac{1}{2}\right) \left(\frac{1}{256}\text{ slugs}\right) (181.87 \text{ ft/sec})^2 \approx 64.6$ ft $\cdot$ lb

31. weight $= 6.5$ oz $= \frac{6.5}{16}$ lb $\Rightarrow$ m $= \frac{6.5}{(16)(32)}$ slugs; W $= \left(\frac{1}{2}\right) \left(\frac{6.5}{(16)(32)}\text{ slugs}\right) (132 \text{ ft/sec})^2 \approx 110.6$ ft $\cdot$ lb

33. We imagine the milkshake divided into thin slabs by planes perpendicular to the y-axis at the points of a partition of the interval $[0, 7]$. The typical slab between the planes at y and $y + \Delta y$ has a volume of about

   $\Delta V = \pi(\text{radius})^2(\text{thickness}) = \pi \left(\frac{y+17.5}{14}\right)^2 \Delta y$ in^3. The force F(y) required to lift this slab is equal to its

   weight: $F(y) = \frac{4}{9} \Delta V = \frac{4\pi}{9} \left(\frac{y+17.5}{14}\right)^2 \Delta y$ oz. The distance through which F(y) must act to lift this slab to

   the level of 1 inch above the top is about $(8 - y)$ in. The work done lifting the slab is about

   $\Delta W = \left(\frac{4\pi}{9}\right) \frac{(y+17.5)^2}{14^2} (8 - y) \Delta y$ in $\cdot$ oz. The work done lifting all the slabs from $y = 0$ to $y = 7$ is

   approximately $W = \sum_0^7 \frac{4\pi}{9 \cdot 14^2} (y + 17.5)^2 (8 - y) \Delta y$ in $\cdot$ oz which is a Riemann sum. The work is the limit of

   these sums as the norm of the partition goes to zero: $W = \int_0^7 \frac{4\pi}{9 \cdot 14^2} (y + 17.5)^2 (8 - y) \, dy$

   $= \frac{4\pi}{9 \cdot 14^2} \int_0^7 (2450 - 26.25y - 27y^2 - y^3) \, dy = \frac{4\pi}{9 \cdot 14^2} \left[-\frac{y^4}{4} - 9y^3 - \frac{26.25}{2} y^2 + 2450y\right]_0^7$

   $= \frac{4\pi}{9 \cdot 14^2} \left[-\frac{7^4}{4} - 9 \cdot 7^3 - \frac{26.25}{2} \cdot 7^2 + 2450 \cdot 7\right] \approx 91.32$ in $\cdot$ oz

35. Work $= \int_{6,370,000}^{35,780,000} \frac{1000\,MG}{r^2} \, dr = 1000\,MG \int_{6,370,000}^{35,780,000} \frac{dr}{r^2} = 1000\,MG \left[-\frac{1}{r}\right]_{6,370,000}^{35,780,000}$

   $= (1000) \left(5.975 \cdot 10^{24}\right) \left(6.672 \cdot 10^{-11}\right) \left(\frac{1}{6,370,000} - \frac{1}{35,780,000}\right) \approx 5.144 \times 10^{10}$ J

6.7 MOMENTS AND CENTERS OF MASS

1. Since the plate is symmetric about the y-axis and its density is constant, the distribution of mass is symmetric about the y-axis and the center of mass lies on the y-axis. This means that $\bar{x} = 0$. It remains to find $\bar{y} = \frac{M_x}{M}$. We model the distribution of mass with *vertical* strips. The typical strip has center of mass:

   $(\tilde{x}, \tilde{y}) = \left(x, \frac{x^2+4}{2}\right)$, length: $4 - x^2$, width: dx, area:

   dA $= (4 - x^2)$ dx, mass: dm $= \delta$ dA $= \delta (4 - x^2)$ dx. The moment of the strip about the x-axis is

   $\tilde{y}$ dm $= \left(\frac{x^2+4}{2}\right) \delta (4 - x^2)$ dx $= \frac{\delta}{2} (16 - x^4)$ dx. The moment of the plate about the x-axis is $M_x = \int \tilde{y}$ dm

   $= \int_{-2}^2 \frac{\delta}{2} (16 - x^4)$ dx $= \frac{\delta}{2} \left[16x - \frac{x^5}{5}\right]_{-2}^2 = \frac{\delta}{2} \left[\left(16 \cdot 2 - \frac{2^5}{5}\right) - \left(-16 \cdot 2 + \frac{2^5}{5}\right)\right] = \frac{\delta \cdot 2}{2} \left(32 - \frac{32}{5}\right) = \frac{128\delta}{5}$. The mass of the

   plate is M $= \int \delta (4 - x^2)$ dx $= \delta \left[4x - \frac{x^3}{3}\right]_{-2}^2 = 2\delta \left(8 - \frac{8}{3}\right) = \frac{32\delta}{3}$. Therefore $\bar{y} = \frac{M_x}{M} = \frac{\left(\frac{128\delta}{5}\right)}{\left(\frac{32\delta}{3}\right)} = \frac{12}{5}$. The plate's center of

   mass is the point $(\bar{x}, \bar{y}) = \left(0, \frac{12}{5}\right)$.

3. Intersection points: $x - x^2 = -x \Rightarrow 2x - x^2 = 0$
   $\Rightarrow x(2 - x) = 0 \Rightarrow x = 0$ or $x = 2$. The typical *vertical*

   strip has center of mass: $(\tilde{x}, \tilde{y}) = \left(x, \frac{(x-x^2) + (-x)}{2}\right)$

   $= \left(x, -\frac{x^2}{2}\right)$, length: $(x - x^2) - (-x) = 2x - x^2$, width: dx,

   area: dA $= (2x - x^2)$ dx, mass: dm $= \delta$ dA $= \delta (2x - x^2)$ dx.
   The moment of the strip about the x-axis is

$\tilde{y}$ dm $= \left(-\frac{x^2}{2}\right) \delta (2x - x^2)$ dx; about the y-axis it is $\tilde{x}$ dm $= x \cdot \delta (2x - x^2)$ dx. Thus, $M_x = \int \tilde{y}$ dm

$= -\int_0^2 \left(\frac{\delta}{2} x^2\right) (2x - x^2)$ dx $= -\frac{\delta}{2} \int_0^2 (2x^3 - x^4)$ dx $= -\frac{\delta}{2} \left[\frac{x^4}{2} - \frac{x^5}{5}\right]_0^2 = -\frac{\delta}{2} \left(2^3 - \frac{2^5}{5}\right) = -\frac{\delta}{2} \cdot 2^3 \left(1 - \frac{4}{5}\right)$

$= -\frac{4\delta}{5}$; $M_y = \int \tilde{x}$ dm $= \int_0^2 x \cdot \delta (2x - x^2)$ dx $= \delta \int_0^2 (2x^2 - x^3) = \delta \left[\frac{2}{3} x^3 - \frac{x^4}{4}\right]_0^2 = \delta \left(2 \cdot \frac{2^3}{3} - \frac{2^4}{4}\right) = \frac{\delta \cdot 2^4}{12} = \frac{4\delta}{3}$;

$M = \int$ dm $= \int_0^2 \delta (2x - x^2)$ dx $= \delta \int_0^2 (2x - x^2)$ dx $= \delta \left[x^2 - \frac{x^3}{3}\right]_0^2 = \delta \left(4 - \frac{8}{3}\right) = \frac{4\delta}{3}$. Therefore, $\bar{x} = \frac{M_y}{M}$

$= \left(\frac{4\delta}{3}\right) \left(\frac{3}{4\delta}\right) = 1$ and $\bar{y} = \frac{M_x}{M} = \left(-\frac{4\delta}{5}\right) \left(\frac{3}{4\delta}\right) = -\frac{3}{5}$ $\Rightarrow$ $(\bar{x}, \bar{y}) = \left(1, -\frac{3}{5}\right)$ is the center of mass.

5. The typical *horizontal* strip has center of mass:

$(\tilde{x}, \tilde{y}) = \left(\frac{y - y^3}{2}, y\right)$, length: $y - y^3$, width: dy,

area: $dA = (y - y^3)$ dy, mass: $dm = \delta dA = \delta (y - y^3)$ dy.
The moment of the strip about the y-axis is

$\tilde{x}$ dm $= \delta \left(\frac{y - y^3}{2}\right) (y - y^3)$ dy $= \frac{\delta}{2} (y - y^3)^2$ dy

$= \frac{\delta}{2} (y^2 - 2y^4 + y^6)$ dy; the moment about the x-axis is

$\tilde{y}$ dm $= \delta y (y - y^3)$ dy $= \delta (y^2 - y^4)$ dy. Thus, $M_x = \int \tilde{y}$ dm $= \delta \int_0^1 (y^2 - y^4)$ dy $= \delta \left[\frac{y^3}{3} - \frac{y^5}{5}\right]_0^1 = \delta \left(\frac{1}{3} - \frac{1}{5}\right) = \frac{2\delta}{15}$;

$M_y = \int \tilde{x}$ dm $= \frac{\delta}{2} \int_0^1 (y^2 - 2y^4 + y^6)$ dy $= \frac{\delta}{2} \left[\frac{y^3}{3} - \frac{2y^5}{5} + \frac{y^7}{7}\right]_0^1 = \frac{\delta}{2} \left(\frac{1}{3} - \frac{2}{5} + \frac{1}{7}\right) = \frac{\delta}{2} \left(\frac{35 - 42 + 15}{3 \cdot 5 \cdot 7}\right) = \frac{4\delta}{105}$; $M = \int$ dm

$= \delta \int_0^1 (y - y^3)$ dy $= \delta \left[\frac{y^2}{2} - \frac{y^4}{4}\right]_0^1 = \delta \left(\frac{1}{2} - \frac{1}{4}\right) = \frac{\delta}{4}$. Therefore, $\bar{x} = \frac{M_y}{M} = \left(\frac{4\delta}{105}\right) \left(\frac{4}{\delta}\right) = \frac{16}{105}$ and $\bar{y} = \frac{M_x}{M} = \left(\frac{2\delta}{15}\right) \left(\frac{4}{\delta}\right)$

$= \frac{8}{15}$ $\Rightarrow$ $(\bar{x}, \bar{y}) = \left(\frac{16}{105}, \frac{8}{15}\right)$ is the center of mass.

7. Applying the symmetry argument analogous to the one used
in Exercise 1, we find $\bar{x} = 0$. The typical *vertical* strip has
center of mass: $(\tilde{x}, \tilde{y}) = \left(x, \frac{\cos x}{2}\right)$, length: cos x, width: dx,
area: $dA = \cos x$ dx, mass: $dm = \delta dA = \delta \cos x$ dx. The
moment of the strip about the x-axis is $\tilde{y}$ dm $= \delta \cdot \frac{\cos x}{2} \cdot \cos x$ dx

$= \frac{\delta}{2} \cos^2 x$ dx $= \frac{\delta}{2} \left(\frac{1 + \cos 2x}{2}\right)$ dx $= \frac{\delta}{4} (1 + \cos 2x)$ dx; thus,

$M_x = \int \tilde{y}$ dm $= \int_{-\pi/2}^{\pi/2} \frac{\delta}{4} (1 + \cos 2x)$ dx $= \frac{\delta}{4} \left[x + \frac{\sin 2x}{2}\right]_{-\pi/2}^{\pi/2} = \frac{\delta}{4} \left[\left(\frac{\pi}{2} + 0\right) - \left(-\frac{\pi}{2}\right)\right] = \frac{\delta \pi}{4}$; $M = \int$ dm $= \delta \int_{-\pi/2}^{\pi/2} \cos x$ dx

$= \delta [\sin x]_{-\pi/2}^{\pi/2} = 2\delta$. Therefore, $\bar{y} = \frac{M_x}{M} = \frac{\delta \pi}{4 \cdot 2\delta} = \frac{\pi}{8}$ $\Rightarrow$ $(\bar{x}, \bar{y}) = \left(0, \frac{\pi}{8}\right)$ is the center of mass.

9. (a) $M_y = \int_1^2 x \left(\frac{1}{x}\right)$ dx $= 1$, $M_x = \int_1^2 \left(\frac{1}{2x}\right) \left(\frac{1}{x}\right)$ dx

$= \frac{1}{2} \int_1^2 \frac{1}{x^2}$ dx $= \left[-\frac{1}{2x}\right]_1^2 = \frac{1}{4}$, $M = \int_1^2 \frac{1}{x}$ dx

$= [\ln |x|]_1^2 = \ln 2 \Rightarrow \bar{x} = \frac{M_y}{M} = \frac{1}{\ln 2} \approx 1.44$ and

$\bar{y} = \frac{M_x}{M} = \frac{\left(\frac{1}{4}\right)}{\ln 2} \approx 0.36$

(b)

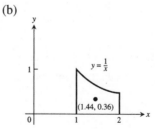

11. $M = \int_0^1 \frac{2}{1 + x^2}$ dx $= 2 [\tan^{-1} x]_0^1 = \frac{\pi}{2}$ and $M_y = \int_0^1 \frac{2x}{1 + x^2}$ dx $= [\ln (1 + x^2)]_0^1 = \ln 2 \Rightarrow \bar{x} = \frac{M_y}{M} = \frac{\ln 2}{\left(\frac{\pi}{2}\right)} = \frac{\ln 4}{\pi}$;
$\bar{y} = 0$ by symmetry

13. $M_y = \int_1^{16} x \left(\frac{1}{\sqrt{x}}\right)$ dx $= \int_1^{16} x^{1/2}$ dx $= \frac{2}{3} [x^{3/2}]_1^{16} = 42$; $M_x = \int_1^{16} \left(\frac{1}{2\sqrt{x}}\right) \left(\frac{1}{\sqrt{x}}\right)$ dx $= \frac{1}{2} \int_1^{16} \frac{1}{x}$ dx

$= \frac{1}{2} [\ln |x|]_1^{16} = \ln 4$, $M = \int_1^{16} \frac{1}{\sqrt{x}}$ dx $= [2x^{1/2}]_1^{16} = 6$ $\Rightarrow \bar{x} = \frac{M_y}{M} = 7$ and $\bar{y} = \frac{M_x}{M} = \frac{\ln 4}{6}$

15. $M_x = \int \tilde{y}\, dm = \int_1^2 \frac{\left(\frac{2}{x^2}\right)}{2} \cdot \delta \cdot \left(\frac{2}{x^2}\right) dx$

$= \int_1^2 \left(\frac{1}{x^2}\right)(x^2)\left(\frac{2}{x^2}\right) dx = \int_1^2 \frac{2}{x^2}\, dx = 2\int_1^2 x^{-2}\, dx$

$= 2\left[-x^{-1}\right]_1^2 = 2\left[\left(-\frac{1}{2}\right) - (-1)\right] = 2\left(\frac{1}{2}\right) = 1;$

$M_y = \int \tilde{x}\, dm = \int_1^2 x \cdot \delta \cdot \left(\frac{2}{x^2}\right) dx$

$= \int_1^2 x(x^2)\left(\frac{2}{x^2}\right) dx = 2\int_1^2 x\, dx = 2\left[\frac{x^2}{2}\right]_1^2$

$= 2\left(2 - \frac{1}{2}\right) = 4 - 1 = 3;\ M = \int dm = \int_1^2 \delta\left(\frac{2}{x^2}\right) dx = \int_1^2 x^2\left(\frac{2}{x^2}\right) dx = 2\int_1^2 dx = 2[x]_1^2 = 2(2-1) = 2.$ So

$\bar{x} = \frac{M_y}{M} = \frac{3}{2}$ and $\bar{y} = \frac{M_x}{M} = \frac{1}{2} \Rightarrow (\bar{x}, \bar{y}) = \left(\frac{3}{2}, \frac{1}{2}\right)$ is the center of mass.

(graph showing $y = 2/x^2$ with $(\bar{x}, \bar{y})$ marked, axes labeled x, y, points 1 and 2)

17. $M_y = \int_1^{16} x\left(\frac{1}{\sqrt{x}}\right)\left(\frac{4}{\sqrt{x}}\right) dx = 4\int_1^{16} dx = 60,\ M_x = \int_1^{16} \left(\frac{1}{2\sqrt{x}}\right)\left(\frac{1}{\sqrt{x}}\right)\left(\frac{4}{\sqrt{x}}\right) dx = 2\int_1^{16} x^{-3/2}\, dx$

$= -4\left[x^{-1/2}\right]_1^{16} = 3,\ M = \int_1^{16} \left(\frac{1}{\sqrt{x}}\right)\left(\frac{4}{\sqrt{x}}\right) dx = 4\int_1^{16} \frac{1}{x}\, dx = [4\ln|x|]_1^{16} = 4\ln 16 \Rightarrow \bar{x} = \frac{M_y}{M} = \frac{15}{\ln 16}$ and

$\bar{y} = \frac{M_x}{M} = \frac{3}{4\ln 16}$

19. (a) We use the shell method: $V = \int_a^b 2\pi\left(\begin{smallmatrix}\text{shell}\\\text{radius}\end{smallmatrix}\right)\left(\begin{smallmatrix}\text{shell}\\\text{height}\end{smallmatrix}\right) dx = \int_1^4 2\pi x\left[\frac{4}{\sqrt{x}} - \left(-\frac{4}{\sqrt{x}}\right)\right] dx$

$= 16\pi\int_1^4 \frac{x}{\sqrt{x}}\, dx = 16\pi\int_1^4 x^{1/2}\, dx = 16\pi\left[\frac{2}{3}x^{3/2}\right]_1^4 = 16\pi\left(\frac{2}{3}\cdot 8 - \frac{2}{3}\right) = \frac{32\pi}{3}(8-1) = \frac{224\pi}{3}$

(b) Since the plate is symmetric about the x-axis and its density $\delta(x) = \frac{1}{x}$ is a function of x alone, the distribution of its mass is symmetric about the x-axis. This means that $\bar{y} = 0$. We use the vertical strip approach to find $\bar{x}$: $M_y = \int \tilde{x}\, dm = \int_1^4 x \cdot \left[\frac{4}{\sqrt{x}} - \left(-\frac{4}{\sqrt{x}}\right)\right] \cdot \delta\, dx = \int_1^4 x \cdot \frac{8}{\sqrt{x}} \cdot \frac{1}{x}\, dx = 8\int_1^4 x^{-1/2}\, dx$

$= 8\left[2x^{1/2}\right]_1^4 = 8(2 \cdot 2 - 2) = 16;\ M = \int dm = \int_1^4 \left[\frac{4}{\sqrt{x}} - \left(\frac{-4}{\sqrt{x}}\right)\right] \cdot \delta\, dx = 8\int_1^4 \left(\frac{1}{\sqrt{x}}\right)\left(\frac{1}{x}\right) dx = 8\int_1^4 x^{-3/2}\, dx$

$= 8\left[-2x^{-1/2}\right]_1^4 = 8[-1 - (-2)] = 8.$ So $\bar{x} = \frac{M_y}{M} = \frac{16}{8} = 2 \Rightarrow (\bar{x}, \bar{y}) = (2, 0)$ is the center of mass.

(c)

21. The mass of a horizontal strip is $dm = \delta\, dA = \delta L\, dy$, where L is the width of the triangle at a distance of y above its base on the x-axis as shown in the figure in the text. Also, by similar triangles we have $\frac{L}{b} = \frac{h-y}{h}$

$\Rightarrow L = \frac{b}{h}(h-y).$ Thus, $M_x = \int \tilde{y}\, dm = \int_0^h \delta y\left(\frac{b}{h}\right)(h-y)\, dy = \frac{\delta b}{h}\int_0^h (hy - y^2)\, dy = \frac{\delta b}{h}\left[\frac{hy^2}{2} - \frac{y^3}{3}\right]_0^h$

$= \frac{\delta b}{h}\left(\frac{h^3}{2} - \frac{h^3}{3}\right) = \delta b h^2\left(\frac{1}{2} - \frac{1}{3}\right) = \frac{\delta b h^2}{6};\ M = \int dm = \int_0^h \delta\left(\frac{b}{h}\right)(h-y)\, dy = \frac{\delta b}{h}\int_0^h (h-y)\, dy = \frac{\delta b}{h}\left[hy - \frac{y^2}{2}\right]_0^h$

$= \frac{\delta b}{h}\left(h^2 - \frac{h^2}{2}\right) = \frac{\delta b h}{2}.$ So $\bar{y} = \frac{M_x}{M} = \left(\frac{\delta b h^2}{6}\right)\left(\frac{2}{\delta b h}\right) = \frac{h}{3} \Rightarrow$ the center of mass lies above the base of the triangle one-third of the way toward the opposite vertex. Similarly the other two sides of the triangle can be placed on the x-axis and the same results will occur. Therefore the centroid does lie at the intersection of the medians, as claimed.

23. From the symmetry about the line x = y it follows that
$\bar{x} = \bar{y}$. It also follows that the line through the points $(0, 0)$
and $\left(\frac{1}{2}, \frac{1}{2}\right)$ is a median $\Rightarrow \bar{y} = \bar{x} = \frac{2}{3} \cdot \left(\frac{1}{2} - 0\right) = \frac{1}{3}$
$\Rightarrow (\bar{x}, \bar{y}) = \left(\frac{1}{3}, \frac{1}{3}\right)$.

25. The point of intersection of the median from the vertex $(0, b)$
to the opposite side has coordinates $\left(0, \frac{a}{2}\right)$
$\Rightarrow \bar{y} = (b - 0) \cdot \frac{1}{3} = \frac{b}{3}$ and $\bar{x} = \left(\frac{a}{2} - 0\right) \cdot \frac{2}{3} = \frac{a}{3}$
$\Rightarrow (\bar{x}, \bar{y}) = \left(\frac{a}{3}, \frac{b}{3}\right)$.

27. $y = x^{1/2} \Rightarrow dy = \frac{1}{2} x^{-1/2} dx$

$\Rightarrow ds = \sqrt{(dx)^2 + (dy)^2} = \sqrt{1 + \frac{1}{4x}} \, dx$;

$M_x = \delta \int_0^2 \sqrt{x} \sqrt{1 + \frac{1}{4x}} \, dx$

$= \delta \int_0^2 \sqrt{x + \frac{1}{4}} \, dx = \frac{2\delta}{3} \left[\left(x + \frac{1}{4}\right)^{3/2}\right]_0^2$

$= \frac{2\delta}{3} \left[\left(2 + \frac{1}{4}\right)^{3/2} - \left(\frac{1}{4}\right)^{3/2}\right]$

$= \frac{2\delta}{3} \left[\left(\frac{9}{4}\right)^{3/2} - \left(\frac{1}{4}\right)^{3/2}\right] = \frac{2\delta}{3} \left(\frac{27}{8} - \frac{1}{8}\right) = \frac{13\delta}{6}$

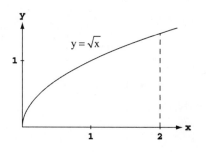

29. From Example 6 we have $M_x = \int_0^\pi a(a \sin \theta)(k \sin \theta) \, d\theta = a^2 k \int_0^\pi \sin^2 \theta \, d\theta = \frac{a^2 k}{2} \int_0^\pi (1 - \cos 2\theta) \, d\theta$

$= \frac{a^2 k}{2} \left[\theta - \frac{\sin 2\theta}{2}\right]_0^\pi = \frac{a^2 k \pi}{2}$; $M_y = \int_0^\pi a(a \cos \theta)(k \sin \theta) \, d\theta = a^2 k \int_0^\pi \sin \theta \cos \theta \, d\theta = \frac{a^2 k}{2} \left[\sin^2 \theta\right]_0^\pi = 0$;

$M = \int_0^\pi ak \sin \theta \, d\theta = ak[- \cos \theta]_0^\pi = 2ak$. Therefore, $\bar{x} = \frac{M_y}{M} = 0$ and $\bar{y} = \frac{M_x}{M} = \left(\frac{a^2 k \pi}{2}\right)\left(\frac{1}{2ak}\right) = \frac{a\pi}{4} \Rightarrow \left(0, \frac{a\pi}{4}\right)$
is the center of mass.

31. Consider the curve as an infinite number of line segments joined together. From the derivation of arc
length we have that the length of a particular segment is $ds = \sqrt{(dx)^2 + (dy)^2}$. This implies that
$M_x = \int \delta y \, ds$, $M_y = \int \delta x \, ds$ and $M = \int \delta \, ds$. If δ is constant, then $\bar{x} = \frac{M_y}{M} = \frac{\int x \, ds}{\int ds} = \frac{\int x \, ds}{length}$ and
$\bar{y} = \frac{M_x}{M} = \frac{\int y \, ds}{\int ds} = \frac{\int y \, ds}{length}$.

CHAPTER 6 PRACTICE EXERCISES

1. $A(x) = \frac{\pi}{4} (diameter)^2 = \frac{\pi}{4} \left(\sqrt{x} - x^2\right)^2$
$= \frac{\pi}{4} \left(x - 2\sqrt{x} \cdot x^2 + x^4\right)$; $a = 0, b = 1$
$\Rightarrow V = \int_a^b A(x) \, dx = \frac{\pi}{4} \int_0^1 \left(x - 2x^{5/2} + x^4\right) dx$
$= \frac{\pi}{4} \left[\frac{x^2}{2} - \frac{4}{7} x^{7/2} + \frac{x^5}{5}\right]_0^1 = \frac{\pi}{4} \left(\frac{1}{2} - \frac{4}{7} + \frac{1}{5}\right)$
$= \frac{\pi}{4 \cdot 70} (35 - 40 + 14) = \frac{9\pi}{280}$

3. $A(x) = \frac{\pi}{4} (\text{diameter})^2 = \frac{\pi}{4} (2 \sin x - 2 \cos x)^2$

$= \frac{\pi}{4} \cdot 4 (\sin^2 x - 2 \sin x \cos x + \cos^2 x)$

$= \pi(1 - \sin 2x); \, a = \frac{\pi}{4}, \, b = \frac{5\pi}{4}$

$\Rightarrow V = \int_a^b A(x) \, dx = \pi \int_{\pi/4}^{5\pi/4} (1 - \sin 2x) \, dx$

$= \pi \left[x + \frac{\cos 2x}{2} \right]_{\pi/4}^{5\pi/4}$

$= \pi \left[\left(\frac{5\pi}{4} + \frac{\cos \frac{5\pi}{2}}{2} \right) - \left(\frac{\pi}{4} - \frac{\cos \frac{\pi}{2}}{2} \right) \right] = \pi^2$

5. $A(x) = \frac{\pi}{4} (\text{diameter})^2 = \frac{\pi}{4} \left(2\sqrt{x} - \frac{x^2}{4} \right)^2 = \frac{\pi}{4} \left(4x - x^{5/2} + \frac{x^4}{16} \right); \, a = 0, \, b = 4 \Rightarrow V = \int_a^b A(x) \, dx$

$= \frac{\pi}{4} \int_0^4 \left(4x - x^{5/2} + \frac{x^4}{16} \right) dx = \frac{\pi}{4} \left[2x^2 - \frac{2}{7} x^{7/2} + \frac{x^5}{5 \cdot 16} \right]_0^4 = \frac{\pi}{4} \left(32 - 32 \cdot \frac{8}{7} + \frac{2}{5} \cdot 32 \right)$

$= \frac{32\pi}{4} \left(1 - \frac{8}{7} + \frac{2}{5} \right) = \frac{8\pi}{35} (35 - 40 + 14) = \frac{72\pi}{35}$

7. (a) *disk method*:

$V = \int_a^b \pi R^2(x) \, dx = \int_{-1}^1 \pi (3x^4)^2 \, dx = \pi \int_{-1}^1 9x^8 \, dx$

$= \pi [x^9]_{-1}^1 = 2\pi$

   (b) *shell method*:

$V = \int_a^b 2\pi \left(\begin{smallmatrix} \text{shell} \\ \text{radius} \end{smallmatrix} \right) \left(\begin{smallmatrix} \text{shell} \\ \text{height} \end{smallmatrix} \right) dx = \int_0^1 2\pi x (3x^4) \, dx = 2\pi \cdot 3 \int_0^1 x^5 \, dx = 2\pi \cdot 3 \left[\frac{x^6}{6} \right]_0^1 = \pi$

Note: The lower limit of integration is 0 rather than -1.

   (c) *shell method*:

$V = \int_a^b 2\pi \left(\begin{smallmatrix} \text{shell} \\ \text{radius} \end{smallmatrix} \right) \left(\begin{smallmatrix} \text{shell} \\ \text{height} \end{smallmatrix} \right) dx = 2\pi \int_{-1}^1 (1 - x) (3x^4) \, dx = 2\pi \left[\frac{3x^5}{5} - \frac{x^6}{2} \right]_{-1}^1 = 2\pi \left[\left(\frac{3}{5} - \frac{1}{2} \right) - \left(-\frac{3}{5} - \frac{1}{2} \right) \right] = \frac{12\pi}{5}$

   (d) *washer method*:

$R(x) = 3, \, r(x) = 3 - 3x^4 = 3(1 - x^4) \Rightarrow V = \int_a^b \pi [R^2(x) - r^2(x)] \, dx = \int_{-1}^1 \pi \left[9 - 9(1 - x^4)^2 \right] dx$

$= 9\pi \int_{-1}^1 [1 - (1 - 2x^4 + x^8)] \, dx = 9\pi \int_{-1}^1 (2x^4 - x^8) \, dx = 9\pi \left[\frac{2x^5}{5} - \frac{x^9}{9} \right]_{-1}^1 = 18\pi \left[\frac{2}{5} - \frac{1}{9} \right] = \frac{2\pi \cdot 13}{5} = \frac{26\pi}{5}$

9. (a) *disk method*:

$V = \pi \int_1^5 \left(\sqrt{x - 1} \right)^2 dx = \pi \int_1^5 (x - 1) \, dx = \pi \left[\frac{x^2}{2} - x \right]_1^5$

$= \pi \left[\left(\frac{25}{2} - 5 \right) - \left(\frac{1}{2} - 1 \right) \right] = \pi \left(\frac{24}{2} - 4 \right) = 8\pi$

   (b) *washer method*:

$R(y) = 5, \, r(y) = y^2 + 1 \Rightarrow V = \int_c^d \pi [R^2(y) - r^2(y)] \, dy = \pi \int_{-2}^2 \left[25 - (y^2 + 1)^2 \right] dy$

$= \pi \int_{-2}^2 (25 - y^4 - 2y^2 - 1) \, dy = \pi \int_{-2}^2 (24 - y^4 - 2y^2) \, dy = \pi \left[24y - \frac{y^5}{5} - \frac{2}{3} y^3 \right]_{-2}^2 = 2\pi \left(24 \cdot 2 - \frac{32}{5} - \frac{2}{3} \cdot 8 \right)$

$= 32\pi \left(3 - \frac{2}{5} - \frac{1}{3} \right) = \frac{32\pi}{15} (45 - 6 - 5) = \frac{1088\pi}{15}$

   (c) *disk method*:

$R(y) = 5 - (y^2 + 1) = 4 - y^2$

$\Rightarrow V = \int_c^d \pi R^2(y) \, dy = \int_{-2}^2 \pi (4 - y^2)^2 \, dy$

$= \pi \int_{-2}^2 (16 - 8y^2 + y^4) \, dy$

$= \pi \left[16y - \frac{8y^3}{3} + \frac{y^5}{5} \right]_{-2}^2 = 2\pi \left(32 - \frac{64}{3} + \frac{32}{5} \right)$

$= 64\pi \left(1 - \frac{2}{3} + \frac{1}{5} \right) = \frac{64\pi}{15} (15 - 10 + 3) = \frac{512\pi}{15}$

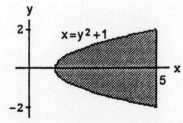

11. *disk method*:

$R(x) = \tan x, a = 0, b = \frac{\pi}{3} \Rightarrow V = \pi \int_0^{\pi/3} \tan^2 x \, dx = \pi \int_0^{\pi/3} (\sec^2 x - 1) \, dx = \pi[\tan x - x]_0^{\pi/3} = \frac{\pi(3\sqrt{3}-\pi)}{3}$

13. *shell method*:

$V = \int_0^1 2\pi \left(\begin{smallmatrix} \text{shell} \\ \text{radius} \end{smallmatrix}\right) \left(\begin{smallmatrix} \text{shell} \\ \text{height} \end{smallmatrix}\right) dx = \int_0^1 2\pi \, y \, e^{y^2} \, dy = \pi \, e^{y^2} \Big|_0^1 = \pi(e-1)$

15. The material removed from the sphere consists of a cylinder and two "caps." From the diagram, the height of the cylinder is 2h, where $h^2 + \left(\sqrt{3}\right)^2 = 2^2$, i.e. $h = 1$. Thus

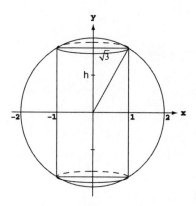

$V_{cyl} = (2h)\pi \left(\sqrt{3}\right)^2 = 6\pi$ ft^3. To get the volume of a cap, use the disk method and $x^2 + y^2 = 2^2$: $V_{cap} = \int_1^2 \pi x^2 dy$

$= \int_1^2 \pi(4 - y^2) dy = \pi\left[4y - \frac{y^3}{3}\right]_1^2$

$= \pi\left[\left(8 - \frac{8}{3}\right) - \left(4 - \frac{1}{3}\right)\right] = \frac{5\pi}{3}$ ft^3. Therefore,

$V_{removed} = V_{cyl} + 2V_{cap} = 6\pi + \frac{10\pi}{3} = \frac{28\pi}{3}$ ft^3.

17. $y = x^{1/2} - \frac{x^{3/2}}{3} \Rightarrow \frac{dy}{dx} = \frac{1}{2}x^{-1/2} - \frac{1}{2}x^{1/2} \Rightarrow \left(\frac{dy}{dx}\right)^2 = \frac{1}{4}\left(\frac{1}{x} - 2 + x\right) \Rightarrow L = \int_1^4 \sqrt{1 + \frac{1}{4}\left(\frac{1}{x} - 2 + x\right)} \, dx$

$\Rightarrow L = \int_1^4 \sqrt{\frac{1}{4}\left(\frac{1}{x} + 2 + x\right)} \, dx = \int_1^4 \sqrt{\frac{1}{4}(x^{-1/2} + x^{1/2})^2} \, dx = \int_1^4 \frac{1}{2}(x^{-1/2} + x^{1/2}) \, dx = \frac{1}{2}\left[2x^{1/2} + \frac{2}{3}x^{3/2}\right]_1^4$

$= \frac{1}{2}\left[\left(4 + \frac{2}{3} \cdot 8\right) - \left(2 + \frac{2}{3}\right)\right] = \frac{1}{2}\left(2 + \frac{14}{3}\right) = \frac{10}{3}$

19. $y = x^2 - \frac{\ln x}{8} \Rightarrow y' = 2x - \frac{1}{8x} \Rightarrow \sqrt{1 + (y')^2} = \sqrt{1 + \left(2x - \frac{1}{8x}\right)^2} = \sqrt{\frac{256x^4 + 32x^2 + 1}{64x^2}} = \sqrt{\frac{(16x^2 + 1)^2}{(8x)^2}} = \frac{16x^2 + 1}{8x}$

$= 2x + \frac{1}{8x} \Rightarrow \text{Length} = \int_1^2 \sqrt{1 + (y')^2} \, dx = \int_1^2 \left(2x + \frac{1}{8x}\right) dx = \left(x^2 + \frac{1}{8}\ln x\right)\Big|_1^2 = \left(4 + \frac{1}{8}\ln 2\right) - \left(1 + \frac{1}{8}\ln 1\right)$

$= 3 + \frac{1}{8}\ln 2$

21. $\frac{dx}{dt} = -5\sin t + 5\sin 5t$ and $\frac{dy}{dt} = 5\cos t - 5\cos 5t \Rightarrow \sqrt{\left(\frac{dx}{dt}\right)^2 + \left(\frac{dy}{dt}\right)^2}$

$= \sqrt{(-5\sin t + 5\sin 5t)^2 + (5\cos t - 5\cos 5t)^2}$

$= 5\sqrt{\sin^2 5t - 2\sin t \sin 5t + \sin^2 t + \cos^2 t - 2\cos t \cos 5t + \cos^2 5t} = 5\sqrt{2 - 2(\sin t \sin 5t + \cos t \cos 5t)}$

$= 5\sqrt{2(1 - \cos 4t)} = 5\sqrt{4\left(\frac{1}{2}\right)(1 - \cos 4t)} = 10\sqrt{\sin^2 2t} = 10|\sin 2t| = 10\sin 2t \text{ (since } 0 \leq t \leq \frac{\pi}{2})$

$\Rightarrow \text{Length} = \int_0^{\pi/2} 10\sin 2t \, dt = [-5\cos 2t]_0^{\pi/2} = (-5)(-1) - (-5)(1) = 10$

23. $\frac{dx}{d\theta} = -3\sin\theta$ and $\frac{dy}{d\theta} = 3\cos\theta \Rightarrow \sqrt{\left(\frac{dx}{d\theta}\right)^2 + \left(\frac{dy}{d\theta}\right)^2} = \sqrt{(-3\sin\theta)^2 + (3\cos\theta)^2} = \sqrt{3(\sin^2\theta + \cos^2\theta)} = 3$

$\Rightarrow \text{Length} = \int_0^{3\pi/2} 3 \, d\theta = 3\int_0^{3\pi/2} d\theta = 3\left(\frac{3\pi}{2} - 0\right) = \frac{9\pi}{2}$

25. $S = \int_a^b 2\pi y \sqrt{1 + \left(\frac{dy}{dx}\right)^2} \, dx; \frac{dy}{dx} = \frac{1}{\sqrt{2x+1}} \Rightarrow \left(\frac{dy}{dx}\right)^2 = \frac{1}{2x+1} \Rightarrow S = \int_0^3 2\pi\sqrt{2x+1}\sqrt{1 + \frac{1}{2x+1}} \, dx$

$= 2\pi\int_0^3 \sqrt{2x+1}\sqrt{\frac{2x+2}{2x+1}} \, dx = 2\sqrt{2}\pi\int_0^3 \sqrt{x+1} \, dx = 2\sqrt{2}\pi\left[\frac{2}{3}(x+1)^{3/2}\right]_0^3 = 2\sqrt{2}\pi \cdot \frac{2}{3}(8 - 1) = \frac{28\pi\sqrt{2}}{3}$

27. $S = \int_c^d 2\pi x \sqrt{1 + \left(\frac{dx}{dy}\right)^2}\, dy$; $\frac{dx}{dy} = \frac{(\frac{1}{2})(4-2y)}{\sqrt{4y-y^2}} = \frac{2-y}{\sqrt{4y-y^2}} \Rightarrow 1 + \left(\frac{dx}{dy}\right)^2 = \frac{4y-y^2+4-4y+y^2}{4y-y^2} = \frac{4}{4y-y^2}$

$\Rightarrow S = \int_1^2 2\pi \sqrt{4y-y^2}\sqrt{\frac{4}{4y-y^2}}\, dy = 4\pi \int_1^2 dx = 4\pi$

29. $x = \frac{t^2}{2}$ and $y = 2t$, $0 \le t \le \sqrt{5} \Rightarrow \frac{dx}{dt} = t$ and $\frac{dy}{dt} = 2 \Rightarrow$ Surface Area $= \int_0^{\sqrt{5}} 2\pi(2t)\sqrt{t^2+4}\, dt = \int_4^9 2\pi u^{1/2}\, du$

$= 2\pi \left[\frac{2}{3}u^{3/2}\right]_4^9 = \frac{76\pi}{3}$, where $u = t^2 + 4 \Rightarrow du = 2t\, dt; t = 0 \Rightarrow u = 4, t = \sqrt{5} \Rightarrow u = 9$

31. $\frac{dy}{dx} = \sqrt{y}\cos^2\sqrt{y} \Rightarrow \frac{dy}{\sqrt{y}\cos^2\sqrt{y}} = dx \Rightarrow 2\tan\sqrt{y} = x + C \Rightarrow y = \left(\tan^{-1}\left(\frac{x+C}{2}\right)\right)^2$

33. $yy' = \sec(y^2)\sec^2 x \Rightarrow \frac{y\, dy}{\sec(y^2)} = \sec^2 x\, dx \Rightarrow \frac{\sin(y^2)}{2} = \tan x + C \Rightarrow \sin(y^2) = 2\tan x + C_1$

35. $\frac{dy}{dx} = e^{-x-y-2} \Rightarrow e^y dy = e^{-(x+2)}dx \Rightarrow e^y = -e^{-(x+2)} + C$. We have $y(0) = -2$, so $e^{-2} = -e^{-2} + C \Rightarrow C = 2e^{-2}$ and

$e^y = -e^{-(x+2)} + 2e^{-2} \Rightarrow y = \ln\left(-e^{-(x+2)} + 2e^{-2}\right)$

37. $x\, dy - (y + \sqrt{y})dx = 0 \Rightarrow \frac{dy}{(y+\sqrt{y})} = \frac{dx}{x} \Rightarrow 2\ln(\sqrt{y}+1) = \ln x + C$. We have $y(1) = 1 \Rightarrow 2\ln\left(\sqrt{1}+1\right) = \ln 1 + C$

$\Rightarrow 2\ln 2 = C = \ln 2^2 = \ln 4$. So $2\ln(\sqrt{y}+1) = \ln x + \ln 4 = \ln(4x) \Rightarrow \ln(\sqrt{y}+1) = \frac{1}{2}\ln(4x) = \ln(4x)^{1/2}$

$\Rightarrow e^{\ln(\sqrt{y}+1)} = e^{\ln(4x)^{1/2}} \Rightarrow \sqrt{y}+1 = 2\sqrt{x} \Rightarrow y = (2\sqrt{x}-1)^2$

39. Since the half life is 5700 years and $A(t) = A_0 e^{kt}$ we have $\frac{A_0}{2} = A_0 e^{5700k} \Rightarrow \frac{1}{2} = e^{5700k} \Rightarrow \ln(0.5) = 5700k$

$\Rightarrow k = \frac{\ln(0.5)}{5700}$. With 10% of the original carbon-14 remaining we have $0.1A_0 = A_0 e^{\frac{\ln(0.5)}{5700}t} \Rightarrow 0.1 = e^{\frac{\ln(0.5)}{5700}t}$

$\Rightarrow \ln(0.1) = \frac{\ln(0.5)}{5700}t \Rightarrow t = \frac{(5700)\ln(0.1)}{\ln(0.5)} \approx 18{,}935$ years (rounded to the nearest year).

41. The equipment alone: the force required to lift the equipment is equal to its weight $\Rightarrow F_1(x) = 100$ N.

The work done is $W_1 = \int_a^b F_1(x)\, dx = \int_0^{40} 100\, dx = [100x]_0^{40} = 4000$ J; the rope alone: the force required

to lift the rope is equal to the weight of the rope paid out at elevation $x \Rightarrow F_2(x) = 0.8(40 - x)$. The work

done is $W_2 = \int_a^b F_2(x)\, dx = \int_0^{40} 0.8(40 - x)\, dx = 0.8\left[40x - \frac{x^2}{2}\right]_0^{40} = 0.8\left(40^2 - \frac{40^2}{2}\right) = \frac{(0.8)(1600)}{2} = 640$ J;

the total work is $W = W_1 + W_2 = 4000 + 640 = 4640$ J

43. Force constant: $F = kx \Rightarrow 20 = k \cdot 1 \Rightarrow k = 20$ lb/ft; the work to stretch the spring 1 ft is

$W = \int_0^1 kx\, dx = k\int_0^1 x\, dx = \left[20\frac{x^2}{2}\right]_0^1 = 10$ ft · lb; the work to stretch the spring an additional foot is

$W = \int_1^2 kx\, dx = k\int_1^2 x\, dx = 20\left[\frac{x^2}{2}\right]_1^2 = 20\left(\frac{4}{2} - \frac{1}{2}\right) = 20\left(\frac{3}{2}\right) = 30$ ft · lb

45. We imagine the water divided into thin slabs by planes
perpendicular to the y-axis at the points of a partition of the
interval $[0, 8]$. The typical slab between the planes at y and
$y + \Delta y$ has a volume of about $\Delta V = \pi(\text{radius})^2(\text{thickness})$
$= \pi\left(\frac{5}{4}y\right)^2\Delta y = \frac{25\pi}{16}y^2\,\Delta y$ ft^3. The force $F(y)$ required to
lift this slab is equal to its weight: $F(y) = 62.4\,\Delta V$
$= \frac{(62.4)(25)}{16}\pi y^2\,\Delta y$ lb. The distance through which $F(y)$

Reservoir's Cross Section

must act to lift this slab to the level 6 ft above the top is

about $(6 + 8 - y)$ ft, so the work done lifting the slab is about $\Delta W = \frac{(62.4)(25)}{16}\pi y^2(14 - y)\,\Delta y$ ft · lb. The work done

lifting all the slabs from $y = 0$ to $y = 8$ to the level 6 ft above the top is approximately

$W \approx \sum_{0}^{8} \frac{(62.4)(25)}{16} \pi y^2 (14 - y) \Delta y$ ft · lb so the work to pump the water is the limit of these Riemann sums as the norm of

the partition goes to zero: $W = \int_0^8 \frac{(62.4)(25)}{(16)} \pi y^2 (14 - y) \, dy = \frac{(62.4)(25)\pi}{16} \int_0^8 (14y^2 - y^3) \, dy = (62.4)\left(\frac{25\pi}{16}\right)\left[\frac{14}{3} y^3 - \frac{y^4}{4}\right]_0^8$

$= (62.4)\left(\frac{25\pi}{16}\right)\left(\frac{14}{3} \cdot 8^3 - \frac{8^4}{4}\right) \approx 418{,}208.81$ ft · lb

47. The tank's cross section looks like the figure in Exercise 45 with right edge given by $x = \frac{5}{10} y = \frac{y}{2}$. A typical

horizontal slab has volume $\Delta V = \pi(\text{radius})^2(\text{thickness}) = \pi\left(\frac{y}{2}\right)^2 \Delta y = \frac{\pi}{4} y^2 \Delta y$. The force required to lift this

slab is its weight: $F(y) = 60 \cdot \frac{\pi}{4} y^2 \Delta y$. The distance through which F(y) must act is $(2 + 10 - y)$ ft, so the

work to pump the liquid is $W = 60 \int_0^{10} \pi(12 - y)\left(\frac{y^2}{4}\right) dy = 15\pi\left[\frac{12y^3}{3} - \frac{y^4}{4}\right]_0^{10} = 22{,}500\pi$ ft · lb; the time needed

to empty the tank is $\frac{22{,}500 \text{ ft·lb}}{275 \text{ ft·lb/sec}} \approx 257$ sec

49. Intersection points: $3 - x^2 = 2x^2 \Rightarrow 3x^2 - 3 = 0$

$\Rightarrow 3(x - 1)(x + 1) = 0 \Rightarrow x = -1$ or $x = 1$. Symmetry

suggests that $\overline{x} = 0$. The typical *vertical* strip has

center of mass: $(\widetilde{x}, \widetilde{y}) = \left(x, \frac{2x^2 + (3 - x^2)}{2}\right) = \left(x, \frac{x^2 + 3}{2}\right)$,

length: $(3 - x^2) - 2x^2 = 3(1 - x^2)$, width: dx,

area: $dA = 3(1 - x^2) dx$, and mass: $dm = \delta \cdot dA$

$= 3\delta(1 - x^2) dx \Rightarrow$ the moment about the x-axis is

$\widetilde{y} \, dm = \frac{3}{2}\delta(x^2 + 3)(1 - x^2) dx = \frac{3}{2}\delta(-x^4 - 2x^2 + 3) dx \Rightarrow M_x = \int \widetilde{y} \, dm = \frac{3}{2}\delta\int_{-1}^1 (-x^4 - 2x^2 + 3) dx$

$= \frac{3}{2}\delta\left[-\frac{x^5}{5} - \frac{2x^3}{3} + 3x\right]_{-1}^1 = 3\delta\left(-\frac{1}{5} - \frac{2}{3} + 3\right) = \frac{3\delta}{15}(-3 - 10 + 45) = \frac{32\delta}{5}; M = \int dm = 3\delta\int_{-1}^1 (1 - x^2) dx$

$= 3\delta\left[x - \frac{x^3}{3}\right]_{-1}^1 = 6\delta\left(1 - \frac{1}{3}\right) = 4\delta \Rightarrow \overline{y} = \frac{M_x}{M} = \frac{\frac{32\delta}{5}}{5 \cdot 4\delta} = \frac{8}{5}$. Therefore, the centroid is $(\overline{x}, \overline{y}) = \left(0, \frac{8}{5}\right)$.

51. The typical *vertical* strip has: center of mass: $(\widetilde{x}, \widetilde{y})$

$= \left(x, \frac{4 + \frac{x^2}{4}}{2}\right)$, length: $4 - \frac{x^2}{4}$, width: dx,

area: $dA = \left(4 - \frac{x^2}{4}\right) dx$, mass: $dm = \delta \cdot dA$

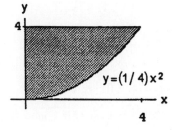

$= \delta\left(4 - \frac{x^2}{4}\right) dx \Rightarrow$ the moment about the x-axis is

$\widetilde{y} \, dm = \delta \cdot \frac{\left(4 + \frac{x^2}{4}\right)}{2}\left(4 - \frac{x^2}{4}\right) dx = \frac{\delta}{2}\left(16 - \frac{x^4}{16}\right) dx$; the

moment about the y-axis is $\widetilde{x} \, dm = \delta\left(4 - \frac{x^2}{4}\right) \cdot x \, dx = \delta\left(4x - \frac{x^3}{4}\right) dx$. Thus, $M_x = \int \widetilde{y} \, dm = \frac{\delta}{2}\int_0^4 \left(16 - \frac{x^4}{16}\right) dx$

$= \frac{\delta}{2}\left[16x - \frac{x^5}{5 \cdot 16}\right]_0^4 = \frac{\delta}{2}\left[64 - \frac{64}{5}\right] = \frac{128\delta}{5}; M_y = \int \widetilde{x} \, dm = \delta\int_0^4 \left(4x - \frac{x^3}{4}\right) dx = \delta\left[2x^2 - \frac{x^4}{16}\right]_0^4$

$= \delta(32 - 16) = 16\delta; M = \int dm = \delta\int_0^4 \left(4 - \frac{x^2}{4}\right) dx = \delta\left[4x - \frac{x^3}{12}\right]_0^4 = \delta\left(16 - \frac{64}{12}\right) = \frac{32\delta}{3}$

$\Rightarrow \overline{x} = \frac{M_y}{M} = \frac{16 \cdot \delta \cdot 3}{32 \cdot \delta} = \frac{3}{2}$ and $\overline{y} = \frac{M_x}{M} = \frac{128 \cdot \delta \cdot 3}{5 \cdot 32 \cdot \delta} = \frac{12}{5}$. Therefore, the centroid is $(\overline{x}, \overline{y}) = \left(\frac{3}{2}, \frac{12}{5}\right)$.

53. A typical horizontal strip has: center of mass: $(\widetilde{x}, \widetilde{y})$

$= \left(\frac{y^2 + 2y}{2}, y\right)$, length: $2y - y^2$, width: dy,

area: $dA = (2y - y^2) dy$, mass: $dm = \delta \cdot dA$

$= (1 + y)(2y - y^2) dy \Rightarrow$ the moment about the

x-axis is $\widetilde{y} \, dm = y(1 + y)(2y - y^2) dy$

$= (2y^2 + 2y^3 - y^3 - y^4) dy$

$= (2y^2 + y^3 - y^4) dy$; the moment about the y-axis is

$$\tilde{x} \, dm = \left(\frac{y^2+2y}{2}\right)(1+y)(2y-y^2)\,dy = \tfrac{1}{2}(4y^2-y^4)(1+y)\,dy = \tfrac{1}{2}(4y^2+4y^3-y^4-y^5)\,dy$$

$$\Rightarrow M_x = \int \tilde{y} \, dm = \int_0^2 (2y^2+y^3-y^4)\,dy = \left[\tfrac{2}{3}y^3+\tfrac{y^4}{4}-\tfrac{y^5}{5}\right]_0^2 = \left(\tfrac{16}{3}+\tfrac{16}{4}-\tfrac{32}{5}\right) = 16\left(\tfrac{1}{3}+\tfrac{1}{4}-\tfrac{2}{5}\right)$$

$$= \tfrac{16}{60}(20+15-24) = \tfrac{4}{15}(11) = \tfrac{44}{15} \,;\, M_y = \int \tilde{x} \, dm = \int_0^2 \tfrac{1}{2}(4y^2+4y^3-y^4-y^5)\,dy = \tfrac{1}{2}\left[\tfrac{4}{3}y^3+y^4-\tfrac{y^5}{5}-\tfrac{y^6}{6}\right]_0^2$$

$$= \tfrac{1}{2}\left(\tfrac{4\cdot2^3}{3}+2^4-\tfrac{2^5}{5}-\tfrac{2^6}{6}\right) = 4\left(\tfrac{4}{3}+2-\tfrac{4}{5}-\tfrac{8}{6}\right) = 4\left(2-\tfrac{4}{5}\right) = \tfrac{24}{5} \,;\, M = \int dm = \int_0^2 (1+y)(2y-y^2)\,dy$$

$$= \int_0^2 (2y+y^2-y^3)\,dy = \left[y^2+\tfrac{y^3}{3}-\tfrac{y^4}{4}\right]_0^2 = \left(4+\tfrac{8}{3}-\tfrac{16}{4}\right) = \tfrac{8}{3} \Rightarrow \bar{x} = \tfrac{M_y}{M} = \left(\tfrac{24}{5}\right)\left(\tfrac{3}{8}\right) = \tfrac{9}{5} \text{ and } \bar{y} = \tfrac{M_x}{M}$$

$$= \left(\tfrac{44}{15}\right)\left(\tfrac{3}{8}\right) = \tfrac{44}{40} = \tfrac{11}{10} \,. \text{ Therefore, the center of mass is } (\bar{x},\bar{y}) = \left(\tfrac{9}{5},\tfrac{11}{10}\right).$$

CHAPTER 6 ADDITIONAL AND ADVANCED EXERCISES

1. $V = \pi \int_a^b [f(x)]^2 \, dx = b^2 - ab \Rightarrow \pi \int_a^x [f(t)]^2 \, dt = x^2 - ax$ for all $x > a \Rightarrow \pi[f(x)]^2 = 2x - a \Rightarrow f(x) = \sqrt{\frac{2x-a}{\pi}}$

3. $s(x) = Cx \Rightarrow \int_0^x \sqrt{1+[f'(t)]^2} \, dt = Cx \Rightarrow \sqrt{1+[f'(x)]^2} = C \Rightarrow f'(x) = \sqrt{C^2-1}$ for $C \geq 1$

   $\Rightarrow f(x) = \int_0^x \sqrt{C^2-1} \, dt + k$. Then $f(0) = a \Rightarrow a = 0 + k \Rightarrow f(x) = \int_0^x \sqrt{C^2-1} \, dt + a \Rightarrow f(x) = x\sqrt{C^2-1} + a$,

   where $C \geq 1$.

5. $y = 2\sqrt{x} \Rightarrow ds = \sqrt{\tfrac{1}{x}+1} \, dx \Rightarrow A = \int_0^3 2\sqrt{x}\sqrt{\tfrac{1}{x}+1} \, dx = \tfrac{4}{3}\left[(1+x)^{3/2}\right]_0^3 = \tfrac{28}{3}$

7. (a) $\frac{dy}{dt} = k\frac{A}{V}(c-y) \Rightarrow dy = -k\frac{A}{V}(y-c)dt \Rightarrow \frac{dy}{y-c} = -k\frac{A}{V}dt \Rightarrow \int\frac{dy}{y-c} = -\int k\frac{A}{V}dt \Rightarrow \ln|y-c| = -k\frac{A}{V}t + C_1$

   $\Rightarrow y - c = \pm e^{C_1}e^{-k\frac{A}{V}t}$. Apply the initial condition, $y(0) = y_0 \Rightarrow y_0 = c + C \Rightarrow C = y_0 - c$

   $\Rightarrow y = c + (y_0-c)e^{-k\frac{A}{V}t}$.

   (b) Steady state solution: $y_\infty = \lim_{t\to\infty} y(t) = \lim_{t\to\infty}\left[c+(y_0-c)e^{-k\frac{A}{V}t}\right] = c + (y_0-c)(0) = c$

9. $F = ma = t^2 \Rightarrow \frac{d^2x}{dt^2} = a = \frac{t^2}{m} \Rightarrow v = \frac{dx}{dt} = \frac{t^3}{3m} + C; v = 0$ when $t = 0 \Rightarrow C = 0 \Rightarrow \frac{dx}{dt} = \frac{t^3}{3m} \Rightarrow x = \frac{t^4}{12m} + C_1;$

   $x = 0$ when $t = 0 \Rightarrow C_1 = 0 \Rightarrow x = \frac{t^4}{12m}$. Then $x = h \Rightarrow t = (12mh)^{1/4}$. The work done is

   $W = \int F \, dx = \int_0^{(12mh)^{1/4}} F(t)\cdot\frac{dx}{dt} \, dt = \int_0^{(12mh)^{1/4}} t^2 \cdot \frac{t^3}{3m} \, dt = \frac{1}{3m}\left[\frac{t^6}{6}\right]_0^{(12mh)^{1/4}} = \left(\frac{1}{18m}\right)(12mh)^{6/4}$

   $= \frac{(12mh)^{3/2}}{18m} = \frac{12mh\cdot\sqrt{12mh}}{18m} = \frac{2h}{3}\cdot2\sqrt{3mh} = \frac{4h}{3}\sqrt{3mh}$

11. From the symmetry of $y = 1 - x^n$, n even, about the y-axis for $-1 \leq x \leq 1$, we have $\bar{x} = 0$. To find $\bar{y} = \frac{M_x}{M}$, we

    use the vertical strips technique. The typical strip has center of mass: $(\tilde{x},\tilde{y}) = \left(x,\frac{1-x^n}{2}\right)$, length: $1 - x^n$,

    width: dx, area: $dA = (1-x^n)\,dx$, mass: $dm = 1\cdot dA = (1-x^n)\,dx$. The moment of the strip about the

    x-axis is $\tilde{y} \, dm = \frac{(1-x^n)^2}{2} \, dx \Rightarrow M_x = \int_{-1}^1 \frac{(1-x^n)^2}{2} \, dx = 2\int_0^1 \tfrac{1}{2}(1-2x^n+x^{2n}) \, dx = \left[x - \frac{2x^{n+1}}{n+1} + \frac{x^{2n+1}}{2n+1}\right]_0^1$

    $= 1 - \frac{2}{n+1} + \frac{1}{2n+1} = \frac{(n+1)(2n+1)-2(2n+1)+(n+1)}{(n+1)(2n+1)} = \frac{2n^2+3n+1-4n-2+n+1}{(n+1)(2n+1)} = \frac{2n^2}{(n+1)(2n+1)} \,.$

    Also, $M = \int_{-1}^1 dA = \int_{-1}^1 (1-x^n) \, dx = 2\int_0^1 (1-x^n) \, dx = 2\left[x - \frac{x^{n+1}}{n+1}\right]_0^1 = 2\left(1-\frac{1}{n+1}\right) = \frac{2n}{n+1}$. Therefore,

    $\bar{y} = \frac{M_x}{M} = \frac{2n^2}{(n+1)(2n+1)} \cdot \frac{(n+1)}{2n} = \frac{n}{2n+1} \Rightarrow \left(0,\frac{n}{2n+1}\right)$ is the location of the centroid. As $n \to \infty$, $\bar{y} \to \frac{1}{2}$ so

    the limiting position of the centroid is $\left(0,\frac{1}{2}\right)$.

13. (a) Consider a single vertical strip with center of mass $(\tilde{x},\tilde{y})$. If the plate lies to the right of the line, then

    the moment of this strip about the line $x = b$ is $(\tilde{x}-b)\,dm = (\tilde{x}-b)\,\delta\,dA \Rightarrow$ the plate's first moment

    about $x = b$ is the integral $\int (x-b)\delta \, dA = \int \delta x \, dA - \int \delta b \, dA = M_y - b\delta A$.

(b) If the plate lies to the left of the line, the moment of a vertical strip about the line $x = b$ is

$(b - \tilde{x}\,)\,dm = (b - \tilde{x}\,)\,\delta\,dA \Rightarrow$ the plate's first moment about $x = b$ is $\int (b - x)\delta\,dA = \int b\delta\,dA - \int \delta x\,dA$
$= b\delta A - M_y$.

15. (a) On $[0, a]$ a typical *vertical* strip has center of mass: $(\tilde{x}\,,\tilde{y}\,) = \left(x, \frac{\sqrt{b^2 - x^2} + \sqrt{a^2 - x^2}}{2}\right)$,

length: $\sqrt{b^2 - x^2} - \sqrt{a^2 - x^2}$, width: dx, area: $dA = \left(\sqrt{b^2 - x^2} - \sqrt{a^2 - x^2}\right) dx$, mass: $dm = \delta\,dA$

$= \delta\left(\sqrt{b^2 - x^2} - \sqrt{a^2 - x^2}\right) dx$. On $[a, b]$ a typical *vertical* strip has center of mass:

$(\tilde{x}\,,\tilde{y}\,) = \left(x, \frac{\sqrt{b^2 - x^2}}{2}\right)$, length: $\sqrt{b^2 - x^2}$, width: dx, area: $dA = \sqrt{b^2 - x^2}\,dx$,

mass: $dm = \delta\,dA = \delta\sqrt{b^2 - x^2}\,dx$. Thus, $M_x = \int \tilde{y}\,dm$

$= \int_0^a \frac{1}{2}\left(\sqrt{b^2 - x^2} + \sqrt{a^2 - x^2}\right)\delta\left(\sqrt{b^2 - x^2} - \sqrt{a^2 - x^2}\right) dx + \int_a^b \frac{1}{2}\sqrt{b^2 - x^2}\,\delta\sqrt{b^2 - x^2}\,dx$

$= \frac{\delta}{2}\int_0^a[(b^2 - x^2) - (a^2 - x^2)]\,dx + \frac{\delta}{2}\int_a^b(b^2 - x^2)\,dx = \frac{\delta}{2}\int_0^a(b^2 - a^2)\,dx + \frac{\delta}{2}\int_a^b(b^2 - x^2)\,dx$

$= \frac{\delta}{2}\left[(b^2 - a^2)\,x\right]_0^a + \frac{\delta}{2}\left[b^2 x - \frac{x^3}{3}\right]_a^b = \frac{\delta}{2}[(b^2 - a^2)\,a] + \frac{\delta}{2}\left[\left(b^3 - \frac{b^3}{3}\right) - \left(b^2 a - \frac{a^3}{3}\right)\right]$

$= \frac{\delta}{2}(ab^2 - a^3) + \frac{\delta}{2}\left(\frac{2}{3}b^3 - ab^2 + \frac{a^3}{3}\right) = \frac{\delta b^3}{3} - \frac{\delta a^3}{3} = \delta\left(\frac{b^3 - a^3}{3}\right); M_y = \int \tilde{x}\,dm$

$= \int_0^a x\delta\left(\sqrt{b^2 - x^2} - \sqrt{a^2 - x^2}\right) dx + \int_a^b x\delta\sqrt{b^2 - x^2}\,dx$

$= \delta\int_0^a x\,(b^2 - x^2)^{1/2}\,dx - \delta\int_0^a x\,(a^2 - x^2)^{1/2}\,dx + \delta\int_a^b x\,(b^2 - x^2)^{1/2}\,dx$

$= \frac{-\delta}{2}\left[\frac{2\,(b^2 - x^2)^{3/2}}{3}\right]_0^a + \frac{\delta}{2}\left[\frac{2\,(a^2 - x^2)^{3/2}}{3}\right]_0^a - \frac{\delta}{2}\left[\frac{2\,(b^2 - x^2)^{3/2}}{3}\right]_a^b$

$= -\frac{\delta}{3}\left[(b^2 - a^2)^{3/2} - (b^2)^{3/2}\right] + \frac{\delta}{3}\left[0 - (a^2)^{3/2}\right] - \frac{\delta}{3}\left[0 - (b^2 - a^2)^{3/2}\right] = \frac{\delta b^3}{3} - \frac{\delta a^3}{3} = \frac{\delta\,(b^3 - a^3)}{3} = M_x;$

We calculate the mass geometrically: $M = \delta A = \delta\left(\frac{\pi b^2}{4}\right) - \delta\left(\frac{\pi a^2}{4}\right) = \frac{\delta\pi}{4}(b^2 - a^2)$. Thus, $\bar{x} = \frac{M_y}{M}$

$= \frac{\delta\,(b^3 - a^3)}{3}\cdot\frac{4}{\delta\pi\,(b^2 - a^2)} = \frac{4}{3\pi}\left(\frac{b^3 - a^3}{b^2 - a^2}\right) = \frac{4}{3\pi}\frac{(b - a)(a^2 + ab + b^2)}{(b - a)(b + a)} = \frac{4\,(a^2 + ab + b^2)}{3\pi(a + b)}$; likewise

$\bar{y} = \frac{M_x}{M} = \frac{4\,(a^2 + ab + b^2)}{3\pi(a + b)}$.

(b) $\displaystyle\lim_{b\to a}\frac{4}{3\pi}\left(\frac{a^2 + ab + b^2}{a + b}\right) = \left(\frac{4}{3\pi}\right)\left(\frac{a^2 + a^2 + a^2}{a + a}\right) = \left(\frac{4}{3\pi}\right)\left(\frac{3a^2}{2a}\right) = \frac{2a}{\pi} \Rightarrow (\bar{x}, \bar{y}) = \left(\frac{2a}{\pi}, \frac{2a}{\pi}\right)$ is the limiting

position of the centroid as $b \to a$. This is the centroid of a circle of radius a (and we note the two circles coincide when $b = a$).

NOTES:

CHAPTER 7 TECHNIQUES OF INTEGRATION

7.1 INTEGRATION BY PARTS

1. $u = x$, $du = dx$; $dv = \sin \frac{x}{2}\, dx$, $v = -2 \cos \frac{x}{2}$;

$\int x \sin \frac{x}{2}\, dx = -2x \cos \frac{x}{2} - \int \left(-2 \cos \frac{x}{2}\right) dx = -2x \cos \left(\frac{x}{2}\right) + 4 \sin \left(\frac{x}{2}\right) + C$

3.

$$
\begin{array}{lll}
 & & \cos t \\[4pt]
t^2 & \xrightarrow{\;(+)\;} & \sin t \\[4pt]
2t & \xrightarrow{\;(-)\;} & -\cos t \\[4pt]
2 & \xrightarrow{\;(+)\;} & -\sin t \\[4pt]
0 & &
\end{array}
$$

$\int t^2 \cos t\, dt = t^2 \sin t + 2t \cos t - 2 \sin t + C$

5. $u = \ln x$, $du = \frac{dx}{x}$; $dv = x\, dx$, $v = \frac{x^2}{2}$;

$\int_1^2 x \ln x\, dx = \left[\frac{x^2}{2} \ln x\right]_1^2 - \int_1^2 \frac{x^2}{2}\, \frac{dx}{x} = 2 \ln 2 - \left[\frac{x^2}{4}\right]_1^2 = 2 \ln 2 - \frac{3}{4} = \ln 4 - \frac{3}{4}$

7. $u = \tan^{-1} y$, $du = \frac{dy}{1+y^2}$; $dv = dy$, $v = y$;

$\int \tan^{-1} y\, dy = y \tan^{-1} y - \int \frac{y\, dy}{(1+y^2)} = y \tan^{-1} y - \frac{1}{2} \ln\left(1 + y^2\right) + C = y \tan^{-1} y - \ln \sqrt{1+y^2} + C$

9. $u = x$, $du = dx$; $dv = \sec^2 x\, dx$, $v = \tan x$;

$\int x \sec^2 x\, dx = x \tan x - \int \tan x\, dx = x \tan x + \ln |\cos x| + C$

11.

$$
\begin{array}{lll}
 & & e^x \\[4pt]
x^3 & \xrightarrow{\;(+)\;} & e^x \\[4pt]
3x^2 & \xrightarrow{\;(-)\;} & e^x \\[4pt]
6x & \xrightarrow{\;(+)\;} & e^x \\[4pt]
6 & \xrightarrow{\;(-)\;} & e^x \\[4pt]
0 & &
\end{array}
$$

$\int x^3 e^x\, dx = x^3 e^x - 3x^2 e^x + 6x e^x - 6 e^x + C = \left(x^3 - 3x^2 + 6x - 6\right) e^x + C$

13.

$$
\begin{array}{lll}
 & & e^x \\[4pt]
x^2 - 5x & \xrightarrow{\;(+)\;} & e^x \\[4pt]
2x - 5 & \xrightarrow{\;(-)\;} & e^x \\[4pt]
2 & \xrightarrow{\;(+)\;} & e^x \\[4pt]
0 & &
\end{array}
$$

$\int (x^2 - 5x)\, e^x\, dx = (x^2 - 5x)\, e^x - (2x - 5) e^x + 2 e^x + C = x^2 e^x - 7x e^x + 7 e^x + C$

$= (x^2 - 7x + 7)\, e^x + C$

15.

$$
\begin{array}{ll}
& e^x \\
x^5 \xrightarrow{(+)} & e^x \\
5x^4 \xrightarrow{(-)} & e^x \\
20x^3 \xrightarrow{(+)} & e^x \\
60x^2 \xrightarrow{(-)} & e^x \\
120x \xrightarrow{(+)} & e^x \\
120 \xrightarrow{(-)} & e^x \\
0 &
\end{array}
$$

$$\int x^5 e^x \, dx = x^5 e^x - 5x^4 e^x + 20x^3 e^x - 60x^2 e^x + 120x e^x - 120 e^x + C$$
$$= \left(x^5 - 5x^4 + 20x^3 - 60x^2 + 120x - 120\right) e^x + C$$

17.

$$
\begin{array}{ll}
& \sin 2\theta \\
\theta^2 \xrightarrow{(+)} & -\tfrac{1}{2}\cos 2\theta \\
2\theta \xrightarrow{(-)} & -\tfrac{1}{4}\sin 2\theta \\
2 \xrightarrow{(+)} & \tfrac{1}{8}\cos 2\theta \\
0 &
\end{array}
$$

$$\int_0^{\pi/2} \theta^2 \sin 2\theta \, d\theta = \left[-\tfrac{\theta^2}{2}\cos 2\theta + \tfrac{\theta}{2}\sin 2\theta + \tfrac{1}{4}\cos 2\theta\right]_0^{\pi/2}$$
$$= \left[-\tfrac{\pi^2}{8}\cdot(-1) + \tfrac{\pi}{4}\cdot 0 + \tfrac{1}{4}\cdot(-1)\right] - \left[0 + 0 + \tfrac{1}{4}\cdot 1\right] = \tfrac{\pi^2}{8} - \tfrac{1}{2} = \tfrac{\pi^2-4}{8}$$

19. $u = \sec^{-1} t, \ du = \dfrac{dt}{t\sqrt{t^2-1}}; \ dv = t \, dt, \ v = \dfrac{t^2}{2};$

$$\int_{2/\sqrt{3}}^2 t \sec^{-1} t \, dt = \left[\tfrac{t^2}{2}\sec^{-1} t\right]_{2/\sqrt{3}}^2 - \int_{2/\sqrt{3}}^2 \left(\tfrac{t^2}{2}\right)\dfrac{dt}{t\sqrt{t^2-1}} = \left(2\cdot\tfrac{\pi}{3} - \tfrac{2}{3}\cdot\tfrac{\pi}{6}\right) - \int_{2/\sqrt{3}}^2 \dfrac{t\,dt}{2\sqrt{t^2-1}}$$
$$= \tfrac{5\pi}{9} - \left[\tfrac{1}{2}\sqrt{t^2-1}\right]_{2/\sqrt{3}}^2 = \tfrac{5\pi}{9} - \tfrac{1}{2}\left(\sqrt{3} - \sqrt{\tfrac{4}{3}-1}\right) = \tfrac{5\pi}{9} - \tfrac{1}{2}\left(\sqrt{3} - \tfrac{\sqrt{3}}{3}\right) = \tfrac{5\pi}{9} - \tfrac{\sqrt{3}}{3} = \tfrac{5\pi - 3\sqrt{3}}{9}$$

21. $I = \int e^\theta \sin\theta \, d\theta; \ [u = \sin\theta, \ du = \cos\theta \, d\theta; \ dv = e^\theta \, d\theta, \ v = e^\theta] \ \Rightarrow \ I = e^\theta \sin\theta - \int e^\theta \cos\theta \, d\theta;$

$[u = \cos\theta, \ du = -\sin\theta \, d\theta; \ dv = e^\theta \, d\theta, \ v = e^\theta] \ \Rightarrow \ I = e^\theta \sin\theta - \left(e^\theta \cos\theta + \int e^\theta \sin\theta \, d\theta\right)$

$= e^\theta \sin\theta - e^\theta \cos\theta - I + C' \ \Rightarrow \ 2I = (e^\theta \sin\theta - e^\theta \cos\theta) + C' \ \Rightarrow \ I = \tfrac{1}{2}(e^\theta \sin\theta - e^\theta \cos\theta) + C$, where $C = \tfrac{C'}{2}$ is another arbitrary constant

23. $I = \int e^{2x} \cos 3x \, dx; \ [u = \cos 3x; \ du = -3\sin 3x \, dx, \ dv = e^{2x} \, dx; \ v = \tfrac{1}{2}e^{2x}]$

$\Rightarrow \ I = \tfrac{1}{2}e^{2x}\cos 3x + \tfrac{3}{2}\int e^{2x}\sin 3x \, dx; \ [u = \sin 3x, \ du = 3\cos 3x, \ dv = e^{2x} \, dx; \ v = \tfrac{1}{2}e^{2x}]$

$\Rightarrow \ I = \tfrac{1}{2}e^{2x}\cos 3x + \tfrac{3}{2}\left(\tfrac{1}{2}e^{2x}\sin 3x - \tfrac{3}{2}\int e^{2x}\cos 3x \, dx\right) = \tfrac{1}{2}e^{2x}\cos 3x + \tfrac{3}{4}e^{2x}\sin 3x - \tfrac{9}{4}I + C'$

$\Rightarrow \ \tfrac{13}{4}I = \tfrac{1}{2}e^{2x}\cos 3x + \tfrac{3}{4}e^{2x}\sin 3x + C' \ \Rightarrow \ \tfrac{e^{2x}}{13}(3\sin 3x + 2\cos 3x) + C$, where $C = \tfrac{4}{13}C'$

25. $\int e^{\sqrt{3s+9}} \, ds; \ \begin{bmatrix} 3s+9 = x^2 \\ ds = \tfrac{2}{3}x \, dx \end{bmatrix} \ \rightarrow \ \int e^x \cdot \tfrac{2}{3}x \, dx = \tfrac{2}{3}\int xe^x \, dx; \ [u = x, \ du = dx; \ dv = e^x \, dx, \ v = e^x];$

$\tfrac{2}{3}\int xe^x \, dx = \tfrac{2}{3}\left(xe^x - \int e^x \, dx\right) = \tfrac{2}{3}(xe^x - e^x) + C = \tfrac{2}{3}\left(\sqrt{3s+9}\,e^{\sqrt{3s+9}} - e^{\sqrt{3s+9}}\right) + C$

27. $u = x$, $du = dx$; $dv = \tan^2 x \, dx$, $v = \int \tan^2 x \, dx = \int \frac{\sin^2 x}{\cos^2 x} \, dx = \int \frac{1 - \cos^2 x}{\cos^2 x} \, dx = \int \frac{dx}{\cos^2 x} - \int dx$

$= \tan x - x; \int_0^{\pi/3} x \tan^2 x \, dx = [x(\tan x - x)]_0^{\pi/3} - \int_0^{\pi/3} (\tan x - x) \, dx = \frac{\pi}{3} \left(\sqrt{3} - \frac{\pi}{3} \right) + \left[\ln |\cos x| + \frac{x^2}{2} \right]_0^{\pi/3}$

$- \frac{\pi}{3} \left(\sqrt{3} - \frac{\pi}{3} \right) + \ln \frac{1}{2} + \frac{\pi^2}{18} = \frac{\pi \sqrt{3}}{3} - \ln 2 - \frac{\pi^2}{18}$

29. $\int \sin(\ln x) \, dx$; $\begin{bmatrix} u = \ln x \\ du = \frac{1}{x} dx \\ dx = e^u \, du \end{bmatrix} \rightarrow \int (\sin u) e^u \, du$. From Exercise 21, $\int (\sin u) e^u \, du = e^u \left(\frac{\sin u - \cos u}{2} \right) + C$

$= \frac{1}{2} [-x \cos(\ln x) + x \sin(\ln x)] + C$

31. (a) $u = x$, $du = dx$; $dv = \sin x \, dx$, $v = -\cos x$;

$S_1 = \int_0^\pi x \sin x \, dx = [-x \cos x]_0^\pi + \int_0^\pi \cos x \, dx = \pi + [\sin x]_0^\pi = \pi$

(b) $S_2 = -\int_\pi^{2\pi} x \sin x \, dx = -\left[[-x \cos x]_\pi^{2\pi} + \int_\pi^{2\pi} \cos x \, dx \right] = -[-3\pi + [\sin x]_\pi^{2\pi}] = 3\pi$

(c) $S_3 = \int_{2\pi}^{3\pi} x \sin x \, dx = [-x \cos x]_{2\pi}^{3\pi} + \int_{2\pi}^{3\pi} \cos x \, dx = 5\pi + [\sin x]_{2\pi}^{3\pi} = 5\pi$

(d) $S_{n+1} = (-1)^{n+1} \int_{n\pi}^{(n+1)\pi} x \sin x \, dx = (-1)^{n+1} \left[[-x \cos x]_{n\pi}^{(n+1)\pi} + [\sin x]_{n\pi}^{(n+1)\pi} \right]$

$= (-1)^{n+1} [-(n+1)\pi(-1)^n + n\pi(-1)^{n+1}] + 0 = (2n+1)\pi$

33. $V = \int_0^{\ln 2} 2\pi(\ln 2 - x) e^x \, dx = 2\pi \ln 2 \int_0^{\ln 2} e^x \, dx - 2\pi \int_0^{\ln 2} x e^x \, dx$

$= (2\pi \ln 2) [e^x]_0^{\ln 2} - 2\pi \left([xe^x]_0^{\ln 2} - \int_0^{\ln 2} e^x \, dx \right)$

$= 2\pi \ln 2 - 2\pi \left(2 \ln 2 - [e^x]_0^{\ln 2} \right) = -2\pi \ln 2 + 2\pi = 2\pi(1 - \ln 2)$

35. (a) $V = \int_0^{\pi/2} 2\pi x \cos x \, dx = 2\pi \left([x \sin x]_0^{\pi/2} - \int_0^{\pi/2} \sin x \, dx \right)$

$= 2\pi \left(\frac{\pi}{2} + [\cos x]_0^{\pi/2} \right) = 2\pi \left(\frac{\pi}{2} + 0 - 1 \right) = \pi(\pi - 2)$

(b) $V = \int_0^{\pi/2} 2\pi \left(\frac{\pi}{2} - x \right) \cos x \, dx$; $u = \frac{\pi}{2} - x$, $du = -dx$; $dv = \cos x \, dx$, $v = \sin x$;

$V = 2\pi \left[\left(\frac{\pi}{2} - x \right) \sin x \right]_0^{\pi/2} + 2\pi \int_0^{\pi/2} \sin x \, dx = 0 + 2\pi [-\cos x]_0^{\pi/2} = 2\pi(0 + 1) = 2\pi$

37. $\text{av}(y) = \frac{1}{2\pi} \int_0^{2\pi} 2e^{-t} \cos t \, dt$

$= \frac{1}{\pi} \left[e^{-t} \left(\frac{\sin t - \cos t}{2} \right) \right]_0^{2\pi}$

(see Exercise 22) $\Rightarrow \text{av}(y) = \frac{1}{2\pi} (1 - e^{-2\pi})$

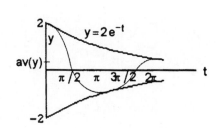

39. $I = \int x^n \cos x \, dx$; $[u = x^n$, $du = nx^{n-1} dx$; $dv = \cos x \, dx$, $v = \sin x]$

$\Rightarrow I = x^n \sin x - \int nx^{n-1} \sin x \, dx$

41. $I = \int x^n e^{ax}\, dx$; $\left[u = x^n,\ du = nx^{n-1}\, dx;\ dv = e^{ax}\, dx,\ v = \frac{1}{a}e^{ax}\right]$

$\Rightarrow I = \frac{x^n e^{ax}}{a}e^{ax} - \frac{n}{a}\int x^{n-1}e^{ax}\, dx,\ a \neq 0$

43. $\int \sin^{-1} x\, dx = x\sin^{-1} x - \int \sin y\, dy = x\sin^{-1} x + \cos y + C = x\sin^{-1} x + \cos(\sin^{-1} x) + C$

45. $\int \sec^{-1} x\, dx = x\sec^{-1} x - \int \sec y\, dy = x\sec^{-1} x - \ln|\sec y + \tan y| + C$

$= x\sec^{-1} x - \ln|\sec(\sec^{-1} x) + \tan(\sec^{-1} x)| + C = x\sec^{-1} x - \ln\left|x + \sqrt{x^2 - 1}\right| + C$

47. Yes, $\cos^{-1} x$ is the angle whose cosine is x which implies $\sin(\cos^{-1} x) = \sqrt{1 - x^2}$.

49. (a) $\int \sinh^{-1} x\, dx = x\sinh^{-1} x - \int \sinh y\, dy = x\sinh^{-1} x - \cosh y + C = x\sinh^{-1} x - \cosh(\sinh^{-1} x) + C$;

check: $d\left[x\sinh^{-1} x - \cosh(\sinh^{-1} x) + C\right] = \left[\sinh^{-1} x + \frac{x}{\sqrt{1+x^2}} - \sinh(\sinh^{-1} x)\frac{1}{\sqrt{1+x^2}}\right] dx$

$= \sinh^{-1} x\, dx$

(b) $\int \sinh^{-1} x\, dx = x\sinh^{-1} x - \int x\left(\frac{1}{\sqrt{1+x^2}}\right) dx = x\sinh^{-1} x - \frac{1}{2}\int (1+x^2)^{-1/2} 2x\, dx$

$= x\sinh^{-1} x - (1+x^2)^{1/2} + C$

check: $d\left[x\sinh^{-1} x - (1+x^2)^{1/2} + C\right] = \left[\sinh^{-1} x + \frac{x}{\sqrt{1+x^2}} - \frac{x}{\sqrt{1+x^2}}\right] dx = \sinh^{-1} x\, dx$

7.2 TRIGONOMETRIC INTEGRALS

1. $\int_0^{\pi/2} \sin^5 x\, dx = \int_0^{\pi/2} (\sin^2 x)^2 \sin x\, dx = \int_0^{\pi/2} (1 - \cos^2 x)^2 \sin x\, dx = \int_0^{\pi/2} (1 - 2\cos^2 x + \cos^4 x)\sin x\, dx$

$= \int_0^{\pi/2} \sin x\, dx - \int_0^{\pi/2} 2\cos^2 x \sin x\, dx + \int_0^{\pi/2} \cos^4 x \sin x\, dx = \left[-\cos x + 2\frac{\cos^3 x}{3} - \frac{\cos^5 x}{5}\right]_0^{\pi/2}$

$= (0) - \left(-1 + \frac{2}{3} - \frac{1}{5}\right) = \frac{8}{15}$

3. $\int_{-\pi/2}^{\pi/2} \cos^3 x\, dx = \int_{-\pi/2}^{\pi/2} (\cos^2 x)\cos x\, dx = \int_{-\pi/2}^{\pi/2} (1 - \sin^2 x)\cos x\, dx = \int_{-\pi/2}^{\pi/2} \cos x\, dx - \int_{-\pi/2}^{\pi/2} \sin^2 x \cos x\, dx$

$= \left[\sin x - \frac{\sin^3 x}{3}\right]_{-\pi/2}^{\pi/2} = \left(1 - \frac{1}{3}\right) - \left(-1 + \frac{1}{3}\right) = \frac{4}{3}$

5. $\int_0^{\pi/2} \sin^7 y\, dy = \int_0^{\pi/2} \sin^6 y \sin y\, dy = \int_0^{\pi/2} (1 - \cos^2 y)^3 \sin y\, dy = \int_0^{\pi/2} \sin y\, dy - 3\int_0^{\pi/2} \cos^2 y \sin y\, dy$

$+ 3\int_0^{\pi/2} \cos^4 y \sin y\, dy - \int_0^{\pi/2} \cos^6 y \sin y\, dy = \left[-\cos y + 3\frac{\cos^3 y}{3} - 3\frac{\cos^5 y}{5} + \frac{\cos^7 y}{7}\right]_0^{\pi/2} = (0) - \left(-1 + 1 - \frac{3}{5} + \frac{1}{7}\right) = \frac{16}{35}$

7. $\int_0^\pi 8\sin^4 x\, dx = 8\int_0^\pi \left(\frac{1 - \cos 2x}{2}\right)^2 dx = 2\int_0^\pi (1 - 2\cos 2x + \cos^2 2x)dx = 2\int_0^\pi dx - 2\int_0^\pi \cos 2x \cdot 2dx + 2\int_0^\pi \frac{1 + \cos 4x}{2}\, dx$

$= [2x - 2\sin 2x]_0^\pi + \int_0^\pi dx + \int_0^\pi \cos 4x\, dx = 2\pi + \left[x + \frac{1}{2}\sin 4x\right]_0^\pi = 2\pi + \pi = 3\pi$

9. $\int_{-\pi/4}^{\pi/4} 16\sin^2 x \cos^2 x\, dx = 16\int_{-\pi/4}^{\pi/4} \left(\frac{1 - \cos 2x}{2}\right)\left(\frac{1 + \cos 2x}{2}\right)dx = 4\int_{-\pi/4}^{\pi/4} (1 - \cos^2 2x)dx = 4\int_{-\pi/4}^{\pi/4} dx - 4\int_{-\pi/4}^{\pi/4} \left(\frac{1 + \cos 4x}{2}\right)dx$

$= [4x]_{-\pi/4}^{\pi/4} - 2\int_{-\pi/4}^{\pi/4} dx - 2\int_{-\pi/4}^{\pi/4} \cos 4x\, dx = \pi + \pi - \left[2x + \frac{\sin 4x}{2}\right]_{-\pi/4}^{\pi/4} = 2\pi - \left(\frac{\pi}{2} - \left(-\frac{\pi}{2}\right)\right) = \pi$

11. $\int_0^{\pi/2} 35\sin^4 x \cos^3 x\, dx = \int_0^{\pi/2} 35\sin^4 x (1 - \sin^2 x)\cos x\, dx = 35\int_0^{\pi/2} \sin^4 x \cos x\, dx - 35\int_0^{\pi/2} \sin^6 x \cos x\, dx$

$= \left[35\frac{\sin^5 x}{5} - 35\frac{\sin^7 x}{7}\right]_0^{\pi/2} = (7 - 5) - (0) = 2$

13. $\int_0^{\pi/4} 8\cos^3 2\theta \sin 2\theta \, d\theta = \left[8\left(-\frac{1}{2}\right)\frac{\cos^4 2\theta}{4} \right]_0^{\pi/4} = \left[-\cos^4 2\theta \right]_0^{\pi/4} = (0) - (-1) = 1$

15. $\int_0^{2\pi} \sqrt{\frac{1-\cos x}{2}} \, dx = \int_0^{2\pi} \left| \sin \frac{x}{2} \right| dx = \int_0^{2\pi} \sin \frac{x}{2} \, dx = \left[-2\cos \frac{x}{2} \right]_0^{2\pi} = 2 + 2 = 4$

17. $\int_0^{\pi} \sqrt{1 - \sin^2 t} \, dt = \int_0^{\pi} |\cos t| \, dt = \int_0^{\pi/2} \cos t \, dt - \int_{\pi/2}^{\pi} \cos t \, dt = [\sin t]_0^{\pi/2} - [\sin t]_{\pi/2}^{\pi} = 1 - 0 - 0 + 1 = 2$

19. $\int_{-\pi/4}^{\pi/4} \sqrt{1 + \tan^2 x} \, dx = \int_{-\pi/4}^{\pi/4} |\sec x| \, dx = \int_{-\pi/4}^{\pi/4} \sec x \, dx = [\ln|\sec x + \tan x|]_{-\pi/4}^{\pi/4} = \ln\left(\sqrt{2} + 1\right) - \ln\left(\sqrt{2} - 1\right)$

    $= \ln\left(\frac{\sqrt{2}+1}{\sqrt{2}-1}\right) = 2\ln\left(1 + \sqrt{2}\right)$

21. $\int_0^{\pi/2} \theta\sqrt{1 - \cos 2\theta} \, d\theta = \int_0^{\pi/2} \theta\sqrt{2} |\sin\theta| \, d\theta = \sqrt{2}\int_0^{\pi/2} \theta \sin\theta \, d\theta = \sqrt{2}\left[-\theta\cos\theta + \sin\theta\right]_0^{\pi/2} = \sqrt{2}(1) = \sqrt{2}$

23. $\int_{-\pi/3}^{0} 2\sec^3 x \, dx; \ u = \sec x, \ du = \sec x \tan x \, dx, \ dv = \sec^2 x \, dx, \ v = \tan x;$

    $\int_{-\pi/3}^{0} 2\sec^3 x \, dx = [2\sec x \tan x]_{-\pi/3}^{0} - 2\int_{-\pi/3}^{0} \sec x \tan^2 x \, dx = 2 \cdot 1 \cdot 0 - 2 \cdot 2 \cdot \sqrt{3} - 2\int_{-\pi/3}^{0} \sec x \, (\sec^2 x - 1)dx$

    $= 4\sqrt{3} - 2\int_{-\pi/3}^{0} \sec^3 x \, dx + 2\int_{-\pi/3}^{0} \sec x \, dx; \ 2\int_{-\pi/3}^{0} 2\sec^3 x \, dx = 4\sqrt{3} + [2\ln|\sec x + \tan x|]_{-\pi/3}^{0}$

    $2\int_{-\pi/3}^{0} 2\sec^3 x \, dx = 4\sqrt{3} + 2\ln|1 + 0| - 2\ln|2 - \sqrt{3}| = 4\sqrt{3} - 2\ln\left(2 - \sqrt{3}\right)$

    $\int_{-\pi/3}^{0} 2\sec^3 x \, dx = 2\sqrt{3} - \ln\left(2 - \sqrt{3}\right)$

25. $\int_0^{\pi/4} \sec^4\theta \, d\theta = \int_0^{\pi/4} (1 + \tan^2\theta)\sec^2\theta \, d\theta = \int_0^{\pi/4} \sec^2\theta \, d\theta + \int_0^{\pi/4} \tan^2\theta \sec^2\theta \, d\theta = \left[\tan\theta + \frac{\tan^3\theta}{3}\right]_0^{\pi/4}$

    $= \left(1 + \frac{1}{3}\right) - (0) = \frac{4}{3}$

27. $\int_{\pi/4}^{\pi/2} \csc^4\theta \, d\theta = \int_{\pi/4}^{\pi/2} (1 + \cot^2\theta)\csc^2\theta \, d\theta = \int_{\pi/4}^{\pi/2} \csc^2\theta \, d\theta + \int_{\pi/4}^{\pi/2} \cot^2\theta \csc^2\theta \, d\theta = \left[-\cot\theta - \frac{\cot^3\theta}{3}\right]_{\pi/4}^{\pi/2}$

    $= (0) - \left(-1 - \frac{1}{3}\right) = \frac{4}{3}$

29. $\int_0^{\pi/4} 4\tan^3 x \, dx = 4\int_0^{\pi/4} (\sec^2 x - 1)\tan x \, dx = 4\int_0^{\pi/4} \sec^2 x \tan x \, dx - 4\int_0^{\pi/4} \tan x \, dx = \left[4\frac{\tan^2 x}{2} - 4\ln|\sec x|\right]_0^{\pi/4}$

    $= 2(1) - 4\ln\sqrt{2} - 2 \cdot 0 + 4\ln 1 = 2 - 2\ln 2$

31. $\int_{\pi/6}^{\pi/3} \cot^3 x \, dx = \int_{\pi/6}^{\pi/3} (\csc^2 x - 1)\cot x \, dx = \int_{\pi/6}^{\pi/3} \csc^2 x \cot x \, dx - \int_{\pi/6}^{\pi/3} \cot x \, dx = \left[-\frac{\cot^2 x}{2} + \ln|\csc x|\right]_{\pi/6}^{\pi/3}$

    $= -\frac{1}{2}\left(\frac{1}{3} - 3\right) + \left(\ln\frac{2}{\sqrt{3}} - \ln 2\right) = \frac{4}{3} - \ln\sqrt{3}$

33. $\int_{-\pi}^{0} \sin 3x \cos 2x \, dx = \frac{1}{2}\int_{-\pi}^{0} (\sin x + \sin 5x) \, dx = \frac{1}{2}\left[-\cos x - \frac{1}{5}\cos 5x\right]_{-\pi}^{0} = \frac{1}{2}\left(-1 - \frac{1}{5} - 1 - \frac{1}{5}\right) = -\frac{6}{5}$

35. $\int_{-\pi}^{\pi} \sin 3x \sin 3x \, dx = \frac{1}{2}\int_{-\pi}^{\pi} (\cos 0 - \cos 6x) \, dx = \frac{1}{2}\int_{-\pi}^{\pi} dx - \frac{1}{2}\int_{-\pi}^{\pi} \cos 6x \, dx = \frac{1}{2}\left[x - \frac{1}{12}\sin 6x\right]_{-\pi}^{\pi} = \frac{\pi}{2} + \frac{\pi}{2} - 0 = \pi$

37. $\int_0^{\pi} \cos 3x \cos 4x \, dx = \frac{1}{2}\int_0^{\pi} (\cos(-x) + \cos 7x) \, dx = \frac{1}{2}\left[-\sin(-x) + \frac{1}{7}\sin 7x\right]_0^{\pi} = \frac{1}{2}(0) = 0$

39. $x = t^{2/3} \Rightarrow t^2 = x^3; y = \frac{t^2}{2} \Rightarrow y = \frac{x^3}{2}; 0 \le t \le 2 \Rightarrow 0 \le x \le 2^{2/3};$

$A = \int_0^{2^{2/3}} 2\pi \left(\frac{x^3}{2}\right) \sqrt{1 + \frac{9}{4}x^4} \, dx; \begin{bmatrix} u = \frac{9}{4}x^4 \\ du = 9x^3 dx \end{bmatrix} \rightarrow \frac{\pi}{9} \int_0^{9(2^{2/3})} \sqrt{1+u} \, du = \left[\frac{\pi}{9} \cdot \frac{2}{3}(1+u)^{3/2}\right]_0^{9(2^{2/3})}$

$= \frac{2\pi}{27}\left[\left(1 + 9(2^{2/3})\right)^{3/2} - 1\right]$

41. $y = \ln(\sec x); y' = \frac{\sec x \tan x}{\sec x} = \tan x; (y')^2 = \tan^2 x; \int_0^{\pi/4} \sqrt{1 + \tan^2 x} \, dx = \int_0^{\pi/4} |\sec x| \, dx = [\ln|\sec x + \tan x|]_0^{\pi/4}$

$= \ln\left(\sqrt{2} + 1\right) - \ln(0 + 1) = \ln\left(\sqrt{2} + 1\right)$

43. $V = \pi \int_0^\pi \sin^2 x \, dx = \pi \int_0^\pi \frac{1 - \cos 2x}{2} \, dx = \frac{\pi}{2}\int_0^\pi dx - \frac{\pi}{2}\int_0^\pi \cos 2x \, dx = \frac{\pi}{2}[x]_0^\pi - \frac{\pi}{4}[\sin 2x]_0^\pi = \frac{\pi}{2}(\pi - 0) - \frac{\pi}{4}(0 - 0) = \frac{\pi^2}{2}$

7.3 TRIGONOMETRIC SUBSTITUTIONS

1. $y = 3 \tan\theta, -\frac{\pi}{2} < \theta < \frac{\pi}{2}, dy = \frac{3 \, d\theta}{\cos^2\theta}, 9 + y^2 = 9(1 + \tan^2\theta) = \frac{9}{\cos^2\theta} \Rightarrow \frac{1}{\sqrt{9 + y^2}} = \frac{|\cos\theta|}{3} = \frac{\cos\theta}{3}$

   $\left(\text{because } \cos\theta > 0 \text{ when } -\frac{\pi}{2} < \theta < \frac{\pi}{2}\right);$

   $\int \frac{dy}{\sqrt{9 + y^2}} = 3 \int \frac{\cos\theta \, d\theta}{3\cos^2\theta} = \int \frac{d\theta}{\cos\theta} = \ln|\sec\theta + \tan\theta| + C' = \ln\left|\frac{\sqrt{9 + y^2}}{3} + \frac{y}{3}\right| + C' = \ln\left|\sqrt{9 + y^2} + y\right| + C$

3. $\int_{-2}^2 \frac{dx}{4 + x^2} = \left[\frac{1}{2} \tan^{-1} \frac{x}{2}\right]_{-2}^2 = \frac{1}{2}\tan^{-1} 1 - \frac{1}{2}\tan^{-1}(-1) = \left(\frac{1}{2}\right)\left(\frac{\pi}{4}\right) - \left(\frac{1}{2}\right)\left(-\frac{\pi}{4}\right) = \frac{\pi}{4}$

5. $\int_0^{3/2} \frac{dx}{\sqrt{9 - x^2}} = \left[\sin^{-1} \frac{x}{3}\right]_0^{3/2} = \sin^{-1} \frac{1}{2} - \sin^{-1} 0 = \frac{\pi}{6} - 0 = \frac{\pi}{6}$

7. $t = 5\sin\theta, -\frac{\pi}{2} < \theta < \frac{\pi}{2}, dt = 5\cos\theta \, d\theta, \sqrt{25 - t^2} = 5\cos\theta;$

   $\int \sqrt{25 - t^2} \, dt = \int (5\cos\theta)(5\cos\theta) \, d\theta = 25\int \cos^2\theta \, d\theta = 25\int \frac{1 + \cos 2\theta}{2} \, d\theta = 25\left(\frac{\theta}{2} + \frac{\sin 2\theta}{4}\right) + C$

   $= \frac{25}{2}(\theta + \sin\theta\cos\theta) + C = \frac{25}{2}\left[\sin^{-1}\left(\frac{t}{5}\right) + \left(\frac{t}{5}\right)\left(\frac{\sqrt{25 - t^2}}{5}\right)\right] + C = \frac{25}{2}\sin^{-1}\left(\frac{t}{5}\right) + \frac{t\sqrt{25 - t^2}}{2} + C$

9. $x = \frac{7}{2}\sec\theta, 0 < \theta < \frac{\pi}{2}, dx = \frac{7}{2}\sec\theta\tan\theta \, d\theta, \sqrt{4x^2 - 49} = \sqrt{49\sec^2\theta - 49} = 7\tan\theta;$

   $\int \frac{dx}{\sqrt{4x^2 - 49}} = \int \frac{\left(\frac{7}{2}\sec\theta\tan\theta\right) d\theta}{7\tan\theta} = \frac{1}{2}\int \sec\theta \, d\theta = \frac{1}{2}\ln|\sec\theta + \tan\theta| + C = \frac{1}{2}\ln\left|\frac{2x}{7} + \frac{\sqrt{4x^2 - 49}}{7}\right| + C$

11. $y = 7\sec\theta, 0 < \theta < \frac{\pi}{2}, dy = 7\sec\theta\tan\theta \, d\theta, \sqrt{y^2 - 49} = 7\tan\theta;$

    $\int \frac{\sqrt{y^2 - 49}}{y} \, dy = \int \frac{(7\tan\theta)(7\sec\theta\tan\theta) \, d\theta}{7\sec\theta} = 7\int \tan^2\theta \, d\theta = 7\int (\sec^2\theta - 1) \, d\theta = 7(\tan\theta - \theta) + C$

    $= 7\left[\frac{\sqrt{y^2 - 49}}{7} - \sec^{-1}\left(\frac{y}{7}\right)\right] + C$

13. $x = \sec\theta, 0 < \theta < \frac{\pi}{2}, dx = \sec\theta\tan\theta \, d\theta, \sqrt{x^2 - 1} = \tan\theta;$

    $\int \frac{dx}{x^2\sqrt{x^2 - 1}} = \int \frac{\sec\theta\tan\theta \, d\theta}{\sec^2\theta\tan\theta} = \int \frac{d\theta}{\sec\theta} = \sin\theta + C = \frac{\sqrt{x^2 - 1}}{x} + C$

15. $x = 2\tan\theta, -\frac{\pi}{2} < \theta < \frac{\pi}{2}, dx = \frac{2 \, d\theta}{\cos^2\theta}, \sqrt{x^2 + 4} = \frac{2}{\cos\theta};$

    $\int \frac{x^3 \, dx}{\sqrt{x^2 + 4}} = \int \frac{(8\tan^3\theta)(\cos\theta) \, d\theta}{\cos^2\theta} = 8\int \frac{\sin^3\theta \, d\theta}{\cos^4\theta} = 8\int \frac{(\cos^2\theta - 1)(-\sin\theta) \, d\theta}{\cos^4\theta};$

    $[t = \cos\theta] \rightarrow 8\int \frac{t^2 - 1}{t^4} \, dt = 8\int \left(\frac{1}{t^2} - \frac{1}{t^4}\right) dt = 8\left(-\frac{1}{t} + \frac{1}{3t^3}\right) + C = 8\left(-\sec\theta + \frac{\sec^3\theta}{3}\right) + C$

    $= 8\left(-\frac{\sqrt{x^2 + 4}}{2} + \frac{(x^2 + 4)^{3/2}}{8\cdot 3}\right) + C = \frac{1}{3}(x^2 + 4)^{3/2} - 4\sqrt{x^2 + 4} + C$

17. $w = 2 \sin \theta, -\frac{\pi}{2} < \theta < \frac{\pi}{2}, dw = 2 \cos \theta \, d\theta, \sqrt{4 - w^2} = 2 \cos \theta;$

$\int \frac{8 \, dw}{w^2 \sqrt{4 - w^2}} = \int \frac{8 \cdot 2 \cos \theta \, d\theta}{4 \sin^2 \theta \cdot 2 \cos \theta} = 2 \int \frac{d\theta}{\sin^2 \theta} = -2 \cot \theta + C = \frac{-2\sqrt{4 - w^2}}{w} + C$

19. $x = \sin \theta, 0 \le \theta \le \frac{\pi}{3}, dx = \cos \theta \, d\theta, (1 - x^2)^{3/2} = \cos^3 \theta;$

$\int_0^{\sqrt{3}/2} \frac{4x^2 \, dx}{(1 - x^2)^{3/2}} = \int_0^{\pi/3} \frac{4 \sin^2 \theta \cos \theta \, d\theta}{\cos^3 \theta} = 4 \int_0^{\pi/3} \left(\frac{1 - \cos^2 \theta}{\cos^2 \theta} \right) d\theta = 4 \int_0^{\pi/3} (\sec^2 \theta - 1) \, d\theta = 4 [\tan \theta - \theta]_0^{\pi/3} = 4\sqrt{3} - \frac{4\pi}{3}$

21. $x = \sec \theta, 0 < \theta < \frac{\pi}{2}, dx = \sec \theta \tan \theta \, d\theta, (x^2 - 1)^{3/2} = \tan^3 \theta;$

$\int \frac{dx}{(x^2 - 1)^{3/2}} = \int \frac{\sec \theta \tan \theta \, d\theta}{\tan^3 \theta} = \int \frac{\cos \theta \, d\theta}{\sin^2 \theta} = -\frac{1}{\sin \theta} + C = -\frac{x}{\sqrt{x^2 - 1}} + C$

23. $x = \sin \theta, -\frac{\pi}{2} < \theta < \frac{\pi}{2}, dx = \cos \theta \, d\theta, (1 - x^2)^{3/2} = \cos^3 \theta;$

$\int \frac{(1 - x^2)^{3/2} \, dx}{x^6} = \int \frac{\cos^3 \theta \cdot \cos \theta \, d\theta}{\sin^6 \theta} = \int \cot^4 \theta \csc^2 \theta \, d\theta = -\frac{\cot^5 \theta}{5} + C = -\frac{1}{5} \left(\frac{\sqrt{1 - x^2}}{x} \right)^5 + C$

25. $x = \frac{1}{2} \tan \theta, -\frac{\pi}{2} < \theta < \frac{\pi}{2}, dx = \frac{1}{2} \sec^2 \theta \, d\theta, (4x^2 + 1)^2 = \sec^4 \theta;$

$\int \frac{8 \, dx}{(4x^2 + 1)^2} = \int \frac{8 \left(\frac{1}{2} \sec^2 \theta \right) d\theta}{\sec^4 \theta} = 4 \int \cos^2 \theta \, d\theta = 2(\theta + \sin \theta \cos \theta) + C = 2 \tan^{-1} 2x + \frac{4x}{(4x^2 + 1)} + C$

27. $v = \sin \theta, -\frac{\pi}{2} < \theta < \frac{\pi}{2}, dv = \cos \theta \, d\theta, (1 - v^2)^{5/2} = \cos^5 \theta;$

$\int \frac{v^2 \, dv}{(1 - v^2)^{5/2}} = \int \frac{\sin^2 \theta \cos \theta \, d\theta}{\cos^5 \theta} = \int \tan^2 \theta \sec^2 \theta \, d\theta = \frac{\tan^3 \theta}{3} + C = \frac{1}{3} \left(\frac{v}{\sqrt{1 - v^2}} \right)^3 + C$

29. Let $e^t = 3 \tan \theta, t = \ln (3 \tan \theta), \tan^{-1} \left(\frac{1}{3} \right) \le \theta \le \tan^{-1} \left(\frac{4}{3} \right), dt = \frac{\sec^2 \theta}{\tan \theta} d\theta, \sqrt{e^{2t} + 9} = \sqrt{9 \tan^2 \theta + 9} = 3 \sec \theta;$

$\int_0^{\ln 4} \frac{e^t \, dt}{\sqrt{e^{2t} + 9}} = \int_{\tan^{-1} (1/3)}^{\tan^{-1} (4/3)} \frac{3 \tan \theta \cdot \sec^2 \theta \, d\theta}{\tan \theta \cdot 3 \sec \theta} = \int_{\tan^{-1} (1/3)}^{\tan^{-1} (4/3)} \sec \theta \, d\theta = [\ln |\sec \theta + \tan \theta|]_{\tan^{-1} (1/3)}^{\tan^{-1} (4/3)}$

$= \ln \left(\frac{5}{3} + \frac{4}{3} \right) - \ln \left(\frac{\sqrt{10}}{3} + \frac{1}{3} \right) = \ln 9 - \ln \left(1 + \sqrt{10} \right)$

31. $\int_{1/12}^{1/4} \frac{2 \, dt}{\sqrt{t} + 4t\sqrt{t}}; \left[u = 2\sqrt{t}, du = \frac{1}{\sqrt{t}} \, dt \right] \rightarrow \int_{1/\sqrt{3}}^1 \frac{2 \, du}{1 + u^2}; u = \tan \theta, \frac{\pi}{6} \le \theta \le \frac{\pi}{4}, du = \sec^2 \theta \, d\theta, 1 + u^2 = \sec^2 \theta;$

$\int_{1/\sqrt{3}}^1 \frac{2 \, du}{1 + u^2} = \int_{\pi/6}^{\pi/4} \frac{2 \sec^2 \theta \, d\theta}{\sec^2 \theta} = [2\theta]_{\pi/6}^{\pi/4} = 2 \left(\frac{\pi}{4} - \frac{\pi}{6} \right) = \frac{\pi}{6}$

33. $x = \sec \theta, 0 < \theta < \frac{\pi}{2}, dx = \sec \theta \tan \theta \, d\theta, \sqrt{x^2 - 1} = \sqrt{\sec^2 \theta - 1} = \tan \theta;$

$\int \frac{dx}{x\sqrt{x^2 - 1}} = \int \frac{\sec \theta \tan \theta \, d\theta}{\sec \theta \tan \theta} = \theta + C = \sec^{-1} x + C$

35. $x = \sec \theta, dx = \sec \theta \tan \theta \, d\theta, \sqrt{x^2 - 1} = \sqrt{\sec^2 \theta - 1} = \tan \theta;$

$\int \frac{x \, dx}{\sqrt{x^2 - 1}} = \int \frac{\sec \theta \cdot \sec \theta \tan \theta \, d\theta}{\tan \theta} = \int \sec^2 \theta \, d\theta = \tan \theta + C = \sqrt{x^2 - 1} + C$

37. $x \frac{dy}{dx} = \sqrt{x^2 - 4}; dy = \sqrt{x^2 - 4} \frac{dx}{x}; y = \int \frac{\sqrt{x^2 - 4}}{x} \, dx;$ $\left[\begin{array}{c} x = 2 \sec \theta, 0 < \theta < \frac{\pi}{2} \\ dx = 2 \sec \theta \tan \theta \, d\theta \\ \sqrt{x^2 - 4} = 2 \tan \theta \end{array} \right]$

$\rightarrow y = \int \frac{(2 \tan \theta)(2 \sec \theta \tan \theta) \, d\theta}{2 \sec \theta} = 2 \int \tan^2 \theta \, d\theta = 2 \int (\sec^2 \theta - 1) \, d\theta = 2(\tan \theta - \theta) + C$

$= 2 \left[\frac{\sqrt{x^2 - 4}}{2} - \sec^{-1} \left(\frac{x}{2} \right) \right] + C; x = 2 \text{ and } y = 0 \Rightarrow 0 = 0 + C \Rightarrow C = 0 \Rightarrow y = 2 \left[\frac{\sqrt{x^2 - 4}}{2} - \sec^{-1} \frac{x}{2} \right]$

39. $(x^2 + 4)\frac{dy}{dx} = 3$, $dy = \frac{3\,dx}{x^2+4}$; $y = 3\int \frac{dx}{x^2+4} = \frac{3}{2}\tan^{-1}\frac{x}{2} + C$; $x = 2$ and $y = 0 \Rightarrow 0 = \frac{3}{2}\tan^{-1}1 + C$

$\Rightarrow C = -\frac{3\pi}{8} \Rightarrow y = \frac{3}{2}\tan^{-1}\left(\frac{x}{2}\right) - \frac{3\pi}{8}$

41. $A = \int_0^3 \frac{\sqrt{9-x^2}}{3}\,dx$; $x = 3\sin\theta$, $0 \le \theta \le \frac{\pi}{2}$, $dx = 3\cos\theta\,d\theta$, $\sqrt{9-x^2} = \sqrt{9 - 9\sin^2\theta} = 3\cos\theta$;

$A = \int_0^{\pi/2} \frac{3\cos\theta \cdot 3\cos\theta\,d\theta}{3} = 3\int_0^{\pi/2}\cos^2\theta\,d\theta = \frac{3}{2}\left[\theta + \sin\theta\cos\theta\right]_0^{\pi/2} = \frac{3\pi}{4}$

7.4 INTEGRATION OF RATIONAL FUNCTIONS BY PARTIAL FRACTIONS

1. $\frac{5x-13}{(x-3)(x-2)} = \frac{A}{x-3} + \frac{B}{x-2} \Rightarrow 5x - 13 = A(x-2) + B(x-3) = (A+B)x - (2A+3B)$

$\Rightarrow \left.\begin{array}{r} A + B = 5 \\ 2A + 3B = 13 \end{array}\right\} \Rightarrow -B = (10-13) \Rightarrow B = 3 \Rightarrow A = 2$; thus, $\frac{5x-13}{(x-3)(x-2)} = \frac{2}{x-3} + \frac{3}{x-2}$

3. $\frac{x+4}{(x+1)^2} = \frac{A}{x+1} + \frac{B}{(x+1)^2} \Rightarrow x + 4 = A(x+1) + B = Ax + (A+B) \Rightarrow \left.\begin{array}{r} A = 1 \\ A + B = 4 \end{array}\right\} \Rightarrow A = 1$ and $B = 3$;

thus, $\frac{x+4}{(x+1)^2} = \frac{1}{x+1} + \frac{3}{(x+1)^2}$

5. $\frac{z+1}{z^2(z-1)} = \frac{A}{z} + \frac{B}{z^2} + \frac{C}{z-1} \Rightarrow z + 1 = Az(z-1) + B(z-1) + Cz^2 \Rightarrow z + 1 = (A+C)z^2 + (-A+B)z - B$

$\Rightarrow \left.\begin{array}{r} A + C = 0 \\ -A + B = 1 \\ -B = 1 \end{array}\right\} \Rightarrow B = -1 \Rightarrow A = -2 \Rightarrow C = 2$; thus, $\frac{z+1}{z^2(z-1)} = \frac{-2}{z} + \frac{-1}{z^2} + \frac{2}{z-1}$

7. $\frac{t^2+8}{t^2-5t+6} = 1 + \frac{5t+2}{t^2-5t+6}$ (after long division); $\frac{5t+2}{t^2-5t+6} = \frac{5t+2}{(t-3)(t-2)} = \frac{A}{t-3} + \frac{B}{t-2}$

$\Rightarrow 5t + 2 = A(t-2) + B(t-3) = (A+B)t + (-2A-3B) \Rightarrow \left.\begin{array}{r} A + B = 5 \\ -2A - 3B = 2 \end{array}\right\} \Rightarrow -B = (10+2) = 12$

$\Rightarrow B = -12 \Rightarrow A = 17$; thus, $\frac{t^2+8}{t^2-5t+6} = 1 + \frac{17}{t-3} + \frac{-12}{t-2}$

9. $\frac{1}{1-x^2} = \frac{A}{1-x} + \frac{B}{1+x} \Rightarrow 1 = A(1+x) + B(1-x)$; $x = 1 \Rightarrow A = \frac{1}{2}$; $x = -1 \Rightarrow B = \frac{1}{2}$;

$\int \frac{dx}{1-x^2} = \frac{1}{2}\int \frac{dx}{1-x} + \frac{1}{2}\int \frac{dx}{1+x} = \frac{1}{2}\left[\ln|1+x| - \ln|1-x|\right] + C$

11. $\frac{x+4}{x^2+5x-6} = \frac{A}{x+6} + \frac{B}{x-1} \Rightarrow x + 4 = A(x-1) + B(x+6)$; $x = 1 \Rightarrow B = \frac{5}{7}$; $x = -6 \Rightarrow A = \frac{-2}{-7} = \frac{2}{7}$;

$\int \frac{x+4}{x^2+5x-6}\,dx = \frac{2}{7}\int \frac{dx}{x+6} + \frac{5}{7}\int \frac{dx}{x-1} = \frac{2}{7}\ln|x+6| + \frac{5}{7}\ln|x-1| + C = \frac{1}{7}\ln|(x+6)^2(x-1)^5| + C$

13. $\frac{y}{y^2-2y-3} = \frac{A}{y-3} + \frac{B}{y+1} \Rightarrow y = A(y+1) + B(y-3)$; $y = -1 \Rightarrow B = \frac{-1}{-4} = \frac{1}{4}$; $y = 3 \Rightarrow A = \frac{3}{4}$;

$\int_4^8 \frac{y\,dy}{y^2-2y-3} = \frac{3}{4}\int_4^8 \frac{dy}{y-3} + \frac{1}{4}\int_4^8 \frac{dy}{y+1} = \left[\frac{3}{4}\ln|y-3| + \frac{1}{4}\ln|y+1|\right]_4^8 = \left(\frac{3}{4}\ln 5 + \frac{1}{4}\ln 9\right) - \left(\frac{3}{4}\ln 1 + \frac{1}{4}\ln 5\right)$

$= \frac{1}{2}\ln 5 + \frac{1}{2}\ln 3 = \frac{\ln 15}{2}$

15. $\frac{1}{t^3+t^2-2t} = \frac{A}{t} + \frac{B}{t+2} + \frac{C}{t-1} \Rightarrow 1 = A(t+2)(t-1) + Bt(t-1) + Ct(t+2)$; $t = 0 \Rightarrow A = -\frac{1}{2}$; $t = -2$

$\Rightarrow B = \frac{1}{6}$; $t = 1 \Rightarrow C = \frac{1}{3}$; $\int \frac{dt}{t^3+t^2-2t} = -\frac{1}{2}\int \frac{dt}{t} + \frac{1}{6}\int \frac{dt}{t+2} + \frac{1}{3}\int \frac{dt}{t-1}$

$= -\frac{1}{2}\ln|t| + \frac{1}{6}\ln|t+2| + \frac{1}{3}\ln|t-1| + C$

17. $\frac{x^3}{x^2+2x+1} = (x-2) + \frac{3x+2}{(x+1)^2}$ (after long division); $\frac{3x+2}{(x+1)^2} = \frac{A}{x+1} + \frac{B}{(x+1)^2} \Rightarrow 3x + 2 = A(x+1) + B$

$= Ax + (A+B) \Rightarrow A = 3, A + B = 2 \Rightarrow A = 3, B = -1$; $\int_0^1 \frac{x^3\,dx}{x^2+2x+1}$

$= \int_0^1 (x-2)\,dx + 3\int_0^1 \frac{dx}{x+1} - \int_0^1 \frac{dx}{(x+1)^2} = \left[\frac{x^2}{2} - 2x + 3\ln|x+1| + \frac{1}{x+1}\right]_0^1 = \left(\frac{1}{2} - 2 + 3\ln 2 + \frac{1}{2}\right) - (1) = 3\ln 2 - 2$

19. $\frac{1}{(x^2-1)^2} = \frac{A}{x+1} + \frac{B}{x-1} + \frac{C}{(x+1)^2} + \frac{D}{(x-1)^2} \Rightarrow 1 = A(x+1)(x-1)^2 + B(x-1)(x+1)^2 + C(x-1)^2 + D(x+1)^2;$

$x = -1 \Rightarrow C = \frac{1}{4}; x = 1 \Rightarrow D = \frac{1}{4};$ coefficient of $x^3 = A + B \Rightarrow A + B = 0;$ constant $= A - B + C + D$

$\Rightarrow A - B + C + D = 1 \Rightarrow A - B = \frac{1}{2};$ thus, $A = \frac{1}{4} \Rightarrow B = -\frac{1}{4}; \int \frac{dx}{(x^2-1)^2}$

$= \frac{1}{4}\int \frac{dx}{x+1} - \frac{1}{4}\int \frac{dx}{x-1} + \frac{1}{4}\int \frac{dx}{(x+1)^2} + \frac{1}{4}\int \frac{dx}{(x-1)^2} = \frac{1}{4}\ln\left|\frac{x+1}{x-1}\right| - \frac{x}{2(x^2-1)} + C$

21. $\frac{1}{(x+1)(x^2+1)} = \frac{A}{x+1} + \frac{Bx+C}{x^2+1} \Rightarrow 1 = A(x^2+1) + (Bx+C)(x+1); x = -1 \Rightarrow A = \frac{1}{2};$ coefficient of x^2

$= A + B \Rightarrow A + B = 0 \Rightarrow B = -\frac{1}{2};$ constant $= A + C \Rightarrow A + C = 1 \Rightarrow C = \frac{1}{2}; \int_0^1 \frac{dx}{(x+1)(x^2+1)}$

$= \frac{1}{2}\int_0^1 \frac{dx}{x+1} + \frac{1}{2}\int_0^1 \frac{(-x+1)}{x^2+1}\,dx = \left[\frac{1}{2}\ln|x+1| - \frac{1}{4}\ln(x^2+1) + \frac{1}{2}\tan^{-1}x\right]_0^1$

$= \left(\frac{1}{2}\ln 2 - \frac{1}{4}\ln 2 + \frac{1}{2}\tan^{-1}1\right) - \left(\frac{1}{2}\ln 1 - \frac{1}{4}\ln 1 + \frac{1}{2}\tan^{-1}0\right) = \frac{1}{4}\ln 2 + \frac{1}{2}\left(\frac{\pi}{4}\right) = \frac{(\pi + 2\ln 2)}{8}$

23. $\frac{y^2+2y+1}{(y^2+1)^2} = \frac{Ay+B}{y^2+1} + \frac{Cy+D}{(y^2+1)^2} \Rightarrow y^2+2y+1 = (Ay+B)(y^2+1) + Cy + D$

$= Ay^3 + By^2 + (A+C)y + (B+D) \Rightarrow A = 0, B = 1; A + C = 2 \Rightarrow C = 2; B + D = 1 \Rightarrow D = 0;$

$\int \frac{y^2+2y+1}{(y^2+1)^2}\,dy = \int \frac{1}{y^2+1}\,dy + 2\int \frac{y}{(y^2+1)^2}\,dy = \tan^{-1}y - \frac{1}{y^2+1} + C$

25. $\frac{2s+2}{(s^2+1)(s-1)^3} = \frac{As+B}{s^2+1} + \frac{C}{s-1} + \frac{D}{(s-1)^2} + \frac{E}{(s-1)^3} \Rightarrow 2s+2$

$= (As+B)(s-1)^3 + C(s^2+1)(s-1)^2 + D(s^2+1)(s-1) + E(s^2+1)$

$= [As^4 + (-3A+B)s^3 + (3A-3B)s^2 + (-A+3B)s - B] + C(s^4 - 2s^3 + 2s^2 - 2s + 1) + D(s^3 - s^2 + s - 1)$

$\quad + E(s^2+1)$

$= (A+C)s^4 + (-3A+B-2C+D)s^3 + (3A-3B+2C-D+E)s^2 + (-A+3B-2C+D)s + (-B+C-D+E)$

$\Rightarrow \left.\begin{array}{rl} A\phantom{{}+3B} + C & = 0 \\ -3A + B - 2C + D & = 0 \\ 3A - 3B + 2C - D + E & = 0 \\ -A + 3B - 2C + D & = 2 \\ -B + C - D + E & = 2 \end{array}\right\}$ summing all equations $\Rightarrow 2E = 4 \Rightarrow E = 2;$

summing eqs (2) and (3) $\Rightarrow -2B + 2 = 0 \Rightarrow B = 1;$ summing eqs (3) and (4) $\Rightarrow 2A + 2 = 2 \Rightarrow A = 0; C = 0$

from eq (1); then $-1 + 0 - D + 2 = 2$ from eq (5) $\Rightarrow D = -1;$

$\int \frac{2s+2}{(s^2+1)(s-1)^3}\,ds = \int \frac{ds}{s^2+1} - \int \frac{ds}{(s-1)^2} + 2\int \frac{ds}{(s-1)^3} = -(s-1)^{-2} + (s-1)^{-1} + \tan^{-1}s + C$

27. $\frac{2\theta^3 + 5\theta^2 + 8\theta + 4}{(\theta^2 + 2\theta + 2)^2} = \frac{A\theta + B}{\theta^2 + 2\theta + 2} + \frac{C\theta + D}{(\theta^2 + 2\theta + 2)^2} \Rightarrow 2\theta^3 + 5\theta^2 + 8\theta + 4 = (A\theta + B)(\theta^2 + 2\theta + 2) + C\theta + D$

$= A\theta^3 + (2A+B)\theta^2 + (2A+2B+C)\theta + (2B+D) \Rightarrow A = 2; 2A + B = 5 \Rightarrow B = 1; 2A + 2B + C = 8 \Rightarrow C = 2;$

$2B + D = 4 \Rightarrow D = 2; \int \frac{2\theta^3 + 5\theta^2 + 8\theta + 4}{(\theta^2 + 2\theta + 2)^2}\,d\theta = \int \frac{2\theta + 1}{(\theta^2 + 2\theta + 2)}\,d\theta + \int \frac{2\theta + 2}{(\theta^2 + 2\theta + 2)^2}\,d\theta$

$= \int \frac{2\theta + 2}{\theta^2 + 2\theta + 2}\,d\theta - \int \frac{d\theta}{\theta^2 + 2\theta + 2} + \int \frac{d(\theta^2 + 2\theta + 2)}{(\theta^2 + 2\theta + 2)^2} = \int \frac{d(\theta^2 + 2\theta + 2)}{\theta^2 + 2\theta + 2} - \int \frac{d\theta}{(\theta+1)^2 + 1} - \frac{1}{\theta^2 + 2\theta + 2}$

$= \frac{-1}{\theta^2 + 2\theta + 2} + \ln(\theta^2 + 2\theta + 2) - \tan^{-1}(\theta + 1) + C$

29. $\frac{2x^3 - 2x^2 + 1}{x^2 - x} = 2x + \frac{1}{x^2 - x} = 2x + \frac{1}{x(x-1)}; \frac{1}{x(x-1)} = \frac{A}{x} + \frac{B}{x-1} \Rightarrow 1 = A(x-1) + Bx; x = 0 \Rightarrow A = -1;$

$x = 1 \Rightarrow B = 1; \int \frac{2x^3 - 2x^2 + 1}{x^2 - x} = \int 2x\,dx - \int \frac{dx}{x} + \int \frac{dx}{x-1} = x^2 - \ln|x| + \ln|x-1| + C = x^2 + \ln\left|\frac{x-1}{x}\right| + C$

31. $\frac{9x^3 - 3x + 1}{x^3 - x^2} = 9 + \frac{9x^2 - 3x + 1}{x^2(x-1)}$ (after long division); $\frac{9x^2 - 3x + 1}{x^2(x-1)} = \frac{A}{x} + \frac{B}{x^2} + \frac{C}{x-1}$

$\Rightarrow 9x^2 - 3x + 1 = Ax(x-1) + B(x-1) + Cx^2; x = 1 \Rightarrow C = 7; x = 0 \Rightarrow B = -1; A + C = 9 \Rightarrow A = 2;$

$\int \frac{9x^3 - 3x + 1}{x^3 - x^2}\,dx = \int 9\,dx + 2\int \frac{dx}{x} - \int \frac{dx}{x^2} + 7\int \frac{dx}{x-1} = 9x + 2\ln|x| + \frac{1}{x} + 7\ln|x-1| + C$

33. $\frac{y^4 + y^2 - 1}{y^3 + y} = y - \frac{1}{y(y^2 + 1)}$; $\frac{1}{y(y^2 + 1)} = \frac{A}{y} + \frac{By + C}{y^2 + 1} \Rightarrow 1 = A(y^2 + 1) + (By + C)y = (A + B)y^2 + Cy + A$

$\Rightarrow A = 1; A + B = 0 \Rightarrow B = -1; C = 0; \int \frac{y^4 + y^2 - 1}{y^3 + y}\, dy = \int y\, dy - \int \frac{dy}{y} + \int \frac{y\, dy}{y^2 + 1} = \frac{y^2}{2} - \ln|y| + \frac{1}{2}\ln(1 + y^2) + C$

35. $\int \frac{e^t\, dt}{e^{2t} + 3e^t + 2} = [e^t = y] \int \frac{dy}{y^2 + 3y + 2} = \int \frac{dy}{y + 1} - \int \frac{dy}{y + 2} = \ln\left|\frac{y + 1}{y + 2}\right| + C = \ln\left(\frac{e^t + 1}{e^t + 2}\right) + C$

37. $\int \frac{\cos y\, dy}{\sin^2 y + \sin y - 6}$; $[\sin y = t, \cos y\, dy = dt] \rightarrow \int \frac{dy}{t^2 + t - 6} = \frac{1}{5}\int\left(\frac{1}{t - 2} - \frac{1}{t + 3}\right) dt = \frac{1}{5}\ln\left|\frac{t - 2}{t + 3}\right| + C$

$= \frac{1}{5}\ln\left|\frac{\sin y - 2}{\sin y + 3}\right| + C$

39. $\int \frac{(x - 2)^2 \tan^{-1}(2x) - 12x^3 - 3x}{(4x^2 + 1)(x - 2)^2}\, dx = \int \frac{\tan^{-1}(2x)}{4x^2 + 1}\, dx - 3\int \frac{x}{(x - 2)^2}\, dx$

$= \frac{1}{2}\int \tan^{-1}(2x)\, d\left(\tan^{-1}(2x)\right) - 3\int \frac{dx}{x - 2} - 6\int \frac{dx}{(x - 2)^2} = \frac{(\tan^{-1} 2x)^2}{4} - 3\ln|x - 2| + \frac{6}{x - 2} + C$

41. $(t^2 - 3t + 2)\frac{dx}{dt} = 1; x = \int \frac{dt}{t^2 - 3t + 2} = \int \frac{dt}{t - 2} - \int \frac{dt}{t - 1} = \ln\left|\frac{t - 2}{t - 1}\right| + C; \frac{t - 2}{t - 1} = Ce^x; t = 3$ and $x = 0$

$\Rightarrow \frac{1}{2} = C \Rightarrow \frac{t - 2}{t - 1} = \frac{1}{2}e^x \Rightarrow x = \ln\left|2\left(\frac{t - 2}{t - 1}\right)\right| = \ln|t - 2| - \ln|t - 1| + \ln 2$

43. $(t^2 + 2t)\frac{dx}{dt} = 2x + 2; \frac{1}{2}\int \frac{dx}{x + 1} = \int \frac{dt}{t^2 + 2t} \Rightarrow \frac{1}{2}\ln|x + 1| = \frac{1}{2}\int \frac{dt}{t} - \frac{1}{2}\int \frac{dt}{t + 2} \Rightarrow \ln|x + 1| = \ln\left|\frac{t}{t + 2}\right| + C$;

$t = 1$ and $x = 1 \Rightarrow \ln 2 = \ln\frac{1}{3} + C \Rightarrow C = \ln 2 + \ln 3 = \ln 6 \Rightarrow \ln|x + 1| = \ln 6\left|\frac{t}{t + 2}\right| \Rightarrow x + 1 = \frac{6t}{t + 2}$

$\Rightarrow x = \frac{6t}{t + 2} - 1, t > 0$

45. $V = \pi \int_{0.5}^{2.5} y^2\, dx = \pi \int_{0.5}^{2.5} \frac{9}{3x - x^2}\, dx = 3\pi \left(\int_{0.5}^{2.5}\left(-\frac{1}{x - 3} + \frac{1}{x}\right)\right) dx = \left[3\pi \ln\left|\frac{x}{x - 3}\right|\right]_{0.5}^{2.5} = 3\pi \ln 25$

47. $A = \int_0^{\sqrt{3}} \tan^{-1} x\, dx = \left[x \tan^{-1} x\right]_0^{\sqrt{3}} - \int_0^{\sqrt{3}} \frac{x}{1 + x^2}\, dx$

$= \frac{\pi\sqrt{3}}{3} - \left[\frac{1}{2}\ln(x^2 + 1)\right]_0^{\sqrt{3}} = \frac{\pi\sqrt{3}}{3} - \ln 2;$

$\bar{x} = \frac{1}{A}\int_0^{\sqrt{3}} x \tan^{-1} x\, dx$

$= \frac{1}{A}\left(\left[\frac{1}{2}x^2 \tan^{-1} x\right]_0^{\sqrt{3}} - \frac{1}{2}\int_0^{\sqrt{3}} \frac{x^2}{1 + x^2}\, dx\right)$

$= \frac{1}{A}\left[\frac{\pi}{2} - \left[\frac{1}{2}(x - \tan^{-1} x)\right]_0^{\sqrt{3}}\right]$

$= \frac{1}{A}\left(\frac{\pi}{2} - \frac{\sqrt{3}}{2} + \frac{\pi}{6}\right) = \frac{1}{A}\left(\frac{2\pi}{3} - \frac{\sqrt{3}}{2}\right) \cong 1.10$

49. (a) $\frac{dx}{dt} = kx(N - x) \Rightarrow \int \frac{dx}{x(N - x)} = \int k\, dt \Rightarrow \frac{1}{N}\int \frac{dx}{x} + \frac{1}{N}\int \frac{dx}{N - x} = \int k\, dt \Rightarrow \frac{1}{N}\ln\left|\frac{x}{N - x}\right| = kt + C$;

$k = \frac{1}{250}, N = 1000, t = 0$ and $x = 2 \Rightarrow \frac{1}{1000}\ln\left|\frac{2}{998}\right| = C \Rightarrow \frac{1}{1000}\ln\left|\frac{x}{1000 - x}\right| = \frac{t}{250} + \frac{1}{1000}\ln\left(\frac{1}{499}\right)$

$\Rightarrow \ln\left|\frac{499x}{1000 - x}\right| = 4t \Rightarrow \frac{499x}{1000 - x} = e^{4t} \Rightarrow 499x = e^{4t}(1000 - x) \Rightarrow (499 + e^{4t})x = 1000e^{4t} \Rightarrow x = \frac{1000e^{4t}}{499 + e^{4t}}$

(b) $x = \frac{1}{2}N = 500 \Rightarrow 500 = \frac{1000e^{4t}}{499 + e^{4t}} \Rightarrow 500 \cdot 499 + 500e^{4t} = 1000e^{4t} \Rightarrow e^{4t} = 499 \Rightarrow t = \frac{1}{4}\ln 499 \approx 1.55$ days

7.5 INTEGRAL TABLES AND COMPUTER ALGEBRA SYSTEMS

1. $\int \frac{dx}{x\sqrt{x - 3}} = \frac{2}{\sqrt{3}}\tan^{-1}\sqrt{\frac{x - 3}{3}} + C$

   (We used FORMULA 13(a) with $a = 1, b = 3$)

3. $\int \frac{x\,dx}{\sqrt{x-2}} = \int \frac{(x-2)\,dx}{\sqrt{x-2}} + 2\int \frac{dx}{\sqrt{x-2}} = \int \left(\sqrt{x-2}\right)^1 dx + 2\int \left(\sqrt{x-2}\right)^{-1} dx$

$= \left(\frac{2}{1}\right) \frac{\left(\sqrt{x-2}\right)^3}{3} + 2\left(\frac{2}{1}\right) \frac{\left(\sqrt{x-2}\right)^1}{1} = \sqrt{x-2}\left[\frac{2(x-2)}{3} + 4\right] + C$

(We used FORMULA 11 with a = 1, b = −2, n = 1 and a = 1, b = −2, n = −1)

5. $\int x\sqrt{2x-3}\,dx = \frac{1}{2}\int (2x-3)\sqrt{2x-3}\,dx + \frac{3}{2}\int \sqrt{2x-3}\,dx = \frac{1}{2}\int \left(\sqrt{2x-3}\right)^3 dx + \frac{3}{2}\int \left(\sqrt{2x-3}\right)^1 dx$

$= \left(\frac{1}{2}\right)\left(\frac{2}{2}\right) \frac{\left(\sqrt{2x-3}\right)^5}{5} + \left(\frac{3}{2}\right)\left(\frac{2}{2}\right) \frac{\left(\sqrt{2x-3}\right)^3}{3} + C = \frac{(2x-3)^{3/2}}{2}\left[\frac{2x-3}{5} + 1\right] + C = \frac{(2x-3)^{3/2}(x+1)}{5} + C$

(We used FORMULA 11 with a = 2, b = −3, n = 3 and a = 2, b = −3, n = 1)

7. $\int \frac{\sqrt{9-4x}}{x^2}\,dx = -\frac{\sqrt{9-4x}}{x} + \frac{(-4)}{2}\int \frac{dx}{x\sqrt{9-4x}} + C$

(We used FORMULA 14 with a = −4, b = 9)

$= -\frac{\sqrt{9-4x}}{x} - 2\left(\frac{1}{\sqrt{9}}\right)\ln\left|\frac{\sqrt{9-4x}-\sqrt{9}}{\sqrt{9-4x}+\sqrt{9}}\right| + C$

(We used FORMULA 13(b) with a = −4, b = 9)

$= \frac{-\sqrt{9-4x}}{x} - \frac{2}{3}\ln\left|\frac{\sqrt{9-4x}-3}{\sqrt{9-4x}+3}\right| + C$

9. $\int x\sqrt{4x-x^2}\,dx = \int x\sqrt{2\cdot 2x - x^2}\,dx = \frac{(x+2)(2x-3\cdot 2)\sqrt{2\cdot 2\cdot x - x^2}}{6} + \frac{2^3}{2}\sin^{-1}\left(\frac{x-2}{2}\right) + C$

$= \frac{(x+2)(2x-6)\sqrt{4x-x^2}}{6} + 4\sin^{-1}\left(\frac{x-2}{2}\right) + C = \frac{(x+2)(x-3)\sqrt{4x-x^2}}{3} + 4\sin^{-1}\left(\frac{x-2}{2}\right) + C$

(We used FORMULA 51 with a = 2)

11. $\int \frac{dx}{x\sqrt{7+x^2}} = \int \frac{dx}{x\sqrt{\left(\sqrt{7}\right)^2 + x^2}} = -\frac{1}{\sqrt{7}}\ln\left|\frac{\sqrt{7}+\sqrt{\left(\sqrt{7}\right)^2+x^2}}{x}\right| + C = -\frac{1}{\sqrt{7}}\ln\left|\frac{\sqrt{7}+\sqrt{7+x^2}}{x}\right| + C$

$\left(\text{We used FORMULA 26 with } a = \sqrt{7}\right)$

13. $\int \frac{\sqrt{4-x^2}}{x}\,dx = \int \frac{\sqrt{2^2-x^2}}{x}\,dx = \sqrt{2^2-x^2} - 2\ln\left|\frac{2+\sqrt{2^2-x^2}}{x}\right| + C = \sqrt{4-x^2} - 2\ln\left|\frac{2+\sqrt{4-x^2}}{x}\right| + C$

(We used FORMULA 31 with a = 2)

15. $\int e^{2t}\cos 3t\,dt = \frac{e^{2t}}{2^2+3^2}(2\cos 3t + 3\sin 3t) + C = \frac{e^{2t}}{13}(2\cos 3t + 3\sin 3t) + C$

(We used FORMULA 108 with a = 2, b = 3)

17. $\int x\cos^{-1}x\,dx = \int x^1\cos^{-1}x\,dx = \frac{x^{1+1}}{1+1}\cos^{-1}x + \frac{1}{1+1}\int \frac{x^{1+1}\,dx}{\sqrt{1-x^2}} = \frac{x^2}{2}\cos^{-1}x + \frac{1}{2}\int \frac{x^2\,dx}{\sqrt{1-x^2}}$

(We used FORMULA 100 with a = 1, n = 1)

$= \frac{x^2}{2}\cos^{-1}x + \frac{1}{2}\left(\frac{1}{2}\sin^{-1}x\right) - \frac{1}{2}\left(\frac{1}{2}x\sqrt{1-x^2}\right) + C = \frac{x^2}{2}\cos^{-1}x + \frac{1}{4}\sin^{-1}x - \frac{1}{4}x\sqrt{1-x^2} + C$

(We used FORMULA 33 with a = 1)

19. $\int x^2\tan^{-1}x\,dx = \frac{x^{2+1}}{2+1}\tan^{-1}x - \frac{1}{2+1}\int \frac{x^{2+1}}{1+x^2}\,dx = \frac{x^3}{3}\tan^{-1}x - \frac{1}{3}\int \frac{x^3}{1+x^2}\,dx$

(We used FORMULA 101 with a = 1, n = 2);

$\int \frac{x^3}{1+x^2}\,dx = \int x\,dx - \int \frac{x\,dx}{1+x^2} = \frac{x^2}{2} - \frac{1}{2}\ln(1+x^2) + C \Rightarrow \int x^2\tan^{-1}x\,dx = \frac{x^3}{3}\tan^{-1}x - \frac{x^2}{6} + \frac{1}{6}\ln(1+x^2) + C$

21. $\int \sin 3x \cos 2x\,dx = -\frac{\cos 5x}{10} - \frac{\cos x}{2} + C$

(We used FORMULA 62(a) with a = 3, b = 2)

23. $\int 8 \sin 4t \sin \frac{t}{2}\, dx = \frac{8}{7} \sin\left(\frac{7t}{2}\right) - \frac{8}{9} \sin\left(\frac{9t}{2}\right) + C = 8\left[\frac{\sin\left(\frac{7t}{2}\right)}{7} - \frac{\sin\left(\frac{9t}{2}\right)}{9}\right] + C$

   (We used FORMULA 62(b) with a = 4, b = $\frac{1}{2}$)

25. $\int \cos \frac{\theta}{3} \cos \frac{\theta}{4}\, d\theta = 6 \sin\left(\frac{\theta}{12}\right) + \frac{6}{7} \sin\left(\frac{7\theta}{12}\right) + C$

   (We used FORMULA 62(c) with a = $\frac{1}{3}$, b = $\frac{1}{4}$)

27. $\int \frac{x^3 + x + 1}{(x^2+1)^2}\, dx = \int \frac{x\, dx}{x^2+1} + \int \frac{dx}{(x^2+1)^2} = \frac{1}{2}\int \frac{d(x^2+1)}{x^2+1} + \int \frac{dx}{(x^2+1)^2} = \frac{1}{2}\ln(x^2+1) + \frac{x}{2(1+x^2)} + \frac{1}{2}\tan^{-1}x + C$

   (For the second integral we used FORMULA 17 with a = 1)

29. $\int \sin^{-1}\sqrt{x}\, dx$; $\begin{bmatrix} u = \sqrt{x} \\ x = u^2 \\ dx = 2u\, du \end{bmatrix} \rightarrow 2\int u^1 \sin^{-1} u\, du = 2\left(\frac{u^{1+1}}{1+1}\sin^{-1}u - \frac{1}{1+1}\int \frac{u^{1+1}}{\sqrt{1-u^2}}\, du\right) = u^2 \sin^{-1}u - \int \frac{u^2\, du}{\sqrt{1-u^2}}$

   (We used FORMULA 99 with a = 1, n = 1)

   $= u^2 \sin^{-1}u - \left(\frac{1}{2}\sin^{-1}u - \frac{1}{2}u\sqrt{1-u^2}\right) + C = \left(u^2 - \frac{1}{2}\right)\sin^{-1}u + \frac{1}{2}u\sqrt{1-u^2} + C$

   (We used FORMULA 33 with a = 1)

   $= \left(x - \frac{1}{2}\right)\sin^{-1}\sqrt{x} + \frac{1}{2}\sqrt{x - x^2} + C$

31. $\int \frac{\sqrt{x}}{\sqrt{1-x}}\, dx$; $\begin{bmatrix} u = \sqrt{x} \\ x = u^2 \\ dx = 2u\, du \end{bmatrix} \rightarrow \int \frac{u \cdot 2u}{\sqrt{1-u^2}}\, du = 2\int \frac{u^2}{\sqrt{1-u^2}}\, du = 2\left(\frac{1}{2}\sin^{-1}u - \frac{1}{2}u\sqrt{1-u^2}\right) + C$

   $= \sin^{-1}u - u\sqrt{1-u^2} + C$

   (We used FORMULA 33 with a = 1)

   $= \sin^{-1}\sqrt{x} - \sqrt{x}\sqrt{1-x} + C = \sin^{-1}\sqrt{x} - \sqrt{x - x^2} + C$

33. $\int (\cot t)\sqrt{1-\sin^2 t}\, dt = \int \frac{\sqrt{1-\sin^2 t}\,(\cos t)\, dt}{\sin t}$; $\begin{bmatrix} u = \sin t \\ du = \cos t\, dt \end{bmatrix} \rightarrow \int \frac{\sqrt{1-u^2}\, du}{u} = \sqrt{1-u^2} - \ln\left|\frac{1+\sqrt{1-u^2}}{u}\right| + C$

   (We used FORMULA 31 with a = 1)

   $= \sqrt{1-\sin^2 t} - \ln\left|\frac{1+\sqrt{1-\sin^2 t}}{\sin t}\right| + C$

35. $\int \frac{dy}{y\sqrt{3 + (\ln y)^2}}$; $\begin{bmatrix} u = \ln y \\ y = e^u \\ dy = e^u\, du \end{bmatrix} \rightarrow \int \frac{e^u\, du}{e^u\sqrt{3+u^2}} = \int \frac{du}{\sqrt{3+u^2}} = \ln\left|u + \sqrt{3+u^2}\right| + C = \ln\left|\ln y + \sqrt{3+(\ln y)^2}\right| + C$

   $\left(\text{We used FORMULA 20 with a} = \sqrt{3}\right)$

37. $\int \sin^5 2x\, dx = -\frac{\sin^4 2x \cos 2x}{5 \cdot 2} + \frac{5-1}{5}\int \sin^3 2x\, dx = -\frac{\sin^4 2x \cos 2x}{10} + \frac{4}{5}\left[-\frac{\sin^2 2x \cos 2x}{3 \cdot 2} + \frac{3-1}{3}\int \sin 2x\, dx\right]$

   (We used FORMULA 60 with a = 2, n = 5 and a = 2, n = 3)

   $= -\frac{\sin^4 2x \cos 2x}{10} - \frac{2}{15}\sin^2 2x \cos 2x + \frac{8}{15}\left(-\frac{1}{2}\right)\cos 2x + C = -\frac{\sin^4 2x \cos 2x}{10} - \frac{2\sin^2 2x \cos 2x}{15} - \frac{4\cos 2x}{15} + C$

39. $\int \sin^2 2\theta \cos^3 2\theta\, d\theta = \frac{\sin^3 2\theta \cos^2 2\theta}{2(2+3)} + \frac{3-1}{3+2}\int \sin^2 2\theta \cos 2\theta\, d\theta$

   (We used FORMULA 69 with a = 2, m = 3, n = 2)

   $= \frac{\sin^3 2\theta \cos^2 2\theta}{10} + \frac{2}{5}\int \sin^2 2\theta \cos 2\theta\, d\theta = \frac{\sin^3 2\theta \cos^2 2\theta}{10} + \frac{2}{5}\left[\frac{1}{2}\int \sin^2 2\theta\, d(\sin 2\theta)\right] = \frac{\sin^3 2\theta \cos^2 2\theta}{10} + \frac{\sin^3 2\theta}{15} + C$

41. $\int 4\tan^3 2x\, dx = 4\left(\frac{\tan^2 2x}{2\cdot 2} - \int \tan 2x\, dx\right) = \tan^2 2x - 4\int \tan 2x\, dx$

   (We used FORMULA 86 with n = 3, a = 2)

   $= \tan^2 2x - \frac{4}{2}\ln|\sec 2x| + C = \tan^2 2x - 2\ln|\sec 2x| + C$

43. $\int 2\sec^3 \pi x\, dx = 2\left[\frac{\sec \pi x \tan \pi x}{\pi(3-1)} + \frac{3-2}{3-1}\int \sec \pi x\, dx\right]$

   (We used FORMULA 92 with n = 3, a = π)

   $= \frac{1}{\pi}\sec \pi x \tan \pi x + \frac{1}{\pi}\ln|\sec \pi x + \tan \pi x| + C$

   (We used FORMULA 88 with a = π)

45. $\int \csc^5 x\, dx = -\frac{\csc^3 x \cot x}{5-1} + \frac{5-2}{5-1}\int \csc^3 x\, dx = -\frac{\csc^3 x \cot x}{4} + \frac{3}{4}\left(-\frac{\csc x \cot x}{3-1} + \frac{3-2}{3-1}\int \csc x\, dx\right)$

   (We used FORMULA 93 with n = 5, a = 1 and n = 3, a = 1)

   $= -\frac{1}{4}\csc^3 x \cot x - \frac{3}{8}\csc x \cot x - \frac{3}{8}\ln|\csc x + \cot x| + C$

   (We used FORMULA 89 with a = 1)

47. $\int e^t \sec^3(e^t - 1)\, dt; \begin{bmatrix} x = e^t - 1 \\ dx = e^t\, dt \end{bmatrix} \rightarrow \int \sec^3 x\, dx = \frac{\sec x \tan x}{3-1} + \frac{3-2}{3-1}\int \sec x\, dx$

   (We used FORMULA 92 with a = 1, n = 3)

   $= \frac{\sec x \tan x}{2} + \frac{1}{2}\ln|\sec x + \tan x| + C = \frac{1}{2}\left[\sec(e^t - 1)\tan(e^t - 1) + \ln|\sec(e^t - 1) + \tan(e^t - 1)|\right] + C$

49. $\int_0^1 2\sqrt{x^2 + 1}\, dx; [x = \tan t] \rightarrow 2\int_0^{\pi/4} \sec t \cdot \sec^2 t\, dt = 2\int_0^{\pi/4} \sec^3 t\, dt = 2\left[\left[\frac{\sec t\cdot \tan t}{3-1}\right]_0^{\pi/4} + \frac{3-2}{3-1}\int_0^{\pi/4}\sec t\, dt\right]$

   (We used FORMULA 92 with n = 3, a = 1)

   $= [\sec t \cdot \tan t + \ln|\sec t + \tan t|]_0^{\pi/4} = \sqrt{2} + \ln\left(\sqrt{2} + 1\right)$

51. $\int_1^2 \frac{(r^2 - 1)^{3/2}}{r}\, dr; [r = \sec \theta] \rightarrow \int_0^{\pi/3} \frac{\tan^3 \theta}{\sec \theta}(\sec \theta \tan \theta)\, d\theta = \int_0^{\pi/3} \tan^4 \theta\, d\theta = \left[\frac{\tan^3 \theta}{4-1}\right]_0^{\pi/3} - \int_0^{\pi/3} \tan^2 \theta\, d\theta$

   $= \left[\frac{\tan^3 \theta}{3} - \tan \theta + \theta\right]_0^{\pi/3} = \frac{3\sqrt{3}}{3} - \sqrt{3} + \frac{\pi}{3} = \frac{\pi}{3}$

   (We used FORMULA 86 with a = 1, n = 4 and FORMULA 84 with a = 1)

53. $S = \int_0^{\sqrt{2}} 2\pi y\sqrt{1 + (y')^2}\, dx$

   $= 2\pi \int_0^{\sqrt{2}} \sqrt{x^2 + 2}\sqrt{1 + \frac{x^2}{x^2 + 2}}\, dx$

   $= 2\sqrt{2}\pi \int_0^{\sqrt{2}} \sqrt{x^2 + 1}\, dx$

   $= 2\sqrt{2}\pi \left[\frac{x\sqrt{x^2 + 1}}{2} + \frac{1}{2}\ln\left|x + \sqrt{x^2 + 1}\right|\right]_0^{\sqrt{2}}$

   (We used FORMULA 21 with a = 1)

   $= \sqrt{2}\pi\left[\sqrt{6} + \ln\left(\sqrt{2} + \sqrt{3}\right)\right] = 2\pi\sqrt{3} + \pi\sqrt{2}\ln\left(\sqrt{2} + \sqrt{3}\right)$

55. $A = \int_0^3 \frac{dx}{\sqrt{x+1}} = \left[2\sqrt{x+1}\right]_0^3 = 2; \bar{x} = \frac{1}{A}\int_0^3 \frac{x\, dx}{\sqrt{x+1}}$ $\bar{y} = \frac{1}{2A}\int_0^3 \frac{dx}{x+1} = \frac{1}{4}[\ln(x+1)]_0^3 = \frac{1}{4}\ln 4 = \frac{1}{2}\ln 2 = \ln\sqrt{2}$

   $= \frac{1}{A}\int_0^3 \sqrt{x+1}\, dx - \frac{1}{A}\int_0^3 \frac{dx}{\sqrt{x+1}}$

   $= \frac{1}{2}\cdot \frac{2}{3}\left[(x+1)^{3/2}\right]_0^3 - 1 = \frac{4}{3};$

   (We used FORMULA 11 with a = 1, b = 1, n = 1 and

   a = 1, b = 1, n = -1)

57. $S = 2\pi \int_{-1}^{1} x^2\sqrt{1 + 4x^2}\, dx;$

$$\begin{bmatrix} u = 2x \\ du = 2\, dx \end{bmatrix} \rightarrow \frac{\pi}{4} \int_{-2}^{2} u^2\sqrt{1 + u^2}\, du$$

$$= \frac{\pi}{4}\left[\frac{u}{8}(1 + 2u^2)\sqrt{1 + u^2} - \frac{1}{8}\ln\left(u + \sqrt{1 + u^2}\right)\right]_{-2}^{2}$$

(We used FORMULA 22 with $a = 1$)

$$= \frac{\pi}{4}\left[\frac{2}{8}(1 + 2\cdot 4)\sqrt{1 + 4} - \frac{1}{8}\ln\left(2 + \sqrt{1 + 4}\right)\right.$$

$$+ \left.\frac{2}{8}(1 + 2\cdot 4)\sqrt{1 + 4} + \frac{1}{8}\ln\left(-2 + \sqrt{1 + 4}\right)\right]$$

$$= \frac{\pi}{4}\left[\frac{9}{2}\sqrt{5} - \frac{1}{8}\ln\left(\frac{2 + \sqrt{5}}{-2 + \sqrt{5}}\right)\right] \approx 7.62$$

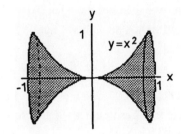

59. The integrand $f(x) = \sqrt{x - x^2}$ is nonnegative, so the integral is maximized by integrating over the function's entire domain, which runs from $x = 0$ to $x = 1$

$$\Rightarrow \int_{0}^{1}\sqrt{x - x^2}\, dx = \int_{0}^{1}\sqrt{2\cdot\frac{1}{2}x - x^2}\, dx = \left[\frac{(x - \frac{1}{2})}{2}\sqrt{2\cdot\frac{1}{2}x - x^2} + \frac{(\frac{1}{2})^2}{2}\sin^{-1}\left(\frac{x - \frac{1}{2}}{\frac{1}{2}}\right)\right]_{0}^{1}$$

(We used FORMULA 48 with $a = \frac{1}{2}$)

$$= \left[\frac{(x - \frac{1}{2})}{2}\sqrt{x - x^2} + \frac{1}{8}\sin^{-1}(2x - 1)\right]_{0}^{1} = \frac{1}{8}\cdot\frac{\pi}{2} - \frac{1}{8}\left(-\frac{\pi}{2}\right) = \frac{\pi}{8}$$

7.6 NUMERICAL INTEGRATION

1. $\int_{1}^{2} x\, dx$

  I. (a) For $n = 4$, $\Delta x = \frac{b - a}{n} = \frac{2 - 1}{4} = \frac{1}{4} \Rightarrow \frac{\Delta x}{2} = \frac{1}{8}$;

  $\sum mf(x_i) = 12 \Rightarrow T = \frac{1}{8}(12) = \frac{3}{2}$;

  $f(x) = x \Rightarrow f'(x) = 1 \Rightarrow f'' = 0 \Rightarrow M = 0$

  $\Rightarrow |E_T| = 0$

	x_i	$f(x_i)$	m	$mf(x_i)$
x_0	1	1	1	1
x_1	5/4	5/4	2	5/2
x_2	3/2	3/2	2	3
x_3	7/4	7/4	2	7/2
x_4	2	2	1	2

  (b) $\int_{1}^{2} x\, dx = \left[\frac{x^2}{2}\right]_{1}^{2} = 2 - \frac{1}{2} = \frac{3}{2} \Rightarrow |E_T| = \int_{1}^{2} x\, dx - T = 0$

  (c) $\frac{|E_T|}{\text{True Value}} \times 100 = 0\%$

  II. (a) For $n = 4$, $\Delta x = \frac{b - a}{n} = \frac{2 - 1}{4} = \frac{1}{4} \Rightarrow \frac{\Delta x}{3} = \frac{1}{12}$;

  $\sum mf(x_i) = 18 \Rightarrow S = \frac{1}{12}(18) = \frac{3}{2}$;

  $f^{(4)}(x) = 0 \Rightarrow M = 0 \Rightarrow |E_S| = 0$

	x_i	$f(x_i)$	m	$mf(x_i)$
x_0	1	1	1	1
x_1	5/4	5/4	4	5
x_2	3/2	3/2	2	3
x_3	7/4	7/4	4	7
x_4	2	2	1	2

  (b) $\int_{1}^{2} x\, dx = \frac{3}{2} \Rightarrow |E_S| = \int_{1}^{2} x\, dx - S = \frac{3}{2} - \frac{3}{2} = 0$

  (c) $\frac{|E_S|}{\text{True Value}} \times 100 = 0\%$

3. $\int_{-1}^{1} (x^2 + 1)\, dx$

   I. (a) For $n = 4$, $\Delta x = \frac{b-a}{n} = \frac{1-(-1)}{4} = \frac{2}{4} = \frac{1}{2} \Rightarrow \frac{\Delta x}{2} = \frac{1}{4}$;

        $\sum mf(x_i) = 11 \Rightarrow T = \frac{1}{4}(11) = 2.75$;

        $f(x) = x^2 + 1 \Rightarrow f'(x) = 2x \Rightarrow f''(x) = 2 \Rightarrow M = 2$

        $\Rightarrow |E_T| \le \frac{1-(-1)}{12}\left(\frac{1}{2}\right)^2(2) = \frac{1}{12}$ or 0.08333

	x_i	$f(x_i)$	m	$mf(x_i)$
x_0	-1	2	1	2
x_1	$-1/2$	$5/4$	2	$5/2$
x_2	0	1	2	2
x_3	$1/2$	$5/4$	2	$5/2$
x_4	1	2	1	2

     (b) $\int_{-1}^{1} (x^2 + 1)\, dx = \left[\frac{x^3}{3} + x\right]_{-1}^{1} = \left(\frac{1}{3} + 1\right) - \left(-\frac{1}{3} - 1\right) = \frac{8}{3} \Rightarrow E_T = \int_{-1}^{1}(x^2 + 1)\,dx - T = \frac{8}{3} - \frac{11}{4} = -\frac{1}{12}$

        $\Rightarrow |E_T| = \left|-\frac{1}{12}\right| \approx 0.08333$

     (c) $\frac{|E_T|}{\text{True Value}} \times 100 = \left(\frac{\frac{1}{12}}{\frac{8}{3}}\right) \times 100 \approx 3\%$

  II. (a) For $n = 4$, $\Delta x = \frac{b-a}{n} = \frac{1-(-1)}{4} = \frac{2}{4} = \frac{1}{2} \Rightarrow \frac{\Delta x}{3} = \frac{1}{6}$;

        $\sum mf(x_i) = 16 \Rightarrow S = \frac{1}{6}(16) = \frac{8}{3} = 2.66667$;

        $f^{(3)}(x) = 0 \Rightarrow f^{(4)}(x) = 0 \Rightarrow M = 0 \Rightarrow |E_S| = 0$

	x_i	$f(x_i)$	m	$mf(x_i)$
x_0	-1	2	1	2
x_1	$-1/2$	$5/4$	4	5
x_2	0	1	2	2
x_3	$1/2$	$5/4$	4	5
x_4	1	2	1	2

     (b) $\int_{-1}^{1}(x^2 + 1)\,dx = \left[\frac{x^3}{3} + x\right]_{-1}^{1} = \frac{8}{3}$

        $\Rightarrow |E_S| = \int_{-1}^{1}(x^2 + 1)\,dx - S = \frac{8}{3} - \frac{8}{3} = 0$

     (c) $\frac{|E_S|}{\text{True Value}} \times 100 = 0\%$

5. $\int_{0}^{2}(t^3 + t)\, dt$

   I. (a) For $n = 4$, $\Delta x = \frac{b-a}{n} = \frac{2-0}{4} = \frac{2}{4} = \frac{1}{2}$

        $\Rightarrow \frac{\Delta x}{2} = \frac{1}{4}$; $\sum mf(t_i) = 25 \Rightarrow T = \frac{1}{4}(25) = \frac{25}{4}$;

        $f(t) = t^3 + t \Rightarrow f'(t) = 3t^2 + 1 \Rightarrow f''(t) = 6t$

        $\Rightarrow M = 12 = f''(2) \Rightarrow |E_T| \le \frac{2-0}{12}\left(\frac{1}{2}\right)^2(12) = \frac{1}{2}$

	t_i	$f(t_i)$	m	$mf(t_i)$
t_0	0	0	1	0
t_1	$1/2$	$5/8$	2	$5/4$
t_2	1	2	2	4
t_3	$3/2$	$39/8$	2	$39/4$
t_4	2	10	1	10

     (b) $\int_{0}^{2}(t^3 + t)\,dt = \left[\frac{t^4}{4} + \frac{t^2}{2}\right]_{0}^{2} = \left(\frac{2^4}{4} + \frac{2^2}{2}\right) - 0 = 6 \Rightarrow |E_T| = \int_{0}^{2}(t^3 + t)\,dt - T = 6 - \frac{25}{4} = -\frac{1}{4} \Rightarrow |E_T| = \frac{1}{4}$

     (c) $\frac{|E_T|}{\text{True Value}} \times 100 = \frac{\left|-\frac{1}{4}\right|}{6} \times 100 \approx 4\%$

  II. (a) For $n = 4$, $\Delta x = \frac{b-a}{n} = \frac{2-0}{4} = \frac{2}{4} = \frac{1}{2} \Rightarrow \frac{\Delta x}{3} = \frac{1}{6}$;

        $\sum mf(t_i) = 36 \Rightarrow S = \frac{1}{6}(36) = 6$;

        $f^{(3)}(t) = 6 \Rightarrow f^{(4)}(t) = 0 \Rightarrow M = 0 \Rightarrow |E_S| = 0$

	t_i	$f(t_i)$	m	$mf(t_i)$
t_0	0	0	1	0
t_1	$1/2$	$5/8$	4	$5/2$
t_2	1	2	2	4
t_3	$3/2$	$39/8$	4	$39/2$
t_4	2	10	1	10

     (b) $\int_{0}^{2}(t^3 + t)\,dt = 6 \Rightarrow |E_S| = \int_{0}^{2}(t^3 + t)\,dt - S$

        $= 6 - 6 = 0$

     (c) $\frac{|E_S|}{\text{True Value}} \times 100 = 0\%$

7. $\int_{1}^{2} \frac{1}{s^2}\, ds$

   I. (a) For $n = 4$, $\Delta x = \frac{b-a}{n} = \frac{2-1}{4} = \frac{1}{4} \Rightarrow \frac{\Delta x}{2} = \frac{1}{8}$;

        $\sum mf(s_i) = \frac{179{,}573}{44{,}100} \Rightarrow T = \frac{1}{8}\left(\frac{179{,}573}{44{,}100}\right) = \frac{179{,}573}{352{,}800}$

        ≈ 0.50899; $f(s) = \frac{1}{s^2} \Rightarrow f'(s) = -\frac{2}{s^3}$

        $\Rightarrow f''(s) = \frac{6}{s^4} \Rightarrow M = 6 = f''(1)$

        $\Rightarrow |E_T| \le \frac{2-1}{12}\left(\frac{1}{4}\right)^2(6) = \frac{1}{32} = 0.03125$

	s_i	$f(s_i)$	m	$mf(s_i)$
s_0	1	1	1	1
s_1	$5/4$	$16/25$	2	$32/25$
s_2	$3/2$	$4/9$	2	$8/9$
s_3	$7/4$	$16/49$	2	$32/49$
s_4	2	$1/4$	1	$1/4$

     (b) $\int_{1}^{2} \frac{1}{s^2}\, ds = \int_{1}^{2} s^{-2}\, ds = \left[-\frac{1}{s}\right]_{1}^{2} = -\frac{1}{2} - \left(-\frac{1}{1}\right) = \frac{1}{2} \Rightarrow E_T = \int_{1}^{2} \frac{1}{s^2}\, ds - T = \frac{1}{2} - 0.50899 = -0.00899$

        $\Rightarrow |E_T| = 0.00899$

     (c) $\frac{|E_T|}{\text{True Value}} \times 100 = \frac{0.00899}{0.5} \times 100 \approx 2\%$

II. (a) For $n = 4$, $\Delta x = \frac{b-a}{n} = \frac{2-1}{4} = \frac{1}{4} \Rightarrow \frac{\Delta x}{3} = \frac{1}{12}$;

	s_i	$f(s_i)$	m	$mf(s_i)$
s_0	1	1	1	1
s_1	5/4	16/25	4	64/25
s_2	3/2	4/9	2	8/9
s_3	7/4	16/49	4	64/49
s_4	2	1/4	1	1/4

$\sum mf(s_i) = \frac{264,821}{44,100} \Rightarrow S = \frac{1}{12}\left(\frac{264,821}{44,100}\right) = \frac{264,821}{529,200}$

≈ 0.50042; $f^{(3)}(s) = -\frac{24}{s^5} \Rightarrow f^{(4)}(s) = \frac{120}{s^6}$

$\Rightarrow M = 120 \Rightarrow |E_S| \leq \left|\frac{2-1}{180}\right|\left(\frac{1}{4}\right)^4(120)$

$= \frac{1}{384} \approx 0.00260$

(b) $\int_1^2 \frac{1}{s^2}\,ds = \frac{1}{2} \Rightarrow E_S = \int_1^2 \frac{1}{s^2}\,ds - S = \frac{1}{2} - 0.50042 = -0.00042 \Rightarrow |E_S| = 0.00042$

(c) $\frac{|E_S|}{\text{True Value}} \times 100 = \frac{0.0004}{0.5} \times 100 \approx 0.08\%$

9. $\int_0^\pi \sin t\,dt$

	t_i	$f(t_i)$	m	$mf(t_i)$
t_0	0	0	1	0
t_1	$\pi/4$	$\sqrt{2}/2$	2	$\sqrt{2}$
t_2	$\pi/2$	1	2	2
t_3	$3\pi/4$	$\sqrt{2}/2$	2	$\sqrt{2}$
t_4	π	0	1	0

I. (a) For $n = 4$, $\Delta x = \frac{b-a}{n} = \frac{\pi-0}{4} = \frac{\pi}{4} \Rightarrow \frac{\Delta x}{2} = \frac{\pi}{8}$;

$\sum mf(t_i) = 2 + 2\sqrt{2} \approx 4.8284$

$\Rightarrow T - \frac{\pi}{8}\left(2 + 2\sqrt{2}\right) \approx 1.89612$;

$f(t) = \sin t \Rightarrow f'(t) = \cos t \Rightarrow f''(t) = -\sin t$

$\Rightarrow M = 1 \Rightarrow |E_T| \leq \frac{\pi-0}{12}\left(\frac{\pi}{4}\right)^2(1) = \frac{\pi^3}{192}$

≈ 0.16149

(b) $\int_0^\pi \sin t\,dt = [-\cos t]_0^\pi = (-\cos \pi) - (-\cos 0) = 2 \Rightarrow |E_T| = \int_0^\pi \sin t\,dt - T \approx 2 - 1.89612 = 0.10388$

(c) $\frac{|E_T|}{\text{True Value}} \times 100 = \frac{0.10388}{2} \times 100 \approx 5\%$

II. (a) For $n = 4$, $\Delta x = \frac{b-a}{n} = \frac{\pi-0}{4} = \frac{\pi}{4} \Rightarrow \frac{\Delta x}{3} = \frac{\pi}{12}$;

	t_i	$f(t_i)$	m	$mf(t_i)$
t_0	0	0	1	0
t_1	$\pi/4$	$\sqrt{2}/2$	4	$2\sqrt{2}$
t_2	$\pi/2$	1	2	2
t_3	$3\pi/4$	$\sqrt{2}/2$	4	$2\sqrt{2}$
t_4	π	0	1	0

$\sum mf(t_i) = 2 + 4\sqrt{2} \approx 7.6569$

$\Rightarrow S = \frac{\pi}{12}\left(2 + 4\sqrt{2}\right) \approx 2.00456$;

$f^{(3)}(t) = -\cos t \Rightarrow f^{(4)}(t) = \sin t$

$\Rightarrow M = 1 \Rightarrow |E_S| \leq \frac{\pi-0}{180}\left(\frac{\pi}{4}\right)^4(1) \approx 0.00664$

(b) $\int_0^\pi \sin t\,dt = 2 \Rightarrow E_S = \int_0^\pi \sin t\,dt - S \approx 2 - 2.00456 = -0.00456 \Rightarrow |E_S| \approx 0.00456$

(c) $\frac{|E_S|}{\text{True Value}} \times 100 = \frac{0.00456}{2} \times 100 \approx 0\%$

11. (a) $M = 0$ (see Exercise 1): Then $n = 1 \Rightarrow \Delta x = 1 \Rightarrow |E_T| = \frac{1}{12}(1)^2(0) = 0 < 10^{-4}$

(b) $M = 0$ (see Exercise 1): Then $n = 2$ (n must be even) $\Rightarrow \Delta x = \frac{1}{2} \Rightarrow |E_S| = \frac{1}{180}\left(\frac{1}{2}\right)^4(0) = 0 < 10^{-4}$

13. (a) $M = 2$ (see Exercise 3): Then $\Delta x = \frac{2}{n} \Rightarrow |E_T| \leq \frac{2}{12}\left(\frac{2}{n}\right)^2(2) = \frac{4}{3n^2} < 10^{-4} \Rightarrow n^2 > \frac{4}{3}(10^4) \Rightarrow n > \sqrt{\frac{4}{3}(10^4)}$

$\Rightarrow n > 115.4$, so let $n = 116$

(b) $M = 0$ (see Exercise 3): Then $n = 2$ (n must be even) $\Rightarrow \Delta x = 1 \Rightarrow |E_S| = \frac{2}{180}(1)^4(0) = 0 < 10^{-4}$

15. (a) $M = 12$ (see Exercise 5): Then $\Delta x = \frac{2}{n} \Rightarrow |E_T| \leq \frac{2}{12}\left(\frac{2}{n}\right)^2(12) = \frac{8}{n^2} < 10^{-4} \Rightarrow n^2 > 8(10^4) \Rightarrow n > \sqrt{8(10^4)}$

$\Rightarrow n > 282.8$, so let $n = 283$

(b) $M = 0$ (see Exercise 5): Then $n = 2$ (n must be even) $\Rightarrow \Delta x = 1 \Rightarrow |E_S| = \frac{2}{180}(1)^4(0) = 0 < 10^{-4}$

17. (a) $M = 6$ (see Exercise 7): Then $\Delta x = \frac{1}{n} \Rightarrow |E_T| \leq \frac{1}{12}\left(\frac{1}{n}\right)^2(6) = \frac{1}{2n^2} < 10^{-4} \Rightarrow n^2 > \frac{1}{2}(10^4) \Rightarrow n > \sqrt{\frac{1}{2}(10^4)}$

$\Rightarrow n > 70.7$, so let $n = 71$

(b) $M = 120$ (see Exercise 7): Then $\Delta x = \frac{1}{n} \Rightarrow |E_S| = \frac{1}{180}\left(\frac{1}{n}\right)^4(120) = \frac{2}{3n^4} < 10^{-4} \Rightarrow n^4 > \frac{2}{3}(10^4)$

$\Rightarrow n > \sqrt[4]{\frac{2}{3}(10^4)} \Rightarrow n > 9.04$, so let $n = 10$ (n must be even)

19. (a) $f(x) = \sqrt{x+1} \Rightarrow f'(x) = \frac{1}{2}(x+1)^{-1/2} \Rightarrow f''(x) = -\frac{1}{4}(x+1)^{-3/2} = -\frac{1}{4\left(\sqrt{x+1}\right)^3} \Rightarrow M = \frac{1}{4\left(\sqrt{1}\right)^3} = \frac{1}{4}.$

Then $\Delta x = \frac{3}{n} \Rightarrow |E_T| \leq \frac{3}{12}\left(\frac{3}{n}\right)^2\left(\frac{1}{4}\right) = \frac{9}{16n^2} < 10^{-4} \Rightarrow n^2 > \frac{9}{16}(10^4) \Rightarrow n > \sqrt{\frac{9}{16}(10^4)} \Rightarrow n > 75,$

so let $n = 76$

(b) $f^{(3)}(x) = \frac{3}{8}(x+1)^{-5/2} \Rightarrow f^{(4)}(x) = -\frac{15}{16}(x+1)^{-7/2} = -\frac{15}{16\left(\sqrt{x+1}\right)^7} \Rightarrow M = \frac{15}{16\left(\sqrt{1}\right)^7} = \frac{15}{16}.$ Then $\Delta x = \frac{3}{n}$

$\Rightarrow |E_S| \leq \frac{3}{180}\left(\frac{3}{n}\right)^4\left(\frac{15}{16}\right) = \frac{3^5(15)}{16(180)n^4} < 10^{-4} \Rightarrow n^4 > \frac{3^5(15)(10^4)}{16(180)} \Rightarrow n > \sqrt[4]{\frac{3^5(15)(10^4)}{16(180)}} \Rightarrow n > 10.6,$ so let

$n = 12$ (n must be even)

21. (a) $f(x) = \sin(x+1) \Rightarrow f'(x) = \cos(x+1) \Rightarrow f''(x) = -\sin(x+1) \Rightarrow M = 1.$ Then $\Delta x = \frac{2}{n} \Rightarrow |E_T| \leq \frac{2}{12}\left(\frac{2}{n}\right)^2(1)$

$= \frac{8}{12n^2} < 10^{-4} \Rightarrow n^2 > \frac{8(10^4)}{12} \Rightarrow n > \sqrt{\frac{8(10^4)}{12}} \Rightarrow n > 81.6,$ so let $n = 82$

(b) $f^{(3)}(x) = -\cos(x+1) \Rightarrow f^{(4)}(x) = \sin(x+1) \Rightarrow M = 1.$ Then $\Delta x = \frac{2}{n} \Rightarrow |E_S| \leq \frac{2}{180}\left(\frac{2}{n}\right)^4(1) = \frac{32}{180n^4} < 10^{-4}$

$\Rightarrow n^4 > \frac{32(10^4)}{180} \Rightarrow n > \sqrt[4]{\frac{32(10^4)}{180}} \Rightarrow n > 6.49,$ so let $n = 8$ (n must be even)

23. $\frac{5}{2}(6.0 + 2(8.2) + 2(9.1)\ldots + 2(12.7) + 13.0)(30) = 15{,}990 \text{ ft}^3.$

25. Using Simpson's Rule, $\Delta x = 1 \Rightarrow \frac{\Delta x}{3} = \frac{1}{3};$

$\sum my_i = 33.6 \Rightarrow$ Cross Section Area $\approx \frac{1}{3}(33.6)$

$= 11.2 \text{ ft}^2.$ Let x be the length of the tank. Then the

Volume V = (Cross Sectional Area) x = 11.2x.

Now 5000 lb of gasoline at 42 lb/ft^3

$\Rightarrow V = \frac{5000}{42} = 119.05 \text{ ft}^3$

$\Rightarrow 119.05 = 11.2x \Rightarrow x \approx 10.63 \text{ ft}$

	x_i	y_i	m	my_i
x_0	0	1.5	1	1.5
x_1	1	1.6	4	6.4
x_2	2	1.8	2	3.6
x_3	3	1.9	4	7.6
x_4	4	2.0	2	4.0
x_5	5	2.1	4	8.4
x_6	6	2.1	1	2.1

27. (a) $|E_S| \leq \frac{b-a}{180}(\Delta x^4)M; \; n = 4 \Rightarrow \Delta x = \frac{\frac{\pi}{2}-0}{4} = \frac{\pi}{8}; \; \left|f^{(4)}\right| \leq 1 \Rightarrow M = 1 \Rightarrow |E_S| \leq \frac{\left(\frac{\pi}{2}-0\right)}{180}\left(\frac{\pi}{8}\right)^4(1) \approx 0.00021$

(b) $\Delta x = \frac{\pi}{8} \Rightarrow \frac{\Delta x}{3} = \frac{\pi}{24};$

$\sum mf(x_i) = 10.47208705$

$\Rightarrow S = \frac{\pi}{24}(10.47208705) \approx 1.37079$

	x_i	$f(x_i)$	m	$mf(x_{1i})$
x_0	0	1	1	1
x_1	$\pi/8$	0.974495358	4	3.897981432
x_2	$\pi/4$	0.900316316	2	1.800632632
x_3	$3\pi/8$	0.784213303	4	3.136853212
x_4	$\pi/2$	0.636619772	1	0.636619772

(c) $\approx \left(\frac{0.00021}{1.37079}\right) \times 100 \approx 0.015\%$

29. $T = \frac{\Delta x}{2}(y_0 + 2y_1 + 2y_2 + 2y_3 + \ldots + 2y_{n-1} + y_n)$ where $\Delta x = \frac{b-a}{n}$ and f is continuous on [a, b]. So

$T = \frac{b-a}{n}\frac{(y_0 + y_1 + y_1 + y_2 + y_2 + \ldots + y_{n-1} + y_{n-1} + y_n)}{2} = \frac{b-a}{n}\left(\frac{f(x_0)+f(x_1)}{2} + \frac{f(x_1)+f(x_2)}{2} + \ldots + \frac{f(x_{n-1})+f(x_n)}{2}\right).$

Since f is continuous on each interval $[x_{k-1}, x_k]$, and $\frac{f(x_{k-1})+f(x_k)}{2}$ is always between $f(x_{k-1})$ and $f(x_k)$, there is a point c_k in

$[x_{k-1}, x_k]$ with $f(c_k) = \frac{f(x_{k-1})+f(x_k)}{2};$ this is a consequence of the Intermediate Value Theorem. Thus our sum is

$\sum_{k=1}^{n}\left(\frac{b-a}{n}\right)f(c_k)$ which has the form $\sum_{k=1}^{n}\Delta x_k f(c_k)$ with $\Delta x_k = \frac{b-a}{n}$ for all k. This is a Riemann Sum for f on [a, b].

31. (a) $a = 1, e = \frac{1}{2} \Rightarrow$ Length $= 4 \int_0^{\pi/2} \sqrt{1 - \frac{1}{4}\cos^2 t}\, dt$

$$= 2 \int_0^{\pi/2} \sqrt{4 - \cos^2 t}\, dt = \int_0^{\pi/2} f(t)\, dt; \text{ use the}$$

Trapezoid Rule with $n = 10 \Rightarrow \Delta t = \frac{b-a}{n} = \frac{\left(\frac{\pi}{2}\right) - 0}{10}$

$$= \frac{\pi}{20} \cdot \int_0^{\pi/2} \sqrt{4 - \cos^2 t}\, dt \approx \sum_{n=0}^{10} mf(x_n) = 37.3686183$$

$$\Rightarrow T = \frac{\Delta t}{2}(37.3686183) = \frac{\pi}{40}(37.3686183)$$

$$= 2.934924419 \Rightarrow \text{Length} = 2(2.934924419) \approx 5.870$$

(b) $|f''(t)| < 1 \Rightarrow M = 1$

$$\Rightarrow |E_T| \le \frac{b-a}{12}\left(\Delta t^2 M\right) \le \frac{\left(\frac{\pi}{2}\right) - 0}{12}\left(\frac{\pi}{20}\right)^2 1 \le 0.0032$$

	x_i	$f(x_i)$	m	$mf(x_i)$
x_0	0	1.732050808	1	1.732050808
x_1	$\pi/20$	1.739100843	2	3.478201686
x_2	$\pi/10$	1.759400893	2	3.518801786
x_3	$3\pi/20$	1.790560631	2	3.581121262
x_4	$\pi/5$	1.82906848	1	3.658136959
x_5	$\pi/4$	1.870828693	1	3.741657387
x_6	$3\pi/10$	1.911676881	2	3.823353762
x_7	$7\pi/20$	1.947791731	2	3.895583461
x_8	$2\pi/5$	1.975982919	2	3.951965839
x_9	$9\pi/20$	1.993872679	2	3.987745357
x_{10}	$\pi/2$	2	1	2

33. The length of the curve $y = \sin\left(\frac{3\pi}{20}x\right)$ from 0 to 20 is: $L = \int_0^{20} \sqrt{1 + \left(\frac{dy}{dx}\right)^2}\, dx;\ \frac{dy}{dx} = \frac{3\pi}{20}\cos\left(\frac{3\pi}{20}x\right) \Rightarrow \left(\frac{dy}{dx}\right)^2$

$$= \frac{9\pi^2}{400}\cos^2\left(\frac{3\pi}{20}x\right) \Rightarrow L = \int_0^{20} \sqrt{1 + \frac{9\pi^2}{400}\cos^2\left(\frac{3\pi}{20}x\right)}\, dx. \text{ Using numerical integration we find } L \approx 21.07 \text{ in}$$

35. $y = \sin x \Rightarrow \frac{dy}{dx} = \cos x \Rightarrow \left(\frac{dy}{dx}\right)^2 = \cos^2 x \Rightarrow S = \int_0^\pi 2\pi(\sin x)\sqrt{1 + \cos^2 x}\, dx;$ a numerical integration gives $S \approx 14.4$

37. A calculator or computer numerical integrator yields $\sin^{-1} 0.6 \approx 0.643501109$.

7.7 IMPROPER INTEGRALS

1. $\int_0^\infty \frac{dx}{x^2 + 1} = \lim_{b \to \infty} \int_0^b \frac{dx}{x^2 + 1} = \lim_{b \to \infty}\left[\tan^{-1} x\right]_0^b = \lim_{b \to \infty}\left(\tan^{-1} b - \tan^{-1} 0\right) = \frac{\pi}{2} - 0 = \frac{\pi}{2}$

3. $\int_0^1 \frac{dx}{\sqrt{x}} = \lim_{b \to 0^+} \int_b^1 x^{-1/2}\, dx = \lim_{b \to 0^+}\left[2x^{1/2}\right]_b^1 = \lim_{b \to 0^+}\left(2 - 2\sqrt{b}\right) = 2 - 0 = 2$

5. $\int_{-1}^1 \frac{dx}{x^{2/3}} = \int_{-1}^0 \frac{dx}{x^{2/3}} + \int_0^1 \frac{dx}{x^{2/3}} = \lim_{b \to 0^-}\left[3x^{1/3}\right]_{-1}^b + \lim_{c \to 0^+}\left[3x^{1/3}\right]_c^1$

$$= \lim_{b \to 0^-}\left[3b^{1/3} - 3(-1)^{1/3}\right] + \lim_{c \to 0^+}\left[3(1)^{1/3} - 3c^{1/3}\right] = (0 + 3) + (3 - 0) = 6$$

7. $\int_0^1 \frac{dx}{\sqrt{1 - x^2}} = \lim_{b \to 1^-}\left[\sin^{-1} x\right]_0^b = \lim_{b \to 1^-}\left(\sin^{-1} b - \sin^{-1} 0\right) = \frac{\pi}{2} - 0 = \frac{\pi}{2}$

9. $\int_{-\infty}^{-2} \frac{2\, dx}{x^2 - 1} = \int_{-\infty}^{-2} \frac{dx}{x - 1} - \int_{-\infty}^{-2} \frac{dx}{x + 1} = \lim_{b \to -\infty}\left[\ln|x - 1|\right]_b^{-2} - \lim_{b \to -\infty}\left[\ln|x + 1|\right]_b^{-2} = \lim_{b \to -\infty}\left[\ln\left|\frac{x - 1}{x + 1}\right|\right]_b^{-2}$

$$= \lim_{b \to -\infty}\left(\ln\left|\frac{-3}{-1}\right| - \ln\left|\frac{b - 1}{b + 1}\right|\right) = \ln 3 - \ln\left(\lim_{b \to -\infty}\frac{b - 1}{b + 1}\right) = \ln 3 - \ln 1 = \ln 3$$

11. $\int_2^\infty \frac{2\, dv}{v^2 - v} = \lim_{b \to \infty}\left[2\ln\left|\frac{v - 1}{v}\right|\right]_2^b = \lim_{b \to \infty}\left(2\ln\left|\frac{b - 1}{b}\right| - 2\ln\left|\frac{2 - 1}{2}\right|\right) = 2\ln(1) - 2\ln\left(\frac{1}{2}\right) = 0 + 2\ln 2 = \ln 4$

13. $\int_{-\infty}^\infty \frac{2x\, dx}{(x^2 + 1)^2} = \int_{-\infty}^0 \frac{2x\, dx}{(x^2 + 1)^2} + \int_0^\infty \frac{2x\, dx}{(x^2 + 1)^2};\ \begin{bmatrix} u = x^2 + 1 \\ du = 2x\, dx \end{bmatrix} \to \int_\infty^1 \frac{du}{u^2} + \int_1^\infty \frac{du}{u^2} = \lim_{b \to \infty}\left[-\frac{1}{u}\right]_b^1 + \lim_{c \to \infty}\left[-\frac{1}{u}\right]_1^c$

$$= \lim_{b \to \infty}\left(-1 + \frac{1}{b}\right) + \lim_{c \to \infty}\left[-\frac{1}{c} - (-1)\right] = (-1 + 0) + (0 + 1) = 0$$

15. $\int_0^1 \frac{\theta + 1}{\sqrt{\theta^2 + 2\theta}}\, d\theta;\ \begin{bmatrix} u = \theta^2 + 2\theta \\ du = 2(\theta + 1)\, d\theta \end{bmatrix} \to \int_0^3 \frac{du}{2\sqrt{u}} = \lim_{b \to 0^+} \int_b^3 \frac{du}{2\sqrt{u}} = \lim_{b \to 0^+}\left[\sqrt{u}\right]_b^3 = \lim_{b \to 0^+}\left(\sqrt{3} - \sqrt{b}\right) = \sqrt{3} - 0$

$$= \sqrt{3}$$

17. $\int_0^\infty \frac{dx}{(1+x)\sqrt{x}}$; $\begin{bmatrix} u = \sqrt{x} \\ du = \frac{dx}{2\sqrt{x}} \end{bmatrix}$ $\rightarrow \int_0^\infty \frac{2\,du}{u^2+1} = \lim_{b \to \infty} \int_0^b \frac{2\,du}{u^2+1} = \lim_{b \to \infty} \left[2\tan^{-1} u\right]_0^b$

$= \lim_{b \to \infty} \left(2\tan^{-1} b - 2\tan^{-1} 0\right) = 2\left(\frac{\pi}{2}\right) - 2(0) = \pi$

19. $\int_0^\infty \frac{dv}{(1+v^2)(1+\tan^{-1} v)} = \lim_{b \to \infty} \left[\ln|1 + \tan^{-1} v|\right]_0^b = \lim_{b \to \infty} \left[\ln|1+\tan^{-1} b|\right] - \ln|1 + \tan^{-1} 0|$

$= \ln\left(1 + \frac{\pi}{2}\right) - \ln(1+0) = \ln\left(1 + \frac{\pi}{2}\right)$

21. $\int_{-\infty}^0 \theta e^\theta \, d\theta = \lim_{b \to -\infty} \left[\theta e^\theta - e^\theta\right]_b^0 = (0 \cdot e^0 - e^0) - \lim_{b \to -\infty} \left[be^b - e^b\right] = -1 - \lim_{b \to -\infty} \left(\frac{b-1}{e^{-b}}\right)$

$= -1 - \lim_{b \to -\infty} \left(\frac{1}{-e^{-b}}\right)$ (l'Hôpital's rule for $\frac{\infty}{\infty}$ form)

$= -1 - 0 = -1$

23. $\int_{-\infty}^0 e^{-|x|} \, dx = \int_{-\infty}^0 e^x \, dx = \lim_{b \to -\infty} \left[e^x\right]_b^0 = \lim_{b \to -\infty} (1 - e^b) = (1 - 0) = 1$

25. $\int_0^1 x \ln x \, dx = \lim_{b \to 0^+} \left[\frac{x^2}{2} \ln x - \frac{x^2}{4}\right]_b^1 = \left(\frac{1}{2} \ln 1 - \frac{1}{4}\right) - \lim_{b \to 0^+} \left(\frac{b^2}{2} \ln b - \frac{b^2}{4}\right) = -\frac{1}{4} - \lim_{b \to 0^+} \frac{\ln b}{\left(\frac{2}{b^2}\right)} + 0$

$= -\frac{1}{4} - \lim_{b \to 0^+} \frac{\left(\frac{1}{b}\right)}{\left(-\frac{4}{b^3}\right)} = -\frac{1}{4} + \lim_{b \to 0^+} \left(\frac{b^2}{4}\right) = -\frac{1}{4} + 0 = -\frac{1}{4}$

27. $\int_0^2 \frac{ds}{\sqrt{4-s^2}} = \lim_{b \to 2^-} \left[\sin^{-1} \frac{s}{2}\right]_0^b = \lim_{b \to 2^-} \left(\sin^{-1} \frac{b}{2}\right) - \sin^{-1} 0 = \frac{\pi}{2} - 0 = \frac{\pi}{2}$

29. $\int_1^2 \frac{ds}{s\sqrt{s^2-1}} = \lim_{b \to 1^+} \left[\sec^{-1} s\right]_b^2 = \sec^{-1} 2 - \lim_{b \to 1^+} \sec^{-1} b = \frac{\pi}{3} - 0 = \frac{\pi}{3}$

31. $\int_{-1}^4 \frac{dx}{\sqrt{|x|}} = \lim_{b \to 0^-} \int_{-1}^b \frac{dx}{\sqrt{-x}} + \lim_{c \to 0^+} \int_c^4 \frac{dx}{\sqrt{x}} = \lim_{b \to 0^-} \left[-2\sqrt{-x}\right]_{-1}^b + \lim_{c \to 0^+} \left[2\sqrt{x}\right]_c^4$

$= \lim_{b \to 0^-} \left(-2\sqrt{-b}\right) - \left(-2\sqrt{-(-1)}\right) + 2\sqrt{4} - \lim_{c \to 0^+} 2\sqrt{c} = 0 + 2 + 2 \cdot 2 - 0 = 6$

33. $\int_{-1}^\infty \frac{d\theta}{\theta^2 + 5\theta + 6} = \lim_{b \to \infty} \left[\ln\left|\frac{\theta+2}{\theta+3}\right|\right]_{-1}^b = \lim_{b \to \infty} \left[\ln\left|\frac{b+2}{b+3}\right|\right] - \ln\left|\frac{-1+2}{-1+3}\right| = 0 - \ln\left(\frac{1}{2}\right) = \ln 2$

35. $\int_0^{\pi/2} \tan \theta \, d\theta = \lim_{b \to \frac{\pi}{2}^-} \left[-\ln|\cos \theta|\right]_0^b = \lim_{b \to \frac{\pi}{2}^-} \left[-\ln|\cos b|\right] + \ln 1 = \lim_{b \to \frac{\pi}{2}^-} \left[-\ln|\cos b|\right] = +\infty$, the integral diverges

37. $\int_0^\pi \frac{\sin \theta \, d\theta}{\sqrt{\pi - \theta}}$; $[\pi - \theta = x]$ $\rightarrow -\int_\pi^0 \frac{\sin x \, dx}{\sqrt{x}} = \int_0^\pi \frac{\sin x \, dx}{\sqrt{x}}$. Since $0 \leq \frac{\sin x}{\sqrt{x}} \leq \frac{1}{\sqrt{x}}$ for all $0 \leq x \leq \pi$ and $\int_0^\pi \frac{dx}{\sqrt{x}}$ converges,

then $\int_0^\pi \frac{\sin x}{\sqrt{x}} \, dx$ converges by the Direct Comparison Test.

39. $\int_0^{\ln 2} x^{-2} e^{-1/x} \, dx$; $\left[\frac{1}{x} = y\right]$ $\rightarrow \int_\infty^{1/\ln 2} \frac{y^2 e^{-y} \, dy}{-y^2} = \int_{1/\ln 2}^\infty e^{-y} \, dy = \lim_{b \to \infty} \left[-e^{-y}\right]_{1/\ln 2}^b = \lim_{b \to \infty} \left[-e^{-b}\right] - \left[-e^{-1/\ln 2}\right]$

$= 0 + e^{-1/\ln 2} = e^{-1/\ln 2}$, so the integral converges.

41. $\int_0^\pi \frac{dt}{\sqrt{t + \sin t}}$. Since for $0 \leq t \leq \pi, 0 \leq \frac{1}{\sqrt{t + \sin t}} \leq \frac{1}{\sqrt{t}}$ and $\int_0^\pi \frac{dt}{\sqrt{t}}$ converges, then the original integral converges as well by the Direct Comparison Test.

43. $\int_0^2 \frac{dx}{1-x^2} = \int_0^1 \frac{dx}{1-x^2} + \int_1^2 \frac{dx}{1-x^2}$ and $\int_0^1 \frac{dx}{1-x^2} = \lim_{b \to 1^-} \left[\frac{1}{2} \ln \left| \frac{1+x}{1-x} \right| \right]_0^b = \lim_{b \to 1^-} \left[\frac{1}{2} \ln \left| \frac{1+b}{1-b} \right| \right] - 0 = \infty$, which

diverges $\Rightarrow \int_0^2 \frac{dx}{1-x^2}$ diverges as well.

45. $\int_{-1}^1 \ln |x| \, dx = \int_{-1}^0 \ln(-x) \, dx + \int_0^1 \ln x \, dx; \int_0^1 \ln x \, dx = \lim_{b \to 0^+} [x \ln x - x]_b^1 = [1 \cdot 0 - 1] - \lim_{b \to 0^+} [b \ln b - b]$

$= -1 - 0 = -1; \int_{-1}^0 \ln(-x) \, dx = -1 \Rightarrow \int_{-1}^1 \ln |x| \, dx = -2$ converges.

47. $\int_1^\infty \frac{dx}{1+x^3}; 0 \le \frac{1}{x^3+1} \le \frac{1}{x^3}$ for $1 \le x < \infty$ and $\int_1^\infty \frac{dx}{x^3}$ converges $\Rightarrow \int_1^\infty \frac{dx}{1+x^3}$ converges by the Direct
Comparison Test.

49. $\int_2^\infty \frac{dv}{\sqrt{v-1}}; \lim_{v \to \infty} \frac{\left(\frac{1}{\sqrt{v-1}} \right)}{\left(\frac{1}{\sqrt{v}} \right)} = \lim_{v \to \infty} \frac{\sqrt{v}}{\sqrt{v-1}} = \lim_{v \to \infty} \frac{1}{\sqrt{1-\frac{1}{v}}} = \frac{1}{\sqrt{1-0}} = 1$ and $\int_2^\infty \frac{dv}{\sqrt{v}} = \lim_{b \to \infty} \left[2\sqrt{v} \right]_2^b = \infty$,

which diverges $\Rightarrow \int_2^\infty \frac{dv}{\sqrt{v-1}}$ diverges by the Limit Comparison Test.

51. $\int_0^\infty \frac{dx}{\sqrt{x^6+1}} = \int_0^1 \frac{dx}{\sqrt{x^6+1}} + \int_1^\infty \frac{dx}{\sqrt{x^6+1}} < \int_0^1 \frac{dx}{\sqrt{x^6+1}} + \int_1^\infty \frac{dx}{x^3}$ and $\int_1^\infty \frac{dx}{x^3} = \lim_{b \to \infty} \left[-\frac{1}{2x^2} \right]_1^b$

$= \lim_{b \to \infty} \left(-\frac{1}{2b^2} + \frac{1}{2} \right) = \frac{1}{2} \Rightarrow \int_0^\infty \frac{dx}{\sqrt{x^6+1}}$ converges by the Direct Comparison Test.

53. $\int_1^\infty \frac{\sqrt{x+1}}{x^2} \, dx; \lim_{x \to \infty} \frac{\left(\frac{\sqrt{x}}{x^2} \right)}{\left(\frac{\sqrt{x+1}}{x^2} \right)} = \lim_{x \to \infty} \frac{\sqrt{x}}{\sqrt{x+1}} = \lim_{x \to \infty} \frac{1}{\sqrt{1+\frac{1}{x}}} = 1; \int_1^\infty \frac{\sqrt{x}}{x^2} \, dx = \int_1^\infty \frac{dx}{x^{3/2}}$

$= \lim_{b \to \infty} \left[-2x^{-1/2} \right]_1^b = \lim_{b \to \infty} \left(\frac{-2}{\sqrt{b}} + 2 \right) = 2 \Rightarrow \int_1^\infty \frac{\sqrt{x+1}}{x^2} \, dx$ converges by the Limit Comparison Test.

55. $\int_\pi^\infty \frac{2+\cos x}{x} \, dx; 0 < \frac{1}{x} \le \frac{2+\cos x}{x}$ for $x \ge \pi$ and $\int_\pi^\infty \frac{dx}{x} = \lim_{b \to \infty} [\ln x]_\pi^b = \infty$, which diverges

$\Rightarrow \int_\pi^\infty \frac{2+\cos x}{x} \, dx$ diverges by the Direct Comparison Test.

57. $\int_4^\infty \frac{2 \, dt}{t^{3/2}-1}; \lim_{t \to \infty} \frac{t^{3/2}}{t^{3/2}-1} = 1$ and $\int_4^\infty \frac{2 \, dt}{t^{3/2}} = \lim_{b \to \infty} \left[-4t^{-1/2} \right]_4^b = \lim_{b \to \infty} \left(\frac{-4}{\sqrt{b}} + 2 \right) = 2 \Rightarrow \int_4^\infty \frac{2 \, dt}{t^{3/2}}$ converges

$\Rightarrow \int_4^\infty \frac{2 \, dt}{t^{3/2}+1}$ converges by the Limit Comparison Test.

59. $\int_1^\infty \frac{e^x}{x} \, dx; 0 < \frac{1}{x} < \frac{e^x}{x}$ for $x > 1$ and $\int_1^\infty \frac{dx}{x}$ diverges $\Rightarrow \int_1^\infty \frac{e^x \, dx}{x}$ diverges by the Direct Comparison Test.

61. $\int_1^\infty \frac{dx}{\sqrt{e^x-x}}; \lim_{x \to \infty} \frac{\left(\frac{1}{\sqrt{e^x-x}} \right)}{\left(\frac{1}{\sqrt{e^x}} \right)} = \lim_{x \to \infty} \frac{\sqrt{e^x}}{\sqrt{e^x-x}} = \lim_{x \to \infty} \frac{1}{\sqrt{1-\frac{x}{e^x}}} = \frac{1}{\sqrt{1-0}} = 1; \int_1^\infty \frac{dx}{\sqrt{e^x}} = \int_1^\infty e^{-x/2} \, dx$

$= \lim_{b \to \infty} \left[-2e^{-x/2} \right]_1^b = \lim_{b \to \infty} \left(-2e^{-b/2} + 2e^{-1/2} \right) = \frac{2}{\sqrt{e}} \Rightarrow \int_1^\infty e^{-x/2} \, dx$ converges $\Rightarrow \int_1^\infty \frac{dx}{\sqrt{e^x-x}}$ converges

by the Limit Comparison Test.

63. $\int_{-\infty}^\infty \frac{dx}{\sqrt{x^4+1}} = 2 \int_0^\infty \frac{dx}{\sqrt{x^4+1}}; \int_0^\infty \frac{dx}{\sqrt{x^4+1}} = \int_0^1 \frac{dx}{\sqrt{x^4+1}} + \int_1^\infty \frac{dx}{\sqrt{x^4+1}} < \int_0^1 \frac{dx}{\sqrt{x^4+1}} + \int_1^\infty \frac{dx}{x^2}$ and

$\int_1^\infty \frac{dx}{x^2} = \lim_{b \to \infty} \left[-\frac{1}{x} \right]_1^b = \lim_{b \to \infty} \left(-\frac{1}{b} + 1 \right) = 1 \Rightarrow \int_{-\infty}^\infty \frac{dx}{\sqrt{x^4+1}}$ converges by the Direct Comparison Test.

65. (a) $\int_1^2 \frac{dx}{x(\ln x)^p}; [t = \ln x] \to \int_0^{\ln 2} \frac{dt}{t^p} = \lim_{b \to 0^+} \left[\frac{1}{-p+1} t^{1-p} \right]_b^{\ln 2} = \lim_{b \to 0^+} \frac{b^{1-p}}{p-1} + \frac{1}{1-p} (\ln 2)^{1-p}$

$\Rightarrow$ the integral converges for $p < 1$ and diverges for $p \ge 1$

(b) $\int_2^\infty \frac{dx}{x(\ln x)^p}$; $[t = \ln x] \rightarrow \int_{\ln 2}^\infty \frac{dt}{t^p}$ and this integral is essentially the same as in Exercise 65(a): it converges
for $p > 1$ and diverges for $p \leq 1$

67. $A = \int_0^\infty e^{-x}\,dx = \lim_{b \to \infty}\ [-e^{-x}]_0^b = \lim_{b \to \infty}\ (-e^{-b}) - (-e^{-0})$

$= 0 + 1 = 1$

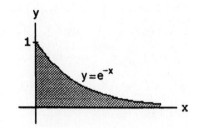

69. $V = \int_0^\infty 2\pi x e^{-x}\,dx = 2\pi \int_0^\infty x e^{-x}\,dx = 2\pi \lim_{b \to \infty}\ [-xe^{-x} - e^{-x}]_0^b = 2\pi \left[\lim_{b \to \infty}\ (-be^{-b} - e^{-b}) - 1\right] = 2\pi$

71. $A = \int_0^{\pi/2} (\sec x - \tan x)\,dx = \lim_{b \to \frac{\pi}{2}^-}\ [\ln|\sec x + \tan x| - \ln|\sec x|]_0^b = \lim_{b \to \frac{\pi}{2}^-}\ \left(\ln\left|1 + \frac{\tan b}{\sec b}\right| - \ln|1 + 0|\right)$

$= \lim_{b \to \frac{\pi}{2}^-}\ \ln|1 + \sin b| = \ln 2$

73. (a) $\int_3^\infty e^{-3x}\,dx = \lim_{b \to \infty}\ \left[-\frac{1}{3}e^{-3x}\right]_3^b = \lim_{b \to \infty}\ \left(-\frac{1}{3}e^{-3b}\right) - \left(-\frac{1}{3}e^{-3\cdot3}\right) = 0 + \frac{1}{3} \cdot e^{-9} = \frac{1}{3}e^{-9}$

$\approx 0.0000411 < 0.000042$. Since $e^{-x^2} \leq e^{-3x}$ for $x > 3$, then $\int_3^\infty e^{-x^2}\,dx < 0.000042$ and therefore

$\int_0^\infty e^{-x^2}\,dx$ can be replaced by $\int_0^3 e^{-x^2}\,dx$ without introducing an error greater than 0.000042.

(b) $\int_0^3 e^{-x^2}\,dx \cong 0.88621$

75. (a)

(b) $>$ int((sin(t))/t, t=0..infinity); $\left(\text{answer is } \frac{\pi}{2}\right)$

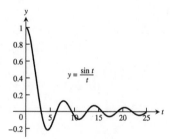

77. (a) $f(x) = \frac{1}{\sqrt{2\pi}}e^{-x^2/2}$

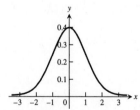

f is increasing on $(-\infty, 0]$. f is decreasing on $[0, \infty)$. f has a local maximum at $(0, f(0)) = \left(0, \frac{1}{\sqrt{2\pi}}\right)$

(b) Maple commands:

$>$f: $= \exp(-x^2/2)(\sqrt{2*pi})$;

$>$int(f, x $= -1..1$); ≈ 0.683

$>$int(f, x $= -2..2$); ≈ 0.954

$>$int(f, x $= -3..3$); ≈ 0.997

(c) Part (b) suggests that as n increases, the integral approaches 1. We can take $\int_{-n}^{n} f(x)\,dx$ as close to 1 as we want by choosing $n > 1$ large enough. Also, we can make $\int_{n}^{\infty} f(x)\,dx$ and $\int_{-\infty}^{-n} f(x)\,dx$ as small as we want by choosing n large enough. This is because $0 < f(x) < e^{-x/2}$ for $x > 1$. (Likewise, $0 < f(x) < e^{x/2}$ for $x < -1$.)

Thus, $\int_{n}^{\infty} f(x)\,dx < \int_{n}^{\infty} e^{-x/2} dx$.

$\int_{n}^{\infty} e^{-x/2} dx = \lim_{c \to \infty} \int_{n}^{c} e^{-x/2} dx = \lim_{c \to \infty} \left[-2e^{-x/2} \right]_{n}^{c} = \lim_{c \to \infty} \left[-2e^{-c/2} + 2e^{-n/2} \right] = 2e^{-n/2}$

As $n \to \infty$, $2e^{-n/2} \to 0$, for large enough n, $\int_{n}^{\infty} f(x)\,dx$ is as small as we want. Likewise for large enough n, $\int_{-\infty}^{-n} f(x)\,dx$ is as small as we want.

CHAPTER 7 PRACTICE EXERCISES

1. $u = \ln(x+1)$, $du = \frac{dx}{x+1}$; $dv = dx$, $v = x$;

$\int \ln(x+1)\,dx = x \ln(x+1) - \int \frac{x}{x+1}\,dx = x \ln(x+1) - \int dx + \int \frac{dx}{x+1} = x \ln(x+1) - x + \ln(x+1) + C_1$
$= (x+1)\ln(x+1) - x + C_1 = (x+1)\ln(x+1) - (x+1) + C$, where $C = C_1 + 1$

3. $u = \tan^{-1} 3x$, $du = \frac{3\,dx}{1+9x^2}$; $dv = dx$, $v = x$;

$\int \tan^{-1} 3x\,dx = x \tan^{-1} 3x - \int \frac{3x\,dx}{1+9x^2}$; $\begin{bmatrix} y = 1+9x^2 \\ dy = 18x\,dx \end{bmatrix} \to x \tan^{-1} 3x - \frac{1}{6} \int \frac{dy}{y}$

$= x \tan^{-1}(3x) - \frac{1}{6} \ln(1+9x^2) + C$

5.

$$
\begin{array}{ll}
 & e^x \\
(x+1)^2 \xrightarrow{(+)} & e^x \\
2(x+1) \xrightarrow{(-)} & e^x \\
2 \xrightarrow{(+)} & e^x \\
0 &
\end{array}
\Rightarrow \int (x+1)^2 e^x\,dx = \left[(x+1)^2 - 2(x+1) + 2 \right] e^x + C
$$

7. $u = \cos 2x$, $du = -2 \sin 2x\,dx$; $dv = e^x\,dx$, $v = e^x$;

$I = \int e^x \cos 2x\,dx = e^x \cos 2x + 2 \int e^x \sin 2x\,dx$;

$u = \sin 2x$, $du = 2 \cos 2x\,dx$; $dv = e^x\,dx$, $v = e^x$;

$I = e^x \cos 2x + 2 \left[e^x \sin 2x - 2 \int e^x \cos 2x\,dx \right] = e^x \cos 2x + 2e^x \sin 2x - 4I \Rightarrow I = \frac{e^x \cos 2x}{5} + \frac{2e^x \sin 2x}{5} + C$

9. $\int \frac{x\,dx}{x^2 - 3x + 2} = \int \frac{2\,dx}{x-2} - \int \frac{dx}{x-1} = 2 \ln|x-2| - \ln|x-1| + C$

11. $\int \frac{dx}{x(x+1)^2} = \int \left(\frac{1}{x} - \frac{1}{x+1} + \frac{-1}{(x+1)^2} \right) dx = \ln|x| - \ln|x+1| + \frac{1}{x+1} + C$

13. $\int \frac{\sin\theta\,d\theta}{\cos^2\theta + \cos\theta - 2}$; $[\cos\theta = y] \to -\int \frac{dy}{y^2 + y - 2} = -\frac{1}{3} \int \frac{dy}{y-1} + \frac{1}{3} \int \frac{dy}{y+2} = \frac{1}{3} \ln\left| \frac{y+2}{y-1} \right| + C$
$= \frac{1}{3} \ln\left| \frac{\cos\theta + 2}{\cos\theta - 1} \right| + C = -\frac{1}{3} \ln\left| \frac{\cos\theta - 1}{\cos\theta + 2} \right| + C$

15. $\int \frac{3x^2 + 4x + 4}{x^3 + x}\,dx = \int \frac{4}{x}\,dx - \int \frac{x-4}{x^2+1}\,dx = 4 \ln|x| - \frac{1}{2} \ln(x^2 + 1) + 4 \tan^{-1} x + C$

17. $\int \frac{(v+3)\,dv}{2v^3 - 8v} = \frac{1}{2} \int \left(-\frac{3}{4v} + \frac{5}{8(v-2)} + \frac{1}{8(v+2)} \right) dv = -\frac{3}{8} \ln|v| + \frac{5}{16} \ln|v-2| + \frac{1}{16} \ln|v+2| + C = \frac{1}{16} \ln\left| \frac{(v-2)^5 (v+2)}{v^6} \right| + C$

19. $\int \frac{dt}{t^4 + 4t^2 + 3} = \frac{1}{2}\int \frac{dt}{t^2+1} - \frac{1}{2}\int \frac{dt}{t^2+3} = \frac{1}{2}\tan^{-1}t - \frac{1}{2\sqrt{3}}\tan^{-1}\left(\frac{t}{\sqrt{3}}\right) + C = \frac{1}{2}\tan^{-1}t - \frac{\sqrt{3}}{6}\tan^{-1}\frac{t}{\sqrt{3}} + C$

21. $\int \frac{x^3+x^2}{x^2+x-2}\,dx = \int\left(x + \frac{2x}{x^2+x-2}\right)dx = \int x\,dx + \frac{2}{3}\int\frac{dx}{x-1} + \frac{4}{3}\int\frac{dx}{x+2} = \frac{x^2}{2} + \frac{4}{3}\ln|x+2| + \frac{2}{3}\ln|x-1| + C$

23. $\int \frac{x^3+4x^2}{x^2+4x+3}\,dx = \int\left(x - \frac{3x}{x^2+4x+3}\right)dx = \int x\,dx + \frac{3}{2}\int\frac{dx}{x+1} - \frac{9}{2}\int\frac{dx}{x+3} = \frac{x^2}{2} - \frac{9}{2}\ln|x+3| + \frac{3}{2}\ln|x+1| + C$

25. $\int \frac{dx}{x\,(3\sqrt{x+1})}$; $\begin{bmatrix} u = \sqrt{x+1} \\ du = \frac{dx}{2\sqrt{x+1}} \\ dx = 2u\,du \end{bmatrix} \to \frac{2}{3}\int\frac{u\,du}{(u^2-1)\,u} = \frac{1}{3}\int\frac{du}{u-1} - \frac{1}{3}\int\frac{du}{u+1} = \frac{1}{3}\ln|u-1| - \frac{1}{3}\ln|u+1| + C$

$= \frac{1}{3}\ln\left|\frac{\sqrt{x+1}-1}{\sqrt{x+1}+1}\right| + C$

27. $\int \frac{ds}{e^s-1}$; $\begin{bmatrix} u = e^s - 1 \\ du = e^s\,ds \\ ds = \frac{du}{u+1} \end{bmatrix} \to \int\frac{du}{u(u+1)} = -\int\frac{du}{u+1} + \int\frac{du}{u} = \ln\left|\frac{u}{u+1}\right| + C = \ln\left|\frac{e^s-1}{e^s}\right| + C = \ln|1 - e^{-s}| + C$

29. (a) $\int \frac{y\,dy}{\sqrt{16-y^2}} = -\frac{1}{2}\int\frac{d(16-y^2)}{\sqrt{16-y^2}} = -\sqrt{16-y^2} + C$

(b) $\int \frac{y\,dy}{\sqrt{16-y^2}}$; $[y = 4\sin x] \to 4\int\frac{\sin x\cos x\,dx}{\cos x} = -4\cos x + C = -\frac{4\sqrt{16-y^2}}{4} + C = -\sqrt{16-y^2} + C$

31. (a) $\int \frac{x\,dx}{4-x^2} = -\frac{1}{2}\int\frac{d(4-x^2)}{4-x^2} = -\frac{1}{2}\ln|4-x^2| + C$

(b) $\int \frac{x\,dx}{4-x^2}$; $[x = 2\sin\theta] \to \int\frac{2\sin\theta\cdot 2\cos\theta\,d\theta}{4\cos^2\theta} = \int\tan\theta\,d\theta = -\ln|\cos\theta| + C = -\ln\left(\frac{\sqrt{4-x^2}}{2}\right) + C$

$= -\frac{1}{2}\ln|4-x^2| + C$

33. $\int \frac{x\,dx}{9-x^2}$; $\begin{bmatrix} u = 9 - x^2 \\ du = -2x\,dx \end{bmatrix} \to -\frac{1}{2}\int\frac{du}{u} = -\frac{1}{2}\ln|u| + C = \ln\frac{1}{\sqrt{u}} + C = \ln\frac{1}{\sqrt{9-x^2}} + C$

35. $\int \frac{dx}{9-x^2} = \frac{1}{6}\int\frac{dx}{3-x} + \frac{1}{6}\int\frac{dx}{3+x} = -\frac{1}{6}\ln|3-x| + \frac{1}{6}\ln|3+x| + C = \frac{1}{6}\ln\left|\frac{x+3}{x-3}\right| + C$

37. $\int \sin^3x\cos^4x\,dx = \int\cos^4x(1-\cos^2x)\sin x\,dx = \int\cos^4x\sin x\,dx - \int\cos^6x\sin x\,dx = -\frac{\cos^5x}{5} + \frac{\cos^7x}{7} + C$

39. $\int \tan^4x\sec^2x\,dx = \frac{\tan^5x}{5} + C$

41. $\int \sin 5\theta\cos 6\theta\,d\theta = \frac{1}{2}\int(\sin(-\theta) + \sin(11\theta))\,d\theta = \frac{1}{2}\int\sin(-\theta)\,d\theta + \frac{1}{2}\int\sin(11\theta)\,d\theta = \frac{1}{2}\cos(-\theta) - \frac{1}{22}\cos 11\theta + C$

$= \frac{1}{2}\cos\theta - \frac{1}{22}\cos 11\theta + C$

43. $\int \sqrt{1 + \cos\left(\frac{t}{2}\right)}\,dt = \int\sqrt{2}\left|\cos\frac{t}{4}\right|\,dt = 4\sqrt{2}\left|\sin\frac{t}{4}\right| + C$

45. $|E_s| \le \frac{3-1}{180}(\triangle x)^4 M$ where $\triangle x = \frac{3-1}{n} = \frac{2}{n}$; $f(x) = \frac{1}{x} = x^{-1} \Rightarrow f'(x) = -x^{-2} \Rightarrow f''(x) = 2x^{-3} \Rightarrow f'''(x) = -6x^{-4}$
$\Rightarrow f^{(4)}(x) = 24x^{-5}$ which is decreasing on $[1, 3] \Rightarrow$ maximum of $f^{(4)}(x)$ on $[1,3]$ is $f^{(4)}(1) = 24 \Rightarrow M = 24$. Then
$|E_s| \le 0.0001 \Rightarrow \left(\frac{3-1}{180}\right)\left(\frac{2}{n}\right)^4(24) \le 0.0001 \Rightarrow \left(\frac{768}{180}\right)\left(\frac{1}{n^4}\right) \le 0.0001 \Rightarrow \frac{1}{n^4} \le (0.0001)\left(\frac{180}{768}\right) \Rightarrow n^4 \ge 10{,}000\left(\frac{768}{180}\right)$
$\Rightarrow n \ge 14.37 \Rightarrow n \ge 16$ (n must be even)

47. $\triangle x = \frac{b-a}{n} = \frac{\pi-0}{6} = \frac{\pi}{6} \Rightarrow \frac{\triangle x}{2} = \frac{\pi}{12}$;

$\sum\limits_{i=0}^{6} mf(x_i) = 12 \Rightarrow T = \left(\frac{\pi}{12}\right)(12) = \pi$;

	x_i	$f(x_i)$	m	$mf(x_i)$
x_0	0	0	1	0
x_1	$\pi/6$	1/2	2	1
x_2	$\pi/3$	3/2	2	3
x_3	$\pi/2$	2	2	4
x_4	$2\pi/3$	3/2	2	3
x_5	$5\pi/6$	1/2	2	1
x_6	π	0	1	0

$\sum\limits_{i=0}^{6} mf(x_i) = 18$ and $\frac{\triangle x}{3} = \frac{\pi}{18} \Rightarrow$

$S = \left(\frac{\pi}{18}\right)(18) = \pi$.

	x_i	$f(x_i)$	m	$mf(x_i)$
x_0	0	0	1	0
x_1	$\pi/6$	1/2	4	2
x_2	$\pi/3$	3/2	2	3
x_3	$\pi/2$	2	4	8
x_4	$2\pi/3$	3/2	2	3
x_5	$5\pi/6$	1/2	4	2
x_6	π	0	1	0

49. $y_{av} = \frac{1}{365-0} \int_0^{365} \left[37 \sin\left(\frac{2\pi}{365}(x-101)\right) + 25\right] dx = \frac{1}{365}\left[-37\left(\frac{365}{2\pi}\right)\cos\left(\frac{2\pi}{365}(x-101)\right) + 25x\right]_0^{365}$

$= \frac{1}{365}\left[\left(-37\left(\frac{365}{2\pi}\right)\cos\left[\frac{2\pi}{365}(365-101)\right] + 25(365)\right) - \left(-37\left(\frac{365}{2\pi}\right)\cos\left[\frac{2\pi}{365}(0-101)\right] + 25(0)\right)\right]$

$= -\frac{37}{2\pi}\cos\left(\frac{2\pi}{365}(264)\right) + 25 + \frac{37}{2\pi}\cos\left(\frac{2\pi}{365}(-101)\right) = -\frac{37}{2\pi}\left(\cos\left(\frac{2\pi}{365}(264)\right) - \cos\left(\frac{2\pi}{365}(-101)\right)\right) + 25$

$\approx -\frac{37}{2\pi}(0.16705 - 0.16705) + 25 = 25°$ F

51. (a) Each interval is 5 min $= \frac{1}{12}$ hour.

$\frac{1}{24}[2.5 + 2(2.4) + 2(2.3) + \ldots + 2(2.4) + 2.3] = \frac{29}{12} \approx 2.42$ gal

(b) $(60 \text{ mph})\left(\frac{12}{29} \text{ hours/gal}\right) \approx 24.83$ mi/gal

53. $\int_0^3 \frac{dx}{\sqrt{9-x^2}} = \lim\limits_{b \to 3^-} \int_0^b \frac{dx}{\sqrt{9-x^2}} = \lim\limits_{b \to 3^-} \left[\sin^{-1}\left(\frac{x}{3}\right)\right]_0^b = \lim\limits_{b \to 3^-} \sin^{-1}\left(\frac{b}{3}\right) - \sin^{-1}\left(\frac{0}{3}\right) = \frac{\pi}{2} - 0 = \frac{\pi}{2}$

55. $\int_{-1}^1 \frac{dy}{y^{2/3}} = \int_{-1}^0 \frac{dy}{y^{2/3}} + \int_0^1 \frac{dy}{y^{2/3}} = 2\int_0^1 \frac{dy}{y^{2/3}} = 2 \cdot 3 \lim\limits_{b \to 0^+} \left[y^{1/3}\right]_b^1 = 6\left(1 - \lim\limits_{b \to 0^+} b^{1/3}\right) = 6$

57. $\int_3^\infty \frac{2\,du}{u^2-2u} = \int_3^\infty \frac{du}{u-2} - \int_3^\infty \frac{du}{u} = \lim\limits_{b \to \infty}\left[\ln\left|\frac{u-2}{u}\right|\right]_3^b = \lim\limits_{b \to \infty}\left[\ln\left|\frac{b-2}{b}\right|\right] - \ln\left|\frac{3-2}{3}\right| = 0 - \ln\left(\frac{1}{3}\right) = \ln 3$

59. $\int_0^\infty x^2 e^{-x}\, dx = \lim\limits_{b \to \infty}\left[-x^2 e^{-x} - 2xe^{-x} - 2e^{-x}\right]_0^b = \lim\limits_{b \to \infty}\left(-b^2 e^{-b} - 2be^{-b} - 2e^{-b}\right) - (-2) = 0 + 2 = 2$

61. $\int_{-\infty}^\infty \frac{dx}{4x^2+9} = 2\int_0^\infty \frac{dx}{4x^2+9} = \frac{1}{2}\int_0^\infty \frac{dx}{x^2+\frac{9}{4}} = \frac{1}{2}\lim\limits_{b \to \infty}\left[\frac{2}{3}\tan^{-1}\left(\frac{2x}{3}\right)\right]_0^b = \frac{1}{2}\lim\limits_{b \to \infty}\left[\frac{2}{3}\tan^{-1}\left(\frac{2b}{3}\right)\right] - \frac{1}{3}\tan^{-1}(0)$

$= \frac{1}{2}\left(\frac{2}{3} \cdot \frac{\pi}{2}\right) - 0 = \frac{\pi}{6}$

63. $\lim\limits_{\theta \to \infty} \frac{\theta}{\sqrt{\theta^2+1}} = 1$ and $\int_6^\infty \frac{d\theta}{\theta}$ diverges $\Rightarrow \int_6^\infty \frac{d\theta}{\sqrt{\theta^2+1}}$ diverges

65. $\int_1^\infty \frac{\ln z}{z}\, dz = \int_1^e \frac{\ln z}{z}\, dz + \int_e^\infty \frac{\ln z}{z}\, dz = \left[\frac{(\ln z)^2}{2}\right]_1^e + \lim\limits_{b \to \infty}\left[\frac{(\ln z)^2}{2}\right]_e^b = \left(\frac{1^2}{2} - 0\right) + \lim\limits_{b \to \infty}\left[\frac{(\ln b)^2}{2} - \frac{1}{2}\right]$

$= \infty \Rightarrow$ diverges

67. $\int_{-\infty}^\infty \frac{2\,dx}{e^x+e^{-x}} = 2\int_0^\infty \frac{2\,dx}{e^x+e^{-x}} < \int_0^\infty \frac{4\,dx}{e^x}$ converges $\Rightarrow \int_{-\infty}^\infty \frac{2\,dx}{e^x+e^{-x}}$ converges

69. $\int \frac{x\,dx}{1+\sqrt{x}}$; $\begin{bmatrix} u = \sqrt{x} \\ du = \frac{dx}{2\sqrt{x}} \end{bmatrix}$ $\rightarrow$ $\int \frac{u^2 \cdot 2u\,du}{1+u} = \int \left(2u^2 - 2u + 2 - \frac{2}{1+u}\right) du = \frac{2}{3}u^3 - u^2 + 2u - 2\ln|1+u| + C$

$= \frac{2x^{3/2}}{3} - x + 2\sqrt{x} - 2\ln\left(1 + \sqrt{x}\right) + C$

71. $\int \frac{dx}{x\,(x^2+1)^2}$; $\begin{bmatrix} x = \tan\theta \\ dx = \sec^2\theta\,d\theta \end{bmatrix}$ $\rightarrow$ $\int \frac{\sec^2\theta\,d\theta}{\tan\theta\,\sec^4\theta} = \int \frac{\cos^3\theta\,d\theta}{\sin\theta} = \int \left(\frac{1-\sin^2\theta}{\sin\theta}\right) d(\sin\theta)$

$= \ln|\sin\theta| - \frac{1}{2}\sin^2\theta + C = \ln\left|\frac{x}{\sqrt{x^2+1}}\right| - \frac{1}{2}\left(\frac{x}{\sqrt{x^2+1}}\right)^2 + C$

73. $\int \frac{2 - \cos x + \sin x}{\sin^2 x}\,dx = \int 2\csc^2 x\,dx - \int \frac{\cos x\,dx}{\sin^2 x} + \int \csc x\,dx = -2\cot x + \frac{1}{\sin x} - \ln|\csc x + \cot x| + C$

$= -2\cot x + \csc x - \ln|\csc x + \cot x| + C$

75. $\int \frac{9\,dv}{81 - v^4} = \frac{1}{2}\int \frac{dv}{v^2+9} + \frac{1}{12}\int \frac{dv}{3-v} + \frac{1}{12}\int \frac{dv}{3+v} = \frac{1}{12}\ln\left|\frac{3+v}{3-v}\right| + \frac{1}{6}\tan^{-1}\frac{v}{3} + C$

77. $\cos(2\theta + 1)$

θ $\xrightarrow{(+)}$ $\frac{1}{2}\sin(2\theta + 1)$

1 $\xrightarrow{(-)}$ $-\frac{1}{4}\cos(2\theta + 1)$

0 $\Rightarrow \int \theta\cos(2\theta + 1)\,d\theta = \frac{\theta}{2}\sin(2\theta + 1) + \frac{1}{4}\cos(2\theta + 1) + C$

79. $\int \frac{\sin 2\theta\,d\theta}{(1 + \cos 2\theta)^2} = -\frac{1}{2}\int \frac{d(1 + \cos 2\theta)}{(1 + \cos 2\theta)^2} = \frac{1}{2(1 + \cos 2\theta)} + C = \frac{1}{4}\sec^2\theta + C$

81. $\int \frac{x\,dx}{\sqrt{2-x}}$; $\begin{bmatrix} y = 2 - x \\ dy = -dx \end{bmatrix}$ $\rightarrow$ $-\int \frac{(2-y)\,dy}{\sqrt{y}} = \frac{2}{3}y^{3/2} - 4y^{1/2} + C = \frac{2}{3}(2-x)^{3/2} - 4(2-x)^{1/2} + C$

$= 2\left[\frac{\left(\sqrt{2-x}\right)^3}{3} - 2\sqrt{2-x}\right] + C$

83. $\int \frac{dy}{y^2 - 2y + 2} = \int \frac{d(y-1)}{(y-1)^2 + 1} = \tan^{-1}(y-1) + C$

85. $\int \frac{z+1}{z^2\,(z^2+4)}\,dz = \frac{1}{4}\int \left(\frac{1}{z} + \frac{1}{z^2} - \frac{z+1}{z^2+4}\right) dz = \frac{1}{4}\ln|z| - \frac{1}{4z} - \frac{1}{8}\ln(z^2+4) - \frac{1}{8}\tan^{-1}\frac{z}{2} + C$

87. $\int \frac{t\,dt}{\sqrt{9 - 4t^2}} = -\frac{1}{8}\int \frac{d(9 - 4t^2)}{\sqrt{9 - 4t^2}} = -\frac{1}{4}\sqrt{9 - 4t^2} + C$

89. $\int \frac{e^t\,dt}{e^{2t} + 3e^t + 2}$; $[e^t = x]$ $\rightarrow$ $\int \frac{dx}{(x+1)(x+2)} = \int \frac{dx}{x+1} - \int \frac{dx}{x+2} = \ln|x+1| - \ln|x+2| + C$

$= \ln\left|\frac{x+1}{x+2}\right| + C = \ln\left(\frac{e^t+1}{e^t+2}\right) + C$

91. $\int_1^\infty \frac{\ln y\,dy}{y^3}$; $\begin{bmatrix} x = \ln y \\ dx = \frac{dy}{y} \\ dy = e^x\,dx \end{bmatrix}$ $\rightarrow$ $\int_0^\infty \frac{x \cdot e^x}{e^{3x}}\,dx = \int_0^\infty xe^{-2x}\,dx = \lim_{b\to\infty}\left[-\frac{x}{2}e^{-2x} - \frac{1}{4}e^{-2x}\right]_0^b$

$= \lim_{b\to\infty}\left(\frac{-b}{2e^{2b}} - \frac{1}{4e^{2b}}\right) - \left(0 - \frac{1}{4}\right) = \frac{1}{4}$

93. $\int e^{\ln\sqrt{x}}\,dx = \int \sqrt{x}\,dx = \frac{2}{3}x^{3/2} + C$

95. $\int \frac{\sin 5t \, dt}{1 + (\cos 5t)^2} ; \begin{bmatrix} u = \cos 5t \\ du = -5 \sin 5t \, dt \end{bmatrix} \rightarrow -\frac{1}{5} \int \frac{du}{1+u^2} = -\frac{1}{5} \tan^{-1} u + C = -\frac{1}{5} \tan^{-1}(\cos 5t) + C$

97. $\int \frac{dr}{1+\sqrt{r}} ; \begin{bmatrix} u = \sqrt{r} \\ du = \frac{dr}{2\sqrt{r}} \end{bmatrix} \rightarrow \int \frac{2u \, du}{1+u} = \int \left(2 - \frac{2}{1+u}\right) du = 2u - 2 \ln|1+u| + C = 2\sqrt{r} - 2 \ln\left(1 + \sqrt{r}\right) + C$

CHAPTER 7 ADDITIONAL AND ADVANCED EXERCISES

1. $u = \left(\sin^{-1} x\right)^2, \, du = \frac{2 \sin^{-1} x \, dx}{\sqrt{1-x^2}} ; \, dv = dx, \, v = x;$

$\int \left(\sin^{-1} x\right)^2 dx = x \left(\sin^{-1} x\right)^2 - \int \frac{2x \sin^{-1} x \, dx}{\sqrt{1-x^2}} ;$

$u = \sin^{-1} x, \, du = \frac{dx}{\sqrt{1-x^2}} ; \, dv = -\frac{2x \, dx}{\sqrt{1-x^2}}, \, v = 2\sqrt{1-x^2};$

$-\int \frac{2x \sin^{-1} x \, dx}{\sqrt{1-x^2}} = 2 \left(\sin^{-1} x\right) \sqrt{1-x^2} - \int 2 \, dx = 2 \left(\sin^{-1} x\right) \sqrt{1-x^2} - 2x + C;$ therefore

$\int \left(\sin^{-1} x\right)^2 dx = x \left(\sin^{-1} x\right)^2 + 2 \left(\sin^{-1} x\right) \sqrt{1-x^2} - 2x + C$

3. $u = \sin^{-1} x, \, du = \frac{dx}{\sqrt{1-x^2}} ; \, dv = x \, dx, \, v = \frac{x^2}{2} ;$

$\int x \sin^{-1} x \, dx = \frac{x^2}{2} \sin^{-1} x - \int \frac{x^2 \, dx}{2\sqrt{1-x^2}} ; \begin{bmatrix} x = \sin \theta \\ dx = \cos \theta \, d\theta \end{bmatrix} \rightarrow \int x \sin^{-1} x \, dx = \frac{x^2}{2} \sin^{-1} x - \int \frac{\sin^2 \theta \cos \theta \, d\theta}{2 \cos \theta}$

$= \frac{x^2}{2} \sin^{-1} x - \frac{1}{2} \int \sin^2 \theta \, d\theta = \frac{x^2}{2} \sin^{-1} x - \frac{1}{2} \left(\frac{\theta}{2} - \frac{\sin 2\theta}{4}\right) + C = \frac{x^2}{2} \sin^{-1} x + \frac{\sin \theta \cos \theta - \theta}{4} + C$

$= \frac{x^2}{2} \sin^{-1} x + \frac{x\sqrt{1-x^2} - \sin^{-1} x}{4} + C$

5. $\int \frac{dt}{t - \sqrt{1-t^2}} ; \begin{bmatrix} t = \sin \theta \\ dt = \cos \theta \, d\theta \end{bmatrix} \rightarrow \int \frac{\cos \theta \, d\theta}{\sin \theta - \cos \theta} = \int \frac{d\theta}{\tan \theta - 1} ; \begin{bmatrix} u = \tan \theta \\ du = \sec^2 \theta \, d\theta \\ d\theta = \frac{du}{u^2+1} \end{bmatrix} \rightarrow \int \frac{du}{(u-1)(u^2+1)}$

$= \frac{1}{2} \int \frac{du}{u-1} - \frac{1}{2} \int \frac{du}{u^2+1} - \frac{1}{2} \int \frac{u \, du}{u^2+1} = \frac{1}{2} \ln \left|\frac{u-1}{\sqrt{u^2+1}}\right| - \frac{1}{2} \tan^{-1} u + C = \frac{1}{2} \ln \left|\frac{\tan \theta - 1}{\sec \theta}\right| - \frac{1}{2} \theta + C$

$= \frac{1}{2} \ln \left(t - \sqrt{1-t^2}\right) - \frac{1}{2} \sin^{-1} t + C$

7. $\lim_{x \to \infty} \int_{-x}^{x} \sin t \, dt = \lim_{x \to \infty} \left[-\cos t\right]_{-x}^{x} = \lim_{x \to \infty} \left[-\cos x + \cos(-x)\right] = \lim_{x \to \infty} \left(-\cos x + \cos x\right) = \lim_{x \to \infty} 0 = 0$

9. $\lim_{n \to \infty} \sum_{k=1}^{n} \ln \sqrt[n]{1 + \frac{k}{n}} = \lim_{n \to \infty} \sum_{k=1}^{n} \ln \left(1 + k \left(\frac{1}{n}\right)\right) \left(\frac{1}{n}\right) = \int_0^1 \ln(1 + x) \, dx; \begin{bmatrix} u = 1 + x, \, du = dx \\ x = 0 \Rightarrow u = 1, \, x = 1 \Rightarrow u = 2 \end{bmatrix}$

$\rightarrow \int_1^2 \ln u \, du = \left[u \ln u - u\right]_1^2 = (2 \ln 2 - 2) - (\ln 1 - 1) = 2 \ln 2 - 1 = \ln 4 - 1$

11. $\frac{dy}{dx} = \sqrt{\cos 2x} \Rightarrow 1 + \left(\frac{dy}{dx}\right)^2 = 1 + \cos 2x = 2 \cos^2 x; \, L = \int_0^{\pi/4} \sqrt{1 + \left(\sqrt{\cos 2t}\right)^2} \, dt = \sqrt{2} \int_0^{\pi/4} \sqrt{\cos^2 t} \, dt$

$= \sqrt{2} \left[\sin t\right]_0^{\pi/4} = 1$

13. $V = \int_a^b 2\pi \left(\begin{smallmatrix}\text{shell}\\\text{radius}\end{smallmatrix}\right)\left(\begin{smallmatrix}\text{shell}\\\text{height}\end{smallmatrix}\right) dx = \int_0^1 2\pi xy \, dx$

$= 6\pi \int_0^1 x^2 \sqrt{1-x} \, dx; \quad \begin{bmatrix} u = 1 - x \\ du = -dx \\ x^2 = (1-u)^2 \end{bmatrix}$

$\rightarrow -6\pi \int_1^0 (1-u)^2 \sqrt{u} \, du$

$= -6\pi \int_1^0 \left(u^{1/2} - 2u^{3/2} + u^{5/2}\right) du$

$= -6\pi \left[\frac{2}{3} u^{3/2} - \frac{4}{5} u^{5/2} + \frac{2}{7} u^{7/2}\right]_1^0 = 6\pi \left(\frac{2}{3} - \frac{4}{5} + \frac{2}{7}\right)$

$= 6\pi \left(\frac{70 - 84 + 30}{105}\right) = 6\pi \left(\frac{16}{105}\right) = \frac{32\pi}{35}$

15. $V = \int_a^b 2\pi \left(\begin{smallmatrix}\text{shell}\\\text{radius}\end{smallmatrix}\right)\left(\begin{smallmatrix}\text{shell}\\\text{height}\end{smallmatrix}\right) dx = \int_0^1 2\pi xe^x \, dx$

$= 2\pi \left[xe^x - e^x\right]_0^1 = 2\pi$

17. (a) $V = \int_1^e \pi \left[1 - (\ln x)^2\right] dx$

$= \pi \left[x - x(\ln x)^2\right]_1^e + 2\pi \int_1^e \ln x \, dx$

(FORMULA 110)

$= \pi \left[x - x(\ln x)^2 + 2(x \ln x - x)\right]_1^e$

$= \pi \left[-x - x(\ln x)^2 + 2x \ln x\right]_1^e$

$= \pi \left[-e - e + 2e - (-1)\right] = \pi$

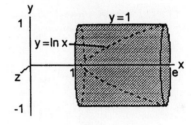

(b) $V = \int_1^e \pi(1 - \ln x)^2 \, dx = \pi \int_1^e \left[1 - 2\ln x + (\ln x)^2\right] dx$

$= \pi \left[x - 2(x \ln x - x) + x(\ln x)^2\right]_1^e - 2\pi \int_1^e \ln x \, dx$

$= \pi \left[x - 2(x \ln x - x) + x(\ln x)^2 - 2(x \ln x - x)\right]_1^e$

$= \pi \left[5x - 4x \ln x + x(\ln x)^2\right]_1^e$

$= \pi \left[(5e - 4e + e) - (5)\right] = \pi(2e - 5)$

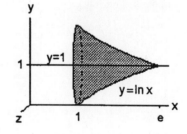

19. (a) $\lim_{x \to 0^+} x \ln x = 0 \Rightarrow \lim_{x \to 0^+} f(x) = 0 = f(0) \Rightarrow f$ is continuous

(b) $V = \int_0^2 \pi x^2 (\ln x)^2 \, dx; \quad \begin{bmatrix} u = (\ln x)^2 \\ du = (2\ln x)\frac{dx}{x} \\ dv = x^2 dx \\ v = \frac{x^3}{3} \end{bmatrix} \rightarrow \pi \left(\lim_{b \to 0^+} \left[\frac{x^3}{3}(\ln x)^2\right]_b^2 - \int_0^2 \left(\frac{x^3}{3}\right)(2 \ln x)\frac{dx}{x}\right)$

$= \pi \left[\left(\frac{8}{3}\right)(\ln 2)^2 - \left(\frac{2}{3}\right)\lim_{b \to 0^+}\left[\frac{x^3}{3}\ln x - \frac{x^3}{9}\right]_b^2\right] = \pi \left[\frac{8(\ln 2)^2}{3} - \frac{16(\ln 2)}{9} + \frac{16}{27}\right]$

21. $M = \int_1^e \ln x \, dx = [x \ln x - x]_1^e = (e - e) - (0 - 1) = 1$;

$M_x = \int_1^e (\ln x)\left(\frac{\ln x}{2}\right) dx = \frac{1}{2} \int_1^e (\ln x)^2 \, dx$

$= \frac{1}{2}\left([x(\ln x)^2]_1^e - 2\int_1^e \ln x \, dx\right) = \frac{1}{2}(e - 2)$;

$M_y = \int_1^e x \ln x \, dx = \left[\frac{x^2 \ln x}{2}\right]_1^e - \frac{1}{2}\int_1^e x \, dx$

$= \frac{1}{2}\left[x^2 \ln x - \frac{x^2}{2}\right]_1^e = \frac{1}{2}\left[\left(e^2 - \frac{e^2}{2}\right) + \frac{1}{2}\right] = \frac{1}{4}(e^2 + 1)$;

therefore, $\bar{x} = \frac{M_y}{M} = \frac{e^2+1}{4}$ and $\bar{y} = \frac{M_x}{M} = \frac{e-2}{2}$

23. $L = \int_1^e \sqrt{1 + \frac{1}{x^2}} \, dx = \int_1^e \frac{\sqrt{x^2+1}}{x} \, dx$; $\begin{bmatrix} x = \tan\theta \\ dx = \sec^2\theta \, d\theta \end{bmatrix} \rightarrow L = \int_{\pi/4}^{\tan^{-1} e} \frac{\sec\theta \cdot \sec^2\theta \, d\theta}{\tan\theta}$

$= \int_{\pi/4}^{\tan^{-1} e} \frac{(\sec\theta)(\tan^2\theta + 1)}{\tan\theta} \, d\theta = \int_{\pi/4}^{\tan^{-1} e} (\tan\theta \sec\theta + \csc\theta) \, d\theta = [\sec\theta - \ln|\csc\theta + \cot\theta|]_{\pi/4}^{\tan^{-1} e}$

$= \left(\sqrt{1 + e^2} - \ln\left|\frac{\sqrt{1+e^2}}{e} + \frac{1}{e}\right|\right) - \left[\sqrt{2} - \ln\left(1 + \sqrt{2}\right)\right] = \sqrt{1 + e^2} - \ln\left(\frac{\sqrt{1+e^2}}{e} + \frac{1}{e}\right) - \sqrt{2} + \ln\left(1 + \sqrt{2}\right)$

25. $L = 4\int_0^1 \sqrt{1 + \left(\frac{dy}{dx}\right)^2} \, dx$; $x^{2/3} + y^{2/3} = 1 \Rightarrow y = \left(1 - x^{2/3}\right)^{3/2} \Rightarrow \frac{dy}{dx} = -\frac{3}{2}\left(1 - x^{2/3}\right)^{1/2}\left(x^{-1/3}\right)\left(\frac{2}{3}\right)$

$\Rightarrow \left(\frac{dy}{dx}\right)^2 = \frac{1-x^{2/3}}{x^{2/3}} \Rightarrow L = 4\int_0^1 \sqrt{1 + \left(\frac{1-x^{2/3}}{x^{2/3}}\right)} \, dx = 4\int_0^1 \frac{dx}{x^{1/3}} = 6\left[x^{2/3}\right]_0^1 = 6$

27. $\int_1^\infty \left(\frac{ax}{x^2+1} - \frac{1}{2x}\right) dx = \lim_{b \to \infty} \int_1^b \left(\frac{ax}{x^2+1} - \frac{1}{2x}\right) dx = \lim_{b \to \infty} \left[\frac{a}{2} \ln(x^2 + 1) - \frac{1}{2} \ln x\right]_1^b = \lim_{b \to \infty} \left[\frac{1}{2} \ln\frac{(x^2+1)^a}{x}\right]_1^b$

$= \lim_{b \to \infty} \frac{1}{2}\left[\ln\frac{(b^2+1)^a}{b} - \ln 2^a\right]$; $\lim_{b \to \infty} \frac{(b^2+1)^a}{b} > \lim_{b \to \infty} \frac{b^{2a}}{b} = \lim_{b \to \infty} b^{2\left(a - \frac{1}{2}\right)} = \infty$ if $a > \frac{1}{2} \Rightarrow$ the improper

integral diverges if $a > \frac{1}{2}$; for $a = \frac{1}{2}$: $\lim_{b \to \infty} \frac{\sqrt{b^2+1}}{b} = \lim_{b \to \infty} \sqrt{1 + \frac{1}{b^2}} = 1 \Rightarrow \lim_{b \to \infty} \frac{1}{2}\left[\ln\frac{(b^2+1)^{1/2}}{b} - \ln 2^{1/2}\right]$

$= \frac{1}{2}\left(\ln 1 - \frac{1}{2} \ln 2\right) = -\frac{\ln 2}{4}$; if $a < \frac{1}{2}$: $0 \le \lim_{b \to \infty} \frac{(b^2+1)^a}{b} < \lim_{b \to \infty} \frac{(b+1)^{2a}}{b+1} = \lim_{b \to \infty} (b+1)^{2a-1} = 0$

$\Rightarrow \lim_{b \to \infty} \ln\frac{(b^2+1)^a}{b} = -\infty \Rightarrow$ the improper integral diverges if $a < \frac{1}{2}$; in summary, the improper integral

$\int_1^\infty \left(\frac{ax}{x^2+1} - \frac{1}{2x}\right) dx$ converges only when $a = \frac{1}{2}$ and has the value $-\frac{\ln 2}{4}$

29. $A = \int_1^\infty \frac{dx}{x^p}$ converges if $p > 1$ and diverges if $p \le 1$. Thus, $p \le 1$ for infinite area. The volume of the solid of revolution

about the x-axis is $V = \int_1^\infty \pi\left(\frac{1}{x^p}\right)^2 dx = \pi \int_1^\infty \frac{dx}{x^{2p}}$ which converges if $2p > 1$ and diverges if $2p \le 1$. Thus we want

$p > \frac{1}{2}$ for finite volume. In conclusion, the curve $y = x^{-p}$ gives infinite area and finite volume for values of p satisfying

$\frac{1}{2} < p \le 1$.

31. (a) $\Gamma(1) = \int_0^\infty e^{-t} \, dt = \lim_{b \to \infty} \int_0^b e^{-t} \, dt = \lim_{b \to \infty} [-e^{-t}]_0^b = \lim_{b \to \infty} \left[-\frac{1}{e^b} - (-1)\right] = 0 + 1 = 1$

(b) $u = t^x$, $du = xt^{x-1} \, dt$; $dv = e^{-t} \, dt$, $v = -e^{-t}$; $x = $ fixed positive real

$\Rightarrow \Gamma(x + 1) = \int_0^\infty t^x e^{-t} \, dt = \lim_{b \to \infty} [-t^x e^{-t}]_0^b + x\int_0^\infty t^{x-1}e^{-t} \, dt = \lim_{b \to \infty} \left(-\frac{b^x}{e^b} + 0^x e^0\right) + x\Gamma(x) = x\Gamma(x)$

(c) $\Gamma(n + 1) = n\Gamma(n) = n!$:

$\quad n = 0$: $\Gamma(0 + 1) = \Gamma(1) = 0!$;

$\quad n = k$: Assume $\Gamma(k + 1) = k!$ for some $k > 0$;

$\quad n = k + 1$: $\Gamma(k + 1 + 1) = (k + 1)\Gamma(k + 1)$ from part (b)

$\quad\quad\quad\quad\quad\quad\quad\quad = (k + 1)k!$ induction hypothesis

$\quad\quad\quad\quad\quad\quad\quad\quad = (k + 1)!$ definition of factorial

Thus, $\Gamma(n + 1) = n\Gamma(n) = n!$ for every positive integer n.

CHAPTER 8 INFINITE SEQUENCES AND SERIES

8.1 SEQUENCES

1. $a_1 = \frac{1-1}{1^2} = 0$, $a_2 = \frac{1-2}{2^2} = -\frac{1}{4}$, $a_3 = \frac{1-3}{3^2} = -\frac{2}{9}$, $a_4 = \frac{1-4}{4^2} = -\frac{3}{16}$

3. $a_1 = \frac{(-1)^2}{2-1} = 1$, $a_2 = \frac{(-1)^3}{4-1} = -\frac{1}{3}$, $a_3 = \frac{(-1)^4}{6-1} = \frac{1}{5}$, $a_4 = \frac{(-1)^5}{8-1} = -\frac{1}{7}$

5. $a_1 = \frac{2}{2^2} = \frac{1}{2}$, $a_2 = \frac{2^2}{2^3} = \frac{1}{2}$, $a_3 = \frac{2^3}{2^4} = \frac{1}{2}$, $a_4 = \frac{2^4}{2^5} = \frac{1}{2}$

7. $a_1 = 1$, $a_2 = 1 + \frac{1}{2} = \frac{3}{2}$, $a_3 = \frac{3}{2} + \frac{1}{2^2} = \frac{7}{4}$, $a_4 = \frac{7}{4} + \frac{1}{2^3} = \frac{15}{8}$, $a_5 = \frac{15}{8} + \frac{1}{2^4} = \frac{31}{16}$, $a_6 = \frac{63}{32}$,
    $a_7 = \frac{127}{64}$, $a_8 = \frac{255}{128}$, $a_9 = \frac{511}{256}$, $a_{10} = \frac{1023}{512}$

9. $a_1 = 2$, $a_2 = \frac{(-1)^2(2)}{2} = 1$, $a_3 = \frac{(-1)^3(1)}{2} = -\frac{1}{2}$, $a_4 = \frac{(-1)^4\left(-\frac{1}{2}\right)}{2} = -\frac{1}{4}$, $a_5 = \frac{(-1)^5\left(-\frac{1}{4}\right)}{2} = \frac{1}{8}$,
    $a_6 = \frac{1}{16}$, $a_7 = -\frac{1}{32}$, $a_8 = -\frac{1}{64}$, $a_9 = \frac{1}{128}$, $a_{10} = \frac{1}{256}$

11. $a_1 = 1$, $a_2 = 1$, $a_3 = 1 + 1 = 2$, $a_4 = 2 + 1 = 3$, $a_5 = 3 + 2 = 5$, $a_6 = 8$, $a_7 = 13$, $a_8 = 21$, $a_9 = 34$, $a_{10} = 55$

13. $a_n = (-1)^{n+1}$, $n = 1, 2, \ldots$

15. $a_n = (-1)^{n+1}n^2$, $n = 1, 2, \ldots$

17. $a_n = n^2 - 1$, $n = 1, 2, \ldots$

19. $a_n = 4n - 3$, $n = 1, 2, \ldots$

21. $a_n = \frac{1 + (-1)^{n+1}}{2}$, $n = 1, 2, \ldots$

23. $\lim\limits_{n \to \infty} 2 + (0.1)^n = 2 \Rightarrow$ converges   (Theorem 5, #4)

25. $\lim\limits_{n \to \infty} \frac{1-2n}{1+2n} = \lim\limits_{n \to \infty} \frac{\left(\frac{1}{n}\right)-2}{\left(\frac{1}{n}\right)+2} = \lim\limits_{n \to \infty} \frac{-2}{2} = -1 \Rightarrow$ converges

27. $\lim\limits_{n \to \infty} \frac{1-5n^4}{n^4+8n^3} = \lim\limits_{n \to \infty} \frac{\left(\frac{1}{n^4}\right)-5}{1+\left(\frac{8}{n}\right)} = -5 \Rightarrow$ converges

29. $\lim\limits_{n \to \infty} \frac{n^2-2n+1}{n-1} = \lim\limits_{n \to \infty} \frac{(n-1)(n-1)}{n-1} = \lim\limits_{n \to \infty} (n-1) = \infty \Rightarrow$ diverges

31. $\lim\limits_{n \to \infty} (1 + (-1)^n)$ does not exist $\Rightarrow$ diverges

33. $\lim\limits_{n \to \infty} \left(\frac{n+1}{2n}\right)\left(1 - \frac{1}{n}\right) = \lim\limits_{n \to \infty} \left(\frac{1}{2} + \frac{1}{2n}\right)\left(1 - \frac{1}{n}\right) = \frac{1}{2} \Rightarrow$ converges

35. $\lim\limits_{n \to \infty} \frac{(-1)^{n+1}}{2n-1} = 0 \Rightarrow$ converges

37. $\lim\limits_{n \to \infty} \sqrt{\frac{2n}{n+1}} = \sqrt{\lim\limits_{n \to \infty} \frac{2n}{n+1}} = \sqrt{\lim\limits_{n \to \infty} \left(\frac{2}{1+\frac{1}{n}}\right)} = \sqrt{2} \Rightarrow$ converges

39. $\lim\limits_{n \to \infty} \sin\left(\frac{\pi}{2} + \frac{1}{n}\right) = \sin\left(\lim\limits_{n \to \infty}\left(\frac{\pi}{2} + \frac{1}{n}\right)\right) = \sin\frac{\pi}{2} = 1 \Rightarrow$ converges

41. $\lim\limits_{n \to \infty} \frac{\sin n}{n} = 0$ because $-\frac{1}{n} \le \frac{\sin n}{n} \le \frac{1}{n} \Rightarrow$ converges by the Sandwich Theorem for sequences

43. $\lim\limits_{n \to \infty} \frac{n}{2^n} = \lim\limits_{n \to \infty} \frac{1}{2^n \ln 2} = 0 \Rightarrow$ converges (using l'Hôpital's rule)

45. $\lim\limits_{n \to \infty} \frac{\ln(n+1)}{\sqrt{n}} = \lim\limits_{n \to \infty} \frac{\left(\frac{1}{n+1}\right)}{\left(\frac{1}{2\sqrt{n}}\right)} = \lim\limits_{n \to \infty} \frac{2\sqrt{n}}{n+1} = \lim\limits_{n \to \infty} \frac{\left(\frac{2}{\sqrt{n}}\right)}{1 + \left(\frac{1}{n}\right)} = 0 \Rightarrow$ converges

47. $\lim\limits_{n \to \infty} 8^{1/n} = 1 \Rightarrow$ converges   (Theorem 5, #3)

49. $\lim\limits_{n \to \infty} \left(1 + \frac{7}{n}\right)^n = e^7 \Rightarrow$ converges   (Theorem 5, #5)

51. $\lim\limits_{n \to \infty} \sqrt[n]{10n} = \lim\limits_{n \to \infty} 10^{1/n} \cdot n^{1/n} = 1 \cdot 1 = 1 \Rightarrow$ converges   (Theorem 5, #3 and #2)

53. $\lim\limits_{n \to \infty} \left(\frac{3}{n}\right)^{1/n} = \frac{\lim\limits_{n \to \infty} 3^{1/n}}{\lim\limits_{n \to \infty} n^{1/n}} = \frac{1}{1} = 1 \Rightarrow$ converges   (Theorem 5, #3 and #2)

55. $\lim\limits_{n \to \infty} \frac{\ln n}{n^{1/n}} = \frac{\lim\limits_{n \to \infty} \ln n}{\lim\limits_{n \to \infty} n^{1/n}} = \frac{\infty}{1} = \infty \Rightarrow$ diverges   (Theorem 5, #2)

57. $\lim\limits_{n \to \infty} \sqrt[n]{4^n\, n} = \lim\limits_{n \to \infty} 4 \sqrt[n]{n} = 4 \cdot 1 = 4 \Rightarrow$ converges   (Theorem 5, #2)

59. $\lim\limits_{n \to \infty} \frac{n!}{n^n} = \lim\limits_{n \to \infty} \frac{1 \cdot 2 \cdot 3 \cdots (n-1)(n)}{n \cdot n \cdot n \cdots n \cdot n} \le \lim\limits_{n \to \infty} \left(\frac{1}{n}\right) = 0$ and $\frac{n!}{n^n} \ge 0 \Rightarrow \lim\limits_{n \to \infty} \frac{n!}{n^n} = 0 \Rightarrow$ converges

61. $\lim\limits_{n \to \infty} \frac{n!}{10^{6n}} = \lim\limits_{n \to \infty} \frac{1}{\left(\frac{(10^6)^n}{n!}\right)} = \infty \Rightarrow$ diverges   (Theorem 5, #6)

63. $\lim\limits_{n \to \infty} \left(\frac{1}{n}\right)^{1/(\ln n)} = \lim\limits_{n \to \infty} \exp\left(\frac{1}{\ln n} \ln\left(\frac{1}{n}\right)\right) = \lim\limits_{n \to \infty} \exp\left(\frac{\ln 1 - \ln n}{\ln n}\right) = e^{-1} \Rightarrow$ converges

65. $\lim\limits_{n \to \infty} \left(\frac{3n+1}{3n-1}\right)^n = \lim\limits_{n \to \infty} \exp\left(n \ln\left(\frac{3n+1}{3n-1}\right)\right) = \lim\limits_{n \to \infty} \exp\left(\frac{\ln(3n+1) - \ln(3n-1)}{\frac{1}{n}}\right)$

$= \lim\limits_{n \to \infty} \exp\left(\frac{\frac{3}{3n+1} - \frac{3}{3n-1}}{\left(-\frac{1}{n^2}\right)}\right) = \lim\limits_{n \to \infty} \exp\left(\frac{6n^2}{(3n+1)(3n-1)}\right) = \exp\left(\frac{6}{9}\right) = e^{2/3} \Rightarrow$ converges

67. $\lim\limits_{n \to \infty} \left(\frac{x^n}{2n+1}\right)^{1/n} = \lim\limits_{n \to \infty} x\left(\frac{1}{2n+1}\right)^{1/n} = x \lim\limits_{n \to \infty} \exp\left(\frac{1}{n} \ln\left(\frac{1}{2n+1}\right)\right) = x \lim\limits_{n \to \infty} \exp\left(\frac{-\ln(2n+1)}{n}\right)$

$= x \lim\limits_{n \to \infty} \exp\left(\frac{-2}{2n+1}\right) = xe^0 = x,\ x > 0 \Rightarrow$ converges

69. $\lim\limits_{n \to \infty} \frac{3^n \cdot 6^n}{2^{-n} \cdot n!} = \lim\limits_{n \to \infty} \frac{36^n}{n!} = 0 \Rightarrow$ converges   (Theorem 5, #6)

71. $\lim\limits_{n \to \infty} \tanh n = \lim\limits_{n \to \infty} \frac{e^n - e^{-n}}{e^n + e^{-n}} = \lim\limits_{n \to \infty} \frac{e^{2n} - 1}{e^{2n} + 1} = \lim\limits_{n \to \infty} \frac{2e^{2n}}{2e^{2n}} = \lim\limits_{n \to \infty} 1 = 1 \Rightarrow$ converges

73. $\lim\limits_{n \to \infty} \frac{n^2 \sin\left(\frac{1}{n}\right)}{2n-1} = \lim\limits_{n \to \infty} \frac{\sin\left(\frac{1}{n}\right)}{\left(\frac{2}{n} - \frac{1}{n^2}\right)} = \lim\limits_{n \to \infty} \frac{-\left(\cos\left(\frac{1}{n}\right)\right)\left(\frac{1}{n^2}\right)}{\left(-\frac{2}{n^2} + \frac{2}{n^3}\right)} = \lim\limits_{n \to \infty} \frac{-\cos\left(\frac{1}{n}\right)}{-2 + \left(\frac{2}{n}\right)} = \frac{1}{2} \Rightarrow$ converges

75. $\lim\limits_{n \to \infty} \tan^{-1} n = \frac{\pi}{2} \Rightarrow$ converges

77. $\lim\limits_{n \to \infty} \left(\frac{1}{3}\right)^n + \frac{1}{\sqrt{2^n}} = \lim\limits_{n \to \infty} \left(\left(\frac{1}{3}\right)^n + \left(\frac{1}{\sqrt{2}}\right)^n\right) = 0 \Rightarrow$ converges (Theorem 5, #4)

79. $\lim\limits_{n \to \infty} \frac{(\ln n)^{200}}{n} = \lim\limits_{n \to \infty} \frac{200\,(\ln n)^{199}}{n} = \lim\limits_{n \to \infty} \frac{200\cdot199\,(\ln n)^{198}}{n} = \cdots = \lim\limits_{n \to \infty} \frac{200!}{n} = 0 \Rightarrow$ converges

81. $\lim\limits_{n \to \infty} \left(n - \sqrt{n^2 - n}\right) = \lim\limits_{n \to \infty} \left(n - \sqrt{n^2 - n}\right)\left(\frac{n + \sqrt{n^2 - n}}{n + \sqrt{n^2 - n}}\right) = \lim\limits_{n \to \infty} \frac{n}{n + \sqrt{n^2 - n}} = \lim\limits_{n \to \infty} \frac{1}{1 + \sqrt{1 - \frac{1}{n}}}$

$= \frac{1}{2} \Rightarrow$ converges

83. $\lim\limits_{n \to \infty} \frac{1}{n}\int_1^n \frac{1}{x}\,dx = \lim\limits_{n \to \infty} \frac{\ln n}{n} = \lim\limits_{n \to \infty} \frac{1}{n} = 0 \Rightarrow$ converges (Theorem 5, #1)

85. $1, 1, 2, 4, 8, 16, 32, \ldots = 1, 2^0, 2^1, 2^2, 2^3, 2^4, 2^5, \ldots \Rightarrow x_1 = 1$ and $x_n = 2^{n-2}$ for $n \geq 2$

87. (a) $f(x) = x^2 - 2$; the sequence converges to $1.414213562 \approx \sqrt{2}$

(b) $f(x) = \tan(x) - 1$; the sequence converges to $0.7853981635 \approx \frac{\pi}{4}$

(c) $f(x) = e^x$; the sequence $1, 0, -1, -2, -3, -4, -5, \ldots$ diverges

89. (a) If $a = 2n + 1$, then $b = \left\lfloor \frac{a^2}{2}\right\rfloor = \left\lfloor \frac{4n^2 + 4n + 1}{2}\right\rfloor = \left\lfloor 2n^2 + 2n + \frac{1}{2}\right\rfloor = 2n^2 + 2n$, $c = \left\lceil \frac{a^2}{2}\right\rceil = \left\lceil 2n^2 + 2n + \frac{1}{2}\right\rceil$

$= 2n^2 + 2n + 1$ and $a^2 + b^2 = (2n + 1)^2 + \left(2n^2 + 2n\right)^2 = 4n^2 + 4n + 1 + 4n^4 + 8n^3 + 4n^2$

$= 4n^4 + 8n^3 + 8n^2 + 4n + 1 = \left(2n^2 + 2n + 1\right)^2 = c^2$.

(b) $\lim\limits_{a \to \infty} \frac{\left\lfloor \frac{a^2}{2}\right\rfloor}{\left\lceil \frac{a^2}{2}\right\rceil} = \lim\limits_{a \to \infty} \frac{2n^2 + 2n}{2n^2 + 2n + 1} = 1$ or $\lim\limits_{a \to \infty} \frac{\left\lfloor \frac{a^2}{2}\right\rfloor}{\left\lceil \frac{a^2}{2}\right\rceil} = \lim\limits_{a \to \infty} \sin\theta = \lim\limits_{\theta \to \pi/2} \sin\theta = 1$

91. (a) $\lim\limits_{n \to \infty} \frac{\ln n}{n^c} = \lim\limits_{n \to \infty} \frac{\left(\frac{1}{n}\right)}{cn^{c-1}} = \lim\limits_{n \to \infty} \frac{1}{cn^c} = 0$

(b) For all $\epsilon > 0$, there exists an $N = e^{-(\ln\epsilon)/c}$ such that $n > e^{-(\ln\epsilon)/c} \Rightarrow \ln n > -\frac{\ln\epsilon}{c} \Rightarrow \ln n^c > \ln\left(\frac{1}{\epsilon}\right)$

$\Rightarrow n^c > \frac{1}{\epsilon} \Rightarrow \frac{1}{n^c} < \epsilon \Rightarrow \left|\frac{1}{n^c} - 0\right| < \epsilon \Rightarrow \lim\limits_{n \to \infty} \frac{1}{n^c} = 0$

93. $\lim\limits_{n \to \infty} n^{1/n} = \lim\limits_{n \to \infty} \exp\left(\frac{1}{n}\ln n\right) = \lim\limits_{n \to \infty} \exp\left(\frac{1}{n}\right) = e^0 = 1$

95. Assume the hypotheses of the theorem and let ϵ be a positive number. For all ϵ there exists a N_1 such that when $n > N_1$ then $|a_n - L| < \epsilon \Rightarrow -\epsilon < a_n - L < \epsilon \Rightarrow L - \epsilon < a_n$, and there exists a N_2 such that when $n > N_2$ then $|c_n - L| < \epsilon \Rightarrow -\epsilon < c_n - L < \epsilon \Rightarrow c_n < L + \epsilon$. If $n > \max\{N_1, N_2\}$, then $L - \epsilon < a_n \leq b_n \leq c_n < L + \epsilon \Rightarrow |b_n - L| < \epsilon \Rightarrow \lim\limits_{n \to \infty} b_n = L$.

97. $a_{n+1} \geq a_n \Rightarrow \frac{3(n+1)+1}{(n+1)+1} > \frac{3n+1}{n+1} \Rightarrow \frac{3n+4}{n+2} > \frac{3n+1}{n+1} \Rightarrow 3n^2 + 3n + 4n + 4 > 3n^2 + 6n + n + 2$

$\Rightarrow 4 > 2$; the steps are reversible so the sequence is nondecreasing; $\frac{3n+1}{n+1} < 3 \Rightarrow 3n + 1 < 3n + 3$

$\Rightarrow 1 < 3$; the steps are reversible so the sequence is bounded above by 3

99. $a_{n+1} \leq a_n \Rightarrow \frac{2^{n+1}3^{n+1}}{(n+1)!} \leq \frac{2^n 3^n}{n!} \Rightarrow \frac{2^{n+1}3^{n+1}}{2^n 3^n} \leq \frac{(n+1)!}{n!} \Rightarrow 2\cdot 3 \leq n + 1$ which is true for $n \geq 5$; the steps are reversible so the sequence is decreasing after a_5, but it is not nondecreasing for all its terms; $a_1 = 6$, $a_2 = 18$, $a_3 = 36$, $a_4 = 54$, $a_5 = \frac{324}{5} = 64.8 \Rightarrow$ the sequence is bounded from above by 64.8

101. $a_n = 1 - \frac{1}{n}$ converges because $\frac{1}{n} \to 0$ by Example 1; also it is a nondecreasing sequence bounded above by 1

103. $a_n = \frac{2^n - 1}{2^n} = 1 - \frac{1}{2^n}$ and $0 < \frac{1}{2^n} < \frac{1}{n}$; since $\frac{1}{n} \to 0$ (by Example 1) $\Rightarrow \frac{1}{2^n} \to 0$, the sequence converges; also it is a nondecreasing sequence bounded above by 1

105. $a_n = ((-1)^n + 1)\left(\frac{n+1}{n}\right)$ diverges because $a_n = 0$ for n odd, while for n even $a_n = 2\left(1 + \frac{1}{n}\right)$ converges to 2; it diverges by definition of divergence

107. If $\{a_n\}$ is nonincreasing with lower bound M, then $\{-a_n\}$ is a nondecreasing sequence with upper bound $-M$. By Theorem 1, $\{-a_n\}$ converges and hence $\{a_n\}$ converges. If $\{a_n\}$ has no lower bound, then $\{-a_n\}$ has no upper bound and therefore diverges. Hence, $\{a_n\}$ also diverges.

109. $a_n \geq a_{n+1} \Leftrightarrow \frac{1 + \sqrt{2n}}{\sqrt{n}} \geq \frac{1 + \sqrt{2(n+1)}}{\sqrt{n+1}} \Leftrightarrow \sqrt{n+1} + \sqrt{2n^2 + 2n} \geq \sqrt{n} + \sqrt{2n^2 + 2n} \Leftrightarrow \sqrt{n+1} \geq \sqrt{n}$

and $\frac{1 + \sqrt{2n}}{\sqrt{n}} \geq \sqrt{2}$; thus the sequence is nonincreasing and bounded below by $\sqrt{2} \Rightarrow$ it converges

111. $\frac{4^{n+1} + 3^n}{4^n} = 4 + \left(\frac{3}{4}\right)^n$ so $a_n \geq a_{n+1} \Leftrightarrow 4 + \left(\frac{3}{4}\right)^n \geq 4 + \left(\frac{3}{4}\right)^{n+1} \Leftrightarrow \left(\frac{3}{4}\right)^n \geq \left(\frac{3}{4}\right)^{n+1} \Leftrightarrow 1 \geq \frac{3}{4}$ and

$4 + \left(\frac{3}{4}\right)^n \geq 4$; thus the sequence is nonincreasing and bounded below by $4 \Rightarrow$ it converges

113. Let $0 < M < 1$ and let N be an integer greater than $\frac{M}{1-M}$. Then $n > N \Rightarrow n > \frac{M}{1-M} \Rightarrow n - nM > M$
$\Rightarrow n > M + nM \Rightarrow n > M(n+1) \Rightarrow \frac{n}{n+1} > M$.

115. The sequence $a_n = 1 + \frac{(-1)^n}{2}$ is the sequence $\frac{1}{2}, \frac{3}{2}, \frac{1}{2}, \frac{3}{2}, \dots$. This sequence is bounded above by $\frac{3}{2}$, but it clearly does not converge, by definition of convergence.

117. Given an $\epsilon > 0$, by definition of convergence there corresponds an N such that for all $n > N$,
$|L_1 - a_n| < \epsilon$ and $|L_2 - a_n| < \epsilon$. Now $|L_2 - L_1| = |L_2 - a_n + a_n - L_1| \leq |L_2 - a_n| + |a_n - L_1| < \epsilon + \epsilon = 2\epsilon$.
$|L_2 - L_1| < 2\epsilon$ says that the difference between two fixed values is smaller than any positive number 2ϵ.
The only nonnegative number smaller than every positive number is 0, so $|L_1 - L_2| = 0$ or $L_1 = L_2$.

119. $a_{2k} \to L \Leftrightarrow$ given an $\epsilon > 0$ there corresponds an N_1 such that $[2k > N_1 \Rightarrow |a_{2k} - L| < \epsilon]$. Similarly,
$a_{2k+1} \to L \Leftrightarrow [2k+1 > N_2 \Rightarrow |a_{2k+1} - L| < \epsilon]$. Let $N = \max\{N_1, N_2\}$. Then $n > N \Rightarrow |a_n - L| < \epsilon$ whether n is even or odd, and hence $a_n \to L$.

121. $\left|\sqrt[n]{0.5} - 1\right| < 10^{-3} \Rightarrow -\frac{1}{1000} < \left(\frac{1}{2}\right)^{1/n} - 1 < \frac{1}{1000} \Rightarrow \left(\frac{999}{1000}\right)^n < \frac{1}{2} < \left(\frac{1001}{1000}\right)^n \Rightarrow n > \frac{\ln\left(\frac{1}{2}\right)}{\ln\left(\frac{999}{1000}\right)} \Rightarrow n > 692.8$
$\Rightarrow N = 692; a_n = \left(\frac{1}{2}\right)^{1/n}$ and $\lim_{n \to \infty} a_n = 1$

123. $(0.9)^n < 10^{-3} \Rightarrow n \ln(0.9) < -3 \ln 10 \Rightarrow n > \frac{-3 \ln 10}{\ln(0.9)} \approx 65.54 \Rightarrow N = 65; a_n = \left(\frac{9}{10}\right)^n$ and $\lim_{n \to \infty} a_n = 0$

125. (a) $f(x) = x^2 - a \Rightarrow f'(x) = 2x \Rightarrow x_{n+1} = x_n - \frac{x_n^2 - a}{2x_n} \Rightarrow x_{n+1} = \frac{2x_n^2 - (x_n^2 - a)}{2x_n} = \frac{x_n^2 + a}{2x_n} = \frac{\left(x_n + \frac{a}{x_n}\right)}{2}$

(b) $x_1 = 2, x_2 = 1.75, x_3 = 1.732142857, x_4 = 1.73205081, x_5 = 1.732050808$; we are finding the positive number where $x^2 - 3 = 0$; that is, where $x^2 = 3, x > 0$, or where $x = \sqrt{3}$.

127. $x_1 = 1, x_2 = 1 + \cos(1) = 1.540302306, x_3 = 1.540302306 + \cos(1 + \cos(1)) = 1.570791601$,
$x_4 = 1.570791601 + \cos(1.570791601) = 1.570796327 = \frac{\pi}{2}$ to 9 decimal places. After a few steps, the arc (x_{n-1}) and line segment $\cos(x_{n-1})$ are nearly the same as the quarter circle.

8.2 INFINITE SERIES

1. $s_n = \frac{a(1-r^n)}{(1-r)} = \frac{2\left(1-\left(\frac{1}{3}\right)^n\right)}{1-\left(\frac{1}{3}\right)} \Rightarrow \lim\limits_{n\to\infty} s_n = \frac{2}{1-\left(\frac{1}{3}\right)} = 3$

3. $s_n = \frac{a(1-r^n)}{(1-r)} = \frac{1-\left(-\frac{1}{2}\right)^n}{1-\left(-\frac{1}{2}\right)} \Rightarrow \lim\limits_{n\to\infty} s_n = \frac{1}{\left(\frac{3}{2}\right)} = \frac{2}{3}$

5. $\frac{1}{(n+1)(n+2)} = \frac{1}{n+1} - \frac{1}{n+2} \Rightarrow s_n = \left(\frac{1}{2} - \frac{1}{3}\right) + \left(\frac{1}{3} - \frac{1}{4}\right) + \ldots + \left(\frac{1}{n+1} - \frac{1}{n+2}\right) = \frac{1}{2} - \frac{1}{n+2} \Rightarrow \lim\limits_{n\to\infty} s_n = \frac{1}{2}$

7. $1 - \frac{1}{4} + \frac{1}{16} - \frac{1}{64} + \ldots$, the sum of this geometric series is $\frac{1}{1-\left(-\frac{1}{4}\right)} = \frac{1}{1+\left(\frac{1}{4}\right)} = \frac{4}{5}$

9. $\frac{7}{4} + \frac{7}{16} + \frac{7}{64} + \ldots$, the sum of this geometric series is $\frac{\left(\frac{7}{4}\right)}{1-\left(\frac{1}{4}\right)} = \frac{7}{3}$

11. $(5+1) + \left(\frac{5}{2} + \frac{1}{3}\right) + \left(\frac{5}{4} + \frac{1}{9}\right) + \left(\frac{5}{8} + \frac{1}{27}\right) + \ldots$, is the sum of two geometric series; the sum is
$\frac{5}{1-\left(\frac{1}{2}\right)} + \frac{1}{1-\left(\frac{1}{3}\right)} = 10 + \frac{3}{2} = \frac{23}{2}$

13. $(1+1) + \left(\frac{1}{2} - \frac{1}{5}\right) + \left(\frac{1}{4} + \frac{1}{25}\right) + \left(\frac{1}{8} - \frac{1}{125}\right) + \ldots$, is the sum of two geometric series; the sum is
$\frac{1}{1-\left(\frac{1}{2}\right)} + \frac{1}{1+\left(\frac{1}{5}\right)} = 2 + \frac{5}{6} = \frac{17}{6}$

15. $\frac{4}{(4n-3)(4n+1)} = \frac{1}{4n-3} - \frac{1}{4n+1} \Rightarrow s_n = \left(1 - \frac{1}{5}\right) + \left(\frac{1}{5} - \frac{1}{9}\right) + \left(\frac{1}{9} - \frac{1}{13}\right) + \ldots + \left(\frac{1}{4n-7} - \frac{1}{4n-3}\right)$
$+ \left(\frac{1}{4n-3} - \frac{1}{4n+1}\right) = 1 - \frac{1}{4n+1} \Rightarrow \lim\limits_{n\to\infty} s_n = \lim\limits_{n\to\infty}\left(1 - \frac{1}{4n+1}\right) = 1$

17. $\frac{40n}{(2n-1)^2(2n+1)^2} = \frac{A}{(2n-1)} + \frac{B}{(2n-1)^2} + \frac{C}{(2n+1)} + \frac{D}{(2n+1)^2} = \frac{A(2n-1)(2n+1)^2 + B(2n+1)^2 + C(2n+1)(2n-1)^2 + D(2n-1)^2}{(2n-1)^2(2n+1)^2}$
$\Rightarrow A(2n-1)(2n+1)^2 + B(2n+1)^2 + C(2n+1)(2n-1)^2 + D(2n-1)^2 = 40n$
$\Rightarrow A\left(8n^3 + 4n^2 - 2n - 1\right) + B\left(4n^2 + 4n + 1\right) + C\left(8n^3 - 4n^2 - 2n + 1\right) = D\left(4n^2 - 4n + 1\right) = 40n$
$\Rightarrow (8A + 8C)n^3 + (4A + 4B - 4C + 4D)n^2 + (-2A + 4B - 2C - 4D)n + (-A + B + C + D) = 40n$
$\Rightarrow \begin{cases} 8A + 8C = 0 \\ 4A + 4B - 4C + 4D = 0 \\ -2A + 4B - 2C - 4D = 40 \\ -A + B + C + D = 0 \end{cases} \Rightarrow \begin{cases} 8A + 8C = 0 \\ A + B - C + D = 0 \\ -A + 2B - C - 2D = 20 \\ -A + B + C + D = 0 \end{cases} \Rightarrow \begin{cases} B + D = 0 \\ 2B - 2D = 20 \end{cases} \Rightarrow 4B = 20 \Rightarrow B = 5$

and $D = -5 \Rightarrow \begin{cases} A + C = 0 \\ -A + 5 + C - 5 = 0 \end{cases} \Rightarrow C = 0$ and $A = 0$. Hence, $\sum\limits_{n=1}^{k}\left[\frac{40n}{(2n-1)^2(2n+1)^2}\right]$

$= 5\sum\limits_{n=1}^{k}\left[\frac{1}{(2n-1)^2} - \frac{1}{(2n+1)^2}\right] = 5\left(\frac{1}{1} - \frac{1}{9} + \frac{1}{9} - \frac{1}{25} + \frac{1}{25} - \ldots - \frac{1}{(2(k-1)+1)^2} + \frac{1}{(2k-1)^2} - \frac{1}{(2k+1)^2}\right)$

$= 5\left(1 - \frac{1}{(2k+1)^2}\right) \Rightarrow$ the sum is $\lim\limits_{n\to\infty} 5\left(1 - \frac{1}{(2k+1)^2}\right) = 5$

19. $s_n = \left(1 - \frac{1}{\sqrt{2}}\right) + \left(\frac{1}{\sqrt{2}} - \frac{1}{\sqrt{3}}\right) + \left(\frac{1}{\sqrt{3}} - \frac{1}{\sqrt{4}}\right) + \ldots + \left(\frac{1}{\sqrt{n-1}} + \frac{1}{\sqrt{n}}\right) + \left(\frac{1}{\sqrt{n}} - \frac{1}{\sqrt{n+1}}\right) = 1 - \frac{1}{\sqrt{n+1}}$
$\Rightarrow \lim\limits_{n\to\infty} s_n = \lim\limits_{n\to\infty}\left(1 - \frac{1}{\sqrt{n+1}}\right) = 1$

21. $s_n = \left(\frac{1}{\ln 3} - \frac{1}{\ln 2}\right) + \left(\frac{1}{\ln 4} - \frac{1}{\ln 3}\right) + \left(\frac{1}{\ln 5} - \frac{1}{\ln 4}\right) + \ldots + \left(\frac{1}{\ln(n+1)} - \frac{1}{\ln n}\right) + \left(\frac{1}{\ln(n+2)} - \frac{1}{\ln(n+1)}\right)$
$= -\frac{1}{\ln 2} + \frac{1}{\ln(n+2)} \Rightarrow \lim\limits_{n\to\infty} s_n = -\frac{1}{\ln 2}$

23. convergent geometric series with sum $\frac{1}{1-\left(\frac{1}{\sqrt{2}}\right)} = \frac{\sqrt{2}}{\sqrt{2}-1} = 2 + \sqrt{2}$

25. convergent geometric series with sum $\frac{\left(\frac{3}{2}\right)}{1 - \left(-\frac{1}{2}\right)} = 1$ 27. $\lim\limits_{n \to \infty} \cos(n\pi) = \lim\limits_{n \to \infty} (-1)^n \neq 0 \Rightarrow$ diverges

29. convergent geometric series with sum $\frac{1}{1 - \left(\frac{1}{e^2}\right)} = \frac{e^2}{e^2 - 1}$

31. convergent geometric series with sum $\frac{2}{1 - \left(\frac{1}{10}\right)} - 2 = \frac{20}{9} - \frac{18}{9} = \frac{2}{9}$

33. difference of two geometric series with sum $\frac{1}{1 - \left(\frac{2}{3}\right)} - \frac{1}{1 - \left(\frac{1}{3}\right)} = 3 - \frac{3}{2} = \frac{3}{2}$

35. $\lim\limits_{n \to \infty} \frac{n!}{1000^n} = \infty \neq 0 \Rightarrow$ diverges

37. $\sum\limits_{n=1}^{\infty} \ln\left(\frac{n}{n+1}\right) = \sum\limits_{n=1}^{\infty} [\ln(n) - \ln(n+1)] \Rightarrow s_n = [\ln(1) - \ln(2)] + [\ln(2) - \ln(3)] + [\ln(3) - \ln(4)] + \ldots$
$+ [\ln(n-1) - \ln(n)] + [\ln(n) - \ln(n+1)] = \ln(1) - \ln(n+1) = -\ln(n+1) \Rightarrow \lim\limits_{n \to \infty} s_n = -\infty, \Rightarrow$ diverges

39. convergent geometric series with sum $\frac{1}{1 - \left(\frac{e}{\pi}\right)} = \frac{\pi}{\pi - e}$

41. $\sum\limits_{n=0}^{\infty} (-1)^n x^n = \sum\limits_{n=0}^{\infty} (-x)^n;\ a = 1, r = -x;$ converges to $\frac{1}{1 - (-x)} = \frac{1}{1 + x}$ for $|x| < 1$

43. $a = 3, r = \frac{x-1}{2};$ converges to $\frac{3}{1 - \left(\frac{x-1}{2}\right)} = \frac{6}{3 - x}$ for $-1 < \frac{x-1}{2} < 1$ or $-1 < x < 3$

45. $a = 1, r = 2x;$ converges to $\frac{1}{1 - 2x}$ for $|2x| < 1$ or $|x| < \frac{1}{2}$

47. $a = 1, r = -(x + 1)^n;$ converges to $\frac{1}{1 + (x + 1)} = \frac{1}{2 + x}$ for $|x + 1| < 1$ or $-2 < x < 0$

49. $a = 1, r = \sin x;$ converges to $\frac{1}{1 - \sin x}$ for $x \neq (2k + 1)\frac{\pi}{2}$, k an integer

51. $0.\overline{23} = \sum\limits_{n=0}^{\infty} \frac{23}{100} \left(\frac{1}{10^2}\right)^n = \frac{\left(\frac{23}{100}\right)}{1 - \left(\frac{1}{100}\right)} = \frac{23}{99}$ 53. $0.\overline{7} = \sum\limits_{n=0}^{\infty} \frac{7}{10} \left(\frac{1}{10}\right)^n = \frac{\left(\frac{7}{10}\right)}{1 - \left(\frac{1}{10}\right)} = \frac{7}{9}$

55. $0.0\overline{6} = \sum\limits_{n=0}^{\infty} \left(\frac{1}{10}\right) \left(\frac{6}{10}\right) \left(\frac{1}{10}\right)^n = \frac{\left(\frac{6}{100}\right)}{1 - \left(\frac{1}{10}\right)} = \frac{6}{90} = \frac{1}{15}$

57. $1.24\overline{123} = \frac{124}{100} + \sum\limits_{n=0}^{\infty} \frac{123}{10^5} \left(\frac{1}{10^3}\right)^n = \frac{124}{100} + \frac{\left(\frac{123}{10^5}\right)}{1 - \left(\frac{1}{10^3}\right)} = \frac{124}{100} + \frac{123}{10^5 - 10^2} = \frac{124}{100} + \frac{123}{99,900} = \frac{123,999}{99,900} = \frac{41,333}{33,300}$

59. (a) $\sum\limits_{n=-2}^{\infty} \frac{1}{(n+4)(n+5)}$ (b) $\sum\limits_{n=0}^{\infty} \frac{1}{(n+2)(n+3)}$ (c) $\sum\limits_{n=5}^{\infty} \frac{1}{(n-3)(n-2)}$

61. (a) one example is $\frac{1}{2} + \frac{1}{4} + \frac{1}{8} + \frac{1}{16} + \ldots = \frac{\left(\frac{1}{2}\right)}{1 - \left(\frac{1}{2}\right)} = 1$

(b) one example is $-\frac{3}{2} - \frac{3}{4} - \frac{3}{8} - \frac{3}{16} - \ldots = \frac{\left(-\frac{3}{2}\right)}{1 - \left(\frac{1}{2}\right)} = -3$

(c) one example is $1 - \frac{1}{2} - \frac{1}{4} - \frac{1}{8} - \frac{1}{16} - \ldots$; the series $\frac{k}{2} + \frac{k}{4} + \frac{k}{8} + \ldots = \frac{\left(\frac{k}{2}\right)}{1 - \left(\frac{1}{2}\right)} = k$ where k is any positive or negative number.

63. Let $a_n = b_n = \left(\frac{1}{2}\right)^n$. Then $\sum\limits_{n=1}^{\infty} a_n = \sum\limits_{n=1}^{\infty} b_n = \sum\limits_{n=1}^{\infty} \left(\frac{1}{2}\right)^n = 1$, while $\sum\limits_{n=1}^{\infty} \left(\frac{a_n}{b_n}\right) = \sum\limits_{n=1}^{\infty} (1)$ diverges.

65. Let $a_n = \left(\frac{1}{4}\right)^n$ and $b_n = \left(\frac{1}{2}\right)^n$. Then $A = \sum\limits_{n=1}^{\infty} a_n = \frac{1}{3}$, $B = \sum\limits_{n=1}^{\infty} b_n = 1$ and $\sum\limits_{n=1}^{\infty} \left(\frac{a_n}{b_n}\right) = \sum\limits_{n=1}^{\infty} \left(\frac{1}{2}\right)^n = 1 \neq \frac{A}{B}$.

67. Since the sum of a finite number of terms is finite, adding or subtracting a finite number of terms from a series that diverges does not change the divergence of the series.

69. (a) $\frac{2}{1-r} = 5 \Rightarrow \frac{2}{5} = 1 - r \Rightarrow r = \frac{3}{5}$; $2 + 2\left(\frac{3}{5}\right) + 2\left(\frac{3}{5}\right)^2 + \ldots$

(b) $\frac{\left(\frac{13}{2}\right)}{1-r} = 5 \Rightarrow \frac{13}{10} = 1 - r \Rightarrow r = -\frac{3}{10}$; $\frac{13}{2} - \frac{13}{2}\left(\frac{3}{10}\right) + \frac{13}{2}\left(\frac{3}{10}\right)^2 - \frac{13}{2}\left(\frac{3}{10}\right)^3 + \ldots$

71. $s_n = 1 + 2r + r^2 + 2r^3 + r^4 + 2r^5 + \ldots + r^{2n} + 2r^{2n+1}, n = 0, 1, \ldots$
$\Rightarrow s_n = (1 + r^2 + r^4 + \ldots + r^{2n}) + (2r + 2r^3 + 2r^5 + \ldots + 2r^{2n+1}) \Rightarrow \lim\limits_{n \to \infty} s_n = \frac{1}{1-r^2} + \frac{2r}{1-r^2}$
$= \frac{1+2r}{1-r^2}$, if $|r^2| < 1$ or $|r| < 1$

8.3 THE INTEGRAL TEST

1. converges; a geometric series with $r = \frac{1}{10} < 1$

3. diverges; by the nth-Term Test for Divergence, $\lim\limits_{n \to \infty} \frac{n}{n+1} = 1 \neq 0$

5. diverges; $\sum\limits_{n=1}^{\infty} \frac{3}{\sqrt{n}} = 3 \sum\limits_{n=1}^{\infty} \frac{1}{\sqrt{n}}$, which is a divergent p-series $\left(p = \frac{1}{2}\right)$

7. converges; a geometric series with $r = \frac{1}{8} < 1$

9. diverges by the Integral Test: $\int_2^n \frac{\ln x}{x} dx = \frac{1}{2}\left(\ln^2 n - \ln 2\right) \Rightarrow \int_2^{\infty} \frac{\ln x}{x} dx \to \infty$

11. converges; a geometric series with $r = \frac{2}{3} < 1$

13. diverges; $\sum\limits_{n=0}^{\infty} \frac{-2}{n+1} = -2 \sum\limits_{n=0}^{\infty} \frac{1}{n+1}$, which diverges by the Integral Test

15. diverges; $\lim\limits_{n \to \infty} a_n = \lim\limits_{n \to \infty} \frac{2^n}{n+1} = \lim\limits_{n \to \infty} \frac{2^n \ln 2}{1} = \infty \neq 0$

17. diverges; $\lim\limits_{n \to \infty} \frac{\sqrt{n}}{\ln n} = \lim\limits_{n \to \infty} \frac{\left(\frac{1}{2\sqrt{n}}\right)}{\left(\frac{1}{n}\right)} = \lim\limits_{n \to \infty} \frac{\sqrt{n}}{2} = \infty \neq 0$

19. diverges; a geometric series with $r = \frac{1}{\ln 2} \approx 1.44 > 1$

21. converges by the Integral Test: $\int_3^\infty \frac{\left(\frac{1}{x}\right)}{(\ln x)\sqrt{(\ln x)^2 - 1}}\, dx;\; \begin{bmatrix} u = \ln x \\ du = \frac{1}{x}\, dx \end{bmatrix} \to \int_{\ln 3}^\infty \frac{1}{u\sqrt{u^2 - 1}}\, du$

$= \lim\limits_{b \to \infty} \left[\sec^{-1} |u|\right]_{\ln 3}^b = \lim\limits_{b \to \infty} \left[\sec^{-1} b - \sec^{-1}(\ln 3)\right] = \lim\limits_{b \to \infty} \left[\cos^{-1}\left(\frac{1}{b}\right) - \sec^{-1}(\ln 3)\right]$

$= \cos^{-1}(0) - \sec^{-1}(\ln 3) = \frac{\pi}{2} - \sec^{-1}(\ln 3) \approx 1.1439$

23. diverges by the nth-Term Test for divergence; $\lim\limits_{n \to \infty} n \sin\left(\frac{1}{n}\right) = \lim\limits_{n \to \infty} \frac{\sin\left(\frac{1}{n}\right)}{\left(\frac{1}{n}\right)} = \lim\limits_{x \to 0} \frac{\sin x}{x} = 1 \neq 0$

25. converges by the Integral Test: $\int_1^\infty \frac{e^x}{1 + e^{2x}}\, dx;\; \begin{bmatrix} u = e^x \\ du = e^x\, dx \end{bmatrix} \to \int_e^\infty \frac{1}{1 + u^2}\, du = \lim\limits_{n \to \infty} \left[\tan^{-1} u\right]_e^b$

$= \lim\limits_{b \to \infty} (\tan^{-1} b - \tan^{-1} e) = \frac{\pi}{2} - \tan^{-1} e \approx 0.35$

27. converges by the Integral Test: $\int_1^\infty \frac{8 \tan^{-1} x}{1 + x^2}\, dx;\; \begin{bmatrix} u = \tan^{-1} x \\ du = \frac{dx}{1 + x^2} \end{bmatrix} \to \int_{\pi/4}^{\pi/2} 8u\, du = \left[4u^2\right]_{\pi/4}^{\pi/2} = 4\left(\frac{\pi^2}{4} - \frac{\pi^2}{16}\right) = \frac{3\pi^2}{4}$

29. converges by the Integral Test: $\int_1^\infty \operatorname{sech} x\, dx = 2 \lim\limits_{b \to \infty} \int_1^b \frac{e^x}{1 + (e^x)^2}\, dx = 2 \lim\limits_{b \to \infty} \left[\tan^{-1} e^x\right]_1^b$

$= 2 \lim\limits_{b \to \infty} (\tan^{-1} e^b - \tan^{-1} e) = \pi - 2 \tan^{-1} e \approx 0.71$

31. $\int_1^\infty \left(\frac{a}{x+2} - \frac{1}{x+4}\right) dx = \lim\limits_{b \to \infty} \left[a \ln |x + 2| - \ln |x + 4|\right]_1^b = \lim\limits_{b \to \infty} \ln \frac{(b+2)^a}{b+4} - \ln \left(\frac{3^a}{5}\right);$

$\lim\limits_{b \to \infty} \frac{(b+2)^a}{b+4} = a \lim\limits_{b \to \infty} (b+2)^{a-1} = \begin{cases} \infty, & a > 1 \\ 1, & a = 1 \end{cases} \Rightarrow$ the series converges to $\ln \left(\frac{5}{3}\right)$ if $a = 1$ and diverges to ∞ if

$a > 1$. If $a < 1$, the terms of the series eventually become negative and the Integral Test does not apply. From

that point on, however, the series behaves like a negative multiple of the harmonic series, and so it diverges.

33. (a)

(b) There are $(13)(365)(24)(60)(60)\,(10^9)$ seconds in 13 billion years; by part (a) $s_n \leq 1 + \ln n$ where

$n = (13)(365)(24)(60)(60)\,(10^9) \Rightarrow s_n \leq 1 + \ln ((13)(365)(24)(60)(60)\,(10^9))$

$= 1 + \ln (13) + \ln (365) + \ln (24) + 2 \ln (60) + 9 \ln (10) \approx 41.55$

35. Yes. If $\sum\limits_{n=1}^\infty a_n$ is a divergent series of positive numbers, then $\left(\frac{1}{2}\right) \sum\limits_{n=1}^\infty a_n = \sum\limits_{n=1}^\infty \left(\frac{a_n}{2}\right)$ also diverges and $\frac{a_n}{2} < a_n$.

There is no "smallest" divergent series of positive numbers: for any divergent series $\sum\limits_{n=1}^\infty a_n$ of positive numbers

$\sum\limits_{n=1}^\infty \left(\frac{a_n}{2}\right)$ has smaller terms and still diverges.

37. Let $A_n = \sum\limits_{k=1}^{n} a_k$ and $B_n = \sum\limits_{k=1}^{n} 2^k a_{(2^k)}$, where $\{a_k\}$ is a nonincreasing sequence of positive terms converging to

0. Note that $\{A_n\}$ and $\{B_n\}$ are nondecreasing sequences of positive terms. Now,

$$B_n = 2a_2 + 4a_4 + 8a_8 + \ldots + 2^n a_{(2^n)} = 2a_2 + (2a_4 + 2a_4) + (2a_8 + 2a_8 + 2a_8 + 2a_8) + \ldots$$

$$+ \underbrace{\left(2a_{(2^n)} + 2a_{(2^n)} + \ldots + 2a_{(2^n)}\right)}_{2^{n-1}\text{ terms}} \le 2a_1 + 2a_2 + (2a_3 + 2a_4) + (2a_5 + 2a_6 + 2a_7 + 2a_8) + \ldots$$

$$+ \left(2a_{(2^{n-1})} + 2a_{(2^{n-1}+1)} + \ldots + 2a_{(2^n)}\right) = 2A_{(2^n)} \le 2\sum\limits_{k=1}^{\infty} a_k. \text{ Therefore if } \sum a_k \text{ converges,}$$

then $\{B_n\}$ is bounded above $\Rightarrow \sum 2^k a_{(2^k)}$ converges. Conversely,

$$A_n = a_1 + (a_2 + a_3) + (a_4 + a_5 + a_6 + a_7) + \ldots + a_n < a_1 + 2a_2 + 4a_4 + \ldots + 2^n a_{(2^n)} = a_1 + B_n < a_1 + \sum\limits_{k=1}^{\infty} 2^k a_{(2^k)}.$$

Therefore, if $\sum\limits_{k=1}^{\infty} 2^k a_{(2^k)}$ converges, then $\{A_n\}$ is bounded above and hence converges.

39. (a) $\int_2^\infty \frac{dx}{x(\ln x)^p} ; \begin{bmatrix} u = \ln x \\ du = \frac{dx}{x} \end{bmatrix} \to \int_{\ln 2}^\infty u^{-p}\, du = \lim\limits_{b \to \infty} \left[\frac{u^{-p+1}}{-p+1}\right]_{\ln 2}^b = \lim\limits_{b \to \infty} \left(\frac{1}{1-p}\right)\left[b^{-p+1} - (\ln 2)^{-p+1}\right]$

$= \begin{cases} \frac{1}{p-1}(\ln 2)^{-p+1}, p > 1 \\ \infty, p < 1 \end{cases} \Rightarrow$ the improper integral converges if $p > 1$ and diverges if $p < 1$.

For $p = 1$: $\int_2^\infty \frac{dx}{x \ln x} = \lim\limits_{b \to \infty} [\ln(\ln x)]_2^b = \lim\limits_{b \to \infty} [\ln(\ln b) - \ln(\ln 2)] = \infty$, so the improper integral

diverges if $p = 1$.

(b) Since the series and the integral converge or diverge together, $\sum\limits_{n=2}^{\infty} \frac{1}{n(\ln n)^p}$ converges if and only if $p > 1$.

41. (a) From Fig. 8.8 in the text with $f(x) = \frac{1}{x}$ and $a_k = \frac{1}{k}$, we have $\int_1^{n+1} \frac{1}{x}\, dx \le 1 + \frac{1}{2} + \frac{1}{3} + \ldots + \frac{1}{n}$

$\le 1 + \int_1^n f(x)\, dx \Rightarrow \ln(n+1) \le 1 + \frac{1}{2} + \frac{1}{3} + \ldots + \frac{1}{n} \le 1 + \ln n \Rightarrow 0 \le \ln(n+1) - \ln n$

$\le \left(1 + \frac{1}{2} + \frac{1}{3} + \ldots + \frac{1}{n}\right) - \ln n \le 1$. Therefore the sequence $\left\{\left(1 + \frac{1}{2} + \frac{1}{3} + \ldots + \frac{1}{n}\right) - \ln n\right\}$ is bounded above

by 1 and below by 0.

(b) From the graph in Fig. 8.8(a) with $f(x) = \frac{1}{x}$, $\frac{1}{n+1} < \int_n^{n+1} \frac{1}{x}\, dx = \ln(n+1) - \ln n$

$\Rightarrow 0 > \frac{1}{n+1} - [\ln(n+1) - \ln n] = \left(1 + \frac{1}{2} + \frac{1}{3} + \ldots + \frac{1}{n+1} - \ln(n+1)\right) - \left(1 + \frac{1}{2} + \frac{1}{3} + \ldots + \frac{1}{n} - \ln n\right)$.

If we define $a_n = 1 + \frac{1}{2} = \frac{1}{3} + \frac{1}{n} - \ln n$, then $0 > a_{n+1} - a_n \Rightarrow a_{n+1} < a_n \Rightarrow \{a_n\}$ is a decreasing sequence of

nonnegative terms.

43. (a) $s_{10} = \sum\limits_{n=1}^{10} \frac{1}{n^3} = 1.97531986$; $\int_{11}^\infty \frac{1}{x^3}\, dx = \lim\limits_{b \to \infty} \int_{11}^b x^{-3}\, dx = \lim\limits_{b \to \infty} \left[-\frac{x^{-2}}{2}\right]_{11}^b = \lim\limits_{b \to \infty} \left(-\frac{1}{2b^2} + \frac{1}{242}\right) = \frac{1}{242}$ and

$\int_{10}^\infty \frac{1}{x^3}\, dx = \lim\limits_{b \to \infty} \int_{10}^b x^{-3}\, dx = \lim\limits_{b \to \infty} \left[-\frac{x^{-2}}{2}\right]_{10}^b = \lim\limits_{b \to \infty} \left(-\frac{1}{2b^2} + \frac{1}{200}\right) = \frac{1}{200}$

$\Rightarrow 1.97531986 + \frac{1}{242} < s < 1.97531986 + \frac{1}{200} \Rightarrow 1.20166 < s < 1.20253$

(b) $s = \sum\limits_{n=1}^{\infty} \frac{1}{n^3} \approx \frac{1.20166 + 1.20253}{2} = 1.202095$; error $\le \frac{1.20253 - 1.20166}{2} = 0.000435$

8.4 COMPARISON TESTS

1. diverges by the Limit Comparison Test (part 1) when compared with $\sum\limits_{n=1}^{\infty} \frac{1}{\sqrt{n}}$, a divergent p-series:

$$\lim\limits_{n \to \infty} \frac{\left(\frac{1}{2\sqrt{n} + \sqrt[3]{n}}\right)}{\left(\frac{1}{\sqrt{n}}\right)} = \lim\limits_{n \to \infty} \frac{\sqrt{n}}{2\sqrt{n} + \sqrt[3]{n}} = \lim\limits_{n \to \infty} \left(\frac{1}{2 + n^{-1/6}}\right) = \frac{1}{2}$$

3. converges by the Direct Comparison Test; $\frac{\sin^2 n}{2^n} \leq \frac{1}{2^n}$, which is the nth term of a convergent geometric series

5. diverges since $\lim\limits_{n \to \infty} \frac{2n}{3n-1} = \frac{2}{3} \neq 0$

7. converges by the Direct Comparison Test; $\left(\frac{n}{3n+1}\right)^n < \left(\frac{n}{3n}\right)^n = \left(\frac{1}{3}\right)^n$, the nth term of a convergent geometric series

9. diverges by the Direct Comparison Test; $n > \ln n \Rightarrow \ln n > \ln \ln n \Rightarrow \frac{1}{n} < \frac{1}{\ln n} < \frac{1}{\ln(\ln n)}$ and $\sum\limits_{n=3}^{\infty} \frac{1}{n}$ diverges

11. converges by the Limit Comparison Test (part 2) when compared with $\sum\limits_{n=1}^{\infty} \frac{1}{n^2}$, a convergent p-series:

$$\lim_{n \to \infty} \frac{\left[\frac{(\ln n)^2}{n^3}\right]}{\left(\frac{1}{n^2}\right)} = \lim_{n \to \infty} \frac{(\ln n)^2}{n} = \lim_{n \to \infty} \frac{2(\ln n)\left(\frac{1}{n}\right)}{1} = 2 \lim_{n \to \infty} \frac{\ln n}{n} = 0$$

13. diverges by the Limit Comparison Test (part 3) with $\frac{1}{n}$, the nth term of the divergent harmonic series:

$$\lim_{n \to \infty} \frac{\left[\frac{1}{\sqrt{n} \ln n}\right]}{\left(\frac{1}{n}\right)} = \lim_{n \to \infty} \frac{\sqrt{n}}{\ln n} = \lim_{n \to \infty} \frac{\left(\frac{1}{2\sqrt{n}}\right)}{\left(\frac{1}{n}\right)} = \lim_{n \to \infty} \frac{\sqrt{n}}{2} = \infty$$

15. diverges by the Limit Comparison Test (part 3) with $\frac{1}{n}$, the nth term of the divergent harmonic series:

$$\lim_{n \to \infty} \frac{\left(\frac{1}{1+\ln n}\right)}{\left(\frac{1}{n}\right)} = \lim_{n \to \infty} \frac{n}{1+\ln n} = \lim_{n \to \infty} \frac{1}{\left(\frac{1}{n}\right)} = \lim_{n \to \infty} n = \infty$$

17. diverges by the Integral Test: $\int_{2}^{\infty} \frac{\ln(x+1)}{x+1} \, dx = \int_{\ln 3}^{\infty} u \, du = \lim_{b \to \infty} \left[\frac{1}{2} u^2\right]_{\ln 3}^{b} = \lim_{b \to \infty} \frac{1}{2}(b^2 - \ln^2 3) = \infty$

19. converges by the Direct Comparison Test with $\frac{1}{n^{3/2}}$, the nth term of a convergent p-series: $n^2 - 1 > n$ for

$n \geq 2 \Rightarrow n^2(n^2 - 1) > n^3 \Rightarrow n\sqrt{n^2 - 1} > n^{3/2} \Rightarrow \frac{1}{n^{3/2}} > \frac{1}{n\sqrt{n^2-1}}$ or use Limit Comparison Test with $\frac{1}{n^2}$.

21. converges because $\sum\limits_{n=1}^{\infty} \frac{1-n}{n 2^n} = \sum\limits_{n=1}^{\infty} \frac{1}{n 2^n} + \sum\limits_{n=1}^{\infty} \frac{-1}{2^n}$ which is the sum of two convergent series:

$\sum\limits_{n=1}^{\infty} \frac{1}{n 2^n}$ converges by the Direct Comparison Test since $\frac{1}{n 2^n} < \frac{1}{2^n}$, and $\sum\limits_{n=1}^{\infty} \frac{-1}{2^n}$ is a convergent geometric series

23. converges by the Direct Comparison Test: $\frac{1}{3^{n-1}+1} < \frac{1}{3^{n-1}}$, which is the nth term of a convergent geometric series

25. diverges by the Limit Comparison Test (part 1) with $\frac{1}{n}$, the nth term of the divergent harmonic series:

$$\lim_{n \to \infty} \frac{\left(\sin \frac{1}{n}\right)}{\left(\frac{1}{n}\right)} = \lim_{x \to 0} \frac{\sin x}{x} = 1$$

27. converges by the Limit Comparison Test (part 1) with $\frac{1}{n^2}$, the nth term of a convergent p-series:

$$\lim_{n \to \infty} \frac{\left(\frac{10n+1}{n(n+1)(n+2)}\right)}{\left(\frac{1}{n^2}\right)} = \lim_{n \to \infty} \frac{10n^2 + n}{n^2 + 3n + 2} = \lim_{n \to \infty} \frac{20n+1}{2n+3} = \lim_{n \to \infty} \frac{20}{2} = 10$$

29. converges by the Direct Comparison Test: $\frac{\tan^{-1} n}{n^{1.1}} < \frac{\frac{\pi}{2}}{n^{1.1}}$ and $\sum\limits_{n=1}^{\infty} \frac{\frac{\pi}{2}}{n^{1.1}} = \frac{\pi}{2} \sum\limits_{n=1}^{\infty} \frac{1}{n^{1.1}}$ is the product of a

convergent p-series and a nonzero constant

31. converges by the Limit Comparison Test (part 1) with $\frac{1}{n^2}$: $\displaystyle\lim_{n \to \infty} \frac{\left(\frac{\coth n}{n^2}\right)}{\left(\frac{1}{n^2}\right)} = \lim_{n \to \infty} \coth n = \lim_{n \to \infty} \frac{e^n + e^{-n}}{e^n - e^{-n}}$

$= \displaystyle\lim_{n \to \infty} \frac{1 + e^{-2n}}{1 - e^{-2n}} = 1$

33. diverges by the Limit Comparison Test (part 1) with $\frac{1}{n}$: $\displaystyle\lim_{n \to \infty} \frac{\left(\frac{1}{n\sqrt[n]{n}}\right)}{\left(\frac{1}{n}\right)} = \lim_{n \to \infty} \frac{1}{\sqrt[n]{n}} = 1$.

35. $\frac{1}{1+2+3+\ldots+n} = \frac{1}{\left(\frac{n(n+1)}{2}\right)} = \frac{2}{n(n+1)}$. The series converges by the Limit Comparison Test (part 1) with $\frac{1}{n^2}$:

$\displaystyle\lim_{n \to \infty} \frac{\left(\frac{2}{n(n+1)}\right)}{\left(\frac{1}{n^2}\right)} = \lim_{n \to \infty} \frac{2n^2}{n^2+n} = \lim_{n \to \infty} \frac{4n}{2n+1} = \lim_{n \to \infty} \frac{4}{2} = 2$.

37. (a) If $\displaystyle\lim_{n \to \infty} \frac{a_n}{b_n} = 0$, then there exists an integer N such that for all $n > N$, $\left|\frac{a_n}{b_n} - 0\right| < 1 \Rightarrow -1 < \frac{a_n}{b_n} < 1$

$\Rightarrow a_n < b_n$. Thus, if $\sum b_n$ converges, then $\sum a_n$ converges by the Direct Comparison Test.

(b) If $\displaystyle\lim_{n \to \infty} \frac{a_n}{b_n} = \infty$, then there exists an integer N such that for all $n > N$, $\frac{a_n}{b_n} > 1 \Rightarrow a_n > b_n$. Thus, if

$\sum b_n$ diverges, then $\sum a_n$ diverges by the Direct Comparison Test.

39. $\displaystyle\lim_{n \to \infty} \frac{a_n}{b_n} = \infty \Rightarrow$ there exists an integer N such that for all $n > N$, $\frac{a_n}{b_n} > 1 \Rightarrow a_n > b_n$. If $\sum a_n$ converges,

then $\sum b_n$ converges by the Direct Comparison Test

8.5 THE RATIO AND ROOT TESTS

1. converges by the Ratio Test: $\displaystyle\lim_{n \to \infty} \frac{a_{n+1}}{a_n} = \lim_{n \to \infty} \frac{\left[\frac{(n+1)^{\sqrt{2}}}{2^{n+1}}\right]}{\left[\frac{n^{\sqrt{2}}}{2^n}\right]} = \lim_{n \to \infty} \frac{(n+1)^{\sqrt{2}}}{2^{n+1}} \cdot \frac{2^n}{n^{\sqrt{2}}} = \lim_{n \to \infty} \left(1 + \frac{1}{n}\right)^{\sqrt{2}} \left(\frac{1}{2}\right) = \frac{1}{2} < 1$

3. diverges by the Ratio Test: $\displaystyle\lim_{n \to \infty} \frac{a_{n+1}}{a_n} = \lim_{n \to \infty} \frac{\left(\frac{(n+1)!}{e^{n+1}}\right)}{\left(\frac{n!}{e^n}\right)} = \lim_{n \to \infty} \frac{(n+1)!}{e^{n+1}} \cdot \frac{e^n}{n!} = \lim_{n \to \infty} \frac{n+1}{e} = \infty$

5. converges by the Ratio Test: $\displaystyle\lim_{n \to \infty} \frac{a_{n+1}}{a_n} = \lim_{n \to \infty} \frac{\left(\frac{(n+1)^{10}}{10^{n+1}}\right)}{\left(\frac{n^{10}}{10^n}\right)} = \lim_{n \to \infty} \frac{(n+1)^{10}}{10^{n+1}} \cdot \frac{10^n}{n^{10}} = \lim_{n \to \infty} \left(1 + \frac{1}{n}\right)^{10} \left(\frac{1}{10}\right) = \frac{1}{10} < 1$

7. converges by the Direct Comparison Test: $\frac{2+(-1)^n}{(1.25)^n} = \left(\frac{4}{5}\right)^n [2 + (-1)^n] \le \left(\frac{4}{5}\right)^n (3)$ which is the nth term of a convergent geometric series

9. diverges; $\displaystyle\lim_{n \to \infty} a_n = \lim_{n \to \infty} \left(1 - \frac{3}{n}\right)^n = \lim_{n \to \infty} \left(1 + \frac{-3}{n}\right)^n = e^{-3} \approx 0.05 \ne 0$

11. converges by the Direct Comparison Test: $\frac{\ln n}{n^3} < \frac{n}{n^3} = \frac{1}{n^2}$ for $n \ge 2$, the nth term of a convergent p-series.

13. diverges by the Direct Comparison Test: $\frac{1}{n} - \frac{1}{n^2} = \frac{n-1}{n^2} > \frac{1}{2}\left(\frac{1}{n}\right)$ for $n > 2$ or by the Limit Comparison Test (part 1) with $\frac{1}{n}$.

15. diverges by the Direct Comparison Test: $\frac{\ln n}{n} > \frac{1}{n}$ for $n \ge 3$

17. converges by the Ratio Test: $\displaystyle\lim_{n \to \infty} \frac{a_{n+1}}{a_n} = \lim_{n \to \infty} \frac{(n+2)(n+3)}{(n+1)!} \cdot \frac{n!}{(n+1)(n+2)} = 0 < 1$

19. converges by the Ratio Test: $\lim\limits_{n \to \infty} \frac{a_{n+1}}{a_n} = \lim\limits_{n \to \infty} \frac{(n+4)!}{3!\,(n+1)!\,3^{n+1}} \cdot \frac{3!\,n!\,3^n}{(n+3)!} = \lim\limits_{n \to \infty} \frac{n+4}{3(n+1)} = \frac{1}{3} < 1$

21. converges by the Ratio Test: $\lim\limits_{n \to \infty} \frac{a_{n+1}}{a_n} = \lim\limits_{n \to \infty} \frac{(n+1)!}{(2n+3)!} \cdot \frac{(2n+1)!}{n!} = \lim\limits_{n \to \infty} \frac{n+1}{(2n+3)(2n+2)} = 0 < 1$

23. converges by the Root Test: $\lim\limits_{n \to \infty} \sqrt[n]{a_n} = \lim\limits_{n \to \infty} \sqrt[n]{\frac{n}{(\ln n)^n}} = \lim\limits_{n \to \infty} \frac{\sqrt[n]{n}}{\ln n} = \lim\limits_{n \to \infty} \frac{1}{\ln n} = 0 < 1$

25. converges by the Direct Comparison Test: $\frac{n!\,\ln n}{n(n+2)!} = \frac{\ln n}{n(n+1)(n+2)} < \frac{n}{n(n+1)(n+2)} = \frac{1}{(n+1)(n+2)} < \frac{1}{n^2}$
    which is the nth-term of a convergent p-series

27. converges by the Ratio Test: $\lim\limits_{n \to \infty} \frac{a_{n+1}}{a_n} = \lim\limits_{n \to \infty} \frac{\left(\frac{1+\sin n}{n}\right) a_n}{a_n} = 0 < 1$

29. diverges by the Ratio Test: $\lim\limits_{n \to \infty} \frac{a_{n+1}}{a_n} = \lim\limits_{n \to \infty} \frac{\left(\frac{3n-1}{2n+5}\right) a_n}{a_n} = \lim\limits_{n \to \infty} \frac{3n-1}{2n+5} = \frac{3}{2} > 1$

31. converges by the Ratio Test: $\lim\limits_{n \to \infty} \frac{a_{n+1}}{a_n} = \lim\limits_{n \to \infty} \frac{\left(\frac{2}{n}\right) a_n}{a_n} = \lim\limits_{n \to \infty} \frac{2}{n} = 0 < 1$

33. converges by the Ratio Test: $\lim\limits_{n \to \infty} \frac{a_{n+1}}{a_n} = \lim\limits_{n \to \infty} \frac{\left(\frac{1+\ln n}{n}\right) a_n}{a_n} = \lim\limits_{n \to \infty} \frac{1+\ln n}{n} = \lim\limits_{n \to \infty} \frac{1}{n} = 0 < 1$

35. diverges by the nth-Term Test: $a_1 = \frac{1}{3}$, $a_2 = \sqrt[2]{\frac{1}{3}}$, $a_3 = \sqrt[3]{\sqrt[2]{\frac{1}{3}}} = \sqrt[6]{\frac{1}{3}}$, $a_4 = \sqrt[4]{\sqrt[3]{\sqrt[2]{\frac{1}{3}}}} = \sqrt[4!]{\frac{1}{3}}, \dots,$

    $a_n = \sqrt[n!]{\frac{1}{3}} \Rightarrow \lim\limits_{n \to \infty} a_n = 1$ because $\left\{\sqrt[n!]{\frac{1}{3}}\right\}$ is a subsequence of $\left\{\sqrt[n]{\frac{1}{3}}\right\}$ whose limit is 1 by Table 8.1

37. converges by the Ratio Test: $\lim\limits_{n \to \infty} \frac{a_{n+1}}{a_n} = \lim\limits_{n \to \infty} \frac{2^{n+1}(n+1)!\,(n+1)!}{(2n+2)!} \cdot \frac{(2n)!}{2^n n!\,n!} = \lim\limits_{n \to \infty} \frac{2(n+1)(n+1)}{(2n+2)(2n+1)}$
    $= \lim\limits_{n \to \infty} \frac{n+1}{2n+1} = \frac{1}{2} < 1$

39. diverges by the Root Test: $\lim\limits_{n \to \infty} \sqrt[n]{a_n} \equiv \lim\limits_{n \to \infty} \sqrt[n]{\frac{(n!)^n}{(n^n)^2}} = \lim\limits_{n \to \infty} \frac{n!}{n^2} = \infty > 1$

41. converges by the Root Test: $\lim\limits_{n \to \infty} \sqrt[n]{a_n} = \lim\limits_{n \to \infty} \sqrt[n]{\frac{n^n}{2^{n^2}}} = \lim\limits_{n \to \infty} \frac{n}{2^n} = \lim\limits_{n \to \infty} \frac{1}{2^n \ln 2} = 0 < 1$

43. converges by the Ratio Test: $\lim\limits_{n \to \infty} \frac{a_{n+1}}{a_n} = \lim\limits_{n \to \infty} \frac{1 \cdot 3 \cdot \cdots \cdot (2n-1)(2n+1)}{4^{n+1} 2^{n+1}(n+1)!} \cdot \frac{4^n 2^n n!}{1 \cdot 3 \cdot \cdots \cdot (2n-1)} = \lim\limits_{n \to \infty} \frac{2n+1}{(4 \cdot 2)(n+1)} = \frac{1}{4} < 1$

45. Ratio: $\lim\limits_{n \to \infty} \frac{a_{n+1}}{a_n} = \lim\limits_{n \to \infty} \frac{1}{(n+1)^p} \cdot \frac{n^p}{1} = \lim\limits_{n \to \infty} \left(\frac{n}{n+1}\right)^p = 1^p = 1 \Rightarrow$ no conclusion

    Root: $\lim\limits_{n \to \infty} \sqrt[n]{a_n} = \lim\limits_{n \to \infty} \sqrt[n]{\frac{1}{n^p}} = \lim\limits_{n \to \infty} \frac{1}{(\sqrt[n]{n})^p} = \frac{1}{(1)^p} = 1 \Rightarrow$ no conclusion

47. $a_n \le \frac{n}{2^n}$ for every n and the series $\sum\limits_{n=1}^{\infty} \frac{n}{2^n}$ converges by the Ratio Test since $\lim\limits_{n \to \infty} \frac{(n+1)}{2^{n+1}} \cdot \frac{2^n}{n} = \frac{1}{2} < 1$

    $\Rightarrow \sum\limits_{n=1}^{\infty} a_n$ converges by the Direct Comparison Test

8.6 ALTERNATING SERIES, ABSOLUTE AND CONDITIONAL CONVERGENCE

1. converges absolutely $\Rightarrow$ converges by the Absolute Convergence Test since $\sum\limits_{n=1}^{\infty} |a_n| = \sum\limits_{n=1}^{\infty} \frac{1}{n^2}$ which is a convergent p-series

3. diverges by the nth-Term Test since for $n > 10 \Rightarrow \frac{n}{10} > 1 \Rightarrow \lim\limits_{n \to \infty} \left(\frac{n}{10}\right)^n \neq 0 \Rightarrow \sum\limits_{n=1}^{\infty} (-1)^{n+1} \left(\frac{n}{10}\right)^n$ diverges

5. converges by the Alternating Series Test because $f(x) = \ln x$ is an increasing function of $x \Rightarrow \frac{1}{\ln x}$ is decreasing
   $\Rightarrow u_n \geq u_{n+1}$ for $n \geq 1$; also $u_n \geq 0$ for $n \geq 1$ and $\lim\limits_{n \to \infty} \frac{1}{\ln n} = 0$

7. diverges by the nth-Term Test since $\lim\limits_{n \to \infty} \frac{\ln n}{\ln n^2} = \lim\limits_{n \to \infty} \frac{\ln n}{2 \ln n} = \lim\limits_{n \to \infty} \frac{1}{2} = \frac{1}{2} \neq 0$

9. converges by the Alternating Series Test since $f(x) = \frac{\sqrt{x}+1}{x+1} \Rightarrow f'(x) = \frac{1-x-2\sqrt{x}}{2\sqrt{x}(x+1)^2} < 0 \Rightarrow f(x)$ is decreasing
   $\Rightarrow u_n \geq u_{n+1}$; also $u_n \geq 0$ for $n \geq 1$ and $\lim\limits_{n \to \infty} u_n = \lim\limits_{n \to \infty} \frac{\sqrt{n}+1}{n+1} = 0$

11. converges absolutely since $\sum\limits_{n=1}^{\infty} |a_n| = \sum\limits_{n=1}^{\infty} \left(\frac{1}{10}\right)^n$ a convergent geometric series

13. converges conditionally since $\frac{1}{\sqrt{n}} > \frac{1}{\sqrt{n+1}} > 0$ and $\lim\limits_{n \to \infty} \frac{1}{\sqrt{n}} = 0 \Rightarrow$ convergence; but $\sum\limits_{n=1}^{\infty} |a_n| = \sum\limits_{n=1}^{\infty} \frac{1}{n^{1/2}}$
    is a divergent p-series

15. converges absolutely since $\sum\limits_{n=1}^{\infty} |a_n| = \sum\limits_{n=1}^{\infty} \frac{n}{n^3+1}$ and $\frac{n}{n^3+1} < \frac{1}{n^2}$ which is the nth-term of a converging p-series

17. converges conditionally since $\frac{1}{n+3} > \frac{1}{(n+1)+3} > 0$ and $\lim\limits_{n \to \infty} \frac{1}{n+3} = 0 \Rightarrow$ convergence; but $\sum\limits_{n=1}^{\infty} |a_n|$
    $= \sum\limits_{n=1}^{\infty} \frac{1}{n+3}$ diverges because $\frac{1}{n+3} \geq \frac{1}{4n}$ and $\sum\limits_{n=1}^{\infty} \frac{1}{n}$ is a divergent series

19. diverges by the nth-Term Test since $\lim\limits_{n \to \infty} \frac{3+n}{5+n} = 1 \neq 0$

21. converges conditionally since $f(x) = \frac{1}{x^2} + \frac{1}{x} \Rightarrow f'(x) = -\left(\frac{2}{x^3} + \frac{1}{x^2}\right) < 0 \Rightarrow f(x)$ is decreasing and hence
    $u_n > u_{n+1} > 0$ for $n \geq 1$ and $\lim\limits_{n \to \infty} \left(\frac{1}{n^2} + \frac{1}{n}\right) = 0 \Rightarrow$ convergence; but $\sum\limits_{n=1}^{\infty} |a_n| = \sum\limits_{n=1}^{\infty} \frac{1+n}{n^2}$
    $= \sum\limits_{n=1}^{\infty} \frac{1}{n^2} + \sum\limits_{n=1}^{\infty} \frac{1}{n}$ is the sum of a convergent and divergent series, and hence diverges

23. converges absolutely by the Ratio Test: $\lim\limits_{n \to \infty} \left(\frac{u_{n+1}}{u_n}\right) = \lim\limits_{n \to \infty} \left[\frac{(n+1)^2 \left(\frac{2}{3}\right)^{n+1}}{n^2 \left(\frac{2}{3}\right)^n}\right] = \frac{2}{3} < 1$

25. converges absolutely by the Integral Test since $\int_1^{\infty} (\tan^{-1} x) \left(\frac{1}{1+x^2}\right) dx = \lim\limits_{b \to \infty} \left[\frac{(\tan^{-1} x)^2}{2}\right]_1^b$
    $= \lim\limits_{b \to \infty} \left[(\tan^{-1} b)^2 - (\tan^{-1} 1)^2\right] = \frac{1}{2}\left[\left(\frac{\pi}{2}\right)^2 - \left(\frac{\pi}{4}\right)^2\right] = \frac{3\pi^2}{32}$

27. diverges by the nth-Term Test since $\lim\limits_{n \to \infty} \frac{n}{n+1} = 1 \neq 0$

29. converges absolutely by the Ratio Test: $\lim\limits_{n \to \infty} \left(\frac{u_{n+1}}{u_n} \right) = \lim\limits_{n \to \infty} \frac{(100)^{n+1}}{(n+1)!} \cdot \frac{n!}{(100)^n} = \lim\limits_{n \to \infty} \frac{100}{n+1} = 0 < 1$

31. converges absolutely by the Direct Comparison Test since $\sum\limits_{n=1}^{\infty} |a_n| = \sum\limits_{n=1}^{\infty} \frac{1}{n^2 + 2n + 1}$ and $\frac{1}{n^2 + 2n + 1} < \frac{1}{n^2}$ which is the

    nth-term of a convergent p-series

33. converges absolutely since $\sum\limits_{n=1}^{\infty} |a_n| = \sum\limits_{n=1}^{\infty} \left| \frac{(-1)^n}{n\sqrt{n}} \right| = \sum\limits_{n=1}^{\infty} \frac{1}{n^{3/2}}$ is a convergent p-series

35. converges absolutely by the Root Test: $\lim\limits_{n \to \infty} \sqrt[n]{|a_n|} = \lim\limits_{n \to \infty} \left(\frac{(n+1)^n}{(2n)^n} \right)^{1/n} = \lim\limits_{n \to \infty} \frac{n+1}{2n} = \frac{1}{2} < 1$

37. diverges by the nth-Term Test since $\lim\limits_{n \to \infty} |a_n| = \lim\limits_{n \to \infty} \frac{(2n)!}{2^n n! \, n} = \lim\limits_{n \to \infty} \frac{(n+1)(n+2)\cdots(2n)}{2^n n}$

    $= \lim\limits_{n \to \infty} \frac{(n+1)(n+2)\cdots(n+(n-1))}{2^{n-1}} > \lim\limits_{n \to \infty} \left(\frac{n+1}{2} \right)^{n-1} = \infty \neq 0$

39. converges conditionally since $\frac{\sqrt{n+1} - \sqrt{n}}{1} \cdot \frac{\sqrt{n+1} + \sqrt{n}}{\sqrt{n+1} + \sqrt{n}} = \frac{1}{\sqrt{n+1} + \sqrt{n}}$ and $\left\{ \frac{1}{\sqrt{n+1} + \sqrt{n}} \right\}$ is a

    decreasing sequence of positive terms which converges to $0 \Rightarrow \sum\limits_{n=1}^{\infty} \frac{(-1)^n}{\sqrt{n+1} + \sqrt{n}}$ converges; but

    $\sum\limits_{n=1}^{\infty} |a_n| = \sum\limits_{n=1}^{\infty} \frac{1}{\sqrt{n+1} + \sqrt{n}}$ diverges by the Limit Comparison Test (part 1) with $\frac{1}{\sqrt{n}}$; a divergent p-series:

    $\lim\limits_{n \to \infty} \left(\frac{\frac{1}{\sqrt{n+1} + \sqrt{n}}}{\frac{1}{\sqrt{n}}} \right) = \lim\limits_{n \to \infty} \frac{\sqrt{n}}{\sqrt{n+1} + \sqrt{n}} = \lim\limits_{n \to \infty} \frac{1}{\sqrt{1 + \frac{1}{n}} + 1} = \frac{1}{2}$

41. diverges by the nth-Term Test since $\lim\limits_{n \to \infty} \left(\sqrt{n + \sqrt{n}} - \sqrt{n} \right) = \lim\limits_{n \to \infty} \left[\left(\sqrt{n + \sqrt{n}} - \sqrt{n} \right) \left(\frac{\sqrt{n + \sqrt{n}} + \sqrt{n}}{\sqrt{n + \sqrt{n}} + \sqrt{n}} \right) \right]$

    $= \lim\limits_{n \to \infty} \frac{\sqrt{n}}{\sqrt{n + \sqrt{n}} + \sqrt{n}} = \lim\limits_{n \to \infty} \frac{1}{\sqrt{1 + \frac{1}{\sqrt{n}}} + 1} = \frac{1}{2} \neq 0$

43. converges absolutely by the Direct Comparison Test since $\text{sech}\,(n) = \frac{2}{e^n + e^{-n}} = \frac{2e^n}{e^{2n} + 1} < \frac{2e^n}{e^{2n}} = \frac{2}{e^n}$ which is the

    nth term of a convergent geometric series

45. $|\text{error}| < \left| (-1)^6 \left(\frac{1}{5} \right) \right| = 0.2$ 47. $|\text{error}| < \left| (-1)^6 \frac{(0.01)^5}{5} \right| = 2 \times 10^{-11}$

49. $\frac{1}{(2n)!} < \frac{5}{10^6} \Rightarrow (2n)! > \frac{10^6}{5} = 200{,}000 \Rightarrow n \geq 5 \Rightarrow 1 - \frac{1}{2!} + \frac{1}{4!} - \frac{1}{6!} + \frac{1}{8!} \approx 0.54030$

51. (a) $a_n \geq a_{n+1}$ fails since $\frac{1}{3} < \frac{1}{2}$

    (b) Since $\sum\limits_{n=1}^{\infty} |a_n| = \sum\limits_{n=1}^{\infty} \left[\left(\frac{1}{3} \right)^n + \left(\frac{1}{2} \right)^n \right] = \sum\limits_{n=1}^{\infty} \left(\frac{1}{3} \right)^n + \sum\limits_{n=1}^{\infty} \left(\frac{1}{2} \right)^n$ is the sum of two absolutely convergent

    series, we can rearrange the terms of the original series to find its sum:

    $\left(\frac{1}{3} + \frac{1}{9} + \frac{1}{27} + \ldots \right) - \left(\frac{1}{2} + \frac{1}{4} + \frac{1}{8} + \ldots \right) = \frac{\left(\frac{1}{3} \right)}{1 - \left(\frac{1}{3} \right)} - \frac{\left(\frac{1}{2} \right)}{1 - \left(\frac{1}{2} \right)} = \frac{1}{2} - 1 = -\frac{1}{2}$

53. The unused terms are $\sum\limits_{j=n+1}^{\infty} (-1)^{j+1}a_j = (-1)^{n+1}\left(a_{n+1} - a_{n+2}\right) + (-1)^{n+3}\left(a_{n+3} - a_{n+4}\right) + \ldots$

$= (-1)^{n+1}\left[(a_{n+1} - a_{n+2}) + (a_{n+3} - a_{n+4}) + \ldots\right]$. Each grouped term is positive, so the remainder

has the same sign as $(-1)^{n+1}$, which is the sign of the first unused term.

55. Theorem 16 states that $\sum\limits_{n=1}^{\infty} |a_n|$ converges $\Rightarrow \sum\limits_{n=1}^{\infty} a_n$ converges. But this is equivalent to $\sum\limits_{n=1}^{\infty} a_n$ diverges $\Rightarrow \sum\limits_{n=1}^{\infty} |a_n|$ diverges

57. (a) $\sum\limits_{n=1}^{\infty} |a_n + b_n|$ converges by the Direct Comparison Test since $|a_n + b_n| \leq |a_n| + |b_n|$ and hence

$\sum\limits_{n=1}^{\infty} (a_n + b_n)$ converges absolutely

(b) $\sum\limits_{n=1}^{\infty} |b_n|$ converges $\Rightarrow \sum\limits_{n=1}^{\infty} -b_n$ converges absolutely; since $\sum\limits_{n=1}^{\infty} a_n$ converges absolutely and

$\sum\limits_{n=1}^{\infty} -b_n$ converges absolutely, we have $\sum\limits_{n=1}^{\infty} [a_n + (-b_n)] = \sum\limits_{n=1}^{\infty} (a_n - b_n)$ converges absolutely by part (a)

(c) $\sum\limits_{n=1}^{\infty} |a_n|$ converges $\Rightarrow |k| \sum\limits_{n=1}^{\infty} |a_n| = \sum\limits_{n=1}^{\infty} |ka_n|$ converges $\Rightarrow \sum\limits_{n=1}^{\infty} ka_n$ converges absolutely

59. $s_1 = -\frac{1}{2}$, $s_2 = -\frac{1}{2} + 1 = \frac{1}{2}$,

$s_3 = -\frac{1}{2} + 1 - \frac{1}{4} - \frac{1}{6} - \frac{1}{8} - \frac{1}{10} - \frac{1}{12} - \frac{1}{14} - \frac{1}{16} - \frac{1}{18} - \frac{1}{20} - \frac{1}{22} \approx -0.5099$,

$s_4 = s_3 + \frac{1}{3} \approx -0.1766$,

$s_5 = s_4 - \frac{1}{24} - \frac{1}{26} - \frac{1}{28} - \frac{1}{30} - \frac{1}{32} - \frac{1}{34} - \frac{1}{36} - \frac{1}{38} - \frac{1}{40} - \frac{1}{42} - \frac{1}{44} \approx -0.512$,

$s_6 = s_5 + \frac{1}{5} \approx -0.312$,

$s_7 = s_6 - \frac{1}{46} - \frac{1}{48} - \frac{1}{50} - \frac{1}{52} - \frac{1}{54} - \frac{1}{56} - \frac{1}{58} - \frac{1}{60} - \frac{1}{62} - \frac{1}{64} - \frac{1}{66} \approx -0.51106$

8.7 POWER SERIES

1. $\lim\limits_{n \to \infty} \left|\frac{u_{n+1}}{u_n}\right| < 1 \Rightarrow \lim\limits_{n \to \infty} \left|\frac{x^{n+1}}{x^n}\right| < 1 \Rightarrow |x| < 1 \Rightarrow -1 < x < 1$; when $x = -1$ we have $\sum\limits_{n=1}^{\infty} (-1)^n$, a divergent

series; when $x = 1$ we have $\sum\limits_{n=1}^{\infty} 1$, a divergent series

(a) the radius is 1; the interval of convergence is $-1 < x < 1$

(b) the interval of absolute convergence is $-1 < x < 1$

(c) there are no values for which the series converges conditionally

3. $\lim\limits_{n \to \infty} \left|\frac{u_{n+1}}{u_n}\right| < 1 \Rightarrow \lim\limits_{n \to \infty} \left|\frac{(4x+1)^{n+1}}{(4x+1)^n}\right| < 1 \Rightarrow |4x+1| < 1 \Rightarrow -1 < 4x + 1 < 1 \Rightarrow -\frac{1}{2} < x < 0$; when $x = -\frac{1}{2}$ we

have $\sum\limits_{n=1}^{\infty} (-1)^n(-1)^n = \sum\limits_{n=1}^{\infty} (-1)^{2n} = \sum\limits_{n=1}^{\infty} 1^n$, a divergent series; when $x = 0$ we have $\sum\limits_{n=1}^{\infty} (-1)^n(1)^n = \sum\limits_{n=1}^{\infty} (-1)^n$,

a divergent series

(a) the radius is $\frac{1}{4}$; the interval of convergence is $-\frac{1}{2} < x < 0$

(b) the interval of absolute convergence is $-\frac{1}{2} < x < 0$

(c) there are no values for which the series converges conditionally

5. $\lim\limits_{n \to \infty} \left| \frac{u_{n+1}}{u_n} \right| < 1 \Rightarrow \lim\limits_{n \to \infty} \left| \frac{(x-2)^{n+1}}{10^{n+1}} \cdot \frac{10^n}{(x-2)^n} \right| < 1 \Rightarrow \frac{|x-2|}{10} < 1 \Rightarrow |x-2| < 10 \Rightarrow -10 < x-2 < 10$

$\Rightarrow -8 < x < 12$; when $x = -8$ we have $\sum\limits_{n=1}^{\infty} (-1)^n$, a divergent series; when $x = 12$ we have $\sum\limits_{n=1}^{\infty} 1$, a divergent series

(a) the radius is 10; the interval of convergence is $-8 < x < 12$

(b) the interval of absolute convergence is $-8 < x < 12$

(c) there are no values for which the series converges conditionally

7. $\lim\limits_{n \to \infty} \left| \frac{u_{n+1}}{u_n} \right| < 1 \Rightarrow \lim\limits_{n \to \infty} \left| \frac{(n+1)x^{n+1}}{(n+3)} \cdot \frac{(n+2)}{nx^n} \right| < 1 \Rightarrow |x| \lim\limits_{n \to \infty} \frac{(n+1)(n+2)}{(n+3)(n)} < 1 \Rightarrow |x| < 1$

$\Rightarrow -1 < x < 1$; when $x = -1$ we have $\sum\limits_{n=1}^{\infty} (-1)^n \frac{n}{n+2}$, a divergent series by the nth-term Test; when $x = 1$ we

have $\sum\limits_{n=1}^{\infty} \frac{n}{n+2}$, a divergent series

(a) the radius is 1; the interval of convergence is $-1 < x < 1$

(b) the interval of absolute convergence is $-1 < x < 1$

(c) there are no values for which the series converges conditionally

9. $\lim\limits_{n \to \infty} \left| \frac{u_{n+1}}{u_n} \right| < 1 \Rightarrow \lim\limits_{n \to \infty} \left| \frac{x^{n+1}}{(n+1)\sqrt{n+1}\, 3^{n+1}} \cdot \frac{n\sqrt{n}\, 3^n}{x^n} \right| < 1 \Rightarrow \frac{|x|}{3} \left(\lim\limits_{n \to \infty} \frac{n}{n+1} \right) \left(\sqrt{\lim\limits_{n \to \infty} \frac{n}{n+1}} \right) < 1$

$\Rightarrow \frac{|x|}{3} (1)(1) < 1 \Rightarrow |x| < 3 \Rightarrow -3 < x < 3$; when $x = -3$ we have $\sum\limits_{n=1}^{\infty} \frac{(-1)^n}{n^{3/2}}$, an absolutely convergent series;

when $x = 3$ we have $\sum\limits_{n=1}^{\infty} \frac{1}{n^{3/2}}$, a convergent p-series

(a) the radius is 3; the interval of convergence is $-3 \le x \le 3$

(b) the interval of absolute convergence is $-3 \le x \le 3$

(c) there are no values for which the series converges conditionally

11. $\lim\limits_{n \to \infty} \left| \frac{u_{n+1}}{u_n} \right| < 1 \Rightarrow \lim\limits_{n \to \infty} \left| \frac{x^{n+1}}{(n+1)!} \cdot \frac{n!}{x^n} \right| < 1 \Rightarrow |x| \lim\limits_{n \to \infty} \left(\frac{1}{n+1} \right) < 1$ for all x

(a) the radius is ∞; the series converges for all x

(b) the series converges absolutely for all x

(c) there are no values for which the series converges conditionally

13. $\lim\limits_{n \to \infty} \left| \frac{u_{n+1}}{u_n} \right| < 1 \Rightarrow \lim\limits_{n \to \infty} \left| \frac{x^{2n+3}}{(n+1)!} \cdot \frac{n!}{x^{2n+1}} \right| < 1 \Rightarrow x^2 \lim\limits_{n \to \infty} \left(\frac{1}{n+1} \right) < 1$ for all x

(a) the radius is ∞; the series converges for all x

(b) the series converges absolutely for all x

(c) there are no values for which the series converges conditionally

15. $\lim\limits_{n \to \infty} \left| \frac{u_{n+1}}{u_n} \right| < 1 \Rightarrow \lim\limits_{n \to \infty} \left| \frac{x^{n+1}}{\sqrt{(n+1)^2+3}} \cdot \frac{\sqrt{n^2+3}}{x^n} \right| < 1 \Rightarrow |x| \sqrt{\lim\limits_{n \to \infty} \frac{n^2+3}{n^2+2n+4}} < 1 \Rightarrow |x| < 1$

$\Rightarrow -1 < x < 1$; when $x = -1$ we have $\sum\limits_{n=1}^{\infty} \frac{(-1)^n}{\sqrt{n^2+3}}$, a conditionally convergent series; when $x = 1$ we have

$\sum\limits_{n=1}^{\infty} \frac{1}{\sqrt{n^2+3}}$, a divergent series

(a) the radius is 1; the interval of convergence is $-1 \le x < 1$

(b) the interval of absolute convergence is $-1 < x < 1$

(c) the series converges conditionally at $x = -1$

17. $\lim\limits_{n \to \infty} \left| \frac{u_{n+1}}{u_n} \right| < 1 \Rightarrow \lim\limits_{n \to \infty} \left| \frac{(n+1)(x+3)^{n+1}}{5^{n+1}} \cdot \frac{5^n}{n(x+3)^n} \right| < 1 \Rightarrow \frac{|x+3|}{5} \lim\limits_{n \to \infty} \left(\frac{n+1}{n} \right) < 1 \Rightarrow \frac{|x+3|}{5} < 1$

$\Rightarrow |x+3| < 5 \Rightarrow -5 < x+3 < 5 \Rightarrow -8 < x < 2$; when $x = -8$ we have $\sum\limits_{n=1}^{\infty} \frac{n(-5)^n}{5^n} = \sum\limits_{n=1}^{\infty} (-1)^n \, n$, a divergent

series; when $x = 2$ we have $\sum\limits_{n=1}^{\infty} \frac{n5^n}{5^n} = \sum\limits_{n=1}^{\infty} n$, a divergent series

(a) the radius is 5; the interval of convergence is $-8 < x < 2$

(b) the interval of absolute convergence is $-8 < x < 2$

(c) there are no values for which the series converges conditionally

19. $\lim\limits_{n \to \infty} \left| \frac{u_{n+1}}{u_n} \right| < 1 \Rightarrow \lim\limits_{n \to \infty} \left| \frac{\sqrt{n+1}\, x^{n+1}}{3^{n+1}} \cdot \frac{3^n}{\sqrt{n}\, x^n} \right| < 1 \Rightarrow \frac{|x|}{3} \sqrt{\lim\limits_{n \to \infty} \left(\frac{n+1}{n} \right)} < 1 \Rightarrow \frac{|x|}{3} < 1 \Rightarrow |x| < 3$

$\Rightarrow -3 < x < 3$; when $x = -3$ we have $\sum\limits_{n=1}^{\infty} (-1)^n \sqrt{n}$, a divergent series; when $x = 3$ we have $\sum\limits_{n=1}^{\infty} \sqrt{n}$, a divergent series

(a) the radius is 3; the interval of convergence is $-3 < x < 3$

(b) the interval of absolute convergence is $-3 < x < 3$

(c) there are no values for which the series converges conditionally

21. $\lim\limits_{n \to \infty} \left| \frac{u_{n+1}}{u_n} \right| < 1 \Rightarrow \lim\limits_{n \to \infty} \left| \frac{\left(1+\frac{1}{n+1}\right)^{n+1} x^{n+1}}{\left(1+\frac{1}{n}\right)^n x^n} \right| < 1 \Rightarrow |x| \left(\frac{\lim\limits_{t \to \infty} \left(1+\frac{1}{t}\right)^t}{\lim\limits_{n \to \infty} \left(1+\frac{1}{n}\right)^n} \right) < 1 \Rightarrow |x| \left(\frac{e}{e} \right) < 1 \Rightarrow |x| < 1$

$\Rightarrow -1 < x < 1$; when $x = -1$ we have $\sum\limits_{n=1}^{\infty} (-1)^n \left(1+\frac{1}{n}\right)^n$, a divergent series by the nth-Term Test since

$\lim\limits_{n \to \infty} \left(1+\frac{1}{n}\right)^n = e \neq 0$; when $x = 1$ we have $\sum\limits_{n=1}^{\infty} \left(1+\frac{1}{n}\right)^n$, a divergent series

(a) the radius is 1; the interval of convergence is $-1 < x < 1$

(b) the interval of absolute convergence is $-1 < x < 1$

(c) there are no values for which the series converges conditionally

23. $\lim\limits_{n \to \infty} \left| \frac{u_{n+1}}{u_n} \right| < 1 \Rightarrow \lim\limits_{n \to \infty} \left| \frac{(n+1)^{n+1} x^{n+1}}{n^n x^n} \right| < 1 \Rightarrow |x| \left(\lim\limits_{n \to \infty} \left(1+\frac{1}{n}\right)^n \right) \left(\lim\limits_{n \to \infty} (n+1) \right) < 1$

$\Rightarrow e \, |x| \lim\limits_{n \to \infty} (n+1) < 1 \Rightarrow$ only $x = 0$ satisfies this inequality

(a) the radius is 0; the series converges only for $x = 0$

(b) the series converges absolutely only for $x = 0$

(c) there are no values for which the series converges conditionally

25. $\lim\limits_{n \to \infty} \left| \frac{u_{n+1}}{u_n} \right| < 1 \Rightarrow \lim\limits_{n \to \infty} \left| \frac{(x+2)^{n+1}}{(n+1)\, 2^{n+1}} \cdot \frac{n 2^n}{(x+2)^n} \right| < 1 \Rightarrow \frac{|x+2|}{2} \lim\limits_{n \to \infty} \left(\frac{n}{n+1} \right) < 1 \Rightarrow \frac{|x+2|}{2} < 1 \Rightarrow |x+2| < 2$

$\Rightarrow -2 < x+2 < 2 \Rightarrow -4 < x < 0$; when $x = -4$ we have $\sum\limits_{n=1}^{\infty} \frac{-1}{n}$, a divergent series; when $x = 0$ we have $\sum\limits_{n=1}^{\infty} \frac{(-1)^{n+1}}{n}$,

the alternating harmonic series which converges conditionally

(a) the radius is 2; the interval of convergence is $-4 < x \leq 0$

(b) the interval of absolute convergence is $-4 < x < 0$

(c) the series converges conditionally at $x = 0$

27. $\lim\limits_{n \to \infty} \left| \frac{u_{n+1}}{u_n} \right| < 1 \Rightarrow \lim\limits_{n \to \infty} \left| \frac{x^{n+1}}{(n+1)\,(\ln(n+1))^2} \cdot \frac{n(\ln n)^2}{x^n} \right| < 1 \Rightarrow |x| \left(\lim\limits_{n \to \infty} \frac{n}{n+1} \right) \left(\lim\limits_{n \to \infty} \frac{\ln n}{\ln(n+1)} \right)^2 < 1$

$\Rightarrow |x| \, (1) \left(\lim\limits_{n \to \infty} \frac{\left(\frac{1}{n}\right)}{\left(\frac{1}{n+1}\right)} \right)^2 < 1 \Rightarrow |x| \left(\lim\limits_{n \to \infty} \frac{n+1}{n} \right)^2 < 1 \Rightarrow |x| < 1 \Rightarrow -1 < x < 1$; when $x = -1$ we have

$\sum\limits_{n=1}^{\infty} \frac{(-1)^n}{n(\ln n)^2}$ which converges absolutely; when x = 1 we have $\sum\limits_{n=1}^{\infty} \frac{1}{n(\ln n)^2}$ which converges

(a) the radius is 1; the interval of convergence is $-1 \leq x \leq 1$

(b) the interval of absolute convergence is $-1 \leq x \leq 1$

(c) there are no values for which the series converges conditionally

29. $\lim\limits_{n \to \infty} \left| \frac{u_{n+1}}{u_n} \right| < 1 \Rightarrow \lim\limits_{n \to \infty} \left| \frac{(4x-5)^{2n+3}}{(n+1)^{3/2}} \cdot \frac{n^{3/2}}{(4x-5)^{2n+1}} \right| < 1 \Rightarrow (4x-5)^2 \left(\lim\limits_{n \to \infty} \frac{n}{n+1} \right)^{3/2} < 1 \Rightarrow (4x-5)^2 < 1$

$\Rightarrow |4x-5| < 1 \Rightarrow -1 < 4x-5 < 1 \Rightarrow 1 < x < \frac{3}{2}$; when x = 1 we have $\sum\limits_{n=1}^{\infty} \frac{(-1)^{2n+1}}{n^{3/2}} = \sum\limits_{n=1}^{\infty} \frac{-1}{n^{3/2}}$ which is

absolutely convergent; when $x = \frac{3}{2}$ we have $\sum\limits_{n=1}^{\infty} \frac{(1)^{2n+1}}{n^{3/2}}$, a convergent p-series

(a) the radius is $\frac{1}{4}$; the interval of convergence is $1 \leq x \leq \frac{3}{2}$

(b) the interval of absolute convergence is $1 \leq x \leq \frac{3}{2}$

(c) there are no values for which the series converges conditionally

31. $\lim\limits_{n \to \infty} \left| \frac{u_{n+1}}{u_n} \right| < 1 \Rightarrow \lim\limits_{n \to \infty} \left| \frac{(x+\pi)^{n+1}}{\sqrt{n+1}} \cdot \frac{\sqrt{n}}{(x+\pi)^n} \right| < 1 \Rightarrow |x+\pi| \lim\limits_{n \to \infty} \left| \sqrt{\frac{n}{n+1}} \right| < 1$

$\Rightarrow |x+\pi| \sqrt{\lim\limits_{n \to \infty} \left(\frac{n}{n+1} \right)} < 1 \Rightarrow |x+\pi| < 1 \Rightarrow -1 < x+\pi < 1 \Rightarrow -1-\pi < x < 1-\pi$;

when $x = -1-\pi$ we have $\sum\limits_{n=1}^{\infty} \frac{(-1)^n}{\sqrt{n}} = \sum\limits_{n=1}^{\infty} \frac{(-1)^n}{n^{1/2}}$, a conditionally convergent series; when $x = 1-\pi$ we have

$\sum\limits_{n=1}^{\infty} \frac{1^n}{\sqrt{n}} = \sum\limits_{n=1}^{\infty} \frac{1}{n^{1/2}}$, a divergent p-series

(a) the radius is 1; the interval of convergence is $(-1-\pi) \leq x < (1-\pi)$

(b) the interval of absolute convergence is $-1-\pi < x < 1-\pi$

(c) the series converges conditionally at $x = -1-\pi$

33. $\lim\limits_{n \to \infty} \left| \frac{u_{n+1}}{u_n} \right| < 1 \Rightarrow \lim\limits_{n \to \infty} \left| \frac{(x-1)^{2n+2}}{4^{n+1}} \cdot \frac{4^n}{(x-1)^{2n}} \right| < 1 \Rightarrow \frac{(x-1)^2}{4} \lim\limits_{n \to \infty} |1| < 1 \Rightarrow (x-1)^2 < 4 \Rightarrow |x-1| < 2$

$\Rightarrow -2 < x-1 < 2 \Rightarrow -1 < x < 3$; at x = -1 we have $\sum\limits_{n=0}^{\infty} \frac{(-2)^{2n}}{4^n} = \sum\limits_{n=0}^{\infty} \frac{4^n}{4^n} = \sum\limits_{n=0}^{\infty} 1$, which diverges; at x = 3

we have $\sum\limits_{n=0}^{\infty} \frac{2^{2n}}{4^n} = \sum\limits_{n=0}^{\infty} \frac{4^n}{4^n} = \sum\limits_{n=0}^{\infty} 1$, a divergent series; the interval of convergence is $-1 < x < 3$; the series

$\sum\limits_{n=0}^{\infty} \frac{(x-1)^{2n}}{4^n} = \sum\limits_{n=0}^{\infty} \left(\left(\frac{x-1}{2} \right)^2 \right)^n$ is a convergent geometric series when $-1 < x < 3$ and the sum is

$\frac{1}{1 - \left(\frac{x-1}{2} \right)^2} = \frac{1}{\left[\frac{4-(x-1)^2}{4} \right]} = \frac{4}{4-x^2+2x-1} = \frac{4}{3+2x-x^2}$

35. $\lim\limits_{n \to \infty} \left| \frac{u_{n+1}}{u_n} \right| < 1 \Rightarrow \lim\limits_{n \to \infty} \left| \frac{(\sqrt{x}-2)^{n+1}}{2^{n+1}} \cdot \frac{2^n}{(\sqrt{x}-2)^n} \right| < 1 \Rightarrow |\sqrt{x}-2| < 2 \Rightarrow -2 < \sqrt{x}-2 < 2 \Rightarrow 0 < \sqrt{x} < 4$

$\Rightarrow 0 < x < 16$; when x = 0 we have $\sum\limits_{n=0}^{\infty} (-1)^n$, a divergent series; when x = 16 we have $\sum\limits_{n=0}^{\infty} (1)^n$, a divergent

series; the interval of convergence is $0 < x < 16$; the series $\sum\limits_{n=0}^{\infty} \left(\frac{\sqrt{x}-2}{2} \right)^n$ is a convergent geometric series when

$0 < x < 16$ and its sum is $\frac{1}{1 - \left(\frac{\sqrt{x}-2}{2} \right)} = \frac{1}{\left(\frac{2-\sqrt{x}+2}{2} \right)} = \frac{2}{4-\sqrt{x}}$

37. $\lim\limits_{n \to \infty} \left| \frac{u_{n+1}}{u_n} \right| < 1 \Rightarrow \lim\limits_{n \to \infty} \left| \left(\frac{x^2+1}{3} \right)^{n+1} \cdot \left(\frac{3}{x^2+1} \right)^n \right| < 1 \Rightarrow \frac{(x^2+1)}{3} \lim\limits_{n \to \infty} |1| < 1 \Rightarrow \frac{x^2+1}{3} < 1 \Rightarrow x^2 < 2$

$\Rightarrow |x| < \sqrt{2} \Rightarrow -\sqrt{2} < x < \sqrt{2}$; at $x = \pm\sqrt{2}$ we have $\sum\limits_{n=0}^{\infty} (1)^n$ which diverges; the interval of convergence is

$-\sqrt{2} < x < \sqrt{2}$; the series $\sum\limits_{n=0}^{\infty} \left(\frac{x^2+1}{3}\right)^n$ is a convergent geometric series when $-\sqrt{2} < x < \sqrt{2}$ and its sum is

$\frac{1}{1 - \left(\frac{x^2+1}{3}\right)} = \frac{1}{\left(\frac{3-x^2-1}{3}\right)} = \frac{3}{2-x^2}$

39. $\lim\limits_{n \to \infty} \left| \frac{(x-3)^{n+1}}{2^{n+1}} \cdot \frac{2^n}{(x-3)^n} \right| < 1 \Rightarrow |x-3| < 2 \Rightarrow 1 < x < 5$; when $x = 1$ we have $\sum\limits_{n=1}^{\infty} (1)^n$ which diverges;

when $x = 5$ we have $\sum\limits_{n=1}^{\infty} (-1)^n$ which also diverges; the interval of convergence is $1 < x < 5$; the sum of this

convergent geometric series is $\frac{1}{1 + \left(\frac{x-3}{2}\right)} = \frac{2}{x-1}$. If $f(x) = 1 - \frac{1}{2}(x-3) + \frac{1}{4}(x-3)^2 + \ldots + \left(-\frac{1}{2}\right)^n (x-3)^n + \ldots$

$= \frac{2}{x-1}$ then $f'(x) = -\frac{1}{2} + \frac{1}{2}(x-3) + \ldots + \left(-\frac{1}{2}\right)^n n(x-3)^{n-1} + \ldots$ is convergent when $1 < x < 5$, and diverges

when $x = 1$ or 5. The sum for $f'(x)$ is $\frac{-2}{(x-1)^2}$, the derivative of $\frac{2}{x-1}$.

41. (a) Differentiate the series for $\sin x$ to get $\cos x = 1 - \frac{3x^2}{3!} + \frac{5x^4}{5!} - \frac{7x^6}{7!} + \frac{9x^8}{9!} - \frac{11x^{10}}{11!} + \ldots$

$= 1 - \frac{x^2}{2!} + \frac{x^4}{4!} - \frac{x^6}{6!} + \frac{x^8}{8!} - \frac{x^{10}}{10!} + \ldots$. The series converges for all values of x since

$\lim\limits_{n \to \infty} \left| \frac{x^{2n+2}}{(2n+2)!} \cdot \frac{(2n)!}{x^{2n}} \right| = x^2 \lim\limits_{n \to \infty} \left(\frac{1}{(2n+1)(2n+2)} \right) = 0 < 1$ for all x.

(b) $\sin 2x = 2x - \frac{2^3 x^3}{3!} + \frac{2^5 x^5}{5!} - \frac{2^7 x^7}{7!} + \frac{2^9 x^9}{9!} - \frac{2^{11} x^{11}}{11!} + \ldots = 2x - \frac{8x^3}{3!} + \frac{32x^5}{5!} - \frac{128x^7}{7!} + \frac{512x^9}{9!} - \frac{2048x^{11}}{11!} + \ldots$

(c) $2 \sin x \cos x = 2 \left[(0 \cdot 1) + (0 \cdot 0 + 1 \cdot 1)x + \left(0 \cdot \frac{-1}{2} + 1 \cdot 0 + 0 \cdot 1\right) x^2 + \left(0 \cdot 0 - 1 \cdot \frac{1}{2} + 0 \cdot 0 - 1 \cdot \frac{1}{3!}\right) x^3 \right.$

$+ \left(0 \cdot \frac{1}{4!} + 1 \cdot 0 - 0 \cdot \frac{1}{2} - 0 \cdot \frac{1}{3!} + 0 \cdot 1\right) x^4 + \left(0 \cdot 0 + 1 \cdot \frac{1}{4!} + 0 \cdot 0 + \frac{1}{2} \cdot \frac{1}{3!} + 0 \cdot 0 + 1 \cdot \frac{1}{5!}\right) x^5$

$+ \left.\left(0 \cdot \frac{1}{6!} + 1 \cdot 0 + 0 \cdot \frac{1}{4!} + 0 \cdot \frac{1}{3!} + 0 \cdot \frac{1}{2} + 0 \cdot \frac{1}{5!} + 0 \cdot 1\right) x^6 + \ldots \right] = 2 \left[x - \frac{4x^3}{3!} + \frac{16x^5}{5!} - \ldots \right]$

$= 2x - \frac{2^3 x^3}{3!} + \frac{2^5 x^5}{5!} - \frac{2^7 x^7}{7!} + \frac{2^9 x^9}{9!} - \frac{2^{11} x^{11}}{11!} + \ldots$

43. (a) $\ln|\sec x| + C = \int \tan x \, dx = \int \left(x + \frac{x^3}{3} + \frac{2x^5}{15} + \frac{17x^7}{315} + \frac{62x^9}{2835} + \ldots \right) dx$

$= \frac{x^2}{2} + \frac{x^4}{12} + \frac{x^6}{45} + \frac{17x^8}{2520} + \frac{31x^{10}}{14,175} + \ldots + C; \; x = 0 \Rightarrow C = 0 \Rightarrow \ln|\sec x| = \frac{x^2}{2} + \frac{x^4}{12} + \frac{x^6}{45} + \frac{17x^8}{2520} + \frac{31x^{10}}{14,175} + \ldots$,

converges when $-\frac{\pi}{2} < x < \frac{\pi}{2}$

(b) $\sec^2 x = \frac{d(\tan x)}{dx} = \frac{d}{dx} \left(x + \frac{x^3}{3} + \frac{2x^5}{15} + \frac{17x^7}{315} + \frac{62x^9}{2835} + \ldots \right) = 1 + x^2 + \frac{2x^4}{3} + \frac{17x^6}{45} + \frac{62x^8}{315} + \ldots$, converges

when $-\frac{\pi}{2} < x < \frac{\pi}{2}$

(c) $\sec^2 x = (\sec x)(\sec x) = \left(1 + \frac{x^2}{2} + \frac{5x^4}{24} + \frac{61x^6}{720} + \ldots \right) \left(1 + \frac{x^2}{2} + \frac{5x^4}{24} + \frac{61x^6}{720} + \ldots \right)$

$= 1 + \left(\frac{1}{2} + \frac{1}{2}\right) x^2 + \left(\frac{5}{24} + \frac{1}{4} + \frac{5}{24}\right) x^4 + \left(\frac{61}{720} + \frac{5}{48} + \frac{5}{48} + \frac{61}{720}\right) x^6 + \ldots$

$= 1 + x^2 + \frac{2x^4}{3} + \frac{17x^6}{45} + \frac{62x^8}{315} + \ldots, \; -\frac{\pi}{2} < x < \frac{\pi}{2}$

45. (a) If $f(x) = \sum\limits_{n=0}^{\infty} a_n x^n$, then $f^{(k)}(x) = \sum\limits_{n=k}^{\infty} n(n-1)(n-2)\cdots(n-(k-1)) a_n x^{n-k}$ and $f^{(k)}(0) = k! a_k$

$\Rightarrow a_k = \frac{f^{(k)}(0)}{k!}$; likewise if $f(x) = \sum\limits_{n=0}^{\infty} b_n x^n$, then $b_k = \frac{f^{(k)}(0)}{k!} \Rightarrow a_k = b_k$ for every nonnegative integer k

(b) If $f(x) = \sum\limits_{n=0}^{\infty} a_n x^n = 0$ for all x, then $f^{(k)}(x) = 0$ for all x $\Rightarrow$ from part (a) that $a_k = 0$ for every nonnegative integer k

47. The series $\sum\limits_{n=1}^{\infty} \frac{x^n}{n}$ converges conditionally at the left-hand endpoint of its interval of convergence $[-1, 1)$; the

series $\sum\limits_{n=1}^{\infty} \frac{x^n}{(n^2)}$ converges absolutely at the left-hand endpoint of its interval of convergence $[-1, 1]$

8.8 TAYLOR AND MACLAURIN SERIES

1. $f(x) = \ln x, f'(x) = \frac{1}{x}, f''(x) = -\frac{1}{x^2}, f'''(x) = \frac{2}{x^3}; f(1) = \ln 1 = 0, f'(1) = 1, f''(1) = -1, f'''(1) = 2 \Rightarrow P_0(x) = 0,$
    $P_1(x) = (x-1), P_2(x) = (x-1) - \frac{1}{2}(x-1)^2, P_3(x) = (x-1) - \frac{1}{2}(x-1)^2 + \frac{1}{3}(x-1)^3$

3. $f(x) = \frac{1}{x} = x^{-1}, f'(x) = -x^{-2}, f''(x) = 2x^{-3}, f'''(x) = -6x^{-4}; f(2) = \frac{1}{2}, f'(2) = -\frac{1}{4}, f''(2) = \frac{1}{4}, f'''(x) = -\frac{3}{8}$
    $\Rightarrow P_0(x) = \frac{1}{2}, P_1(x) = \frac{1}{2} - \frac{1}{4}(x-2), P_2(x) = \frac{1}{2} - \frac{1}{4}(x-2) + \frac{1}{8}(x-2)^2,$
    $P_3(x) = \frac{1}{2} - \frac{1}{4}(x-2) + \frac{1}{8}(x-2)^2 - \frac{1}{16}(x-2)^3$

5. $f(x) = \sin x, f'(x) = \cos x, f''(x) = -\sin x, f'''(x) = -\cos x; f\left(\frac{\pi}{4}\right) = \sin\frac{\pi}{4} = \frac{\sqrt{2}}{2}, f'\left(\frac{\pi}{4}\right) = \cos\frac{\pi}{4} = \frac{\sqrt{2}}{2},$
    $f''\left(\frac{\pi}{4}\right) = -\sin\frac{\pi}{4} = -\frac{\sqrt{2}}{2}, f'''\left(\frac{\pi}{4}\right) = -\cos\frac{\pi}{4} = -\frac{\sqrt{2}}{2} \Rightarrow P_0 = \frac{\sqrt{2}}{2}, P_1(x) = \frac{\sqrt{2}}{2} + \frac{\sqrt{2}}{2}\left(x - \frac{\pi}{4}\right),$
    $P_2(x) = \frac{\sqrt{2}}{2} + \frac{\sqrt{2}}{2}\left(x - \frac{\pi}{4}\right) - \frac{\sqrt{2}}{4}\left(x - \frac{\pi}{4}\right)^2, P_3(x) = \frac{\sqrt{2}}{2} + \frac{\sqrt{2}}{2}\left(x - \frac{\pi}{4}\right) - \frac{\sqrt{2}}{4}\left(x - \frac{\pi}{4}\right)^2 - \frac{\sqrt{2}}{12}\left(x - \frac{\pi}{4}\right)^3$

7. $f(x) = \sqrt{x} = x^{1/2}, f'(x) = \left(\frac{1}{2}\right)x^{-1/2}, f''(x) = \left(-\frac{1}{4}\right)x^{-3/2}, f'''(x) = \left(\frac{3}{8}\right)x^{-5/2}; f(4) = \sqrt{4} = 2,$
    $f'(4) = \left(\frac{1}{2}\right)4^{-1/2} = \frac{1}{4}, f''(4) = \left(-\frac{1}{4}\right)4^{-3/2} = -\frac{1}{32}, f'''(4) = \left(\frac{3}{8}\right)4^{-5/2} = \frac{3}{256} \Rightarrow P_0(x) = 2, P_1(x) = 2 + \frac{1}{4}(x-4),$
    $P_2(x) = 2 + \frac{1}{4}(x-4) - \frac{1}{64}(x-4)^2, P_3(x) = 2 + \frac{1}{4}(x-4) - \frac{1}{64}(x-4)^2 + \frac{1}{512}(x-4)^3$

9. $e^x = \sum_{n=0}^{\infty} \frac{x^n}{n!} \Rightarrow e^{-x} = \sum_{n=0}^{\infty} \frac{(-x)^n}{n!} = 1 - x + \frac{x^2}{2!} - \frac{x^3}{3!} + \frac{x^4}{4!} - \cdots$

11. $f(x) = (1+x)^{-1} \Rightarrow f'(x) = -(1+x)^{-2}, f''(x) = 2(1+x)^{-3}, f'''(x) = -3!(1+x)^{-4} \Rightarrow \cdots f^{(k)}(x)$
    $= (-1)^k k!(1+x)^{-k-1}; f(0) = 1, f'(0) = -1, f''(0) = 2, f'''(0) = -3!, \ldots, f^{(k)}(0) = (-1)^k k!$
    $\Rightarrow \frac{1}{1+x} = 1 - x + x^2 - x^3 + \cdots = \sum_{n=0}^{\infty}(-x)^n = \sum_{n=0}^{\infty}(-1)^n x^n$

13. $\sin x = \sum_{n=0}^{\infty} \frac{(-1)^n x^{2n+1}}{(2n+1)!} \Rightarrow \sin 3x = \sum_{n=0}^{\infty} \frac{(-1)^n (3x)^{2n+1}}{(2n+1)!} = \sum_{n=0}^{\infty} \frac{(-1)^n 3^{2n+1} x^{2n+1}}{(2n+1)!} = 3x - \frac{3^3 x^3}{3!} + \frac{3^5 x^5}{5!} - \cdots$

15. $7\cos(-x) = 7\cos x = 7\sum_{n=0}^{\infty} \frac{(-1)^n x^{2n}}{(2n)!} = 7 - \frac{7x^2}{2!} + \frac{7x^4}{4!} - \frac{7x^6}{6!} + \cdots,$ since the cosine is an even function

17. $\cosh x = \frac{e^x + e^{-x}}{2} = \frac{1}{2}\left[\left(1 + x^2 + \frac{x^2}{2!} + \frac{x^3}{3!} + \frac{x^4}{4!} + \cdots\right) + \left(1 - x + \frac{x^2}{2!} - \frac{x^3}{3!} + \frac{x^4}{4!} - \cdots\right)\right] = 1 + \frac{x^2}{2!} + \frac{x^4}{4!} + \frac{x^6}{6!} + \cdots$
    $= \sum_{n=0}^{\infty} \frac{x^{2n}}{(2n)!}$

19. $f(x) = x^4 - 2x^3 - 5x + 4 \Rightarrow f'(x) = 4x^3 - 6x^2 - 5, f''(x) = 12x^2 - 12x, f'''(x) = 24x - 12, f^{(4)}(x) = 24$
    $\Rightarrow f^{(n)}(x) = 0$ if $n \geq 5; f(0) = 4, f'(0) = -5, f''(0) = 0, f'''(0) = -12, f^{(4)}(0) = 24, f^{(n)}(0) = 0$ if $n \geq 5$
    $\Rightarrow x^4 - 2x^3 - 5x + 4 = 4 - 5x - \frac{12}{3!}x^3 + \frac{24}{4!}x^4 = x^4 - 2x^3 - 5x + 4$ itself

21. $f(x) = x^3 - 2x + 4 \Rightarrow f'(x) = 3x^2 - 2, f''(x) = 6x, f'''(x) = 6 \Rightarrow f^{(n)}(x) = 0$ if $n \geq 4; f(2) = 8, f'(2) = 10,$
    $f''(2) = 12, f'''(2) = 6, f^{(n)}(2) = 0$ if $n \geq 4 \Rightarrow x^3 - 2x + 4 = 8 + 10(x-2) + \frac{12}{2!}(x-2)^2 + \frac{6}{3!}(x-2)^3$
    $= 8 + 10(x-2) + 6(x-2)^2 + (x-2)^3$

23. $f(x) = x^4 + x^2 + 1 \Rightarrow f'(x) = 4x^3 + 2x, f''(x) = 12x^2 + 2, f'''(x) = 24x, f^{(4)}(x) = 24, f^{(n)}(x) = 0$ if $n \geq 5;$
    $f(-2) = 21, f'(-2) = -36, f''(-2) = 50, f'''(-2) = -48, f^{(4)}(-2) = 24, f^{(n)}(-2) = 0$ if $n \geq 5 \Rightarrow x^4 + x^2 + 1$
    $= 21 - 36(x+2) + \frac{50}{2!}(x+2)^2 - \frac{48}{3!}(x+2)^3 + \frac{24}{4!}(x+2)^4 = 21 - 36(x+2) + 25(x+2)^2 - 8(x+2)^3 + (x+2)^4$

25. $f(x) = x^{-2} \Rightarrow f'(x) = -2x^{-3}, f''(x) = 3! x^{-4}, f'''(x) = -4! x^{-5} \Rightarrow f^{(n)}(x) = (-1)^n(n+1)! x^{-n-2};$
$f(1) = 1, f'(1) = -2, f''(1) = 3!, f'''(1) = -4!, f^{(n)}(1) = (-1)^n(n+1)! \Rightarrow \frac{1}{x^2}$

$= 1 - 2(x-1) + 3(x-1)^2 - 4(x-1)^3 + \ldots = \sum\limits_{n=0}^{\infty} (-1)^n(n+1)(x-1)^n$

27. $f(x) = e^x \Rightarrow f'(x) = e^x, f''(x) = e^x \Rightarrow f^{(n)}(x) = e^x; f(2) = e^2, f'(2) = e^2, \ldots f^{(n)}(2) = e^2$

$\Rightarrow e^x = e^2 + e^2(x-2) + \frac{e^2}{2}(x-2)^2 + \frac{e^3}{3!}(x-2)^3 + \ldots = \sum\limits_{n=0}^{\infty} \frac{e^2}{n!}(x-2)^n$

29. If $e^x = \sum\limits_{n=0}^{\infty} \frac{f^{(n)}(a)}{n!}(x-a)^n$ and $f(x) = e^x$, we have $f^{(n)}(a) = e^a$ for all $n = 0, 1, 2, 3, \ldots$

$\Rightarrow e^x = e^a \left[\frac{(x-a)^0}{0!} + \frac{(x-a)^1}{1!} + \frac{(x-a)^2}{2!} + \ldots \right] = e^a \left[1 + (x-a) + \frac{(x-a)^2}{2!} + \ldots \right]$ at $x = a$

31. $f(x) = f(a) + f'(a)(x-a) + \frac{f''(a)}{2}(x-a)^2 + \frac{f'''(a)}{3!}(x-a)^3 + \ldots \Rightarrow f'(x)$

$= f'(a) + f''(a)(x-a) + \frac{f'''(a)}{3!} 3(x-a)^2 + \ldots \Rightarrow f''(x) = f''(a) + f'''(a)(x-a) + \frac{f^{(4)}(a)}{4!} 4 \cdot 3(x-a)^2 + \ldots$

$\Rightarrow f^{(n)}(x) = f^{(n)}(a) + f^{(n+1)}(a)(x-a) + \frac{f^{(n+2)}(a)}{2}(x-a)^2 + \ldots$

$\Rightarrow f(a) = f(a) + 0, f'(a) = f'(a) + 0, \ldots, f^{(n)}(a) = f^{(n)}(a) + 0$

33. $f(x) = \ln(\cos x) \Rightarrow f'(x) = -\tan x$ and $f''(x) = -\sec^2 x; f(0) = 0, f'(0) = 0, f''(0) = -1 \Rightarrow L(x) = 0$ and $Q(x) = -\frac{x^2}{2}$

35. $f(x) = (1-x^2)^{-1/2} \Rightarrow f'(x) = x(1-x^2)^{-3/2}$ and $f''(x) = (1-x^2)^{-3/2} + 3x^2(1-x^2)^{-5/2}; f(0) = 1, f'(0) = 0,$
$f''(0) = 1 \Rightarrow L(x) = 1$ and $Q(x) = 1 + \frac{x^2}{2}$

37. $f(x) = \sin x \Rightarrow f'(x) = \cos x$ and $f''(x) = -\sin x; f(0) = 0, f'(0) = 1, f''(0) = 0 \Rightarrow L(x) = x$ and $Q(x) = x$

8.9 CONVERGENCE OF TAYLOR SERIES

1. $e^x = 1 + x + \frac{x^2}{2!} + \ldots = \sum\limits_{n=0}^{\infty} \frac{x^n}{n!} \Rightarrow e^{-5x} = 1 + (-5x) + \frac{(-5x)^2}{2!} + \ldots = 1 - 5x + \frac{5^2 x^2}{2!} - \frac{5^3 x^3}{3!} + \ldots = \sum\limits_{n=0}^{\infty} \frac{(-1)^n 5^n x^n}{n!}$

3. $\sin x = x - \frac{x^3}{3!} + \frac{x^5}{5!} - \ldots = \sum\limits_{n=0}^{\infty} \frac{(-1)^n x^{2n+1}}{(2n+1)!} \Rightarrow 5\sin(-x) = 5\left[(-x) - \frac{(-x)^3}{3!} + \frac{(-x)^5}{5!} - \ldots \right] = \sum\limits_{n=0}^{\infty} \frac{5(-1)^{n+1} x^{2n+1}}{(2n+1)!}$

5. $\cos x = \sum\limits_{n=0}^{\infty} \frac{(-1)^n x^{2n}}{(2n)!} \Rightarrow \cos\sqrt{x+1} = \sum\limits_{n=0}^{\infty} \frac{(-1)^n \left[(x+1)^{1/2} \right]^{2n}}{(2n)!} = \sum\limits_{n=0}^{\infty} \frac{(-1)^n(x+1)^n}{(2n)!} = 1 - \frac{x+1}{2!} + \frac{(x+1)^2}{4!} - \frac{(x+1)^3}{6!} + \ldots$

7. $e^x = \sum\limits_{n=0}^{\infty} \frac{x^n}{n!} \Rightarrow xe^x = x\left(\sum\limits_{n=0}^{\infty} \frac{x^n}{n!} \right) = \sum\limits_{n=0}^{\infty} \frac{x^{n+1}}{n!} = x + x^2 + \frac{x^3}{2!} + \frac{x^4}{3!} + \frac{x^5}{4!} + \ldots$

9. $\cos x = \sum\limits_{n=0}^{\infty} \frac{(-1)^n x^{2n}}{(2n)!} \Rightarrow \frac{x^2}{2} - 1 + \cos x = \frac{x^2}{2} - 1 + \sum\limits_{n=0}^{\infty} \frac{(-1)^n x^{2n}}{(2n)!} = \frac{x^2}{2} - 1 + 1 - \frac{x^2}{2} + \frac{x^4}{4!} - \frac{x^6}{6!} + \frac{x^8}{8!} - \frac{x^{10}}{10!} + \ldots$

$= \frac{x^4}{4!} - \frac{x^6}{6!} + \frac{x^8}{8!} - \frac{x^{10}}{10!} + \ldots = \sum\limits_{n=2}^{\infty} \frac{(-1)^n x^{2n}}{(2n)!}$

11. $\cos x = \sum\limits_{n=0}^{\infty} \frac{(-1)^n x^{2n}}{(2n)!} \Rightarrow x\cos\pi x = x\sum\limits_{n=0}^{\infty} \frac{(-1)^n(\pi x)^{2n}}{(2n)!} = \sum\limits_{n=0}^{\infty} \frac{(-1)^n \pi^{2n} x^{2n+1}}{(2n)!} = x - \frac{\pi^2 x^3}{2!} + \frac{\pi^4 x^5}{4!} - \frac{\pi^6 x^7}{6!} + \ldots$

13. $\cos^2 x = \frac{1}{2} + \frac{\cos 2x}{2} = \frac{1}{2} + \frac{1}{2} \sum_{n=0}^{\infty} \frac{(-1)^n (2x)^{2n}}{(2n)!} = \frac{1}{2} + \frac{1}{2} \left[1 - \frac{(2x)^2}{2!} + \frac{(2x)^4}{4!} - \frac{(2x)^6}{6!} + \frac{(2x)^8}{8!} - \dots \right]$

$= 1 - \frac{(2x)^2}{2 \cdot 2!} + \frac{(2x)^4}{2 \cdot 4!} - \frac{(2x)^6}{2 \cdot 6!} + \frac{(2x)^8}{2 \cdot 8!} - \dots = 1 + \sum_{n=1}^{\infty} \frac{(-1)^n (2x)^{2n}}{2 \cdot (2n)!} = 1 + \sum_{n=1}^{\infty} \frac{(-1)^n \, 2^{2n-1} \, x^{2n}}{(2n)!}$

15. $\frac{x^2}{1-2x} = x^2 \left(\frac{1}{1-2x} \right) = x^2 \sum_{n=0}^{\infty} (2x)^n = \sum_{n=0}^{\infty} 2^n x^{n+2} = x^2 + 2x^3 + 2^2 x^4 + 2^3 x^5 + \dots$

17. $\frac{1}{1-x} = \sum_{n=0}^{\infty} x^n = 1 + x + x^2 + x^3 + \dots \;\Rightarrow\; \frac{d}{dx}\left(\frac{1}{1-x} \right) = \frac{1}{(1-x)^2} = 1 + 2x + 3x^2 + \dots = \sum_{n=1}^{\infty} nx^{n-1} = \sum_{n=0}^{\infty} (n+1)x^n$

19. By the Alternating Series Estimation Theorem, the error is less than $\frac{|x|^5}{5!} \;\Rightarrow\; |x|^5 < (5!)\,(5 \times 10^{-4}) \;\Rightarrow\; |x|^5 < 600 \times 10^{-4}$

$\Rightarrow\; |x| < \sqrt[5]{6 \times 10^{-2}} \approx 0.56968$

21. If $\sin x = x$ and $|x| < 10^{-3}$, then the error is less than $\frac{(10^{-3})^3}{3!} \approx 1.67 \times 10^{-10}$, by Alternating Series Estimation Theorem;

The Alternating Series Estimation Theorem says $R_2(x)$ has the same sign as $-\frac{x^3}{3!}$. Moreover, $x < \sin x$

$\Rightarrow\; 0 < \sin x - x = R_2(x) \;\Rightarrow\; x < 0 \;\Rightarrow\; -10^{-3} < x < 0.$

23. $|R_2(x)| = \left| \frac{e^c x^3}{3!} \right| < \frac{3^{(0.1)}(0.1)^3}{3!} < 1.87 \times 10^{-4}$, where c is between 0 and x

25. $F(x) = \int_0^x \left(t^2 - \frac{t^6}{3!} + \frac{t^{10}}{5!} - \frac{t^{14}}{7!} + \dots \right) dt = \left[\frac{t^3}{3} - \frac{t^7}{7 \cdot 3!} + \frac{t^{11}}{11 \cdot 5!} - \frac{t^{15}}{15 \cdot 7!} + \dots \right]_0^x \approx \frac{x^3}{3} - \frac{x^7}{7 \cdot 3!} + \frac{x^{11}}{11 \cdot 5!}$

$\Rightarrow\; |\text{error}| < \frac{1}{15 \cdot 7!} \approx 0.000013$

27. (a) $F(x) = \int_0^x \left(t - \frac{t^3}{3} + \frac{t^5}{5} - \frac{t^7}{7} + \dots \right) dt = \left[\frac{t^2}{2} - \frac{t^4}{12} + \frac{t^6}{30} - \dots \right]_0^x \approx \frac{x^2}{2} - \frac{x^4}{12} \;\Rightarrow\; |\text{error}| < \frac{(0.5)^6}{30} \approx .00052$

(b) $|\text{error}| < \frac{1}{33 \cdot 34} \approx .00089$ when $F(x) \approx \frac{x^2}{2} - \frac{x^4}{3 \cdot 4} + \frac{x^6}{5 \cdot 6} - \frac{x^8}{7 \cdot 8} + \dots + (-1)^{15} \frac{x^{32}}{31 \cdot 32}$

29. $\frac{1}{x^2}\left(e^x - (1+x) \right) = \frac{1}{x^2}\left(\left(1 + x + \frac{x^2}{2} + \frac{x^3}{3!} + \dots \right) - 1 - x \right) = \frac{1}{2} + \frac{x}{3!} + \frac{x^2}{4!} + \dots \;\Rightarrow\; \lim_{x \to 0} \frac{e^x - (1+x)}{x^2}$

$= \lim_{x \to 0} \left(\frac{1}{2} + \frac{x}{3!} + \frac{x^2}{4!} + \dots \right) = \frac{1}{2}$

31. $\frac{1}{t^4}\left(1 - \cos t - \frac{t^2}{2} \right) = \frac{1}{t^4}\left[1 - \frac{t^2}{2} - \left(1 - \frac{t^2}{2} + \frac{t^4}{4!} - \frac{t^6}{6!} + \dots \right) \right] = -\frac{1}{4!} + \frac{t^2}{6!} - \frac{t^4}{8!} + \dots \;\Rightarrow\; \lim_{t \to 0} \frac{1 - \cos t - \left(\frac{t^2}{2} \right)}{t^4}$

$= \lim_{t \to 0} \left(-\frac{1}{4!} + \frac{t^2}{6!} - \frac{t^4}{8!} + \dots \right) = -\frac{1}{24}$

33. $\frac{1}{y^3}\left(y - \tan^{-1} y \right) = \frac{1}{y^3}\left[y - \left(y - \frac{y^3}{3} + \frac{y^5}{5} - \dots \right) \right] = \frac{1}{3} - \frac{y^2}{5} + \frac{y^4}{7} - \dots \;\Rightarrow\; \lim_{y \to 0} \frac{y - \tan^{-1} y}{y^3} = \lim_{y \to 0} \left(\frac{1}{3} - \frac{y^2}{5} + \frac{y^4}{7} - \dots \right)$

$= \frac{1}{3}$

35. $e^x \sin x = 0 + x + x^2 + x^3 \left(-\frac{1}{3!} + \frac{1}{2!} \right) + x^4 \left(-\frac{1}{3!} + \frac{1}{3!} \right) + x^5 \left(\frac{1}{5!} - \frac{1}{2!}\frac{1}{3!} + \frac{1}{4!} \right) + x^6 \left(\frac{1}{5!} - \frac{1}{3!}\frac{1}{3!} + \frac{1}{5!} \right) + \dots$

$= x + x^2 + \frac{1}{3} x^3 - \frac{1}{30} x^5 - \frac{1}{90} x^6 + \dots$

37. $\sin^2 x = \left(\frac{1 - \cos 2x}{2} \right) = \frac{1}{2} - \frac{1}{2}\cos 2x = \frac{1}{2} - \frac{1}{2}\left(1 - \frac{(2x)^2}{2!} + \frac{(2x)^4}{4!} - \frac{(2x)^6}{6!} + \dots \right) = \frac{2x^2}{2!} - \frac{2^3 x^4}{4!} + \frac{2^5 x^6}{6!} - \dots$

$\Rightarrow\; \frac{d}{dx}\left(\sin^2 x \right) = \frac{d}{dx}\left(\frac{2x^2}{2!} - \frac{2^3 x^4}{4!} + \frac{2^5 x^6}{6!} - \dots \right) = 2x - \frac{(2x)^3}{3!} + \frac{(2x)^5}{5!} - \frac{(2x)^7}{7!} + \dots \;\Rightarrow\; 2\sin x \cos x$

$= 2x - \frac{(2x)^3}{3!} + \frac{(2x)^5}{5!} - \frac{(2x)^7}{7!} + \dots = \sin 2x$, which checks

39. A special case of Taylor's Theorem is $f(b) = f(a) + f'(c)(b - a)$, where c is between a and b $\Rightarrow$ $f(b) - f(a) = f'(c)(b - a)$, the Mean Value Theorem.

41. (a) $f'' \leq 0$, $f'(a) = 0$ and $x = a$ interior to the interval I $\Rightarrow$ $f(x) - f(a) = \frac{f''(c_2)}{2}(x - a)^2 \leq 0$ throughout I
$\Rightarrow$ $f(x) \leq f(a)$ throughout I $\Rightarrow$ f has a local maximum at $x = a$

   (b) similar reasoning gives $f(x) - f(a) = \frac{f''(c_2)}{2}(x - a)^2 \geq 0$ throughout I $\Rightarrow$ $f(x) \geq f(a)$ throughout I $\Rightarrow$ f has a local minimum at $x = a$

43. (a) $f(x) = (1 + x)^k \Rightarrow f'(x) = k(1 + x)^{k-1} \Rightarrow f''(x) = k(k - 1)(1 + x)^{k-2}$; $f(0) = 1$, $f'(0) = k$, and $f''(0) = k(k - 1)$
$\Rightarrow Q(x) = 1 + kx + \frac{k(k - 1)}{2}x^2$

   (b) $|R_2(x)| = \left|\frac{3 \cdot 2 \cdot 1}{3!}x^3\right| < \frac{1}{100} \Rightarrow |x^3| < \frac{1}{100} \Rightarrow 0 < x < \frac{1}{100^{1/3}}$ or $0 < x < .21544$

45. If $f(x) = \sum_{n=0}^{\infty} a_n x^n$, then $f^{(k)}(x) = \sum_{n=k}^{\infty} n(n - 1)(n - 2) \cdots (n - k + 1) a_n x^{n-k}$ and $f^{(k)}(0) = k! \, a_k$

$\Rightarrow a_k = \frac{f^{(k)}(0)}{k!}$ for k a nonnegative integer. Therefore, the coefficients of $f(x)$ are identical with the corresponding coefficients in the Maclaurin series of $f(x)$ and the statement follows.

47. (a) Suppose $f(x)$ is a continuous periodic function with period p. Let x_0 be an arbitrary real number. Then f assumes a minimum m_1 and a maximum m_2 in the interval $[x_0, x_0 + p]$; i.e., $m_1 \leq f(x) \leq m_2$ for all x in $[x_0, x_0 + p]$. Since f is periodic it has exactly the same values on all other intervals $[x_0 + p, x_0 + 2p]$, $[x_0 + 2p, x_0 + 3p]$, ... , and $[x_0 - p, x_0]$, $[x_0 - 2p, x_0 - p]$, ... , and so forth. That is, for all real numbers $-\infty < x < \infty$ we have $m_1 \leq f(x) \leq m_2$. Now choose $M = \max\{|m_1|, |m_2|\}$. Then $-M \leq -|m_1| \leq m_1 \leq f(x) \leq m_2 \leq |m_2| \leq M \Rightarrow |f(x)| \leq M$ for all x.

   (b) The dominate term in the nth order Taylor polynomial generated by cos x about $x = a$ is $\frac{\sin(a)}{n!}(x - a)^n$ or $\frac{\cos(a)}{n!}(x - a)^n$. In both cases, as $|x|$ increases the absolute value of these dominate terms tends to ∞, causing the graph of $P_n(x)$ to move away from cos x.

49. (a) $e^{-i\pi} = \cos(-\pi) + i\sin(-\pi) = -1 + i(0) = -1$

   (b) $e^{i\pi/4} = \cos\left(\frac{\pi}{4}\right) + i\sin\left(\frac{\pi}{4}\right) = \frac{1}{\sqrt{2}} + \frac{i}{\sqrt{2}} = \left(\frac{1}{\sqrt{2}}\right)(1 + i)$

   (c) $e^{-i\pi/2} = \cos\left(-\frac{\pi}{2}\right) + i\sin\left(-\frac{\pi}{2}\right) = 0 + i(-1) = -i$

51. $e^x = 1 + x + \frac{x^2}{2!} + \frac{x^3}{3!} + \frac{x^4}{4!} + \ldots \Rightarrow e^{i\theta} = 1 + i\theta + \frac{(i\theta)^2}{2!} + \frac{(i\theta)^3}{3!} + \frac{(i\theta)^4}{4!} + \ldots$ and
$e^{-i\theta} = 1 - i\theta + \frac{(-i\theta)^2}{2!} + \frac{(-i\theta)^3}{3!} + \frac{(-i\theta)^4}{4!} + \ldots = 1 - i\theta + \frac{(i\theta)^2}{2!} - \frac{(i\theta)^3}{3!} + \frac{(i\theta)^4}{4!} - \ldots$

$\Rightarrow \frac{e^{i\theta} + e^{-i\theta}}{2} = \frac{\left(1 + i\theta + \frac{(i\theta)^2}{2!} + \frac{(i\theta)^3}{3!} + \frac{(i\theta)^4}{4!} + \ldots\right) + \left(1 - i\theta + \frac{(i\theta)^2}{2!} - \frac{(i\theta)^3}{3!} + \frac{(i\theta)^4}{4!} - \ldots\right)}{2}$

$= 1 - \frac{\theta^2}{2!} + \frac{\theta^4}{4!} - \frac{\theta^6}{6!} + \ldots = \cos\theta$;

$\frac{e^{i\theta} - e^{-i\theta}}{2i} = \frac{\left(1 + i\theta + \frac{(i\theta)^2}{2!} + \frac{(i\theta)^3}{3!} + \frac{(i\theta)^4}{4!} + \ldots\right) - \left(1 - i\theta + \frac{(i\theta)^2}{2!} - \frac{(i\theta)^3}{3!} + \frac{(i\theta)^4}{4!} - \ldots\right)}{2i}$

$= \theta - \frac{\theta^3}{3!} + \frac{\theta^5}{5!} - \frac{\theta^7}{7!} + \ldots = \sin\theta$

53. $e^x \sin x = \left(1 + x + \frac{x^2}{2!} + \frac{x^3}{3!} + \frac{x^4}{4!} + \ldots\right)\left(x - \frac{x^3}{3!} + \frac{x^5}{5!} - \frac{x^7}{7!} + \ldots\right)$

$= (1)x + (1)x^2 + \left(-\frac{1}{6} + \frac{1}{2}\right)x^3 + \left(-\frac{1}{6} + \frac{1}{6}\right)x^4 + \left(\frac{1}{120} - \frac{1}{12} + \frac{1}{24}\right)x^5 + \ldots = x + x^2 + \frac{1}{3}x^3 - \frac{1}{30}x^5 + \ldots$;

$e^x \cdot e^{ix} = e^{(1+i)x} = e^x(\cos x + i\sin x) = e^x \cos x + i(e^x \sin x) \Rightarrow e^x \sin x$ is the series of the imaginary part of $e^{(1+i)x}$ which we calculate next; $e^{(1+i)x} = \sum_{n=0}^{\infty} \frac{(x+ix)^n}{n!} = 1 + (x + ix) + \frac{(x+ix)^2}{2!} + \frac{(x+ix)^3}{3!} + \frac{(x+ix)^4}{4!} + \ldots$

$= 1 + x + ix + \frac{1}{2!}(2ix^2) + \frac{1}{3!}(2ix^3 - 2x^3) + \frac{1}{4!}(-4x^4) + \frac{1}{5!}(-4x^5 - 4ix^5) + \frac{1}{6!}(-8ix^6) + \ldots \Rightarrow$ the imaginary part of $e^{(1+i)x}$ is $x + \frac{2}{2!}x^2 + \frac{2}{3!}x^3 - \frac{4}{5!}x^5 - \frac{8}{6!}x^6 + \ldots = x + x^2 + \frac{1}{3}x^3 - \frac{1}{30}x^5 - \frac{1}{90}x^6 + \ldots$ in agreement with our product calculation. The series for $e^x \sin x$ converges for all values of x.

55. (a) $e^{i\theta_1}e^{i\theta_2} = (\cos\theta_1 + i\sin\theta_1)(\cos\theta_2 + i\sin\theta_2) = (\cos\theta_1\cos\theta_2 - \sin\theta_1\sin\theta_2) + i(\sin\theta_1\cos\theta_2 + \sin\theta_2\cos\theta_1)$

$= \cos(\theta_1 + \theta_2) + i\sin(\theta_1 + \theta_2) = e^{i(\theta_1 + \theta_2)}$

(b) $e^{-i\theta} = \cos(-\theta) + i\sin(-\theta) = \cos\theta - i\sin\theta = (\cos\theta - i\sin\theta)\left(\frac{\cos\theta + i\sin\theta}{\cos\theta + i\sin\theta}\right) = \frac{1}{\cos\theta + i\sin\theta} = \frac{1}{e^{i\theta}}$

8.10 THE BINOMIAL SERIES

1. $(1+x)^{1/2} = 1 + \frac{1}{2}x + \frac{\left(\frac{1}{2}\right)\left(-\frac{1}{2}\right)x^2}{2!} + \frac{\left(\frac{1}{2}\right)\left(-\frac{1}{2}\right)\left(-\frac{3}{2}\right)x^3}{3!} + \ldots = 1 + \frac{1}{2}x - \frac{1}{8}x^2 + \frac{1}{16}x^3 - \ldots$

3. $(1-x)^{-1/2} = 1 - \frac{1}{2}(-x) + \frac{\left(-\frac{1}{2}\right)\left(-\frac{3}{2}\right)(-x)^2}{2!} + \frac{\left(-\frac{1}{2}\right)\left(-\frac{3}{2}\right)\left(-\frac{5}{2}\right)(-x)^3}{3!} + \ldots = 1 + \frac{1}{2}x + \frac{3}{8}x^2 + \frac{5}{16}x^3 + \ldots$

5. $\left(1 + \frac{x}{2}\right)^{-2} = 1 - 2\left(\frac{x}{2}\right) + \frac{(-2)(-3)\left(\frac{x}{2}\right)^2}{2!} + \frac{(-2)(-3)(-4)\left(\frac{x}{2}\right)^3}{3!} + \ldots = 1 - x + \frac{3}{4}x^2 - \frac{1}{2}x^3$

7. $(1+x^3)^{-1/2} = 1 - \frac{1}{2}x^3 + \frac{\left(-\frac{1}{2}\right)\left(-\frac{3}{2}\right)(x^3)^2}{2!} + \frac{\left(-\frac{1}{2}\right)\left(-\frac{3}{2}\right)\left(-\frac{5}{2}\right)(x^3)^3}{3!} + \ldots = 1 - \frac{1}{2}x^3 + \frac{3}{8}x^6 - \frac{5}{16}x^9 + \ldots$

9. $\left(1 + \frac{1}{x}\right)^{1/2} = 1 + \frac{1}{2}\left(\frac{1}{x}\right) + \frac{\left(\frac{1}{2}\right)\left(-\frac{1}{2}\right)\left(\frac{1}{x}\right)^2}{2!} + \frac{\left(\frac{1}{2}\right)\left(-\frac{1}{2}\right)\left(-\frac{3}{2}\right)\left(\frac{1}{x}\right)^3}{3!} + \ldots = 1 + \frac{1}{2x} - \frac{1}{8x^2} + \frac{1}{16x^3} + \ldots$

11. $(1+x)^4 = 1 + 4x + \frac{(4)(3)x^2}{2!} + \frac{(4)(3)(2)x^3}{3!} + \frac{(4)(3)(2)x^4}{4!} = 1 + 4x + 6x^2 + 4x^3 + x^4$

13. $(1-2x)^3 = 1 + 3(-2x) + \frac{(3)(2)(-2x)^2}{2!} + \frac{(3)(2)(1)(-2x)^3}{3!} = 1 - 6x + 12x^2 - 8x^3$

15. (a) $(1-x^2)^{-1/2} \approx 1 + \frac{x^2}{2} + \frac{3x^4}{8} + \frac{5x^6}{16} \Rightarrow \sin^{-1}x \approx x + \frac{x^3}{6} + \frac{3x^5}{40} + \frac{5x^7}{112}$; Using the Ratio Test:

$\lim\limits_{n \to \infty}\left|\frac{1\cdot3\cdot5\cdots(2n-1)(2n+1)x^{2n+3}}{2\cdot4\cdot6\cdots(2n)(2n+2)(2n+3)} \cdot \frac{2\cdot4\cdot6\cdots(2n)(2n+1)}{1\cdot3\cdot5\cdots(2n-1)x^{2n+1}}\right| < 1 \Rightarrow x^2 \lim\limits_{n \to \infty}\left|\frac{(2n+1)(2n+1)}{(2n+2)(2n+3)}\right| < 1$

$\Rightarrow |x| < 1 \Rightarrow$ the radius of convergence is 1. See Exercise 19.

(b) $\frac{d}{dx}(\cos^{-1}x) = -(1-x^2)^{-1/2} \Rightarrow \cos^{-1}x = \frac{\pi}{2} - \sin^{-1}x \approx \frac{\pi}{2} - \left(x + \frac{x^3}{6} + \frac{3x^5}{40} + \frac{5x^7}{112}\right) \approx \frac{\pi}{2} - x - \frac{x^3}{6} - \frac{3x^5}{40} - \frac{5x^7}{112}$

17. $\frac{-1}{1+x} = -\frac{1}{1-(-x)} = -1 + x - x^2 + x^3 - \ldots \Rightarrow \frac{d}{dx}\left(\frac{-1}{1+x}\right) = \frac{1}{1+x^2} = \frac{d}{dx}(-1 + x - x^2 + x^3 - \ldots)$

$= 1 - 2x + 3x^2 - 4x^3 + \ldots$

19. $(1-x^2)^{-1/2} = (1+(-x^2))^{-1/2} = (1)^{-1/2} + \left(-\frac{1}{2}\right)(1)^{-3/2}(-x^2) + \frac{\left(-\frac{1}{2}\right)\left(-\frac{3}{2}\right)(1)^{-5/2}(-x^2)^2}{2!}$

$+ \frac{\left(-\frac{1}{2}\right)\left(-\frac{3}{2}\right)\left(-\frac{5}{2}\right)(1)^{-7/2}(-x^2)^3}{3!} + \ldots = 1 + \frac{x^2}{2} + \frac{1\cdot3x^4}{2^2\cdot2!} + \frac{1\cdot3\cdot5x^6}{2^3\cdot3!} + \ldots = 1 + \sum\limits_{n=1}^{\infty}\frac{1\cdot3\cdot5\cdots(2n-1)x^{2n}}{2^n\cdot n!}$

$\Rightarrow \sin^{-1}x = \int_0^x (1-t^2)^{-1/2}\,dt = \int_0^x\left(1 + \sum\limits_{n=1}^{\infty}\frac{1\cdot3\cdot5\cdots(2n-1)x^{2n}}{2^n\cdot n!}\right)dt = x + \sum\limits_{n=1}^{\infty}\frac{1\cdot3\cdot5\cdots(2n-1)x^{2n+1}}{2\cdot4\cdots(2n)(2n+1)}$,

where $|x| < 1$

CHAPTER 8 PRACTICE EXERCISES

1. converges to 1, since $\lim\limits_{n \to \infty} a_n = \lim\limits_{n \to \infty} \left(1 + \frac{(-1)^n}{n}\right) = 1$

3. converges to -1, since $\lim\limits_{n \to \infty} a_n = \lim\limits_{n \to \infty} \left(\frac{1 - 2^n}{2^n}\right) = \lim\limits_{n \to \infty} \left(\frac{1}{2^n} - 1\right) = -1$

5. diverges, since $\left\{\sin \frac{n\pi}{2}\right\} = \{0, 1, 0, -1, 0, 1, \dots\}$

7. converges to 0, since $\lim\limits_{n \to \infty} a_n = \lim\limits_{n \to \infty} \frac{\ln n^2}{n} = 2 \lim\limits_{n \to \infty} \frac{\left(\frac{1}{n}\right)}{1} = 0$

9. converges to 1, since $\lim\limits_{n \to \infty} a_n = \lim\limits_{n \to \infty} \left(\frac{n + \ln n}{n}\right) = \lim\limits_{n \to \infty} \frac{1 + \left(\frac{1}{n}\right)}{1} = 1$

11. converges to e^{-5}, since $\lim\limits_{n \to \infty} a_n = \lim\limits_{n \to \infty} \left(\frac{n - 5}{n}\right)^n = \lim\limits_{n \to \infty} \left(1 + \frac{(-5)}{n}\right)^n = e^{-5}$ by Theorem 5

13. converges to 3, since $\lim\limits_{n \to \infty} a_n = \lim\limits_{n \to \infty} \left(\frac{3^n}{n}\right)^{1/n} = \lim\limits_{n \to \infty} \frac{3}{n^{1/n}} = \frac{3}{1} = 3$ by Theorem 5

15. converges to $\ln 2$, since $\lim\limits_{n \to \infty} a_n = \lim\limits_{n \to \infty} n\left(2^{1/n} - 1\right) = \lim\limits_{n \to \infty} \frac{2^{1/n} - 1}{\left(\frac{1}{n}\right)} = \lim\limits_{n \to \infty} \frac{\left[\frac{\left(-2^{1/n} \ln 2\right)}{n^2}\right]}{\left(\frac{-1}{n^2}\right)} = \lim\limits_{n \to \infty} 2^{1/n} \ln 2$
$= 2^0 \cdot \ln 2 = \ln 2$

17. diverges, since $\lim\limits_{n \to \infty} a_n = \lim\limits_{n \to \infty} \frac{(n + 1)!}{n!} = \lim\limits_{n \to \infty} (n + 1) = \infty$

19. $\frac{1}{(2n - 3)(2n - 1)} = \frac{\left(\frac{1}{2}\right)}{2n - 3} - \frac{\left(\frac{1}{2}\right)}{2n - 1} \Rightarrow s_n = \left[\frac{\left(\frac{1}{2}\right)}{3} - \frac{\left(\frac{1}{2}\right)}{5}\right] + \left[\frac{\left(\frac{1}{2}\right)}{5} - \frac{\left(\frac{1}{2}\right)}{7}\right] + \dots + \left[\frac{\left(\frac{1}{2}\right)}{2n - 3} - \frac{\left(\frac{1}{2}\right)}{2n - 1}\right] = \frac{\left(\frac{1}{2}\right)}{3} - \frac{\left(\frac{1}{2}\right)}{2n - 1}$

$\Rightarrow \lim\limits_{n \to \infty} s_n = \lim\limits_{n \to \infty} \left[\frac{1}{6} - \frac{\left(\frac{1}{2}\right)}{2n - 1}\right] = \frac{1}{6}$

21. $\frac{9}{(3n - 1)(3n + 2)} = \frac{3}{3n - 1} - \frac{3}{3n + 2} \Rightarrow s_n = \left(\frac{3}{2} - \frac{3}{5}\right) + \left(\frac{3}{5} - \frac{3}{8}\right) + \left(\frac{3}{8} - \frac{3}{11}\right) + \dots + \left(\frac{3}{3n - 1} - \frac{3}{3n + 2}\right)$
$= \frac{3}{2} - \frac{3}{3n + 2} \Rightarrow \lim\limits_{n \to \infty} s_n = \lim\limits_{n \to \infty} \left(\frac{3}{2} - \frac{3}{3n + 2}\right) = \frac{3}{2}$

23. $\sum\limits_{n=0}^{\infty} e^{-n} = \sum\limits_{n=0}^{\infty} \frac{1}{e^n}$, a convergent geometric series with $r = \frac{1}{e}$ and $a = 1 \Rightarrow$ the sum is $\frac{1}{1 - \left(\frac{1}{e}\right)} = \frac{e}{e - 1}$

25. diverges, a p-series with $p = \frac{1}{2}$

27. Since $f(x) = \frac{1}{x^{1/2}} \Rightarrow f'(x) = -\frac{1}{2x^{3/2}} < 0 \Rightarrow f(x)$ is decreasing $\Rightarrow a_{n+1} < a_n$, and $\lim\limits_{n \to \infty} a_n = \lim\limits_{n \to \infty} \frac{1}{\sqrt{n}} = 0$, the

series $\sum\limits_{n=1}^{\infty} \frac{(-1)^n}{\sqrt{n}}$ converges by the Alternating Series Test. Since $\sum\limits_{n=1}^{\infty} \frac{1}{\sqrt{n}}$ diverges, the given series converges conditionally.

29. The given series does not converge absolutely by the Direct Comparison Test since $\frac{1}{\ln(n + 1)} > \frac{1}{n + 1}$, which is

the nth term of a divergent series. Since $f(x) = \frac{1}{\ln(x + 1)} \Rightarrow f'(x) = -\frac{1}{(\ln(x + 1))^2(x + 1)} < 0 \Rightarrow f(x)$ is decreasing

$\Rightarrow a_{n+1} < a_n$, and $\lim\limits_{n \to \infty} a_n = \lim\limits_{n \to \infty} \frac{1}{\ln(n + 1)} = 0$, the given series converges conditionally by the Alternating

Series Test.

31. converges absolutely by the Direct Comparison Test since $\frac{\ln n}{n^3} < \frac{n}{n^3} = \frac{1}{n^2}$, the nth term of a convergent p-series

33. $\lim\limits_{n \to \infty} \frac{\left(\frac{1}{n\sqrt{n^2+1}}\right)}{\left(\frac{1}{n^2}\right)} = \sqrt{\lim\limits_{n \to \infty} \frac{n^2}{n^2+1}} = \sqrt{1} = 1 \Rightarrow$ converges absolutely by the Limit Comparison Test

35. converges absolutely by the Ratio Test since $\lim\limits_{n \to \infty} \left[\frac{n+2}{(n+1)!} \cdot \frac{n!}{n+1}\right] = \lim\limits_{n \to \infty} \frac{n+2}{(n+1)^2} = 0 < 1$

37. converges absolutely by the Ratio Test since $\lim\limits_{n \to \infty} \left[\frac{3^{n+1}}{(n+1)!} \cdot \frac{n!}{3^n}\right] = \lim\limits_{n \to \infty} \frac{3}{n+1} = 0 < 1$

39. converges absolutely by the Limit Comparison Test since $\lim\limits_{n \to \infty} \frac{\left(\frac{1}{n^{3/2}}\right)}{\left(\frac{1}{\sqrt{n(n+1)(n+2)}}\right)} = \sqrt{\lim\limits_{n \to \infty} \frac{n(n+1)(n+2)}{n^3}} = 1$

41. $\lim\limits_{n \to \infty} \left|\frac{u_{n+1}}{u_n}\right| < 1 \Rightarrow \lim\limits_{n \to \infty} \left|\frac{(x+4)^{n+1}}{(n+1)3^{n+1}} \cdot \frac{n3^n}{(x+4)^n}\right| < 1 \Rightarrow \frac{|x+4|}{3} \lim\limits_{n \to \infty} \left(\frac{n}{n+1}\right) < 1 \Rightarrow \frac{|x+4|}{3} < 1$

$\Rightarrow |x+4| < 3 \Rightarrow -3 < x+4 < 3 \Rightarrow -7 < x < -1$; at $x = -7$ we have $\sum\limits_{n=1}^{\infty} \frac{(-1)^n 3^n}{n3^n} = \sum\limits_{n=1}^{\infty} \frac{(-1)^n}{n}$, the alternating

harmonic series, which converges conditionally; at $x = -1$ we have $\sum\limits_{n=1}^{\infty} \frac{3^n}{n3^n} = \sum\limits_{n=1}^{\infty} \frac{1}{n}$, the divergent harmonic series

(a) the radius is 3; the interval of convergence is $-7 \le x < -1$

(b) the interval of absolute convergence is $-7 < x < -1$

(c) the series converges conditionally at $x = -7$

43. $\lim\limits_{n \to \infty} \left|\frac{u_{n+1}}{u_n}\right| < 1 \Rightarrow \lim\limits_{n \to \infty} \left|\frac{(3x-1)^{n+1}}{(n+1)^2} \cdot \frac{n^2}{(3x-1)^n}\right| < 1 \Rightarrow |3x-1| \lim\limits_{n \to \infty} \frac{n^2}{(n+1)^2} < 1 \Rightarrow |3x-1| < 1$

$\Rightarrow -1 < 3x - 1 < 1 \Rightarrow 0 < 3x < 2 \Rightarrow 0 < x < \frac{2}{3}$; at $x = 0$ we have $\sum\limits_{n=1}^{\infty} \frac{(-1)^{n-1}(-1)^n}{n^2} = \sum\limits_{n=1}^{\infty} \frac{(-1)^{2n-1}}{n^2}$

$= -\sum\limits_{n=1}^{\infty} \frac{1}{n^2}$, a nonzero constant multiple of a convergent p-series, which is absolutely convergent; at $x = \frac{2}{3}$ we

have $\sum\limits_{n=1}^{\infty} \frac{(-1)^{n-1}(1)^n}{n^2} = \sum\limits_{n=1}^{\infty} \frac{(-1)^{n-1}}{n^2}$, which converges absolutely

(a) the radius is $\frac{1}{3}$; the interval of convergence is $0 \le x \le \frac{2}{3}$

(b) the interval of absolute convergence is $0 \le x \le \frac{2}{3}$

(c) there are no values for which the series converges conditionally

45. $\lim\limits_{n \to \infty} \left|\frac{u_{n+1}}{u_n}\right| < 1 \Rightarrow \lim\limits_{n \to \infty} \left|\frac{x^{n+1}}{(n+1)^{n+1}} \cdot \frac{n^n}{x^n}\right| < 1 \Rightarrow |x| \lim\limits_{n \to \infty} \left|\left(\frac{n}{n+1}\right)^n \left(\frac{1}{n+1}\right)\right| < 1 \Rightarrow \frac{|x|}{e} \lim\limits_{n \to \infty} \left(\frac{1}{n+1}\right) < 1$

$\Rightarrow \frac{|x|}{e} \cdot 0 < 1$, which holds for all x

(a) the radius is ∞; the series converges for all x

(b) the series converges absolutely for all x

(c) there are no values for which the series converges conditionally

47. $\lim\limits_{n \to \infty} \left|\frac{u_{n+1}}{u_n}\right| < 1 \Rightarrow \lim\limits_{n \to \infty} \left|\frac{(n+2)x^{2n+1}}{3^{n+1}} \cdot \frac{3^n}{(n+1)x^{2n-1}}\right| < 1 \Rightarrow \frac{x^2}{3} \lim\limits_{n \to \infty} \left(\frac{n+2}{n+1}\right) < 1 \Rightarrow -\sqrt{3} < x < \sqrt{3}$;

the series $\sum\limits_{n=1}^{\infty} -\frac{n+1}{\sqrt{3}}$ and $\sum\limits_{n=1}^{\infty} \frac{n+1}{\sqrt{3}}$, obtained with $x = \pm\sqrt{3}$, both diverge

(a) the radius is $\sqrt{3}$; the interval of convergence is $-\sqrt{3} < x < \sqrt{3}$

(b) the interval of absolute convergence is $-\sqrt{3} < x < \sqrt{3}$

(c) there are no values for which the series converges conditionally

49. $\lim\limits_{n \to \infty} \left| \frac{u_{n+1}}{u_n} \right| < 1 \Rightarrow \lim\limits_{n \to \infty} \left| \frac{\text{csch}\,(n+1)x^{n+1}}{\text{csch}\,(n)x^n} \right| < 1 \Rightarrow |x| \lim\limits_{n \to \infty} \left| \frac{\left(\frac{2}{e^{n+1} - e^{-n-1}} \right)}{\left(\frac{2}{e^n - e^{-n}} \right)} \right| < 1$

$\Rightarrow |x| \lim\limits_{n \to \infty} \left| \frac{e^{-1} - e^{-2n-1}}{1 - e^{-2n-2}} \right| < 1 \Rightarrow \frac{|x|}{e} < 1 \Rightarrow -e < x < e$; the series $\sum\limits_{n=1}^{\infty} (\pm e)^n \, \text{csch}\, n$, obtained with $x = \pm e$,

both diverge since $\lim\limits_{n \to \infty} (\pm e)^n \, \text{csch}\, n \neq 0$

(a) the radius is e; the interval of convergence is $-e < x < e$

(b) the interval of absolute convergence is $-e < x < e$

(c) there are no values for which the series converges conditionally

51. The given series has the form $1 - x + x^2 - x^3 + \ldots + (-x)^n + \ldots = \frac{1}{1+x}$, where $x = \frac{1}{4}$; the sum is $\frac{1}{1 + \left(\frac{1}{4} \right)} = \frac{4}{5}$

53. The given series has the form $x - \frac{x^3}{3!} + \frac{x^5}{5!} - \ldots + (-1)^n \frac{x^{2n+1}}{(2n+1)!} + \ldots = \sin x$, where $x = \pi$; the sum is $\sin \pi = 0$

55. The given series has the form $1 + x + \frac{x^2}{2!} + \frac{x^2}{3!} + \ldots + \frac{x^n}{n!} + \ldots = e^x$, where $x = \ln 2$; the sum is $e^{\ln(2)} = 2$

57. Consider $\frac{1}{1-2x}$ as the sum of a convergent geometric series with $a = 1$ and $r = 2x \Rightarrow \frac{1}{1-2x}$

$= 1 + (2x) + (2x)^2 + (2x)^3 + \ldots = \sum\limits_{n=0}^{\infty} (2x)^n = \sum\limits_{n=0}^{\infty} 2^n x^n$ where $|2x| < 1 \Rightarrow |x| < \frac{1}{2}$

59. $\sin x = \sum\limits_{n=0}^{\infty} \frac{(-1)^n x^{2n+1}}{(2n+1)!} \Rightarrow \sin \pi x = \sum\limits_{n=0}^{\infty} \frac{(-1)^n (\pi x)^{2n+1}}{(2n+1)!} = \sum\limits_{n=0}^{\infty} \frac{(-1)^n \pi^{2n+1} x^{2n+1}}{(2n+1)!}$

61. $\cos x = \sum\limits_{n=0}^{\infty} \frac{(-1)^n x^{2n}}{(2n)!} \Rightarrow \cos \left(x^{5/2} \right) = \sum\limits_{n-0}^{\infty} \frac{(-1)^n \left(x^{5/2} \right)^{2n}}{(2n)!} = \sum\limits_{n=0}^{\infty} \frac{(-1)^n x^{5n}}{(2n)!}$

63. $e^x = \sum\limits_{n=0}^{\infty} \frac{x^n}{n!} \Rightarrow e^{(\pi x/2)} = \sum\limits_{n=0}^{\infty} \frac{\left(\frac{\pi x}{2} \right)^n}{n!} = \sum\limits_{n=0}^{\infty} \frac{\pi^n x^n}{2^n n!}$

65. $f(x) = \sqrt{3 + x^2} = (3 + x^2)^{1/2} \Rightarrow f'(x) = x(3 + x^2)^{-1/2} \Rightarrow f''(x) = -x^2(3 + x^2)^{-3/2} + (3 + x^2)^{-1/2}$

$\Rightarrow f'''(x) = 3x^3(3 + x^2)^{-5/2} - 3x(3 + x^2)^{-3/2};\ f(-1) = 2,\ f'(-1) = -\frac{1}{2},\ f''(-1) = -\frac{1}{8} + \frac{1}{2} = \frac{3}{8}$,

$f'''(-1) = -\frac{3}{32} + \frac{3}{8} = \frac{9}{32} \Rightarrow \sqrt{3 + x^2} = 2 - \frac{(x+1)}{2 \cdot 1!} + \frac{3(x+1)^2}{2^3 \cdot 2!} + \frac{9(x+1)^3}{2^5 \cdot 3!} + \ldots$

67. $f(x) = \frac{1}{x+1} = (x+1)^{-1} \Rightarrow f'(x) = -(x+1)^{-2} \Rightarrow f''(x) = 2(x+1)^{-3} \Rightarrow f'''(x) = -6(x+1)^{-4};\ f(3) = \frac{1}{4}$,

$f'(3) = -\frac{1}{4^2},\ f''(3) = \frac{2}{4^3},\ f'''(2) = \frac{-6}{4^4} \Rightarrow \frac{1}{x+1} = \frac{1}{4} - \frac{1}{4^2}(x-3) + \frac{1}{4^3}(x-3)^2 - \frac{1}{4^4}(x-3)^3 + \ldots$

69. $\int_0^{1/2} \exp(-x^3)\, dx = \int_0^{1/2} \left(1 - x^3 + \frac{x^6}{2!} - \frac{x^9}{3!} + \frac{x^{12}}{4!} + \ldots \right) dx = \left[x - \frac{x^4}{4} + \frac{x^7}{7 \cdot 2!} - \frac{x^{10}}{10 \cdot 3!} + \frac{x^{13}}{13 \cdot 4!} - \ldots \right]_0^{1/2}$

$\approx \frac{1}{2} - \frac{1}{2^4 \cdot 4} + \frac{1}{2^7 \cdot 7 \cdot 2!} - \frac{1}{2^{10} \cdot 10 \cdot 3!} + \frac{1}{2^{13} \cdot 13 \cdot 4!} - \frac{1}{2^{16} \cdot 16 \cdot 5!} \approx 0.484917143$

71. $\int_1^{1/2} \frac{\tan^{-1} x}{x}\, dx = \int_1^{1/2} \left(1 - \frac{x^2}{3} + \frac{x^4}{5} - \frac{x^6}{7} + \frac{x^8}{9} - \frac{x^{10}}{11} + \ldots \right) dx = \left[x - \frac{x^3}{9} + \frac{x^5}{25} - \frac{x^7}{49} + \frac{x^9}{81} - \frac{x^{11}}{121} + \ldots \right]_0^{1/2}$

$\approx \frac{1}{2} - \frac{1}{9 \cdot 2^3} + \frac{1}{5^2 \cdot 2^5} - \frac{1}{7^2 \cdot 2^7} + \frac{1}{9^2 \cdot 2^9} - \frac{1}{11^2 \cdot 2^{11}} + \frac{1}{13^2 \cdot 2^{13}} - \frac{1}{15^2 \cdot 2^{15}} + \frac{1}{17^2 \cdot 2^{17}} - \frac{1}{19^2 \cdot 2^{19}} + \frac{1}{21^2 \cdot 2^{21}}$

≈ 0.4872223583

73. $\lim\limits_{x \to 0} \frac{7 \sin x}{e^{2x} - 1} = \lim\limits_{x \to 0} \frac{7 \left(x - \frac{x^3}{3!} + \frac{x^5}{5!} - \ldots \right)}{\left(2x + \frac{2^2 x^2}{2!} + \frac{2^3 x^3}{3!} + \ldots \right)} = \lim\limits_{x \to 0} \frac{7 \left(1 - \frac{x^2}{3!} + \frac{x^4}{5!} - \ldots \right)}{\left(2 + \frac{2^2 x}{2!} + \frac{2^3 x^2}{3!} + \ldots \right)} = \frac{7}{2}$

75. $\lim\limits_{t \to 0} \left(\frac{1}{2 - 2\cos t} - \frac{1}{t^2}\right) = \lim\limits_{t \to 0} \frac{t^2 - 2 + 2\cos t}{2t^2(1 - \cos t)} = \lim\limits_{t \to 0} \frac{t^2 - 2 + 2\left(1 - \frac{t^2}{2} + \frac{t^4}{4!} - \cdots\right)}{2t^2\left(1 - 1 + \frac{t^2}{2} - \frac{t^4}{4!} + \cdots\right)} = \lim\limits_{t \to 0} \frac{2\left(\frac{t^4}{4!} - \frac{t^6}{6!} + \cdots\right)}{\left(t^4 - \frac{2t^6}{4!} + \cdots\right)}$

$= \lim\limits_{t \to 0} \frac{2\left(\frac{1}{4!} - \frac{t^2}{6!} + \cdots\right)}{\left(1 - \frac{2t^2}{4!} + \cdots\right)} = \frac{1}{12}$

77. $\lim\limits_{z \to 0} \frac{1 - \cos^2 z}{\ln(1 - z) + \sin z} = \lim\limits_{z \to 0} \frac{1 - \left(1 - z^2 + \frac{z^4}{3} - \cdots\right)}{\left(-z - \frac{z^2}{2} - \frac{z^3}{3} - \cdots\right) + \left(z - \frac{z^3}{3!} + \frac{z^5}{5!} - \cdots\right)} = \lim\limits_{z \to 0} \frac{\left(z^2 - \frac{z^4}{3} + \cdots\right)}{\left(-\frac{z^2}{2} - \frac{2z^3}{3} - \frac{z^4}{4} - \cdots\right)}$

$= \lim\limits_{z \to 0} \frac{\left(1 - \frac{z^2}{3} + \cdots\right)}{\left(-\frac{1}{2} - \frac{2z}{3} - \frac{z^2}{4} - \cdots\right)} = -2$

79. $\lim\limits_{x \to 0} \left(\frac{\sin 3x}{x^3} + \frac{r}{x^2} + s\right) = \lim\limits_{x \to 0} \left[\frac{\left(3x - \frac{(3x)^3}{6} + \frac{(3x)^5}{120} - \cdots\right)}{x^3} + \frac{r}{x^2} + s\right] = \lim\limits_{x \to 0} \left(\frac{3}{x^2} - \frac{9}{2} + \frac{81x^2}{40} + \cdots + \frac{r}{x^2} + s\right) = 0$

$\Rightarrow \frac{r}{x^2} + \frac{3}{x^2} = 0$ and $s - \frac{9}{2} = 0 \Rightarrow r = -3$ and $s = \frac{9}{2}$

81. $\lim\limits_{n \to \infty} \left|\frac{2 \cdot 5 \cdot 8 \cdots (3n - 1)(3n + 2)x^{n+1}}{2 \cdot 4 \cdot 6 \cdots (2n)(2n + 2)} \cdot \frac{2 \cdot 4 \cdot 6 \cdots (2n)}{2 \cdot 5 \cdot 8 \cdots (3n - 1)x^n}\right| < 1 \Rightarrow |x| \lim\limits_{n \to \infty} \left|\frac{3n + 2}{2n + 2}\right| < 1 \Rightarrow |x| < \frac{2}{3}$

$\Rightarrow$ the radius of convergence is $\frac{2}{3}$

83. $\sum\limits_{k=2}^{n} \ln\left(1 - \frac{1}{k^2}\right) = \sum\limits_{k=2}^{n} \left[\ln\left(1 + \frac{1}{k}\right) + \ln\left(1 - \frac{1}{k}\right)\right] = \sum\limits_{k=2}^{n} \left[\ln(k + 1) - \ln k + \ln(k - 1) - \ln k\right]$

$= [\ln 3 - \ln 2 + \ln 1 - \ln 2] + [\ln 4 - \ln 3 + \ln 2 - \ln 3] + [\ln 5 - \ln 4 + \ln 3 - \ln 4] + [\ln 6 - \ln 5 + \ln 4 - \ln 5]$

$+ \cdots + [\ln(n + 1) - \ln n + \ln(n - 1) - \ln n] = [\ln 1 - \ln 2] + [\ln(n + 1) - \ln n]$ after cancellation

$\Rightarrow \sum\limits_{k=2}^{n} \ln\left(1 - \frac{1}{k^2}\right) = \ln\left(\frac{n+1}{2n}\right) \Rightarrow \sum\limits_{k=2}^{\infty} \ln\left(1 - \frac{1}{k^2}\right) = \lim\limits_{n \to \infty} \ln\left(\frac{n+1}{2n}\right) = \ln\frac{1}{2}$ is the sum

85. (a) $\lim\limits_{n \to \infty} \left|\frac{1 \cdot 4 \cdot 7 \cdots (3n - 2)(3n + 1)x^{3n+3}}{(3n + 3)!} \cdot \frac{(3n)!}{1 \cdot 4 \cdot 7 \cdots (3n - 2)x^{3n}}\right| < 1 \Rightarrow |x^3| \lim\limits_{n \to \infty} \frac{(3n + 1)}{(3n + 1)(3n + 2)(3n + 3)}$

$= |x^3| \cdot 0 < 1 \Rightarrow$ the radius of convergence is ∞

(b) $y = 1 + \sum\limits_{n=1}^{\infty} \frac{1 \cdot 4 \cdot 7 \cdots (3n - 2)}{(3n)!} x^{3n} \Rightarrow \frac{dy}{dx} = \sum\limits_{n=1}^{\infty} \frac{1 \cdot 4 \cdot 7 \cdots (3n - 2)}{(3n - 1)!} x^{3n-1}$

$\Rightarrow \frac{d^2y}{dx^2} = \sum\limits_{n=1}^{\infty} \frac{1 \cdot 4 \cdot 7 \cdots (3n - 2)}{(3n - 2)!} x^{3n-2} = x + \sum\limits_{n=2}^{\infty} \frac{1 \cdot 4 \cdot 7 \cdots (3n-5)}{(3n-3)!} x^{3n-2}$

$= x\left(1 + \sum\limits_{n=1}^{\infty} \frac{1 \cdot 4 \cdot 7 \cdots (3n-2)}{(3n)!} x^{3n}\right) = xy + 0 \Rightarrow a = 1$ and $b = 0$

87. Yes, the series $\sum\limits_{n=1}^{\infty} a_n b_n$ converges as we now show. Since $\sum\limits_{n=1}^{\infty} a_n$ converges it follows that $a_n \to 0 \Rightarrow a_n < 1$

for $n >$ some index $N \Rightarrow a_n b_n < b_n$ for $n > N \Rightarrow \sum\limits_{n=1}^{\infty} a_n b_n$ converges by the Direct Comparison Test with $\sum\limits_{n=1}^{\infty} b_n$

89. $\sum\limits_{n=1}^{\infty} (x_{n+1} - x_n) = \lim\limits_{n \to \infty} \sum\limits_{k=1}^{\infty}(x_{k+1} - x_k) = \lim\limits_{n \to \infty} (x_{n+1} - x_1) = \lim\limits_{n \to \infty} (x_{n+1}) - x_1 \Rightarrow$ both the series and

sequence must either converge or diverge.

91. $\sum\limits_{n=1}^{\infty} \frac{a_n}{n} = a_1 + \frac{a_2}{2} + \frac{a_3}{3} + \frac{a_4}{4} + \cdots \geq a_1 + \left(\frac{1}{2}\right)a_2 + \left(\frac{1}{3} + \frac{1}{4}\right)a_4 + \left(\frac{1}{5} + \frac{1}{6} + \frac{1}{7} + \frac{1}{8}\right)a_8$

$+ \left(\frac{1}{9} + \frac{1}{10} + \frac{1}{11} + \cdots + \frac{1}{16}\right)a_{16} + \cdots \geq \frac{1}{2}(a_2 + a_4 + a_8 + a_{16} + \cdots)$ which is a divergent series

CHAPTER 8 ADDITIONAL AND ADVANCED EXERCISES

1. converges since $\frac{1}{(3n-2)^{(2n+1)/2}} < \frac{1}{(3n-2)^{3/2}}$ and $\sum_{n=1}^{\infty} \frac{1}{(3n-2)^{3/2}}$ converges by the Limit Comparison Test:

$$\lim_{n \to \infty} \frac{\left(\frac{1}{n^{3/2}}\right)}{\left(\frac{1}{(3n-2)^{3/2}}\right)} = \lim_{n \to \infty} \left(\frac{3n-2}{n}\right)^{3/2} = 3^{3/2}$$

3. diverges by the nth-Term Test since $\lim_{n \to \infty} a_n = \lim_{n \to \infty} (-1)^n \tanh n = \lim_{b \to \infty} (-1)^n \left(\frac{1-e^{-2n}}{1+e^{-2n}}\right) = \lim_{n \to \infty} (-1)^n$ does not exist

5. converges by the Direct Comparison Test: $a_1 = 1 = \frac{12}{(1)(3)(2)^2}$, $a_2 = \frac{1 \cdot 2}{3 \cdot 4} = \frac{12}{(2)(4)(3)^2}$, $a_3 = \left(\frac{2 \cdot 3}{4 \cdot 5}\right)\left(\frac{1 \cdot 2}{3 \cdot 4}\right)$

$= \frac{12}{(3)(5)(4)^2}$, $a_4 = \left(\frac{3 \cdot 4}{5 \cdot 6}\right)\left(\frac{2 \cdot 3}{4 \cdot 5}\right)\left(\frac{1 \cdot 2}{3 \cdot 4}\right) = \frac{12}{(4)(6)(5)^2}$, $\dots$ $\Rightarrow 1 + \sum_{n=1}^{\infty} \frac{12}{(n+1)(n+3)(n+2)^2}$ represents the

given series and $\frac{12}{(n+1)(n+3)(n+2)^2} < \frac{12}{n^4}$, which is the nth-term of a convergent p-series

7. diverges by the nth-Term Test since if $a_n \to L$ as $n \to \infty$, then $L = \frac{1}{1+L} \Rightarrow L^2 + L - 1 = 0 \Rightarrow L = \frac{-1 \pm \sqrt{5}}{2} \neq 0$

9. $f(x) = \cos x$ with $a = \frac{\pi}{3} \Rightarrow f\left(\frac{\pi}{3}\right) = 0.5, f'\left(\frac{\pi}{3}\right) = -\frac{\sqrt{3}}{2}, f''\left(\frac{\pi}{3}\right) = -0.5, f'''\left(\frac{\pi}{3}\right) = \frac{\sqrt{3}}{2}, f^{(4)}\left(\frac{\pi}{3}\right) = 0.5;$

$\cos x = \frac{1}{2} - \frac{\sqrt{3}}{2}\left(x - \frac{\pi}{3}\right) - \frac{1}{4}\left(x - \frac{\pi}{3}\right)^2 + \frac{\sqrt{3}}{12}\left(x - \frac{\pi}{3}\right)^3 + \dots$

11. $e^x = 1 + x + \frac{x^2}{2!} + \frac{x^3}{3!} + \dots$ with $a = 0$

13. $f(x) = \cos x$ with $a = 22\pi \Rightarrow f(22\pi) = 1, f'(22\pi) = 0, f''(22\pi) = -1, f'''(22\pi) = 0, f^{(4)}(22\pi) = 1,$

$f^{(5)}(22\pi) = 0, f^{(6)}(22\pi) = -1; \cos x = 1 - \frac{1}{2}(x - 22\pi)^2 + \frac{1}{4!}(x - 22\pi)^4 - \frac{1}{6!}(x - 22\pi)^6 + \dots$

15. Yes, the sequence converges: $c_n = (a^n + b^n)^{1/n} \Rightarrow c_n = b\left(\left(\frac{a}{b}\right)^n + 1\right)^{1/n} \Rightarrow \lim_{n \to \infty} c_n = \ln b + \lim_{n \to \infty} \frac{\ln\left(\left(\frac{a}{b}\right)^n + 1\right)}{n}$

$= \ln b + \lim_{n \to \infty} \frac{\left(\frac{a}{b}\right)^n \ln\left(\frac{a}{b}\right)}{\left(\frac{a}{b}\right)^n + 1} = \ln b + \frac{0 \cdot \ln\left(\frac{a}{b}\right)}{0 + 1} = \ln b$ since $0 < a < b$. Thus, $\lim_{n \to \infty} c_n = e^{\ln b} = b$.

17. $s_n = \sum_{k=0}^{n-1} \int_k^{k+1} \frac{dx}{1+x^2} \Rightarrow s_n = \int_0^1 \frac{dx}{1+x^2} + \int_1^2 \frac{dx}{1+x^2} + \dots + \int_{n-1}^n \frac{dx}{1+x^2} \Rightarrow s_n = \int_0^n \frac{dx}{1+x^2}$

$\Rightarrow \lim_{n \to \infty} s_n = \lim_{n \to \infty} (\tan^{-1} n - \tan^{-1} 0) = \frac{\pi}{2}$

19. (a) No, the limit does not appear to depend on the value of the constant a
    (b) Yes, the limit depends on the value of b

    (c) $s = \left(1 - \frac{\cos\left(\frac{a}{n}\right)}{n}\right)^n \Rightarrow \ln s = \frac{\ln\left(1 - \frac{\cos\left(\frac{a}{n}\right)}{n}\right)}{\left(\frac{1}{n}\right)} \Rightarrow \lim_{n \to \infty} \ln s = \frac{\left(\frac{1}{1 - \frac{\cos\left(\frac{a}{n}\right)}{n}}\right)\left(\frac{-\frac{a}{n}\sin\left(\frac{a}{n}\right) + \cos\left(\frac{a}{n}\right)}{n^2}\right)}{\left(-\frac{1}{n^2}\right)}$

    $= \lim_{n \to \infty} \frac{\frac{a}{n}\sin\left(\frac{a}{n}\right) - \cos\left(\frac{a}{n}\right)}{1 - \frac{\cos\left(\frac{a}{n}\right)}{n}} = \frac{0-1}{1-0} = -1 \Rightarrow \lim_{n \to \infty} s = e^{-1} \approx 0.3678794412;$ similarly,

    $\lim_{n \to \infty} \left(1 - \frac{\cos\left(\frac{a}{n}\right)}{bn}\right)^n = e^{-1/b}$

21. $\lim_{n \to \infty} \left|\frac{u_{n+1}}{u_n}\right| < 1 \Rightarrow \lim_{n \to \infty} \left|\frac{b^{n+1}x^{n+1}}{\ln(n+1)} \cdot \frac{\ln n}{b^n x^n}\right| < 1 \Rightarrow |bx| < 1 \Rightarrow -\frac{1}{b} < x < \frac{1}{b} = 5 \Rightarrow b = \pm\frac{1}{5}$

23. $\lim\limits_{x \to 0} \frac{\sin(ax) - \sin x - x}{x^3} = \lim\limits_{x \to 0} \frac{\left(ax - \frac{a^3x^3}{3!} + \dots\right) - \left(x - \frac{x^3}{3!} + \dots\right) - x}{x^3} = \lim\limits_{x \to 0} \left[\frac{a-2}{x^2} - \frac{a^3}{3!} + \frac{1}{3!} - \left(\frac{a^5}{5!} - \frac{1}{5!}\right)x^2 + \dots\right]$

is finite if $a - 2 = 0 \Rightarrow a = 2$; $\lim\limits_{x \to 0} \frac{\sin 2x - \sin x - x}{x^3} = -\frac{2^3}{3!} + \frac{1}{3!} = -\frac{7}{6}$

25. (a) $\frac{u_n}{u_{n+1}} = \frac{(n+1)^2}{n^2} = 1 + \frac{2}{n} + \frac{1}{n^2} \Rightarrow C = 2 > 1$ and $\sum\limits_{n=1}^{\infty} \frac{1}{n^2}$ converges

   (b) $\frac{u_n}{u_{n+1}} = \frac{n+1}{n} = 1 + \frac{1}{n} + \frac{0}{n^2} \Rightarrow C = 1 \le 1$ and $\sum\limits_{n=1}^{\infty} \frac{1}{n}$ diverges

27. (a) $\sum\limits_{n=1}^{\infty} a_n = L \Rightarrow a_n^2 \le a_n \sum\limits_{n=1}^{\infty} a_n = a_n L \Rightarrow \sum\limits_{n=1}^{\infty} a_n^2$ converges by the Direct Comparison Test

   (b) converges by the Limit Comparison Test: $\lim\limits_{n \to \infty} \frac{\left(\frac{a_n}{1 - a_n}\right)}{a_n} = \lim\limits_{n \to \infty} \frac{1}{1 - a_n} = 1$ since $\sum\limits_{n=1}^{\infty} a_n$ converges and

   therefore $\lim\limits_{x \to \infty} a_n = 0$

29. $(1 - x)^{-1} = 1 + \sum\limits_{n=1}^{\infty} x^n$ where $|x| < 1 \Rightarrow \frac{1}{(1-x)^2} = \frac{d}{dx}(1 - x)^{-1} = \sum\limits_{n=1}^{\infty} nx^{n-1}$ and when $x = \frac{1}{2}$ we have

   $4 = 1 + 2\left(\frac{1}{2}\right) + 3\left(\frac{1}{2}\right)^2 + 4\left(\frac{1}{2}\right)^3 + \dots + n\left(\frac{1}{2}\right)^{n-1} + \dots$

31. (a) $\frac{1}{(1-x)^2} = \frac{d}{dx}\left(\frac{1}{1-x}\right) = \frac{d}{dx}(1 + x + x^2 + x^3 + \dots) = 1 + 2x + 3x^2 + 4x^3 + \dots = \sum\limits_{n=1}^{\infty} nx^{n-1}$

   (b) from part (a) we have $\sum\limits_{n=1}^{\infty} n\left(\frac{5}{6}\right)^{n-1}\left(\frac{1}{6}\right) = \left(\frac{1}{6}\right)\left[\frac{1}{1 - \left(\frac{5}{6}\right)}\right]^2 = 6$

   (c) from part (a) we have $\sum\limits_{n=1}^{\infty} np^{n-1}q = \frac{q}{(1-p)^2} = \frac{q}{q^2} = \frac{1}{q}$

CHAPTER 9 POLAR COORDINATES AND CONICS

9.1 POLAR COORDINATES

1. a, e; b, g; c, h; d, f

3. (a) $\left(2, \frac{\pi}{2} + 2n\pi\right)$ and $\left(-2, \frac{\pi}{2} + (2n+1)\pi\right)$, n an integer
    (b) $(2, 2n\pi)$ and $(-2, (2n+1)\pi)$, n an integer
    (c) $\left(2, \frac{3\pi}{2} + 2n\pi\right)$ and $\left(-2, \frac{3\pi}{2} + (2n+1)\pi\right)$, n an integer
    (d) $(2, (2n+1)\pi)$ and $(-2, 2n\pi)$, n an integer

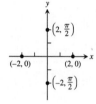

5. (a) $x = r\cos\theta = 3\cos 0 = 3, y = r\sin\theta = 3\sin 0 = 0 \Rightarrow$ Cartesian coordinates are $(3, 0)$
    (b) $x = r\cos\theta = -3\cos 0 = -3, y = r\sin\theta = -3\sin 0 = 0 \Rightarrow$ Cartesian coordinates are $(-3, 0)$
    (c) $x = r\cos\theta = 2\cos \frac{2\pi}{3} = -1, y = r\sin\theta = 2\sin \frac{2\pi}{3} = \sqrt{3} \Rightarrow$ Cartesian coordinates are $\left(-1, \sqrt{3}\right)$
    (d) $x = r\cos\theta = 2\cos \frac{7\pi}{3} = 1, y = r\sin\theta = 2\sin \frac{7\pi}{3} = \sqrt{3} \Rightarrow$ Cartesian coordinates are $\left(1, \sqrt{3}\right)$
    (e) $x = r\cos\theta = -3\cos \pi = 3, y = r\sin\theta = -3\sin \pi = 0 \Rightarrow$ Cartesian coordinates are $(3, 0)$
    (f) $x = r\cos\theta = 2\cos \frac{\pi}{3} = 1, y = r\sin\theta = 2\sin \frac{\pi}{3} = \sqrt{3} \Rightarrow$ Cartesian coordinates are $\left(1, \sqrt{3}\right)$
    (g) $x = r\cos\theta = -3\cos 2\pi = -3, y = r\sin\theta = -3\sin 2\pi = 0 \Rightarrow$ Cartesian coordinates are $(-3, 0)$
    (h) $x = r\cos\theta = -2\cos \left(-\frac{\pi}{3}\right) = -1, y = r\sin\theta = -2\sin \left(-\frac{\pi}{3}\right) = \sqrt{3} \Rightarrow$ Cartesian coordinates are $\left(-1, \sqrt{3}\right)$

7.

9.

11.

13.

15.

17.

19.

21.

23. $r \cos \theta = 2 \Rightarrow x = 2$, vertical line through $(2, 0)$ 25. $r \sin \theta = 0 \Rightarrow y = 0$, the x-axis

27. $r = 4 \csc \theta \Rightarrow r = \frac{4}{\sin \theta} \Rightarrow r \sin \theta = 4 \Rightarrow y = 4$, a horizontal line through $(0, 4)$

29. $r \cos \theta + r \sin \theta = 1 \Rightarrow x + y = 1$, line with slope $m = -1$ and intercept $b = 1$

31. $r^2 = 1 \Rightarrow x^2 + y^2 = 1$, circle with center $C = (0, 0)$ and radius 1

33. $r = \frac{5}{\sin \theta - 2 \cos \theta} \Rightarrow r \sin \theta - 2r \cos \theta = 5 \Rightarrow y - 2x = 5$, line with slope $m = 2$ and intercept $b = 5$

35. $r = \cot \theta \csc \theta = \left(\frac{\cos \theta}{\sin \theta} \right) \left(\frac{1}{\sin \theta} \right) \Rightarrow r \sin^2 \theta = \cos \theta \Rightarrow r^2 \sin^2 \theta = r \cos \theta \Rightarrow y^2 = x$, parabola with vertex $(0, 0)$
    which opens to the right

37. $r = (\csc \theta) e^{r \cos \theta} \Rightarrow r \sin \theta = e^{r \cos \theta} \Rightarrow y = e^x$, graph of the natural exponential function

39. $r^2 + 2r^2 \cos \theta \sin \theta = 1 \Rightarrow x^2 + y^2 + 2xy = 1 \Rightarrow x^2 + 2xy + y^2 = 1 \Rightarrow (x + y)^2 = 1 \Rightarrow x + y = \pm 1$, two parallel
    straight lines of slope -1 and y-intercepts $b = \pm 1$

41. $r^2 = -4r \cos \theta \Rightarrow x^2 + y^2 = -4x \Rightarrow x^2 + 4x + y^2 = 0 \Rightarrow x^2 + 4x + 4 + y^2 = 4 \Rightarrow (x + 2)^2 + y^2 = 4$, a circle with
    center $C(-2, 0)$ and radius 2

43. $r = 8 \sin \theta \Rightarrow r^2 = 8r \sin \theta \Rightarrow x^2 + y^2 = 8y \Rightarrow x^2 + y^2 - 8y = 0 \Rightarrow x^2 + y^2 - 8y + 16 = 16$
    $\Rightarrow x^2 + (y - 4)^2 = 16$, a circle with center $C(0, 4)$ and radius 4

45. $r = 2 \cos \theta + 2 \sin \theta \Rightarrow r^2 = 2r \cos \theta + 2r \sin \theta \Rightarrow x^2 + y^2 = 2x + 2y \Rightarrow x^2 - 2x + y^2 - 2y = 0$
    $\Rightarrow (x - 1)^2 + (y - 1)^2 = 2$, a circle with center $C(1, 1)$ and radius $\sqrt{2}$

47. $r \sin \left(\theta + \frac{\pi}{6} \right) = 2 \Rightarrow r \left(\sin \theta \cos \frac{\pi}{6} + \cos \theta \sin \frac{\pi}{6} \right) = 2 \Rightarrow \frac{\sqrt{3}}{2} r \sin \theta + \frac{1}{2} r \cos \theta = 2 \Rightarrow \frac{\sqrt{3}}{2} y + \frac{1}{2} x = 2$
    $\Rightarrow \sqrt{3} y + x = 4$, line with slope $m = -\frac{1}{\sqrt{3}}$ and intercept $b = \frac{4}{\sqrt{3}}$

49. $x = 7 \Rightarrow r \cos \theta = 7$ 51. $x = y \Rightarrow r \cos \theta = r \sin \theta \Rightarrow \theta = \frac{\pi}{4}$

53. $x^2 + y^2 = 4 \Rightarrow r^2 = 4 \Rightarrow r = 2$ or $r = -2$

55. $\frac{x^2}{9} + \frac{y^2}{4} = 1 \Rightarrow 4x^2 + 9y^2 = 36 \Rightarrow 4r^2 \cos^2 \theta + 9r^2 \sin^2 \theta = 36$

57. $y^2 = 4x \Rightarrow r^2 \sin^2 \theta = 4r \cos \theta \Rightarrow r \sin^2 \theta = 4 \cos \theta$

59. $x^2 + (y - 2)^2 = 4 \Rightarrow x^2 + y^2 - 4y + 4 = 4 \Rightarrow x^2 + y^2 = 4y \Rightarrow r^2 = 4r \sin \theta \Rightarrow r = 4 \sin \theta$

61. $(x - 3)^2 + (y + 1)^2 = 4 \Rightarrow x^2 - 6x + 9 + y^2 + 2y + 1 = 4 \Rightarrow x^2 + y^2 = 6x - 2y - 6 \Rightarrow r^2 = 6r \cos \theta - 2r \sin \theta - 6$

63. $(0, \theta)$ where θ is any angle

9.2 GRAPHING IN POLAR COORDINATES

1. $1 + \cos(-\theta) = 1 + \cos\theta = r \Rightarrow$ symmetric about the
    x-axis; $1 + \cos(-\theta) \neq -r$ and $1 + \cos(\pi - \theta)$
    $= 1 - \cos\theta \neq r \Rightarrow$ not symmetric about the y-axis;
    therefore not symmetric about the origin

3. $1 - \sin(-\theta) = 1 + \sin\theta \neq r$ and $1 - \sin(\pi - \theta)$
    $= 1 - \sin\theta \neq -r \Rightarrow$ not symmetric about the x-axis;
    $1 - \sin(\pi - \theta) = 1 - \sin\theta = r \Rightarrow$ symmetric about
    the y-axis; therefore not symmetric about the origin

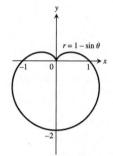

5. $2 + \sin(-\theta) = 2 - \sin\theta \neq r$ and $2 + \sin(\pi - \theta)$
    $= 2 + \sin\theta \neq -r \Rightarrow$ not symmetric about the x-axis;
    $2 + \sin(\pi - \theta) = 2 + \sin\theta = r \Rightarrow$ symmetric about the
    y-axis; therefore not symmetric about the origin

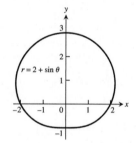

7. $\sin\left(-\frac{\theta}{2}\right) = -\sin\left(\frac{\theta}{2}\right) = -r \Rightarrow$ symmetric about the y-axis;
    $\sin\left(\frac{2\pi - \theta}{2}\right) = \sin\left(\frac{\theta}{2}\right)$, so the graph is symmetric about the
    x-axis, and hence the origin.

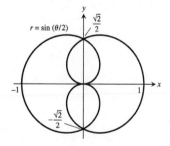

9. $\cos(-\theta) = \cos\theta = r^2 \Rightarrow (r, -\theta)$ and $(-r, -\theta)$ are on the
    graph when (r, θ) is on the graph $\Rightarrow$ symmetric about the
    x-axis and the y-axis; therefore symmetric about the origin

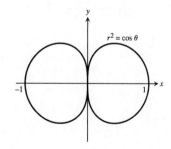

11. $-\sin(\pi-\theta)=-\sin\theta=r^2 \Rightarrow (r,\pi-\theta)$ and $(-r,\pi-\theta)$
are on the graph when (r,θ) is on the graph $\Rightarrow$ symmetric
about the y-axis and the x-axis; therefore symmetric about
the origin

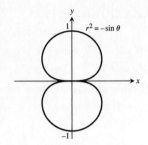

13. Since $(\pm r,-\theta)$ are on the graph when (r,θ) is on the graph
$\left((\pm r)^2=4\cos 2(-\theta) \Rightarrow r^2=4\cos 2\theta\right)$, the graph is
symmetric about the x-axis and the y-axis $\Rightarrow$ the graph is
symmetric about the origin

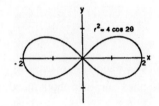

15. Since (r,θ) on the graph $\Rightarrow (-r,\theta)$ is on the graph
$\left((\pm r)^2=-\sin 2\theta \Rightarrow r^2=-\sin 2\theta\right)$, the graph is
symmetric about the origin. But $-\sin 2(-\theta)=-(-\sin 2\theta)$
$\sin 2\theta\neq r^2$ and $-\sin 2(\pi-\theta)=-\sin(2\pi-2\theta)$
$=-\sin(-2\theta)=-(-\sin 2\theta)=\sin 2\theta\neq r^2 \Rightarrow$ the graph
is not symmetric about the x-axis; therefore the graph is
not symmetric about the y-axis

17. $\theta=\frac{\pi}{2} \Rightarrow r=-1 \Rightarrow \left(-1,\frac{\pi}{2}\right)$, and $\theta=-\frac{\pi}{2} \Rightarrow r=-1$
$\Rightarrow \left(-1,-\frac{\pi}{2}\right)$; $r'=\frac{dr}{d\theta}=-\sin\theta$; Slope $=\frac{r'\sin\theta+r\cos\theta}{r'\cos\theta-r\sin\theta}$
$=\frac{-\sin^2\theta+r\cos\theta}{-\sin\theta\cos\theta-r\sin\theta} \Rightarrow$ Slope at $\left(-1,\frac{\pi}{2}\right)$ is
$\frac{-\sin^2\left(\frac{\pi}{2}\right)+(-1)\cos\frac{\pi}{2}}{-\sin\frac{\pi}{2}\cos\frac{\pi}{2}-(-1)\sin\frac{\pi}{2}}=-1$; Slope at $\left(-1,-\frac{\pi}{2}\right)$ is
$\frac{-\sin^2\left(-\frac{\pi}{2}\right)+(-1)\cos\left(-\frac{\pi}{2}\right)}{-\sin\left(-\frac{\pi}{2}\right)\cos\left(-\frac{\pi}{2}\right)-(-1)\sin\left(-\frac{\pi}{2}\right)}=1$

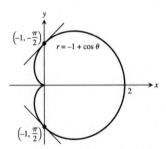

19. $\theta=\frac{\pi}{4} \Rightarrow r=1 \Rightarrow \left(1,\frac{\pi}{4}\right)$; $\theta=-\frac{\pi}{4} \Rightarrow r=-1$
$\Rightarrow \left(-1,-\frac{\pi}{4}\right)$; $\theta=\frac{3\pi}{4} \Rightarrow r=-1 \Rightarrow \left(-1,\frac{3\pi}{4}\right)$;
$\theta=-\frac{3\pi}{4} \Rightarrow r=1 \Rightarrow \left(1,-\frac{3\pi}{4}\right)$;
$r'=\frac{dr}{d\theta}=2\cos 2\theta$;
Slope $=\frac{r'\sin\theta+r\cos\theta}{r'\cos\theta-r\sin\theta}=\frac{2\cos 2\theta\sin\theta+r\cos\theta}{2\cos 2\theta\cos\theta-r\sin\theta}$
$\Rightarrow$ Slope at $\left(1,\frac{\pi}{4}\right)$ is $\frac{2\cos\left(\frac{\pi}{2}\right)\sin\left(\frac{\pi}{4}\right)+(1)\cos\left(\frac{\pi}{4}\right)}{2\cos\left(\frac{\pi}{2}\right)\cos\left(\frac{\pi}{4}\right)-(1)\sin\left(\frac{\pi}{4}\right)}=-1$;

Slope at $\left(-1,-\frac{\pi}{4}\right)$ is $\frac{2\cos\left(-\frac{\pi}{2}\right)\sin\left(-\frac{\pi}{4}\right)+(-1)\cos\left(-\frac{\pi}{4}\right)}{2\cos\left(-\frac{\pi}{2}\right)\cos\left(-\frac{\pi}{4}\right)-(-1)\sin\left(-\frac{\pi}{4}\right)}=1$;

Slope at $\left(-1,\frac{3\pi}{4}\right)$ is $\frac{2\cos\left(\frac{3\pi}{2}\right)\sin\left(\frac{3\pi}{4}\right)+(-1)\cos\left(\frac{3\pi}{4}\right)}{2\cos\left(\frac{3\pi}{2}\right)\cos\left(\frac{3\pi}{4}\right)-(-1)\sin\left(\frac{3\pi}{4}\right)}=1$;

Slope at $\left(1,-\frac{3\pi}{4}\right)$ is $\frac{2\cos\left(-\frac{3\pi}{2}\right)\sin\left(-\frac{3\pi}{4}\right)+(1)\cos\left(-\frac{3\pi}{4}\right)}{2\cos\left(-\frac{3\pi}{2}\right)\cos\left(-\frac{3\pi}{4}\right)-(1)\sin\left(-\frac{3\pi}{4}\right)}=-1$

21. (a)

(b)

23. (a)

(b)

25.

27.

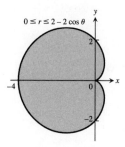

29. Note that (r, θ) and $(-r, \theta + \pi)$ describe the same point in the plane. Then $r = 1 - \cos\theta \Leftrightarrow -1 - \cos(\theta + \pi)$
$= -1 - (\cos\theta\cos\pi - \sin\theta\sin\pi) = -1 + \cos\theta = -(1 - \cos\theta) = -r$; therefore (r, θ) is on the graph of
$r = 1 - \cos\theta \Leftrightarrow (-r, \theta + \pi)$ is on the graph of $r = -1 - \cos\theta \Rightarrow$ the answer is (a).

31.

33. (a)   (b)   (c)   (d)

9.3 AREA AND LENGTHS IN POLAR COORDINATES

1. $A = \int_0^{2\pi} \frac{1}{2}(4 + 2\cos\theta)^2 \, d\theta = \int_0^{2\pi} \frac{1}{2}(16 + 16\cos\theta + 4\cos^2\theta) \, d\theta = \int_0^{2\pi} \left[8 + 8\cos\theta + 2\left(\frac{1+\cos 2\theta}{2}\right)\right] d\theta$

$= \int_0^{2\pi} (9 + 8\cos\theta + \cos 2\theta) \, d\theta = \left[9\theta + 8\sin\theta + \frac{1}{2}\sin 2\theta\right]_0^{2\pi} = 18\pi$

3. $A = 2\int_0^{\pi/4} \frac{1}{2}\cos^2 2\theta \, d\theta = \int_0^{\pi/4} \frac{1+\cos 4\theta}{2} \, d\theta = \frac{1}{2}\left[\theta + \frac{\sin 4\theta}{4}\right]_0^{\pi/4} = \frac{\pi}{8}$

5. $A = \int_0^{\pi/2} \frac{1}{2}(4\sin 2\theta) \, d\theta = \int_0^{\pi/2} 2\sin 2\theta \, d\theta = [-\cos 2\theta]_0^{\pi/2} = 2$

7. $r = 2\cos\theta$ and $r = 2\sin\theta \Rightarrow 2\cos\theta = 2\sin\theta$

$\Rightarrow \cos\theta = \sin\theta \Rightarrow \theta = \frac{\pi}{4}$; therefore

$A = 2\int_0^{\pi/4} \frac{1}{2}(2\sin\theta)^2 \, d\theta = \int_0^{\pi/4} 4\sin^2\theta \, d\theta$

$= \int_0^{\pi/4} 4\left(\frac{1-\cos 2\theta}{2}\right) d\theta = \int_0^{\pi/4} (2 - 2\cos 2\theta) \, d\theta$

$= [2\theta - \sin 2\theta]_0^{\pi/4} = \frac{\pi}{2} - 1$

9. $r = 2$ and $r = 2(1 - \cos\theta) \Rightarrow 2 = 2(1 - \cos\theta)$

$\Rightarrow \cos\theta = 0 \Rightarrow \theta = \pm\frac{\pi}{2}$; therefore

$A = 2\int_0^{\pi/2} \frac{1}{2}[2(1 - \cos\theta)]^2 \, d\theta + \frac{1}{2}$ area of the circle

$= \int_0^{\pi/2} 4(1 - 2\cos\theta + \cos^2\theta) \, d\theta + \left(\frac{1}{2}\pi\right)(2)^2$

$= \int_0^{\pi/2} 4\left(1 - 2\cos\theta + \frac{1+\cos 2\theta}{2}\right) d\theta + 2\pi$

$= \int_0^{\pi/2} (4 - 8\cos\theta + 2 + 2\cos 2\theta) \, d\theta + 2\pi$

$= [6\theta - 8\sin\theta + \sin 2\theta]_0^{\pi/2} + 2\pi = 5\pi - 8$

11. $r = \sqrt{3}$ and $r^2 = 6\cos 2\theta \Rightarrow 3 = 6\cos 2\theta \Rightarrow \cos 2\theta = \frac{1}{2}$

$\Rightarrow \theta = \frac{\pi}{6}$ (in the 1st quadrant); we use symmetry of the

graph to find the area, so

$A = 4\int_0^{\pi/6} \left[\frac{1}{2}(6\cos 2\theta) - \frac{1}{2}\left(\sqrt{3}\right)^2\right] d\theta$

$= 2\int_0^{\pi/6} (6\cos 2\theta - 3) \, d\theta = 2[3\sin 2\theta - 3\theta]_0^{\pi/6}$

$= 3\sqrt{3} - \pi$

13. $r = 1$ and $r = -2 \cos \theta \Rightarrow 1 = -2 \cos \theta \Rightarrow \cos \theta = -\frac{1}{2}$
$\Rightarrow \theta = \frac{2\pi}{3}$ in quadrant II; therefore

$A = 2 \int_{2\pi/3}^{\pi} \frac{1}{2} \left[(-2 \cos \theta)^2 - 1^2 \right] d\theta = \int_{2\pi/3}^{\pi} (4 \cos^2 \theta - 1) \, d\theta$

$= \int_{2\pi/3}^{\pi} [2(1 + \cos 2\theta) - 1] \, d\theta = \int_{2\pi/3}^{\pi} (1 + 2 \cos 2\theta) \, d\theta$

$= [\theta + \sin 2\theta]_{2\pi/3}^{\pi} = \frac{\pi}{3} + \frac{\sqrt{3}}{2}$

15. (a) $r = \tan \theta$ and $r = \left(\frac{\sqrt{2}}{2} \right) \csc \theta \Rightarrow \tan \theta = \left(\frac{\sqrt{2}}{2} \right) \csc \theta$

$\Rightarrow \sin^2 \theta = \left(\frac{\sqrt{2}}{2} \right) \cos \theta \Rightarrow 1 - \cos^2 \theta = \left(\frac{\sqrt{2}}{2} \right) \cos \theta$

$\Rightarrow \cos^2 \theta + \left(\frac{\sqrt{2}}{2} \right) \cos \theta - 1 = 0 \Rightarrow \cos \theta = -\sqrt{2}$ or

$\frac{\sqrt{2}}{2}$ (use the quadratic formula) $\Rightarrow \theta = \frac{\pi}{4}$ (the solution

in the first quadrant); therefore the area of R_1 is

$A_1 = \int_0^{\pi/4} \frac{1}{2} \tan^2 \theta \, d\theta = \frac{1}{2} \int_0^{\pi/4} (\sec^2 \theta - 1) \, d\theta = \frac{1}{2} [\tan \theta - \theta]_0^{\pi/4} = \frac{1}{2} \left(\tan \frac{\pi}{4} - \frac{\pi}{4} \right) = \frac{1}{2} - \frac{\pi}{8}$;

$AO = \left(\frac{\sqrt{2}}{2} \right) \csc \frac{\pi}{2} = \frac{\sqrt{2}}{2}$ and $OB = \left(\frac{\sqrt{2}}{2} \right) \csc \frac{\pi}{4} = 1 \Rightarrow AB = \sqrt{1^2 - \left(\frac{\sqrt{2}}{2} \right)^2} = \frac{\sqrt{2}}{2}$

$\Rightarrow$ the area of R_2 is $A_2 = \frac{1}{2} \left(\frac{\sqrt{2}}{2} \right) \left(\frac{\sqrt{2}}{2} \right) = \frac{1}{4}$; therefore the area of the region shaded in the text is

$2 \left(\frac{1}{2} - \frac{\pi}{8} + \frac{1}{4} \right) = \frac{3}{2} - \frac{\pi}{4}$. Note: The area must be found this way since no common interval generates the region. For

example, the interval $0 \le \theta \le \frac{\pi}{4}$ generates the arc OB of $r = \tan \theta$ but does not generate the segment AB of the line

$r = \frac{\sqrt{2}}{2} \csc \theta$. Instead the interval generates the half-line from B to $+\infty$ on the line $r = \frac{\sqrt{2}}{2} \csc \theta$.

(b) $\lim_{\theta \to \pi/2^-} \tan \theta = \infty$ and the line $x = 1$ is $r = \sec \theta$ in polar coordinates; then $\lim_{\theta \to \pi/2^-} (\tan \theta - \sec \theta)$

$= \lim_{\theta \to \pi/2^-} \left(\frac{\sin \theta}{\cos \theta} - \frac{1}{\cos \theta} \right) = \lim_{\theta \to \pi/2^-} \left(\frac{\sin \theta - 1}{\cos \theta} \right) = \lim_{\theta \to \pi/2^-} \left(\frac{\cos \theta}{-\sin \theta} \right) = 0 \Rightarrow r = \tan \theta$ approaches

$r = \sec \theta$ as $\theta \to \frac{\pi}{2}^- \Rightarrow r = \sec \theta$ (or $x = 1$) is a vertical asymptote of $r = \tan \theta$. Similarly, $r = -\sec \theta$

(or $x = -1$) is a vertical asymptote of $r = \tan \theta$.

17. $r = \theta^2, 0 \le \theta \le \sqrt{5} \Rightarrow \frac{dr}{d\theta} = 2\theta$; therefore Length $= \int_0^{\sqrt{5}} \sqrt{(\theta^2)^2 + (2\theta)^2} \, d\theta = \int_0^{\sqrt{5}} \sqrt{\theta^4 + 4\theta^2} \, d\theta$

$= \int_0^{\sqrt{5}} |\theta| \sqrt{\theta^2 + 4} \, d\theta = $ (since $\theta \ge 0$) $\int_0^{\sqrt{5}} \theta \sqrt{\theta^2 + 4} \, d\theta$; $\left[u = \theta^2 + 4 \Rightarrow \frac{1}{2} du = \theta \, d\theta; \theta = 0 \Rightarrow u = 4, \right.$

$\left. \theta = \sqrt{5} \Rightarrow u = 9 \right] \to \int_4^9 \frac{1}{2} \sqrt{u} \, du = \frac{1}{2} \left[\frac{2}{3} u^{3/2} \right]_4^9 = \frac{19}{3}$

19. $r = 1 + \cos \theta \Rightarrow \frac{dr}{d\theta} = -\sin \theta$; therefore Length $= \int_0^{2\pi} \sqrt{(1 + \cos \theta)^2 + (-\sin \theta)^2} \, d\theta$

$= 2 \int_0^{\pi} \sqrt{2 + 2 \cos \theta} \, d\theta = 2 \int_0^{\pi} \sqrt{\frac{4(1 + \cos \theta)}{2}} \, d\theta = 4 \int_0^{\pi} \sqrt{\frac{1 + \cos \theta}{2}} \, d\theta = 4 \int_0^{\pi} \cos \left(\frac{\theta}{2} \right) \, d\theta = 4 \left[2 \sin \frac{\theta}{2} \right]_0^{\pi} = 8$

21. $r = \frac{6}{1 + \cos \theta}, 0 \le \theta \le \frac{\pi}{2} \Rightarrow \frac{dr}{d\theta} = \frac{6 \sin \theta}{(1 + \cos \theta)^2}$; therefore Length $= \int_0^{\pi/2} \sqrt{\left(\frac{6}{1 + \cos \theta} \right)^2 + \left(\frac{6 \sin \theta}{(1 + \cos \theta)^2} \right)^2} \, d\theta$

$= \int_0^{\pi/2} \sqrt{\frac{36}{(1 + \cos \theta)^2} + \frac{36 \sin^2 \theta}{(1 + \cos \theta)^4}} \, d\theta = 6 \int_0^{\pi/2} \left| \frac{1}{1 + \cos \theta} \right| \sqrt{1 + \frac{\sin^2 \theta}{(1 + \cos \theta)^2}} \, d\theta$

$= \left(\text{since } \frac{1}{1 + \cos \theta} > 0 \text{ on } 0 \le \theta \le \frac{\pi}{2} \right) 6 \int_0^{\pi/2} \left(\frac{1}{1 + \cos \theta} \right) \sqrt{\frac{1 + 2 \cos \theta + \cos^2 \theta + \sin^2 \theta}{(1 + \cos \theta)^2}} \, d\theta$

$= 6 \int_0^{\pi/2} \left(\frac{1}{1 + \cos \theta} \right) \sqrt{\frac{2 + 2 \cos \theta}{(1 + \cos \theta)^2}} \, d\theta = 6\sqrt{2} \int_0^{\pi/2} \frac{d\theta}{(1 + \cos \theta)^{3/2}} = 6\sqrt{2} \int_0^{\pi/2} \frac{d\theta}{\left(2 \cos^2 \frac{\theta}{2} \right)^{3/2}} = 3 \int_0^{\pi/2} \left| \sec^3 \frac{\theta}{2} \right| \, d\theta$

$$= 3 \int_0^{\pi/2} \sec^3 \tfrac{\theta}{2} \, d\theta = 6 \int_0^{\pi/4} \sec^3 u \, du = \text{(use tables)} \; 6 \left(\left[\tfrac{\sec u \tan u}{2} \right]_0^{\pi/4} + \tfrac{1}{2} \int_0^{\pi/4} \sec u \, du \right)$$

$$= 6 \left(\tfrac{1}{\sqrt{2}} + \left[\tfrac{1}{2} \ln |\sec u + \tan u| \right]_0^{\pi/4} \right) = 3 \left[\sqrt{2} + \ln \left(1 + \sqrt{2} \right) \right]$$

23. $r = \cos^3 \tfrac{\theta}{3} \Rightarrow \tfrac{dr}{d\theta} = -\sin \tfrac{\theta}{3} \cos^2 \tfrac{\theta}{3}$; therefore Length $= \int_0^{\pi/4} \sqrt{\left(\cos^3 \tfrac{\theta}{3} \right)^2 + \left(-\sin \tfrac{\theta}{3} \cos^2 \tfrac{\theta}{3} \right)^2} \, d\theta$

$$= \int_0^{\pi/4} \sqrt{\cos^6 \left(\tfrac{\theta}{3} \right) + \sin^2 \left(\tfrac{\theta}{3} \right) \cos^4 \left(\tfrac{\theta}{3} \right)} \, d\theta = \int_0^{\pi/4} \left(\cos^2 \tfrac{\theta}{3} \right) \sqrt{\cos^2 \left(\tfrac{\theta}{3} \right) + \sin^2 \left(\tfrac{\theta}{3} \right)} \, d\theta = \int_0^{\pi/4} \cos^2 \left(\tfrac{\theta}{3} \right) \, d\theta$$

$$= \int_0^{\pi/4} \tfrac{1 + \cos \left(\frac{2\theta}{3} \right)}{2} \, d\theta = \tfrac{1}{2} \left[\theta + \tfrac{3}{2} \sin \tfrac{2\theta}{3} \right]_0^{\pi/4} = \tfrac{\pi}{8} + \tfrac{3}{8}$$

25. $r = \sqrt{1 + \cos 2\theta} \Rightarrow \tfrac{dr}{d\theta} = \tfrac{1}{2} (1 + \cos 2\theta)^{-1/2} (-2 \sin 2\theta)$; therefore Length $= \int_0^{\pi\sqrt{2}} \sqrt{(1 + \cos 2\theta) + \tfrac{\sin^2 2\theta}{(1 + \cos 2\theta)}} \, d\theta$

$$= \int_0^{\pi\sqrt{2}} \sqrt{\tfrac{1 + 2 \cos 2\theta + \cos^2 2\theta + \sin^2 2\theta}{1 + \cos 2\theta}} \, d\theta = \int_0^{\pi\sqrt{2}} \sqrt{\tfrac{2 + 2 \cos 2\theta}{1 + \cos 2\theta}} \, d\theta = \int_0^{\pi\sqrt{2}} \sqrt{2} \, d\theta = \left[\sqrt{2} \, \theta \right]_0^{\pi\sqrt{2}} = 2\pi$$

27. Let $r = f(\theta)$. Then $x = f(\theta) \cos \theta \Rightarrow \tfrac{dx}{d\theta} = f'(\theta) \cos \theta - f(\theta) \sin \theta \Rightarrow \left(\tfrac{dx}{d\theta} \right)^2 = [f'(\theta) \cos \theta - f(\theta) \sin \theta]^2$

$$= [f'(\theta)]^2 \cos^2 \theta - 2f'(\theta) f(\theta) \sin \theta \cos \theta + [f(\theta)]^2 \sin^2 \theta; \; y = f(\theta) \sin \theta \Rightarrow \tfrac{dy}{d\theta} = f'(\theta) \sin \theta + f(\theta) \cos \theta$$

$$\Rightarrow \left(\tfrac{dy}{d\theta} \right)^2 = [f'(\theta) \sin \theta + f(\theta) \cos \theta]^2 = [f'(\theta)]^2 \sin^2 \theta + 2f'(\theta) f(\theta) \sin \theta \cos \theta + [f(\theta)]^2 \cos^2 \theta. \; \text{Therefore}$$

$$\left(\tfrac{dx}{d\theta} \right)^2 + \left(\tfrac{dy}{d\theta} \right)^2 = [f'(\theta)]^2 (\cos^2 \theta + \sin^2 \theta) + [f(\theta)]^2 (\cos^2 \theta + \sin^2 \theta) = [f'(\theta)]^2 + [f(\theta)]^2 = r^2 + \left(\tfrac{dr}{d\theta} \right)^2.$$

Thus, $L = \int_\alpha^\beta \sqrt{\left(\tfrac{dx}{d\theta} \right)^2 + \left(\tfrac{dy}{d\theta} \right)^2} \, d\theta = \int_\alpha^\beta \sqrt{r^2 + \left(\tfrac{dr}{d\theta} \right)^2} \, d\theta.$

29. $r = 2f(\theta), \; \alpha \le \theta \le \beta \Rightarrow \tfrac{dr}{d\theta} = 2f'(\theta) \Rightarrow r^2 + \left(\tfrac{dr}{d\theta} \right)^2 = [2f(\theta)]^2 + [2f'(\theta)]^2 \Rightarrow \text{Length} = \int_\alpha^\beta \sqrt{4[f(\theta)]^2 + 4[f'(\theta)]^2} \, d\theta$

$$= 2 \int_\alpha^\beta \sqrt{[f(\theta)]^2 + [f'(\theta)]^2} \, d\theta \; \text{which is twice the length of the curve } r = f(\theta) \text{ for } \alpha \le \theta \le \beta.$$

9.4 CONIC SECTIONS

1. $x = \tfrac{y^2}{8} \Rightarrow 4p = 8 \Rightarrow p = 2$; focus is $(2, 0)$, directrix is $x = -2$

3. $y = -\tfrac{x^2}{6} \Rightarrow 4p = 6 \Rightarrow p = \tfrac{3}{2}$; focus is $\left(0, -\tfrac{3}{2} \right)$, directrix is $y = \tfrac{3}{2}$

5. $\tfrac{x^2}{4} - \tfrac{y^2}{9} = 1 \Rightarrow c = \sqrt{4 + 9} = \sqrt{13} \Rightarrow$ foci are $\left(\pm \sqrt{13}, 0 \right)$; vertices are $(\pm 2, 0)$; asymptotes are $y = \pm \tfrac{3}{2} x$

7. $\tfrac{x^2}{2} + y^2 = 1 \Rightarrow c = \sqrt{2 - 1} = 1 \Rightarrow$ foci are $(\pm 1, 0)$; vertices are $\left(\pm \sqrt{2}, 0 \right)$

9. $y^2 = 12x \Rightarrow x = \tfrac{y^2}{12} \Rightarrow 4p = 12 \Rightarrow p = 3$;   11. $x^2 = -8y \Rightarrow y = \tfrac{x^2}{-8} \Rightarrow 4p = 8 \Rightarrow p = 2$;
   focus is $(3, 0)$, directrix is $x = -3$ focus is $(0, -2)$, directrix is $y = 2$

13. $y = 4x^2 \Rightarrow y = \frac{x^2}{\left(\frac{1}{4}\right)} \Rightarrow 4p = \frac{1}{4} \Rightarrow p = \frac{1}{16}$;

  focus is $\left(0, \frac{1}{16}\right)$, directrix is $y = -\frac{1}{16}$

15. $x = -3y^2 \Rightarrow x = -\frac{y^2}{\left(\frac{1}{3}\right)} \Rightarrow 4p = \frac{1}{3} \Rightarrow p = \frac{1}{12}$;

  focus is $\left(-\frac{1}{12}, 0\right)$, directrix is $x = \frac{1}{12}$

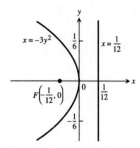

17. $16x^2 + 25y^2 = 400 \Rightarrow \frac{x^2}{25} + \frac{y^2}{16} = 1$

  $\Rightarrow c = \sqrt{a^2 - b^2} = \sqrt{25 - 16} = 3$

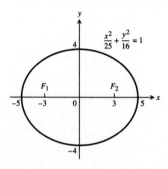

19. $2x^2 + y^2 = 2 \Rightarrow x^2 + \frac{y^2}{2} = 1$

  $\Rightarrow c = \sqrt{a^2 - b^2} = \sqrt{2 - 1} = 1$

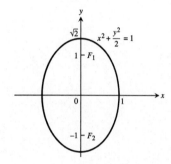

21. $3x^2 + 2y^2 = 6 \Rightarrow \frac{x^2}{2} + \frac{y^2}{3} = 1$

  $\Rightarrow c = \sqrt{a^2 - b^2} = \sqrt{3 - 2} = 1$

23. $6x^2 + 9y^2 = 54 \Rightarrow \frac{x^2}{9} + \frac{y^2}{6} = 1$

  $\Rightarrow c = \sqrt{a^2 - b^2} = \sqrt{9 - 6} = \sqrt{3}$

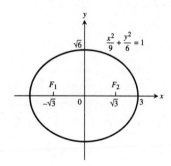

25. Foci: $\left(\pm\sqrt{2}, 0\right)$, Vertices: $(\pm 2, 0) \Rightarrow a = 2, c = \sqrt{2} \Rightarrow b^2 = a^2 - c^2 = 4 - \left(\sqrt{2}\right)^2 = 2 \Rightarrow \frac{x^2}{4} + \frac{y^2}{2} = 1$

27. $x^2 - y^2 = 1 \Rightarrow c = \sqrt{a^2 + b^2} = \sqrt{1 + 1} = \sqrt{2}$;
asymptotes are $y = \pm x$

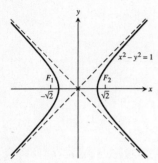

29. $y^2 - x^2 = 8 \Rightarrow \frac{y^2}{8} - \frac{x^2}{8} = 1 \Rightarrow c = \sqrt{a^2 + b^2}$
$= \sqrt{8 + 8} = 4$; asymptotes are $y = \pm x$

31. $8x^2 - 2y^2 = 16 \Rightarrow \frac{x^2}{2} - \frac{y^2}{8} = 1 \Rightarrow c = \sqrt{a^2 + b^2}$
$= \sqrt{2 + 8} = \sqrt{10}$; asymptotes are $y = \pm 2x$

33. $8y^2 - 2x^2 = 16 \Rightarrow \frac{y^2}{2} - \frac{x^2}{8} = 1 \Rightarrow c = \sqrt{a^2 + b^2}$
$= \sqrt{2 + 8} = \sqrt{10}$; asymptotes are $y = \pm \frac{x}{2}$

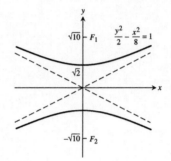

35. Foci: $\left(0, \pm \sqrt{2}\right)$, Asymptotes: $y = \pm x \Rightarrow c = \sqrt{2}$ and $\frac{a}{b} = 1 \Rightarrow a = b \Rightarrow c^2 = a^2 + b^2 = 2a^2 \Rightarrow 2 = 2a^2$
$\Rightarrow a = 1 \Rightarrow b = 1 \Rightarrow y^2 - x^2 = 1$

37. Vertices: $(\pm 3, 0)$, Asymptotes: $y = \pm \frac{4}{3}x \Rightarrow a = 3$ and $\frac{b}{a} = \frac{4}{3} \Rightarrow b = \frac{4}{3}(3) = 4 \Rightarrow \frac{x^2}{9} - \frac{y^2}{16} = 1$

39. (a) $y^2 = 8x \Rightarrow 4p = 8 \Rightarrow p = 2 \Rightarrow$ directrix is $x = -2$,
focus is $(2, 0)$, and vertex is $(0, 0)$; therefore the new
directrix is $x = -1$, the new focus is $(3, -2)$, and the
new vertex is $(1, -2)$

41. (a) $\frac{x^2}{16} + \frac{y^2}{9} = 1 \Rightarrow$ center is $(0, 0)$, vertices are $(-4, 0)$
and $(4, 0)$; $c = \sqrt{a^2 - b^2} = \sqrt{7} \Rightarrow$ foci are $\left(\sqrt{7}, 0\right)$
and $\left(-\sqrt{7}, 0\right)$; therefore the new center is $(4, 3)$, the
new vertices are $(0, 3)$ and $(8, 3)$, and the new foci are
$\left(4 \pm \sqrt{7}, 3\right)$

(b)

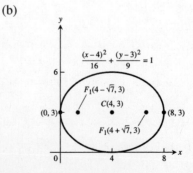

43. (a) $\frac{x^2}{16} - \frac{y^2}{9} = 1 \Rightarrow$ center is $(0,0)$, vertices are $(-4,0)$

and $(4,0)$, and the asymptotes are $\frac{x}{4} = \pm\frac{y}{3}$ or

$y = \pm\frac{3x}{4}$; $c = \sqrt{a^2 + b^2} = \sqrt{25} = 5 \Rightarrow$ foci are

$(-5,0)$ and $(5,0)$; therefore the new center is $(2,0)$, the

new vertices are $(-2,0)$ and $(6,0)$, the new foci

are $(-3,0)$ and $(7,0)$, and the new asymptotes are

$y = \pm\frac{3(x-2)}{4}$

(b)

45. $y^2 = 4x \Rightarrow 4p = 4 \Rightarrow p = 1 \Rightarrow$ focus is $(1,0)$, directrix is $x = -1$, and vertex is $(0,0)$; therefore the new
vertex is $(-2,-3)$, the new focus is $(-1,-3)$, and the new directrix is $x = -3$; the new equation is
$(y+3)^2 = 4(x+2)$

47. $x^2 = 8y \Rightarrow 4p = 8 \Rightarrow p = 2 \Rightarrow$ focus is $(0,2)$, directrix is $y = -2$, and vertex is $(0,0)$; therefore the new
vertex is $(1,-7)$, the new focus is $(1,-5)$, and the new directrix is $y = -9$; the new equation is
$(x-1)^2 = 8(y+7)$

49. $\frac{x^2}{6} + \frac{y^2}{9} = 1 \Rightarrow$ center is $(0,0)$, vertices are $(0,3)$ and $(0,-3)$; $c = \sqrt{a^2 - b^2} = \sqrt{9-6} = \sqrt{3} \Rightarrow$ foci are $\left(0, \sqrt{3}\right)$

and $\left(0, -\sqrt{3}\right)$; therefore the new center is $(-2,-1)$, the new vertices are $(-2,2)$ and $(-2,-4)$, and the new foci

are $\left(-2, -1 \pm \sqrt{3}\right)$; the new equation is $\frac{(x+2)^2}{6} + \frac{(y+1)^2}{9} = 1$

51. $\frac{x^2}{3} + \frac{y^2}{2} = 1 \Rightarrow$ center is $(0,0)$, vertices are $\left(\sqrt{3},0\right)$ and $\left(-\sqrt{3},0\right)$; $c = \sqrt{a^2 - b^2} = \sqrt{3-2} = 1 \Rightarrow$ foci are

$(-1,0)$ and $(1,0)$; therefore the new center is $(2,3)$, the new vertices are $\left(2 \pm \sqrt{3}, 3\right)$, and the new foci are $(1,3)$

and $(3,3)$; the new equation is $\frac{(x-2)^2}{3} + \frac{(y-3)^2}{2} = 1$

53. $\frac{x^2}{4} - \frac{y^2}{5} = 1 \Rightarrow$ center is $(0,0)$, vertices are $(2,0)$ and $(-2,0)$; $c = \sqrt{a^2 + b^2} = \sqrt{4+5} = 3 \Rightarrow$ foci are $(3,0)$ and

$(-3,0)$; the asymptotes are $\pm\frac{x}{2} = \frac{y}{\sqrt{5}} \Rightarrow y = \pm\frac{\sqrt{5}x}{2}$; therefore the new center is $(2,2)$, the new vertices are

$(4,2)$ and $(0,2)$, and the new foci are $(5,2)$ and $(-1,2)$; the new asymptotes are $y - 2 = \pm\frac{\sqrt{5}(x-2)}{2}$; the new

equation is $\frac{(x-2)^2}{4} - \frac{(y-2)^2}{5} = 1$

55. $y^2 - x^2 = 1 \Rightarrow$ center is $(0,0)$, vertices are $(0,1)$ and $(0,-1)$; $c = \sqrt{a^2 + b^2} = \sqrt{1+1} = \sqrt{2} \Rightarrow$ foci are

$\left(0, \pm\sqrt{2}\right)$; the asymptotes are $y = \pm x$; therefore the new center is $(-1,-1)$, the new vertices are $(-1,0)$ and

$(-1,-2)$, and the new foci are $\left(-1, -1 \pm \sqrt{2}\right)$; the new asymptotes are $y + 1 = \pm(x+1)$; the new equation is

$(y+1)^2 - (x+1)^2 = 1$

57. $x^2 + 4x + y^2 = 12 \Rightarrow x^2 + 4x + 4 + y^2 = 12 + 4 \Rightarrow (x+2)^2 + y^2 = 16$; this is a circle: center at $C(-2,0)$, $a = 4$

59. $x^2 + 2x + 4y - 3 = 0 \Rightarrow x^2 + 2x + 1 = -4y + 3 + 1 \Rightarrow (x+1)^2 = -4(y-1)$; this is a parabola: $V(-1,1)$, $F(-1,0)$

61. $x^2 + 5y^2 + 4x = 1 \Rightarrow x^2 + 4x + 4 + 5y^2 = 5 \Rightarrow (x+2)^2 + 5y^2 = 5 \Rightarrow \frac{(x+2)^2}{5} + y^2 = 1$; this is an ellipse: the

center is $(-2,0)$, the vertices are $\left(-2 \pm \sqrt{5}, 0\right)$; $c = \sqrt{a^2 - b^2} = \sqrt{5-1} = 2 \Rightarrow$ the foci are $(-4,0)$ and $(0,0)$

63. $x^2 + 2y^2 - 2x - 4y = -1 \Rightarrow x^2 - 2x + 1 + 2\left(y^2 - 2y + 1\right) = 2 \Rightarrow (x-1)^2 + 2(y-1)^2 = 2$

$\Rightarrow \frac{(x-1)^2}{2} + (y-1)^2 = 1$; this is an ellipse: the center is $(1, 1)$, the vertices are $\left(1 \pm \sqrt{2}, 1\right)$;

$c = \sqrt{a^2 - b^2} = \sqrt{2-1} = 1 \Rightarrow$ the foci are $(2, 1)$ and $(0, 1)$

65. $x^2 - y^2 - 2x + 4y = 4 \Rightarrow x^2 - 2x + 1 - \left(y^2 - 4y + 4\right) = 1 \Rightarrow (x-1)^2 - (y-2)^2 = 1$; this is a hyperbola:

the center is $(1, 2)$, the vertices are $(2, 2)$ and $(0, 2)$; $c = \sqrt{a^2 + b^2} = \sqrt{1+1} = \sqrt{2} \Rightarrow$ the foci are $\left(1 \pm \sqrt{2}, 2\right)$;

the asymptotes are $y - 2 = \pm(x-1)$

67. $2x^2 - y^2 + 6y = 3 \Rightarrow 2x^2 - \left(y^2 - 6y + 9\right) = -6 \Rightarrow \frac{(y-3)^2}{6} - \frac{x^2}{3} = 1$; this is a hyperbola: the center is $(0, 3)$,

the vertices are $\left(0, 3 \pm \sqrt{6}\right)$; $c = \sqrt{a^2 + b^2} = \sqrt{6+3} = 3 \Rightarrow$ the foci are $(0, 6)$ and $(0, 0)$; the asymptotes are

$\frac{y-3}{\sqrt{6}} = \pm \frac{x}{\sqrt{3}} \Rightarrow y = \pm\sqrt{2}x + 3$

69. (a) $y^2 = kx \Rightarrow x = \frac{y^2}{k}$; the volume of the solid formed by

revolving R_1 about the y-axis is $V_1 = \int_0^{\sqrt{kx}} \pi \left(\frac{y^2}{k}\right)^2 dy$

$= \frac{\pi}{k^2} \int_0^{\sqrt{kx}} y^4\, dy = \frac{\pi x^2 \sqrt{kx}}{5}$; the volume of the right

circular cylinder formed by revolving PQ about the

y-axis is $V_2 = \pi x^2 \sqrt{kx} \Rightarrow$ the volume of the solid

formed by revolving R_2 about the y-axis is

$V_3 = V_2 - V_1 = \frac{4\pi x^2 \sqrt{kx}}{5}$. Therefore we can see the

ratio of V_3 to V_1 is 4:1.

(b) The volume of the solid formed by revolving R_2 about the x-axis is $V_1 = \int_0^x \pi \left(\sqrt{kt}\right)^2 dt = \pi k \int_0^x t\, dt$

$= \frac{\pi k x^2}{2}$. The volume of the right circular cylinder formed by revolving PS about the x-axis is

$V_2 = \pi \left(\sqrt{kx}\right)^2 x = \pi k x^2 \Rightarrow$ the volume of the solid formed by revolving R_1 about the x-axis is

$V_3 = V_2 - V_1 = \pi k x^2 - \frac{\pi k x^2}{2} = \frac{\pi k x^2}{2}$. Therefore the ratio of V_3 to V_1 is 1:1.

71. $x^2 = 4py$ and $y = p \Rightarrow x^2 = 4p^2 \Rightarrow x = \pm 2p$. Therefore the line $y = p$ cuts the parabola at points $(-2p, p)$ and

$(2p, p)$, and these points are $\sqrt{[2p - (-2p)]^2 + (p-p)^2} = 4p$ units apart.

73. Let $y = \sqrt{1 - \frac{x^2}{4}}$ on the interval $0 \le x \le 2$. The area of the inscribed rectangle is given by

$A(x) = 2x \left(2\sqrt{1 - \frac{x^2}{4}}\right) = 4x\sqrt{1 - \frac{x^2}{4}}$ (since the length is $2x$ and the height is $2y$)

$\Rightarrow A'(x) = 4\sqrt{1 - \frac{x^2}{4}} - \frac{x^2}{\sqrt{1 - \frac{x^2}{4}}}$. Thus $A'(x) = 0 \Rightarrow 4\sqrt{1 - \frac{x^2}{4}} - \frac{x^2}{\sqrt{1 - \frac{x^2}{4}}} = 0 \Rightarrow 4\left(1 - \frac{x^2}{4}\right) - x^2 = 0 \Rightarrow x^2 = 2$

$\Rightarrow x = \sqrt{2}$ (only the positive square root lies in the interval). Since $A(0) = A(2) = 0$ we have that $A\left(\sqrt{2}\right) = 4$

is the maximum area when the length is $2\sqrt{2}$ and the height is $\sqrt{2}$.

75. $9x^2 - 4y^2 = 36 \Rightarrow y^2 = \frac{9x^2 - 36}{4} \Rightarrow y = \pm\frac{3}{2}\sqrt{x^2 - 4}$ on the interval $2 \le x \le 4 \Rightarrow V = \int_2^4 \pi \left(\frac{3}{2}\sqrt{x^2 - 4}\right)^2 dx$

$= \frac{9\pi}{4} \int_2^4 (x^2 - 4)\, dx = \frac{9\pi}{4}\left[\frac{x^3}{3} - 4x\right]_2^4 = \frac{9\pi}{4}\left[\left(\frac{64}{3} - 16\right) - \left(\frac{8}{3} - 8\right)\right] = \frac{9\pi}{4}\left(\frac{56}{3} - 8\right) = \frac{3\pi}{4}(56 - 24) = 24\pi$

9.5 CONICS IN POLAR COORDINATES

1. $16x^2 + 25y^2 = 400 \Rightarrow \frac{x^2}{25} + \frac{y^2}{16} = 1 \Rightarrow c = \sqrt{a^2 - b^2}$
   $= \sqrt{25 - 16} = 3 \Rightarrow e = \frac{c}{a} = \frac{3}{5}$; $F(\pm 3, 0)$;
   directrices are $x = 0 \pm \frac{a}{e} = \pm \frac{5}{\left(\frac{3}{5}\right)} = \pm \frac{25}{3}$

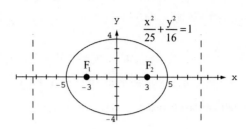

3. $2x^2 + y^2 = 2 \Rightarrow x^2 + \frac{y^2}{2} = 1 \Rightarrow c = \sqrt{a^2 - b^2}$
   $= \sqrt{2 - 1} = 1 \Rightarrow e = \frac{c}{a} = \frac{1}{\sqrt{2}}$; $F(0, \pm 1)$;
   directrices are $y = 0 \pm \frac{a}{e} = \pm \frac{\sqrt{2}}{\left(\frac{1}{\sqrt{2}}\right)} = \pm 2$

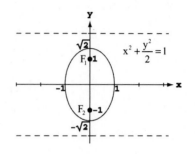

5. $3x^2 + 2y^2 = 6 \Rightarrow \frac{x^2}{2} + \frac{y^2}{3} = 1 \Rightarrow c = \sqrt{a^2 - b^2}$
   $= \sqrt{3 - 2} = 1 \Rightarrow e = \frac{c}{a} = \frac{1}{\sqrt{3}}$; $F(0, \pm 1)$;
   directrices are $y = 0 \pm \frac{a}{e} = \pm \frac{\sqrt{3}}{\left(\frac{1}{\sqrt{3}}\right)} = \pm 3$

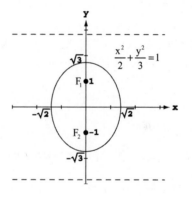

7. $6x^2 + 9y^2 = 54 \Rightarrow \frac{x^2}{9} + \frac{y^2}{6} = 1 \Rightarrow c = \sqrt{a^2 - b^2}$
   $= \sqrt{9 - 6} = \sqrt{3} \Rightarrow e = \frac{c}{a} = \frac{\sqrt{3}}{3}$; $F\left(\pm \sqrt{3}, 0\right)$;
   directrices are $x = 0 \pm \frac{a}{e} = \pm \frac{3}{\left(\frac{\sqrt{3}}{3}\right)} = \pm 3\sqrt{3}$

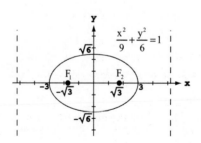

9. Foci: $(0, \pm 3)$, $e = 0.5 \Rightarrow c = 3$ and $a = \frac{c}{e} = \frac{3}{0.5} = 6 \Rightarrow b^2 = 36 - 9 = 27 \Rightarrow \frac{x^2}{27} + \frac{y^2}{36} = 1$

11. Vertices: $(0, \pm 70)$, $e = 0.1 \Rightarrow a = 70$ and $c = ae = 70(0.1) = 7 \Rightarrow b^2 = 4900 - 49 = 4851 \Rightarrow \frac{x^2}{4851} + \frac{y^2}{4900} = 1$

13. Focus: $\left(\sqrt{5}, 0\right)$, Directrix: $x = \frac{9}{\sqrt{5}} \Rightarrow c = ae = \sqrt{5}$ and $\frac{a}{e} = \frac{9}{\sqrt{5}} \Rightarrow \frac{ae}{e^2} = \frac{9}{\sqrt{5}} \Rightarrow \frac{\sqrt{5}}{e^2} = \frac{9}{\sqrt{5}} \Rightarrow e^2 = \frac{5}{9}$
    $\Rightarrow e = \frac{\sqrt{5}}{3}$. Then $PF = \frac{\sqrt{5}}{3} PD \Rightarrow \sqrt{\left(x - \sqrt{5}\right)^2 + (y - 0)^2} = \frac{\sqrt{5}}{3}\left|x - \frac{9}{\sqrt{5}}\right| \Rightarrow \left(x - \sqrt{5}\right)^2 + y^2 = \frac{5}{9}\left(x - \frac{9}{\sqrt{5}}\right)^2$
    $\Rightarrow x^2 - 2\sqrt{5}x + 5 + y^2 = \frac{5}{9}\left(x^2 - \frac{18}{\sqrt{5}}x + \frac{81}{5}\right) \Rightarrow \frac{4}{9}x^2 + y^2 = 4 \Rightarrow \frac{x^2}{9} + \frac{y^2}{4} = 1$

15. Focus: $(-4, 0)$, Directrix: $x = -16 \Rightarrow c = ae = 4$ and $\frac{a}{e} = 16 \Rightarrow \frac{ae}{e^2} = 16 \Rightarrow \frac{4}{e^2} = 16 \Rightarrow e^2 = \frac{1}{4} \Rightarrow e = \frac{1}{2}$. Then

$PF = \frac{1}{2} PD \Rightarrow \sqrt{(x+4)^2 + (y-0)^2} = \frac{1}{2} |x + 16| \Rightarrow (x+4)^2 + y^2 = \frac{1}{4}(x+16)^2 \Rightarrow x^2 + 8x + 16 + y^2$

$= \frac{1}{4}(x^2 + 32x + 256) \Rightarrow \frac{3}{4}x^2 + y^2 = 48 \Rightarrow \frac{x^2}{64} + \frac{y^2}{48} = 1$

17. $x^2 - y^2 = 1 \Rightarrow c = \sqrt{a^2 + b^2} = \sqrt{1 + 1} = \sqrt{2} \Rightarrow e = \frac{c}{a}$

$= \frac{\sqrt{2}}{1} = \sqrt{2}$; asymptotes are $y = \pm x$; $F\left(\pm\sqrt{2}, 0\right)$;

directrices are $x = 0 \pm \frac{a}{e} = \pm\frac{1}{\sqrt{2}}$

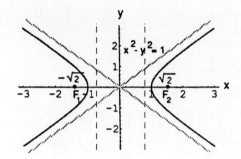

19. $y^2 - x^2 = 8 \Rightarrow \frac{y^2}{8} - \frac{x^2}{8} = 1 \Rightarrow c = \sqrt{a^2 + b^2}$

$= \sqrt{8 + 8} = 4 \Rightarrow e = \frac{c}{a} = \frac{4}{\sqrt{8}} = \sqrt{2}$; asymptotes are

$y = \pm x$; $F(0, \pm 4)$; directrices are $y = 0 \pm \frac{a}{e}$

$= \pm\frac{\sqrt{8}}{\sqrt{2}} = \pm 2$

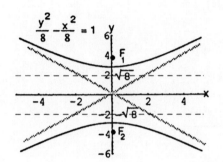

21. $8x^2 - 2y^2 = 16 \Rightarrow \frac{x^2}{2} - \frac{y^2}{8} = 1 \Rightarrow c = \sqrt{a^2 + b^2}$

$= \sqrt{2 + 8} = \sqrt{10} \Rightarrow e = \frac{c}{a} = \frac{\sqrt{10}}{\sqrt{2}} = \sqrt{5}$; asymptotes

are $y = \pm 2x$; $F\left(\pm\sqrt{10}, 0\right)$; directrices are $x = 0 \pm \frac{a}{e}$

$= \pm\frac{\sqrt{2}}{\sqrt{5}} = \pm\frac{2}{\sqrt{10}}$

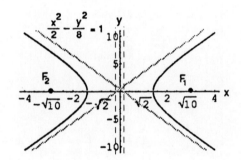

23. $8y^2 - 2x^2 = 16 \Rightarrow \frac{y^2}{2} - \frac{x^2}{8} = 1 \Rightarrow c = \sqrt{a^2 + b^2}$

$= \sqrt{2 + 8} = \sqrt{10} \Rightarrow e = \frac{c}{a} = \frac{\sqrt{10}}{\sqrt{2}} = \sqrt{5}$; asymptotes

are $y = \pm\frac{x}{2}$; $F\left(0, \pm\sqrt{10}\right)$; directrices are $y = 0 \pm \frac{a}{e}$

$= \pm\frac{\sqrt{2}}{\sqrt{5}} = \pm\frac{2}{\sqrt{10}}$

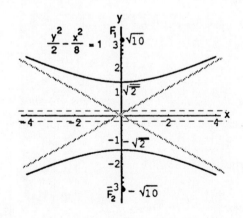

25. Vertices $(0, \pm 1)$ and $e = 3 \Rightarrow a = 1$ and $e = \frac{c}{a} = 3 \Rightarrow c = 3a = 3 \Rightarrow b^2 = c^2 - a^2 = 9 - 1 = 8 \Rightarrow y^2 - \frac{x^2}{8} = 1$

27. Foci $(\pm 3, 0)$ and $e = 3 \Rightarrow c = 3$ and $e = \frac{c}{a} = 3 \Rightarrow c = 3a \Rightarrow a = 1 \Rightarrow b^2 = c^2 - a^2 = 9 - 1 = 8 \Rightarrow x^2 - \frac{y^2}{8} = 1$

29. $e = 1, x = 2 \Rightarrow k = 2 \Rightarrow r = \frac{2(1)}{1 + (1)\cos\theta} = \frac{2}{1 + \cos\theta}$

31. $e = 5, y = -6 \Rightarrow k = 6 \Rightarrow r = \frac{6(5)}{1 - 5 \sin \theta} = \frac{30}{1 - 5 \sin \theta}$

33. $e = \frac{1}{2}, x = 1 \Rightarrow k = 1 \Rightarrow r = \frac{\left(\frac{1}{2}\right)(1)}{1 + \left(\frac{1}{2}\right) \cos \theta} = \frac{1}{2 + \cos \theta}$

35. $e = \frac{1}{5}, x = -10 \Rightarrow k = 10 \Rightarrow r = \frac{\left(\frac{1}{5}\right)(10)}{1 - \left(\frac{1}{5}\right) \sin \theta} = \frac{10}{5 - \sin \theta}$

37. $r = \frac{1}{1 + \cos \theta} \Rightarrow e = 1, k = 1 \Rightarrow x = 1$

39. $r = \frac{25}{10 - 5 \cos \theta} \Rightarrow r = \frac{\left(\frac{25}{10}\right)}{1 - \left(\frac{5}{10}\right) \cos \theta} = \frac{\left(\frac{5}{2}\right)}{1 - \left(\frac{1}{2}\right) \cos \theta}$

$\Rightarrow e = \frac{1}{2}, k = 5 \Rightarrow x = -5; a\left(1 - e^2\right) = ke$

$\Rightarrow a\left[1 - \left(\frac{1}{2}\right)^2\right] = \frac{5}{2} \Rightarrow \frac{3}{4}a = \frac{5}{2} \Rightarrow a = \frac{10}{3} \Rightarrow ea = \frac{5}{3}$

41. $r = \frac{400}{16 + 8 \sin \theta} \Rightarrow r = \frac{\left(\frac{400}{16}\right)}{1 + \left(\frac{8}{16}\right) \sin \theta} \Rightarrow r = \frac{25}{1 + \left(\frac{1}{2}\right) \sin \theta}$

$e = \frac{1}{2}, k = 50 \Rightarrow y = 50; a\left(1 - e^2\right) = ke$

$\Rightarrow a\left[1 - \left(\frac{1}{2}\right)^2\right] = 25 \Rightarrow \frac{3}{4}a = 25 \Rightarrow a = \frac{100}{3}$

$\Rightarrow ea = \frac{50}{3}$

43. $r = \frac{8}{2 - 2 \sin \theta} \Rightarrow r = \frac{4}{1 - \sin \theta} \Rightarrow e = 1,$
    $k = 4 \Rightarrow y = -4$

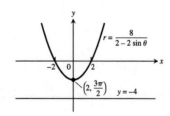

45. $r \cos\left(\theta - \frac{\pi}{4}\right) = \sqrt{2} \Rightarrow r\left(\cos \theta \cos \frac{\pi}{4} + \sin \theta \sin \frac{\pi}{4}\right)$

$= \sqrt{2} \Rightarrow \frac{1}{\sqrt{2}} r \cos \theta + \frac{1}{\sqrt{2}} r \sin \theta = \sqrt{2} \Rightarrow \frac{1}{\sqrt{2}} x + \frac{1}{\sqrt{2}} y$

$= \sqrt{2} \Rightarrow x + y = 2 \Rightarrow y = 2 - x$

47. $r \cos\left(\theta - \frac{2\pi}{3}\right) = 3 \Rightarrow r\left(\cos\theta \cos\frac{2\pi}{3} + \sin\theta \sin\frac{2\pi}{3}\right) = 3$

$\Rightarrow -\frac{1}{2}r\cos\theta + \frac{\sqrt{3}}{2}r\sin\theta = 3 \Rightarrow -\frac{1}{2}x + \frac{\sqrt{3}}{2}y = 3$

$\Rightarrow -x + \sqrt{3}y = 6 \Rightarrow y = \frac{\sqrt{3}}{3}x + 2\sqrt{3}$

49. $\sqrt{2}x + \sqrt{2}y = 6 \Rightarrow \sqrt{2}r\cos\theta + \sqrt{2}r\sin\theta = 6 \Rightarrow r\left(\frac{\sqrt{2}}{2}\cos\theta + \frac{\sqrt{2}}{2}\sin\theta\right) = 3 \Rightarrow r\left(\cos\frac{\pi}{4}\cos\theta + \sin\frac{\pi}{4}\sin\theta\right)$

$= 3 \Rightarrow r\cos\left(\theta - \frac{\pi}{4}\right) = 3$

51. $y = -5 \Rightarrow r\sin\theta = -5 \Rightarrow -r\sin\theta = 5 \Rightarrow r\sin(-\theta) = 5 \Rightarrow r\cos\left(\frac{\pi}{2} - (-\theta)\right) = 5 \Rightarrow r\cos\left(\theta + \frac{\pi}{2}\right) = 5$

53.

55.

57. $(x - 6)^2 + y^2 = 36 \Rightarrow C = (6, 0), a = 6$

$\Rightarrow r = 12\cos\theta$ is the polar equation

59. $x^2 + (y - 5)^2 = 25 \Rightarrow C = (0, 5), a = 5$

$\Rightarrow r = 10\sin\theta$ is the polar equation

61. $x^2 + 2x + y^2 = 0 \Rightarrow (x + 1)^2 + y^2 = 1$

$\Rightarrow C = (-1, 0), a = 1 \Rightarrow r = -2\cos\theta$ is

the polar equation

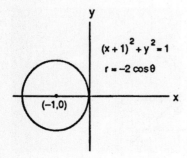

63. $x^2 + y^2 + y = 0 \Rightarrow x^2 + \left(y + \frac{1}{2}\right)^2 = \frac{1}{4}$

$\Rightarrow C = \left(0, -\frac{1}{2}\right), a = \frac{1}{2} \Rightarrow r = -\sin\theta$ is the

polar equation

65.

67.

69.

71.

73.

75. (a) Perihelion $= a - ae = a(1 - e)$, Aphelion $= ea + a = a(1 + e)$

(b)

Planet	Perihelion	Aphelion
Mercury	0.3075 AU	0.4667 AU
Venus	0.7184 AU	0.7282 AU
Earth	0.9833 AU	1.0167 AU
Mars	1.3817 AU	1.6663 AU
Jupiter	4.9512 AU	5.4548 AU
Saturn	9.0210 AU	10.0570 AU
Uranus	18.2977 AU	20.0623 AU
Neptune	29.8135 AU	30.3065 AU
Pluto	29.6549 AU	49.2251 AU

77. $x^2 + y^2 - 2ay = 0 \Rightarrow (r\cos\theta)^2 + (r\sin\theta)^2 - 2ar\sin\theta = 0$
$\Rightarrow r^2\cos^2\theta + r^2\sin^2\theta - 2ar\sin\theta = 0 \Rightarrow r^2 = 2ar\sin\theta$
$\Rightarrow r = 2a\sin\theta$

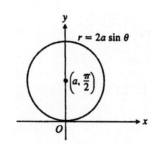

79. $x \cos \alpha + y \sin \alpha = p \Rightarrow r \cos \theta \cos \alpha + r \sin \theta \sin \alpha = p$
    $\Rightarrow r(\cos \theta \cos \alpha + \sin \theta \sin \alpha) = p \Rightarrow r \cos (\theta - \alpha) = p$

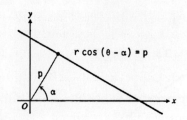

9.6 CONICS AND PARAMETRIC EQUATIONS; THE CYCLOID

1. $x = \cos t, y = \sin t, 0 \le t \le \pi$
   $\Rightarrow \cos^2 t + \sin^2 t = 1 \Rightarrow x^2 + y^2 = 1$

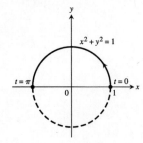

3. $x = 4 \cos t, y = 5 \sin t, 0 \le t \le \pi$
   $\Rightarrow \frac{16 \cos^2 t}{16} + \frac{25 \sin^2 t}{25} = 1 \Rightarrow \frac{x^2}{16} + \frac{y^2}{25} = 1$

5. $x = t, y = \sqrt{t}, t \ge 0 \Rightarrow y = \sqrt{x}$

7. $x = - \sec t, y = \tan t, - \frac{\pi}{2} < t < \frac{\pi}{2}$
   $\Rightarrow \sec^2 t - \tan^2 t = 1 \Rightarrow x^2 - y^2 = 1$

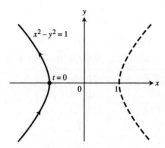

9. $x = t, y = \sqrt{4 - t^2}, 0 \le t \le 2$
   $\Rightarrow y = \sqrt{4 - x^2}$

11. $x = - \cosh t, y = \sinh t, -\infty < 1 < \infty$
    $\Rightarrow \cosh^2 t - \sinh^2 t = 1 \Rightarrow x^2 - y^2 = 1$

13. Arc PF = Arc AF since each is the distance rolled and

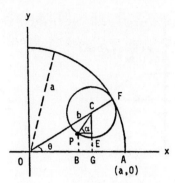

$\frac{\text{Arc PF}}{b} = \angle FCP \Rightarrow \text{Arc PF} = b(\angle FCP); \frac{\text{Arc AF}}{a} = \theta$

$\Rightarrow \text{Arc AF} = a\theta \Rightarrow a\theta = b(\angle FCP) \Rightarrow \angle FCP = \frac{a}{b}\theta;$

$\angle OCG = \frac{\pi}{2} - \theta; \angle OCG = \angle OCP + \angle PCE$

$= \angle OCP + \left(\frac{\pi}{2} - \alpha\right).$ Now $\angle OCP = \pi - \angle FCP$

$= \pi - \frac{a}{b}\theta.$ Thus $\angle OCG = \pi - \frac{a}{b}\theta + \frac{\pi}{2} - \alpha \Rightarrow \frac{\pi}{2} - \theta$

$= \pi - \frac{a}{b}\theta + \frac{\pi}{2} - \alpha \Rightarrow \alpha = \pi - \frac{a}{b}\theta + \theta = \pi - \left(\frac{a-b}{b}\theta\right).$

Then $x = OG - BG = OG - PE = (a - b)\cos\theta - b\cos\alpha = (a - b)\cos\theta - b\cos\left(\pi - \frac{a-b}{b}\theta\right)$

$= (a - b)\cos\theta + b\cos\left(\frac{a-b}{b}\theta\right).$ Also $y = EG = CG - CE = (a - b)\sin\theta - b\sin\alpha$

$= (a - b)\sin\theta - b\sin\left(\pi - \frac{a-b}{b}\theta\right) = (a - b)\sin\theta - b\sin\left(\frac{a-b}{b}\theta\right).$ Therefore

$x = (a - b)\cos\theta + b\cos\left(\frac{a-b}{b}\theta\right)$ and $y = (a - b)\sin\theta - b\sin\left(\frac{a-b}{b}\theta\right).$

If $b = \frac{a}{4},$ then $x = \left(a - \frac{a}{4}\right)\cos\theta + \frac{a}{4}\cos\left(\frac{a - \left(\frac{a}{4}\right)}{\left(\frac{a}{4}\right)}\theta\right)$

$= \frac{3a}{4}\cos\theta + \frac{a}{4}\cos 3\theta = \frac{3a}{4}\cos\theta + \frac{a}{4}(\cos\theta\cos 2\theta - \sin\theta\sin 2\theta)$

$= \frac{3a}{4}\cos\theta + \frac{a}{4}((\cos\theta)(\cos^2\theta - \sin^2\theta) - (\sin\theta)(2\sin\theta\cos\theta))$

$= \frac{3a}{4}\cos\theta + \frac{a}{4}\cos^3\theta - \frac{a}{4}\cos\theta\sin^2\theta - \frac{2a}{4}\sin^2\theta\cos\theta$

$= \frac{3a}{4}\cos\theta + \frac{a}{4}\cos^3\theta - \frac{3a}{4}(\cos\theta)(1 - \cos^2\theta) = a\cos^3\theta;$

$y = \left(a - \frac{a}{4}\right)\sin\theta - \frac{a}{4}\sin\left(\frac{a - \left(\frac{a}{4}\right)}{\left(\frac{a}{4}\right)}\theta\right) = \frac{3a}{4}\sin\theta - \frac{a}{4}\sin 3\theta = \frac{3a}{4}\sin\theta - \frac{a}{4}(\sin\theta\cos 2\theta + \cos\theta\sin 2\theta)$

$= \frac{3a}{4}\sin\theta - \frac{a}{4}((\sin\theta)(\cos^2\theta - \sin^2\theta) + (\cos\theta)(2\sin\theta\cos\theta))$

$= \frac{3a}{4}\sin\theta - \frac{a}{4}\sin\theta\cos^2\theta + \frac{a}{4}\sin^3\theta - \frac{2a}{4}\cos^2\theta\sin\theta$

$= \frac{3a}{4}\sin\theta - \frac{3a}{4}\sin\theta\cos^2\theta + \frac{a}{4}\sin^3\theta$

$= \frac{3a}{4}\sin\theta - \frac{3a}{4}(\sin\theta)(1 - \sin^2\theta) + \frac{a}{4}\sin^3\theta = a\sin^3\theta.$

15. Draw line AM in the figure and note that $\angle AMO$ is a right angle since it is an inscribed angle which spans the diameter of a circle. Then $AN^2 = MN^2 + AM^2.$ Now, $OA = a,$ $\frac{AN}{a} = \tan t,$ and $\frac{AM}{a} = \sin t.$ Next $MN = OP$

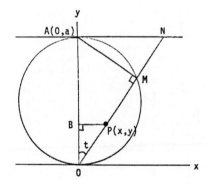

$\Rightarrow OP^2 = AN^2 - AM^2 = a^2\tan^2 t - a^2\sin^2 t$

$\Rightarrow OP = \sqrt{a^2\tan^2 t - a^2\sin^2 t}$

$= (a\sin t)\sqrt{\sec^2 t - 1} = \frac{a\sin^2 t}{\cos t}.$ In triangle BPO,

$x = OP\sin t = \frac{a\sin^3 t}{\cos t} = a\sin^2 t\tan t$ and

$y = OP\cos t = a\sin^2 t \Rightarrow x = a\sin^2 t\tan t$ and $y = a\sin^2 t.$

17. $D = \sqrt{(x - 2)^2 + \left(y - \frac{1}{2}\right)^2} \Rightarrow D^2 = (x - 2)^2 + \left(y - \frac{1}{2}\right)^2 = (t - 2)^2 + \left(t^2 - \frac{1}{2}\right)^2 \Rightarrow D^2 = t^4 - 4t + \frac{17}{4}$

$\Rightarrow \frac{d(D^2)}{dt} = 4t^3 - 4 = 0 \Rightarrow t = 1.$ The second derivative is always positive for $t \neq 0 \Rightarrow t = 1$ gives a local minimum for D^2 (and hence D) which is an absolute minimum since it is the only extremum $\Rightarrow$ the closest point on the parabola is $(1, 1).$

19. (a)

(b)

(c)

21.

23. (a)

(b)

25. (a)

(b)

(c)

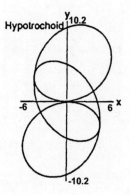

CHAPTER 9 PRACTICE EXERCISES

1. $r \cos\left(\theta + \frac{\pi}{3}\right) = 2\sqrt{3} \Rightarrow r\left(\cos\theta \cos\frac{\pi}{3} - \sin\theta \sin\frac{\pi}{3}\right)$
    $= 2\sqrt{3} \Rightarrow \frac{1}{2} r \cos\theta - \frac{\sqrt{3}}{2} r \sin\theta = 2\sqrt{3}$
    $\Rightarrow r \cos\theta - \sqrt{3} r \sin\theta = 4\sqrt{3} \Rightarrow x - \sqrt{3}y = 4\sqrt{3}$
    $\Rightarrow y = \frac{\sqrt{3}}{3} x - 4$

3. $r = 2\sec\theta \Rightarrow r = \frac{2}{\cos\theta} \Rightarrow r\cos\theta = 2 \Rightarrow x = 2$

5. $r = -\frac{3}{2}\csc\theta \Rightarrow r\sin\theta = -\frac{3}{2} \Rightarrow y = -\frac{3}{2}$

7. $r = -4\sin\theta \Rightarrow r^2 = -4r\sin\theta \Rightarrow x^2 + y^2 + 4y = 0$
    $\Rightarrow x^2 + (y+2)^2 = 4$; circle with center $(0, -2)$ and
    radius 2.

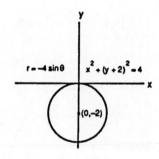

9. $r = 2\sqrt{2}\cos\theta \Rightarrow r^2 = 2\sqrt{2}\, r\cos\theta$
    $\Rightarrow x^2 + y^2 - 2\sqrt{2}x = 0 \Rightarrow \left(x - \sqrt{2}\right)^2 + y^2 = 2$;
    circle with center $\left(\sqrt{2}, 0\right)$ and radius $\sqrt{2}$

11. $x^2 + y^2 + 5y = 0 \Rightarrow x^2 + \left(y + \frac{5}{2}\right)^2 = \frac{25}{4} \Rightarrow C = \left(0, -\frac{5}{2}\right)$

    and $a = \frac{5}{2}$; $r^2 + 5r \sin \theta = 0 \Rightarrow r = -5 \sin \theta$

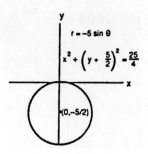

13. $x^2 + y^2 - 3x = 0 \Rightarrow \left(x - \frac{3}{2}\right)^2 + y^2 = \frac{9}{4} \Rightarrow C = \left(\frac{3}{2}, 0\right)$

    and $a = \frac{3}{2}$; $r^2 - 3r \cos \theta = 0 \Rightarrow r = 3 \cos \theta$

15.

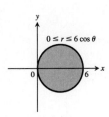

17. d   19. l   21. k   23. i

25. $A = 2 \int_0^\pi \frac{1}{2} r^2 \, d\theta = \int_0^\pi (2 - \cos \theta)^2 \, d\theta = \int_0^\pi (4 - 4 \cos \theta + \cos^2 \theta) \, d\theta = \int_0^\pi \left(4 - 4 \cos \theta + \frac{1 + \cos 2\theta}{2}\right) d\theta$

    $= \int_0^\pi \left(\frac{9}{2} - 4 \cos \theta + \frac{\cos 2\theta}{2}\right) d\theta = \left[\frac{9}{2} \theta - 4 \sin \theta + \frac{\sin 2\theta}{4}\right]_0^\pi = \frac{9}{2} \pi$

27. $r = 1 + \cos 2\theta$ and $r = 1 \Rightarrow 1 = 1 + \cos 2\theta \Rightarrow 0 = \cos 2\theta \Rightarrow 2\theta = \frac{\pi}{2} \Rightarrow \theta = \frac{\pi}{4}$; therefore

    $A = 4 \int_0^{\pi/4} \frac{1}{2} \left[(1 + \cos 2\theta)^2 - 1^2\right] d\theta = 2 \int_0^{\pi/4} (1 + 2 \cos 2\theta + \cos^2 2\theta - 1) \, d\theta$

    $= 2 \int_0^{\pi/4} \left(2 \cos 2\theta + \frac{1}{2} + \frac{\cos 4\theta}{2}\right) d\theta = 2 \left[\sin 2\theta + \frac{1}{2} \theta + \frac{\sin 4\theta}{8}\right]_0^{\pi/4} = 2 \left(1 + \frac{\pi}{8} + 0\right) = 2 + \frac{\pi}{4}$

29. $r = -1 + \cos \theta \Rightarrow \frac{dr}{d\theta} = - \sin \theta$; Length $= \int_0^{2\pi} \sqrt{(-1 + \cos \theta)^2 + (- \sin \theta)^2} \, d\theta = \int_0^{2\pi} \sqrt{2 - 2 \cos \theta} \, d\theta$

    $= \int_0^{2\pi} \sqrt{\frac{4(1 - \cos \theta)}{2}} \, d\theta = \int_0^{2\pi} 2 \sin \frac{\theta}{2} \, d\theta = \left[-4 \cos \frac{\theta}{2}\right]_0^{2\pi} = (-4)(-1) - (-4)(1) = 8$

31. $r = 8 \sin^3 \left(\frac{\theta}{3}\right), 0 \le \theta \le \frac{\pi}{4} \Rightarrow \frac{dr}{d\theta} = 8 \sin^2 \left(\frac{\theta}{3}\right) \cos \left(\frac{\theta}{3}\right)$; $r^2 + \left(\frac{dr}{d\theta}\right)^2 = \left[8 \sin^3 \left(\frac{\theta}{3}\right)\right]^2 + \left[8 \sin^2 \left(\frac{\theta}{3}\right) \cos \left(\frac{\theta}{3}\right)\right]^2$

    $= 64 \sin^4 \left(\frac{\theta}{3}\right) \Rightarrow L = \int_0^{\pi/4} \sqrt{64 \sin^4 \left(\frac{\theta}{3}\right)} \, d\theta = \int_0^{\pi/4} 8 \sin^2 \left(\frac{\theta}{3}\right) d\theta = \int_0^{\pi/4} 8 \left[\frac{1 - \cos \left(\frac{2\theta}{3}\right)}{2}\right] d\theta$

    $= \int_0^{\pi/4} \left[4 - 4 \cos \left(\frac{2\theta}{3}\right)\right] d\theta = \left[4\theta - 6 \sin \left(\frac{2\theta}{3}\right)\right]_0^{\pi/4} = 4 \left(\frac{\pi}{4}\right) - 6 \sin \left(\frac{\pi}{6}\right) - 0 = \pi - 3$

…

33. $x^2 = -4y \Rightarrow y = -\frac{x^2}{4} \Rightarrow 4p = 4 \Rightarrow p = 1;$
therefore Focus is $(0, -1)$, Directrix is $y = 1$

35. $y^2 = 3x \Rightarrow x = \frac{y^2}{3} \Rightarrow 4p = 3 \Rightarrow p = \frac{3}{4};$
therefore Focus is $\left(\frac{3}{4}, 0\right)$, Directrix is $x = -\frac{3}{4}$

37. $16x^2 + 7y^2 = 112 \Rightarrow \frac{x^2}{7} + \frac{y^2}{16} = 1$
$\Rightarrow c^2 = 16 - 7 = 9 \Rightarrow c = 3; e = \frac{c}{a} = \frac{3}{4}$

39. $3x^2 - y^2 = 3 \Rightarrow x^2 - \frac{y^2}{3} = 1 \Rightarrow c^2 = 1 + 3 = 4$
$\Rightarrow c = 2; e = \frac{c}{a} = \frac{2}{1} = 2;$ the asymptotes are $y = \pm \sqrt{3}\, x$

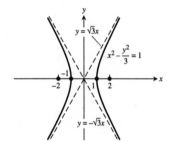

41. $x^2 = -12y \Rightarrow -\frac{x^2}{12} = y \Rightarrow 4p = 12 \Rightarrow p = 3 \Rightarrow$ focus is $(0, -3)$, directrix is $y = 3$, vertex is $(0, 0)$; therefore new vertex is $(2, 3)$, new focus is $(2, 0)$, new directrix is $y = 6$, and the new equation is $(x - 2)^2 = -12(y - 3)$

43. $\frac{x^2}{9} + \frac{y^2}{25} = 1 \Rightarrow a = 5$ and $b = 3 \Rightarrow c = \sqrt{25 - 9} = 4 \Rightarrow$ foci are $(0, \pm 4)$, vertices are $(0, \pm 5)$, center is $(0, 0)$; therefore the new center is $(-3, -5)$, new foci are $(-3, -1)$ and $(-3, -9)$, new vertices are $(-3, -10)$ and $(-3, 0)$, and the new equation is $\frac{(x + 3)^2}{9} + \frac{(y + 5)^2}{25} = 1$

45. $\frac{y^2}{8} - \frac{x^2}{2} = 1 \Rightarrow a = 2\sqrt{2}$ and $b = \sqrt{2} \Rightarrow c = \sqrt{8 + 2} = \sqrt{10} \Rightarrow$ foci are $\left(0, \pm \sqrt{10}\right)$, vertices are $\left(0, \pm 2\sqrt{2}\right)$, center is $(0, 0)$, and the asymptotes are $y = \pm 2x$; therefore the new center is $\left(2, 2\sqrt{2}\right)$, new foci are $\left(2, 2\sqrt{2} \pm \sqrt{10}\right)$, new vertices are $\left(2, 4\sqrt{2}\right)$ and $(2, 0)$, the new asymptotes are $y = 2x - 4 + 2\sqrt{2}$ and $y = -2x + 4 + 2\sqrt{2}$; the new equation is $\frac{\left(y - 2\sqrt{2}\right)^2}{8} - \frac{(x - 2)^2}{2} = 1$

47. $x^2 - 4x - 4y^2 = 0 \Rightarrow x^2 - 4x + 4 - 4y^2 = 4 \Rightarrow (x - 2)^2 - 4y^2 = 4 \Rightarrow \frac{(x - 2)^2}{4} - y^2 = 1,$ a hyperbola; $a = 2$ and $b = 1 \Rightarrow c = \sqrt{1 + 4} = \sqrt{5}$; the center is $(2, 0)$, the vertices are $(0, 0)$ and $(4, 0)$; the foci are $\left(2 \pm \sqrt{5}, 0\right)$ and the asymptotes are $y = \pm \frac{x - 2}{2}$

49. $y^2 - 2y + 16x = -49 \Rightarrow y^2 - 2y + 1 = -16x - 48 \Rightarrow (y - 1)^2 = -16(x + 3),$ a parabola; the vertex is $(-3, 1)$; $4p = 16 \Rightarrow p = 4 \Rightarrow$ the focus is $(-7, 1)$ and the directrix is $x = 1$

51. $9x^2 + 16y^2 + 54x - 64y = -1 \Rightarrow 9\left(x^2 + 6x\right) + 16\left(y^2 - 4y\right) = -1 \Rightarrow 9\left(x^2 + 6x + 9\right) + 16\left(y^2 - 4y + 4\right) = 144$
$\Rightarrow 9(x + 3)^2 + 16(y - 2)^2 = 144 \Rightarrow \frac{(x + 3)^2}{16} + \frac{(y - 2)^2}{9} = 1,$ an ellipse; the center is $(-3, 2)$; $a = 4$ and $b = 3$
$\Rightarrow c = \sqrt{16 - 9} = \sqrt{7}$; the foci are $\left(-3 \pm \sqrt{7}, 2\right)$; the vertices are $(1, 2)$ and $(-7, 2)$

53. $x^2 + y^2 - 2x - 2y = 0 \Rightarrow x^2 - 2x + 1 + y^2 - 2y + 1 = 2 \Rightarrow (x - 1)^2 + (y - 1)^2 = 2$, a circle with center $(1, 1)$ and radius $= \sqrt{2}$

55. $r = \frac{2}{1 + \cos \theta} \Rightarrow e = 1 \Rightarrow$ parabola with vertex at $(1, 0)$

57. $r = \frac{6}{1 - 2 \cos \theta} \Rightarrow e = 2 \Rightarrow$ hyperbola; $ke = 6 \Rightarrow 2k = 6$
    $\Rightarrow k = 3 \Rightarrow$ vertices are $(2, \pi)$ and $(6, \pi)$

59. $e = 2$ and $r \cos \theta = 2 \Rightarrow x = 2$ is directrix $\Rightarrow k = 2$; the conic is a hyperbola; $r = \frac{ke}{1 + e \cos \theta} \Rightarrow r = \frac{(2)(2)}{1 + 2 \cos \theta}$
    $\Rightarrow r = \frac{4}{1 + 2 \cos \theta}$

61. $e = \frac{1}{2}$ and $r \sin \theta = 2 \Rightarrow y = 2$ is directrix $\Rightarrow k = 2$; the conic is an ellipse; $r = \frac{ke}{1 + e \sin \theta} \Rightarrow r = \frac{(2)\left(\frac{1}{2}\right)}{1 + \left(\frac{1}{2}\right) \sin \theta}$
    $\Rightarrow r = \frac{2}{2 + \sin \theta}$

63. $x = \frac{1}{2} \tan t$ and $y = \frac{1}{2} \sec t \Rightarrow x^2 = \frac{1}{4} \tan^2 t$ 65. $x = -\cos t$ and $y = \cos^2 t \Rightarrow y = (-x)^2 = x^2$
    and $y^2 = \frac{1}{4} \sec^2 t \Rightarrow 4x^2 = \tan^2 t$ and
    $4y^2 = \sec^2 t \Rightarrow 4x^2 + 1 = 4y^2 \Rightarrow 4y^2 - 4x^2 = 1$

       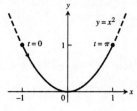

67. (a) Around the x-axis: $9x^2 + 4y^2 = 36 \Rightarrow y^2 = 9 - \frac{9}{4} x^2 \Rightarrow y = \pm \sqrt{9 - \frac{9}{4} x^2}$ and we use the positive root:
    $$V = 2 \int_0^2 \pi \left(\sqrt{9 - \frac{9}{4} x^2}\right)^2 dx = 2 \int_0^2 \pi \left(9 - \frac{9}{4} x^2\right) dx = 2\pi \left[9x - \frac{3}{4} x^3\right]_0^2 = 24\pi$$

    (b) Around the y-axis: $9x^2 + 4y^2 = 36 \Rightarrow x^2 = 4 - \frac{4}{9} y^2 \Rightarrow x = \pm \sqrt{4 - \frac{4}{9} y^2}$ and we use the positive root:
    $$V = 2 \int_0^3 \pi \left(\sqrt{4 - \frac{4}{9} y^2}\right)^2 dy = 2 \int_0^3 \pi \left(4 - \frac{4}{9} y^2\right) dy = 2\pi \left[4y - \frac{4}{27} y^3\right]_0^3 = 16\pi$$

69. (a) $r = \frac{k}{1 + e \cos \theta} \Rightarrow r + er \cos \theta = k \Rightarrow \sqrt{x^2 + y^2} + ex = k \Rightarrow \sqrt{x^2 + y^2} = k - ex \Rightarrow x^2 + y^2$
    $= k^2 - 2kex + e^2 x^2 \Rightarrow x^2 - e^2 x^2 + y^2 + 2kex - k^2 = 0 \Rightarrow (1 - e^2) x^2 + y^2 + 2kex - k^2 = 0$

(b) $e = 0 \Rightarrow x^2 + y^2 - k^2 = 0 \Rightarrow x^2 + y^2 = k^2 \Rightarrow$ circle;

$0 < e < 1 \Rightarrow e^2 < 1 \Rightarrow e^2 - 1 < 0 \Rightarrow B^2 - 4AC = 0^2 - 4\,(1 - e^2)\,(1) = 4\,(e^2 - 1) < 0 \Rightarrow$ ellipse;

$e = 1 \Rightarrow B^2 - 4AC = 0^2 - 4(0)(1) = 0 \Rightarrow$ parabola;

$e > 1 \Rightarrow e^2 > 1 \Rightarrow B^2 - 4AC = 0^2 - 4\,(1 - e^2)\,(1) = 4e^2 - 4 > 0 \Rightarrow$ hyperbola

CHAPTER 9 ADDITIONAL AND ADVANCED EXERCISES

1. Directrix $x = 3$ and focus $(4, 0) \Rightarrow$ vertex is $\left(\frac{7}{2}, 0\right)$

    $\Rightarrow p = \frac{1}{2} \Rightarrow$ the equation is $x - \frac{7}{2} = \frac{y^2}{2}$

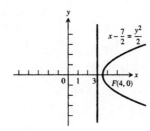

3. $x^2 = 4y \Rightarrow$ vertex is $(0, 0)$ and $p = 1 \Rightarrow$ focus is $(0, 1)$; thus the distance from $P(x, y)$ to the vertex is $\sqrt{x^2 + y^2}$

    and the distance from P to the focus is $\sqrt{x^2 + (y - 1)^2} \Rightarrow \sqrt{x^2 + y^2} = 2\sqrt{x^2 + (y - 1)^2}$

    $\Rightarrow x^2 + y^2 = 4\,[x^2 + (y - 1)^2] \Rightarrow x^2 + y^2 = 4x^2 + 4y^2 - 8y + 4 \Rightarrow 3x^2 + 3y^2 - 8y + 4 = 0$, which is a circle

5. Vertices are $(0, \pm 2) \Rightarrow a = 2; e = \frac{c}{a} \Rightarrow 0.5 = \frac{c}{2} \Rightarrow c = 1 \Rightarrow$ foci are $(0, \pm 1)$

7. Let the center of the hyperbola be $(0, y)$.

    (a) Directrix $y = -1$, focus $(0, -7)$ and $e = 2 \Rightarrow c - \frac{a}{e} = 6 \Rightarrow \frac{a}{e} = c - 6 \Rightarrow a = 2c - 12$. Also $c = ae = 2a$

    $\Rightarrow a = 2(2a) - 12 \Rightarrow a = 4 \Rightarrow c = 8; y - (-1) = \frac{a}{e} = \frac{4}{2} = 2 \Rightarrow y = 1 \Rightarrow$ the center is $(0, 1); c^2 = a^2 + b^2$

    $\Rightarrow b^2 = c^2 - a^2 = 64 - 16 = 48$; therefore the equation is $\frac{(y - 1)^2}{16} - \frac{x^2}{48} = 1$

    (b) $e = 5 \Rightarrow c - \frac{a}{e} = 6 \Rightarrow \frac{a}{e} = c - 6 \Rightarrow a = 5c - 30$. Also, $c = ae = 5a \Rightarrow a = 5(5a) - 30 \Rightarrow 24a = 30 \Rightarrow a = \frac{5}{4}$

    $\Rightarrow c = \frac{25}{4}; y - (-1) = \frac{a}{e} = \frac{\left(\frac{5}{4}\right)}{5} = \frac{1}{4} \Rightarrow y = -\frac{3}{4} \Rightarrow$ the center is $\left(0, -\frac{3}{4}\right); c^2 = a^2 + b^2 \Rightarrow b^2 = c^2 - a^2$

    $= \frac{625}{16} - \frac{25}{16} = \frac{75}{2}$; therefore the equation is $\frac{\left(y + \frac{3}{4}\right)^2}{\left(\frac{25}{16}\right)} - \frac{x^2}{\left(\frac{75}{2}\right)} = 1$ or $\frac{16\left(y + \frac{3}{4}\right)^2}{25} - \frac{2x^2}{75} = 1$

9. $b^2 x^2 + a^2 y^2 = a^2 b^2 \Rightarrow \frac{dy}{dx} = -\frac{b^2 x}{a^2 y}$; at (x_1, y_1) the tangent line is $y - y_1 = \left(-\frac{b^2 x_1}{a^2 y_1}\right)(x - x_1)$

    $\Rightarrow a^2 y y_1 + b^2 x x_1 = b^2 x_1^2 + a^2 y_1^2 = a^2 b^2 \Rightarrow b^2 x x_1 + a^2 y y_1 - a^2 b^2 = 0$

11.

13.

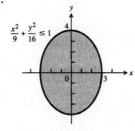

15. $(9x^2 + 4y^2 - 36)(4x^2 + 9y^2 - 16) \le 0$

$\Rightarrow 9x^2 + 4y^2 - 36 \le 0$ and $4x^2 + 9y^2 - 16 \ge 0$

or $9x^2 + 4y^2 - 36 \ge 0$ and $4x^2 + 9y^2 - 16 \le 0$

17. (a) $x = e^{2t} \cos t$ and $y = e^{2t} \sin t \Rightarrow x^2 + y^2 = e^{4t} \cos^2 t + e^{4t} \sin^2 t = e^{4t}$. Also $\frac{y}{x} = \frac{e^{2t} \sin t}{e^{2t} \cos t} = \tan t$

$\Rightarrow t = \tan^{-1}\left(\frac{y}{x}\right) \Rightarrow x^2 + y^2 = e^{4\tan^{-1}(y/x)}$ is the Cartesian equation. Since $r^2 = x^2 + y^2$ and

$\theta = \tan^{-1}\left(\frac{y}{x}\right)$, the polar equation is $r^2 = e^{4\theta}$ or $r = e^{2\theta}$ for $r > 0$

(b) $ds^2 = r^2\, d\theta^2 + dr^2; r = e^{2\theta} \Rightarrow dr = 2e^{2\theta}\, d\theta$

$\Rightarrow ds^2 = r^2\, d\theta^2 + \left(2e^{2\theta}\, d\theta\right)^2 = \left(e^{2\theta}\right)^2 d\theta^2 + 4e^{4\theta}\, d\theta^2$

$= 5e^{4\theta}\, d\theta^2 \Rightarrow ds = \sqrt{5}\, e^{2\theta}\, d\theta \Rightarrow L = \int_0^{2\pi} \sqrt{5}\, e^{2\theta}\, d\theta$

$= \left[\frac{\sqrt{5}\, e^{2\theta}}{2}\right]_0^{2\pi} = \frac{\sqrt{5}}{2}\left(e^{4\pi} - 1\right)$

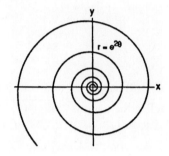

19. $e = 2$ and $r \cos \theta = 2 \Rightarrow x = 2$ is the directrix $\Rightarrow k = 2$; the conic is a hyperbola with $r = \frac{ke}{1 + e \cos \theta}$

$\Rightarrow r = \frac{(2)(2)}{1 + 2 \cos \theta} = \frac{4}{1 + 2 \cos \theta}$

21. $e = \frac{1}{2}$ and $r \sin \theta = 2 \Rightarrow y = 2$ is the directrix $\Rightarrow k = 2$; the conic is an ellipse with $r = \frac{ke}{1 + e \sin \theta}$

$\Rightarrow r = \frac{2\left(\frac{1}{2}\right)}{1 + \left(\frac{1}{2}\right) \sin \theta} = \frac{2}{2 + \sin \theta}$

23. Arc PF = Arc AF since each is the distance rolled;

$\angle PCF = \frac{\text{Arc } PF}{b} \Rightarrow \text{Arc } PF = b(\angle PCF); \theta = \frac{\text{Arc } AF}{a}$

$\Rightarrow \text{Arc } AF = a\theta \Rightarrow a\theta = b(\angle PCF) \Rightarrow \angle PCF = \left(\frac{a}{b}\right)\theta;$

$\angle OCB = \frac{\pi}{2} - \theta$ and $\angle OCB = \angle PCF - \angle PCE$

$= \angle PCF - \left(\frac{\pi}{2} - \alpha\right) = \left(\frac{a}{b}\right)\theta - \left(\frac{\pi}{2} - \alpha\right) \Rightarrow \frac{\pi}{2} - \theta$

$= \left(\frac{a}{b}\right)\theta - \left(\frac{\pi}{2} - \alpha\right) \Rightarrow \frac{\pi}{2} - \theta = \left(\frac{a}{b}\right)\theta - \frac{\pi}{2} + \alpha$

$\Rightarrow \alpha = \pi - \theta - \left(\frac{a}{b}\right)\theta \Rightarrow \alpha = \pi - \left(\frac{a+b}{b}\right)\theta.$

Now $x = OB + BD = OB + EP = (a + b) \cos \theta + b \cos \alpha = (a + b) \cos \theta + b \cos \left(\pi - \left(\frac{a+b}{b}\right)\theta\right)$

$= (a + b) \cos \theta + b \cos \pi \cos \left(\left(\frac{a+b}{b}\right)\theta\right) + b \sin \pi \sin \left(\left(\frac{a+b}{b}\right)\theta\right) = (a + b) \cos \theta - b \cos \left(\left(\frac{a+b}{b}\right)\theta\right)$ and

$y = PD = CB - CE = (a + b) \sin \theta - b \sin \alpha = (a + b) \sin \theta - b \sin \left(\left(\frac{a+b}{b}\right)\theta\right)$

$= (a + b) \sin \theta - b \sin \pi \cos \left(\left(\frac{a+b}{b}\right)\theta\right) + b \cos \pi \sin \left(\left(\frac{a+b}{b}\right)\theta\right) = (a + b) \sin \theta - b \sin \left(\left(\frac{a+b}{b}\right)\theta\right);$

therefore $x = (a + b) \cos \theta - b \cos \left(\left(\frac{a+b}{b}\right)\theta\right)$ and $y = (a + b) \sin \theta - b \sin \left(\left(\frac{a+b}{b}\right)\theta\right)$

25. $\beta = \psi_2 - \psi_1 \Rightarrow \tan \beta = \tan(\psi_2 - \psi_1) = \frac{\tan \psi_2 - \tan \psi_1}{1 + \tan \psi_2 \tan \psi_1};$

the curves will be orthogonal when $\tan \beta$ is undefined, or

when $\tan \psi_2 = \frac{-1}{\tan \psi_1} \Rightarrow \frac{r}{g'(\theta)} = \frac{-1}{\left[\frac{r}{f'(\theta)}\right]}$

$\Rightarrow r^2 = -f'(\theta)\, g'(\theta)$

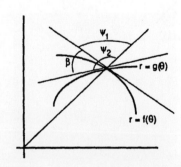

27. $r = 2a \sin 3\theta \Rightarrow \frac{dr}{d\theta} = 6a \cos 3\theta \Rightarrow \tan \psi = \frac{r}{\left(\frac{dr}{d\theta}\right)} = \frac{2a \sin 3\theta}{6a \cos 3\theta} = \frac{1}{3} \tan 3\theta$; when $\theta = \frac{\pi}{6}$, $\tan \psi = \frac{1}{3} \tan \frac{\pi}{2}$

$\Rightarrow \psi = \frac{\pi}{2}$

29. $\tan \psi_1 = \frac{\sqrt{3} \cos \theta}{-\sqrt{3} \sin \theta} = -\cot \theta$ is $-\frac{1}{\sqrt{3}}$ at $\theta = \frac{\pi}{3}$; $\tan \psi_2 = \frac{\sin \theta}{\cos \theta} = \tan \theta$ is $\sqrt{3}$ at $\theta = \frac{\pi}{3}$; since the product of these slopes is -1, the tangents are perpendicular

NOTES: